About the Authors

Omri Rand is a Professor of Aerospace Engineering at the Technion – Israel Institute of Technology. He has been involved in research on theoretical modeling and analysis in the area of anisotropic elasticity for the last fifteen years, he is the author of many journal papers and conference presentations in this area. Dr. Rand has been extensively active in composite rotor blade analysis, and established many well recognized analytical and numerical approaches. He teaches graduate courses in the area of anisotropic elasticity, serves as the Editor-in-Chief of *Science and Engineering of Composite Materials,* as a reviewer for leading professional journals, and as a consultant to various research and development organizations.

Vladimir Rovenski is a Professor of Mathematics and a well known researcher in the area of Riemannian and computational geometry. He is a corresponding member of the Natural Science Academy of Russia, a member of the American Mathematical Society, and serves as a reviewer of *Zentralblatt für Mathematik.* He is the author of many journal papers and books, including *Foliations on Riemannian Manifolds and Submanifolds* (Birkhäuser, 1997), and *Geometry of Curves and Surfaces with MAPLE* (Birkhäuser, 2000). Since 1999, Dr. Rovenski has been a senior scientist at the faculty of Aerospace Engineering at the Technion – Israel Institute of Technology, and a lecturer at Haifa University.

Omri Rand
Vladimir Rovenski

Analytical Methods in Anisotropic Elasticity

with Symbolic Computational Tools

Birkhäuser
Boston • Basel • Berlin

Omri Rand
Technion — Israel Institute of Technology
Faculty of Aerospace Engineering
Haifa 32000
Israel

Vladimir Rovenski
Technion — Israel Institute of Technology
Faculty of Aerospace Engineering
Haifa 32000
Israel

AMS Subject Classifications: 74E10, 74Bxx, 74Sxx, 65C20, 65Z05, 68W30, 74-XX, 74A10, 74A40, 74Axx, 74Fxx, 74Gxx, 74H10, 74Kxx, 74N15, 68W05, 65Nxx, 35J55 (Primary); 74-01, 74-04, 65-XX, 68Uxx, 68-XX (Secondary)

Library of Congress Cataloging-in-Publication Data
Rand, Omri.
Analytical methods in anisotropic elasticity : with symbolic computational tools / Omri Rand, Vladimir Rovenski.
p. cm.
Includes bibliographical references and index.
ISBN 0-8176-4372-2 (alk. paper)
1. Elasticity. 2. Anisotropy. 3. Anisotropy—Mathematical models. 4. Inhomogeneous materials. I. Rovenskii, Vladimir Y, 1953- II. Title
QA931.R36 2004
531′.382–dc22 2004054558

Additional material to this book can be downloaded from http:// extras .springer .com .

ISBN 0-8176-4272-2 Printed on acid-free paper.
ISBN 978-0-8176-4272-3

9 8 7 6 5 4 3 2 1 SPIN 10855936

To my family, **Ora**, **Shahar**, **Tal** and **Boaz**,

Omri Rand

To my teacher, Professor **Victor Toponogov**,

Vladimir Rovenski

Preface

Prior to the computer era, analytical methods in elasticity had already been developed and improved up to impressive levels. Relevant mathematical techniques were extensively exploited, contributing significantly to the understanding of physical phenomena. In recent decades, numerical computerized techniques have been refined and modernized, and have reached high levels of capabilities, standardization and automation. This trend, accompanied by convenient and high resolution graphical visualization capability, has made analytical methods less attractive, and the amount of effort devoted to them has become substantially smaller. Yet, with some tenacity, the tremendous advances in computerized tools have yielded various mature programs for symbolic manipulation. Such tools have revived many abandoned analytical methodologies by easing the tedious effort that was previously required, and by providing additional capabilities to perform complex derivation processes that were once considered impractical.

Generally speaking, it is well recognized that analytical solutions should be applied to relatively simple problems, while numerical techniques may handle more complex cases. However, it is also agreed that analytical solutions provide better insight and improved understanding of the involved physical phenomena, and enable a clear representation of the role taken by each of the problem parameters. Nowadays, analytical and numerical methods are considered as complementary: that is, while analytical methods provide the required understanding, numerical solutions provide accuracy and the capability to deal with cases where the geometry and other characteristics impose relatively complex solutions.

Nevertheless, from a practical point of view, analytic solutions are still considered as "art", while numerical codes (such as codes that are based on the finite-element method) seem to offer a "straightforward" solution for any type and geometry of a new problem. One of the reasons for this view emerges from the variety of techniques that are used for analytical solutions. For example, one has the option to select either the deformation field or the stress field to construct the initial solution hypothesis, or, one has the option to formulate the governing equations using differential equilibrium, or by employing more integral energy methodologies for the same task. Hence, the main obstacle to using analytical approaches seems to be the fact that many researchers and engineers tend to believe that, as far as analytic solutions are considered, each problem is associated with a specific solution type and that a different solution methodology has to be tailored for every new problem.

In light of the above, the objective of this book is twofold. First, it brings together and refreshes the fundamentals of anisotropic elasticity and reviews various mathematical tools and analytical solution trails that are encountered in this area. Then, it presents a collection of classical and advanced problems in anisotropic elasticity that encompasses various two-dimensional problems and different types of three-dimensional beam models. The book includes models of various mathematical complexity and physical accuracy levels, and provides the theoretical background for composite material analysis. One of the most advanced formulations presented is a complete analytical model and solution scheme for an arbitrarily loaded non-homogeneous beam structure of generic anisotropy.

All classic and modern analytic solutions are derived using symbolic computational techniques. Emphasis is put on the basic principles of the analytic approach (problem statement, setting of simplifying assumptions, satisfying the field and boundary conditions, proof of solution, etc.), and their implementation using symbolic computational tools, so that the reader will be able to employ the relevant approach to new problems that frequently arise. Discussions are devoted to the physical interpretation of the presented mathematical solutions.

From a format point of view, the book provides the background and mathematical formulation for each problem or topic. The main steps of the analytical solution and the graphical results are discussed as well, while the complete system of symbolic codes (written in `Maple`) are available on the enclosed disc.

A unique characteristic of this book is the fact that the *entire* analytical derivation and *all* solution expressions are *symbolically proved by suitable (computerized) codes*. Hence, the chance for (human) error or typographical mistake is eliminated. The symbolic worksheets are therefore absolute and firm testimony to the exactness of the presented expressions. For that reason, the specific solutions included in the text should be viewed as *illustrative examples* only, while the solution exactness and its generic applicability are proved symbolically in the most generic manner.

The book is aimed at graduate and senior undergraduate students, professors, engineers, applied mathematicians, numerical analysis experts, mechanics researchers and composite materials scientists.

Chapter description:

The first part of the book (Chapters 1–4) contains the fundamentals of anisotropic elasticity. The second part (Chapters 5–10) is devoted to various beam analyses and contains recent and advanced models developed by the authors.

Chapter 1 addresses *fundamental issues of anisotropic elasticity and analytical methodologies*. It provides a review of deformation measures and strain in generic orthogonal curvilinear coordinates, and reaches the complete nonlinear compatibility equations in such systems. It then introduces fundamental stress measures and the associated equilibrium equations. Later on, energy theorems and variational analyses are derived, followed by a general discussion of analytical methodologies and typical solution trails.

Chapter 2 reviews the mathematical representation of general *anisotropic materials,* including the special cases of Monoclinic, Orthotropic, Tetragonal, Transversely Isotropic, Cubic and Isotropic materials. Later, transformations between coordinate systems of the compliance and stiffness matrices (or tensors) are presented. The chapter also addresses issues such as planes of elastic symmetry, principal directions of anisotropy and non-Cartesian anisotropy.

Chapter 3 defines two-dimensional homogeneous and non-homogeneous domain topologies, and presents various *plane deformation* problems and analyses, including detailed formulation of plane-strain/stress and plane-shear states. The derivation yields formal definitions of generalized Neumann/Dirichlet and biharmonic boundary value problems (BVPs). The chapter also addresses Coupled-Plane BVP for materials of general anisotropy. Along the same lines, the classical anisotropic laminated plate theory is then presented.

Chapter 4 presents various *solution methodologies* for the BVPs derived in Chapter 3, and establishes solution schemes that facilitate applications presented later on. Explicit analytic expressions for low-order exact/conditional polynomial solutions, and approximate high-order polynomial solutions in a homogeneous simply connected domain are derived and illustrated. A formulation based on complex potentials is also thoroughly derived and demonstrated by Fourier series solutions.

Chapter 5 reviews some basic aspects and general definitions of *anisotropic beam analysis*, approximate analysis techniques and relevant literature. It discusses the associated coupling characteristics at both the material and structural levels.

Chapter 6 presents an analysis of *general anisotropic beams* that may be viewed as a level-based extension of the classical Lekhnitskii formulation, and is capable of handling beams of general anisotropy and cross-section geometry that undergo generic distribution of surface, body-force and tip loading. The derivation is founded on the BVPs deduced in Chapter 3, and despite its complexity, it provides a clear insight into the associated structural behavior and coupling mechanisms.

Chapter 7 contains a closed-form formulation for *uncoupled monoclinic homogeneous beams*. It first presents solutions for tip loads, and then a generic formulation for axially non-uniform distribution of surface and body loads. Later on, analysis and examples of beams of cylindrical anisotropy are presented.

The entire reasoning of the approach in this chapter is founded on St. Venant's semi-inverse method of solution and may be considered as dual (though less generic) to the method presented in Chapter 6.

Chapter 8 is focused on problems in various *non-homogeneous domains*. It first reviews generic formulations of *plane BVPs*, and then extends the analysis of Chapter 7 to the case of *monoclinic non-homogeneous beams under tip loading*, which is founded on extending the classical definition of the auxiliary problems of plane deformation to the anisotropic case. The discussion encompasses the determination of the principal axis of extension, principal planes of bending and shear center. The chapter also presents a generalization of the derivation in Chapter 7 to the case of uncoupled non-homogeneous beams that undergo *generic distributed loading*.

Chapter 9 discusses *coupled solid monoclinic beams*. The analysis presents an approximate model that provides insight into and fundamental understanding of the coupling mechanisms within anisotropic beams at the structural level. The model also supplies a simplified but relatively accurate tool for quantitative estimation of coupled beam behavior.

In addition, the chapter presents an exact, level-based solution scheme for coupled beams. The derivation employs a series of properly interconnected solution levels and reaches the exact solution in an iterative manner.

Chapter 10 handles *coupled thin-wall monoclinic beams* in a similar (approximate) manner to Chapter 9. The analysis encompasses beams having either *multiply connected domain* ("closed") or *simply connected domain* ("open") cross-sections.

Chapter 11 presents instructions for the symbolic and illustrative programs included in this book (implemented in `Maple`).

General Style Clarification Notes:

(1) As a general rule, we use a "tilde" (e.g. $\widetilde{A}$), for "temporary" variables that have no meaningful role, and are introduced for the sake of clarification and analytic convenience. Such variables are valid "locally" within the immediate paragraphs in which they appear in. Hence, if such a notation is repeated elsewhere, it stands for a different "local" meaning; likewise, the superscript $()'$ may have different meanings in various contexts.

(2) Due to the dependency of most involved functions on many parameters, *both ordinary and partial derivatives*, say $d(\ )/d\alpha$ or $\partial(\ )/\partial\alpha$, are abbreviated as $(\ )_{,\alpha}$. Similarly, $d^2(\ )/d\alpha^2$ or $\partial^2(\ )/\partial\alpha^2$, are abbreviated as $(\ )_{,\alpha\alpha}$.

(3) Integrals appear in a short notation by omitting the explicit indication of the integration variables. Two examples are an integration along a closed loop with a circumferential coordinate, s, i.e., $\oint_{\partial\Omega}\widetilde{\mathrm{F}}ds$, which is written simply as $\oint_{\partial\Omega}\widetilde{\mathrm{F}}$, and the area integration in the xy-plane, i.e., $\iint_{\Omega}\widetilde{\mathrm{F}}dxdy$, which is written as $\iint_{\Omega}\widetilde{\mathrm{F}}$.

(4) Within the equation notation, e.g. (1.3), the first digits stand for the chapter in which it appears, while (1.15a) is an example for an equation in a group of (sub-)equations. By a notation like (1.9a:b) we refer to the second equation of a group of equations that appears in one line that is collectively denoted (1.9a).

(5) Within the Section, Program, Remark, Example, Figure and Table notation, e.g. S.1.2, **P.1.2**, Remark 1.2, Example 1.2, Fig. 1.2, Table 1.2, the first digit stands for the chapter in which it appears.

Acknowledgements

We wish to acknowledge the great help of Dr. Michael Kazar (Kezerashvili) who had a unique role in exposing us to some great contributions to this science made by the Eastern academia discussed in Chapters 7,8.

We are also thankful to the Ph.D. student Michael Grebshtein who made a tremendous contribution to the rigor, the analytical uniformity and the symbolic verification of the derivation in Chapters 4,7,8.

We warmly thank Ann Kostant, Executive Editor of Mathematics and Physics at Birkhäuser Boston, for her support during the publishing process.

Omri Rand
Vladimir Rovenski

Haifa, Israel

Contents

Analytical Methods in Anisotropic Elasticity

1

Fundamentals of Anisotropic Elasticity and Analytical Methodologies

This chapter is devoted to fundamental issues in anisotropic elasticity that should be defined and reviewed before specific problems are tackled. Hence, the main purpose of this chapter is to provide a common and general background and terminology that will allow further development of analytical tools.

In contrast with traditional and other modern textbooks in the general area of elasticity, see e.g. (Muskhelishvili, 1953), (Milne-Thomson, 1960), (Novozhilov, 1961), (Hearmon, 1961), (Filonenko-Borodich, 1965), (Steeds, 1973), (Sokolnikoff, 1983), (Parton and Perlin, 1984), (Landau and Lifschitz, 1986), (Ciarlet, 1988), (Reismann and Pawlik, 1991), (Barber, 1992), (Green and Zerna, 1992), (Chou and Pagano, 1992), (Wu *et al.*, 1992), (Saada, 1993), (Gould, 1994), (Ting, 1996), (Chernykh and Kulman, 1998), (Soutas-Little, 1999), (Boresi and Chong, 1999), (Doghri, 2000), (Atanackovic and Guran, 2000), (Slaughter, 2001), the fundamental issues presented in what follows are fully backed by symbolic codes that testify for the exactness of the derivation, and may be employed to produce an enormous amount of additional information and results, in a clear, complete and analytically accurate manner.

1.1 Deformation Measures and Strain

The mathematical representation of deformation of an *elastic* body is under discussion in this section. Here and throughout this book, the notion *elastic* stands for a non-rigid solid medium that is deformed under external loading and fully recovers its size and shape when loading is removed. Hence, we shall use the term "elastic" as equivalent to "non-rigid". The change in the relative position of points is generally termed *deformation*, and the study of deformations is the province of the *analysis of strain*, (Sokolnikoff, 1983).

The discussion presented in what follows supplies measures mainly for small and finite deformations. Such expressions are essential for proper modeling of the linear behavior of anisotropic elastic media. Under certain assumptions, these expressions may also be interpreted to provide an insight into the physics of the deformation.

Unless stated differently, we shall assume that the deformation throughout the elastic body is *continuous*, while mathematically, we presume that all deformation expressions are differentiable with a number of continuous derivatives as required. Cases where the media consists of different materials, such as laminated composite structures and other non-homogeneous domains will be treated as an extension of the homogeneous approach in further chapters.

To analytically describe the deformation, we shall adopt a suitable system of curvilinear orthogonal coordinates in a Euclidean space, E^3, or in a plane, E^2. The analysis of non-orthogonal curvilinear coordinates is much more complicated, while yielding less practical advantage in the context of this book.

In essence, the selection of a coordinate system for a specific problem is immaterial, since clearly, any given physical deformation may be described in many and various coordinate systems. However, in the course of any search for analytical solutions, it becomes obvious that the selection of a suitable coordinate system may have a tremendous effect on the ability to derive a solution, and on the effort that is required to achieve such a solution.

Traditionally, when isotropic elasticity has been under discussion, selection of a proper coordinate system was primarily based on the geometry of the problem and the complexity involved within the fulfillment of the boundary conditions. As will become clear in further chapters, when an anisotropic elastic body is under discussion, material type and direction should also be taken into account while selecting a coordinate system, and in many cases, material anisotropy has a predominant influence on this selection. In what follows, we shall therefore establish and review the mathematical definition of deformation for a variety of coordinate systems. For the sake of clarity, we will first deal with Cartesian coordinates, and then generalize the approach for other orthogonal coordinate systems.

1.1.1 Displacements in Cartesian Coordinates

One of the ways to present *deformation* in elastic media is based on specifying the *displacement components*, which for the sake of brevity will also be denoted as *displacements*. By definition, the displacements of a material particle are determined by its initial and final locations, while the path of the particle between these two points is irrelevant.

Suppose now that due to the deformation described by the *displacement vector* $\mathbf{u} = (u, v, w)$, a material particle, M, located at $\mathbf{x} = (x, y, z)$ in the elastic media before deformation, has moved to a new location, $M^*(\xi, \eta, \zeta)$, with position vector $\mathbf{x} + \mathbf{u}$, Fig. 1.1(a).

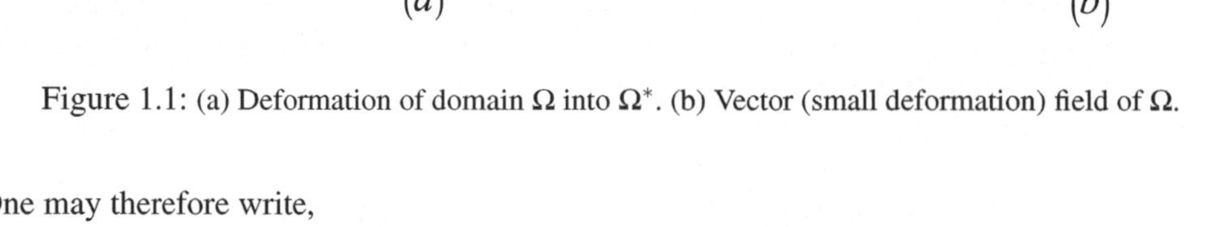

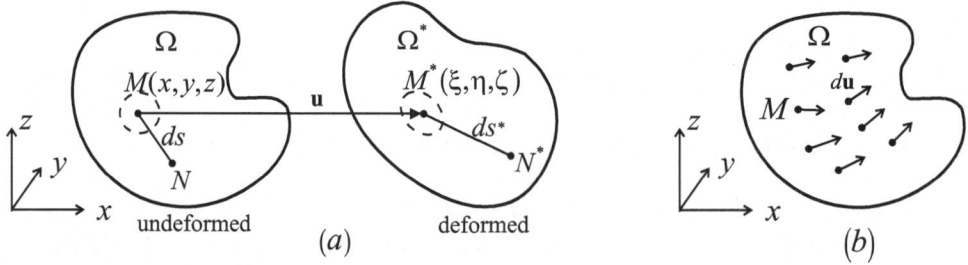

Figure 1.1: (a) Deformation of domain Ω into Ω^*. (b) Vector (small deformation) field of Ω.

One may therefore write,

$$\xi = x + u, \qquad \eta = y + v, \qquad \zeta = z + w, \tag{1.1}$$

where $u = u(x, y, z)$, $v = v(x, y, z)$, $w = w(x, y, z)$. Differentiating the above equations yields

$$d\xi = (1 + u_{,x})\,dx + u_{,y}\,dy + u_{,z}\,dz,$$

$$d\eta = v_{,x}dx + (1+v_{,y})\,dy + v_{,z}dz,$$
$$d\zeta = w_{,x}dx + w_{,y}dy + (1+w_{,z})\,dz. \tag{1.2}$$

A necessary and sufficient condition for a continuous deformation $\mathbf{u}$ to be physically possible (i.e. with locally single-valued inverse) is that the Jacobian of (1.2) is greater than zero, (Boresi and Chong, 1999),

$$J = \begin{vmatrix} 1+u_{,x} & u_{,y} & u_{,z} \\ v_{,x} & 1+v_{,y} & v_{,z} \\ w_{,x} & w_{,y} & 1+w_{,z} \end{vmatrix} > 0. \tag{1.3}$$

The simplest types of displacement vector are *translation* (i.e. $\mathbf{u} = \mathbf{u}_0$, where $\mathbf{u}_0 = (u^0, v^0, w^0)$ is a constant vector) and *small rotation* about the origin (i.e. $\mathbf{u} = \bar{\omega}_\mathbf{0} \times \mathbf{x}$, where $\bar{\omega}_\mathbf{0} = (\omega_1^0, \omega_2^0, \omega_3^0)$ is a constant vector, see also Remark 1.1). These may be combined to form the basic type of deformation known as *rigid body displacements*. In fact, all bodies are to some extent deformable, and a rigid body deformation (or a "non-deformable" case) stands for an ideal case where the distance between every pair of points of the body remains invariant throughout the history of the body. In practice, any rigid deformation (that includes no relative displacement of material points) may be composed of three translation components and three rotation components, which are uniformly applied to all material particles. For example, in Cartesian coordinates where the point $\mathbf{x}_0 = (x_0, y_0, z_0)$ is fixed (i.e. belongs to the axis of rotation), the rigid body displacements u_r, v_r and w_r are expressed using linear functions of x, y, z as

$$u_r = \omega_2^0(z-z_0) - \omega_3^0(y-y_0) + u^0,$$
$$v_r = \omega_3^0(x-x_0) - \omega_1^0(z-z_0) + v^0,$$
$$w_r = \omega_1^0(y-y_0) - \omega_2^0(x-x_0) + w^0, \tag{1.4}$$

i.e., $\mathbf{u} = \bar{\omega}_\mathbf{0} \times (\mathbf{x} - \mathbf{x}_0) + \mathbf{u}_0$. Note that the Jacobian of the above rigid body deformation is equal to or greater than 1, since

$$J_r = \begin{vmatrix} 1 & -\omega_3^0 & \omega_2^0 \\ \omega_3^0 & 1 & -\omega_1^0 \\ -\omega_2^0 & \omega_1^0 & 1 \end{vmatrix} = 1 + \sum_{i=1}^{3} (\omega_i^0)^2. \tag{1.5}$$

In general, (1.1) in which u, v, w are linear functions of the coordinates constitute an *affine deformation*, see S.1.7.1.

A common way to present a field of small deformation in the elastic media is the so-called *vector field description*. Within this technique, at each point a vector that represents the direction and relative magnitude of the displacement is drawn, see Fig. 1.1(b) and examples in Chapters 7, 8.

Remark 1.1 Note that the $\mathbf{u} = \bar{\omega}_\mathbf{0} \times \mathbf{x}$ definition of displacement due to rotation holds for small rotation only. Finite rotation should be treated by transformation matrices as discussed in S.1.7.1.

1.1.2 Strain in Cartesian Coordinates

In order to derive proper deformation measures and define the strain components in a *Cartesian* coordinate system x, y, z, in addition to the material points $M(x,y,z)$, $M^*(\xi, \eta, \zeta)$ discussed earlier, we assume that a point N, which is infinitesimally close to M and located before deformation at $(x+dx, y+dy, z+dz)$ has moved to $N^*(\xi + d\xi, \eta + d\eta, \zeta + d\zeta)$, see Fig. 1.1(a). Let $ds = |MN|$ be the distance between the two points before deformation, and $ds^* = |M^*N^*|$ be the distance between the same material points after deformation. The squares of distances are therefore given by

$$(ds)^2 = (dx)^2 + (dy)^2 + (dz)^2, \qquad (ds^*)^2 = (d\xi)^2 + (d\eta)^2 + (d\zeta)^2. \tag{1.6}$$

We shall now establish a definition for suitable measures that will properly describe the deformation at the vicinity of the point M. The common *measure of deformation* is mathematically defined as $\frac{1}{2}\left[(ds^*)^2 - (ds)^2\right]$, and may be expressed as a sum of its components according to the resulting six different combinations of products of infinitesimal distances, namely,

$$\frac{1}{2}[(ds^*)^2 - (ds)^2] = \varepsilon_{xx}(dx)^2 + \varepsilon_{yy}(dy)^2 + \varepsilon_{zz}(dz)^2 + \varepsilon_{yz}\,dy\,dz + \varepsilon_{xz}\,dx\,dz + \varepsilon_{xy}\,dx\,dy. \tag{1.7}$$

The coefficients $\varepsilon_{\alpha\beta}$ are traditionally referred to as the *engineering* strain components at point M, and are given by

$$\begin{aligned}
&\varepsilon_{xx} = u_{,x} + \frac{1}{2}\left(u_{,x}^2 + v_{,x}^2 + w_{,x}^2\right), \quad \varepsilon_{yy} = v_{,y} + \frac{1}{2}\left(u_{,y}^2 + v_{,y}^2 + w_{,y}^2\right), \quad \varepsilon_{zz} = w_{,z} + \frac{1}{2}\left(u_{,z}^2 + v_{,z}^2 + w_{,z}^2\right),\\
&\varepsilon_{yz} = v_{,z} + w_{,y} + u_{,y}u_{,z} + v_{,y}v_{,z} + w_{,y}w_{,z}, \qquad \varepsilon_{xz} = u_{,z} + w_{,x} + u_{,x}u_{,z} + v_{,x}v_{,z} + w_{,x}w_{,z},\\
&\varepsilon_{xy} = u_{,y} + v_{,x} + u_{,y}u_{,x} + v_{,y}v_{,x} + w_{,y}w_{,x}.
\end{aligned} \tag{1.8}$$

A reduction of the above expressions leads to their linear version, $e_{\alpha\beta}$, namely,

$$e_{xx} = u_{,x}, \qquad e_{yy} = v_{,y}, \qquad e_{zz} = w_{,z}, \tag{1.9a}$$

$$e_{yz} = v_{,z} + w_{,y}, \qquad e_{xz} = u_{,z} + w_{,x}, \qquad e_{xy} = u_{,y} + v_{,x}. \tag{1.9b}$$

Occasionally, it is convenient to replace $\varepsilon_{\alpha\beta}$ $(\alpha \neq \beta)$ with $\gamma_{\alpha\beta}$ or to use a numerical index notation in which the strain vector $\{\varepsilon_{xx}, \varepsilon_{yy}, \varepsilon_{zz}, \varepsilon_{yz}, \varepsilon_{xz}, \varepsilon_{xy}\}$ is written as $\{\varepsilon_1, \varepsilon_2, \varepsilon_3, \varepsilon_4, \varepsilon_5, \varepsilon_6\}$.

The nonlinear version of $\varepsilon_{\alpha\beta}$ will be dealt with in what follows as we introduce a more consistent analysis of strain using tensor ("index") notation. For that purpose, we define $\{\mathbf{k}_i\}_{i=1,2,3}$ as the orthogonal unit vectors of the Cartesian basis (i.e. in the x, y and z directions, respectively). Here and in the following derivation, we shall use the index notation $x_1 \equiv x$, $x_2 \equiv y$, $x_3 \equiv z$. The position vector $\mathbf{r}(\mathbf{x}) = \mathbf{x}$ of the point M is written as $\mathbf{r} = \sum_{i=1}^{3} x_i \mathbf{k}_i$, while the displacement vector $\mathbf{u}(\mathbf{x})$ in the static (i.e. a time independent) case takes the form

$$\mathbf{u} = \sum\nolimits_{i=1}^{3} u_i(x_1, x_2, x_3)\,\mathbf{k}_i. \tag{1.10}$$

As already indicated, due to the elastic deformation, the point M is relocated to M^*, and the position vector of which $\mathbf{r}^*(\mathbf{x}) = \mathbf{x} + \mathbf{u}$ is expressed as $\mathbf{r}^* = \sum_{i=1}^{3}(x_i + u_i)\,\mathbf{k}_i$.

To determine the associated metric tensors, $\mathbf{g} = \{g_{ij}\}$ and $\mathbf{g}^* = \{g_{ij}^*\}$ before and after deformation, respectively, given by

$$g_{ij} = \mathbf{r}_{,i} \cdot \mathbf{r}_{,j}, \qquad g_{ij}^* = \mathbf{r}_{,i}^* \cdot \mathbf{r}_{,j}^*, \tag{1.11}$$

we expand the partial derivatives of $\mathbf{r}$ and $\mathbf{r}^*$ with respect to the coordinate system base $\mathbf{r}_{,i} = \mathbf{k}_i$, and similarly to (1.2),

$$\mathbf{r}_{,i}^* = \sum\nolimits_{j=1}^{3}(\delta_{ij} + u_{j,i})\,\mathbf{k}_j, \tag{1.12}$$

where δ_{ij} is Kronecker's symbol. Hence, in the Cartesian case under discussion, $g_{ij} = \delta_{ij}$, while

$$g_{ij}^* = \delta_{ij} + \underline{u_{i,j} + u_{j,i}} + \sum\nolimits_{m=1}^{3} u_{m,i}u_{m,j}. \tag{1.13}$$

By definition, the strain tensor components of the Lagrange-Green deformation measures are

$$\varepsilon_{ij} = \frac{1}{2}(g^*_{ij} - g_{ij}). \tag{1.14}$$

Thus, the linear (the underlined terms of (1.13)) and nonlinear expressions for strains in *tensor notation* are

$$e_{ij} = \frac{1}{2}(u_{i,j} + u_{j,i}), \tag{1.15a}$$

$$\varepsilon_{ij} = \frac{1}{2}(u_{i,j} + u_{j,i}) + \frac{1}{2}\sum_{m=1}^{3} u_{m,i}u_{m,j}. \tag{1.15b}$$

In particular one obtains $e_{ii} = u_{i,i}$ and $\varepsilon_{ii} = u_{i,i} + \frac{1}{2}\sum_{m=1}^{3}(u_{m,i})^2$. The resulting nonlinear strain tensor components may be also written as

$$\varepsilon_{ij} = e_{ij} + \frac{1}{2}\sum_{m=1}^{3}(e_{im} + \omega_{im})(e_{jm} + \omega_{jm}), \tag{1.16}$$

where ω_{ij} are the components of the antisymmetric ($\omega_{ji} = -\omega_{ij}$) *rotation tensor*, ω, while $\mathbf{e}$ and ω are commonly put in a matrix form as

$$\mathbf{e} = \begin{bmatrix} e_{11} & e_{12} & e_{13} \\ e_{12} & e_{22} & e_{23} \\ e_{13} & e_{23} & e_{33} \end{bmatrix}, \qquad \omega = \begin{bmatrix} 0 & \omega_{12} & \omega_{13} \\ -\omega_{12} & 0 & \omega_{23} \\ -\omega_{13} & -\omega_{23} & 0 \end{bmatrix}. \tag{1.17}$$

The rotation tensor components may also be defined by the rotation vector $\bar{\omega}$ as

$$\bar{\omega} = \frac{1}{2}\nabla \times \mathbf{u} = \omega_{23}\mathbf{k}_1 + \omega_{31}\mathbf{k}_2 + \omega_{12}\mathbf{k}_3. \tag{1.18}$$

In the above equation, the "nabla" operator is written as $\nabla = \sum_i \frac{\partial}{\partial x_i}\mathbf{k}_i$, and the vector cross-product operation yields $\omega_{ij} = \frac{1}{2}(u_{j,i} - u_{i,j})$, namely,

$$\omega_{23} = \frac{1}{2}(u_{3,2} - u_{2,3}), \qquad \omega_{31} = \frac{1}{2}(u_{1,3} - u_{3,1}), \qquad \omega_{12} = \frac{1}{2}(u_{2,1} - u_{1,2}). \tag{1.19}$$

By comparison with the expressions presented by (1.9a,b), we conclude that the strain components in tensor notation (which are written by indices as ε_{ij}, $i,j = 1,2,3$) and the strain components in engineering notation are related as the symmetric matrices

$$\begin{bmatrix} \varepsilon_{11} & \varepsilon_{12} & \varepsilon_{13} \\ \varepsilon_{12} & \varepsilon_{22} & \varepsilon_{23} \\ \varepsilon_{13} & \varepsilon_{23} & \varepsilon_{33} \end{bmatrix} = \begin{bmatrix} \varepsilon_x & \frac{1}{2}\varepsilon_{xy} & \frac{1}{2}\varepsilon_{xz} \\ \frac{1}{2}\varepsilon_{xy} & \varepsilon_y & \frac{1}{2}\varepsilon_{yz} \\ \frac{1}{2}\varepsilon_{xz} & \frac{1}{2}\varepsilon_{yz} & \varepsilon_z \end{bmatrix}. \tag{1.20}$$

The profound advantages of using the tensor notation will become clearer within the coordinate transformation techniques developed in S.1.3.3, S.1.3.4. In addition, it should be indicated that the above tensorial definition of the strain is very attractive in many analytical applications, as it only requires the ability to define the position vectors of a material point before and after deformation. Then, (1.11), (1.14) may be directly applied.

For the sake of clarification, we may now summarize all notation forms mentioned above for strain in Cartesian coordinates. These forms will be exploited as convenience requires throughout this book:

$$\begin{aligned} &\varepsilon_1 \equiv \varepsilon_{11} \equiv \varepsilon_{xx} \equiv \varepsilon_x, && \varepsilon_2 \equiv \varepsilon_{22} \equiv \varepsilon_{yy} \equiv \varepsilon_y, && \varepsilon_3 \equiv \varepsilon_{33} \equiv \varepsilon_{zz} \equiv \varepsilon_z, \\ &\varepsilon_4 \equiv 2\varepsilon_{23} \equiv \varepsilon_{yz} \equiv \gamma_{yz}, && \varepsilon_5 \equiv 2\varepsilon_{13} \equiv \varepsilon_{xz} \equiv \gamma_{xz}, && \varepsilon_6 \equiv 2\varepsilon_{12} \equiv \varepsilon_{xy} \equiv \gamma_{xy}, \\ &\omega_x \equiv \omega_1 \equiv \omega_{23}, && \omega_y \equiv \omega_2 \equiv \omega_{31}, && \omega_z \equiv \omega_3 \equiv \omega_{12}. \end{aligned} \tag{1.21}$$

P.1.1, P.1.2 (with $s = 0$) demonstrate a derivation of the strain components in Cartesian coordinates.

1.1.3 Strain in Orthogonal Curvilinear Coordinates

The derivation presented in this section for deformation measures of orthogonal curvilinear coordinates in Euclidean space E^3 is founded on the generic discussion of coordinate systems presented in S.1.7. The reader is therefore advised to become familiar with the mathematical aspects involved and the notation described in S.1.7. We define the deformation in orthogonal curvilinear coordinates using three functions $\{u_i(\alpha_1, \alpha_2, \alpha_3)\}_{i=1,2,3}$ and a local basis $\{\hat{\mathbf{k}}_i\}$, see (1.215), as

$$\mathbf{u} = \sum_i u_i(\alpha_1, \alpha_2, \alpha_3)\,\hat{\mathbf{k}}_i\,. \tag{1.22}$$

We also define in S.1.7 three functions, f_1, f_2, f_3, that convert curvilinear coordinates of E^3 space into Cartesian ones, see (1.210). By substituting $\hat{\mathbf{k}}_i$ of (1.215), one rewrites (1.22) as

$$\mathbf{u} = u_x\mathbf{k}_1 + u_y\mathbf{k}_2 + u_z\mathbf{k}_3, \tag{1.23}$$

where $u_\beta = u_\beta(u_1, u_2, u_3, \alpha_1, \alpha_2, \alpha_3)$, $\beta = x, y, z$. Since the displacement components u_1, u_2, u_3 are functions of $\alpha_1, \alpha_2, \alpha_3$ by themselves, one may also consider the form $u_\beta = u_\beta(\alpha_1, \alpha_2, \alpha_3)$. Subsequently, the position vectors of a point before and after deformation, $\mathbf{r}$ and $\mathbf{r}^* = \mathbf{r} + \mathbf{u}$, respectively, may be exploited to construct the metric tensors, see (1.11),

$$g_{ij} = \mathbf{r}_{,\alpha_i} \cdot \mathbf{r}_{,\alpha_j}, \qquad g^*_{ij} = \mathbf{r}^*_{,\alpha_i} \cdot \mathbf{r}^*_{,\alpha_j}, \tag{1.24}$$

for which $(ds)^2 = \sum_{ij} g_{ij}\,d\alpha_i\,d\alpha_j$ and $(ds^*)^2 = \sum_{ij} g^*_{ij}\,d\alpha_i\,d\alpha_j$. The Lagrange-Green definition of the strain tensor in this case becomes

$$\varepsilon_{ij} = \frac{g^*_{ij} - g_{ij}}{2H_iH_j}. \tag{1.25}$$

Note that $\mathbf{g}$ is always a diagonal matrix while *Lamé parameters*, H_i are defined as $H_i = \sqrt{g_{ii}}$. As an example, in cylindrical coordinates, where $(\alpha_1, \alpha_2, \alpha_3) = (\rho, \theta^c, z)$, in view of (1.218a) one obtains $g_{11} = 1$, $g_{22} = \rho^2$, $g_{33} = 1$, and $H_1 = 1$, $H_2 = \rho$, $H_3 = 1$, while

$$\begin{aligned} g^*_{\rho\rho} &= 1 + 2u_{\rho,\rho} + u^2_{\rho,\rho} + u^2_{\theta,\rho} + u^2_{z,\rho}, \\ g^*_{\rho\theta} &= u_{\theta,\rho}(\rho + u_{\theta,\theta} + u_\rho) + u_{z,\rho}u_{z,\theta} + (u_{\rho,\theta} - u_\theta)(1 + u_{\rho,\rho}), \qquad \text{etc.,} \end{aligned} \tag{1.26}$$

and hence,

$$\begin{aligned} \varepsilon_{11} &= \underline{u_{\rho,\rho}} + \frac{1}{2}(u^2_{\rho,\rho} + u^2_{\theta,\rho} + u^2_{z,\rho}), \\ \varepsilon_{12} &= \frac{1}{2\rho}(\underline{u_{\rho,\theta} - u_\theta + \rho\,u_{\theta,\rho}} + u_{z,\rho}\,u_{z,\theta} - u_{\rho,\rho}\,u_\theta + u_{\rho,\rho}\,u_{\rho,\theta} + u_{\theta,\rho}\,u_{\theta,\theta} + u_{\theta,\rho}\,u_\rho), \\ &\text{etc.,} \end{aligned} \tag{1.27}$$

where the linear terms are underlined. Any orthogonal coordinates may be directly incorporated into **P.1.1, P.1.2** to produce the associated strain expressions (in addition to the built-in systems in these programs).

Note that although originally derived for the Cartesian system, one may use (1.16) to express the nonlinear strain components in terms of their linear parts and the rotation vector components in the case under discussion as well.

To express e_{ij} for the present case (i.e. the linear parts of ε_{ij}), we define an operation, which will be denoted "cyc-ijk", as *the operation in which we create two additional equations out of a given equation by "forward" replacement of the indices for the first equation, namely,* $i \rightarrow j$,

$j \to k$, $k \to i$, and "backwards" replacement for the second equation, $i \to k$, $j \to i$, $k \to j$. Hence, by applying cyc-123 to the following two equations, one obtains the required six e_{ij} components

$$e_{11} = \frac{1}{H_1} u_{1,\alpha_1} + \frac{1}{H_1 H_2} H_{1,\alpha_2} u_2 + \frac{1}{H_1 H_3} H_{1,\alpha_3} u_3,$$
$$e_{12} = \frac{H_2}{H_1} \left(\frac{u_2}{H_2} \right)_{,\alpha_1} + \frac{H_1}{H_2} \left(\frac{u_1}{H_1} \right)_{,\alpha_2}. \tag{1.28}$$

The rotation tensor components are extracted from

$$\bar{\omega} = \frac{1}{2} \nabla \times \mathbf{u} = \frac{1}{2 H_1 H_2 H_3} \begin{vmatrix} H_1 \hat{\mathbf{k}}_1 & H_2 \hat{\mathbf{k}}_2 & H_3 \hat{\mathbf{k}}_3 \\ \frac{\partial}{\partial \alpha_1} & \frac{\partial}{\partial \alpha_2} & \frac{\partial}{\partial \alpha_3} \\ H_1 u_1 & H_2 u_2 & H_3 u_3 \end{vmatrix}, \tag{1.29}$$

(note that the "nabla" operator in this case is $\nabla = \sum_i \frac{1}{H_i} \frac{\partial}{\partial \alpha_i} \hat{\mathbf{k}}_i$), which in view of (1.18) yields (apply cyc-123)

$$\omega_{ij} = \frac{1}{2 H_i H_j} [(H_j u_j)_{,\alpha_i} - (H_i u_i)_{,\alpha_j}]. \tag{1.30}$$

In some applications, it is useful to define the unit vectors in the coordinate line directions of the deformed state, namely

$$\hat{\mathbf{k}}_i^* = \mathbf{r}^*_{,\alpha_i} / |\mathbf{r}^*_{,\alpha_i}|. \tag{1.31}$$

1.1.4 Physical Interpretation of Strain Components

1.1.4.1 Relative Extension and Angle Change

So far, the deformation measures ε_{ij} have been defined mathematically, and were shown to provide a set of six parameters that reflect the deformation at a given material point. To gain some physical insight and clearer interpretation of the above discussed measures, we will first define a more physically-based deformation measure known as the "relative extension", which is essentially the ratio of the change of distances between two adjoint points to the initial distance, namely: $E_{MN} = \frac{ds^* - ds}{ds}$, see Fig. 1.1(a). While trying to relate the above expression to the previously derived strain components we first note that

$$\frac{(ds^*)^2 - (ds)^2}{2} = (\frac{1}{2} E_{MN}^2 + E_{MN})(ds)^2. \tag{1.32}$$

We subsequently substitute the r.h.s. of (1.7) in the above equation and express the relative extension as

$$\underline{\frac{1}{2} E_{MN}^2} + E_{MN} = \varepsilon_x \cos(\bar{\mathbf{s}}, x)^2 + \varepsilon_y \cos(\bar{\mathbf{s}}, y)^2 + \varepsilon_z \cos(\bar{\mathbf{s}}, z)^2$$
$$+ \varepsilon_{yz} \cos(\bar{\mathbf{s}}, y) \cos(\bar{\mathbf{s}}, z) + \varepsilon_{xz} \cos(\bar{\mathbf{s}}, x) \cos(\bar{\mathbf{s}}, z) + \varepsilon_{xy} \cos(\bar{\mathbf{s}}, x) \cos(\bar{\mathbf{s}}, y), \tag{1.33}$$

where the underlined term may be neglected for small strain. Note that the direction cosines between a material element $\bar{\mathbf{s}} = \vec{MN}$ that is placed at a generic orientation and the (say Cartesian) axes are defined as $\cos(\bar{\mathbf{s}}, \alpha) = d\alpha/ds$ ($\alpha = x, y, z$). As a special case, the above result shows that the relative extensions (elongations at the point M) in the x, y, z directions are given by (apply cyc-xyz)

$$\frac{1}{2} E_x^2 + E_x = \varepsilon_x \qquad \text{or} \qquad E_x = \sqrt{1 + 2\varepsilon_x} - 1, \tag{1.34}$$

while as already indicated, for the small strain case, $E_x \cong \varepsilon_x$. In addition, one may use (1.12), (1.15b), (1.34) to show that (apply cyc-xyz)

$$\left|\mathbf{r}^*_{,x}\right| = 1 + E_x. \tag{1.35}$$

Thus, we may draw two important conclusions from the above discussion. First, as shown by (1.33), all strain components play a role in the determination of the relative extension in an arbitrary direction. Secondly, and a more specific conclusion, is that the $\varepsilon_{\alpha\alpha}$ strain components, may be viewed as *the relative extension in* α *direction* for small strain values ($\varepsilon_{\alpha\alpha} \gg \varepsilon^2_{\alpha\alpha}/2$), while for the large strain case, the relative extension of the coordinate axes is given by (1.34).

The above is completely applicable for curvilinear coordinates as well. Thus, (1.33) may be rewritten by replacing x, y, z with $1, 2, 3$ and $\varepsilon_{\alpha\beta}$ with $2\varepsilon_{ij}$. The direction cosines will now represent the angles from the element to the curvilinear directions, namely, $\cos(\bar{\mathbf{s}}, \mathbf{k}_i)$. Subsequently, the relative extensions along the coordinate lines are

$$E_i = \sqrt{1 + 2\varepsilon_{ii}} - 1. \tag{1.36}$$

We continue the physical interpretation of the strain component by looking at the angle change (caused by the deformation), $\Delta\varphi_{\alpha\beta}$, between two unit vectors, $\mathbf{k}^*_\alpha$ and the $\mathbf{k}^*_\beta$, which clearly were perpendicular before deformation (where they were denoted $\mathbf{k}_\alpha$ and the $\mathbf{k}_\beta$, respectively). For example, in Cartesian coordinates, the angle between $\mathbf{k}^*_x$ and $\mathbf{k}^*_y$ is given by

$$\mathbf{k}^*_x \cdot \mathbf{k}^*_y = \cos(\mathbf{k}^*_x, \mathbf{k}^*_y) = \cos\left(\frac{\pi}{2} - \Delta\varphi_{xy}\right) = \sin(\Delta\varphi_{xy}). \tag{1.37}$$

Therefore, with the aid of (1.12), (1.34), (1.35) one may write

$$\mathbf{k}^*_x = \frac{(1 + u_{,x})\,\mathbf{k}_x + v_{,x}\mathbf{k}_y + w_{,x}\mathbf{k}_z}{\sqrt{1 + 2\varepsilon_x}}, \qquad \mathbf{k}^*_y = \frac{u_{,y}\mathbf{k}_x + (1 + v_{,y})\,\mathbf{k}_y + w_{,y}\mathbf{k}_z}{\sqrt{1 + 2\varepsilon_y}}. \tag{1.38}$$

Hence, in view of (1.15b) in a tensorial notation we obtain

$$\sin\Delta\varphi_{ij} = \frac{2\varepsilon_{ij}}{\sqrt{1 + 2\varepsilon_{ii}}\sqrt{1 + 2\varepsilon_{jj}}}, \tag{1.39}$$

which is applicable for curvilinear coordinates as well. It is therefore shown that for small strain $\Delta\varphi_{\alpha\beta} \cong \varepsilon_{\alpha\beta}$, or more generally, $\Delta\varphi_{ij} \cong 2\varepsilon_{ij}$. Thus, *in the linear case,* ε_{ij} *may be viewed as half the change of angle in the i, j-plane caused by the deformation.*

1.1.4.2 Relative Change in Volume

The relative change in volume is an additional measure that may be expressed by the strain components and therefore serves also as a physical interpretation of these components.

The volume of an infinitesimal cubic element before deformation is $dV = dx\,dy\,dz$. Due to the deformation, an infinitesimal cubic element is deformed by (1.2) into a parallelepiped, the volume of which is given by $dV^* = J\,dx\,dy\,dz$, see (1.3). Using (1.15b) with engineering notation for Cartesian coordinates, one may write

$$\left(\frac{dV^*}{dV}\right)^2 = \begin{vmatrix} 1 + 2\varepsilon_x & \varepsilon_{xy} & \varepsilon_{xz} \\ \varepsilon_{xy} & 1 + 2\varepsilon_y & \varepsilon_{yz} \\ \varepsilon_{xz} & \varepsilon_{yz} & 1 + 2\varepsilon_z \end{vmatrix} = 1 + 2\varepsilon_x + 2\varepsilon_y + 2\varepsilon_z + 4\varepsilon_y\varepsilon_z + 4\varepsilon_x\varepsilon_z$$
$$+ 4\varepsilon_x\varepsilon_y + 8\varepsilon_x\varepsilon_y\varepsilon_z + \underline{2\varepsilon_{xy}\varepsilon_{xz}\varepsilon_{yz} - \varepsilon_{yz}{}^2 - 2\varepsilon_{yz}{}^2\varepsilon_x - \varepsilon_{xz}{}^2 - 2\varepsilon_{xz}{}^2\varepsilon_y - \varepsilon_{xy}{}^2 - 2\varepsilon_{xy}{}^2\varepsilon_z}. \tag{1.40}$$

Clearly, if one uses the above result for the state of principal strain (that will be derived within S.1.3.4), the underlined terms in (1.40) vanish.

Alternatively, by employing the strain tensor invariants Ξ_1, Ξ_2, Ξ_3 of S.1.3.4, one may write

$$\left(\frac{dV^*}{dV}\right)^2 = 1 + 2\Xi_1 \underline{+ 4\Xi_2 + 8\Xi_3}. \tag{1.41}$$

Hence, for small strain, the underlined terms in the above equation may be neglected, and the relative change in volume, $dV^*/dV - 1$, becomes a simple invariant of the coordinate system orientation, namely,

$$\frac{dV^* - dV}{dV} \cong \varepsilon_x + \varepsilon_y + \varepsilon_z = \Xi_1. \tag{1.42}$$

For curvilinear coordinates, (1.40) should be written by replacing $\varepsilon_{\alpha\alpha}$ with ε_{ii} and $\varepsilon_{\alpha\beta}$ with $2\varepsilon_{ij}$ where $\alpha, \beta \in \{x,y,z\}$, $i, j \in \{1,2,3\}$.

1.2 Displacement by Strain Integration

In many occasions, analytical solution methods yield expressions of the strain distributions with no previous use or inclusion of the displacement field. When such distributions are known, various "integration" steps should be carried out in order to create the displacement components. Two approaches to the general derivation of these steps will be developed in what follows. Yet, before any integration steps are taken, one needs to verify that the equations are integrable. For that purpose we shall first develop the compatibility equations.

1.2.1 Compatibility Equations

One of the fundamental sets of governing equations in the theory of elasticity is known as the *compatibility equations*. To clarify the role and origin of these equations, we shall first restrict ourselves to small displacements in Cartesian coordinates, where in order to solve a given problem, one presumes a set of analytical forms for the six strain components. Clearly, the six strain components $\varepsilon = \{\varepsilon_{ij}\}_{j \geq i = 1,2,3}$ cannot be selected arbitrarily as functions of the Cartesian coordinates $\mathbf{x} = \{x_i\}$, as they are determined completely by only three displacement components $\mathbf{u} = \{u_i\}_{i=1,2,3}$, see (1.15b). The required additional relations are known as the *compatibility equations*, and, as already mentioned, in some mathematical contexts they are also referred to as the "integrability conditions".

Prior to the general case discussion, we shall present the linear reduction of these equations, which is founded on the six independent components of the linear strain tensor, $\mathbf{e} = \{e_{ij}\}_{i,j=1,2,3}$, of (1.15a). By taking second derivatives of $\mathbf{e}$ in Cartesian coordinates one may write the following identities:

$$e_{mn,ij} + e_{ij,mn} = e_{im,jn} + e_{jn,im}, \qquad i, j, m, n \in \{1,2,3\}. \tag{1.43}$$

Only six out of the above 81 equations are independent, for example, those obtained by the following index sets:

$$(mnij) = (1212),\ (2323),\ (3131),\ (1213),\ (2321),\ (3132). \tag{1.44}$$

Subsequently, following (1.43), in engineering notation, the (linear) compatibility equations in Cartesian coordinates are

$$\gamma_{xy,xy} = \varepsilon_{x,yy} + \varepsilon_{y,xx}, \tag{1.45a}$$

$$\gamma_{yz,yz} = \varepsilon_{y,zz} + \varepsilon_{z,yy}, \tag{1.45b}$$

$$\gamma_{xz,xz} = \varepsilon_{x,zz} + \varepsilon_{z,xx}, \tag{1.45c}$$

$$2\varepsilon_{x,yz} = \gamma_{xz,xy} + \gamma_{xy,xz} - \gamma_{yz,xx}, \tag{1.45d}$$

$$2\varepsilon_{y,xz} = \gamma_{xy,yz} + \gamma_{yz,xy} - \gamma_{xz,yy}, \tag{1.45e}$$

$$2\varepsilon_{z,xy} = \gamma_{yz,xz} + \gamma_{xz,yz} - \gamma_{xy,zz}. \tag{1.45f}$$

For an x, y-plane deformation, only one independent equation, (1.45a), is obtained and may be written in index notation with $i = j = 2$, $m = n = 1$ (the first case of (1.44)) as

$$e_{11,22} + e_{22,11} = 2e_{12,12}. \tag{1.46}$$

To facilitate the discussion of the complete nonlinear strain analysis case, we shall define the compatibility equations as *the conditions imposed on the Lagrange-Green tensor, that guarantee the existence of a unique displacement solution when the fully nonlinear strain expressions are utilized.*

To derive the desired conditions for curvilinear coordinates we shall employ the metric tensors in the undeformed and deformed configuration, previously denoted as $\mathbf{g}$ and $\mathbf{g}^*$, respectively. We shall also make use of (1.25), which shows that, for generic orthogonal curvilinear coordinates (in Euclidean space), the metric tensor after deformation is expressed as

$$g^*_{ij} = g_{ij} + 2\varepsilon_{ij} H_i H_j. \tag{1.47}$$

The key to the development of the compatibility conditions is the fact that, similar to the undeformed configuration that occupies a part of a Euclidean space of a given topology while its *curvature tensor* vanishes, the curvature tensor of the deformed configuration must vanish as well *since the deformed body occupies (again) a part of a Euclidean space while preserving the domain topology* (namely, a multiply connected domain of any level will be preserved as such). We therefore adopt the Riemann-Christoffel curvature tensor of (1.213), see S.1.7.2.2, for $\mathbf{g}^*$, and the associated 81 complete compatibility equations become

$$R^*_{mnij} = 0, \qquad i, j, m, n \in \{1, 2, 3\}, \tag{1.48}$$

where $R^*_{mnij} = \frac{1}{2}(g^*_{mj,ni} + g^*_{in,mj} - g^*_{mi,nj} - g^*_{jn,im}) - g^{*fh}(\Gamma^*_{f,im}\Gamma^*_{h,jn} - \Gamma^*_{f,ni}\Gamma^*_{h,jm})$ and $\Gamma^*_{p,ms} = \frac{1}{2}(g^*_{mp,s} + g^*_{ps,m} - g^*_{ms,p})$. Note that (1.47) enables us to write the above equations in terms of strain components. As already indicated, only six of these equations are independent, see (1.44).

P.1.3, P.1.4 are capable of producing the fully *nonlinear compatibility equations* for both three- and two-dimensional deformation fields. For example, the compatibility equation for a two-dimensional case in Cartesian coordinates is

$$\begin{aligned} 2\varepsilon_{12,12} - \varepsilon_{11,22} - \varepsilon_{22,11} = {} & g^{*11}[\varepsilon_{11,1}(\varepsilon_{12,2} - \varepsilon_{22,1}) - \varepsilon^2_{11,2}] + g^{*22}(\varepsilon_{22,2}(\varepsilon_{12,1} - \varepsilon_{11,2}) - \varepsilon^2_{22,1}) \\ & + g^{*12}[(\varepsilon_{12,1} - \varepsilon_{11,2})(\varepsilon_{11,2} - \varepsilon_{22,1}) + \varepsilon_{11,1}\varepsilon_{22,2} - 2\varepsilon_{11,2}\varepsilon_{22,1}], \end{aligned} \tag{1.49}$$

where

$$g^{*11} = \frac{1 + 2\varepsilon_{22}}{\widetilde{D}}, \qquad g^{*12} = -\frac{2\varepsilon_{12}}{\widetilde{D}}, \qquad g^{*22} = \frac{1 + 2\varepsilon_{11}}{\widetilde{D}}, \tag{1.50}$$

and $\widetilde{D} = 1 + 2\varepsilon_{22} + 2\varepsilon_{11} + 4\varepsilon_{11}\varepsilon_{22} - 4\varepsilon^2_{12}$. Assuming that all strains and their derivatives are small compared with unity, one may linearize (1.49) and reach the compatibility equation, (1.46). Note that for this level of accuracy, the expressions for the strain components must be linearized as well, i.e. only the substitution of $\varepsilon_{ij} \approx e_{ij}$ would be consistent. As another example, we present the linearized compatibility equation in polar coordinates, (ρ, θ)

$$\varepsilon_{\rho\rho,\theta\theta} - 2\rho\,\varepsilon_{\rho\theta,\rho\theta} + 2\rho\,\varepsilon_{\theta\theta,\rho} + \rho^2\varepsilon_{\theta\theta,\rho\rho} - \rho\,\varepsilon_{\rho\rho,\rho} - 2\varepsilon_{\rho\theta,\theta} = 0. \tag{1.51}$$

The fully nonlinear compatibility equations for other cases (including three-dimensional cases of generic orthogonal curvilinear coordinates) are very lengthy, and as already indicated may be obtained in full by activating **P.1.3, P.1.4**.

1.2.2 Continuous Approach

To derive continuous expressions for the strain components integration, we shall be focused on the linear strain components in a Cartesian coordinate system (i.e. the case of $\varepsilon_{ij} = e_{ij}$, see S.1.1.2), as no general analysis may be drawn for a general nonlinear case. Although similar procedures may be carried out for any curvilinear coordinates along the same lines, yet in practice, it is more convenient to transform the strain components into Cartesian coordinates (see S.1.3.4) and integrate them there (and if necessary, transform the resulting displacements back to the curvilinear coordinates).

It is important to reiterate and state that the underlying assumption in the following derivation is the fulfillment of the compatibility equations by the strain components. Otherwise, the system is not integrable, and the three displacement components u, v and w can not be consistently extracted from the six strain components (see discussion in S.1.2.1).

In what follows, we shall assume that all strain components, in engineering notation as described by (1.9a,b), namely, ε_x, ε_y, ε_z, γ_{yz}, γ_{xz} and γ_{xy}, are known as general functions of x, y and z. In addition, the rigid body displacements and rotation components are given as the values of the three displacement components $u = u^0$, $v = v^0$, $w = w^0$ and the three rotation components $\omega_x = \omega_x^0$, $\omega_y = \omega_y^0$ and $\omega_z = \omega_z^0$ at the system origin point, P_0 (i.e. at $x = y = z = 0$). Modifying the resulting solution for a case where the rigid body components are defined at other locations is simple.

As a *first step* we determine each component of the rotation vector from its three given partial derivatives

$$\omega_{i,j} = f_{ij}(x,y,z), \qquad i,j \in \{x,y,z\}, \tag{1.52}$$

where

$$f_{xx} = \frac{1}{2}(\gamma_{xz,y} - \gamma_{xy,z}), \qquad f_{xy} = \frac{1}{2}\gamma_{yz,y} - \varepsilon_{y,z}, \qquad f_{xz} = \varepsilon_{z,y} - \frac{1}{2}\gamma_{yz,z}, \tag{1.53a}$$

$$f_{yx} = \varepsilon_{x,z} - \frac{1}{2}\gamma_{xz,x}, \qquad f_{yy} = \frac{1}{2}(\gamma_{xy,z} - \gamma_{yz,x}), \quad f_{yz} = \frac{1}{2}\gamma_{xz,z} - \varepsilon_{z,x}, \tag{1.53b}$$

$$f_{zx} = \frac{1}{2}\gamma_{xy,x} - \varepsilon_{x,y}, \qquad f_{zy} = \varepsilon_{y,x} - \frac{1}{2}\gamma_{xy,y}, \qquad f_{zz} = \frac{1}{2}(\gamma_{yz,x} - \gamma_{xz,y}). \tag{1.53c}$$

Hence, since the rotations at $P_0(0,0,0)$ are known, those of another point, say $P(x,y,z)$, may be presented as

$$\omega_i = \omega_i^0 + \int_{P_0}^{P} (f_{ix}\,dx + f_{iy}\,dy + f_{iz}\,dz), \qquad i \in \{x,y,z\}. \tag{1.54}$$

Note that the expressions under the above integrals are complete differentials in view of

$$f_{ij,k} = f_{ik,j}, \qquad i,j,k \in \{x,y,z\}, \tag{1.55}$$

which are essentially equivalent to the compatibility equations, (1.45a–f).

As a *second step* we determine each component of displacement from its three given partial derivatives

$$u_{,x}=\varepsilon_x, \qquad u_{,y}=\frac{1}{2}\gamma_{xy}-\omega_z, \qquad u_{,z}=\frac{1}{2}\gamma_{xz}+\omega_y, \tag{1.56a}$$

$$v_{,y}=\varepsilon_y, \qquad v_{,z}=\frac{1}{2}\gamma_{yz}-\omega_x, \qquad v_{,x}=\frac{1}{2}\gamma_{xy}+\omega_z, \tag{1.56b}$$

$$w_{,z}=\varepsilon_z, \qquad w_{,x}=\frac{1}{2}\gamma_{xz}-\omega_y, \qquad w_{,y}=\frac{1}{2}\gamma_{yz}+\omega_x. \tag{1.56c}$$

Similar to the rotation case, when the displacement components at P_0 are known, those of point P may be expressed as

$$u=u^0+\int_{P_0}^{P}[\varepsilon_x\,dx+(\frac{1}{2}\gamma_{xy}-\omega_z)\,dy+(\frac{1}{2}\gamma_{xz}+\omega_y)\,dz], \tag{1.57a}$$

$$v=v^0+\int_{P_0}^{P}[(\frac{1}{2}\gamma_{xy}+\omega_z)\,dx+\varepsilon_y\,dy+(\frac{1}{2}\gamma_{yz}-\omega_x)\,dz], \tag{1.57b}$$

$$w=w^0+\int_{P_0}^{P}[(\frac{1}{2}\gamma_{xz}-\omega_y)\,dx+(\frac{1}{2}\gamma_{yz}+\omega_x)\,dy+\varepsilon_z\,dz]. \tag{1.57c}$$

Again, the expressions under the integrals are complete differentials in view of the compatibility equations. The above procedure may be executed for a given consistent set of strain functions by activating **P.1.5**.

Remark 1.2 The procedure described above may be reduced to the two-dimensional case, where $w=0$ and the displacements u, v are functions of x,y only, all strain components that include the z index vanish (namely, $\gamma_{xz}=\gamma_{yz}=\varepsilon_z=0$), and the remaining components $\varepsilon_x(x,y)$, $\varepsilon_y(x,y)$, $\gamma_{xy}(x,y)$ satisfy compatibility (1.45a). As a first step, we determine the z component of the rotation vector, $\omega_z(x,y)$ (obviously, $\omega_x=\omega_y=0$ in this case), from its two given partial derivatives, see (1.54) with $i=3$,

$$\begin{aligned}\omega_z&=\omega_z^0+\int_{P_0}^{P}f_{zx}\,dx+f_{zy}\,dy\\&=\omega_z^0-\int_0^x\varepsilon_{x,y}\,dx+\int_0^y\varepsilon_{y,x}(0,y)\,dy+\frac{1}{2}\gamma_{xy}-\gamma_{xy}(0,y)+\frac{1}{2}\gamma_{xy}(0,0).\end{aligned} \tag{1.58}$$

As a second step, we determine each component of the displacements $u(x,y)$ and $v(x,y)$ from their two given x- and y- partial derivatives, see (1.56a,b), which yields

$$u=u_0+\int_{P_0}^{P}[\varepsilon_x\,dx+(\frac{1}{2}\gamma_{xy}-\omega_z)\,dy], \qquad v=v_0+\int_{P_0}^{P}[\varepsilon_y\,dy+(\frac{1}{2}\gamma_{xy}+\omega_z)\,dx]. \tag{1.59}$$

This procedure may be executed for a given consistent set of strain functions by activating **P.1.6**.

1.2.3 Level Approach

Many problems in elasticity may be analyzed by series expansion of the involved expressions with respect to one of the coordinate systems. Beam analyses, see Chapters 6–9, are classic examples for schemes where all quantities may be expanded as Taylor series of the longitudinal coordinate, say z. Such representation of the involved expressions may be exploited to substantially simplify the integration process described in S.1.2.2.

In general, we shall expand all spatial functions as *truncated* polynomials of degree $K_m \geq 0$ of the axial variable, z, and as continuous functions of the cross-section variables x and y. For any generic function $\widetilde{G}(x,y,z)$ written as

$$\widetilde{G}(x,y,z) = \sum_{k=0}^{K_m} \widetilde{G}^{(k)}(x,y)z^k, \tag{1.60}$$

we shall refer to K_m as the *expansion degree*, and to $\widetilde{G}^{(k)}$ as the $k\underline{th}$ (level) *component* of $\widetilde{G}$. Clearly, $\widetilde{G}$ may be integrated and differentiated with respect to z as

$$\int_0^z \widetilde{G}dz = \sum_{k=0}^{K_m} \widetilde{G}^{(k)} \frac{z^{k+1}}{k+1}, \qquad \widetilde{G}_{,z} = \sum_{k=0}^{K_m-1} (k+1)\widetilde{G}^{(k+1)} z^k. \tag{1.61}$$

To derive the present approach, we also assume that the strain components are truncated as

$$\varepsilon_i = \sum_{k=0}^{K_m} \varepsilon_i^{(k)}(x,y)z^k. \tag{1.62}$$

In cases where the strain components are not given as in (1.62), one may employ a standard Taylor series expansion, and therefore, in some cases the above truncated form represents an approximation. We subsequently expect the displacements and rotations to be expressed in levels as well, namely

$$u = \sum_{k=0}^{K_m} u^{(k)}(x,y)z^k, \qquad v = \sum_{k=0}^{K_m} v^{(k)}(x,y)z^k, \qquad w = \sum_{k=0}^{K_m} w^{(k)}(x,y)z^k, \tag{1.63a}$$

$$\omega_i = \sum_{k=0}^{K_m} \omega_i^{(k)}(x,y)z^k, \qquad i = 1,2,3. \tag{1.63b}$$

In what follows, K_m stands for the maximal expansion degree of all analysis components (clearly, the expansion degree of the strain components will be less than those of the displacement components). The f_{ij} functions of (1.53a–c) are also written in levels as $f_{ij} = \sum_{k=0}^{K_m} f_{ij}^{(k)}(x,y)z^k$, while

$$f_{xx}^{(k)} = \frac{1}{2}[\gamma_{xz,y}^{(k)} - (k+1)\gamma_{xy}^{(k+1)}], \; f_{xy}^{(k)} = \frac{1}{2}\gamma_{yz,y}^{(k)} - (k+1)\varepsilon_y^{(k+1)}, \; f_{xz}^{(k)} = \varepsilon_{z,y}^{(k)} - \frac{1}{2}(k+1)\gamma_{yz}^{(k+1)},$$

$$f_{yx}^{(k)} = (k+1)\varepsilon_x^{(k+1)} - \frac{1}{2}\gamma_{xz,x}^{(k)}, \; f_{yy}^{(k)} = \frac{1}{2}[(k+1)\gamma_{xy}^{(k+1)} - \gamma_{yz,x}^{(k)}], \; f_{yz}^{(k)} = \frac{1}{2}(k+1)\gamma_{xz}^{(k+1)} - \varepsilon_{z,x}^{(k)},$$

$$f_{zx}^{(k)} = \frac{1}{2}\gamma_{xy,x}^{(k)} - \varepsilon_{x,y}^{(k)}, \qquad f_{zy}^{(k)} = \varepsilon_{y,x}^{(k)} - \frac{1}{2}\gamma_{xy,y}^{(k)}, \qquad f_{zz}^{(k)} = \frac{1}{2}(\gamma_{yz,x}^{(k)} - \gamma_{xz,y}^{(k)}). \tag{1.64}$$

By selecting the polygonal integration trajectory $(0,0,0) \to (0,y,0) \to (x,y,0) \to (x,y,z)$, we write (1.54) as

$$\omega_i = \omega_i^0 + \int_0^x f_{ix}^{(0)}(x,y)dx + \int_0^y f_{iy}^{(0)}(0,y)dy + \sum_{k=0}^{K_m} f_{xz}^{(k)}(x,y)\frac{z^{k+1}}{k+1}, \qquad i \in \{x,y,z\}. \tag{1.65}$$

Hence, the level components of the rotations are

$$\omega_i^{(0)} = \omega_i^0 + \int_0^x f_{ix}^{(0)}(x,y)\,dx + \int_0^y f_{iy}^{(0)}(0,y)\,dy, \qquad \omega_i^{(k)} = \frac{1}{k} f_{iz}^{(k-1)}(x,y) \tag{1.66}$$

where $k \geq 1$, $i \in \{x,y,z\}$. At this stage, (1.56a–c) show that

$$u_{,x}^{(k)} = \varepsilon_{xx}^{(k)}, \qquad u_{,y}^{(k)} = \frac{1}{2}\gamma_{xy}^{(k)} - \omega_z^{(k)}, \qquad u_{,z}^{(k)} = \frac{1}{2}\gamma_{xz}^{(k)} + \omega_y^{(k)}, \tag{1.67a}$$

$$v_{,y}^{(k)} = \varepsilon_{yy}^{(k)}, \qquad v_{,z}^{(k)} = \frac{1}{2}\gamma_{yz}^{(k)} - \omega_x^{(k)}, \qquad v_{,x}^{(k)} = \frac{1}{2}\gamma_{xy}^{(k)} + \omega_z^{(k)}, \tag{1.67b}$$

$$w_{,z}^{(k)} = \varepsilon_{zz}^{(k)}, \qquad w_{,x}^{(k)} = \frac{1}{2}\gamma_{xz}^{(k)} - \omega_y^{(k)}, \qquad w_{,y}^{(k)} = \frac{1}{2}\gamma_{yz}^{(k)} + \omega_x^{(k)}, \tag{1.67c}$$

and therefore, by analogy with the rotation integration process, one may write

$$u^{(0)} = u^0 + \int_0^x \varepsilon_{xx}^{(0)}\,dx + \int_0^y (\frac{1}{2}\gamma_{xy}^{(0)} - \omega_z^{(0)})(0,y)\,dy, \qquad u^{(k)} = \frac{1}{k}\frac{1}{2}(\frac{1}{2}\gamma_{xz}^{(k-1)} + \omega_y^{(k-1)}), \quad (1.68)$$

$$v^{(0)} = v^0 + \int_0^x [\frac{1}{2}\gamma_{xy}^{(0)} + \omega_z^{(0)}]\,dx + \int_0^y \varepsilon_{yy}^{(0)}(0,y)\,dy, \qquad v^{(k)} = \frac{1}{k}(\frac{1}{2}\gamma_{yz}^{(k-1)} - \omega_x^{(k-1)}),$$

$$w^{(0)} = w^0 + \int_0^x [\frac{1}{2}\gamma_{xz}^{(0)} - \omega_y^{(0)}]\,dx + \int_0^y (\frac{1}{2}\gamma_{yz}^{(0)} + \omega_x^{(0)})(0,y)\,dy, \qquad w^{(k)} = \frac{1}{k}\varepsilon_{zz}^{(k-1)}$$

where $k \geq 1$.

Remark 1.3 An alternative way to handle the strain integration for level-based solution calls for exploiting the strain-displacement relations of (1.9a,b) to show that

$$\varepsilon_x^{(k)} = u_{,x}^{(k)}, \qquad \varepsilon_y^{(k)} = v_{,y}^{(k)}, \qquad \gamma_{xy}^{(k)} = u_{,y}^{(k)} + v_{,x}^{(k)}, \quad (1.69a)$$

$$\varepsilon_z^{(k)} = (k+1)w^{(k+1)}, \quad \gamma_{yz}^{(k)} = (k+1)v^{(k+1)} + w_{,y}^{(k)}, \quad \gamma_{xz}^{(k)} = (k+1)u^{(k+1)} + w_{,x}^{(k)}. \quad (1.69b)$$

Thus, one can extract the displacement components from (1.69b) as

$$w^{(k+1)} = \frac{1}{k+1}\varepsilon_z^{(k)}, \quad (1.70a)$$

$$u^{(k+2)} = \frac{1}{k+2}(\gamma_{xz}^{(k+1)} - \frac{1}{k+1}\varepsilon_{z,x}^{(k)}), \qquad v^{(k+2)} = \frac{1}{k+2}(\gamma_{yz}^{(k+1)} - \frac{1}{k+1}\varepsilon_{z,y}^{(k)}), \quad (1.70b)$$

$$u^{(1)} = \gamma_{xz}^{(0)} - w_{,x}^{(0)}, \qquad v^{(1)} = \gamma_{yz}^{(0)} - w_{,y}^{(0)}, \quad (1.70c)$$

using the known strain components and the zero level of displacements obtained from (1.69a) with $k = 0$. In view of (1.19) one may also extract the rotations components as

$$\omega_1^{(k)} = \frac{1}{2}(w_{,y}^{(k)} - (k+1)v^{(k+1)}), \quad \omega_2^{(k)} = \frac{1}{2}((k+1)u^{(k+1)} - w_{,x}^{(k)}), \quad \omega_3^{(k)} = \frac{1}{2}(v_{,x}^{(k)} - u_{,y}^{(k)}). \quad (1.71)$$

To execute the above approach one should pursue the following steps:

(1) Calculate $w^{(k)}$ for all $k > 0$ levels by (1.70a).

(2) Calculate $u^{(k)}$, $v^{(k)}$ for all $k > 0$ levels by (1.70b).

(3) Calculate $\{\omega_i^{(0)}\}_{,i=1,2,3}$ from (1.66), including the introduction of the rigid body rotations $\{\omega_i^0\}_{i=1,2,3}$. This step is analogous to (1.54) with $f_{xz} = f_{yz} = f_{zz} = 0$.

(4) Calculate $u^{(0)}$, $v^{(0)}$, $w^{(0)}$ from (1.68), including the introduction of the rigid body displacements u^0, v^0, w^0. This step is analogous to (1.57a–c) with $u_{,z} = v_{,z} = w_{,z} = 0$.

(5) Calculate $u^{(1)}$, $v^{(1)}$ by (1.70c).

(6) Calculate $\{\omega_i^{(k)}\}_{i=1,2,3}$ for all $k > 0$ levels from (1.71).

1.3 Stress Measures

In this section we shall introduce the notion of *stress* and derive different forms of its expression. We shall also discuss the equilibrium equations that are most commonly associated to and written by the stress components. The derivation is largely founded on the coordinate systems analysis of S.1.7.

1.3.1 Definition of Stress

To define the stress tensor at a point, it is worthwhile to first examine the mathematical definition of stress in a simple linear case, which may be easily interpreted and associated by complementary physical quantities. As will be shown later on, when all nonlinear effects are included, one is forced to work with "generalized stresses", which are mathematical measures of the actual stress components and are more difficult to be physically interpreted.

We shall first examine a three-dimensional body by virtually cutting it over an interior plane in which we define a small area A (say, a small circle with a center located at P) as shown in Fig. 1.2. The force acting over A normal to the plane will be denoted "**N**" while the tangent

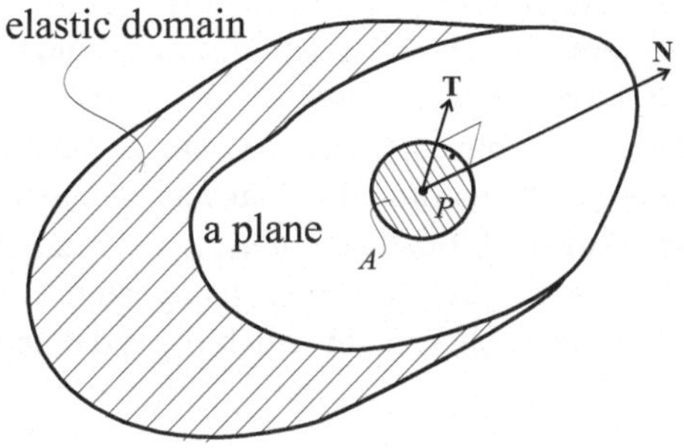

Figure 1.2: Normal and tangential loads over a small area in an elastic domain.

(in-plane) force component will be denoted "**T**". Both **N** and **T** are functions of the location of the circle over the plane (or essentially the location of its center point, P in the plane). By narrowing the area A, the point P and the area A collide, and the normal and tangential stress components at P are defined as

$$\sigma_N(P) = \lim_{A\to 0} \frac{\mathbf{N}}{A}, \qquad \sigma_T(P) = \lim_{A\to 0} \frac{\mathbf{T}}{A}. \tag{1.72}$$

Hence, the dimension of the stress components is *force per unit area.*

At each point, one may examine an infinitesimally small material element, as shown in Fig. 1.3. When described in the *coordinate space* (as opposed to the Euclidean space), be-

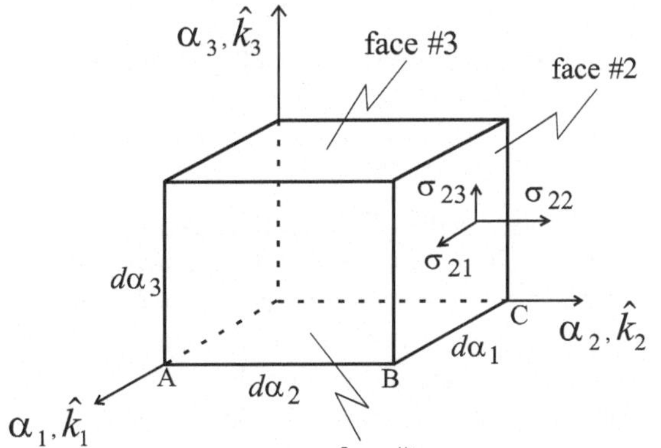

Figure 1.3: An infinitesimal element in its coordinate space.

fore deformation, the above element may be viewed as an infinitesimal cube regardless of the specific system employed. When described in a Euclidean space in the undeformed state, an infinitesimal material element of general curvilinear coordinates may be described by a cubic topology while all faces are different four-edges-polygons. Hence, its six faces are different quadrangles (this description is usually termed a "rectangular parallelepiped" since all considered coordinate systems are orthogonal). For example, the x coordinates of the A and B vertices (see Fig. 1.3) are, respectively, $f_{1,\alpha_1}d\alpha_1$, $f_{1,\alpha_1}d\alpha_1 + f_{1,\alpha_2}d\alpha_2$. When the same material element after deformation is examined, it may be no longer described as a cubic even in its coordinate space. The material element in the deformed state in both the coordinate space and the Euclidean space is generally called an "oblique angle parallelepiped", and may be generally viewed as a cubic, the corners of which have been displaced differently, so its six faces are now different quadrangles (in essence, as previously discussed, this general description holds for the undeformed case in Euclidean space as well).

We initially consider the stress components as an asymmetric second-order tensor. By definition, σ_{ij} is the stress component in the $\hat{\mathbf{k}}_j$-direction that acts on a plane, which is perpendicular to the $\hat{\mathbf{k}}_i$-direction — e.g. σ_{1j} in Fig. 1.3. When the same material element after deformation is examined, one may decompose the stress components by using various coordinate systems. However, we preserve the same notation logic in which σ_{ij} is defined as the stress component in the deformed $j^{\underline{th}}$ direction of *a given coordinate system* that acts on a plane, which before deformation was perpendicular to the $\hat{\mathbf{k}}_i$-direction.

On each of the deformed faces one may define stress vectors. We will denote by $\vec{\sigma}_1$, $\vec{\sigma}_2$ and $\vec{\sigma}_3$ the stress vectors over faces #1,2 and 3, see Fig. 1.3. Before deformation the edges' lengths are $H_i\,d\alpha_i$ (see (1.215)), and their respective areas are $A_i = H_j H_k\,d\alpha_j\,d\alpha_k$ (apply cyc-ijk). The areas of the corresponding faces after deformation are denoted A_i^*.

Based on the above definitions, the forces that act over faces #1,2 and 3 may be written as $A_i^*\vec{\sigma}_i$ $(i \in \{1,3\})$, or, equivalently, $\frac{A_i^*}{A_i}\vec{\sigma}_i H_j H_k\,d\alpha_j\,d\alpha_k$. We will now decompose $\vec{\sigma}_i$ along the deformed and the undeformed directions, respectively, as

$$\vec{\sigma}_i = \sum_{j=1}^{3} \sigma_{ij}\hat{\mathbf{k}}_j^*, \qquad \frac{A_i^*}{A_i}\vec{\sigma}_i = \sum_{j=1}^{3} s_{ij}\hat{\mathbf{k}}_j. \tag{1.73}$$

It may be verified that

$$\begin{Bmatrix} s_{i1} \\ s_{i2} \\ s_{i3} \end{Bmatrix} = [\mathbf{I}+\mathbf{e}+\omega]\begin{Bmatrix} \bar{\bar{\sigma}}_{i1} \\ \bar{\bar{\sigma}}_{i2} \\ \bar{\bar{\sigma}}_{i3} \end{Bmatrix}, \qquad i = 1,2,3. \tag{1.74}$$

In the above, $\mathbf{I}$ is a unit matrix, $\mathbf{e}$ and ω are given by (1.17), and

$$\bar{\bar{\sigma}}_{ij} = \frac{A_i^*}{A_i}\frac{\sigma_{ij}}{1+E_j}, \tag{1.75}$$

where E_j is the relative extension in the $j^{\underline{th}}$ direction, see (1.36). $\bar{\bar{\sigma}}_{ij}$ are usually referred to as the"generalized stresses" and are not stresses in the strict sense. As shown by (1.75), these quantities are based on the element volume before deformation. However, they have important symmetry characteristics that are missing in σ_{ij}, namely, $\bar{\bar{\sigma}}_{ij} = \bar{\bar{\sigma}}_{ji}$ (while $\sigma_{ij} \neq \sigma_{ji}$). Therefore, under a coordinate system transformation the tensor $\bar{\bar{\sigma}} = \{\bar{\bar{\sigma}}_{ij}\}$ acts like a second-order symmetric tensor. Subsequently, this tensor should be regarded as the stress tensor when a fully nonlinear analysis is employed. Once a solution of a specific problem is carried out, and the components of $\bar{\bar{\sigma}}$ are derived, the values of σ_{ij} may be recovered by (1.75). Note that boundary conditions should be imposed on σ_{ij} (and not on $\bar{\bar{\sigma}}_{ij}$), as they represent the physical stress components in the deformed state.

At this stage, the only missing component in the scheme is the area ratio A_i^*/A_i, which may be written as (apply cyc-ijk)

$$\frac{A_i^*}{A_i} = (1+E_j)(1+E_k)\sin(\hat{\mathbf{k}}_j^*, \hat{\mathbf{k}}_k^*). \tag{1.76}$$

Here E_i are given in (1.36) and $\sin(\mathbf{k}_j^*, \mathbf{k}_k^*)$ may be evaluated using (1.39), which yields (apply cyc-ijk)

$$\frac{A_i^*}{A_i} = \sqrt{(1+2\varepsilon_{jj})(1+2\varepsilon_{kk}) - 4\varepsilon_{jk}^2}, \qquad i \neq j \neq k. \tag{1.77}$$

Transformation of stress components between curvilinear and Cartesian coordinates appears in S.1.3.3. As previously mentioned, when a nonlinear analysis is under discussion, the tensor $\overline{\sigma}$ should be transformed between coordinate systems.

Note that in the linear case, we pay no attention to the difference between the element shape before and after deformation. Subsequently, $A_i^* \cong A_i$, $E_j \ll 1$, $\mathbf{e} \ll \mathbf{I}$, $\omega \ll \mathbf{I}$, and therefore, $s_{ij} \cong \overline{\overline{\sigma}}_{ij} \cong \sigma_{ij}$, and σ_{ij} becomes a symmetric tensor as well.

1.3.2 Equilibrium Equations

An extremely important and necessary ingredient in the set of governing equations for each problem in the theory of elasticity are the *equilibrium equations*. To express these equations using the above derived stress definitions, the body forces acting over *the unit volume* at each material point should also be considered. The body forces are described by their components in the undeformed directions as

$$\mathbf{F}_b = \sum\nolimits_{j=1}^{3} F_{bj}\mathbf{k}_j, \tag{1.78}$$

while the total force action on a volume element is given by $\mathbf{F}_b dV = \mathbf{F}_b H_1 H_2 H_3 d\alpha_1 d\alpha_2 d\alpha_3$. Subsequently, equilibrium may be imposed by equating the resultant vector of all forces acting on the material element to zero, namely,

$$\left(H_2H_3\frac{A_1^*}{A_1}\vec{\sigma}_1\right)_{,\alpha_1} + \left(H_1H_3\frac{A_2^*}{A_2}\vec{\sigma}_2\right)_{,\alpha_2} + \left(H_1H_2\frac{A_3^*}{A_3}\vec{\sigma}_3\right)_{,\alpha_3} + H_1H_2H_3\vec{\mathbf{F}}_\mathbf{b} = 0. \tag{1.79}$$

The above vector equation may be now decomposed into its components. This yields three equations of equilibrium that may be written as (apply cyc-123)

$$\begin{aligned}(H_2H_3s_{11})_{,\alpha_1} + (H_1H_3s_{21})_{,\alpha_2} + (H_1H_2s_{31})_{,\alpha_3} + H_3H_{1,\alpha_2}s_{12} + H_2H_{1,\alpha_3}s_{13}\\ - H_3H_{2,\alpha_1}s_{22} - H_2H_{3,\alpha_1}s_{33} + H_1H_2H_3F_{b1} = 0,\end{aligned} \tag{1.80}$$

while s_{ij} are given in (1.74). As already indicated, in the linear case, $s_{ij} \cong \sigma_{ij}$.

P.1.7, P.1.8 are capable of producing equilibrium equations for various orthogonal coordinates in E^3 or E^2. These equations are written in terms of σ_{ij}. When nonlinear analysis is required, σ_{ij} should be replaced by s_{ij}.

We shall present here some illustrative examples of the linear case. For *Cartesian coordinates* we denote the body-force components in the x, y, z directions, as

$$\mathbf{F}_b = X_b\mathbf{k}_1 + Y_b\mathbf{k}_2 + Z_b\mathbf{k}_3, \tag{1.81}$$

and write

$$\sigma_{x,x} + \tau_{xy,y} + \tau_{xz,z} + X_b = 0, \tag{1.82a}$$
$$\tau_{xy,x} + \sigma_{y,y} + \tau_{yz,z} + Y_b = 0, \tag{1.82b}$$
$$\tau_{xz,x} + \tau_{yz,y} + \sigma_{z,z} + Z_b = 0. \tag{1.82c}$$

Note that for the present linear Cartesian case, moment differential equilibrium may be easily seen as a direct consequence of the stress tensor symmetry.

It should be noted that the differential equilibrium equations of (1.82a–c) may be derived from an integral ("static") equilibrium that is written with the aid of the body and the surface loads that act over the *volume* of each material point, and over the outer surface of the body. Similar to (1.78), surface loads are defined as forces per *the unit area* at each boundary material point and described by their components in the undeformed directions as

$$\mathbf{F}_s = \sum_{j=1}^{3} F_{sj}\mathbf{k}_j. \tag{1.83}$$

In Cartesian coordinates we write

$$\mathbf{F}_s = X_s\mathbf{k}_1 + Y_s\mathbf{k}_2 + Z_s\mathbf{k}_3, \tag{1.84}$$

where

$$X_s = \sigma_x \cos(\bar{\mathbf{n}}, x) + \tau_{xy}\cos(\bar{\mathbf{n}}, y) + \tau_{xz}\cos(\bar{\mathbf{n}}, z), \tag{1.85a}$$
$$Y_s = \tau_{xy}\cos(\bar{\mathbf{n}}, x) + \sigma_y \cos(\bar{\mathbf{n}}, y) + \tau_{yz}\cos(\bar{\mathbf{n}}, z), \tag{1.85b}$$
$$Z_s = \tau_{xz}\cos(\bar{\mathbf{n}}, x) + \tau_{yz}\cos(\bar{\mathbf{n}}, y) + \sigma_z \cos(\bar{\mathbf{n}}, z), \tag{1.85c}$$

and $\cos(\bar{\mathbf{n}}, x)$, $\cos(\bar{\mathbf{n}}, y)$ and $\cos(\bar{\mathbf{n}}, z)$ are angle cosines between the normal to the surface and the x, y, z directions, respectively.

At this stage, we express integral force equilibrium as

$$\iint_S \mathbf{F}_s + \iiint_V \mathbf{F}_b = 0. \tag{1.86}$$

Hence, by substituting (1.81, 1.84, 1.85a-c) in (1.86) and applying the Divergence Theorem, we reach three integral equations (over the entire body volume) the integrands of which are (1.82a–c). Since these equations apply to each infintisimal volume as well, (1.82a–c) are re-established.

For *cylindrical coordinates* we denote by R_b, Θ_b, Z_b the body-force components in the ρ, θ^c, z directions, respectively, and write

$$\frac{1}{\rho}\left(\sigma_{\rho\rho} - \sigma_{\theta\theta} + \sigma_{\rho\theta,\theta}\right) + \sigma_{\rho\rho,\rho} + \sigma_{\rho z,z} + R_b = 0, \tag{1.87a}$$
$$\frac{1}{\rho}\left(\sigma_{\theta\theta,\theta} + 2\sigma_{\rho\theta}\right) + \sigma_{\theta z,z} + \sigma_{\rho\theta,\rho} + \Theta_b = 0, \tag{1.87b}$$
$$\frac{1}{\rho}\left(\sigma_{\theta z,\theta} + \sigma_{\rho z}\right) + \sigma_{\rho z,\rho} + \sigma_{zz,z} + Z_b = 0. \tag{1.87c}$$

For *spherical coordinates* we denote by R_b, Θ_b and Φ_b the body-force components in the ρ, θ^s, ϕ^s directions, respectively, and write

$$\frac{1}{\rho}\left(\sigma_{\rho\phi,\phi} + 2\sigma_{\rho\rho} - \sigma_{\theta\theta} - \sigma_{\phi\phi} + \cot\phi\,\sigma_{\rho\phi}\right) + \frac{1}{\rho\sin\phi}\sigma_{\rho\theta,\theta} + \sigma_{\rho\rho,\rho} + R_b = 0, \tag{1.88a}$$
$$\frac{1}{\rho}\left(\sigma_{\theta\phi,\phi} + 3\sigma_{\rho\theta}\right) + \frac{1}{\rho\sin\phi}\left(\sigma_{\theta\theta,\theta} + 2\cos\phi\,\sigma_{\theta\phi}\right) + \sigma_{\rho\theta,\rho} + \Theta_b = 0, \tag{1.88b}$$
$$\frac{1}{\rho}\left(\sigma_{\phi\phi,\phi} + 3\sigma_{\rho\phi}\right) + \frac{1}{\rho\sin\phi}\left[\sigma_{\theta\phi,\theta} + \cos\phi\left(\sigma_{\phi\phi} - \sigma_{\theta\theta}\right)\right] + \sigma_{\rho\phi,\rho} + \Phi_b = 0. \tag{1.88c}$$

For *elliptical-cylindrical coordinates* (with $\widetilde{a} = 1$, see (1.218b)), one obtains

$$(\sigma_{11}-\sigma_{22})\cosh(\alpha_1)\sinh(\alpha_1)+\widetilde{A}\sigma_{11,1}+2\sigma_{12}\cos(\alpha_2)\sin(\alpha_2)+\widetilde{A}\sigma_{12,2}+\widetilde{A}^{\frac{3}{2}}\sigma_{13,3}+F_{b1}\widetilde{A}^{\frac{3}{2}}=0,$$
$$2\sigma_{12}\cosh(\alpha_1)\sinh(\alpha_1)+\widetilde{A}\sigma_{12,1}+(\sigma_{22}-\sigma_{11})\cos(\alpha_2)\sin(\alpha_2)+\widetilde{A}\sigma_{22,2}+\widetilde{A}^{\frac{3}{2}}\sigma_{23,3}+F_{b2}\widetilde{A}^{\frac{3}{2}}=0,$$
$$\sigma_{13}\cosh(\alpha_1)\sinh(\alpha_1)+\widetilde{A}\sigma_{13,1}+\sigma_{23}\cos(\alpha_2)\sin(\alpha_2)+\widetilde{A}\sigma_{23,2}+\widetilde{A}^{\frac{3}{2}}\sigma_{33,3}+F_{b3}\widetilde{A}^{\frac{3}{2}}=0 \quad (1.89)$$

where $\widetilde{A} = \cosh^2(\alpha_1) - \cos^2(\alpha_2)$, and F_{b1}, F_{b2} and F_{b3} are the body-force components in the α_1, α_2 and α_3 directions, respectively.

1.3.3 Stress Tensor Transformation due to Coordinate System Rotation

We shall now exploit the general derivation for coordinate systems presented in S.1.7 to transform the stress components at a point. More specifically, for a given stress tensor in one coordinate system, we wish to determine the six independent stress components of the same tensor as seen by another coordinate system. For that purpose, we shall consider the elements of the transformation matrix $\mathbf{T}$ (which are functions of the rotation angles ψ, θ and ϕ, see (1.203), (1.206)) as the T_{ij} element of the corresponding transformation tensor, and define the (symmetric) second-order stress tensor as

$$\sigma = \begin{bmatrix} \sigma_{11} & \sigma_{12} & \sigma_{13} \\ \sigma_{12} & \sigma_{22} & \sigma_{23} \\ \sigma_{13} & \sigma_{23} & \sigma_{33} \end{bmatrix}. \quad (1.90)$$

Hence, the components of the stress tensor in the new system, $\overline{\sigma} = \{\overline{\sigma}_{ij}\}$, are obtained by the standard tensor transformation

$$\overline{\sigma}_{ij} = \sigma_{ab} T_{ia} T_{jb}. \quad (1.91)$$

This operation may be expressed using matrix notation as well, as

$$\overline{\sigma} = \mathbf{T} \cdot \sigma \cdot \mathbf{T}^T. \quad (1.92)$$

To simplify the relations between the stress tensor components before and after transformation, we shall look at the above formula using the vectors σ and $\overline{\sigma}$, which contain the stress components before and after transformation, namely,

$$\sigma = [\sigma_{xx}, \sigma_{yy}, \sigma_{zz}, \sigma_{yz}, \sigma_{xz}, \sigma_{xy}]^T, \qquad \overline{\sigma} = [\overline{\sigma}_{xx}, \overline{\sigma}_{yy}, \overline{\sigma}_{zz}, \overline{\sigma}_{yz}, \overline{\sigma}_{xz}, \overline{\sigma}_{xy}]^T. \quad (1.93)$$

These vectors may be related as $\overline{\sigma} = \mathbf{M}_\sigma \cdot \sigma$, where $\mathbf{M}_\sigma$ is a (non-symmetric) 6×6 matrix, which is clearly a function of the transformation (Euler's) rotation angles ψ, θ and ϕ. Example terms are

$$\mathbf{M}_\sigma(3,2) = (\cos\phi \sin\theta \sin\psi - \sin\phi \cos\psi)^2, \qquad \mathbf{M}_\sigma(2,3) = \sin^2\phi \cos^2\theta. \quad (1.94)$$

The reader may activate **P.1.9** to generate all terms of $\mathbf{M}_\sigma$ symbolically. Figure 1.4 presents an example of axis rotation and the associated $\mathbf{M}_\sigma$ matrix. Such a view for different sets of rotation angles may be obtained by activating **P.1.10**.

To transform a stress tensor given in orthogonal curvilinear coordinates to Cartesian coordinates, we select the transformation tensor of (1.216). The result is a function of the specific point under discussion, and may be written as a function of the curvilinear coordinates $\alpha_1, \alpha_2, \alpha_3$. The corresponding transformation angles may be determined by (1.208).

P.1.11 executes stress tensor transformations between curvilinear and Cartesian coordinates.

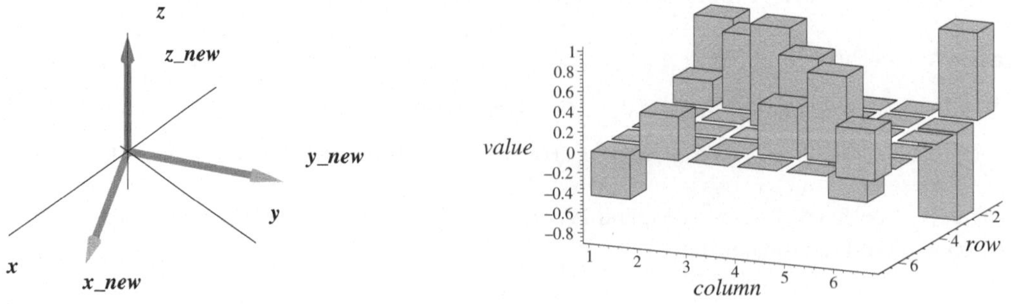

(a) Axis orientation. (b) The matrix $\mathbf{M}_\sigma$.

Figure 1.4: Example of axis orientation and the resulting $\mathbf{M}_\sigma$ for $\psi = 30°$, $\theta = \phi = 0$.

1.3.3.1 Principal Stresses

We shall now seek expressions for the *principal stresses* at a point. In essence, we are looking for a set of rotation angles that will define a new system orientation, in which only the normal stress components σ_α ($\alpha = x, y, z$) are nonzero at a point. Clearly, these values of rotation angles will ensure that the vector

$$\sigma_{\mathbf{L}} = [\sigma_{yz}, \sigma_{xz}, \sigma_{xy}]^T \tag{1.95}$$

will vanish, or in other words, the state where σ is diagonal, and its elements are called *the principal stresses* at a point. One way to carry out this task is to set the following set of three equations:

$$\overline{\sigma}_{\mathbf{L}} = \mathbf{L}_\sigma \cdot \sigma = 0, \tag{1.96}$$

where $\mathbf{L}_\sigma$ is a 3×6 matrix (essentially the last three lines of the matrix $\mathbf{M}_\sigma$). For a given stress tensor, these are three equations in the three unknown orientation angles ψ, θ, and ϕ. The solution of such a system is not trivial, and it is much more convenient to adopt a more standard way. In this new course, we carry out an eigenvalue analysis for σ and obtain its eigenvalues $\{\sigma_i^P\}_{i=1,2,3}$ and eigenvectors $\{\mathbf{v}_i\}_{i=1,2,3}$ so that

$$\sigma \cdot \mathbf{v}_i = \sigma^P \cdot \mathbf{v}_i. \tag{1.97}$$

Hence, in the principal stress state, $\sigma_{ii} = \sigma_i^P$ and $\sigma_{ij} = 0$ ($i \neq j$).

Implementation of the above eigenvalue analysis requires that the system matrix determinant will vanish, namely,

$$\begin{vmatrix} \sigma_{11} - \sigma^P & \sigma_{12} & \sigma_{13} \\ \sigma_{12} & \sigma_{22} - \sigma^P & \sigma_{23} \\ \sigma_{13} & \sigma_{23} & \sigma_{33} - \sigma^P \end{vmatrix} = 0, \tag{1.98}$$

which leads to the cubic polynomial equation

$$\sigma^3 - \Theta_1 \sigma^2 + \Theta_2 \sigma - \Theta_3 = 0. \tag{1.99}$$

Θ_i are usually referred to as the *stress invariants* and are given by the matrix minors as

$$\Theta_1 = tr\,\sigma = \sigma_{11} + \sigma_{22} + \sigma_{33}, \tag{1.100a}$$

$$\Theta_2 = \frac{1}{2}\left[(tr\,\sigma)^2 - tr(\sigma^2)\right] = \sigma_{11}\sigma_{22} + \sigma_{22}\sigma_{33} + \sigma_{33}\sigma_{11} \underline{- \sigma_{12}^2 - \sigma_{13}^2 - \sigma_{23}^2}, \tag{1.100b}$$

$$\Theta_3 = \det\sigma = \sigma_{11}\sigma_{22}\sigma_{33} \underline{+ 2\sigma_{12}\sigma_{13}\sigma_{23} - \sigma_{11}\sigma_{23}^2 - \sigma_{22}\sigma_{13}^2 - \sigma_{33}\sigma_{12}^2}. \tag{1.100c}$$

Since Θ_i are invariants, the resulting eigenvalues are invariants as well. Therefore, the above invariants may be expressed in a more compact form using the principal stresses by ignoring the underlined terms (1.100a–c) and replacing σ_{ii} with σ_i^P. From this point on, we shall assume that the three eigenvalues obtained by solving (1.99) are put in a decreasing order, $\sigma_1^P \geq \sigma_2^P \geq \sigma_3^P$, and their eigenvectors $\mathbf{v}_i = \{v_i(1), v_i(2), v_i(3)\}$ are put in the same order.

To determine the above stress eigenvalues, it is convenient to define the *stress deviator tensor* $\sigma_{ij}^D = \sigma_{ij} - \frac{1}{3}\Theta_1\delta_{ij}$, which is a second-order tensor as well, and its eigenvalues will be denoted $\sigma_1^{DP} \geq \sigma_2^{DP} \geq \sigma_3^{DP}$. It is therefore clear that (apply cyc-123)

$$\sigma_1^{DP} = \frac{1}{3}\left(2\sigma_1^P - \sigma_2^P - \sigma_3^P\right). \tag{1.101}$$

In addition, the invariants of the stress deviator tensor may be expressed as functions of the stress tensor invariants, namely,

$$\Theta_1^D = \sigma_1^{DP} + \sigma_2^{DP} + \sigma_3^{DP} = 0, \tag{1.102a}$$

$$\Theta_2^D = \sigma_1^{DP}\sigma_2^{DP} + \sigma_1^{DP}\sigma_3^{DP} + \sigma_2^{DP}\sigma_3^{DP} = \Theta_2 - \frac{1}{3}\Theta_1^2, \tag{1.102b}$$

$$\Theta_3^D = \sigma_1^{DP}\sigma_2^{DP}\sigma_3^{DP} = \frac{2}{27}\Theta_1^3 - \frac{1}{3}\Theta_1\Theta_2 + \Theta_3. \tag{1.102c}$$

Hence, to determine σ_i^{DP}, one needs to solve the cubic equation

$$(\sigma^D)^3 + \Theta_2^D\sigma^D - \Theta_3^D = 0, \tag{1.103}$$

which is simpler than (1.99). Once σ_i^{DP} are known, σ_i^P are directly obtained via $\sigma_i^P = \sigma_i^{DP} + \frac{1}{3}\Theta_1$, see also Remark 1.4.

To determine the transformation matrix, $\mathbf{T_E}$, from a given state to the principal directions along with the corresponding transformation angles ψ, θ and ϕ, we use the above derived eigenvectors $\mathbf{v}_i$ as

$$\mathbf{T_E} = \begin{bmatrix} v_1(1) & v_1(2) & v_1(3) \\ v_2(1) & v_2(2) & v_2(3) \\ v_3(1) & v_3(2) & v_3(3) \end{bmatrix}. \tag{1.104}$$

To preserve orientation (the transformation matrix determinant must be equal to a unit), one should change the sign of one of the eigenvectors if required. Also, by simple eigenvalue analysis argumentation we obtain

$$\mathbf{T_E} \cdot \sigma \cdot (\mathbf{T_E})^T = \begin{bmatrix} \sigma_1^P & 0 & 0 \\ 0 & \sigma_2^P & 0 \\ 0 & 0 & \sigma_3^P \end{bmatrix}. \tag{1.105}$$

Example 1.1 *A Generic Stress State.*

To demonstrate the above derivation, we shall examine a generic stress state as shown in Fig. 1.5(a). This state serves as an example only (while units will not be indicated). The above procedure shows that the eigenvalues are $\sigma_1^P = .0823$, $\sigma_2^P = -.0187$, $\sigma_3^P = -.0937$ and that

$$\mathbf{T_E} = \begin{bmatrix} -0.138 & 0.982 & -0.194 \\ -0.446 & -0.237 & -0.860 \\ 0.882 & 0.0288 & -0.464 \end{bmatrix}. \tag{1.106}$$

Thus, according to (1.208), the rotation angles are $\psi = -81.99°$, $\theta = -11.19°$ and $\phi = 61.65°$. The corresponding principal stress state is shown in Fig. 1.5(b). The principal stress directions at a point may be derived by **P.1.12**.

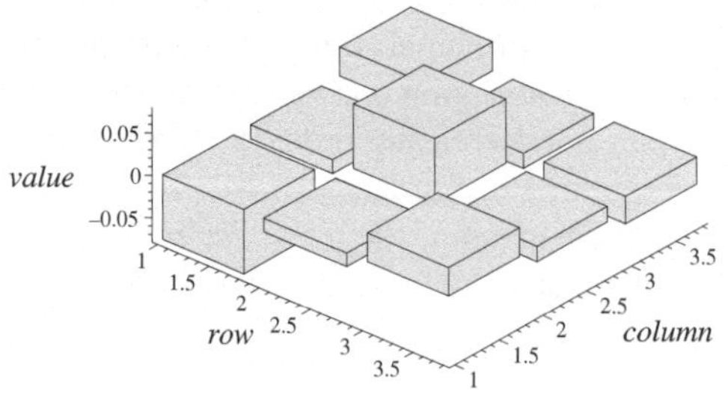

(a) A generic stress tensor.

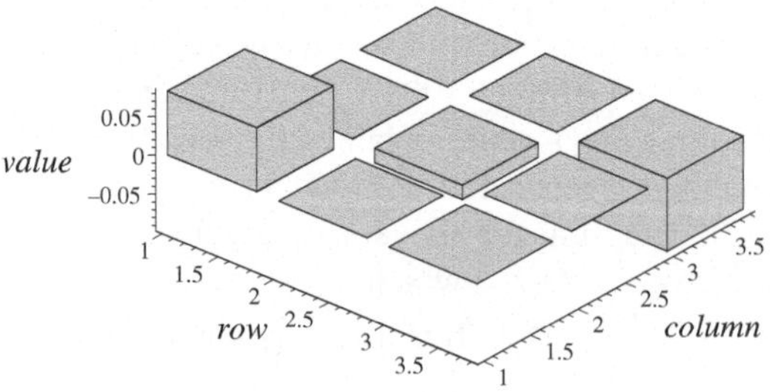

(b) The principal stress tensor.

Figure 1.5: Example of a stress tensor and its principal state.

Remark 1.4 Two additional stress invariants, Θ_4 and Θ_5 (which are identical for both the stress and the stress deviator tensors) may be defined. The first one is

$$\Theta_4 = \frac{1}{\sqrt{6}}\sqrt{\left(\sigma_1^P-\sigma_2^P\right)^2+\left(\sigma_2^P-\sigma_3^P\right)^2+\left(\sigma_3^P-\sigma_1^P\right)^2}, \tag{1.107}$$

while it is easy to verify that $\Theta_4 = \sqrt{\frac{1}{3}\Theta_1^2-\Theta_2} = \sqrt{-\Theta_2^D}$. The second invariant is defined by

$$\tan(\Theta_5) = \frac{1}{\sqrt{3}}\frac{2\sigma_2^P-\sigma_1^P-\sigma_3^P}{\sigma_1^P-\sigma_3^P} = \sqrt{3}\frac{\sigma_2^{DP}}{\sigma_1^{DP}-\sigma_3^{DP}}. \tag{1.108}$$

Due to the order of the principal stresses, i.e. $\sigma_1^P \geq \sigma_2^P \geq \sigma_3^P$, this invariant is bounded as $|\Theta_5| \leq \pi/6$. The above definitions enable us to write

$$\sigma_1^{DP}, \sigma_3^{DP} = \frac{2\Theta_4}{\sqrt{3}}\sin(\Theta_5 \pm \frac{2\pi}{3}), \qquad \sigma_2^{DP} = \frac{2\Theta_4}{\sqrt{3}}\sin(\Theta_5), \tag{1.109}$$

which allows graphical interpretation of Θ_4 and Θ_5 as presented in Fig. 1.6(a).

1.3.3.2 Visualizing the State of Stress at a Point

Many visualization methods of the state of stress at a point have been discussed extensively in the literature. In view of the powerful modern visualization tools, the classical methods seem less attractive and important. We will describe the main ideas in this area briefly.

A good starting point is the examination of the stresses over a face of an infinitesimal cube having general orientation so that the normal to the face under discussion is oriented at the $\bar{x}$ direction as shown in Fig. 1.6(b). Also, the normal stress and the resultant shear stress over this face are denoted as

$$\sigma_N = \bar{\sigma}_{11}, \tag{1.110a}$$

$$\sigma_T = \sqrt{\bar{\sigma}_{12}^2+\bar{\sigma}_{13}^2}. \tag{1.110b}$$

We shall now assume that the original x-, y-, z-axes are the directions of the principal stress state, and write the above two equations using (1.91), in addition to (1.204) (with $i = 1$). This

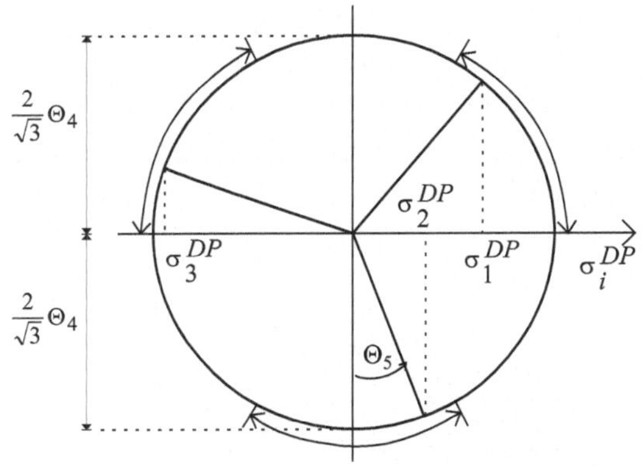

(a) Geometrical relations between the Θ_4, Θ_5 invariants and the principal σ_i^{DP} eigenvalue.

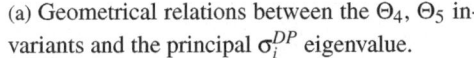

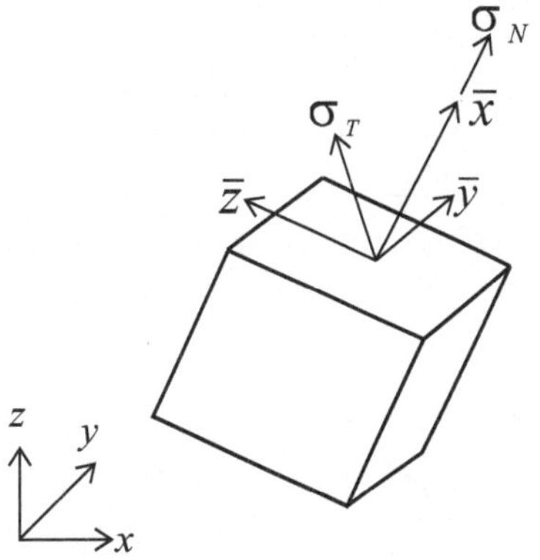

(b) An infinitesimal cube at an arbitrary orientation.

Figure 1.6: Stress invariants and transformation notation.

procedure constitutes the following set of three equations:

$$\sigma_N = \sigma_1^P T_{11}^2 + \sigma_2^P T_{12}^2 + \sigma_3^P T_{13}^2, \tag{1.111a}$$
$$\sigma_T^2 = (\sigma_1^P)^2 T_{11}^2 + (\sigma_2^P)^2 T_{12}^2 + (\sigma_3^P)^2 T_{13}^2 - \sigma_N^2, \tag{1.111b}$$
$$1 = T_{11}^2 + T_{12}^2 + T_{13}^2, \tag{1.111c}$$

which may be solved for T_{11}^2, T_{12}^2 and T_{13}^2 as

$$T_{11}^2 = \frac{\sigma_T^2 + (\sigma_N - \sigma_2^P)(\sigma_N - \sigma_3^P)}{(\sigma_2^P - \sigma_1^P)(\sigma_3^P - \sigma_1^P)}, \tag{1.112a}$$
$$T_{12}^2 = \frac{\sigma_T^2 + (\sigma_N - \sigma_3^P)(\sigma_N - \sigma_1^P)}{(\sigma_3^P - \sigma_2^P)(\sigma_1^P - \sigma_2^P)}, \tag{1.112b}$$
$$T_{13}^2 = \frac{\sigma_T^2 + (\sigma_N - \sigma_1^P)(\sigma_N - \sigma_2^P)}{(\sigma_1^P - \sigma_3^P)(\sigma_2^P - \sigma_3^P)}. \tag{1.112c}$$

In view of the eigenvalues' order ($\sigma_1^P \geq \sigma_2^P \geq \sigma_3^P$) and the fact that the l.h.s. of the above equations are non-negative, the following inequalities are obtained:

$$\sigma_T^2 + (\sigma_N - \sigma_2^P)(\sigma_N - \sigma_3^P) \geq 0, \tag{1.113a}$$
$$\sigma_T^2 + (\sigma_N - \sigma_3^P)(\sigma_N - \sigma_1^P) \leq 0, \tag{1.113b}$$
$$\sigma_T^2 + (\sigma_N - \sigma_1^P)(\sigma_N - \sigma_2^P) \geq 0. \tag{1.113c}$$

We may now plot the above conditions in the $\sigma_N - \sigma_T$ plane, in order to find a region where all inequalities are satisfied, which will yield the valid combinations of normal and shear stress at the point under discussion. This graphical analysis is shown in Fig. 1.7(a) and is traditionally referred to as *Mohr's diagram*. The first inequality shows a region outside a circle, the diameter of which is located over the σ_N-axis between $\sigma_N = \sigma_2^P$ and $\sigma_N = \sigma_3^P$. The second inequality shows a region inside a circle, the diameter of which is located over the σ_N-axis between

$\sigma_N = \sigma_3^P$ and $\sigma_N = \sigma_1^P$. The third inequality shows a region outside a circle, the diameter of which is located over the σ_N-axis between $\sigma_N = \sigma_1^P$ and $\sigma_N = \sigma_2^P$. Note that only the upper part of the σ_N-σ_T plane is of interest in this case since by definition $\sigma_T \geq 0$, see (1.110b). An immediate and clear result of Mohr's diagram is that the maximal shear stress is given by $\sigma_T^m = \frac{\sigma_1^P - \sigma_3^P}{2}$, which occurs at $\sigma_N^m = \frac{\sigma_1^P + \sigma_3^P}{2}$. For the state of stress described in Example 1.1, **P.1.12** has produced Mohr's diagram shown in Fig. 1.7(b).

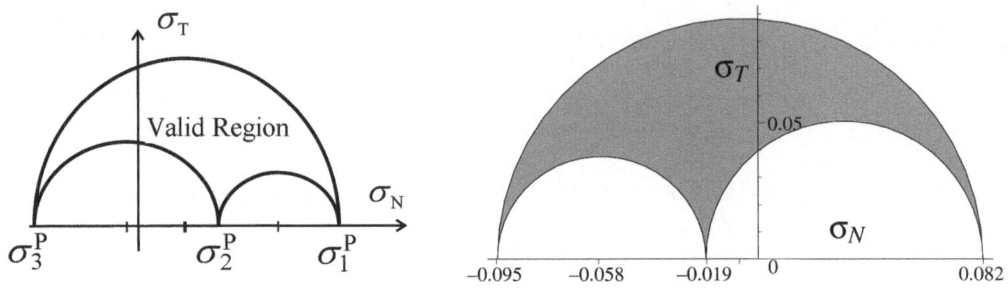

(a) General notation. (b) The values of Example 1.1.

Figure 1.7: Mohr's diagram.

For each combination of σ_N and σ_T in Mohr's diagram corresponds a pair of angles (ψ, θ). Note that the value of ϕ has no importance in this case, since it represents a rotation about the $\bar{x}$-direction (as it will change neither σ_N nor σ_T). To determine the rotation angles between the principal directions and the coordinate systems of the point under discussion for a given set of σ_N and σ_T, we evaluate T_{11}, T_{12} by (1.112a,b) and use (1.208:a,b). The inverse problem is of course much easier to solve since for each set of two rotation angles ψ and θ, one can calculate σ_N and σ_T directly from (1.111a,b), (1.207), see also Remark 1.5.

Suppose now that we wish to calculate the angles ψ and θ that are required to reach a given set of σ_N and σ_T from a state of stress, which is not principal. To do that, we substitute σ_N and σ_T in the l.h.s. of (1.110a,b) while in the r.h.s. we substitute the expressions obtained from (1.92) with $\phi = 0$, namely

$$\begin{aligned}\bar{\sigma}_{11} =& \cos^2\theta\cos^2\psi\,\sigma_{11} + 2\cos^2\theta\cos\psi\sin\psi\,\sigma_{12} - 2\cos\theta\cos\psi\sin\theta\,\sigma_{13}\\ &+ \cos^2\theta\sin^2\psi\,\sigma_{22} - 2\cos\theta\sin\psi\sin\theta\,\sigma_{23} + \sin^2\theta\,\sigma_{33}, \qquad (1.114a)\\ \bar{\sigma}_{12} =& -\cos\theta\cos\psi\sin\psi\,\sigma_{11} + \cos\theta(\cos^2\psi - \sin^2\psi)\,\sigma_{12}\\ &+ \sin\theta\sin\psi\,\sigma_{13} + \cos\theta\sin\psi\cos\psi\,\sigma_{22} - \sin\theta\cos\psi\,\sigma_{23}, \qquad (1.114b)\\ \bar{\sigma}_{13} =& \sin\theta\cos^2\psi\cos\theta\,\sigma_{11} + 2\sin\theta\cos\psi\cos\theta\sin\psi\,\sigma_{12} + (\cos^2\theta - \sin^2\theta)\cos\psi\,\sigma_{13}\\ &+ \sin\theta\sin^2\psi\cos\theta\,\sigma_{22} + (\cos^2\theta - \sin^2\theta)\sin\psi\,\sigma_{23} - \cos\theta\sin\theta\,\sigma_{33}. \qquad (1.114c)\end{aligned}$$

Substitutions of the above in (1.110a,b) converts these equations into a system of two equations that should be simultaneously solved for ψ and θ,

Since in the general case, the expressions are not simple enough for analytic solution, a graphical representation of these equations is given in Fig. 1.8 for two cases (produced by **P.1.12**). In Fig. 1.8(a), the angles are measured from the non-principal stress state of Example 1.1, while in Fig. 1.8(b), the angles are measured from the corresponding principal stress state. In these figures, the thick line shows solutions for (1.110a), while the thin line shows solutions for (1.110b) (angles are given in radians). The desired solutions are therefore the points where the above lines coincide.

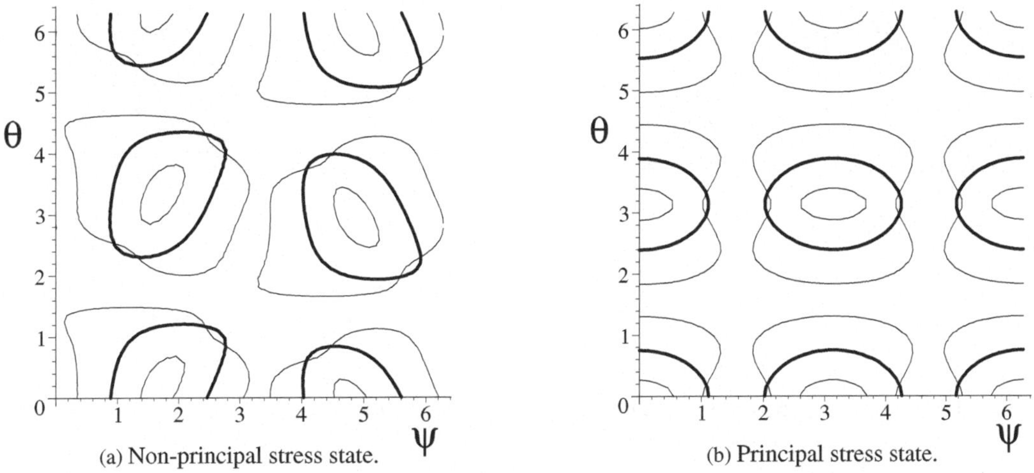

(a) Non-principal stress state. (b) Principal stress state.

Figure 1.8: Graphical solution of (1.110a,b) for $\sigma_N = 0$, $\tau_T = 0.045$.

As shown, there are a number of solutions in the region under discussion, and the picture has a period of π in θ and 2π in ψ. To visualize the above solution, one may also plot a graph

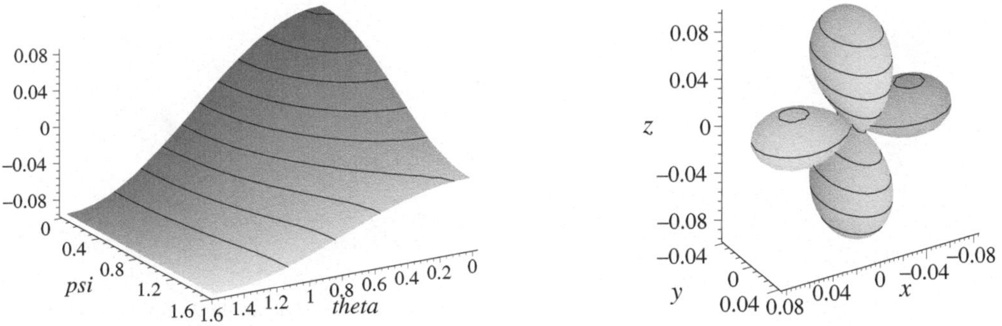

(a) $|\sigma_N|$ over $\psi - \theta$ plane. (b) 3D surface of $|\sigma_N|$.

Figure 1.9: σ_N as function of ψ and θ measured from the (principal) state of stress, see Example 1.1.

of $\sigma_N = \overline{\sigma}_{11}$ over the ψ-θ plane, see Fig. 1.9(a), and similarly, by substituting (1.114b,c) in (1.110b), σ_T as described by the surface in Fig. 1.10(a) is obtained.

One may also create a three-dimensional (spherical) surface of $\sigma_N = \overline{\sigma}_{11}(\theta, \psi)$, and $\sigma_T = \sqrt{\overline{\sigma}_{12}^2(\theta, \psi) + \overline{\sigma}_{13}^2(\theta, \psi)}$. To do that we use the results of Remark 1.12 and replace ψ by θ^s and θ by $\phi^s - \frac{\pi}{2}$, respectively, where θ^s and ϕ^s are spherical angles, see Fig. 1.20(b). These spherical plots are shown in Figs. 1.9(b), 1.10(b) where each point on the surface represents an orientation of the $\overline{x}$-axis of the transformed system (by connecting the origin with it). Thus, σ_N and σ_T are directly proportional to the distance that is measured along the $\overline{x}$-axis between the origin and the corresponding surfaces.

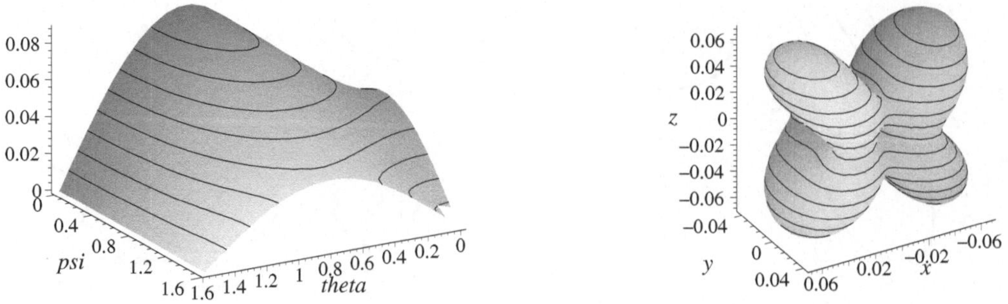

(a) σ_T over $\psi - \theta$ plane.

(b) 3D surface of σ_T.

Figure 1.10: σ_T as a function of ψ and θ measured from the (principal) state of stress, see Example 1.1.

Further on, we may also plot graphs of ψ and θ as functions of σ_N and σ_T over the valid region of Mohr's diagram. From (1.112a–c), for each (σ_N, σ_T) point we calculate T_{11}, T_{12} and T_{13}, and then use (1.208) to determine ψ and θ, respectively. The resulting diagrams are shown in Fig. 1.11 produced by **P.1.12**. Note that each surface is plotted over the valid area of Mohr's diagram only. See also Remark 1.6.

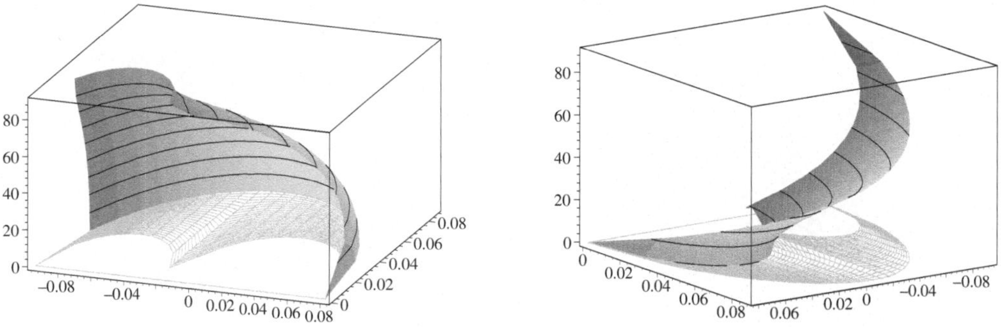

(a) The angle ψ (deg.).

(b) The angle θ (deg.).

Figure 1.11: The angles ψ and θ that transform a principal stress state for each σ_N, σ_T point.

Remark 1.5 One may define the *mean shear stress*, σ_T^0, by integrating τ_T over the surface, ω, of an infinitesimally small sphere, the center of which is located at the point under discussion. This operation may be expressed as

$$\sigma_T^0 = \lim_{\omega \to 0} \sqrt{\frac{1}{S(\omega)} \iint_\omega \sigma_T^2} = \sqrt{\frac{2}{5}} \Theta_4. \tag{1.115}$$

The above relation connects the mean shear stress with the principal stresses. According to (1.109), the maximal shear stress may be written as $\sigma_T^m = \frac{1}{2}(\sigma_1^P - \sigma_3^P) = \frac{1}{2}(\sigma_1^{DP} - \sigma_3^{DP}) = 2\Theta_4 \cos(\Theta_5)$, and thus, $\sigma_T^0 / \sigma_T^m = \frac{\sqrt{2/5}}{\cos(\Theta_5)}$ which, by taking into account the valid range of Θ_5

in (1.109), yields

$$\sqrt{\frac{8}{15}} \geq \frac{\sigma_T^0}{\sigma_T^m} \geq \sqrt{\frac{2}{5}}. \tag{1.116}$$

Remark 1.6 An additional, rather classical visualization tool is based on the well-known *Stress Quadric of Cauchy*. To derive the related equations, we start from the principal stress directions (namely, assume that the x, y, z system origin is located at the point under discussion while the coordinate lines coincide with the principal directions at that point). As a first step, we look at a point located at a distance $\widetilde{A}$ and arbitrary direction, which may be expressed by ψ and θ. The coordinates of this point are, see Remark 1.12,

$$x = \widetilde{A}\cos\theta\,\cos\psi, \qquad y = \widetilde{A}\cos\theta\,\sin\psi, \qquad z = -\widetilde{A}\sin\theta. \tag{1.117}$$

As a second step, we define $\sigma_N = \overline{\sigma}_{11}$ as the normal stress obtained in the $\overline{x}$-direction of the new system by rotating the coordinate system with the above angles. This stress component may be easily obtained from (1.114a) by setting $\sigma_{ii} = \sigma_i^P$ and $\sigma_{ij} = 0\,(i \neq j)$, as

$$\overline{\sigma}_{11} = \cos^2\theta\,\cos^2\psi\,\sigma_1^P + \cos^2\theta\,\sin^2\psi\,\sigma_2^P + \sin^2\theta\,\sigma_3^P, \tag{1.118}$$

and by using (1.117)

$$\widetilde{A}^2\overline{\sigma}_{11} = \sigma_1^P x^2 + \sigma_2^P y^2 + \sigma_3^P z^2. \tag{1.119}$$

We shall now call for $\overline{\sigma}_{11}$ to be proportional to the inverse of $\widetilde{A}$, namely, $\overline{\sigma}_{11} = c/\widetilde{A}^2$ where c is a normalization constant that may take both positive and negative values. Such a requirement yields a relatively simple quadratic surface, which is given by

$$\sigma_1^P x^2 + \sigma_2^P y^2 + \sigma_3^P z^2 = c. \tag{1.120}$$

The above surface may be classified by the four types: (1) ellipsoid, (2) unparted and biparted hyperboloids, (3) cylinder over ellipse, (4) hyperbola and parallel planes. Figure 1.12 (produced by **P.1.12**) presents an illustrative case where the value $c = 1$ yields the unparted hyper-

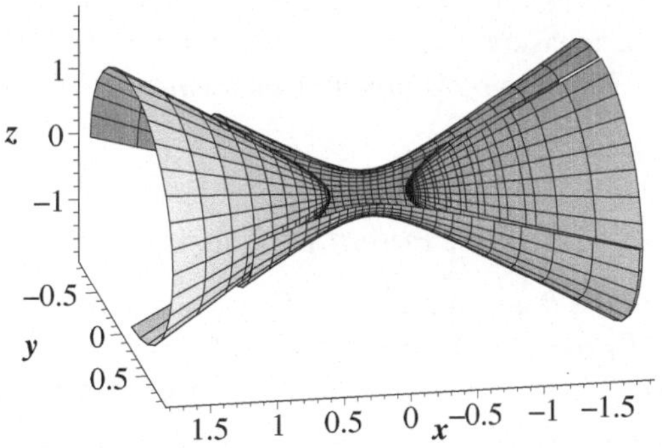

Figure 1.12: Three-dimensional plot of (1.120) for the state of stress described in Example 1.1.

boloid (i.e. the outer surface), and the value $c = -1$ yields the biparted hyperboloid (i.e. the inner surface). Hence, when applying stress transformation from a given system to a "new" system (denoted by a bar), the interpretation of the surface in Fig. 1.12 is as follows: *the value of $\overline{\sigma}_N$ is inversely proportional to the square of the distance that is measured along the $\overline{x}$-axis between the origin and the surface.*

1.3.4 Strain Tensor Transformation due to Coordinate System Rotation

Since both the strain and the stress at a point are described as symmetric second-order tensors, their transformation, invariants and principal axes are derived by the same rules.

Hence, similar to the discussion of S.1.3.3, one may transform the strain components at a point by applying the tensor transformation or by using the matrix notation

$$\bar{\varepsilon}_{ab} = \varepsilon_{ab} T_{ia} T_{jb}, \qquad \Leftrightarrow \qquad \bar{\varepsilon} = \mathbf{T} \cdot \varepsilon \cdot \mathbf{T}^T. \tag{1.121}$$

Compared with S.1.3.3, the discussion regarding the strain tensor transformation will be brief in view of the above similarities between the stress and the strain tensor transformation, the associated eigenvalue analysis, the principal axis determination, etc., Hence, the discussion regarding the stress tensor is applicable in a direct manner by replacing σ_{ij} with ε_{ij}, etc. Subsequently, a matrix $\mathbf{M}_\varepsilon = \mathbf{M}_\sigma$ may be defined, and the strain tensor invariants Ξ_1, Ξ_2, Ξ_3, the eigenvalues ε_i^P and the principal axes may be calculated analogously.

One may also draw a Mohr's diagram in the ε_N-ε_T plane (where $\varepsilon_N = \bar{\varepsilon}_{11}$, $\varepsilon_T = \sqrt{\bar{\varepsilon}_{12}^2 + \bar{\varepsilon}_{13}^2}$) using the same technique that has been described for the stress description in the σ_N-σ_T plane (see S.1.3.3), and similarly, all other visualization methods are applicable in this case as well, see also discussion in Remark 2.3.

The reader may activate **P.1.9, P.1.10** to transform strain coefficients by coordinate system rotation and **P.1.12** for visualization of a state of strain at a point. **P.1.11** executes transformations of a strain tensor between curvilinear and Cartesian coordinates.

1.4 Energy Theorems

In this section we shall review some work- and energy-based measures that are encountered in the theory of elasticity. These measures may be further exploited to create functionals that may be combined with *variational analysis* to derive the associated *Euler's equations*, see S.1.5, and other solutions in this area.

In general, two main measures, the *"Potential Energy"* and the *"Complementary Energy"*, should be addressed. These measures serve as the basis to the well known *Theorem of Minimum Potential Energy*, the *Theorem of Minimum Complementary Energy* and the *Theorem of Reciprocity.*

1.4.1 The Theorem of Minimum Potential Energy

Suppose that a body of volume B and surface $S = S_L + S_D$, has reached equilibrium under the action of distributed load $\mathbf{F}_s$ over the S_L part of its surface, and body force, $\mathbf{F}_b$, that applies at each material point. The deformation $\mathbf{u} = (u_1, u_2, u_3)$ over the remaining S_D part of the surface is known as well. We introduce a variation of them, $\delta\mathbf{u}$ (i.e. "virtual displacements"), that vanish over S_D. Since the body and the surface loads are fixed during the application of the virtual displacements, and $\delta\mathbf{u} = 0$ on S_D, one can write the "virtual work" as

$$\delta U = \iint_{S_L} \mathbf{F}_s \cdot \delta\mathbf{u} + \iiint_B \mathbf{F}_b \cdot \delta\mathbf{u} = \delta\Big(\iint_{S_L} \mathbf{F}_s \cdot \mathbf{u} + \iiint_B \mathbf{F}_b \cdot \mathbf{u}\Big) \tag{1.122}$$

where " $\cdot$ " is a scalar product of vectors. Further on, the strain energy is given by

$$U \equiv \iiint_B W, \tag{1.123}$$

where W is the *volume density of the strain energy*, see S.2.10. Equating the variation of the external load potential and the strain energy, may be written as $\delta V = 0$, where

$$V = \iiint_B W - \iint_{S_L} \mathbf{F}_s \cdot \mathbf{u} - \iiint_B \mathbf{F}_b \cdot \mathbf{u}. \tag{1.124}$$

This enables us to conclude that *the potential energy, V, has a stationary value among all admissible variations of the displacements u_i from the equilibrium state.*

Since $V(\mathbf{u} + \delta\mathbf{u}) \geq 0$ for any variation $\delta\mathbf{u}$, the Theorem of Minimum Potential Energy may be expressed as:

Among all displacements that satisfy the boundary conditions, those that satisfy the equilibrium equations as well make the potential energy an absolute minimum.

The converse theorem is true as well, see (Sokolnikoff, 1983).

Remark 1.7 In the time-dependent case, the above statement may be extended to the well-known *Hamilton's principle* for the time-dependent virtual displacements that satisfy all geometrical boundary conditions at all times, see e.g. (Langhaar, 1962):

$$\int_{t_1}^{t_2} \delta(T - V)\, dt = 0, \tag{1.125}$$

where T stands for the *total kinetic energy* stored in the system, and t_1 and t_2 are times where the deformation is known. This principle may be invoked to derive the equations of motion of an elastic body, or in other words, the resulting Euler's equations of such functional are time-dependent. For time-dependent cases, one should also consider employing Ritz's method (see (Sokolnikoff, 1983) and further on in S.1.6.5.4), which leads to Lagrange's equations, see Examples 1.6, 1.8.

1.4.2 The Theorem of Minimum Complementary Energy

In contrast with the previous case where displacement components variations were dealt with, we shall now employ small variations of the stress distribution.

Let σ_{ij} be a known set of stress distributions satisfying the equilibrium and compatibility equations, the natural boundary conditions on S_L, and the geometrical boundary conditions on S_D as required. We introduce a perturbation, or a "virtual" (small) system of stress $\delta\sigma_{ij}$, see (1.85a–c), as

$$\delta X_s = \delta\sigma_x \cos(\bar{\mathbf{n}}, x) + \delta\tau_{xy} \cos(\bar{\mathbf{n}}, y) + \delta\tau_{xz} \cos(\bar{\mathbf{n}}, z), \tag{1.126a}$$
$$\delta Y_s = \delta\tau_{xy} \cos(\bar{\mathbf{n}}, x) + \delta\sigma_y \cos(\bar{\mathbf{n}}, y) + \delta\tau_{yz} \cos(\bar{\mathbf{n}}, z), \tag{1.126b}$$
$$\delta Z_s = \delta\tau_{xz} \cos(\bar{\mathbf{n}}, x) + \delta\tau_{yz} \cos(\bar{\mathbf{n}}, y) + \delta\sigma_{zz} \cos(\bar{\mathbf{n}}, z). \tag{1.126c}$$

Further on, we assume that:

(a) $\delta\sigma_{ij}$ satisfy equilibrium (when body forces are ignored);

(b) $\delta\mathbf{F}_s$ vanish on S_L;

(c) $\delta\sigma_{ij}$ are arbitrary on S_D (i.e. not constrained).

Note that $\delta\sigma_{ij}$ do not necessary satisfy the compatibility equations. Subsequently, for a given $\delta\sigma_{ij}$, the variation of the strain energy is defined as

$$\delta U = \iiint_B \delta W = \underline{\iint_{S_D} \mathbf{u} \cdot \delta\mathbf{F}_s} + \underline{\underline{\iiint_B W(\delta\sigma_{ij})}}, \tag{1.127}$$

where $\mathbf{u}$ are the displacements associated with σ_{ij}. The once underlined term on the r.h.s. of (1.127) represents the work done by the virtual stresses on the displacements over S_D. The twice underlined term is the strain energy associated with $\delta\sigma_{ij}$ (which is a function of specific constitutive relations) and is always positive (see (2.25) and the discussion in S.2.10 regarding the positive-definite stress-strain law). Thus, one may define a quantity called "complementary energy" as

$$V^* = U - \iint_{S_D} \mathbf{u} \cdot \mathbf{F}_s, \tag{1.128}$$

and constitute the *Theorem of Minimum Complementary Energy* by stating that:

Among all stress distributions that satisfy the equilibrium equations and the boundary conditions, those that satisfy the compatibility equations as well make the complementary energy an absolute minimum.

The converse theorem is true as well, see also (Sokolnikoff, 1983) and Example 7.2.

We may now utilize the Theorem of Minimum Complementary Energy and confine the discussion to the case of $S_D = 0$ (i.e. $S_L = S$) or the case where $\mathbf{u}$ vanish on S_D, which formally yields $V^* = U$. In words, in the case where no geometrical constraints exist (i.e. only natural boundary conditions are imposed), or in the case where zero displacements are forced over S_D, one may state that:

Among all stress distributions that satisfy the equilibrium equations and the natural boundary condition, those that satisfy the compatibility equations as well make the strain energy an absolute minimum.

This theorem is frequently referred to as *The Principle of Least Work.*

Remark 1.8 In a state of equilibrium under the action of surface loads $\mathbf{F}_s$ and body force $\mathbf{F}_b$, the strain energy, U, is equal to half of the work that would be done by surface and body loads (of magnitude equal to their values in the equilibrium state) by acting through the displacements from the unstressed state to the equilibrium state, namely,

$$2U = \iint_{S_L} \mathbf{F}_s \cdot \mathbf{u} + \iiint_B \mathbf{F}_b \cdot \mathbf{u}. \tag{1.129}$$

This *Clapeyron's Theorem* does not conflict with (1.124), which is set for a *small virtual perturbation* of the displacements and not *their magnitude in the equilibrium state.*

1.4.3 Theorem of Reciprocity

Consider two observations of the response of a given elastic body to two different systems of loads. First, the loading system $\mathbf{F}_s^{(1)}$, $\mathbf{F}_b^{(1)}$ has been applied and created a deformation $\mathbf{u}^{(1)}$ and strain energy $U^{(1)}$. Then, (after the first loading system has been removed) a loading system $\mathbf{F}_s^{(2)}$, $\mathbf{F}_b^{(2)}$ has created the deformation $\mathbf{u}^{(2)}$ and the associated strain energy $U^{(2)}$. Applying now the first system loading and then the second one (without removing the first one), we get the deformation $\mathbf{u}^{(1)} + \mathbf{u}^{(2)}$ and strain energy $U^{(1)} + U^{(2)} + U^{(12)}$. Inverting the order of loading we get the deformation $\mathbf{u}^{(2)} + \mathbf{u}^{(1)}$ and strain energy $U^{(2)} + U^{(1)} + U^{(21)}$. Since we are reaching the same state in both experiments, $U^{(21)} = U^{(12)}$. This result constitutes the *Reciprocal Theorem* of Betti and Rayleigh:

If an elastic body is subjected to two systems of body and surface loads, the work that would be done by the first system in acting through the displacements that were created by the action of the second system is equal to the work that would be done by the second system in acting through the displacements that were created by the action of the first system.

This result may be written as

$$\iint_S \mathbf{F}_s^{(1)} \cdot \mathbf{u}_i^{(2)} + \iiint_B \mathbf{F}_b^{(1)} \cdot \mathbf{u}_i^{(2)} = \iint_S \mathbf{F}_s^{(2)} \cdot \mathbf{u}_i^{(1)} + \iiint_B \mathbf{F}_b^{(2)} \cdot \mathbf{u}_i^{(1)}. \tag{1.130}$$

As a special case of the above theorem, we may consider two discrete surface loads acting over points A and B of an elastic body, and suppose that no body forces are applied. When the force $\mathbf{F}_s(A)$ (at A) is applied, the displacements $\mathbf{u}_A^A$ at A and $\mathbf{u}_B^A$ at B are obtained. When the force $\mathbf{F}_s(B)$ (at B) is applied, the displacements $\mathbf{u}_A^B$ at A and $\mathbf{u}_B^B$ at B are obtained. For this case (1.130) yields

$$\mathbf{F}_s(A) \cdot \mathbf{u}_A^B = \mathbf{F}_s(B) \cdot \mathbf{u}_B^A. \tag{1.131}$$

If we now adopt the notation $\mathbf{u}_A^B = \alpha_{AB}\,\mathbf{F}_s(B)$ and $\mathbf{u}_B^A = \alpha_{BA}\,\mathbf{F}_s(A)$ (where the matrices α_{AB} and α_{BA} are usually referred to as the *influence coefficients*), (1.131) shows that $\alpha_{AB} = \alpha_{BA}$.

1.4.4 Castigliano's Theorems

We shall now examine the case of the Theorem of Minimum Potential Energy, where body forces are absent and the surface loading $\mathbf{F}_{sk}$ $(k = 1, \ldots, K)$ are discrete. For such a case

$$V = \iiint_B W(\mathbf{u}_1, \ldots, \mathbf{u}_K) - \sum_{k=1}^{K} \mathbf{F}_{sk} \cdot \mathbf{u}_k, \tag{1.132}$$

where $\mathbf{F}_{sk}$ are surface loading and $\mathbf{u}_k$ are displacements of the $k^{\underline{th}}$ point (of the application of the force $\mathbf{F}_{sk}$). Variation of V leads to $\sum_{k=1}^{K} \frac{\partial V}{\partial \mathbf{u}_k} \delta \mathbf{u}_k = 0$, while (1.123) together with the fact that the variations $\delta \mathbf{u}_k$ are arbitrary, yield the system $\mathbf{F}_{sk} = \frac{\partial U}{\partial \mathbf{u}_k}$. This result, widely known as *Castigliano's First Theorem*, may be interpreted as follows: *the partial derivative of the strain energy with respect to generalized displacements is equal to the corresponding generalized force*.

Note that this result stands for small perturbations, and thus should be applied either to the case of relatively small loads, or to the case where the system is linear in the sense that the resulting displacements vary linearly with the loads (in such a case, $U_{,u_k}$ is a linear function of u_k). To develop *Castigliano's Second Theorem* we write the potential energy as

$$V = \iiint_B W(\mathbf{F}_{s1}, \ldots, \mathbf{F}_{sK}) - \sum_{k=1}^{K} \mathbf{F}_{sk} \cdot \mathbf{u}_k, \tag{1.133}$$

and again the minimization leads to equations $\sum_{k=1}^{K} \frac{\partial V}{\partial \mathbf{F}_{sk}} \delta \mathbf{F}_{sk} = 0$, which for arbitrary variations $\delta \mathbf{F}_{sk}$ yield $\mathbf{u}_k = \frac{\partial U}{\partial \mathbf{F}_{sk}}$. The above may put in words as:

The partial derivative of the strain energy with respect to a generalized force is equal to the corresponding generalized displacements.

The reservation mentioned above regarding the system linearity holds in this case as well.

1.5 Euler's Equations

In this section we shall employ *variational analysis* (or *calculus of variations*) techniques that deal with minimization of functionals. In the present context, a functional is an operator that converts a set of functions to a number. A fundamental result of the calculus of variations is that the extreme values of a functional must satisfy an associated differential equation (or a set of differential equations) over the discussed domain that are generally termed *Euler's equations*.

The notion “extreme values” stands for local minima, maxima or inflection points. Hence, the underlying idea is founded on the existence of a physical global quantity that remains maximal or minimal at all times, regardless of the nature (i.e. stationary or time-dependent) of the problem. Note that variational calculus is a fundamental analytical tool in many other areas of general physics and engineering, and a review of the mathematical methods associated with it may be found, e.g., in (Sagan, 1969).

Two general assumptions are typically associated with the analytical methodologies applied in calculus of variations. First, we generally assume that all functions and functionals are continuous and have continuous derivatives as required. In addition, the functional values are assumed to be positive.

This section contains a brief survey of the techniques associated with deriving Euler’s equations out of a given functional, and examples for use of these equations in the area of elasticity.

For the sake of abbreviating, in this section, derivatives of functions of one variable, for example $y(x)$, are denoted as $\frac{dy}{dx} = y'$, $\frac{d^2y}{dx^2} = y''$, $\frac{d^my}{dx^m} = y^{(m)}$.

1.5.1 Functional Based on Functions of One Variable

We shall first calculate extreme values of integral functional, J, whose integrand, F, contains one or several functions associated with the admissible function $y(x)$ of the C^2 class on the interval $[x_0,x_1]$. As an example, consider the problem

$$J(y) = \int_{x_0}^{x_1} F(x,y,y')\,dx \rightarrow \min, \tag{1.134}$$

where F is a continuous function of three arguments (the problem of determining a maximum may be dispensed with F replaced by $-F$). The boundary values of $y(x)$ are generally given as

$$y(x_0) = y_0, \qquad y(x_1) = y_1. \tag{1.135}$$

The minimization of the functional $J(y)$ leads to the *Euler’s equation* for its integrand, see (Sokolnikoff, 1983),

$$F_{,y} - \frac{d}{dx}(F_{,y'}) = 0, \tag{1.136}$$

where, obviously,

$$\frac{d}{dx}(F_{,y'}) = F_{,y'y'}y'' + F_{,y'y}y' + F_{,y'x}. \tag{1.137}$$

Equation (1.136) is a necessary condition that $J(y)$ possess a stationary value.

A similar derivation may be carrying out for the extreme problem

$$J(y) = \int_{x_0}^{x_1} F(x,y,y',y'',\ldots,y^{(m)})\,dx \rightarrow \min, \tag{1.138}$$

where $m \geq 1$, and the admissible function $y(x)$ belongs to the C^{m+1} class on the interval $[x_0,x_1]$, and satisfies the boundary conditions

$$\begin{aligned} y(x_0) &= y_0, \quad y'(x_0) = y_0{}', \quad \ldots \quad y^{(m)}(x_0) = y_0, \\ y(x_1) &= y_1, \quad y'(x_1) = y_1{}', \quad \ldots \quad y^{(m)}(x_1) = y_1. \end{aligned} \tag{1.139}$$

The minimization in this case leads to the following *Euler’s equation*:

$$F_{,y} - \frac{d}{dx}(F_{,y'}) + \frac{d^2}{dx^2}(F_{,y''}) -, \cdots, +(-1)^m \frac{d^m}{dx^m}(F_{,y^{(m)}}) = 0, \tag{1.140}$$

while $\frac{d^i}{dx^i}F_{,y^{(i)}}$ are derived analogously to (1.137).

To generalize the previous case of (1.134), consider the problem defined in a different way by the functional

$$J(y) = \int_{x_0}^{x_1} F(x, y_1, \dots, y_n, y_1{}', \dots, y_n{}')\, dx \to \min, \tag{1.141}$$

where $n \geq 1$, and F is a continuous function of $2n+1$ arguments. We suppose that the admissible functions $y_i(x)$ of one variable belong to the C^2 class on the interval $[x_0, x_1]$, and that the boundary values are defined as

$$y_i(x_0) = y_{i0}, \qquad y_i(x_1) = y_{i1}, \qquad 1 \leq i \leq n. \tag{1.142}$$

The minimization leads to the following system of Euler's equations:

$$F_{,y_i} - \frac{d}{dx}(F_{,y_i{}'}) = 0, \qquad 1 \leq i \leq n. \tag{1.143}$$

To generalize the case of (1.141) furthermore, we note that similar calculations performed on the extreme problem

$$J(y) = \int_{x_0}^{x_1} F(x, y_1, \dots, y_n, y_1{}' \dots, y_n{}', \dots, y_1^{(m)}, \dots, y_n^{(m)})\, dx \to \min \tag{1.144}$$

lead to the following system of Euler's equations:

$$F_{,y_i} - \frac{d}{dx}(F_{,y_i{}'}) + \frac{d^2}{dx^2}(F_{,y_i{}''}) -, \cdots, +(-1)^m \frac{d^m}{dx^m}(F_{,y_i^{(m)}}) = 0, \qquad 1 \leq i \leq n. \tag{1.145}$$

The **Euler Lagrange**$(F, x, y(x))$ and **Euler Lagrange**$(F, x, [y_1(x), \dots, y_n(x)])$ commands from the *Variational Calculus* package of (Maple, 2003), compute the Euler's equations of the functionals (1.134), (1.141). The higher-order functionals (1.138), (1.144) may be reduced to these forms as well.

Remark 1.9 Consider the variational problem (1.141) when for each function, only one of the boundary conditions (1.142) is given. For example, $y_i(x_0) = y_{i0}$, $1 \leq i \leq n$. Clearly, the admissible functions $y_i(x)$ in this case form a larger class. The minimizing process shows that Euler's equations given in (1.143) are obtained only if the following boundary conditions (usually called "natural") are satisfied:

$$F_{,y_i{}'}(x_1) = 0, \qquad 1 \leq i \leq n. \tag{1.146}$$

This characteristic of the variational process is one of its profound advantages in the area of elasticity, as it is capable of providing part of the boundary conditions as well.

Example 1.2 *Rotating Beam.*

This example demonstrates the application of the Theorem of Minimum Potential Energy to derive Euler's equation associated with rotating isotropic beam. In this case, there are three components that contribute to the potential energy: the strain energy, the rotational tension, and the surface loads. Let z be a coordinate along the beam axis, while we are looking for the beam axis deflection in the y direction $v(z)$, $0 \leq z \leq l$, see Fig. 1.13. The (bending) strain energy U_B in this example will be expressed by the bending curvature v'' and Young's modulus, E (see (5.44) with $1/a_{33} = E$), namely

$$U_B = \iiint_B W = \frac{1}{2}\iiint_B \sigma_z \varepsilon_z = \frac{1}{2}\int_0^l EI_x \left(v''\right)^2, \tag{1.147}$$

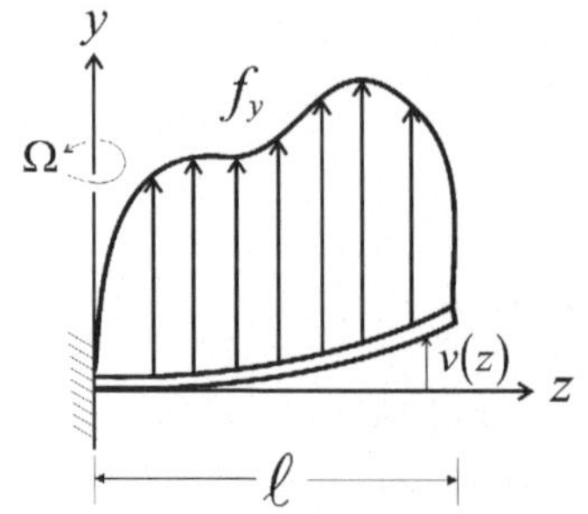

Figure 1.13: Rotating beam notation in Example 1.2.

where $I_x = \iint_\Omega y^2$ is the cross-section moment (of inertia) about the x-axis. The rotational tension may be treated in two different ways. We first may consider this effect as a contributor to the strain energy. Its contribution equals to the product of the (given) tensile force, $T(z)$, and the extension created by the bending $v(z)$, which is written as $\sqrt{1+v'^2}-1 \cong \frac{1}{2}(v')^2$. Hence,

$$U_T = \int_0^l \frac{1}{2} T(z) \left(v'\right)^2 dz. \tag{1.148}$$

Alternatively, the tension effect may be viewed as a transverse distributed loading of $(T(z)\,v')'$. Since this loading is not constant but *depends on the displacements* (unlike the body loads in (1.124)), its potential should be therefore written as

$$V_T = -\frac{1}{2}\int_0^l (T(z)\,v')'\,v\,dz. \tag{1.149}$$

The given surface loads in the y direction, $f_y(z)$, do not depend on the deformation. Therefore, their potential is $-\int_0^l f_y(z) y\,dz$. Overall, the functional F of (1.138) may be written as

$$F = \frac{1}{2} EI_x \left(v''\right)^2 - \frac{1}{2}(T(z)\,v')'\,v - f_y(z) v. \tag{1.150}$$

By employing the minimization results of (1.140) with $F = F(z, v, v', v'')$, see also **P.1.13**, we obtain the Euler's equation

$$(EI_x v'')'' - (T(z) v')' - f_y(z) = 0. \tag{1.151}$$

For a rotation about the y-axis as shown in Fig. 1.13, the tension force is $T = \frac{1}{2} m(z) \Omega^2 (l^2 - z^2)$, where $m(z)$ is the mass distribution per unit length and Ω is the rotational velocity.

1.5.2 Variational Problems with Constraints

Consider a case where *Lagrange multipliers* should be employed within the variational problem (1.134) with the *isoperimeter conditions*

$$J_k(y) = \int_{x_0}^{x_1} F_k(x, y, y')\,dx = g_k, \qquad 1 \le k \le n, \tag{1.152}$$

where g_k are constants and F_k are continuous functions of three arguments. The corresponding Lagrange functional has the form

$$J_L(y) = \int_{x_0}^{x_1} [F(x, y, y') + \textstyle\sum_{k=1}^{n} \lambda_k F_k(x, y, y')]\,dx. \tag{1.153}$$

In this case we solve the variation problem $J_L(y) \to \min$, by considering Lagrange multipliers, λ_k, as constants.

Example 1.3 *Elastica.*

The analysis described in this example deals with large deformations of an *elastic rod.* This kind of problems are traditionally termed "Elastica". For further reading see (Frisch-Fay, 1962), (Stronge and Yu, 1993). The results presented below are documented in **P.1.14**.

The *bending energy* of an elastic rod, which is deformed as the plane curve $\mathbf{r}(s) = [x(s), y(s)]$ of length l is assumed, for simplicity, to be proportional to the integral of the squared curvature over the length of the curve, i.e. $\int_0^l \kappa^2(s)\,ds$, see Example 1.2. Here $s \in [0,l]$ is the natural length parameter of the curve. Recall that the *curvature* $\kappa(s)$ of a plane curve is given as $\frac{d\theta}{ds}$ where $\theta(s)$ is the angle between the local tangent line and the x-axis. One may therefore ask the following question: *what shape will the curve take if the total turning of its tangent is given and the turning is zero and* θ_e *at the endpoints*. As a constrained variational problem we write

$$J = \int_0^l (\frac{d\theta}{ds})^2\,ds \to \min, \quad J_1 = \int_0^l \theta\,ds = \tilde{g}, \quad \text{with} \quad \theta(0) = 0, \quad \theta(l) = \theta_e. \tag{1.154}$$

The Lagrange functional of (1.153) takes the form $J_L(y) = \int_0^l [(\frac{d\theta}{ds})^2 + \lambda\theta]\,ds$ and the function F depends on two of the three variables $s, \theta, \frac{d\theta}{ds}$ and on the parameter λ. The Euler's equation for the constrained problem is of the form of (1.136),

$$\lambda - 2\frac{d^2\theta}{ds^2} = 0. \tag{1.155}$$

Using the boundary conditions one may find the solution $\theta = \frac{\lambda}{4}s^2 - \frac{\lambda l^2 - 4\theta_e}{4l}s$, that depends on the parameter λ, which with the constraint $J_1 = \tilde{g}$ gives

$$\tilde{g} = (\frac{\lambda}{12}s^3 - \frac{\lambda l^2 - 4\theta_e}{8l}s^2)|_0^l = \frac{l}{24}\left(12\theta_e - \lambda l^2\right). \tag{1.156}$$

Hence, $\lambda = \frac{12}{l^3}\left(\theta_e l - 2\tilde{g}\right)$ and

$$\theta = (3\theta_e - \frac{6\tilde{g}}{l})\left(\frac{s}{l}\right)^2 + (-2\theta_e + \frac{6\tilde{g}}{l})\frac{s}{l}. \tag{1.157}$$

To present the above result we note that a plane curve $\mathbf{r}(s) = [x(s), y(s)]$ with a given curvature function $\kappa(s) = \frac{d\theta}{ds}$, may be reconstructed in view of

$$\frac{dx}{ds}(s) = \cos\theta, \qquad \frac{dy}{ds}(s) = \sin\theta, \tag{1.158}$$

by the formulas

$$x(s) = x_0 + \int_0^s \cos\theta\,ds, \qquad y(s) = y_0 + \int_0^s \sin\theta\,ds, \qquad \theta = \int_0^s \kappa\,ds. \tag{1.159}$$

The resulting curve is known as *Euler's spiral* (or *the spiral of Cornu*), and an example of it is shown in Fig. 1.14.

Another example in the same class deals with an elastic rod, the shape of which is expressed as $x = x(s), y = y(s)$ while its angles at the end points (x_1, y_1) and (x_2, y_2) are θ_1 and θ_2, respectively. The constraints of this problem are written by the curve projections onto the x- and y- axes as, see (1.158),

$$\int_0^l \cos\theta\,ds = x_2 - x_1, \qquad \int_0^l \sin\theta\,ds = y_2 - y_1. \tag{1.160}$$

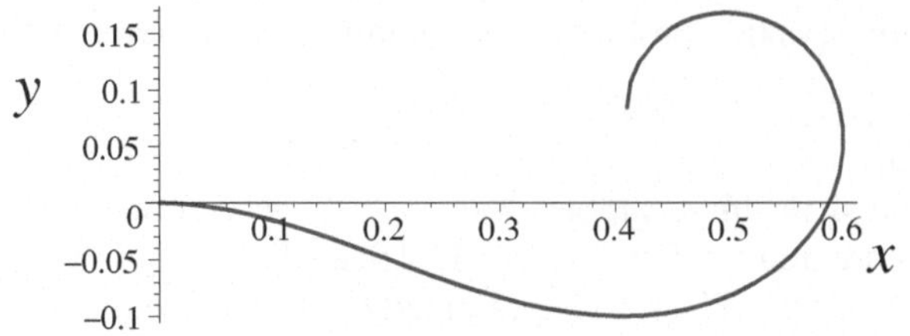

Figure 1.14: Euler's spiral for $l = 1,\ \theta_e = \frac{3}{2}\pi,\ \widetilde{g} = 1$.

Using two Lagrange multipliers, we write the (1.153) type integral

$$J_L = \int_0^l [(\frac{d\theta}{ds})^2 + \lambda_1 \cos\theta + \lambda_2 \sin\theta]\, ds \rightarrow min. \tag{1.161}$$

The resulting Euler's equation (1.136) depends on the parameters λ_1, λ_2 and has the form of

$$-2\frac{d^2\theta}{ds^2} + \lambda_2 \cos\theta - \lambda_1 \sin\theta = 0. \tag{1.162}$$

One may now move to a new coordinate system, say $\widetilde{x}$, $\widetilde{y}$, by rotating the plane by an angle, say $\Delta\theta$, so that the curve of (1.162) will be described as

$$\frac{d^2\widetilde{\theta}}{ds^2} = \mu \cos\widetilde{\theta}, \tag{1.163}$$

where $\widetilde{\theta} = \theta + \Delta\theta$ and $\mu = \frac{1}{2}\sqrt{\lambda_1^2 + \lambda_2^2}$. This new problem is now associated with the modified boundary conditions and constraints, which are written with the transformed values $(\widetilde{x}_1, \widetilde{y}_1)$, $(\widetilde{x}_2, \widetilde{y}_2)$. Subsequently, the above problem should be solved under the conditions of $\widetilde{\theta}_1 = \theta_1 + \Delta\theta$ at $(\widetilde{x}_1, \widetilde{y}_1)$ and $\widetilde{\theta}_2 = \theta_2 + \Delta\theta$ at $(\widetilde{x}_2, \widetilde{y}_2)$. The solution, namely, $\widetilde{\theta} = \widetilde{\theta}(s, \lambda_1, \lambda_2)$, should also fulfill the modified constraints

$$\int_0^l \cos\widetilde{\theta}\, ds = \widetilde{x}_2 - \widetilde{x}_1, \qquad \int_0^l \sin\widetilde{\theta}\, ds = \widetilde{y}_2 - \widetilde{y}_1, \tag{1.164}$$

which enables the determination of λ_1 and λ_2. For more details regarding the above solution, see (Oprea, 1997).

1.5.3 Functional Based on Function of Several Variables

We next consider the problem

$$J(u) = \iint_\Omega F(x, y, u, u_x, u_y) \rightarrow \text{min}, \tag{1.165}$$

where F is a continuous function of five arguments. We suppose that the admissible functions $u(x,y)$ belong to the C^2 class on the two-dimensional domain Ω, while the boundary condition is formulated in terms of a given continuous function $\varphi(x,y)$ along the contour

$$u = \varphi \qquad \text{on} \quad \partial\Omega. \tag{1.166}$$

The minimization leads to the Euler's equation

$$F_{,u} - \frac{\partial}{\partial x}(F_{,u_x}) - \frac{\partial}{\partial y}(F_{,u_y}) = 0, \tag{1.167}$$

where, obviously,

$$\frac{\partial}{\partial x}(F_{,u_x}) = F_{,u_x u_x} u_{,xx} + F_{,u_x u} u_{,x} + F_{,u_x x}, \qquad \frac{\partial}{\partial y}(F_{,u_y}) = F_{,u_y u_y} u_{,yy} + F_{,u_y u} u_{,y} + F_{,u_y y}. \quad (1.168)$$

More generally, if the functional $F(x_1, \ldots, x_n, u, u_{x_1}, \ldots, u_{x_n})$ depends on $2n+1$ $(n > 1)$ variables including the function $u(x_1, \ldots, x_n)$ of n variables and its first derivatives, then the resulting Euler's equation takes the form

$$F_{,u} - \sum_{i=1}^{n} \frac{\partial}{\partial x_i}(F_{,u_{x_i}}) = 0. \quad (1.169)$$

Similar to (1.165) case, when higher (second-order) derivatives are included the extreme problem becomes

$$J(u) = \iint_G F(x, y, u, u_x, u_y, u_{xx}, u_{xy}, u_{yy}) \;\rightarrow\; \min, \quad (1.170)$$

where the admissible functions $u(x,y)$ belong to the C^3 class on the domain Ω, and take specified continuous values on the boundary. These calculations lead to the following *Euler's equation*

$$F_{,u} - \frac{\partial}{\partial x}(F_{,u_x}) - \frac{\partial}{\partial y}(F_{,u_y}) + \frac{\partial^2}{\partial x^2}(F_{,u_{xx}}) + \frac{\partial^2}{\partial x \partial y}(F_{,u_{xy}}) + \frac{\partial^2}{\partial y^2}(F_{,u_{yy}}) = 0. \quad (1.171)$$

We shall not treat here explicitly the general Euler's equation, where the functional F depends on the function $u(x_1, \ldots, x_n)$ (of n variables) and its higher derivatives (of order $\leq m$). Such derivation is implemented in **P.1.13**, which was used to create the following examples.

Example 1.4 *Variational Problem Related to Poisson's Equation.*
Of a particular interest is the special class of Euler's equations that emerge from the functional

$$F = a_{44}(u_x)^2 - 2a_{45} u_x u_y + a_{55}(u_y)^2 + 2uf(x,y), \quad (1.172)$$

where f is a given function, and the real numbers $a_{ij},\ i,j = 4,5$ satisfy $a_{44} > 0,\ a_{55} > 0,\ a_{44}a_{55} - a_{45}^2 > 0$. In this particular case of (1.165), the minimizing process yields the Euler's equation known as the generalized *Poisson's equation* in Ω:

$$a_{44} u_{,xx} - 2a_{45} u_{,xy} + a_{55} u_{,yy} = f(x,y). \quad (1.173)$$

In the isotropic case, where $a_{44} = a_{55},\ a_{45} = 0$, the differential operator of (1.173) is proportional to the classic *Laplacian* $\nabla^{(2)} = \frac{\partial^2}{\partial x^2} + \frac{\partial^2}{\partial y^2}$ (which is a scalar square of the "nabla" operator $\nabla = \{\frac{\partial}{\partial x}, \frac{\partial}{\partial y}\}$), see also (Sokolnikoff, 1983). More general versions of this operator appear in Chapter 3.

Example 1.5 *Variational Problem Related to the Biharmonic problem.*
One may examine the variational problem of the functional

$$F = (u_{xx} + u_{yy})^2 - 2f(x,y)u, \quad (1.174)$$

where f is a given function, and the admissible functions, $u(x,y)$, belong to the C^4 class on the two-dimensional domain, Ω, and are subjected to the boundary conditions on the contour, $\partial\Omega$, given by

$$u = \varphi, \qquad \frac{d}{dn}u = h \qquad \text{on} \quad \partial\Omega, \quad (1.175)$$

where $\frac{d}{dn}$ is a normal derivative and $\varphi(x,y)$, $h(x,y)$ are given functions on $\partial\Omega$. In this case of (1.170), we find that the minimization process yields the Euler's equation known as the fundamental form of the *biharmonic equation* in Ω:

$$\nabla^{(4)} u = f, \tag{1.176}$$

where $\nabla^{(4)} = \frac{\partial^4}{\partial x^4} + 2\frac{\partial^4}{\partial x^2 \partial y^2} + \frac{\partial^4}{\partial x^4}$ is the *biharmonic operator*, see (Sokolnikoff, 1983). Replacing the functional of (1.174), for example, by

$$F = c_0(u_{xx})^2 + c_1 u_{xx} u_{xy} + c_2 u_{xx} u_{yy} + c_3 u_{xy} u_{yy} + c_4(u_{yy})^2 + 2f(x,y)u, \tag{1.177}$$

where c_i are real numbers satisfying certain "positiveness" conditions, one obtains (1.176) with a more general version of the biharmonic operator $\nabla^{(4)}$, see **P.1.13**,

$$c_0\frac{\partial^4}{\partial x^4} + c_1\frac{\partial^4}{\partial x^3 \partial y} + c_2\frac{\partial^4}{\partial x^2 \partial y^2} + c_3\frac{\partial^4}{\partial x \partial y^3} + c_4\frac{\partial^4}{\partial x^4}. \tag{1.178}$$

This operator is extensively treated in Chapter 3.

Example 1.6 *Dynamic Torsion of a Beam.*
Consider an isotropic beam of circular cross-section, the axis of which is stretched along the z-axis and is acted upon by a torsional moment distribution $m_z(z,t)$. The beam is characterized by a torsional rigidity, $D(z)$, and a polar moment of inertia, $\mathrm{I}_p(z)$. The twist angle is denoted $\phi(z,t)$. In this case V and T may be simplified to (see also Remark 7.5)

$$V = \frac{1}{2}\int_0^l D(z)\,(\phi_{,z})^2 dz - \int_0^l m_z(z,t)\,\phi\, dz, \qquad T = \frac{1}{2}\int_0^l \mathrm{I}_p(z)\,(\phi_{,t})^2 dz, \tag{1.179}$$

and therefore, according to (1.125)

$$J(\phi) = \int_{t_1}^{t_2} (T - V)\, dt. \tag{1.180}$$

We then arrive at a functional $F(z,t,\phi,\phi_z,\phi_t)$ defined in (1.165) that takes the form

$$F = \frac{1}{2}\mathrm{I}_p(z)\,(\phi_t)^2 - \frac{1}{2}D(z)\,(\phi_z)^2 + m_z(z,t)\,\phi. \tag{1.181}$$

Subsequently, (1.167) yields with the aid of (1.168) the governing Euler's equation, which may be titled as the "equation of motion" in this case, see **P.1.13**,

$$[D(z)\,\phi_{,z}]_{,z} - \mathrm{I}_p(z)\,\phi_{,tt} + m_z(z,t) = 0. \tag{1.182}$$

1.6 Analytical Methodologies

Strictly speaking, there is no rigorous set of rules that allows one to follow and create analytical solutions for any new problem. For that reason, analytic solutions are still considered as "art", while numerical codes (such as the finite-element method, etc.) seem to offer a more "straightforward" solution for any type and geometry of a new problem. This fact emerges from the variety of techniques that are used for analytical solutions, and from the almost "uniform" approach that may be exploited when numerical tools are utilized (see also the preface to this book). Hence, with that respect, analytical methods are less "direct" compared with numerical methods, and a bit of guessing and intuition are always helpful. Yet, there are few standard paths that are commonly used to create analytical solutions, and it may be stated that most of the existing analytical solutions make use of one of these paths. It is therefore beneficial to classify and discuss these groups of methodologies.

In what follows, we shall present the most useful range of analytical methodologies in anisotropic elasticity. Bearing in mind that each of the relevant methodologies is a discipline by itself, we shall confine the discussion here to the fundamentals of each method.

Prior to the description of the above paths, we shall summarize and review some basic definitions and the various systems of governing equations that are encountered in the field of elasticity, and the specific role of each one of them.

Similar to the previous section, we shall adopt here the notation for derivatives of functions of one variable as $\frac{dy}{dx} = y'$, $\frac{d^2y}{dx^2} = y''$, $\frac{d^my}{dx^m} = y^{(m)}$.

1.6.1 *The Fundamental Problems of Elasticity*

We shall first review the three basic fundamental problems that are generally treated within the theory of elasticity, while it is a common practice to try and describe all problems encountered as special cases of these three problems. We shall present here only the steady state version of these problems and for simplicity, work in Cartesian coordinates.

The first fundamental problem considers a body that undergoes a given distribution of loads over its boundary surface, S. This problem may be expressed as: *find the deformation (displacement) functions $u(x,y,z)$, $v(x,y,z)$ and $w(x,y,z)$ that satisfy the equilibrium equations (1.82a–c) and the boundary conditions*

$$X_s = f_x, \qquad Y_s = f_y, \qquad Z_s = f_z \qquad on \quad S, \tag{1.183}$$

where X_s, Y_s, Z_s are the stress resultants of (1.85a–c), and f_x, f_y, f_z are known functions on S.

The second fundamental problem considers a body that undergoes a given distribution of boundary surface deformation. It may be expressed as: *find the deformation (displacement) functions $u(x,y,z)$, $v(x,y,z)$ and $w(x,y,z)$ that satisfy the equilibrium equations (1.82a–c) and the boundary conditions*

$$u = u^*, \qquad v = v^*, \qquad w = w^* \qquad on \quad S, \tag{1.184}$$

where u^*, v^*, w^* are known functions on S.

The third fundamental problem (sometimes referred to as *the mixed problem*, see (Muskhelishvili, 1953)), considers a body that undergoes a given distribution of surface loads on part of its boundary surface, say S_L, and a given distribution of surface deformation on the remaining boundary surface, say S_D (i.e. $S = S_L \bigcup S_D$). It may be expressed as: *find the deformation (displacement) functions $u(x,y,z)$, $v(x,y,z)$ and $w(x,y,z)$ that satisfy the equilibrium equations (1.82a–c) and the boundary conditions*

$$X_s = f_x, \qquad Y_s = f_y, \qquad Z_s = f_z \qquad on \quad S_L, \tag{1.185a}$$

$$u = u^*, \qquad v = v^*, \qquad w = w^* \qquad on \quad S_D, \tag{1.185b}$$

where f_x, f_y, f_z and u^*, v^*, w^* are known functions on S_L and S_D, respectively.

Note that in all cases, the equilibrium equations of (1.82a–c) contain the body-force components as well.

1.6.2 *Fundamental Ingredients of Analytical Solutions*

Equilibrium Equations: The first system of equations that should be discussed are the equilibrium equations in their generic form (1.79), or in their linear version for Cartesian coordinates as presented by (1.82a–c). These equations were generally derived for elastic media regardless of its characteristics (i.e., isotropic or anisotropic) in terms of the stress components and the body force at a point. By employing the appropriate stress-strain relations, one may express these equations in terms of the strain components, which will make the resulting equations material-dependent (and may complicate their exploitation in many cases). Therefore, the equilibrium equations are usually kept in terms of stresses. When applied, the differential equilibrium equations are also referred to as *field equations* as they should be fulfilled at each material point.

Note that some analytical solutions also utilize various integral forms of the equilibrium equations, which are obtained by integrating the stresses over some definite volume portion, V, bounded by a surface S, of the elastic body. Adopting the notation of (1.81), (1.84) for the Cartesian case, simple static force and moment equilibrium show that

$$\iint_S \{X_s, Y_s, Z_s, xY_s - yX_s, xZ_s - zX_s, yZ_s - zY_s\}$$
$$+ \iiint_V \{X_b, Y_b, Z_b, xY_b - yX_b, xZ_b - zX_b, yZ_b - zY_b\} = \{0, 0, 0, 0, 0, 0\}. \quad (1.186)$$

The first three of the above equations are identical to (1.86).

Stress-Strain Relationships: For materials that exhibit linear stress-strain relationships, this system is basically the generalized Hook's law, that will be derived in S.2.1. For nonlinear materials, such relationships is typically based on power or exponential expressions. Most of the existing analytical solutions for *elastic* media (as opposed to solutions in plasticity), are derived for linear stress-strain relationships. This is also true for solutions that include geometrically nonlinear effects.

Compatibility Equations: The system of six compatibility equations derived in S.1.2.1 and represented by (1.48) (while the linear version for Cartesian coordinates is given by (1.45a–f)), is initially written in terms of the strain components. Clearly, using the appropriate stress-strain relations, the compatibility equations may be expressed in terms of stresses as well, but they will include material properties, which will make them less attractive in generic analytical solutions for anisotropic elasticity.

The linear and nonlinear version of the compatibility equations shown in S.1.2.1 should not be confused with linear and nonlinear expressions of displacement derivatives in the strain components, see S.1.1.2. However, for consistency, linear strain-displacements relations should be invoked when linear compatibility equations are employed, while nonlinear strain-displacements relations should be invoked when nonlinear compatibility equations are employed. Nevertheless, nonlinear analyses typically bypass the need for nonlinear compatibility equations by selecting solution trail "A" discussed in what follows within S.1.6.5.1.

Boundary Conditions: There are two main types of boundary conditions that should be discussed here, and they are traditionally referred to as *"natural"* and *"geometrical" (or "essential")* boundary conditions.

The *natural* boundary conditions are those of (1.85a–c), and are based on equating the internal stress components over the boundaries to the external surface load that acts there. These boundary conditions may sometimes be applied in an integral form. For example, at the "free-end" of a slender beam, one may require that instead of demanding that all six σ_i components will vanish at each point over the end cross-section, only the net (integral) resultant loads will vanish there (see further discussion in S.5.1.3).

The *geometrical* boundary conditions are restrictions posed on the displacements (and/or its derivatives) at some portion of the outer surface of the elastic media as shown by (1.184) or

(1.185b). This type of boundary conditions may also have an integral version. For example, for a clamp slender beam (cantilever), one may require that the mean displacements over the root cross-section will vanish as opposed to the requirement of zero displacements at each point over that cross-section. In this case, the requirement of zero displacement field yields three integral equations. A similar approach may be adopted for the root rotations (see details in S.5.1.3).

1.6.3 St. Venant's Semi-Inverse Method of Solution

The so-called "St. Venant's Semi-Inverse Method of Solution" serves as a basic solution philosophy in the theory of elasticity although, fundamentally, it is a general method that has no specific relation to a specific field or theory, and it may be applied elsewhere. Therefore, the method deserves special attention.

The St. Venant's Semi-Inverse Method of Solution consists of a set of equations that represent the *solution "mathematical structure"* or the *"solution hypothesis"*. These are sometimes called "assumptions", but not in the sense that some parts of the solution or some details of the problem's physics are ignored or overlooked. Essentially, these "assumptions" are just a presumed structure of the final mathematical solution, and typically include various parameters to be determined. Hence, the initial solution structure is an assumption, which may be based on a "guess", "intuition", "past experience", etc. It is important to stress the point that the source of these preliminary assumptions *need not be proved as long as they fulfill all relevant governing conditions once all unknown parameters are determined.* Thus, once a solution has been successfully reached, the above "assumptions" are converted into a valid solution.

When nonlinear problems are under discussion, one may raise the question of solution uniqueness. Typically, this question remains open, and the solution justification should be based on additional physical arguments.

1.6.4 Variational Analysis of Energy Based Functionals

Following the discussion of S.1.5, it is worth reiterating here that within the context of energy consideration in the theory of elasticity, the notion *variational analysis* stands for the class of methodologies that determine, reach and view the state of equilibrium in elastic media from a unique point of view. Such methods, utilizing the well-established theorems derived in S.1.4 that employ energy-related measures, show that equilibrium in elastic bodies is governed by global minimum principles, and combine it with the *Calculus of variations* as demonstrated by the elementary examples of S.1.5. The introduction of these methods into the theory of elasticity was first proposed by Ritz and Rayleigh, see (Sokolnikoff, 1983), and developed further by many others, see e.g. (Courant and Hilbert, 1989).

When variational tools are employed, two types of outcome should be expected. The first one includes the differential governing equations of the physical process, and is generally termed as the corresponding "Euler's Equations", already discussed in S.1.5. However, the second type of outcome is the actual (static or time-dependent) response of the elastic body. This type of outcome will be discussed further on within S.1.6.5.4.

1.6.5 Typical Solution Trails

As a general conclusion from the discussion so far, it may be stated that the governing equations of elasticity for an anisotropic body may be expressed using two different approaches. First, one may employ the approach that is generally referred to as *"Differential Equilibrium"*,

"Newton's laws" or *"Vector Mechanics"*, which all stand for the creation of a set of equations that will assure the fulfillment of the differential equilibrium equations, the differential compatibility equations, and the local (natural and geometrical) boundary conditions. Alternatively, one may employ the variational analysis of energy based functionals discussed in S.1.5, S.1.6.4 to either derive the governing equations and boundary conditions, or to directly derive a specific solution.

Strictly speaking, the two methods described above are equivalent. One may claim that the former approach that utilizes equilibrium considerations is more "physical oriented", while the concept of the latter is less intuitive in that sense. Generally speaking, while both approaches are suitable to accommodate closed-form or numerical solutions, variational analysis is sometimes superior as it is capable of supplying, in addition to a specific solution, the governing equations, and sometimes part of the boundary conditions (see Remark 1.9), as well.

In what follows, we shall review the basic solution trails that are typically encountered within the theory of anisotropic elasticity.

1.6.5.1 Solution Trail "A": Deformation Hypothesis

In view of S.1.6.3, Trail "A" consists of prescribing the expressions for the deformation components (or their measures) as presented by Fig. 1.15. Subsequently, the strain components

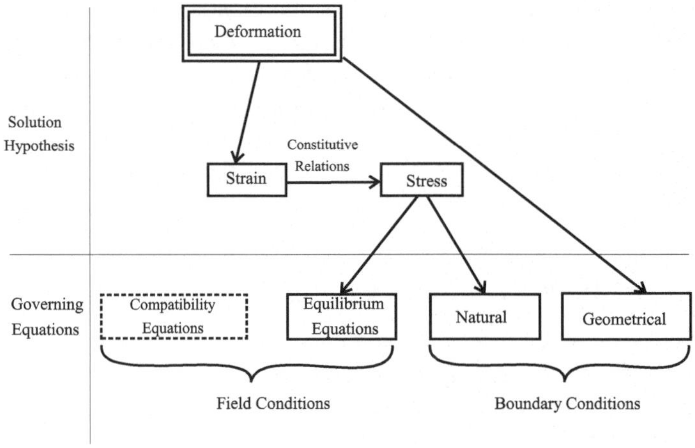

Figure 1.15: Solution Trail "A".

are determined, see S.1.1, and the stress components are derived using the constitutive relations thereafter, see S.2.1. The stress components enable the construction of the equilibrium equations, see S.1.3.2, and the natural boundary conditions, (1.185a), while the geometrical boundary conditions, (1.185b), are directly expressed in terms of the assumed deformation.

1.6.5.2 Solution Trail "B": Stress/Strain Hypothesis

Trail "B" consists of prescribing the expressions for the stress *or* the strain components as presented by Fig. 1.16. Since the strain components may always be obtained by the stress components via the constitutive relations, see S.2.1, and vice versa, the following procedure is identical to both cases. Subsequently, the strain components enable us to construct the compatibility equations, see S.1.2.1, while the stress components enable the construction of the equilibrium equations, see S.1.3.2, and the natural boundary conditions (1.185a). The displacements are then derived by strain integration (see S.1.2) and so, the geometrical boundary conditions (1.185b) may also be expressed.

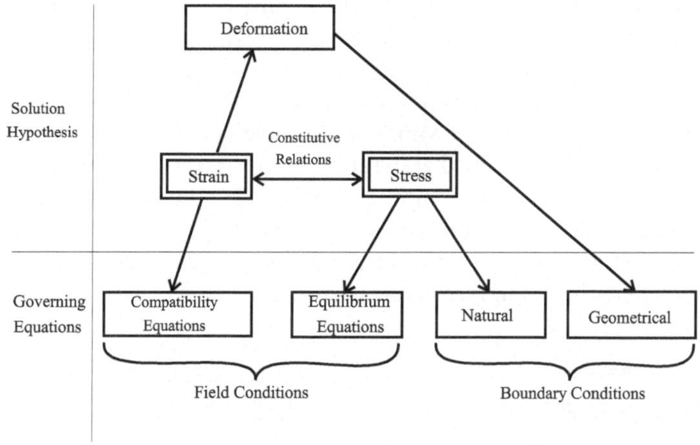

Figure 1.16: Solution Trail “B”.

1.6.5.3 Solution Trail “C”: Stress Functions Hypothesis

Trail “C” consists of prescribing the stress functions expressions as presented by Fig. 1.17. In this case, the stress components are determined according to a specific kind of the stress

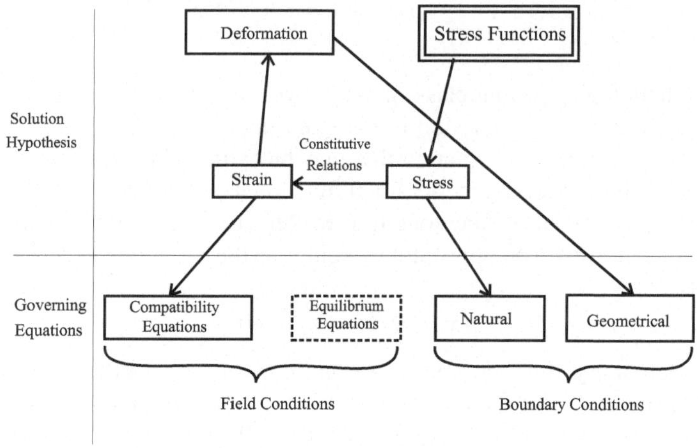

Figure 1.17: Solution Trail “C”.

function/s employed and therefore satisfy equilibrium by definition. The strain components are then derived using the constitutive relations, see S.2.1. The strain components enable us to construct the compatibility equations, see S.1.2.1, while the stress components enable the construction of the natural boundary conditions (1.185a). In addition, the displacements are derived by strain integration (see S.1.2) so that the geometrical boundary conditions (1.185b) may also be expressed.

1.6.5.4 Solution Trails "D" and "E": Energy Considerations

In this section, a generic methodology that employs energy theorems and provides the actual response *without* explicit formulation of Euler's equations will be discussed.

In trail "D" we apply the Theorem of Minimum Potential Energy. In such a case, one is required to prescribe the displacements (in such a way that the geometrical boundary conditions are satisfied), see Fig. 1.18(a), and to minimize V of (1.124).

In trail "E" we apply the Theorem of Minimum Complementary Energy, and one is required to prescribe the stresses (in such a way that the equilibrium and the natural boundary conditions are satisfied), see Fig. 1.18(b), and to minimize V^* of (1.128). In the latter case, displacements may be obtained by suitable integration, see S.1.2. The variational minimization process takes care of the missing part, i.e. the equilibrium and natural boundary conditions in the first case, and the compatibility conditions in the second case.

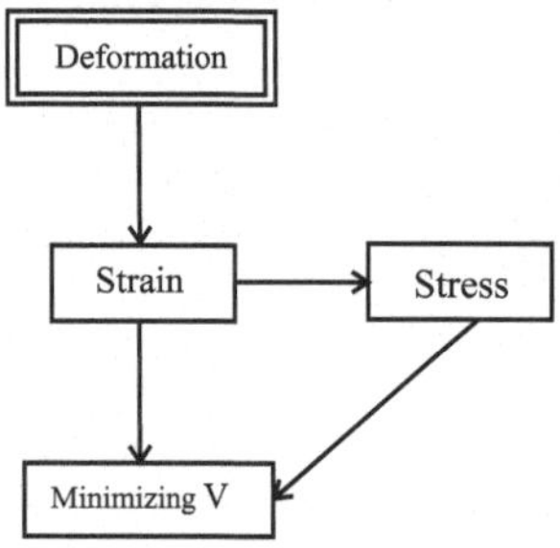

(a) Solution Trail "D".

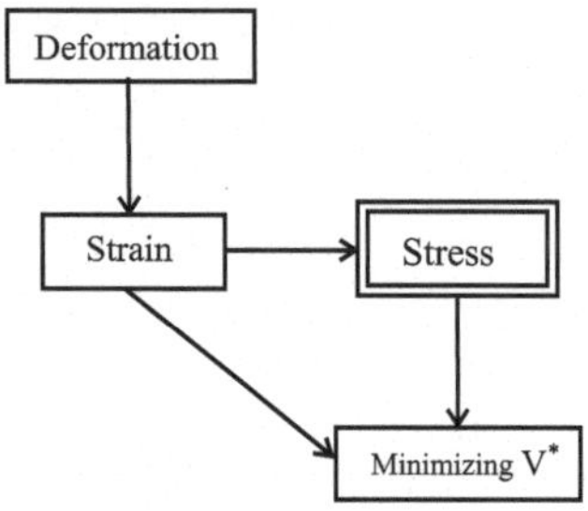

(b) Solution Trail "E".

Figure 1.18: Energy based solution trails "D" and "E" .

Within the methodologies in this class of trails, we shall provide some more attention to *Ritz's Method* and confine the discussion to the one-dimensional case where we seek an approximate solution of a governing equation that may be extracted from a functional, J.

To execute Ritz's Method for a generic $J > 0$ functional, see e.g. (1.134), one first needs to adopt a *relatively complete* set of functions η_i (see Remark 1.10), each of which satisfies the boundary conditions (1.135). The approximate solution (the exact one is denoted by $y^*(x)$) is therefore expressed as

$$y_N^*(x) = \bar{y}(x) + \sum_{i=1}^{N} a_i \eta_i \tag{1.187}$$

where $\bar{y}(x)$ is an auxiliary function and a_i are unknown (real) coefficients (which actually depend on N) that should be determined by the method. One may therefore examine the value of J when N terms are employed in (1.187), which will be denoted J_N. Subsequently, one may substitute $y_N^*(x)$ in the functional of (1.124) and express it as $J_N(a_1, a_2, \ldots, a_N) = J(y_N^*)$. Minimization of J_N with respect to $\{a_i\}$ yields the following linear system of N unknowns and equations:

$$\frac{\partial J_N}{\partial a_i} = 0, \qquad i = 1, \ldots, N. \tag{1.188}$$

Each approximation level includes the previous one, and a non-increasing series $\min J_N \leq \min J_{N-1} \leq \cdots \leq \min J_1$ is obtained. Thus, this approximation yields an upper bound for $\min J$. Since F is a continuous function, the condition $\left| F(x, y_N^*, y_{N,x}^*) - F(x, y^*, y_{,x}^*) \right| < \varepsilon_1$ may be satisfied for any (small) ε_1 by a appropriate value of N, and since $J = \int_{x_0}^{x_1} F\, dx$, it will be also bounded by (small) $\varepsilon_2 = (x_1 - x_0)\varepsilon_1$ as

$$|J_N^*(y_N^*) - J(y)| = \int_{x_0}^{x_1} |F^*(x, y_N^*, y_{N,x}^*) - F(x, y^*, y_{,x}^*)|\, dx < \varepsilon_2 . \tag{1.189}$$

In two (or more) dimensions we analogously write

$$u^*(x,y) = \overline{u}(x,y) + \sum_{i=1}^{N} a_i \eta_i(x,y), \tag{1.190}$$

where $\overline{u}(x,y)$ is an auxiliary function and a_i are unknown (real) coefficients that should be determined by the method. In some, mainly "dynamics oriented" analyses, the coefficients a_i are referred to as "Generalized Coordinates". As shown in Example 1.8, when a time-dependent problem is under discussion, this concept assigns a_i to be time-dependent while the "shape functions" η_i remain spatial functions only.

Remark 1.10 In the one-dimensional case, a set of functions $\{\eta_i(x)\}$ on $[x_0, x_1]$ is said to be *relatively complete* if for every $\varepsilon > 0$ and $y(x)$ there are N and y_N^* as defined by (1.187) such that $|y_N^* - y| < \varepsilon$ and $\left|y_{N,x}^* - y_{,x}\right| < \varepsilon$ for all $x \in [x_0, x_1]$. In two dimensions, a set of functions $\{\eta_i(x,y)\}$ on Ω is said to be *relatively complete* if for every $\varepsilon > 0$ and $u(x,y)$ there are N and u_N^* as defined by (1.190) such that $|u_N^* - u| < \varepsilon$, $\left|u_{N,x}^* - u_{,x}\right| < \varepsilon$ and $\left|u_{N,y}^* - u_{,y}\right| < \varepsilon$ for all $(x,y) \in \Omega$. The extension of the above for higher dimensions is clear.

Example 1.7 *Bending of a Beam.*

We shall demonstrate here the usage of Ritz's Method with the Theorem of Minimum Potential Energy, i.e. $J = V$, see (1.124) for a "simply-supported" uniform isotropic beam of length l, and Young's modulus E that undergoes a transverse loading $f_y(z)$ as shown in Fig. 1.19(a). In the absence of body loads, (1.124) yields for this case (see also Example 1.2)

$$V = \int_0^l [\frac{1}{2} E I_x \left(v^{*\prime\prime}\right)^2 - f_y(z) v^*] dz . \tag{1.191}$$

To employ Ritz's method for determining $v(z)$ we substitute $v = v^*$ in (1.191), where

$$v^* = \sum_{i=1}^{N} a_i \sin(\frac{i\pi z}{l}), \tag{1.192}$$

which clearly satisfies the geometrical boundary conditions $v^*(0) = v^*(l) = 0$ as required.

For uniform loading, i.e. $f_y = f_0 = const.$ and $N = 3$, one obtains

$$V = -\frac{2 l f_0}{\pi}(a_1 + \frac{1}{3} a_3) + \frac{E I_x \pi^4}{l^3}(\frac{1}{4} a_1^2 + 4 a_2^2 + \frac{81}{4} a_3^2) . \tag{1.193}$$

The solution of the three equations, $V_{,a_i} = 0$, yields $a_1 = \frac{4 l^4 f_0}{E I_x \pi^5}$, $a_2 = 0$ and $a_3 = \frac{4 l^4 f_0}{243 E I_x \pi^5}$. Even with only the first term, the above approximation, i.e. $v = 4 \frac{l^4 f_0}{E I_x \pi^5} \sin(\pi\zeta)$ where $\zeta = \frac{z}{l}$, is in a satisfactory agreement with the exact one given by $v = \frac{l^4 f_0}{24 E I_x} \zeta(\zeta^3 - 2\zeta^2 + 1)$.

In addition, for linear loading distribution of $f_y = f_0(z - \frac{l}{2})$ we obtain

$$V = a_2 \frac{l^2 f_0}{2\pi} + \frac{E I_x \pi^4}{l^3}(\frac{1}{4} a_1^2 + 4 a_2^2 + \frac{81}{4} a_3^2) . \tag{1.194}$$

The solution of the three equations, $V_{,a_i} = 0$, in this case yields $a_1 = a_3 = 0$, $a_2 = -\frac{l^5 f_0}{16 E I_x \pi^5}$.

When the above beam is "clamped-free" (cantilever) as shown in Fig. 1.19(b), one may alternatively use the approximation

$$v^* = \sum_{i=2}^{N} a_i z^i, \tag{1.195}$$

which satisfies the geometrical boundary conditions $v = v' = 0$ at $z = 0$. For $f_y = f_0 = const.$ such approximations yield the exact one, $v = \frac{l^4 f_0}{24EI_x}\zeta^2\left(\zeta^2 - 4\zeta + 6\right)$, for $N = 4$ (i.e. three terms).

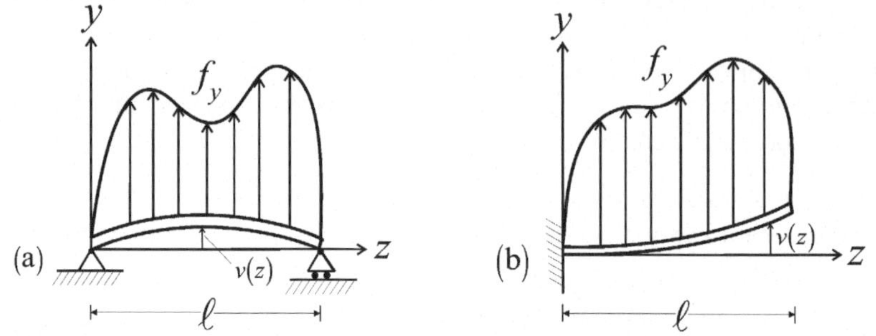

Figure 1.19: Isotropic beam under transverse loading: (a) Simply-supported, (b) Cantilever.

Example 1.8 *Lagrange's Equations.*

We adopt here the concept of expanding each displacement component as a relatively complete set of shape functions as in Example 1.7, but we consider a time-dependent problem by setting the generalized coordinates of the problem, $\mathbf{a} = (a_1, \ldots, a_{3N})$, to be functions of time, t. Hence, in the general case

$$u_i^*(x,y,z,t) = \bar{u}_i(x,y,z,t) + \sum_{j=1}^{N} a_m(t)\eta_j(x,y,z), \qquad i = 1,2,3, \tag{1.196}$$

where $\bar{u}_i$ are auxiliary (known) functions and $m = (i-1)N + j$. We now employ the Theorem of Minimum Potential Energy by writing the variation of V in (1.124) as

$$\delta V = (\text{grad}_a U - Q^S - Q^B)\cdot\delta\mathbf{a} = 0 \tag{1.197}$$

where $\text{grad}_a = (\frac{\partial}{\partial a_1}, \ldots, \frac{\partial}{\partial a_{3N}})$. In the above we have employed the notation

$$\text{grad}_a U = \text{grad}_a\left(\iiint_B W\right), \qquad Q^S = (Q_1^S, \ldots, Q_N^S), \quad Q_j^S = \iint_S \mathbf{F}_s\cdot\mathbf{u}_{,a_j},$$
$$Q^B = (Q_1^B, \ldots, Q_N^B), \quad Q_j^B = \iiint_B \mathbf{F}_b\cdot\mathbf{u}_{,a_j}. \tag{1.198}$$

By assuming that in a time-dependent case, the dominant portion of the body force is $\mathbf{F}_b = -\rho\ddot{\mathbf{u}}$ (where ρ is the material density and a dot stands for a time derivative) we write

$$Q_j^B = -\iiint_B \rho\ddot{\mathbf{u}}\cdot\mathbf{u}_{,a_j} = -\frac{\partial}{\partial t}\underline{\left(\iiint_B \rho\dot{\mathbf{u}}\cdot\mathbf{u}_{,a_j}\right)} + \underline{\underline{\iiint_B \rho\dot{\mathbf{u}}\cdot\dot{\mathbf{u}}_{,a_j}}}. \tag{1.199}$$

We further note that the kinetic energy is given by $T = \frac{1}{2}\iiint\limits_B (\rho\sum_{i=1}^3 \dot{u}_i^2)$, and therefore, the once underlined term in (1.199) is $\partial T/\partial\dot{a}_j$ (note that $\partial\dot{u}_i/\partial\dot{a}_j = \partial U_i/\partial a_j$ in view of (1.196)), and the twice underlined term is $\partial T/\partial a_j$. Hence, (1.197) become

$$\frac{\partial}{\partial t}\left(\frac{\partial T}{\partial \dot{a}_j}\right) - \frac{\partial T}{\partial a_j} + \frac{\partial U}{\partial a_j} - Q_j^S = 0, \qquad j = 1,\ldots,3N, \tag{1.200}$$

which are well known as *Lagrange's equations*.

1.6.5.5 Solution Trail "F": Galerkin's Method

Galerkin's method is a general powerful method for solving boundary value problems, and there are many applications related to it in the theory of elasticity. It resembles Ritz's method in the sense that it is also based on expanding an approximate solution into a series of relatively complete sets of "shape" functions, but it is founded on a different functional type. In Galerkin's method, we select the functional to be the error $L(u^*)$, but instead of employing the operation

$$J = \iint_\Omega |L(u^*)|^2 \to \min, \tag{1.201}$$

we set the requirement

$$J_i = \iint_\Omega L(u^*)\eta_i = 0, \tag{1.202}$$

by which we force the error function to be orthogonal to each one of the shape functions.

Since J is based on the differential operator and not on the potential energy, each of the shape functions must satisfy *all* boundary conditions (natural and geometrical). This may pose substantial constraints on the admissible families of shape functions. For example, the series used in (1.195) for a cantilever beam is not applicable here since the η_i functions do not satisfy the natural boundary conditions (one by one) at the free tip.

1.7 Appendix: Coordinate Systems

1.7.1 Transformation Between Coordinate Systems

There are many occasions when one wishes to change the coordinate system for which the problem is expressed. As already discussed in the introduction of S.1.1, from an analytical point of view, such a change may be crucial for the existence and complexity of a closed-form solution. In this book we deal with transformations of three types of mathematical identities. The first one is the vector space transformation (which is equivalent to coordinate transformation of a point in Euclidean space). The second one is the (second-order) tensor transformation, which should be applied to the strain and stress components, see S.1.3.3, S.1.3.4, and the third one is the (fourth-order) tensor transformation of the constitutive relations, namely, the compliance and the stiffness matrices, which are dealt with in Chapter 2.

As a first step we shall define a generic linear transformation in Euclidean space E^3 (with Cartesian metric), by a transformation matrix, $\mathbf{T}$, written as

$$\mathbf{T} = \begin{bmatrix} T_{11} & T_{12} & T_{13} \\ T_{21} & T_{22} & T_{23} \\ T_{31} & T_{32} & T_{33} \end{bmatrix}. \tag{1.203}$$

$\mathbf{T}$ is a 3×3 matrix that transforms a position vector $\mathbf{v} = (v_1, v_2, v_3)$ given in a certain coordinate system, S, to another vector $\overline{\mathbf{v}} = (\overline{v}_1, \overline{v}_2, \overline{v}_3)$, which defines the same point in a new system, $\overline{S}$, by $\overline{\mathbf{v}} = \mathbf{T} \cdot \mathbf{v}$. The matrix $\mathbf{T}$ is *orthogonal*, and its transposed and inverse matrices coincide, i.e., $\mathbf{T}^T = \mathbf{T}^{-1}$, or

$$\sum_{k=1}^{3} T_{ik} T_{jk} = \delta_{ij}, \qquad i,j \in \{1,2,3\}. \tag{1.204}$$

Hence, $det(\mathbf{T}) = \pm 1$ (while in order to preserve orientation, the $+1$ is selected).

The orthogonal matrix, $\mathbf{T}$, may be composed in various ways. Here, we shall adopt the usage of three Euler's angles ϕ, θ, ψ, that represent rotations about the x, y, z directions, respectively. For any single rotation about the x, y or z directions, one may assign as $\mathbf{T}$ the following $\mathbf{T}_x(\phi)$, $\mathbf{T}_y(\theta)$ or $\mathbf{T}_z(\psi)$ matrices, respectively:

$$\mathbf{T}_x(\phi) = \begin{bmatrix} 1 & 0 & 0 \\ 0 & \cos\phi & \sin\phi \\ 0 & -\sin\phi & \cos\phi \end{bmatrix}, \qquad \mathbf{T}_y(\theta) = \begin{bmatrix} \cos\theta & 0 & -\sin\theta \\ 0 & 1 & 0 \\ \sin\theta & 0 & \cos\theta \end{bmatrix},$$

$$\mathbf{T}_z(\psi) = \begin{bmatrix} \cos\psi & \sin\psi & 0 \\ -\sin\psi & \cos\psi & 0 \\ 0 & 0 & 1 \end{bmatrix}. \tag{1.205}$$

However, a generic orthogonal transformation will be composed as a rotation by the angle ψ about the z-axis, followed by a rotation by the angle θ about the y-axis (in its new position x', y', z'), followed by a rotation by the angle ϕ about the x-axis (in its new position x'', y'', z''), see Fig. 1.20(a). Thus, the final resulting system is $\overline{x}, \overline{y}, \overline{z}$. In this case, the orthogonal matrix

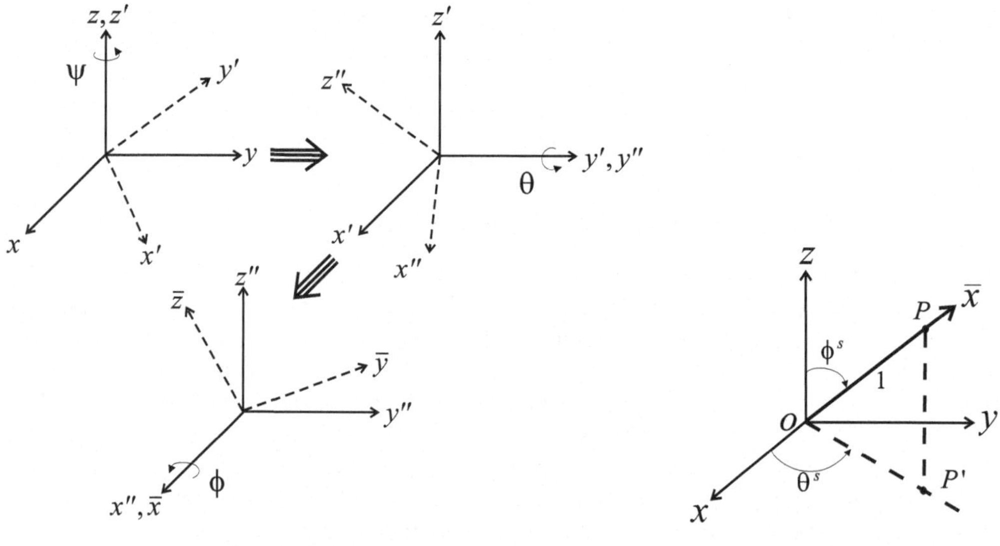

(a) Generic Euler's angles. (b) Spherical angles.

Figure 1.20: Definition of Euler's and spherical angles.

$\mathbf{T} = \{T_{ij}\}$ takes the form

$$\mathbf{T} = \mathbf{T}_x(\phi) \cdot \mathbf{T}_y(\theta) \cdot \mathbf{T}_z(\psi), \tag{1.206}$$

where its elements are, see **P.1.15**,

$$T_{11} = \cos\theta \cos\psi, \quad T_{12} = \cos\theta \sin\psi, \quad T_{13} = -\sin\theta, \tag{1.207}$$

$$T_{21} = \sin\phi\,\sin\theta\,\cos\psi - \cos\phi\,\sin\psi, \quad T_{22} = \sin\phi\,\sin\theta\,\sin\psi + \cos\phi\,\cos\psi, \quad T_{23} = \sin\phi\,\cos\theta,$$
$$T_{31} = \cos\phi\,\sin\theta\,\cos\psi + \sin\phi\,\sin\psi, \quad T_{32} = \cos\phi\,\sin\theta\,\sin\psi - \sin\phi\,\cos\psi, \quad T_{33} = \cos\phi\,\cos\theta.$$

As will be discussed further on, $\mathbf{T}$ will also be utilized as a second-order tensor. As a general rule, we shall carry on the notation $\overline{(\,)}$ for the value of $(\,)$ in the transformed ("new") coordinate system. Note that for a given transformation tensor, the corresponding transformation angles may be determined by five terms of (1.207), as

$$\tan\psi = \frac{T_{12}}{T_{11}}, \qquad \sin\theta = -T_{13}, \qquad \tan\phi = \frac{T_{23}}{T_{33}}. \tag{1.208}$$

Special attention should be devoted to the angles' sign in the case where $T_{11} = 0$ and/or $T_{33} = 0$.

Remark 1.11 While utilizing the procedure described above for constructing the transformation matrix by Euler's angles, it should be noted that the order of rotation angles has a crucial importance. For example, the reader may verify that a transformation based on $\psi = \pi/2,\ \theta = \pi/2,\ \phi = 0$, is inverted by $\psi = -\pi/2,\ \theta = 0,\ \phi = \pi/2$. There are six different ways of selecting the order of the three different axes. The selection made above (z,y,x) is convenient for stress and strain analysis (see S.1.3.3). However, a different order may be easily constructed by changing the order in the matrix product of (1.206) (see also **P.1.15**). Note that there are similar methods where the transformation may be defined by rotation about repeated axes like $\mathbf{T} = \mathbf{T}_z(\phi)\cdot\mathbf{T}_y(\theta)\cdot\mathbf{T}_z(\psi)$.

Remark 1.12 It is important to distinguish between the above Euler's angles, ϕ, θ and ψ and the "standard" spherical angles, $\theta^s = \angle(\vec{OP}',x) \in [0,2\pi]$ and $\phi^s = \angle(\vec{OP},z) \in [0,\pi]$ used (in analytical geometry) to locate a point in space. As shown in Fig. 1.20(b), one may define the x,y,z coordinates of a point, P, in space using the standard spherical angles θ^s and ϕ^s (P' is the orthogonal projection of P onto the x,y-plane). By setting the distance of P from the origin to be a unit ($|\vec{OP}| = 1$) we write

$$x = \sin\phi^s\cos\theta^s, \qquad y = \sin\phi^s\sin\theta^s, \qquad z = \cos\phi^s. \tag{1.209}$$

However, using Euler's angles with $\phi = 0$ (see Fig. 1.20(a)), we may write the coordinates of point P assuming that it is placed on the $\bar{x}$ coordinate line (of the new system) by setting $\bar{\mathbf{v}} = (1,0,0)$, and calculating $\mathbf{v} = \mathbf{T}^{-1}\cdot\bar{\mathbf{v}}$, which yields $x = \cos\theta\cos\psi,\ y = \cos\theta\sin\psi,\ z = -\sin\theta$. Therefore, $\theta^s = \psi,\ \phi^s = \frac{\pi}{2} + \theta$. Note that while the spherical angles are commutative, as indicated in Remark 1.11, Euler's angles are not, and the above holds for the order of angles adopted in this section only.

1.7.2 Curvilinear Coordinate Systems

1.7.2.1 Definition

We shall assign $(\alpha_1,\alpha_2,\alpha_3) \in A$ to be the *curvilinear coordinates* in a domain C of Euclidean space E^3 (with orthonormal basis $\{\mathbf{k}_i\}$), see Fig. 1.21, and therefore, the position vector of each point may be expressed by three functions that relate α_i to the Cartesian coordinates by a topological transformation, $\mathbf{r}: A \to C$, as

$$x_i = f_i(\alpha_1,\alpha_2,\alpha_3), \quad i = 1,2,3, \quad \Leftrightarrow \quad \mathbf{r} = \textstyle\sum_{i=1}^{3} f_i(\alpha_1,\alpha_2,\alpha_3)\,\mathbf{k}_i. \tag{1.210}$$

Similarly, two functions of α_1,α_2 are used for curvilinear coordinates in a plane domain. Curvilinear coordinates are called *regular* if the Jacobian, J_r, of (1.210) is nonzero, namely,

$$J_r = \det(\mathbf{dr}) \neq 0, \qquad \mathbf{dr} = \begin{bmatrix} f_{1,\alpha_1} & f_{1,\alpha_2} & f_{1,\alpha_3} \\ f_{2,\alpha_1} & f_{2,\alpha_2} & f_{2,\alpha_3} \\ f_{3,\alpha_1} & f_{3,\alpha_2} & f_{3,\alpha_3} \end{bmatrix}. \tag{1.211}$$

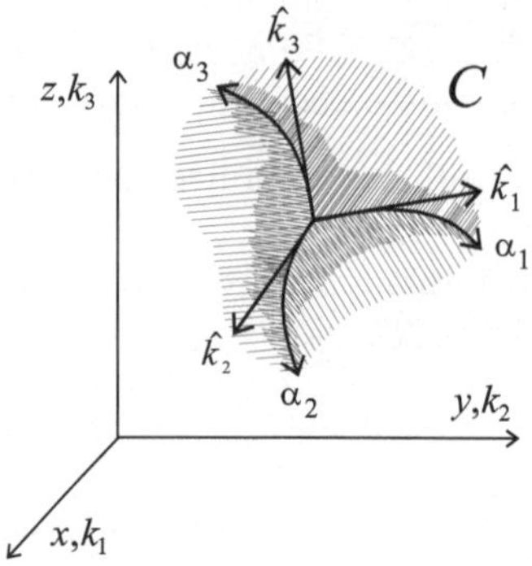

Figure 1.21: Orthogonal curvilinear coordinates in domain C of E^3.

In the general case of curvilinear coordinates in E^3 as described by (1.210), one may construct the *metric differential quadratic form* and the *metric tensor*

$$ds^2 = \sum_{ij} g_{ij}\, d\alpha_i\, d\alpha_j, \qquad \mathbf{g} = \{g_{ij}\}, \tag{1.212}$$

whose coefficients $g_{ij} = \mathbf{r}_{,\alpha_i} \cdot \mathbf{r}_{,\alpha_j}$ $(i, j = 1,2,3)$ are functions of α_i.

In the particular case of Cartesian coordinates, $g_{ij} = \delta_{ij}$ and $ds^2 = dx^2 + dy^2 + dz^2$. Another special case is the *affine* coordinate system which is defined by linear transformation functions $f_i = \overline{f}_{i1}\alpha_1 + \overline{f}_{i2}\alpha_2 + \overline{f}_{i3}\alpha_3$ $(i = 1,2,3)$, where $\overline{f}_i$ are constants. For this kind of coordinates, both the matrix $\mathbf{dr}$ of (1.211) and the metric tensor $\mathbf{g} = \{g_{ij}\}$ are constants.

1.7.2.2 The Riemann-Christoffel Curvature Tensor

The following *two problems*, that were first formulated and solved by Lamé for a particular case, see (Cartan, 1946), are essential in the theory of curvilinear coordinates.

First: for a given metric form ds^2 or metric tensor $\mathbf{g} = \{g_{ij}\}$ (see (1.212)), find conditions on the coefficients g_{ij} so they may represent the metric of some curvilinear coordinates α_j in a domain A of E^3 (or E^2).

Second: for a given metric form ds^2 or metric tensor $\mathbf{g} = \{g_{ij}\}$ (see (1.212)), of some curvilinear coordinates α_j in a domain A of E^3 (or E^2), find a transformation in the form of (1.210), i.e. find the associated functions $f_i(\alpha_1, \alpha_2, \alpha_3)$.

We shall elaborate here on the first problem, as the second one finds its answer in the strain integration schemes presented within S.1.2.2, S.1.2.3.

The discussion below is founded on the well-known Riemann Theorem, see (Oprea, 1997), that may be put as: *A symmetric tensor* $\mathbf{g} = \{g_{ij}\}$, $i, j = 1,2,3$ *is a metric tensor for a Euclidean space if it is non-singular (*$\det(\mathbf{g}) \neq 0$*), positive-definite (*$\mathbf{g} > 0$*), and the Riemann-Christoffel curvature tensor* $\mathbf{R} = \{R_{mnij}\}$ *that is formed from it, vanishes identically.*

The components of the Riemann-Christoffel curvature tensor are known to be

$$R_{mnij} = \frac{1}{2}\left(g_{mj,ni} + g_{in,mj} - g_{mi,nj} - g_{jn,im}\right) - g^{fh}\left(\Gamma_{f,im}\Gamma_{h,jn} - \Gamma_{f,ni}\Gamma_{h,jm}\right), \tag{1.213}$$

while $\mathbf{g}^{-1} = \{g^{qs}\}$ is the inverse matrix of $\mathbf{g}$. Here we have used the notation $\Gamma_{p,ms}$ for *Christoffel symbols of the first kind*, (not a tensor), which are based on the first-order derivatives of the metric g_{ij}, namely,

$$\Gamma_{p,ms} = \frac{1}{2}\left(g_{mp,s} + g_{ps,m} - g_{ms,p}\right). \tag{1.214}$$

Only six of the 81 curvature tensor **R** components are independent, and are obtained by the same set of (1.44). In a two-dimensional case, there is only one (out of 16) independent components of curvature tensor, for example, R_{1221}.

1.7.2.3 Orthogonal Curvilinear Coordinate Systems

When the matrix **g** is diagonal (i.e. $g_{ij} = 0$ for $i \neq j$), the coordinates (1.210) are said to be *orthogonal*, and one may define the well-known *Lamé's parameters*, which are essentially the absolute values of $\mathbf{r}_{,\alpha_i}$, namely, $H_i = |\mathbf{r}_{,\alpha_i}|$. Subsequently, as shown in Fig. 1.21, we may define unit vectors $\hat{\mathbf{k}}_i$ in the α_i directions as

$$\hat{\mathbf{k}}_i = \frac{1}{H_i}\mathbf{r}_{,\alpha_i}, \qquad i = 1,2,3. \tag{1.215}$$

The arc-lengths of elements in the $\hat{\mathbf{k}}_i$ directions are $ds_i = H_i d\alpha_i$. Hence, in view of (1.211), a transformation matrix that converts a vector in Cartesian coordinates into a vector in orthogonal curvilinear coordinates is given by

$$\mathbf{T} = \begin{bmatrix} \frac{1}{H_1} f_{1,\alpha_1} & \frac{1}{H_1} f_{2,\alpha_1} & \frac{1}{H_1} f_{3,\alpha_1} \\ \frac{1}{H_2} f_{1,\alpha_2} & \frac{1}{H_2} f_{2,\alpha_2} & \frac{1}{H_2} f_{3,\alpha_2} \\ \frac{1}{H_3} f_{1,\alpha_3} & \frac{1}{H_3} f_{2,\alpha_3} & \frac{1}{H_3} f_{3,\alpha_3} \end{bmatrix}. \tag{1.216}$$

The derivatives of the unit vectors $\hat{\mathbf{k}}_i$ with respect to the curvilinear coordinates are, see (Novozhilov, 1961),

$$\begin{Bmatrix} \hat{\mathbf{k}}_{1,\alpha_1} \\ \hat{\mathbf{k}}_{1,\alpha_2} \\ \hat{\mathbf{k}}_{1,\alpha_3} \\ \hat{\mathbf{k}}_{2,\alpha_1} \\ \hat{\mathbf{k}}_{2,\alpha_2} \\ \hat{\mathbf{k}}_{2,\alpha_3} \\ \hat{\mathbf{k}}_{3,\alpha_1} \\ \hat{\mathbf{k}}_{3,\alpha_2} \\ \hat{\mathbf{k}}_{3,\alpha_3} \end{Bmatrix} = \begin{bmatrix} 0 & -\frac{1}{H_2}H_{1,\alpha_2} & -\frac{1}{H_3}H_{1,\alpha_3} \\ 0 & \frac{1}{H_1}H_{2,\alpha_1} & 0 \\ 0 & 0 & \frac{1}{H_1}H_{3,\alpha_1} \\ \frac{1}{H_2}H_{1,\alpha_2} & 0 & 0 \\ -\frac{1}{H_1}H_{2,\alpha_1} & 0 & -\frac{1}{H_3}H_{2,\alpha_3} \\ 0 & 0 & \frac{1}{H_2}H_{3,\alpha_2} \\ \frac{1}{H_3}H_{1,\alpha_3} & 0 & 0 \\ 0 & \frac{1}{H_3}H_{2,\alpha_3} & 0 \\ -\frac{1}{H_1}H_{3,\alpha_1} & -\frac{1}{H_2}H_{3,\alpha_2} & 0 \end{bmatrix} \begin{Bmatrix} \hat{\mathbf{k}}_1 \\ \hat{\mathbf{k}}_2 \\ \hat{\mathbf{k}}_3 \end{Bmatrix}. \tag{1.217}$$

Example 1.9 *Standard Orthogonal Coordinate Systems.*

In the few illustrative orthogonal coordinate systems listed below the coordinate parameters are denoted by $\alpha_1, \alpha_2, \alpha_3$. The expressions for f_i (see (1.210)) that appear in (1.218a–d) are for: (a) *Cylindrical coordinates* ($\alpha_1 = \rho, \alpha_2 = \theta^c, \alpha_3 = z$), (b) *Elliptical-cylindrical coordinates* ($\alpha_3 = z$), (c) *Spherical coordinates* ($\alpha_1 = \rho, \alpha_2 = \theta^s, \alpha_3 = \phi^s$), (d) *Bipolar-cylindrical coordinates* ($\alpha_3 = z$),

$$f_1 = \alpha_1 \cos\alpha_2, \qquad f_2 = \alpha_1 \sin\alpha_2, \qquad f_3 = \alpha_3, \tag{1.218a}$$

$$f_1 = \tilde{a}\cosh\alpha_1 \cos\alpha_2, \qquad f_2 = \tilde{a}\sinh\alpha_1 \sin\alpha_2, \qquad f_3 = \alpha_3, \tag{1.218b}$$

$$f_1 = \alpha_1 \cos\alpha_2 \sin\alpha_3, \qquad f_2 = \alpha_1 \sin\alpha_2 \sin\alpha_3, \qquad f_3 = \alpha_1 \cos\alpha_3, \tag{1.218c}$$

$$f_1 = \frac{\tilde{a}\sinh\alpha_2}{\cosh\alpha_2 - \cos\alpha_1}, \qquad f_2 = \frac{\tilde{a}\sin\alpha_1}{\cosh\alpha_2 - \cos\alpha_1}, \qquad f_3 = \alpha_3 \tag{1.218d}$$

where $\widetilde{a} > 0$. Additional example is the *Ellipsoidal coordinate system*, which is defined by

$$f_1 = \frac{\alpha_1\alpha_2\alpha_3}{\widetilde{a}\widetilde{b}},\ f_2 = \frac{\sqrt{(\alpha_1^2-\widetilde{b}^2)(\alpha_2^2-\widetilde{b}^2)(\widetilde{b}^2-\alpha_3^2)}}{\widetilde{b}\sqrt{\widetilde{a}^2-\widetilde{b}^2}},\ f_3 = \frac{\sqrt{(\alpha_1^2-\widetilde{a}^2)(\widetilde{a}^2-\alpha_2^2)(\widetilde{a}^2-\alpha_3^2)}}{\widetilde{a}\sqrt{\widetilde{a}^2-\widetilde{b}^2}}, \tag{1.219}$$

where $\widetilde{a} \geq \widetilde{b} > 0$. Symbolic computational tools (e.g. (Maple, 2003)) support many other three-

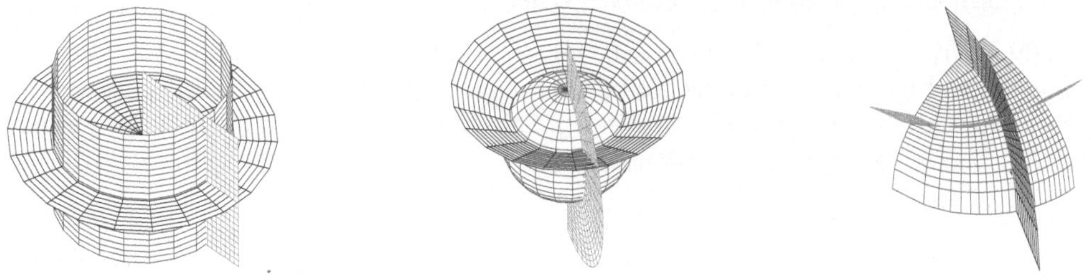

(a) Cylindrical. (b) Spherical. (c) Ellipsoidal.

Figure 1.22: Image of various curvilinear coordinate systems.

dimensional orthogonal systems. By activating **P.1.16, P.1.17**, the reader may examine additional curvilinear coordinates, see Fig. 1.22 for illustrative examples.

It is interesting to note that for orthogonal curvilinear coordinates, one may write the six independent components of (1.213) as the following six differential relations between Lamé parameters (apply cyc-123):

$$\underline{(\frac{H_{2,\alpha_1}}{H_1})_{,\alpha_1} + (\frac{H_{1,\alpha_2}}{H_2})_{,\alpha_2}} + \frac{H_{1,\alpha_2}H_{1,\alpha_3}}{H_2^2} \underline{= 0}, \qquad (\frac{H_{1,\alpha_3}}{H_3})_{,\alpha_2} - \frac{H_{1,\alpha_2}H_{2,\alpha_3}}{H_2H_3} = 0. \tag{1.220}$$

The above six conditions may therefore be titled as the integrability conditions of a given metric tensor $g = \Sigma_{i=1}^{3} H_i^2\,(d\alpha_i)^2$. One may also reach (1.220) by considering the identities

$$(\hat{\mathbf{k}}_{i,\alpha_i})_{,\alpha_j} = (\hat{\mathbf{k}}_{j,\alpha_j})_{,\alpha_i}, \tag{1.221}$$

and substitute in it the $\hat{\mathbf{k}}_{i,\alpha_j}$ of (1.217). Then, each one of the three basic sets $(i,j) = (1,2)$, $(1,3)$, $(2,3)$ yields two equations. These are obtained by the corresponding (two) $\hat{\mathbf{k}}_i$-components that appear in each equation.

For orthogonal coordinates in plane (*polar*, *bipolar*, *elliptical*, etc.) the equality $R_{1221} = 0$ coincides with the underlined terms in (1.220:a).

2
Anisotropic Materials

This chapter reviews the mathematical representation of anisotropic materials and their characteristics as observed by various coordinate systems. Generally speaking, the elastic properties are characterized by certain functional relationships between loads and deformation, and the nature of these relationships is in the focus of this chapter.

Prior to the detailed consideration, it is worth clarifying the notions "anisotropy" and "non-homogeneous", which are highly relevant for the type of boundary value problems discussed in this book. For that purpose, suppose that we wish to measure the properties of a given elastic medium using a given coordinate system. We shall define the medium as "non-homogeneous" if we find that the material properties are functions of the coordinate system *location*. On the other hand, we shall define the medium as "anisotropic" if we find that the material properties are functions of the coordinate system *orientation*. This dependence of properties on their measuring axes orientation enables us to define the notion "anisotropy" also as *the tendency of the material to react differently to stresses applied in different directions*.

It is therefore possible to have all four combinations of "isotropic and homogeneous", "anisotropic and homogeneous", "isotropic and non-homogeneous" and "anisotropic and non-homogeneous" materials (or elastic domains).

This chapter is devoted to the anisotropic material properties, while the homogeneity issue will be considered within specific applications in further chapters. The mathematical representation of anisotropic materials is closely related to the type of analysis and the class of problems that are under discussion. Within this book, we will be focused on the class of problems in anisotropic elasticity, which are characterized by typical physical dimensions that justify the employment of *effective elastic properties*. In general, the concept of working with effective elastic properties implies that emphasis is put on the macro-mechanical behavior of the involved phenomena, while paying much less attention to the exact micro-structure of the material. From an analytical methodologies point of view, this enables the utilization of continuous functions for describing the deformation (and likewise other physical measures) over finite volumes, the typical dimensions of which are on the order of those of the structure under discussion.

Although the above approach applies to isotropic materials as well, it is of special importance for anisotropic materials due to their special micro-structure. For anisotropic materials, the micro-structure usually consists of different components or layers that are tied together in "natural" materials, e.g. crystals, rocks, wood, etc., or a number of ingredients that are molded or bonded together in "man-made" materials, such as *composite materials* that typically consist of a matrix, which is reinforced by high-strength fibers. In the latter case, the concept of working with effective elastic properties, which are determined by various "homogenization methods", completely bypasses the physical mechanisms that describe the fiber-matrix relations, and treats the effective combination as if the material is homogeneous.

The above way of working with effective material properties should not be confused with "smearing" and "averaging" methods that are frequently encountered in laminated composite analysis, see S.3.5.2, S.5.3.1. In such cases, laminated stacks of composite plies are handled in a simpler way by treating them as an "averaged" homogeneous material, see e.g. (Ochoa and Reddy, 1992).

For further reading on anisotropic and composite materials systems see (Vinson and Chou, 1975), (Halpin, 1992), (Mallick, 1993), (Parton and Kudryavtsev, 1993), (Daniel and Ishai, 1994), (Gibson, 1994), (Bull, 1995), (Hull and Clyne, 1996), (Mallick, 1997), (Jones, 1999), (Kelly and Zweben, 2000), (Vinson and Sierakowski, 2002), (Reifsnider and Case, 2002), (Kollár and Springer, 2003).

For further reading on homogenization methods and effective properties see (Devries *et al.*, 1989), (Tarnopol'skii *et al.*, 1992), (Ashbee, 1993), (Ghosh *et al.*, 1995), (Ponte-Castaneda and Willis, 1995), (Ponte-Castaneda, 1996), (Kalamkarov and Kolpakov, 1997), (Kolpakov, 2004).

2.1 The Generalized Hook's Law

The definition of a generic anisotropic material is founded on the so-called "Generalized Hooke's Law" (GHL), which in fact establishes the material "constitutive relations" or the "stress-strain relations". At this stage we shall be focused on the case of Cartesian anisotropy where the GHL linearly connects the engineering strain and stress vectors, ε , and, σ , respectively. These vectors are written as (see S.1.1.2)

$$\varepsilon = [\varepsilon_{xx}, \varepsilon_{yy}, \varepsilon_{zz}, \gamma_{yz}, \gamma_{xz}, \gamma_{xy}]^T, \qquad \sigma = [\sigma_{xx}, \sigma_{yy}, \sigma_{zz}, \tau_{yz}, \tau_{xz}, \tau_{xy}]^T. \tag{2.1}$$

Subsequently, the GHL is typically written in one of the following forms:

$$\varepsilon = \mathbf{a} \cdot \sigma\,, \qquad \sigma = \mathbf{A} \cdot \varepsilon\,, \tag{2.2}$$

where clearly, the above sixth-order symmetric matrices are related by $\mathbf{A} = \mathbf{a}^{-1}$. We shall denote $\mathbf{a}$ and $\mathbf{A}$ as the *compliance matrix* and the *stiffness matrix*, respectively.

In what follows the discussion is confined to the case of linear ("engineering") strain expressions, which are based on derivatives of the displacements as shown by (1.9a,b), namely, $\varepsilon_{\alpha\alpha} = u_{\alpha,\alpha}$ and $\gamma_{\alpha\beta} = u_{\alpha,\beta} + u_{\beta,\alpha}$ for $\alpha, \beta = x, y, z$.

For the present purpose of defining the material properties, we shall make two additional assumptions. First, we shall restrict our attention to the range of relatively small strain where most materials exhibit linear stress-strain relations, and completely disregard materials that do not exhibit linear stress-strain relations at all. In addition, we shall ignore the fact that many materials exhibit linear behavior that should be quantified by different elastic moduli over the two sides of the zero-strain point (e.g., various Graphite fabrics show different elastic moduli for tension and compression, see (Chen and Saleeb, 1994)). Programs **P.2.1, P.2.2, P.2.3** derive the compliance and stiffness matrices in both symbolic and numerical manners.

Remark 2.1 The strain vector of (2.2), i.e., $\varepsilon = [\varepsilon_1, \ldots, \varepsilon_6]^T$ (see notation in (1.21)), is the *mechanical strain* vector that is induced by the stress vector. When the body is subjected to a change of temperature, this strain vector should be replaced by $\varepsilon - \varepsilon^\circ$ where ε will now stand for the *total strain* vector and $\varepsilon^\circ = [\varepsilon_1^\circ, \ldots, \varepsilon_6^\circ]^T$ is the *thermal strain* vector, which is induced by the change in temperature. For general anisotropic materials (see S.2.2) one may write $\varepsilon_i^\circ = \alpha_i^\circ (T - T_r)$, where α_i° are thermal expansion coefficients (of dimension 1/per degree), T is the actual temperature and T_r is a reference temperature for which the *thermal strain* components are defined as zero. Equation (2.2) is therefore converted to $\varepsilon - \varepsilon^\circ = \mathbf{a} \cdot \sigma$ and $\sigma = \mathbf{A} \cdot (\varepsilon - \varepsilon^\circ)$. For isotropic materials (see S.2.8) it is customary to write $\varepsilon_i^\circ = \alpha^\circ (T - T_r)$ for $i = 1,2,3$ and $\varepsilon_i^\circ = 0$ for $i = 4,5,6$, where α° is a constant.

2.2 General Anisotropic Materials

For the most general anisotropic materials, which are also referred to as *triclinic materials*, the compliance matrix $\mathbf{a} = \{a_{ij}\}$ contains 21 independent elastic coefficients. For further use and for obvious reasons, this matrix will be denoted GEN21 and is written as (note that as a general rule, symmetric matrices will be left blank in their lower left part)

$$\mathbf{a} = \begin{bmatrix} a_{11} & a_{12} & a_{13} & a_{14} & a_{15} & a_{16} \\ & a_{22} & a_{23} & a_{24} & a_{25} & a_{26} \\ & & a_{33} & a_{34} & a_{35} & a_{36} \\ & & & a_{44} & a_{45} & a_{46} \\ & Sym. & & & a_{55} & a_{56} \\ & & & & & a_{66} \end{bmatrix} . \tag{2.3}$$

Similarly, the stiffness matrix of this general anisotropic material, $\mathbf{A} = \{A_{ij}\}$, may be written by its 21 components as

$$\mathbf{A} = \begin{bmatrix} A_{11} & A_{12} & A_{13} & A_{14} & A_{15} & A_{16} \\ & A_{22} & A_{23} & A_{24} & A_{25} & A_{26} \\ & & A_{33} & A_{34} & A_{35} & A_{36} \\ & & & A_{44} & A_{45} & A_{46} \\ & Sym. & & & A_{55} & A_{56} \\ & & & & & A_{66} \end{bmatrix} . \tag{2.4}$$

Hence, by adopting the above GHL representation of the material constitutive relations, under the above mentioned provisions, all materials may be defined by numerical assignment of all terms in their matrices a_{ij} and/or A_{ij}. These terms will be further referred to as "elastic moduli". Later on, when the transformation of these matrices due to a rotation of the coordinate system will be discussed, we shall be obliged to define the compliance and stiffness tensors as well.

Man-made materials and many natural materials exhibit some symmetry and micro-structure simplification that lead to GHL that consists of less than 21 independent elastic moduli. For example, (Lekhnitskii, 1981) presents the properties of quartz (rock crystal) in which nine of the a_{ij} moduli vanish, and there are six additional relations between the remaining moduli, and hence, overall, there are only six independent moduli that characterize the GHL in that case. Subsequently, the following sections will be devoted to materials that are less general than the above GEN21 material. Most of the materials below are named by their internal micro-structure, and we shall present their effective elastic properties only as reflected by the above GHL.

2.3 Monoclinic Materials

A first type of materials that may be defined by less than 21 independent parameters are *monoclinic materials* that are defined by 13 independent coefficients and, as will become clearer later on (see S.2.16), contain one plane of elastic symmetry. There are three main kinds of materials in this category, and their classification is based on the type of population of their compliance and stiffness matrices. The first kind is the material described by

$$\mathbf{a} = \begin{bmatrix} a_{11} & a_{12} & a_{13} & 0 & 0 & a_{16} \\ & a_{22} & a_{23} & 0 & 0 & a_{26} \\ & & a_{33} & 0 & 0 & a_{36} \\ & & & a_{44} & a_{45} & 0 \\ & Sym. & & & a_{55} & 0 \\ & & & & & a_{66} \end{bmatrix}, \tag{2.5}$$

which for future use will be denoted MON13z. The numerical part of this notation stands for the number of independent coefficients, and the index "z" is added to indicate that the x,y-plane is a plane of elastic symmetry. A similar population of the compliance (or the stiffness) matrix may be obtained when orthotropic material is rotated about the z-axis, as will be shown later on within S.2.4. The second and the third kinds of materials in this family are described by the compliance matrices

$$\begin{bmatrix} a_{11} & a_{12} & a_{13} & a_{14} & 0 & 0 \\ & a_{22} & a_{23} & a_{24} & 0 & 0 \\ & & a_{33} & a_{34} & 0 & 0 \\ & & & a_{44} & 0 & 0 \\ Sym. & & & & a_{55} & a_{56} \\ & & & & & a_{66} \end{bmatrix}, \quad \begin{bmatrix} a_{11} & a_{12} & a_{13} & 0 & a_{15} & 0 \\ & a_{22} & a_{23} & 0 & a_{25} & 0 \\ & & a_{33} & 0 & a_{35} & 0 \\ & & & a_{44} & 0 & a_{46} \\ & Sym. & & & a_{55} & 0 \\ & & & & & a_{66} \end{bmatrix}, \tag{2.6}$$

and for similar reasons will be denoted MON13x and MON13y materials, respectively. The same matrix population is obtained for the corresponding stiffness matrices (**A**) of monoclinic materials.

2.4 Orthotropic Materials

A simpler class of materials known as *orthotropic materials*, is defined by nine independent moduli only. These materials have three orthogonal planes of elastic symmetry and are usually defined by the so-called *engineering constants* E_{ii}, G_{ij} and ν_{ij}, as

$$\mathbf{a} = \begin{bmatrix} \frac{1}{E_{11}} & -\frac{\nu_{21}}{E_{22}} & -\frac{\nu_{31}}{E_{33}} & 0 & 0 & 0 \\ -\frac{\nu_{12}}{E_{11}} & \frac{1}{E_{22}} & -\frac{\nu_{32}}{E_{33}} & 0 & 0 & 0 \\ -\frac{\nu_{13}}{E_{11}} & -\frac{\nu_{23}}{E_{22}} & \frac{1}{E_{33}} & 0 & 0 & 0 \\ 0 & 0 & 0 & \frac{1}{G_{23}} & 0 & 0 \\ 0 & 0 & 0 & 0 & \frac{1}{G_{13}} & 0 \\ 0 & 0 & 0 & 0 & 0 & \frac{1}{G_{12}} \end{bmatrix}. \tag{2.7}$$

For further use, this matrix form will also be denoted as ORT9. The above matrix is written in full to emphasize its symmetry properties $\frac{\nu_{12}}{E_{11}} = \frac{\nu_{21}}{E_{22}}$, $\frac{\nu_{13}}{E_{11}} = \frac{\nu_{31}}{E_{33}}$, $\frac{\nu_{23}}{E_{22}} = \frac{\nu_{32}}{E_{33}}$. One may gain some physical interpretation of the above notation by noticing first that $\frac{1}{E_{ii}}$ is the amount of ε_{ii}

per unit σ_{ii}. Also, $-\nu_{12}$ is the ratio $\varepsilon_{yy}/\varepsilon_{xx}$ when σ_{xx} is applied, $-\nu_{13}$ is the ratio $\varepsilon_{zz}/\varepsilon_{xx}$ when σ_{xx} is applied, etc. In addition, $\frac{1}{G_{ij}}$ is the amount of γ_{ij} per unit σ_{ij} for $(i,j) = (1,2),(1,3)$ and $(2,3)$. This physical interpretation serves also as the basis for experimental procedures that establish the physical magnitude of the elastic moduli for materials in this class, see e.g. (Jenkins, 1998), (Jones, 1999). As already indicated, simple inversion of the compliance matrix, **a**, yields the stiffness matrix **A** that exhibits population and symmetry, which are identical to those of **a**. The non-vanishing terms of the stiffness matrix are given by

$$\begin{aligned}
&A_{11} = \frac{E_{11}}{D_0}(1-\nu_{23}\nu_{32}), \qquad A_{12} = \frac{E_{11}}{D_0}(\nu_{21}+\nu_{23}\nu_{31}), \qquad A_{13} = \frac{E_{11}}{D_0}(\nu_{21}\nu_{32}+\nu_{31}), \\
&A_{22} = \frac{E_{22}}{D_0}(1-\nu_{13}\nu_{31}), \qquad A_{23} = \frac{E_{22}}{D_0}(\nu_{32}+\nu_{12}\nu_{31}), \qquad A_{33} = \frac{E_{33}}{D_0}(1-\nu_{12}\nu_{21}), \\
&A_{44} = G_{23}, \qquad A_{55} = G_{13}, \qquad A_{66} = G_{12},
\end{aligned} \tag{2.8}$$

where $D_0/(E_{11}E_{22}E_{33})$ is the determinant of the 3×3 upper-left minor of the **a** matrix of (2.7). D_0 may be written as

$$D_0 = 1-\nu_{12}\nu_{21}-\nu_{13}\nu_{31}-\nu_{23}\nu_{32}-\nu_{12}\nu_{23}\nu_{31}-\nu_{13}\nu_{21}\nu_{32}\,. \tag{2.9}$$

The same stiffness matrix may be written using the nine parameters of the lower part of **a** only (i.e. E_{ii}, ν_{12}, ν_{13}, ν_{23} and G_{ij}) as

$$\begin{aligned}
&A_{11} = \frac{E_{11}}{D_0}\Big(1-\nu_{23}^2\frac{E_{33}}{E_{22}}\Big), \quad A_{12} = \frac{E_{22}}{D_0}\Big(\nu_{12}+\nu_{23}\nu_{13}\frac{E_{33}}{E_{22}}\Big), \quad A_{13} = \frac{E_{33}}{D_0}(\nu_{12}\nu_{23}+\nu_{13}), \\
&A_{22} = \frac{E_{22}}{D_0}\Big(1-\nu_{13}^2\frac{E_{33}}{E_{11}}\Big), \quad A_{23} = \frac{E_{33}}{D_0}\Big(\nu_{23}+\nu_{12}\nu_{13}\frac{E_{22}}{E_{11}}\Big), \quad A_{33} = \frac{E_{33}}{D_0}\Big(1-\nu_{12}^2\frac{E_{22}}{E_{11}}\Big),
\end{aligned} \tag{2.10}$$

and

$$D_0 = 1-\nu_{23}^2\frac{E_{33}}{E_{22}}-\nu_{12}^2\frac{E_{22}}{E_{11}}-2\nu_{12}\nu_{23}\nu_{13}\frac{E_{33}}{E_{11}}-\nu_{13}^2\frac{E_{33}}{E_{11}}. \tag{2.11}$$

Clearly, if required, the stiffness matrix may also be expressed using the nine coefficients E_{11}, E_{22}, E_{33}, ν_{21}, ν_{31}, ν_{32} and G_{12}, G_{13}, G_{23} only.

It is sometimes useful to visualize the population and the relative magnitude of the various coefficients in the compliance and stiffness matrices, as shown in Fig. 2.1 for a typical orthotropic material.

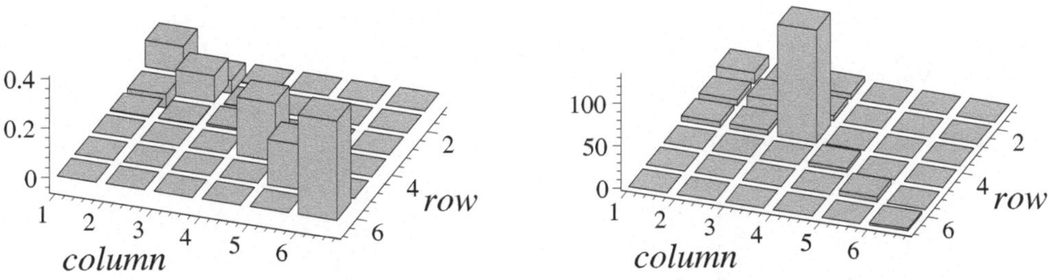

(a) The compliance matrix **a** (1/GPa). (b) The stiffness matrix **A** (GPa).

Figure 2.1: The compliance and stiffness matrices for typical orthotropic Graphite/Epoxy material.

As already indicated in S.2.3, when an orthotropic material is rotated about the x-, y- or z- directions, the resulting compliance/stiffness matrix population is identical to the form of

MON13x, MON13y and MON13z, respectively. Such materials are sometimes referred to as *Generally Orthotropic*. However, it should be noted that their matrices are based on 10 independent parameters only (nine of the orthotropic material plus one angle of rotation), while generic monoclinic materials are based on 13 independent parameters.

For example, Fig. 2.2, presents a beam of a MON13y material that was created from orthotropic material by rotating it by the angle $-\theta_y$ about the y-axis (x, y, z are the beam axes, x', y', z' are the orthotropic material principal axes). As shown, this rotation may also be viewed

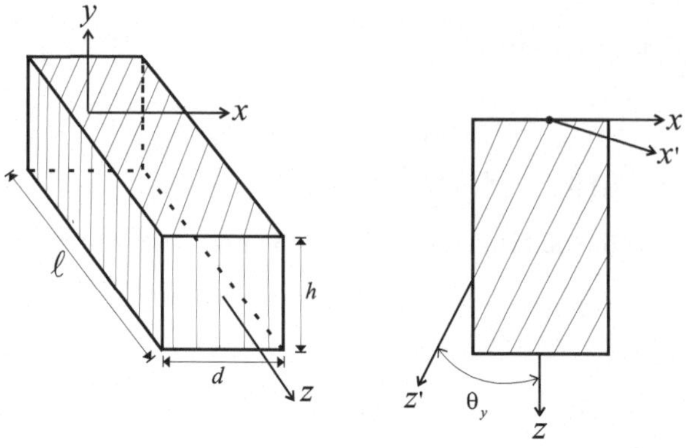

Figure 2.2: A MON13y homogeneous beam made of rotated orthotropic material.

as a rotation of the x', y', z' coordinate system by the angle θ_y.

2.5 Tetragonal Materials

Tetragonal material is an orthotropic one that has a pair of coordinate directions along which the elastic properties are identical, and different properties along the third coordinate axis. Hence, there are three fundamental types of this material. For example, when the (x, y) axes are selected, the material remains unchanged under the transformation $x \to -y$, $y \to x$ (where the "minus" has been introduced just for preserving the system orientation). Such a material may therefore be defined by the *six* independent parameters E, E', G, G', ν, ν', so that

$$E_{11} = E_{22} = E, \qquad E_{33} = E', \qquad G_{23} = G_{13} = G', \qquad G_{12} = G,$$
$$\nu_{12} = \nu_{21} = \nu, \qquad \nu_{13} = \nu_{23} = \nu'\frac{E}{E'}, \qquad \nu_{31} = \nu_{32} = \nu', \tag{2.12}$$

which yields the following compliance matrix:

$$\mathbf{a} = \begin{bmatrix} \frac{1}{E} & -\frac{\nu}{E} & -\frac{\nu'}{E'} & 0 & 0 & 0 \\ & \frac{1}{E} & -\frac{\nu'}{E'} & 0 & 0 & 0 \\ & & \frac{1}{E'} & 0 & 0 & 0 \\ & Sym. & & \frac{1}{G'} & 0 & 0 \\ & & & & \frac{1}{G'} & 0 \\ & & & & & \frac{1}{G} \end{bmatrix}. \tag{2.13}$$

For further use, tetragonal materials where $x-y$ directions are interchangeable will be denoted TRG6z, while similarly, the other two kinds will be denoted TRG6x and TRG6y. The population of the above compliance matrix for all three pairs of identical directions is summarized in

Material → Interchangeable Directions → Module ↓	TRG6x $y-z$	TRG6y $x-z$	TRG6z $x-y$
a_{11}	$\frac{1}{E'}$	$\frac{1}{E}$	$\frac{1}{E}$
a_{22}	$\frac{1}{E}$	$\frac{1}{E'}$	$\frac{1}{E}$
a_{33}	$\frac{1}{E}$	$\frac{1}{E}$	$\frac{1}{E'}$
a_{12}	$-\frac{\nu'}{E'}$	$-\frac{\nu'}{E'}$	$-\frac{\nu}{E}$
a_{13}	$-\frac{\nu'}{E'}$	$-\frac{\nu}{E}$	$-\frac{\nu'}{E'}$
a_{23}	$-\frac{\nu}{E}$	$-\frac{\nu'}{E'}$	$-\frac{\nu'}{E'}$
a_{44}	$\frac{1}{G}$	$\frac{1}{G'}$	$\frac{1}{G'}$
a_{55}	$\frac{1}{G'}$	$\frac{1}{G}$	$\frac{1}{G'}$
a_{66}	$\frac{1}{G'}$	$\frac{1}{G'}$	$\frac{1}{G}$

Table 2.1: The compliance matrix moduli for the three fundamental tetragonal materials.

Table 2.1. The stiffness matrix of tetragonal materials may be defined using the following Q_i quantities:

$$Q_1 = \frac{E}{1-\nu-2\nu'^2\frac{E}{E'}} \cdot \frac{1-\nu'^2\frac{E}{E'}}{1+\nu}, \qquad Q_2 = \frac{E'(1-\nu)}{1-\nu-2\nu'^2\frac{E}{E'}},$$
$$Q_3 = \frac{E}{1-\nu-2\nu'^2\frac{E}{E'}} \cdot \frac{\nu+\nu'^2\frac{E}{E'}}{1+\nu}, \qquad Q_4 = \frac{E\nu'}{1-\nu-2\nu'^2\frac{E}{E'}}, \tag{2.14}$$

while for the case where the $x-y$ directions are interchangeable,

$$\mathbf{A} = \begin{bmatrix} Q_1 & Q_3 & Q_4 & 0 & 0 & 0 \\ & Q_1 & Q_4 & 0 & 0 & 0 \\ & & Q_2 & 0 & 0 & 0 \\ & & & G' & 0 & 0 \\ & Sym. & & & G' & 0 \\ & & & & & G \end{bmatrix}. \tag{2.15}$$

The population of the above stiffness matrix for all three types of tetragonal materials is summarized in Table 2.2. Note that Q_i depend only on the four parameters E, E', ν, and ν', which together with G', G constitute a set of six independent parameters.

2.6 Transversely Isotropic (Hexagonal) Materials

Many natural and man-made materials are classified as *transversely isotropic* (*transtropic* for short, or *hexagonal*). This definition stands for the case when one can find a line that allows a rotation of the material about it without changing its properties. The plane, which is perpendicular to this line (which is in fact an axis of rotational symmetry) is called a *plane of elastic*

Material → Interchangeable Directions → Module ↓	TRG6x $y-z$	TRG6y $x-z$	TRG6z $x-y$
A_{11}	Q_2	Q_1	Q_1
A_{22}	Q_1	Q_2	Q_1
A_{33}	Q_1	Q_1	Q_2
A_{12}	Q_4	Q_4	Q_3
A_{13}	Q_4	Q_3	Q_4
A_{23}	Q_3	Q_4	Q_4
A_{44}	G	G'	G'
A_{55}	G'	G	G'
A_{66}	G'	G'	G

Table 2.2: The stiffness matrix moduli for the three fundamental tetragonal materials.

symmetry or *plane of isotropy*. Hence, each plane that contains the axis of rotation is a plane of symmetry, and therefore, transversely isotropic material admits an infinite number of elastic symmetries. A modern example for such a material are laminates made of randomly oriented chopped fibers that are in general placed in a certain plane. The effective material properties have no profound direction in that plane, which then becomes a plane of elastic symmetry, see (Giurgiutiu and Reifsnider, 1994).

Thus, transversely isotropic materials have *one* isotropic plane, and are defined by *five* independent parameters: E, E' — Young's moduli for tension or compression in the plane of isotropy and in a direction normal to it, respectively, ν' — Poisson's ratio characterizing transverse contraction in the plane of isotropy when tension is applied in a direction normal to it, respectively, G', G — shear moduli for the plane of isotropy and any plane perpendicular to it, respectively, see (Lekhnitskii, 1981). Note the notion "Young's modulus" and "Poisson's ratio" are adopted from pure isotropic terminology, and used here as analogy only.

Consider for example, the case when the x,y-plane is isotropic, namely, where a rotation of the material about the z-axis has no influence on its properties. The compliance matrix of this transversely isotropic material is identical to that of the tetragonal material TRG6z, while in addition, a dependency of G on E and ν should be introduced. The source of this dependency is the "isotropic relation" $A_{66} = (A_{11} - A_{12})/2$ or $a_{66} = 2(a_{11} - a_{12})$ that will be clarified by the discussion of isotropic materials within S.2.8. Hence, in tetragonal material terminology for TRG6z, the above isotropic relation yields $G = (Q_1 - Q_3)/2$, which may be simplified as

$$G = \frac{E}{2(1+\nu)}. \tag{2.16}$$

An identical expression is deduced for the other two material versions. Subsequently, to obtain the three kinds of transversely isotropic materials, TI5x, TI5y, TI5z, from the tetragonal materials TRG6x, TRG6y, TRG6z, respectively, one should introduce $\nu = \nu(G,E)$ as shown by (2.16) directly into Table 2.1, and into the Q_i terms in Table 2.2.

2.7 Cubic Materials

A sub-group of the above discussed materials are the so-called *cubic materials* which are based on three independent parameters only, and may be defined as orthotropic materials with $E_{11} = E_{22} = E_{33} = E$, $\nu_{12} = \nu_{13} = \nu_{23} = \nu$, and $G_{12} = G_{13} = G_{23} = G$. Hence, the compliance and the stiffness matrices for cubic material may be written as

$$\begin{bmatrix} \frac{1}{E} & -\frac{\nu}{E} & -\frac{\nu}{E} & 0 & 0 & 0 \\ & \frac{1}{E} & -\frac{\nu}{E} & 0 & 0 & 0 \\ & & \frac{1}{E} & 0 & 0 & 0 \\ & & & \frac{1}{G} & 0 & 0 \\ & Sym. & & & \frac{1}{G} & 0 \\ & & & & & \frac{1}{G} \end{bmatrix}, \begin{bmatrix} \frac{E(1-\nu)}{(1+\nu)(1-2\nu)} & \frac{E\nu}{(1+\nu)(1-2\nu)} & \frac{E\nu}{(1+\nu)(1-2\nu)} & 0 & 0 & 0 \\ & \frac{E(1-\nu)}{(1+\nu)(1-2\nu)} & \frac{E\nu}{(1+\nu)(1-2\nu)} & 0 & 0 & 0 \\ & & \frac{E(1-\nu)}{(1+\nu)(1-2\nu)} & 0 & 0 & 0 \\ & & & G & 0 & 0 \\ & Sym. & & & G & 0 \\ & & & & & G \end{bmatrix}. \tag{2.17}$$

For further use, these materials will be denoted CUB3, and clearly, they may also be defined as special transversely isotropic or tetragonal materials with $E' = E$, $G' = G$ and $\nu' = \nu$.

2.8 Isotropic Materials

To derive the stress-strain relationships for isotropic materials, we make use of the stress and strain deviator tensors $\sigma_{ij}^D = \sigma_{ij} - \frac{1}{3}\Theta_1\delta_{ij}$ and $\varepsilon_{ij}^D = \varepsilon_{ij} - \frac{1}{3}\Xi_1\delta_{ij}$. Here we have also invoked the stress and strain first linear invariants, namely, the sum of normal stresses $\Theta_1 = \sigma_{11} + \sigma_{22} + \sigma_{33}$, see (1.100a), and the sum of normal strains $\Xi_1 = \varepsilon_{11} + \varepsilon_{22} + \varepsilon_{33}$, which is the relative change in volume for the linear theory, see S.1.1.4.2.

The uniqueness of isotropic stress-strain relationships emerges from the fact that for these materials, *the components of the stress deviator are proportional to the components of the strain deviator*. For the moment, we shall denote this proportion constant (which is clearly a material property) as $2G$, and therefore $\sigma_{ij}^D = 2G\varepsilon_{ij}^D$. Thus, by substituting the above mentioned stress and strain deviators one obtains

$$\sigma_{ij} = 2G\varepsilon_{ij} + (K - \frac{2G}{3})\Xi_1\delta_{ij}, \tag{2.18}$$

where $K = \Theta_1/3\Xi_1$ is another independent material property, the *bulk modulus*. The physical interpretation of K is clear as it is *the ratio of the mean normal stress ($\Theta_1/3$) to the relative change in volume (Ξ_1)*, and therefore may be viewed as a measure or *the material resistance to a change in volume*. Hence, overall, we have now two material independent properties, K and G, which for the sake of analytical uniformity will be replaced by two other independent constants, the modulus of elasticity E, and the non-dimensional ratio ν, so that

$$K = \frac{E}{3(1-2\nu)}, \qquad G = \frac{E}{2(1+\nu)}. \tag{2.19}$$

For $\nu = \frac{1}{2}$ we have $K = \infty$, which characterizes an *incompressible elastic material* (i.e. a material that does not change in volume when subjected to tensile or compressive strains). The above notation allows one to consider isotropic material as a special orthotropic material with $E_{11} = E_{22} = E_{33} = E$, $\nu_{12} = \nu_{13} = \nu_{23} = \nu$ and $G_{12} = G_{13} = G_{23} = \frac{E}{2(1+\nu)}$ (see (2.7)), and therefore

$$\mathbf{a} = \frac{1}{E}\begin{bmatrix} 1 & -\nu & -\nu & 0 & 0 & 0 \\ & 1 & -\nu & 0 & 0 & 0 \\ & & 1 & 0 & 0 & 0 \\ & & & 2(1+\nu) & 0 & 0 \\ & & & & 2(1+\nu) & 0 \\ & & & & & 2(1+\nu) \end{bmatrix}, \tag{2.20a}$$

$$\mathbf{A} = \frac{E}{(1+\nu)(1-2\nu)}\begin{bmatrix} 1-\nu & \nu & \nu & 0 & 0 & 0 \\ & 1-\nu & \nu & 0 & 0 & 0 \\ & & 1-\nu & 0 & 0 & 0 \\ & & & \frac{1}{2}-\nu & 0 & 0 \\ & & & & \frac{1}{2}-\nu & 0 \\ & & & & & \frac{1}{2}-\nu \end{bmatrix}. \tag{2.20b}$$

The physical interpretation of G as the *elastic shear modulus* becomes now clear as it relates the shear and strain components $\sigma_{ij} = 2G\varepsilon_{ij}$ $(ij = 12, 13, 23)$ or $\sigma_{\alpha\beta} = G\varepsilon_{\alpha\beta}$ $(\alpha\beta = xy, xz, yz)$. Traditionally, E is referred to as *Young's modulus*, and ν as *Poisson's Ratio*. Note that for $\nu = 1/2$, the r.h.s. of (1.42) vanishes, which again testifies for the material incompressibility.

For further use, isotropic materials will be denoted ISO2 which clearly reflects the fact that they are fully characterized by two independent parameters only.

Remark 2.2 It is worth mentioning some classical notation used for isotropic materials. By defining *Lamé's constants* as $\mu = G$, $\lambda = \frac{\nu E}{(1+\nu)(1-2\nu)}$, one may write the constitutive relations using Kronecker's symbol δ_{ij} as

$$\sigma_{ij} = \lambda\delta_{ij}\Xi_1 + 2\mu\varepsilon_{ij}, \qquad \varepsilon_{ij} = -\frac{\lambda}{2\mu}\delta_{ij}\Xi_1 + \frac{1}{2\mu}\sigma_{ij}. \tag{2.21}$$

2.9 Engineering Notation of Composites

Engineering notation of (anisotropic) composite materials is usually founded on slightly different (and approximate) convention. Typically, *four* parameters, E_1, E_2, ν_{12} and G_{12} are defined, where the subscript 1 stands for the "fiber" (or the "stiff") direction, and the subscript 2 stands for the "perpendicular to the fiber" (or the "soft") direction. In some cases, 1 and 2 are replaced by L ("Longitudinal") and T ("Transverse"), respectively. This notation is used mainly for unidirectional materials, or for materials that have a profound stiff direction. Such materials may be viewed as transversely isotropic materials (that should be defined by *five* independent parameters, see S.2.6), and in order to examine these engineering parameters the reader is referred to Fig. 2.3. In this figure, the material is transversely isotropic where the x, y-plane is an isotropic plane, and the stiff direction is stretched along the z-axis. Hence, in engineering notation, the x_1- and x_2- directions coincide with the z- and x- directions, respectively. E_1 and E_2 are the moduli in these directions and therefore, based on (2.12), (2.16), $E_1 = E'$ and $E_2 = E$. ν_{12} is the major Poisson's ratio defined as a contraction in the x-direction due to unit extension in the z-direction, and therefore $\nu_{12} = \nu'$. G_{12} is the shear modulus in the x_1, x_2-plane, and therefore, $G_{12} = G'$. Hence, data regarding G (the shear module in the x, y-plane) or ν (the minor Poisson's ratio that represents, for example, a contraction in the y-direction due to unit extension in the x-direction) is usually not measured nor reported. Note that only one of these parameters is required as they are related by (2.16). It should also be noted that for typical transversely isotropic materials, the variation of ν will have only negligible influence on A_{33}

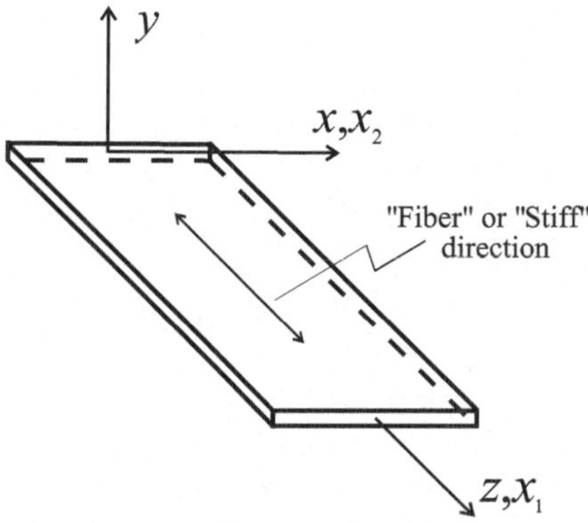

Figure 2.3: Engineering notation of transversely isotropic material (TI5z).

which is the main driver of the structural behavior in slender structures, where unidirectional materials are typically placed (beams, slender plates, etc., see Chapter 5). To show that, we may expand A_{33} (Q_2 in this case, see Table 2.2 and (2.14)) into Taylor series around $\nu = 0$, and examine the case of $\nu' < 1$ and $E/E' \ll 1$, namely

$$A_{33} = \frac{E'^2}{E' - 2\nu'^2 E}[1 + \underline{\frac{2\nu'^2 E}{E' - 2\nu'^2 E}}\nu + O(\nu^2)] \qquad \text{or} \qquad \lim_{\frac{E}{E'}\to 0} A_{33} = E'. \tag{2.22}$$

Hence, the above underlined term is negligible compared with the unit. Thus, A_{33} is only a weak function of ν, and lack of data in this case is of minor importance. On the other hand, expanding A_{12} (Q_3 in this case, see Table 2.2 and (2.14)) yields

$$A_{12} = \frac{\nu'^2 E^2}{E' - 2\nu'^2 E}[1 + \underline{\frac{E'^2 - 2\nu'^2 EE' + 2\nu'^4 E^2}{E\nu'^2 (E' - 2\nu'^2 E)}}\nu + O(\nu^2)] \text{ or } \lim_{\frac{E}{E'}\to 0} A_{12} = \nu\frac{E}{1-\nu^2}. \tag{2.23}$$

Examination of the underlined term under the same arguments shows that A_{12} is a strong function of ν, and in cases where this module is required and important, knowledge of the value of ν is inevitable.

Remark 2.3 The reader should be aware of the fact that unlike the isotropic case, for anisotropic materials, the stress principal directions are not identical to the strain principal directions, see S.1.3.3, S.1.3.4. This fact becomes clearer if one examines the normal strain, $\varepsilon_N = \varepsilon_{11}$, and the shear strain, $\varepsilon_T = \sqrt{\varepsilon_{12}^2 + \varepsilon_{13}^2}$, in the stress principal directions

$$\varepsilon_N = a_{11}\sigma_1^P + a_{12}\sigma_2^P + a_{13}\sigma_3^P, \tag{2.24a}$$

$$\varepsilon_T^2 = \left(a_{51}\sigma_1^P + a_{52}\sigma_2^P + a_{53}\sigma_3^P\right)^2 + \left(a_{61}\sigma_1^P + a_{62}\sigma_2^P + a_{63}\sigma_3^P\right)^2, \tag{2.24b}$$

from which follows that in the general case, ε_T is not zero. However, orthotropic (and simpler) materials do show coincidence of stress and strain principal directions.

2.10 Positive-Definite Stress-Strain Law

So far, the above discussed stress-strain laws were presented as real symmetric 6×6 matrices of different types of population. However, energy considerations show that there are certain positive definiteness constraints on the coefficients in these matrices, which mathematically may be put as a system of inequalities.

To derive the above inequalities, we should first review the expression for W, the *volume density of the strain energy* or the *elastic potential* that is stored in an elastic body, which is deformed by external loads. For a *linear, anisotropic* elastic media, W is put in the following quadratic form:

$$W = \frac{1}{2}\sum_{i,j} \widetilde{\alpha}_{ij}\varepsilon_i\varepsilon_j, \tag{2.25}$$

where $\widetilde{\alpha}_{ij}$ are related to the elements of the stiffness matrix as $A_{ij} = \frac{1}{2}(\widetilde{\alpha}_{ij} + \widetilde{\alpha}_{ji})$. The above expression for W emerges from various constraints. The first one is the requirement that W will be a function of the strain components that vanishes for $\varepsilon_i = 0$, and thus, a constant value (that reflects a pre-stressed state) should not be incorporated. In addition, W should reflect a linear stress-strain relation, namely,

$$\sigma_i = W_{,\varepsilon_i} = \frac{1}{2}\sum_j (\widetilde{\alpha}_{ij} + \widetilde{\alpha}_{ji})\,\varepsilon_j. \tag{2.26}$$

The above shows that the stress-strain matrix is symmetric (i.e. $A_{ij} = (\widetilde{\alpha}_{ij} + \widetilde{\alpha}_{ji})/2$). Higher-order dependency of W in ε_i would have caused a nonlinear stress-strain relation. Subsequently, W may also be written as

$$W = \frac{1}{2}\sum_i \sigma_i\varepsilon_i. \tag{2.27}$$

Equation (2.27) shows that *in order to ensure that W will attain positive values for all nonzero strain states (i.e. energy will not be created), the stress-strain relations matrix should be positive-definite*. Recall that a matrix, $\widetilde{\mathbf{M}}$, is said to be *positive-definite* if the inequality $\widetilde{\mathbf{z}}^T \cdot \widetilde{\mathbf{M}} \cdot \widetilde{\mathbf{z}} > 0$ holds for any nonzero "test vector" $\widetilde{\mathbf{z}}$. The actual implementation of a positive-definite check of a matrix may be easily performed via its eigenvalue analysis.

Thus, in the general case, there are six inequalities that should be simultaneously fulfilled, see also **P.2.4**. For example, examination of the compliance matrix of a MON13z material, the number of inequalities may be reduced by the obvious preliminary assumption that considers all the diagonal terms, a_{ii}, as positive (as clearly demanded by the requirement $\varepsilon_i/\sigma_i > 0$). In such a case, one obtains an additional four inequalities, the simplest two of which are $a_{11}a_{22} - a_{12}^2 > 0$ and $a_{44}a_{55} - a_{45}^2 > 0$. For orthotropic materials (see (2.7)), by considering the stiffness matrix and assuming positive E_{11}, E_{22}, E_{33}, G_{23}, G_{13}, G_{12}, the stress-strain matrix becomes positive-definite if

$$\frac{E_{11}}{E_{22}} > \nu_{12}^2 \quad \text{and} \quad D_0 > 0, \tag{2.28}$$

where D_0 is given by (2.9) or (2.11). Simple argumentation of coordinate interchanging reveals that the first inequality of (2.28) may be replaced by one of the following inequalities:

$$\frac{E_{11}}{E_{33}} > \nu_{13}^2, \qquad \frac{E_{22}}{E_{33}} > \nu_{23}^2. \tag{2.29}$$

For transversely isotropic materials we assume that E, E' and G are all positive. Hence, for the typical case where the x,y-plane is isotropic, one obtains the inequalities $(\nu - 1)E' + 2\nu'^2 E < 0,\ \nu > -1$, which may also be written as

$$\nu'^2 < \frac{E'}{E}, \qquad |\nu| < 1. \tag{2.30}$$

For isotropic materials, by assuming positive E, one obtains the conditions $(2\nu - 1)(\nu + 1) < 0,\ \nu > -1$, which may be solved for ν as

$$-1 < \nu < \frac{1}{2}. \tag{2.31}$$

2.11 Typical Material Characteristics

There is a wide range of properties for anisotropic/composite materials, and the tables in this section are provided for orientation purposes only, while exact values should be carefully extracted from proper sources, see e.g. (Dostal, 1987), (Stuart, 1996), (Akovali, 2001), (Adams *et al.*, 2002).

The tables below are for typical properties of unidirectional composites (mainly commercial fabrics), and isotropic materials. Yet, it should be noted that characterization of composite materials encompasses, in addition to the classical strength values, many other considerations, such as specific strength (i.e. strength to weight ratio), thermal expansion coefficients, damage tolerance and fatigue characteristics, production aspects, etc., which are not treated in the present context.

Anisotropic natural materials are mainly various kinds of rocks and woods. In some cases, naturally laminated rock structures are found, and for the purpose of macro-analysis these may be viewed as transversely isotropic materials. In addition, many kinds of woods may be regarded as reinforced material as they are characterized by a natural lay-up where the relatively stiff fibers are oriented in a certain direction, and may be considered as transversely isotropic materials as well.

Man-made materials are usually produced by various mixture rules of stiff reinforced fibers layered in a relatively soft "matrix", see e.g. (Mallick, 1993). In such cases, the material properties strongly depend on the volume fracture of each component and on the ratio of the amount of fibers layered in the principal direction and perpendicular to it. When the fibers are placed along the principal direction only, a pure *unidirectional lamina* is obtained.

In practices, composite materials are typically used as "prepregs". The engineering term "prepreg" stands for reinforced fabrics that are pre-impregnated with resin and partially cured. Such a material may be used by manufacturers to mold various parts, even without adding any resin, which allows avoiding of wet lay-up. Table 2.3 presents typical properties of unidirectional prepregs, where ρ stands for the laminate density, and V_f is the fiber volume fraction.

Property	Graphite/ Epoxy	Boron/ Epoxy	Carbon/ Epoxy	Kevlar/ Epoxy	E-Glass/ Epoxy	S-Glass/ Epoxy
E' ("E_{11}") (GPa)	290.	204.	140.–180.	88.	39.	43.
E ("E_{22}") (GPa)	6.	20.	9.–10.	5.6	8.6	8.9
G' ("G_{12}") (GPa)	5.	9.	4.–7.	2.2	3.8	4.5
ν' ("ν_{12}")	0.23	0.23	0.28–0.30	0.34	0.28	0.27
ρ (g/cm^3)	1.6	2.0	1.5–1.6	1.4	2.1	2.0
V_f	0.57	0.67	0.6–0.7	0.60	0.55	0.50

Table 2.3: Typical properties of unidirectional prepregs.

For reference purposes, Table 2.4 provides typical properties of three isotropic metals.

Property	Aluminum	Steel	Titanium
E (GPa)	73.	210.	110.
ν	0.33	0.30	0.30

Table 2.4: Typical properties of isotropic metals.

Remark 2.4 As shown by the above tables, Poisson's ratios, ν (or ν'), are typically positive. Rubber-type materials are examples of relatively high Poisson's ratio that is close to 0.5. Yet, some materials of internal "foam structures" exhibit negative Poisson's ratios (i.e. expand laterally when stretched), see e.g. (Lakes, 1987).

2.12 Compliance Matrix Transformation

To create a consistent transformation for the compliance matrix due to spatial rotation, one should first develop the following fourth-order *compliance tensor*, **C**, that plays the role of the constitutive relations and consistently connects the strain and stress tensors as (note that summation is carried out for identical indices)

$$\varepsilon_{ab} = C_{abcd}\sigma_{cd}. \tag{2.32}$$

Since the tensor **C** plays the role of the matrix **a** in the generalized Hooke's law discussed in S.2.1, it exhibits triple symmetry, namely

$$C_{abcd} = C_{bacd} = C_{abdc} = C_{cdab}. \tag{2.33}$$

To support the discussion regarding the transformation of **C**, two typical strain components, ε_{xx} and γ_{xz} are examined. We first write them in the matrix notation of (2.3)

$$\varepsilon_{xx} = a_{11}\sigma_{xx} + a_{12}\sigma_{yy} + a_{13}\sigma_{zz} + a_{14}\sigma_{yz} + a_{15}\sigma_{xz} + a_{16}\sigma_{xy}, \tag{2.34a}$$

$$\gamma_{xz} = a_{51}\sigma_{xx} + a_{52}\sigma_{yy} + a_{53}\sigma_{zz} + a_{54}\sigma_{yz} + a_{55}\sigma_{xz} + a_{56}\sigma_{xy}. \tag{2.34b}$$

Subsequently, we express the same terms in tensor notation. In view of $\varepsilon_{xx} = \varepsilon_{11}$ we write

$$\varepsilon_{xx} = C_{1111}\sigma_{11} + C_{1122}\sigma_{22} + C_{1133}\sigma_{33} + 2C_{1123}\sigma_{23} + 2C_{1113}\sigma_{13} + 2C_{1112}\sigma_{12}, \tag{2.35}$$

and in view of $\gamma_{xz} = 2\varepsilon_{13}$ we write

$$\gamma_{xz} = 2C_{1311}\sigma_{11} + 2C_{1322}\sigma_{22} + 2C_{1333}\sigma_{33} + 4C_{1323}\sigma_{23} + 4C_{1313}\sigma_{13} + 4C_{1312}\sigma_{12}. \tag{2.36}$$

The above form is obtained due to the double summation in (2.32). As an example, for $c = 2$ and $d = 3$, one gets the term $C_{ab23}\sigma_{23}$, while for $c = 3$ and $d = 2$, one gets the term $C_{ab32}\sigma_{32}$, which is identical to the previous one. Therefore, it may concluded that $C_{1111} = a_{11}$, $2C_{1123} = a_{14}$, $2C_{1311} = a_{15}$, $4C_{1323} = a_{45}$, etc. To take the above considerations into account, we define a modified compliance matrix that connects the vectors $\varepsilon = [\varepsilon_{11}, \varepsilon_{22}, \varepsilon_{33}, \varepsilon_{23}, \varepsilon_{13}, \varepsilon_{12}]^T$ and $\sigma = [\sigma_{11}, \sigma_{22}, \sigma_{33}, \sigma_{23}, \sigma_{13}, \sigma_{12}]^T$, as $\varepsilon = \mathbf{C}\cdot\sigma$. Hence, the matrix $\mathbf{C} = \{c_{ij}\}$ is given by

$$\mathbf{C} = \begin{bmatrix} a_{11} & a_{12} & a_{13} & a_{14}/2 & a_{15}/2 & a_{16}/2 \\ & a_{22} & a_{23} & a_{24}/2 & a_{25}/2 & a_{26}/2 \\ & & a_{33} & a_{34}/2 & a_{35}/2 & a_{36}/2 \\ & & & a_{44}/4 & a_{45}/4 & a_{46}/4 \\ & & & & a_{55}/4 & a_{56}/4 \\ & & & & & a_{66}/4 \end{bmatrix}. \tag{2.37}$$

Analogously to the notation in (1.21), we also establish the following transformation (known as Voigt's contracted indicial notation) between pairs of indices, say (i, j) $(i, j = 1, 2, 3)$ and a single index, say k $(k = 1, \ldots, 6)$

$$(1,1) \to 1, \quad (2,2) \to 2, \quad (3,3) \to 3; \quad (2,3) \to 4, \quad (1,3) \to 5, \quad (1,2) \to 6, \tag{2.38}$$

while an (i,j) pair is treated as a (j,i) pair. Subsequently, we use the above index transformation for the pair of indices (i_1,i_2) to determine a single index i, and for the pair of indices (j_1,j_2) to determine a single j, and construct the tensor $\mathbf{C}$ out of the matrix $\mathbf{C}$ as

$$C_{i_1i_2j_1j_2} = c_{ij}. \tag{2.39}$$

We now apply a tensor transformation to $\mathbf{C}$ based on the transformation tensor $\mathbf{T}$ developed in S.1.7.1 and obtain the corresponding tensor in the new system

$$\overline{C}_{klst} = C_{mnpr}T_{km}T_{ln}T_{sp}T_{tr}. \tag{2.40}$$

Based on the indices notation of (2.38), the modified compliance matrix $\bar{\mathbf{c}} = \{\bar{c}_{ij}\}$ in the new system becomes

$$\bar{c}_{ij} = \overline{C}_{i_1i_2j_1j_2}. \tag{2.41}$$

Finally, one may construct the desired new compliance matrix $\bar{\mathbf{a}} = \{\bar{a}_{ij}\}$ according to the relations shown in (2.37) (namely, the relations between $\bar{c}_{ij}$ and $\bar{a}_{ij}$ are identical to those between c_{ij} and a_{ij})

$$\bar{\mathbf{a}} = \begin{bmatrix} \bar{c}_{11} & \bar{c}_{12} & \bar{c}_{13} & 2\bar{c}_{14} & 2\bar{c}_{15} & 2\bar{c}_{16} \\ & \bar{c}_{22} & \bar{c}_{23} & 2\bar{c}_{24} & 2\bar{c}_{25} & 2\bar{c}_{26} \\ & & \bar{c}_{33} & 2\bar{c}_{34} & 2\bar{c}_{35} & 2\bar{c}_{36} \\ & & & 4\bar{c}_{44} & 4\bar{c}_{45} & 4\bar{c}_{46} \\ & & & & 4\bar{c}_{55} & 4\bar{c}_{56} \\ & & & & & 4\bar{c}_{66} \end{bmatrix}. \tag{2.42}$$

The expressions for generic rotation that include all three angles are huge, and the reader may examine all of them symbolically by activating **P.2.5**. In what follows, only representative examples are given. For that purpose, we shall define a 21 terms vector $\mathbf{a_V}$ as

$$\mathbf{a_V} = \left[a_{11},\ldots,a_{16},a_{22},\ldots,a_{26},a_{33},\ldots,a_{36},a_{44},\ldots,a_{46},a_{55},a_{56},a_{66}\right]^T, \tag{2.43}$$

while its version after rotation will be denoted analogously as $\bar{\mathbf{a}}_\mathbf{V} = \{\bar{a}_{ij}\}$. These vectors are related by $\bar{\mathbf{a}}_\mathbf{V} = \mathbf{M_a} \cdot \mathbf{a_V}$, where $\mathbf{M_a}$ is a (non-symmetric) 21×21 matrix. We shall now look at the coefficient of a_{12} in the expression for $\bar{a}_{16}$, which according to the above notation is given by $M_a(6,2)$. To clarify that, we write $M_a(6,2) = 2\widetilde{A}_1\widetilde{B}_1$, where

$$\widetilde{A}_1 = \cos\phi\cos^2\psi + 2\,\sin\theta\sin\phi\sin\psi\cos\psi - \cos\phi\sin^2\psi, \qquad \widetilde{B}_1 = \cos\psi\cos^3\theta\sin\psi.$$

As another example, we examine the coefficient of a_{44} in the expression for $\bar{a}_{25}$, which is given by $M_a(10,16) = \widetilde{A}_2\widetilde{B}_2\widetilde{C}_2$, where

$$\begin{aligned} \widetilde{A}_2 &= \sin\phi\cos\theta, \qquad \widetilde{B}_2 = \sin\phi\sin\theta\sin\psi + \cos\phi\cos\psi, \\ \widetilde{C}_2 &= \sin\phi\sin\theta\cos\psi - \sin\psi\cos\phi\sin^2\theta + \cos^2\theta\cos\phi\sin\psi. \end{aligned} \tag{2.44}$$

Again, all $441(=21^2)$ terms of $\mathbf{M_a}$ may be easily produced by **P.2.5**, and the reader may create compliance matrices for specific materials and rotation angles by activating **P.2.6**.

Due to the symmetry characteristics of the tensor $\mathbf{C}$, see (2.33), there are two quantities that are invariants under the most general transformation (that includes all three rotation angles ϕ, θ and ψ). These are defined as (see linear invariants in **P.2.5**)

$$J_1^a = C_{iklm}\delta_{ik}\delta_{lm} = C_{1111} + C_{2222} + C_{3333} + 2C_{1122} + 2C_{1133} + 2C_{2233}, \tag{2.45a}$$

$$J_2^a = C_{iklm}\delta_{il}\delta_{km} = C_{1111} + C_{2222} + C_{3333} + 2C_{1212} + 2C_{2323} + 2C_{1313}, \tag{2.45b}$$

where δ_{ab} is Kronecker's symbol. In terms of a_{ij}, these invariants become

$$J_1^a = a_{11} + a_{22} + a_{33} + 2(a_{12} + a_{13} + a_{23}), \quad J_2^a = a_{11} + a_{22} + a_{33} + \frac{a_{44} + a_{55} + a_{66}}{2}. \tag{2.46}$$

In addition, there are invariants for each direction, namely, quantities that remain unchanged when the rotation is carried out about a single axis only. Four of these invariants, which are denoted I_i^a, are given in Table 2.5 (may also be produced by **P.2.5** by setting any two rotation angles to zero). The role played by those and other invariants will be discussed later on within S.2.15. As indicated, the explicit expressions for $\overline{a}_{ij}$ of GEN21 materials are lengthy. Yet, for

Rotation Axis → Invariant ↓	x	y	z
I_1^a	$a_{33} + a_{22} + 2a_{32}$	$a_{11} + a_{33} + 2a_{13}$	$a_{11} + a_{22} + 2a_{12}$
I_2^a	$a_{44} - 4a_{32}$	$a_{55} - 4a_{13}$	$a_{66} - 4a_{12}$
I_3^a	$a_{66} + a_{55}$	$a_{44} + a_{66}$	$a_{44} + a_{55}$
I_4^a	$a_{13} + a_{21}$	$a_{12} + a_{23}$	$a_{13} + a_{23}$
I_1^A	$A_{33} + A_{22} + 2A_{32}$	$A_{11} + A_{33} + 2A_{13}$	$A_{11} + A_{22} + 2A_{12}$
I_2^A	$A_{44} - A_{32}$	$A_{55} - A_{13}$	$A_{66} - A_{12}$
I_3^A	$A_{66} + A_{55}$	$A_{44} + A_{66}$	$A_{44} + A_{55}$
I_4^A	$A_{13} + A_{21}$	$A_{12} + A_{23}$	$A_{13} + A_{23}$

Table 2.5: Invariants for a single axis transformation.

orthotropic materials one may use **P.2.7**, which yields, for example,

$$\begin{aligned} \overline{a}_{11} = {} & \frac{1}{E_{22}} \cos^4\theta \sin^4\psi + (\widetilde{A}\sin^2\theta + \widetilde{B}\cos^2\theta\cos^2\psi)\cos^2\theta\sin^2\psi + \frac{1}{E_{33}}\sin^4\theta \\ & + \widetilde{C}\cos^2\theta\cos^2\psi\sin^2\theta + \frac{1}{E_{11}}\cos^4\theta\cos^4\psi, \end{aligned} \tag{2.47}$$

which as expected, does not depend on ϕ, while

$$\widetilde{A} = -2\frac{\nu_{23}}{E_{22}} + \frac{1}{G_{23}}, \qquad \widetilde{B} = -2\frac{\nu_{12}}{E_{11}} + \frac{1}{G_{12}}, \qquad \widetilde{C} = -2\frac{\nu_{13}}{E_{11}} + \frac{1}{G_{13}}. \tag{2.48}$$

The general transformation invariants in this case take the form

$$J_1^a = -2\frac{\nu_{12}}{E_{11}} - 2\frac{\nu_{13}}{E_{11}} - 2\frac{\nu_{23}}{E_{22}} + \frac{1}{E_{22}} + \frac{1}{E_{33}} + \frac{1}{E_{11}}, \tag{2.49a}$$

$$J_2^a = \frac{1}{E_{22}} + \frac{1}{E_{33}} + \frac{1}{2G_{13}} + \frac{1}{2G_{23}} + \frac{1}{E_{11}} + \frac{1}{2G_{12}}. \tag{2.49b}$$

P.2.8 provides two kinds of visualization techniques. First, as shown in Fig. 2.4, each term of the compliance matrix may be plotted over the ψ,θ-plane, while for the terms that are functions of ϕ, the latter should be fixed. In addition, spherical plots of each $\overline{a}_{ij}$ term may be of interest. To do that, we use the results of Remark 1.12 and replace ψ by θ^s and θ by $\phi^s - \frac{\pi}{2}$, respectively (where θ^s and ϕ^s are spherical angles, see Fig. 1.20(b)), and set a value for ϕ when

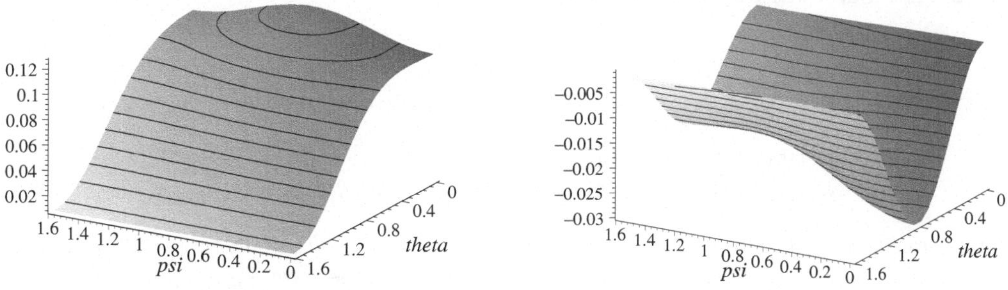

(a) a_{11}.

(b) a_{13} for $\phi = 0$.

Figure 2.4: Two terms of the compliance matrix over the $\psi - \theta$ plane ($\times 10^9$ 1/GPa).

required. We then create a surface where each point of it represents an orientation of the $\overline{x}$-axis of the transformed system (by connecting the origin with it), and the $\overline{a}_{ij}$ under discussion are directly proportional to the distance that is measured along the $\overline{x}$-axis between the origin and the surface. Examples of two terms of the compliance matrix are given in Fig. 2.5.

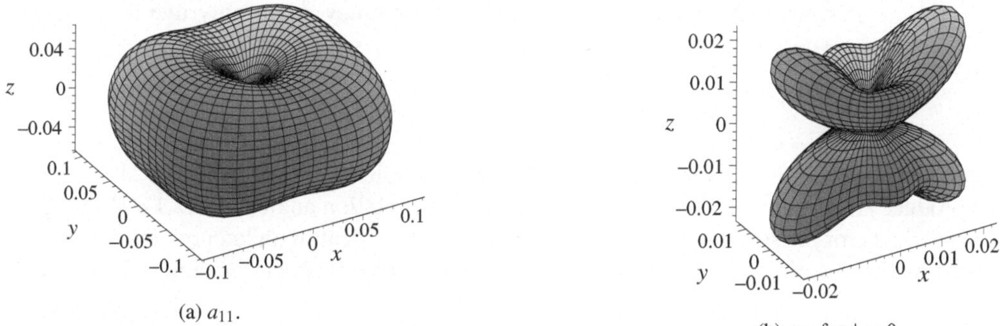

(a) a_{11}.

(b) a_{13} for $\phi = 0$.

Figure 2.5: Spherical plot of two terms of the compliance matrix ($\times 10^9$ 1/GPa).

2.13 Stiffness Matrix Transformation

The derivation of the stiffness matrix transformation, **A**, see (2.4), is similar to the one presented in S.2.12 for the compliance matrix, but involves some very important modifications due to the different matrix definitions in this case. Again, we first write two typical stress-strain relations using the stiffness matrix

$$\sigma_{xx} = A_{11}\varepsilon_{xx} + A_{12}\varepsilon_{yy} + A_{13}\varepsilon_{zz} + A_{14}\varepsilon_{yz} + A_{15}\varepsilon_{xz} + A_{16}\varepsilon_{xy}, \tag{2.50a}$$

$$\sigma_{xz} = A_{51}\varepsilon_{xx} + A_{52}\varepsilon_{yy} + A_{53}\varepsilon_{zz} + A_{54}\varepsilon_{yz} + A_{55}\varepsilon_{xz} + A_{56}\varepsilon_{xy}. \tag{2.50b}$$

In tensor notation, the stiffness tensor is defined as

$$\sigma_{ab} = S_{abcd}\varepsilon_{cd}, \tag{2.51}$$

where $\mathbf{S}$ is a fourth-order tensor with triple symmetry,

$$S_{abcd} = S_{bacd} = S_{abdc} = S_{badc}. \tag{2.52}$$

In view of $\sigma_{xx} = \sigma_{11}$, $\sigma_{xz} = \sigma_{13}$, the above two representative relations may be written as

$$\sigma_{xx} = S_{1111}\varepsilon_{11} + S_{1122}\varepsilon_{22} + S_{1133}\varepsilon_{33} + 2S_{1123}\varepsilon_{13} + 2S_{1113}\varepsilon_{13} + 2S_{1112}\varepsilon_{12}, \tag{2.53a}$$

$$\sigma_{xz} = S_{1311}\varepsilon_{11} + S_{1322}\varepsilon_{22} + S_{1333}\varepsilon_{33} + 2S_{1323}\varepsilon_{23} + 2S_{1313}\varepsilon_{13} + 2S_{1312}\varepsilon_{12}. \tag{2.53b}$$

Similar to the previous case of the compliance matrix, due to the double summation, some terms appear twice (for example, for $c = 2$, $d = 3$ one gets the term $S_{ab23}\varepsilon_{23}$ and for $c = 3$, $d = 2$ one gets the term $S_{ab32}\varepsilon_{32}$, which are identical). Noting that $\varepsilon_{13} = \gamma_{xz}/2$, $\varepsilon_{xx} = \varepsilon_{11}$, etc., one may conclude that $S_{1111} = A_{11}$, $S_{1123} = A_{14}$, $S_{1311} = A_{51}$, $S_{1323} = A_{54}$, etc. Hence, in this case, a modified stiffness matrix is not required, and using the index transformation of (2.38), we determine the tensor $\mathbf{S}$ as $S_{i_1 i_2 j_1 j_2} = A_{ij}$, while the corresponding stress tensor transformation yields

$$\overline{S}_{klst} = S_{mnpr} T_{km} T_{ln} T_{sp} T_{tr}. \tag{2.54}$$

The modified compliance matrix, $\overline{\mathbf{A}}$, in the new coordinate system is extracted from $\overline{\mathbf{S}}$, as $\overline{A}_{ij} = \overline{S}_{i_1 i_2 j_1 j_2}$. To present some examples in this case, we may define, analogously to (2.43), two vectors $\mathbf{A_V}$ and $\overline{\mathbf{A}}_\mathbf{V}$, and a matrix $\mathbf{M_A}$. It turns out that the coefficient of A_{12} in the expression for $\overline{A}_{16}$ is given by $M_A(6,2) = M_a(6,2)/2$, while the coefficient of A_{44} in the expression for $\overline{A}_{25}$ is $M_A(10,16) = 2M_a(10,16)$ (see S.2.12). In general, it is interesting to note that $M_A(i,j) = D_A^a(i,j) M_a(i,j)$ where the elements of the matrix $D_A^a(i,j)$ may attain only five discrete real values (which are not functions of the rotation angles),

$$D_A^a(i,j) \in \{\frac{1}{4}, \frac{1}{2}, 1, 2, 4\}. \tag{2.55}$$

The reader may examine all the terms of $\overline{\mathbf{A}}$ using the symbolic results of **P.2.9** (see also **P.2.10**), and produce stiffness matrices for specific materials and rotation angles by **P.2.11**. While general invariants for the stiffness transformation matrix will be dealt with later on, the single axis transformation invariants for the present case are denoted I_i^A and appear in Table 2.5, see **P.2.9**.

2.14 Compliance and Stiffness Matrix Transformation to Curvilinear Coordinates

To transform the constitutive relations from Cartesian coordinates into orthogonal curvilinear coordinates, one may use the techniques presented in S.2.12, S.2.13, but replace the transformation matrix $\mathbf{T}$ with the one given by (1.216). In the general case, the resulting compliance and stiffness matrices become functions of the α_1, α_2 and α_3 coordinates. Such a case may also be termed as a case of *non-uniform constitutive relations*, i.e. where the material properties depend on the coordinates.

P.2.12 executes transformations of a compliance matrix given in Cartesian coordinates into its form in various orthogonal curvilinear coordinates.

2.15 Principal Directions of Anisotropy

In this section we shall develop the analytical background required for the determination of the principal directions of anisotropy. These directions are useful for general material classification, and in particular for the task of comparing anisotropic materials. Conceptually, the following discussion is of "local nature" and deals with the "local principal directions of anisotropy", i.e. at a given point.

To illustrate the main issue under discussion, suppose that two GEN21 anisotropic materials given by two sets of 21 independent coefficients, need to be compared. Clearly, two materials are identical if all their 21 coefficients are equal, see also Remark 2.5, but there is a possibility, that two materials that apparently seem different, differ only by rigid body rotation (i.e. a rotation that may depend on one, two or three Euler angles). To clarify and study this possibility, there is a need to define *unique canonical principal directions of anisotropy*, in which there is only one way to present a given material regardless of its orientation with respect to the coordinate system. The common criteria for defining the principal directions of anisotropy is based on the following statement: *A coordinate system is said to coincide with the principal directions of anisotropy if the material, when subjected to "all-around uniform pure extension state", forms a "pure tension state".*

To develop the required transformation of a given stiffness matrix to the principal directions of anisotropy, we first introduce the symmetric *bulk modulus tensor*

$$\mathbf{K} = \begin{bmatrix} K_{11} & K_{12} & K_{13} \\ K_{12} & K_{22} & K_{23} \\ K_{13} & K_{23} & K_{33} \end{bmatrix}. \tag{2.56}$$

$\mathbf{K}$ is one of two tensors that may be obtained by contraction of the stiffness tensor $\mathbf{S}$, and its elements are written as

$$K_{ij} = \sum\nolimits_{k=1}^{3} S_{ijkk} = \sum\nolimits_{k=1}^{3} A_{mk}, \tag{2.57}$$

where the index m corresponds to the pair (i,j) by the index transformation (2.38).

We may now examine the special case of strain principal directions and write the state of "all-round uniform extension" as $\varepsilon_{ij} = \widetilde{\varepsilon}\delta_{ij}$ $(i,j = 1,2,3)$, where $\widetilde{\varepsilon}$ is a reference (constant) strain. According to (2.57), (2.2:b), the stress components are given by $\sigma_{ij} = K_{ij}\widetilde{\varepsilon}$. Hence, the principal directions of the tensor $\mathbf{K}$ coincide with the stress principal directions for the case under discussion, which as stated above, constitute the *principal directions of anisotropy*. Quantitatively, in these principal directions, $\sigma_{ij} = K_i^P \widetilde{\varepsilon}\delta_{ij}$, where $\{K_i^P\}$ are the three eigenvalues of the tensor $\mathbf{K}$. Extracting the rotation angles out of the eigenvalue analysis of the bulk modulus tensor $\mathbf{K}$ is identical to that discussed for the eigenvalue analysis of the stress tensor σ — see Example 1.1 of S.1.3.3.1.

Strictly speaking, to ensure a *canonical* (unique) principal system of anisotropy, the eigenvalues of $\mathbf{K}$ should always be put in a certain order before calculating the resulting transformation matrix. In the present case we select the order $K_3^P \geq K_2^P \geq K_1^P$ and for the moment assume that all eigenvalues of $\mathbf{K}$ are different (cases with multiple eigenvalues are discussed in Remark 2.6). This order force the material to orient its "strongest" direction in the z-direction and the "weakest" one in the x-direction. In such a case, the transformation is completely unique since the eigenvalue analysis itself is unique.

To illustrate the above concept, suppose that we deal with an orthotropic material that has been rotated so its stiffness matrix is populated as MON13z material (see Fig. 2.8), and we wish to restore the material principal directions of anisotropy, in which $\overline{A}_{16} = \overline{A}_{26} = \overline{A}_{36} = \overline{A}_{45} = 0$. The condition $K_{12} = 0$ only guarantees that $\overline{A}_{16} + \overline{A}_{26} + \overline{A}_{36} = 0$, see (2.57) (also note that $K_{13} = K_{23} = 0$ in this case). However, since the transformation is unique, there is no more than one solution to the equation $K_{12} = 0$, and thus, this solution must yield a rotation that will bring the material back to its orthotropic population. Furthermore, in this case we employ the angle ψ only. The relevant terms in the transformed system may be written as functions of the coefficients in the original system as

$$\begin{aligned}
\overline{A}_{16} &= -A_{26}\sin^4\psi + (A_{22} - 2A_{66} - A_{12})\cos\psi\sin^3\psi + 3(A_{26} - A_{16})\cos^2\psi\sin^2\psi \\
&\quad + (2A_{66} - A_{11} + A_{12})\cos^3\psi\sin\psi + A_{16}\cos^4\psi, \\
\overline{A}_{26} &= -A_{16}\sin^4\psi + (2A_{66} - A_{11} + A_{12})\cos\psi\sin^3\psi + 3(A_{16} - A_{26})\cos^2\psi\sin^2\psi \\
&\quad + (A_{22} - 2A_{66} - A_{12})\cos^3\psi\sin\psi + A_{26}\cos^4\psi, \\
\overline{A}_{36} &= A_{36}\cos 2\psi + (A_{23} - A_{13})\cos\psi\sin\psi, \quad \overline{A}_{45} = A_{45}\cos 2\psi + (A_{44} - A_{55})\cos\psi\sin\psi. \quad (2.58)
\end{aligned}$$

Thus, by substitution of the expressions of (2.58) in $K_{12} = \overline{A}_{16} + \overline{A}_{26} + \overline{A}_{36}$ and utilization of elementary trigonometric identities, one obtains

$$K_{12} = (A_{16} + A_{26} + A_{36})\cos 2\psi + \frac{1}{2}(A_{22} - A_{11} - A_{13} + A_{23})\sin 2\psi, \tag{2.59}$$

and the condition $K_{12}(\psi) = 0$ yields

$$\psi = \frac{1}{2}\arctan\left(\frac{2(A_{36} + A_{26} + A_{16})}{A_{13} - A_{23} - A_{22} + A_{11}}\right). \tag{2.60}$$

Hence, the coordinate system should be rotated about the z-axis in the amount given above in order to convert the coordinate system into its principal directions of anisotropy, see also Remark 2.7. Note that a generic MON13z material will remain of MON13z-type despite the fulfillment of the condition $K_{12} = 0$.

As already indicated, due to its symmetry properties, the fourth-order stiffness tensor **S** has another (in addition to **K**) symmetric second-order contracted tensor which is given by

$$\mathbf{L} = \begin{bmatrix} L_{11} & L_{12} & L_{13} \\ L_{21} & L_{22} & L_{23} \\ L_{31} & L_{32} & L_{33} \end{bmatrix}, \tag{2.61}$$

where

$$L_{ij} = \sum_{k=1}^{3} S_{ikjk} = \sum_{k=1}^{3} A_{mn}, \tag{2.62}$$

and again, the indices m and n correspond to the pairs (i,k) and (j,k), respectively, by the index transformation of (2.38).

K and **L**, and hence their six invariants may be used for "fast" comparisons between two sets of stiffness matrices. Identical invariants is of course a necessary (but not sufficient) condition for declaring identity between two matrices. The invariants may be expressed (analogously to the stress tensor invariants) by their tensor trace and determinant as

$$I_1^K = tr\mathbf{K}, \qquad I_2^K = \frac{1}{2}\left[(tr\mathbf{K})^2 - tr\left(\mathbf{K}^2\right)\right], \qquad I_3^K = \det\mathbf{K}, \tag{2.63}$$

while those of **L** are obtained analogously. Hence I_i^K, I_i^L $(i = 1,2,3)$, are the six invariants of the stiffness matrix, while explicit expressions for the first two linear invariants are

$$I_1^K = A_{11} + A_{22} + A_{33} + 2(A_{12} + A_{13} + A_{23}), \quad I_1^L = A_{11} + A_{22} + A_{33} + 2(A_{44} + A_{55} + A_{66}). \tag{2.64}$$

Note that in the isotropic case the bulk modulus, see S.2.8, is $K = \frac{1}{9}I_1^K$ or $\frac{1}{3}I_2^K/I_1^K$ or I_3^K/I_2^K. The above invariants could also be obtained by replacing **C** with **S** in (2.45a,b), see also **P.2.9**.

Remark 2.5 Strictly speaking, two materials behave in an identical manner if the expression for W, the *volume density of the strain energy* or the *elastic potential*, that is stored in the elastic body as expressed by the strain component (see S.2.10) is identical for both materials. In view of (2.25), this requirement is equivalent to equality of all their 21 elastic components.

Example 2.1 *Principal Directions of Anisotropy.*

The illustrative transformation of the stiffness matrix to its principal directions of anisotropy, shown in Fig. 2.6(a), may be produced by **P.2.13**. The **K** tensor eigenvalues in this case are 0.225×10^{11}, 0.234×10^{11} and 1.386×10^{11}, and the resulting transformation matrix is (note that each line of $\mathbf{T}$ is an eigenvector of $\mathbf{K}$)

$$\mathbf{T} = \begin{bmatrix} -0.612 & -0.148 & -0.777 \\ -0.354 & -0.828 & 0.436 \\ -0.707 & 0.542 & 0.455 \end{bmatrix}. \tag{2.65}$$

As demonstrated by Fig. 2.6(b), and although not seen from Fig. 2.6(a), the original matrix has three planes of elastic symmetry since its population is identical to that of an orthotropic (ORT9) material.

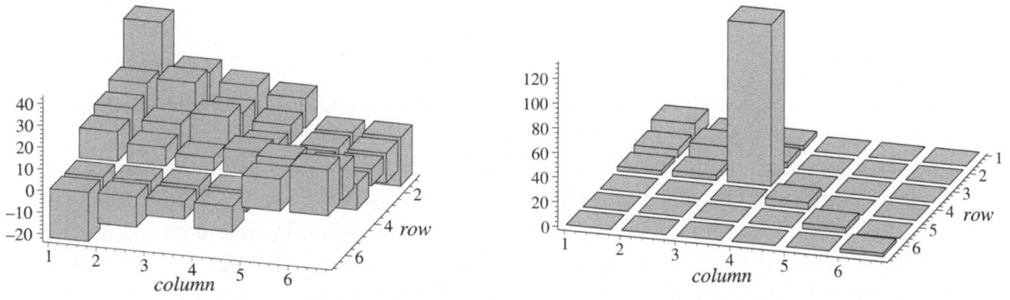

(a) A generic stiffness matrix.

(b) Canonical-principal system.

Figure 2.6: A stiffness matrix and its image in a canonical-principal coordinate system (GPa).

Remark 2.6 When the material poses a state of multiple eigenvalues of **K**, the canonical transformation might not be unique, since there are an infinite number of sets of principal directions of anisotropy (trivial example for that are isotropic materials for which **K** is diagonal with equal elements $K_{ii} = E/(1-2\nu)$, but there may be other special cases of stiffness matrices of multiple eigenvalues of **K**). In most cases, multiple eigenvalues indicate equivalence of stiffness in two or more directions and pose no difficulty.

It should be noted that such situations are not "stable" in the sense that even a slightly small change in the numerical value of the material characteristics will destroy this type of symmetry and yield different eigenvalues of **K**.

Remark 2.7 In addition to K_{12} discussed above, the remaining non-diagonal terms of **K**,

$$K_{13} = \overline{A}_{15} + \overline{A}_{25} + \overline{A}_{35}, \qquad K_{23} = \overline{A}_{14} + \overline{A}_{24} + \overline{A}_{34}, \tag{2.66}$$

enable us to determine the following additional *single angle* transformations:

$$\theta = \frac{1}{2} \arctan\left(\frac{2(A_{25}+A_{15}+A_{35})}{A_{33}-A_{11}+A_{23}-A_{12}}\right), \qquad \phi = \frac{1}{2} \arctan\left(\frac{2(A_{14}+A_{34}+A_{24})}{A_{12}-A_{13}-A_{33}+A_{22}}\right). \tag{2.67}$$

Analogously to (2.60), the above may be used to determine rotations required to set to zero the respective terms of the *bulk modulus tensor*, and are valid for single angle rotation only.

2.16 Planes of Elastic Symmetry

One of the interesting characteristics of anisotropic materials is the existence and the number of *planes of elastic symmetry* created by their micro-structure. This notion may be examined from both mathematical and physical points of view. Similar to the case of local principal directions of anisotropy, planes of elastic symmetry are defined locally, i.e. at a given point.

The transformation techniques of S.2.15 will be adopted for the task of proving the existence of a symmetry of a given material with respect to a given (or presumed) plane. From its constitutive relations point of view, a material is said to be *symmetric with respect to the* $x_i - x_j$ *plane*, if its stiffness (or compliance) matrix remains unchanged (invariant) under a transformation, in which the sign of the x_k-direction ($k \neq i, j$) is changed.

For example, a material that remains unaltered by the coordinate transformation matrix

$$\mathbf{T}_{xy} = \begin{bmatrix} 1 & 0 & 0 \\ 0 & 1 & 0 \\ 0 & 0 & -1 \end{bmatrix} \tag{2.68}$$

is symmetric with respect to the x, y-plane, and an analogous technique may be used for other planes (including those that are not perpendicular to one of the coordinate lines).

Another mathematical point of view is based on the transformation techniques discussed in S.2.13, which clarify the fact that a stiffness matrix transformation is not sensitive to a change of sign of the transformation matrix, see (2.54). Hence, the above $\mathbf{T}_{xy}$ could be replaced with $\mathbf{T}'_{xy} = -\mathbf{T}_{xy}$, which is the transformation matrix obtained for $\psi = \pi$ and $\theta = \phi = 0$ (the same is of course true for the compliance matrix). From a physical point of view, the above transformation shows that one may rotate the material by 180° about an axis that is perpendicular to a plane of elastic symmetry without changing the material characteristics such as the transformation of the $x - y$ system into the $x' - y'$ system shown in Fig. 2.7. This kind of symmetry is usually termed a "symmetry of the second-order" (as a symmetry of the $n^{\underline{th}}$ -order is the case where a rotation of $2\pi/n$ does not alter the material).

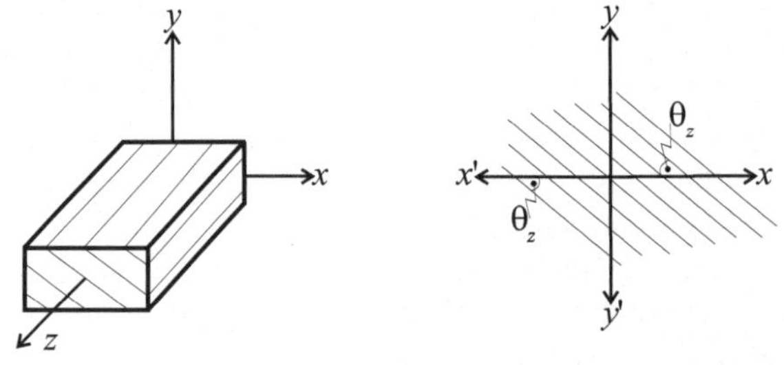

Figure 2.7: An example of a beam with x, y-plane of elastic symmetry. Rotating the coordinate system by 180° about the z-axis does not alter the stiffness and compliance matrices.

As far as the number of planes of elastic symmetry is concerned, for a given GEN21 anisotropic material (see S.2.1), we may set the following question: *are there one or more planes that could be virtually drawn inside the material with respect to which the material is symmetric ?*

In principle, one may adopt the transformation techniques of S.2.15 to rotate the system of coordinates to a presumed new spatial state, and then check if one of the new coordinate planes is a plane of elastic symmetry. In the general case, where the existence and number of the planes of elastic symmetry is not known, this process becomes quite tedious. A much more efficient way is offered in what follows.

The reasoning of the following method is founded on the fact that a transformation to the principal system of anisotropy ensures that materials with one plane of elastic symmetry will be transformed into one of the monoclinic forms, and materials with three orthogonal planes of elastic symmetry will be transformed into the orthotropic form, see S.2.15.

Hence, to evaluate the existence and number of the planes of elastic symmetry, we first rotate the material to its principal directions of anisotropy x_A^P, y_A^P, z_A^P. Once the stiffness matrix in the principal directions of anisotropy is obtained, one may examine the matrix population, and compare it to that of orthotropic and (the three fundamental kinds of) monoclinic materials as shown in Fig. 2.8. This comparison indicates that if the matrix is populated as the MON13x material, the $y_A^P - z_A^P$ plane is an elastic symmetry plane. Similarly, if the matrix is populated as the MON13y or MON13z materials, the planes of elastic symmetry are $x_A^P - z_A^P$ or $x_A^P - y_A^P$, respectively. If the matrix is populated as the ORT9 material, all three $y_A^P - z_A^P$, $x_A^P - y_A^P$ and $x_A^P - z_A^P$ planes are elastic symmetry planes. From the above, it is clear that no case of (only) two planes of elastic symmetry exists.

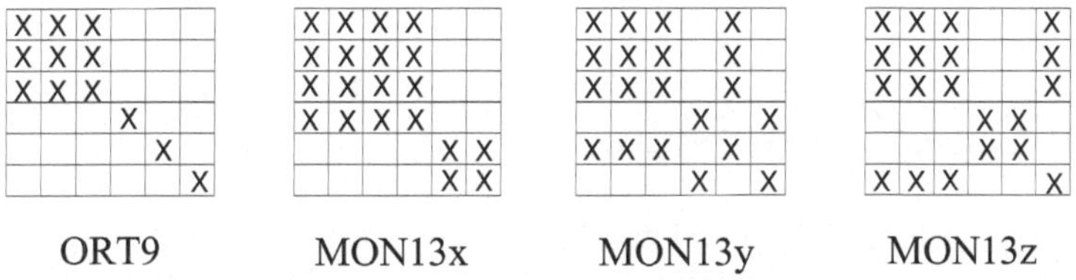

Figure 2.8: Complience/Stiffness matrix populations of orthotropic and monoclinic materials.

We may now summarize our findings regarding the number of planes of elastic symmetry in various material types in a form of (N_i, N_p) pairs where N_i stands for the *minimal* number of nonzero independent elastic moduli, and N_p stands for the *maximal* number of planes of symmetry. For that purpose, we first observe all materials in their principal directions of anisotropy, see S.2.15. We subsequently denote the case of general anisotropic material as $(18,0)$ since it may be rotated by three Euler angles that will zero out (up to) three of its elastic moduli.

Similarly, monoclinic material will be denoted $(12,1)$ as it may be rotated by one angle to zero out (at most) one of its elastic moduli (see also Remark 2.8). Orthotropic material will be denoted as $(9,3)$, and tetragonal materials as $(6,\infty)$ (both may not be rotated to zero out one or more elastic moduli). Heuristically, and without a proof, it is interesting to note that all of the above cases obey a simple rule given by

$$N_p = \frac{18 - N_i}{N_i - 6} \qquad (18 \geq N_i \geq 6). \tag{2.69}$$

All cases of $N_i < 6$, such as the transversely isotropic materials $(5,\infty)$, cubic materials $(3,\infty)$, and the isotropic materials $(2,\infty)$, yield infinite N_p as well. Equation (2.69) is presented in Fig. 2.9, which generally illustrates the increasing number of planes of symmetry with the decreasing number of independent unknowns used to characterize the material.

It should be noted again that from the (2.69) and Fig. 2.9 point of view, N_p stands for *the maximal number* of planes of elastic symmetry of each material type, since clearly, for a given N_i, examples of materials with fewer planes of elastic symmetry may be easily created.

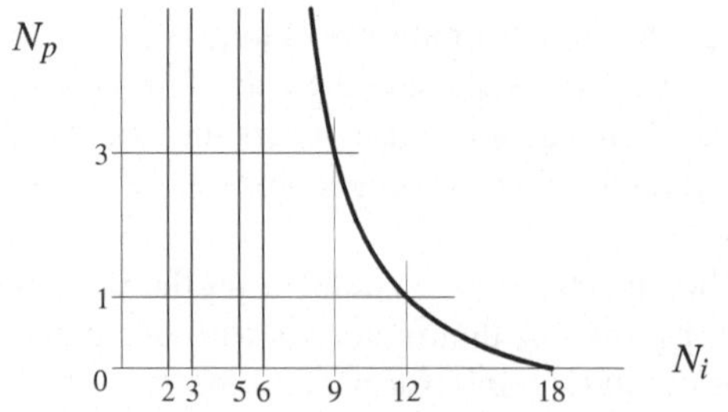

Figure 2.9: Number of symmetry planes vs. number of independent unknowns.

Remark 2.8 In rare and special cases, the procedure discussed in this section may mislead by its conclusion regarding the number of planes of elastic symmetry. An example is the stiffness matrix populated as ORT9 in addition to $A_{16} = \widetilde{A}$, $A_{26} = -\widetilde{A}$, while $\widetilde{A} \neq 0$ and $A_{11} = A_{22}$, $A_{13} = A_{23}$, $A_{44} = A_{55}$. In such a case, the criteria for principal directions are fulfilled and (2.59) shows that no rotation is required (and the material seems to be of MON13z-type that has only one plane of elastic symmetry). Yet, ψ may be selected so that both $\overline{A}_{16}$ and $\overline{A}_{26}$ vanish (which will yield an ORT9 material that has three planes of elastic symmetry). Such cases do not pose any difficulty as they are not "stable" in the sense discussed in Remark 2.6.

2.17 Non-Cartesian Anisotropy

In S.2.1, we have presented the generalized Hook's law for the case where the strain and the stress components are described in Cartesian coordinates and for material that poses Cartesian anisotropy. Yet, one also ought to deal with cases where the material directions are not parallel (or perpendicular) to the axes of a Cartesian system. Such cases are generally referred to as a *non-Cartesian anisotropy*. A common example for a non-Cartesian anisotropy is the so-called *cylindrical anisotropy*, where two of the material axes coincide with the tangent and the normal to concentric circles in a given plane, and the third one coincides with the direction perpendicular to that plane. In composite material terminology, this is the type of material lay-up produced by *filament winding* of a circular shell, and it may also be viewed locally as a thin layer of Cartesian anisotropy materials that are laid over the surface of a prismatic circular tube.

To study the case of *cylindrical anisotropy*, we note that $\alpha_1 = r(=\rho)$, $\alpha_2 = \theta^c$, $\alpha_3 = z$ (see Example 1.9) and the corresponding Lamé's parameters are $H_1 = 1$, $H_2 = r$, $H_3 = 1$. Thus, $\mathbf{T}$ of (1.216) becomes a function of θ^c only,

$$\mathbf{T} = \begin{bmatrix} \cos\theta^c & \sin\theta^c & 0 \\ -\sin\theta^c & \cos\theta^c & 0 \\ 0 & 0 & 1 \end{bmatrix}. \tag{2.70}$$

We now define the *cylindrical anisotropy* as the case where at each point, the material has the properties of *a reference material of Cartesian anisotropy so that its coordinate lines x, y and z coincide with the local r, θ^c and z directions, respectively* — see Fig. 2.10. Hence, the material properties become independent of θ^c, and we therefore conclude that the stiffness or compliance matrices of a material of *cylindrical anisotropy* are *constants*. By selecting $\theta^c = 0$, $\mathbf{T}$ becomes a unit matrix and the above matrices become identical to those of *Cartesian anisotropy*. To demonstrate that, consider Fig. 2.11(a) that presents a cylinder with orthotropic

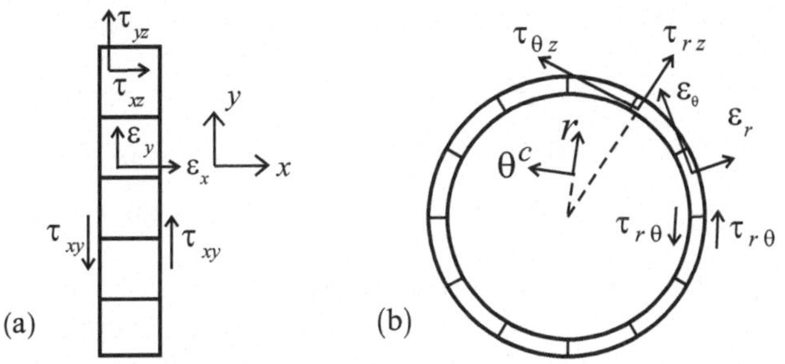

Figure 2.10: A thin "material strip" of (a) Cartesian anisotropy, (b) cylindrical anisotropy.

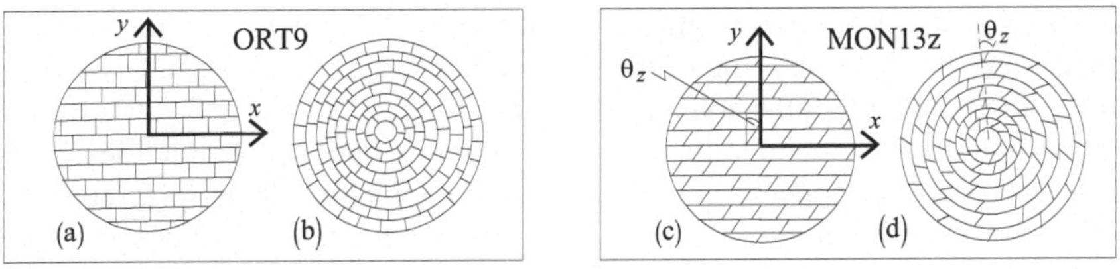

Figure 2.11: Circular domains of Cartesian, (a), (c), and cylindrical, (b), (d), anisotropy.

laminae having the compliance matrix, **a**, of (2.7). When these laminae are wrapped over a circle, see Fig. 2.11(b), a material of *cylindrical anisotropy* is obtained and may be written as

$$\begin{Bmatrix} \varepsilon_r \\ \varepsilon_\theta \\ \varepsilon_z \\ \gamma_{\theta z} \\ \gamma_{rz} \\ \gamma_{r\theta} \end{Bmatrix} = \begin{bmatrix} \frac{1}{E_{11}} & -\frac{\nu_{21}}{E_{22}} & -\frac{\nu_{31}}{E_{33}} & 0 & 0 & 0 \\ -\frac{\nu_{12}}{E_{11}} & \frac{1}{E_{22}} & -\frac{\nu_{32}}{E_{33}} & 0 & 0 & 0 \\ -\frac{\nu_{13}}{E_{11}} & -\frac{\nu_{23}}{E_{22}} & \frac{1}{E_{33}} & 0 & 0 & 0 \\ 0 & 0 & 0 & \frac{1}{G_{23}} & 0 & 0 \\ 0 & 0 & 0 & 0 & \frac{1}{G_{13}} & 0 \\ 0 & 0 & 0 & 0 & 0 & \frac{1}{G_{12}} \end{bmatrix} \begin{Bmatrix} \sigma_r \\ \sigma_\theta \\ \sigma_z \\ \tau_{\theta z} \\ \tau_{rz} \\ \tau_{r\theta} \end{Bmatrix}. \tag{2.71}$$

Similarly, when the above orthotropic material is rotated by an angle θ_z and yields a MON13z as shown in Fig. 2.11(c), one may visualize placing the laminae as shown in Fig. 2.11(d) to form the following constitutive relations:

$$\begin{Bmatrix} \varepsilon_r \\ \varepsilon_\theta \\ \varepsilon_z \\ \gamma_{\theta z} \\ \gamma_{rz} \\ \gamma_{r\theta} \end{Bmatrix} = \begin{bmatrix} a_{11} & a_{12} & a_{13} & 0 & 0 & a_{16} \\ & a_{22} & a_{23} & 0 & 0 & a_{26} \\ & & a_{33} & 0 & 0 & a_{36} \\ & & & a_{44} & a_{45} & 0 \\ & Sym. & & & a_{55} & 0 \\ & & & & & a_{66} \end{bmatrix} \begin{Bmatrix} \sigma_r \\ \sigma_\theta \\ \sigma_z \\ \tau_{\theta z} \\ \tau_{rz} \\ \tau_{r\theta} \end{Bmatrix}. \tag{2.72}$$

See further use of cylindrical anisotropy in S.7.5 and Example 10.1. Along the same lines, other cases of non-Cartesian anisotropy may be defined. To form a material with a *spherical anisotropy* we examine the transformation matrix **T** for a spherical coordinate system (see

(1.206), Remark 1.12 and Example 1.9), which is a function of θ^s and ϕ^s, namely,

$$\mathbf{T} = \begin{bmatrix} \sin(\phi^s)\cos(\theta^s) & \sin(\theta^s)\sin(\phi^s) & \cos(\phi^s) \\ -\sin(\theta^s) & \cos(\theta^s) & 0 \\ \cos(\phi^s)\cos(\theta^s) & \sin(\theta^s)\cos(\phi^s) & -\sin(\phi^s) \end{bmatrix}. \tag{2.73}$$

Here, we define a material with a *spherical anisotropy*, as the case where at each point, the material has the properties of *a reference material of Cartesian anisotropy so that its coordinate lines x, y and z are taken to coincide with the local r,* θ^s *and* $-\phi^s$ *directions, respectively.*

Hence, the transformation becomes independent of θ^s and ϕ^s in this case as well, and $\mathbf{T}$ may be taken as the constant matrix that corresponds to $\theta^s = 0$, $\phi^s = \frac{\pi}{2}$.

As indicated in S.1.1, S.1.7, the description of the strain measures and the stress components in generic curvilinear coordinates might have a tremendous influence on the complexity of the analytical solution. Typically, for isotropic materials, specific curvilinear coordinates are selected in a way that simplifies the boundary conditions. Yet, as shown in S.2.17, the *directions of anisotropy* might have a crucial influence on the resulting constitutive relations.

To clarify the above, consider a given homogeneous domain (of uniform elastic properties), which is described by uniform (constant) constitutive relations in Cartesian coordinates. When the same domain is described by another coordinate system, the constitutive relations become coordinate-dependent. As examples, a homogeneous domain of Cartesian anisotropy will be described as a coordinate-dependent material in cylindrical coordinates, and homogeneous domain of cylindrical anisotropy will be described as a coordinate-dependent material in Cartesian coordinates. This phenomena may cause considerable complexity that should be considered against the simplification gained by using the curvilinear coordinates.

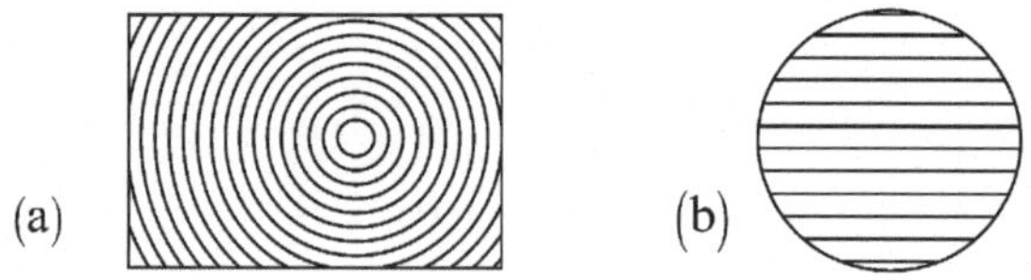

Figure 2.12: A rectangle of cylindrical anisotropy and a circle of Cartesian anisotropy.

As a result of the above discussed phenomena, in most cases, the selection of the coordinate system in anisotropic elasticity is determined by the material properties solely and *not* by the boundary shape. For example, Fig. 2.12 shows a rectangular domain of cylindrical anisotropy (a "wood-type" problem), which should be analyzed by cylindrical coordinates to prevent a non-uniform constitutive relation, and a circular domain of Cartesian anisotropy that should be analyzed using Cartesian coordinates for the same reason.

It is therefore clear that employing Cartesian coordinates for Cartesian anisotropy is ideal for "rectangular" domains, and likewise, employing cylindrical coordinates for cylindrical anisotropy is ideal for tubes and shells (provided that the origin of the cylindrical anisotropy coincides with the origin of the cylindrical coordinates, in which the problem is solved, see also S.7.5 and Fig. 7.14). In the above cases, the presentation of both the material and the boundary conditions is "natural".

3
Plane Deformation Analysis

This chapter reviews various types of plane deformation analyses encountered in diversified problems within the area of anisotropic elasticity, and presents them in a uniform mathematical format. The associated *boundary value problems* (BVPs) are classified and reduced to certain types, which are commonly expressed using the *generalized Laplace*, the *biharmonic* and the third-order differential operators.

Needless to say, any kind of a three-dimensional problem reduction into a two-dimensional one is accompanied by appropriate approximation level, which reflects a neglect of some components of the strain and/or stress tensors. The problems derived in this chapter are classified according to the components that are neglected in each case. Table 3.1 presents a summary of the discussed cases where neglected components are denoted by "0" (while ~ 0 stands for a case where consistent formulation yields a zero value, but a suitable approach may yield an estimation of this small quantity).

	Plane Strain	Plane Stress	Plane Shear	Coupled plane problem	Plates under Bending
$\varepsilon_x, \varepsilon_y, \gamma_{xy}$ $\sigma_x, \sigma_y, \tau_{xy}$			0		
ε_z	0		0	0	
$\gamma_{yz}, \gamma_{xz}, \tau_{yz}, \tau_{xz}$	0	0			~ 0
σ_z		0	0		0
Material	MON13z	MON13z	MON13z	GEN21	MON13z
Derived in	S.3.2.1	S.3.2.2	S.3.3	S.3.4	S.3.5.2

Table 3.1: Classification of plane problems.

One should realize that various problems encountered in the mechanics of anisotropic elasticity may pose identical (or similar) BVP. As an example, it will be shown that the three problems of plane-strain, plane-stress and bending of plates are mathematically identical as they require a solution of a biharmonic BVP of the same type.

For each BVP, we shall be focused mainly on the *necessary* existence conditions associated with its mathematical presentation correctness, while assuming that their sufficiency (similar to the solution uniqueness) emerges from the mechanical sense of the (linear) problems.

3.1 Plane Domain Definition and Contour Parametrization

3.1.1 Plane Domain Topology

Figure 3.1(a) presents a simply connected plane domain, Ω, its circumference $\partial\Omega$, and the circumferential arc-length coordinate, s. The entire circumference length is denoted by L, and therefore $L \geq s \geq 0$ (a domain, Ω, is said to be *simply connected*, if any closed curve inside it may be shrunk into a point by means of continuous deformation during which it always remains in Ω. Otherwise, Ω is said to be *multiply connected*).

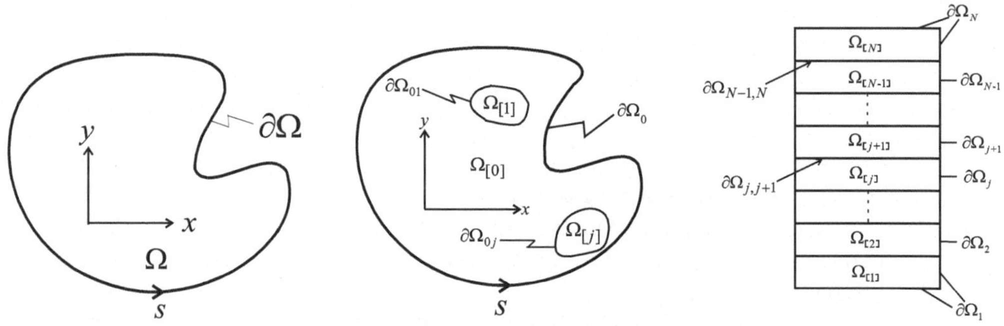

(a) Simply connected. (b) Multiply connected and non-homogeneous. (c) Laminated.

Figure 3.1: Plane domain geometry notation.

A *non-homogeneous domain* consists of several simply connected domains of different "materials", $\Omega_{[1]},\ldots,\Omega_{[N]}$, see Fig. 3.1(b), while the exterior domain, $\Omega_{[0]}$, serves as a "surrounding material": $\Omega = \bigcup_{j=0}^{N}\Omega_{[j]}$. The boundary contours of the domains are denoted by $\partial\Omega_j$, while the contour $\partial\Omega_0$ contains all those with $j > 0$, see Fig. 3.1(b). Topologies where more closed boundaries encircle other closed boundaries are also possible (see e.g. Fig. 3.2). Multiply connected domains are similar in essence, and may be generally described by Fig. 3.1(b) where the material of one (or more) of the domains $\Omega_{[1]},\ldots,\Omega_{[N]}$ is removed.

Many applications pose non-homogeneous *laminated* domains, and so many analyses are focused on modern composites. In such cases, Ω is divided into $N > 1$ (rectangular) domains of different materials $\Omega_{[j]}$ $(j = 1,\ldots,N)$ as shown in Fig. 3.1(c), and hence, $\Omega_{[0]} = \emptyset$.

In both Figs. 3.1(b),(c), the boundaries of the domains $\Omega_{[j]}$ $(j = 1,\ldots,N)$ are separated into the "dividing curves" and the "free curves". The dividing curves between two different domains of materials $\Omega_{[i]}$ and $\Omega_{[j]}$ are denoted by $\partial\Omega_{ij}$ (or $\partial\Omega_{ji}$). The "free" curves of each domain (i.e. the curves that are free of contact with other domains) are denoted by $\partial\Omega_j$. As shown, there are cases where the free curve of a domain consists of different segments. Yet, $\partial\Omega_j$ serves as a common indication for all these free segments of the $\Omega_{[j]}$ domain.

The above general topologies may be formatted as shown in Fig. 3.2. In these symmetric tables, we indicate on the diagonal the number of free surface segments of each domain, while

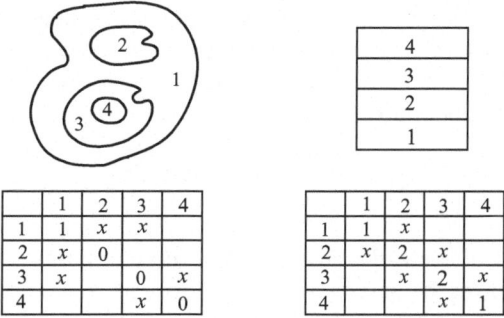

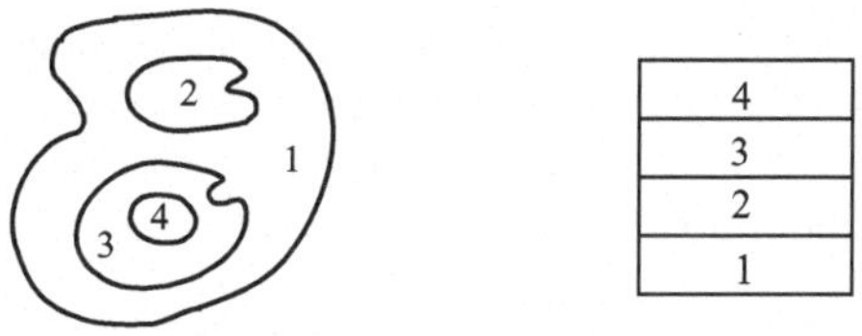

	1	2	3	4
1	1	x	x	
2	x	0		
3	x		0	x
4			x	0

	1	2	3	4
1	1	x		
2	x	2	x	
3		x	2	x
4			x	1

Figure 3.2: Topology tables of two non-homogeneous cross-sections: (a) Generic, (b) Laminated.

for the off diagonal terms, "x" stands for a mutual dividing curve between the "row region" and the "column region". The overall number of boundary contours in a given domain is therefore equal to the sum of the diagonal terms and the number of "x" signs over the upper (or lower) triangles of these topology tables.

A few words should be devoted to the notation convention that applies in what follows. A superscript in square brackets, say $\widetilde{\mathbf{H}}^{[j]}$, indicates that the value of $\widetilde{\mathbf{H}}$ is calculated for (or is taken for) the material of domain $\Omega_{[j]}$. In what follows, the reader will also find functions of x, y that contain characteristics (such as elastic moduli, or generalized harmonic functions) that appear without domain index. The value of these characteristics is determined *locally* according to the domain in the (x, y) location under discussion.

For future purposes (see S.3.2.6, S.3.3.5, S.3.4, Chapter 8, etc.) we define for a general quantity $\widetilde{\mathbf{H}}$,

$$\widetilde{\mathbf{H}}_{[i]}^{[j]} = \widetilde{\mathbf{H}}^{[j]} - \widetilde{\mathbf{H}}^{[i]}. \tag{3.1}$$

Hence, the above equation may be interpreted as subtraction of $\widetilde{\mathbf{H}}$ that has been calculated for domain $\Omega_{[i]}$ from $\widetilde{\mathbf{H}}$ that has been calculated for domain $\Omega_{[j]}$.

Remark 3.1 The nature of the non-homogeneity presented above may be defined as *piecewise constant non-homogeneity* since it basically assumes that each region is homogeneous within itself and material properties "jump" between regions. Such a non-homogeneity dominates many applications of man-made materials and structures, Yet, non-homogeneity may be also defined by materials with *continuously varying elastic moduli.* Only a few examples will be devoted to such cases (see e.g. Remark 8.3).

The underlying assumption that remains valid for all two-dimensional analyses and all subsequent beam analyses, is that all sub-domains involved are perfectly bonded such that no slippage occurs, and all in-plane and out-of-plane displacements at the interfaces are continuous.

3.1.2 Contour Parametrization and Directional Cosines

Many solution methods require parametrizations of plane region contours. The contour of a simply connected shape (see e.g. Fig. 3.1(a)) is a *simple closed curve*, that may be generally defined by periodic functions of parameter t as

$$x = x(t), \qquad y = y(t), \qquad t \in [0, T]. \tag{3.2}$$

Deriving the functions $x(t)$ and $y(t)$ for general shapes of the domain Ω may be carried out in various ways as will be described shortly. **P.3.1** enables the examination of some prescribed parametrization functions. Once the parametrization functions are determined, the *arc-length function*, $s(t)$, is derived by the following integration:

$$s(t) = \int_0^t \sqrt{(\frac{dx}{dt})^2 + (\frac{dy}{dt})^2}\, dt. \tag{3.3}$$

From an analytical point of view, the above integral may impose a relatively difficult task. Therefore, ways to minimize the participation of an explicit $s(t)$ expression will be explored within the solution methodologies described further on in Chapter 4.

Note that the above parametrization technique may be also applied to any other non-Cartesian coordinates, where α_i of S.1.7.2 are defined as functions of t.

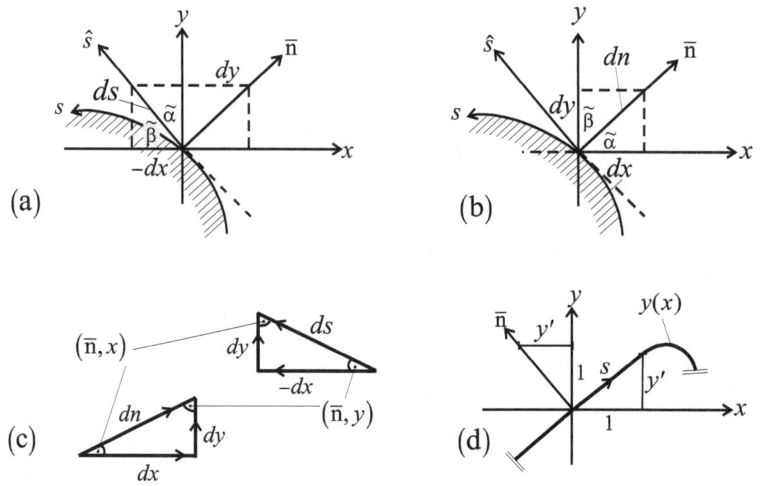

Figure 3.3: Directional cosine notation: $\tilde{\alpha} \equiv (\bar{\mathbf{n}}, x)$, $\tilde{\beta} \equiv (\bar{\mathbf{n}}, y)$.

To define the directional cosines, we first examine Figs. 3.3(a),(b), where the x, y coordinates, the normal to the contour, $\bar{\mathbf{n}}$, and the circumferential direction, s, are shown. Accordingly, the angle cosines between the normal to the contour and the x- and y- axes, see also Fig. 3.3(c), are given by

$$\cos(\bar{\mathbf{n}}, x) = \frac{dy}{ds} = \frac{dx}{dn}, \qquad \cos(\bar{\mathbf{n}}, y) = -\frac{dx}{ds} = \frac{dy}{dn} \tag{3.4}$$

and the following identity holds:

$$\cos(\bar{\mathbf{n}}, x)^2 + \cos(\bar{\mathbf{n}}, y)^2 = 1. \tag{3.5}$$

When the domain contour is presented as a C^1-regular function $y(x)$, see Fig. 3.3(d), $\cos(\bar{\mathbf{n}}, x) = -\frac{dx/dt}{\sqrt{1+(dy/dt)^2}}$, and $\cos(\bar{\mathbf{n}}, y) = \frac{1}{\sqrt{1+(dy/dt)^2}}$. Note that signs should be changed if the direction of s in Fig. 3.3(d) is reversed.

Example 3.1 *Directional Cosines in an Ellipse.*

For the case of an ellipse of semi-axes $a \geq b > 0$, one may adopt the geometrical angle $t \in [0, 2\pi]$ as the contour's parameter and write $x = a\cos t$, $y = b\sin t$. Hence the cosines of the angles between the normal to the ellipse and the x- and y- axes are given as (see (3.4))

$$\cos(\bar{\mathbf{n}},x) = \frac{b}{\widetilde{\lambda}} \cos t = \frac{b}{\widetilde{\lambda}} \frac{x}{a}, \qquad \cos(\bar{\mathbf{n}},y) = \frac{a}{\widetilde{\lambda}} \sin t = \frac{a}{\widetilde{\lambda}} \frac{y}{b}, \tag{3.6}$$

where $\widetilde{\lambda} = \frac{ds}{dt} = \sqrt{a^2 \sin^2 t + b^2 \cos^2 t} = \sqrt{\frac{b^2}{a^2}x^2 + \frac{a^2}{b^2}y^2}$ is the *"parametrization velocity"*.

Remark 3.2 The above definitions are valuable for common integral evaluations, which are founded on Green's Theorem, $\iint_\Omega (N_{,x} + M_{,y}) = \oint_{\partial\Omega} N\cos(\bar{\mathbf{n}},x) + M\cos(\bar{\mathbf{n}},y)$, as

$$\iint_\Omega \Lambda_{,y} = \oint_{\partial\Omega} \Lambda \cos(\bar{\mathbf{n}},y), \qquad \iint_\Omega \Lambda_{,x} = \oint_{\partial\Omega} \Lambda \cos(\bar{\mathbf{n}},x), \tag{3.7}$$

or the corollaries when $\Lambda = 1$, $\Lambda = x$ or $\Lambda = y$

$$\oint_{\partial\Omega} \{1, x, y\} \cos(\bar{\mathbf{n}},x) = \{0, S_\Omega, 0\}, \qquad \oint_{\partial\Omega} \{1, x, y\} \cos(\bar{\mathbf{n}},y) = \{0, 0, S_\Omega\}. \tag{3.8}$$

For further use, we shall document here some integral identities that contain the function Λ of a constant value Λ_0 over a contour $\partial\Omega$. Using identities $(x\Lambda)_{,x} = \Lambda + x\Lambda_{,x}$, $(y\Lambda)_{,y} = \Lambda + y\Lambda_{,y}$ and (3.7), (3.8) we obtain

$$\iint_\Omega \{x, y\}\Lambda_{,y} = \{0,\ S_\Omega \Lambda_0 - \iint_\Omega \Lambda\}, \tag{3.9a}$$

$$\iint_\Omega \{x, y\}\Lambda_{,x} = \{S_\Omega \Lambda_0 - \iint_\Omega \Lambda,\ 0\}. \tag{3.9b}$$

3.1.3 Parametrization by Conformal Mapping

The *conformal mapping* technique is a powerful analytical tool that may be invoked for contour parametrization. Within this approach, we look at each point on the contour as a complex number $z = x + \mathbf{i}y$ ($\mathbf{i} = \sqrt{-1}$), and employ the appropriate conformal mapping to transform a domain of a unit circle to the simple connected domain under discussion.

The reader should also be aware of the fact that unlike the classical techniques (see e.g. (Sokolnikoff, 1983)), we exploit the conformal mapping only for deriving the parametrization functions $x(t)$ and $y(t)$, and hence, within the present context, we care for the contour points only.

Transformations of a unit circle to various simply connected domains are widely discussed in the literature. **P.3.1** enables the examination of some prescribed $w(z)$ transformation functions.

The Schwarz-Christoffel transformation methodology may be further invoked to map a unit circle exterior onto the exterior of a general *polygon* of n vertices, see (Muskhelishvili, 1953). By denoting the complex location at the origin plane, where the unit circle is defined, as ζ $(= \xi + \mathbf{i}\eta)$, this transformation is given by

$$w(z) = \widetilde{A} \int_1^z P(\zeta)\, d\zeta + \widetilde{B}, \qquad P(\zeta) = \frac{1}{\zeta^2} \prod_{k=1}^n (\zeta - Z_k)^{\alpha_k}, \qquad n \geq 3, \quad |\zeta| \geq 1, \tag{3.10}$$

where the polygon vertices w_k correspond to the points Z_k, $k = 1, \ldots, n$ on the unit circle. The interior angles at the polygon vertices, φ_k, are determined by the parameters $\alpha_k \in (-1, 1)$ so that $\varphi_k = \pi(1 - \alpha_k)$. The complex constants $\widetilde{A}$ and $\widetilde{B}$ represent "stretching" and "rigid" rotation and translation of the domain. In the simple case of a regular n-polygon, $\alpha_k = \frac{2}{n}$, one may select $Z_k = \exp(\frac{2\pi(k-1)}{n}\mathbf{i})$, and the transformation (3.10) is written as, see Example 3.2,

$$w_n = \widetilde{A} \int_1^z (1 - \zeta^n)^{\frac{2}{n}} \frac{d\zeta}{\zeta^2} + \widetilde{B}, \qquad n \geq 3. \tag{3.11}$$

For a generic n-vertices polygon, the sum of internal angles is given by $\sum_k \varphi_k = \pi(n-2)$, and therefore $\sum_{k=1}^n \alpha_k = 2$. Hence, $P(\zeta)$ of (3.10) may be written as $P(\zeta) = \prod_{k=1}^n (1-\frac{Z_k}{\zeta})^{\alpha_k}$. Using the product of a binomial z-series $(1-\frac{Z_k}{\zeta})^{\alpha_k} = 1-\alpha_k\frac{Z_k}{\zeta}+\cdots$, one may write $P(\zeta) = 1-\frac{1}{\zeta}\sum_{k=1}^n \alpha_k Z_k+\cdots$ (i.e. a series expansion about infinity). To ensure that $w(z)$ is a single-valued function on the unit circle exterior (i.e. contains no logarithmic singularity at the origin), one must require vanishing of the $\frac{1}{\zeta}$ coefficient in this series, i.e.,

$$\sum_{k=1}^n \alpha_k Z_k = 0. \tag{3.12}$$

Obviously, the position of a point, Z_k, on a unit circle is defined by the value of its central angle (i.e. a scalar). Therefore, for a given n-vertices polygon one needs to determine n such central angles and the complex constants $\widetilde{A}$ and $\widetilde{B}$, i.e. a total of $n+4$ scalars. These have to satisfy the following $n+4$ equations/conditions: given location of one vertix, given orientation of one edge, $n-1$ given edge lengths, and (3.12). This constitutes an implicit procedure that may be derived in a closed form only for a few simple cases.

Since in the general case, the integrals of (3.10) may not be carried out analytically in a closed form, we expand them into Laurent-Taylor power series and practically use only a truncated version of it, namely, $w \approx \frac{C_0}{\zeta}+\sum_{n=1}^{k_p} C_n\zeta^n$. Here, C_n are known constants ($C_{k_p} \neq 0$), and $k_p > 0$ will be referred to as the *parametrization level*. The parametrization of an ellipse in Example 3.2 is therefore characterized by $k_p = 1$, as the corresponding conformal mapping is $w = \frac{a-b}{2}\frac{1}{\zeta}+\frac{a+b}{2}\zeta$. The image of the unit circle $|\zeta| = 1$ by such truncated parametrization is a simple closed curve that approximates the polygon, given accurately by (3.10).

Example 3.2 *Regular n-Polygon Parametrization.*

To clarify the discussion in this section regarding a regular n-polygon parametrization, three transformations, for a triangle, $w_3(z)$, a square, $w_4(z)$, and a pentagon $w_5(z)$, will be presented here. These, and additional results for *any* regular polygon may be derived by activating **P.3.2.** The transformations map the circumference of a unit circle (i.e., $|z| = 1$) directly onto the desired contour and are given using (3.11) as

$$w_3 = -\int_1^z \frac{(1-\zeta^3)^{\frac{2}{3}}}{\zeta^2}d\zeta, \quad w_4 = -\int_1^z \frac{(1-\zeta^4)^{\frac{1}{2}}}{\zeta^2}d\zeta, \quad w_5 = -\int_1^z \frac{(1-\zeta^5)^{\frac{2}{5}}}{\zeta^2}d\zeta. \tag{3.13}$$

By expanding each one of (3.13) into a series, we get

$$w_3 \approx -\frac{1}{z}-\frac{z^2}{3}-\frac{z^5}{45}+\cdots, \quad w_4 \approx -\frac{1}{z}-\frac{z^3}{6}-\frac{z^7}{56}+\cdots, \quad w_5 \approx -\frac{1}{z}-\frac{z^4}{10}-\frac{z^9}{75}+\cdots. \tag{3.14}$$

While transforming a unit circle circumference, we usually assign the parameter $t \in [0, 2\pi]$ to be the central angle of this circle, namely, $z = \cos t+\mathbf{i}\sin t$, and thus, the contours $w_n(t) = x_n(t)+\mathbf{i}y_n(t)$ are approximated by polynomials of $\cos t$ and $\sin t$ of order k_p. Note that there are minimal values of $k_p > 1$ for each configuration (except an ellipse with $k_p = 1$) that make the truncated series valid. These are $k_p = 2$ for a triangle, $k_p = 3$ for a square and $k_p = 4$ for a pentagon, etc. For a triangle, a square and pentagon, $x_n(t)$ and $y_n(t)$, are approximated as

$$\begin{aligned} x_3(t) &\approx -\frac{1}{3}\cos t-\frac{2}{3}\cos^2 t, & y_3(t) &\approx \sin t-\frac{2}{3}\cos t\sin t,\\ x_4(t) &\approx -\frac{1}{2}\cos t-\frac{2}{3}\cos^3 t, & y_4(t) &\approx \frac{7}{6}\sin t-\frac{2}{3}\cos^2 t\sin t,\\ x_5(t) &\approx -\frac{1}{10}\cos t-\frac{4}{5}\cos^4 t+\frac{4}{5}\cos^2 t, & y_5(t) &\approx \sin t+\frac{2}{5}\cos t\sin t-\frac{4}{5}\cos^3 t\sin t. \end{aligned} \tag{3.15}$$

One may easily convert the above expressions into Fourier series form as

$$\begin{aligned} x_3(t) &\approx -\cos t - \frac{1}{3}\cos(2t), & y_3(t) &\approx \sin t - \frac{1}{3}\sin(2t), \\ x_4(t) &\approx -\cos t - \frac{1}{6}\cos(3t), & y_4(t) &\approx \sin t - \frac{1}{6}\sin(3t), \\ x_5(t) &\approx -\cos t - \frac{1}{10}\cos(4t), & y_5(t) &\approx \sin t - \frac{1}{10}\sin(4t). \end{aligned} \tag{3.16}$$

P.3.3 produces generic polygon parametrizations. Figure 3.4 presents an example with vertices $Z_i = [1, .306 + .952\mathbf{i}, -.807 + .590\mathbf{i}, -.508 - .862\mathbf{i}, .187 - .982\mathbf{i}]$ where the corresponding angles are $\varphi_i = [98°, 118°, 88°, 128°, 108°]$.

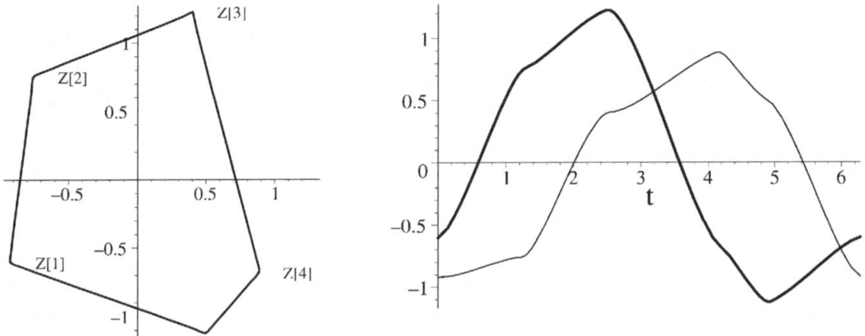

(a) The polygon.

(b) The parametrization functions $x(t)$ (thin) and $y(t)$ (thick).

Figure 3.4: Parametrization by conformal mapping of a generic pentagon.

3.1.4 Parametrization by Piecewise Linear Functions

An alternative and very robust method to parameterize an n–vertices polygon by trigonometric polynomials is founded on expressing the functions $x(t)$ and $y(t)$ in a piecewise linear manner. Within this method, $0 \le t \le 2\pi$ is assumed, and the parametrization functions are written as

$$x(t) = \begin{cases} a_1 t + b_1, & t \in [0, t_1], \\ a_2 t + b_2, & t \in [t_1, t_2], \\ & \dots \\ a_n t + b_n, & t \in [t_{n-1}, 2\pi], \end{cases} \qquad y(t) = \begin{cases} c_1 t + d_1, & t \in [0, t_1], \\ c_2 t + d_2, & t \in [t_1, t_2], \\ & \dots \\ c_n t + d_n, & t \in [t_{n-1}, 2\pi]. \end{cases} \tag{3.17}$$

Two trails to employ the above piecewise linear functions $x(t)$ and $y(t)$ may be applied.

The first one (denoted "Method A") uses a non-dimensional parameter defined as $t = 2\pi \frac{s}{l}$, where l is the circumferential length. In this case, the integration of (3.3) is not required, and the "length" devoted in terms of t for each edge is proportional to its actual length, namely, $\Delta t = 2\pi \frac{\Delta s}{l}$.

Another way (denoted "Method B") is based on dividing the range of the parameter $0 \le t \le 2\pi$ into n *equal* segments for a generic polygon of n edges, so that each segment, regardless of its length, occupies an equal "length" in terms of t, namely, $\Delta t = \frac{2\pi}{n}$.

If necessary, a Fourier series expansion is then carried out for $x(t)$ and $y(t)$ separately, and their truncated versions, up to k_p harmonics, are obtained as the parametrization functions.

P.3.4 describes various polygon parametrizations by the above two methods, see also Example 3.3. Method B has some advantages over Method A when a polygon with large *edge length diversity* is under discussion. Qualitatively, in such a case, when Method A is employed, the functions $x(t)$ and $y(t)$ contain sharp turning points and abrupt changes that result in "Gibbs-type" phenomena. These phenomena are much less sharp when Method B is used, see Example 3.3.

Example 3.3 *Generic (Non-Convex) Polygon Parametrization.*

By activating **P.3.4**, a series of parametrizations for a non-convex polygon were obtained, see Fig. 3.5(a). The thin and thick lines represent the (actual) piecewise and the trigonometric

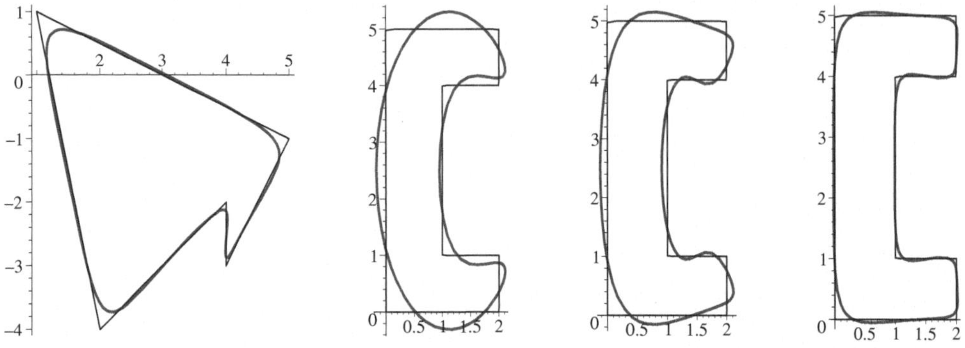

(a) $k_p = 4$. (b) $k_p = 3$. (c) $k_p = 5$. (d) $k_p = 7$.

Figure 3.5: Parametrization by a piecewise linear and trigonometric functions (Method B). (a): a generic pentagon, (b)–(d): A "C" domain approximated by various parametrization levels.

(approximate) curves, respectively. For illustration, the constant and the first four harmonics of $y(t)$ of Fig. 3.5(a) are written as

$$\begin{aligned} y \cong & -1.800 - 1.225\cos t + 0.802\cos(2t) - 0.356\cos(3t) + 0.0766\cos(4t) \\ & + 0.793\sin t - 0.622\sin(2t) - 0.276\sin(3t) + 0.0496\sin(4t) + \cdots . \end{aligned} \tag{3.18}$$

In addition, Figs. 3.5(b)–(d) show a series of approximations of a "C"–domain (with eight vertices).

To demonstrate the difference between Methods A and B, Fig. 3.6 presents a slender triangle and the quality of approximation obtained by $k_p = 3$ for its $x(t)$ function.

3.2 Plane-Strain and Plane-Stress

There are plenty of physical situations, where complete three-dimensional considerations are not necessary, and various kinds of two-dimensional reductions of the problem are (fully or partially) acceptable. The fundamental two-dimensional states are widely known as "Plane-Strain" and the "Plane-Stress". Each of the above cases leads to a different analysis, which is formulated by a different version of the *biharmonic BVP*. We shall first consider the plane-

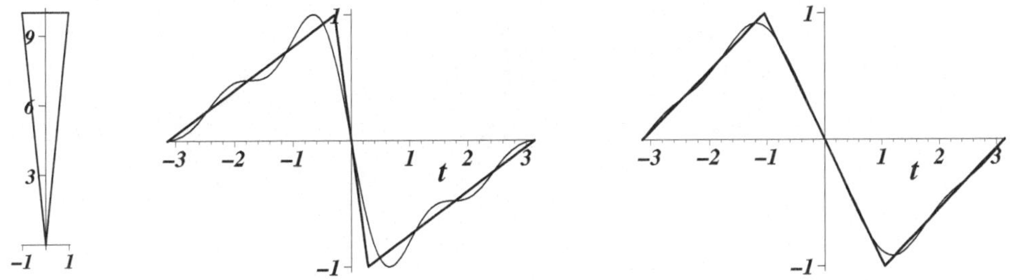

(a)

(b) Method A.

(c) Method B.

Figure 3.6: The function $x(t)$ for the slender triangle shown in (a) as approximated by Methods A, B.

strain case, since, in contrast to only an approximate reduction in the plane-stress case, it does constitute a complete and exact presentation of a special three-dimensional state.

To support the following discussion, it is convenient to express the generalized Hook's law (GHL) for a linear material (see S.2.1) as

$$\bar{\varepsilon} = \mathbf{A}_\varepsilon \cdot \bar{\sigma} + \mathbf{B}_\varepsilon \cdot \bar{\bar{\sigma}}, \qquad \bar{\bar{\varepsilon}} = \mathbf{C}_\varepsilon \cdot \bar{\sigma} + \mathbf{D}_\varepsilon \cdot \bar{\bar{\sigma}}, \tag{3.19}$$

where $\bar{\varepsilon} = [\varepsilon_x, \varepsilon_y, \gamma_{xy}]$, $\bar{\bar{\varepsilon}} = [\varepsilon_z, \gamma_{yz}, \gamma_{xz}]$, $\bar{\sigma} = [\sigma_x, \sigma_y, \tau_{xy}]$, $\bar{\bar{\sigma}} = [\sigma_z, \tau_{yz}, \tau_{xz}]$. The involved matrices are

$$\mathbf{A}_\varepsilon = \begin{bmatrix} a_{11} & a_{12} & a_{16} \\ a_{12} & a_{22} & a_{26} \\ a_{16} & a_{26} & a_{66} \end{bmatrix}, \quad \mathbf{B}_\varepsilon = \begin{bmatrix} a_{13} & a_{14} & a_{15} \\ a_{23} & a_{24} & a_{25} \\ a_{36} & a_{46} & a_{56} \end{bmatrix}, \quad \mathbf{D}_\varepsilon = \begin{bmatrix} a_{33} & a_{34} & a_{35} \\ a_{34} & a_{44} & a_{45} \\ a_{35} & a_{45} & a_{55} \end{bmatrix}, \tag{3.20}$$

and $\mathbf{C}_\varepsilon = \mathbf{B}_\varepsilon^T$. Similarly, the GHL may be inverted and expressed by the stiffness matrix as

$$\bar{\sigma} = \mathbf{A}_\sigma \cdot \bar{\varepsilon} + \mathbf{B}_\sigma \cdot \bar{\bar{\varepsilon}}, \qquad \bar{\bar{\sigma}} = \mathbf{C}_\sigma \cdot \bar{\varepsilon} + \mathbf{D}_\sigma \cdot \bar{\bar{\varepsilon}}, \tag{3.21}$$

where

$$\mathbf{A}_\sigma = \begin{bmatrix} A_{11} & A_{12} & A_{16} \\ A_{12} & A_{22} & A_{26} \\ A_{16} & A_{26} & A_{66} \end{bmatrix}, \quad \mathbf{B}_\sigma = \begin{bmatrix} A_{13} & A_{14} & A_{15} \\ A_{23} & A_{24} & A_{25} \\ A_{36} & A_{46} & A_{56} \end{bmatrix}, \quad \mathbf{D}_\sigma = \begin{bmatrix} A_{33} & A_{34} & A_{35} \\ A_{34} & A_{44} & A_{45} \\ A_{35} & A_{45} & A_{55} \end{bmatrix}, \tag{3.22}$$

and $\mathbf{C}_\sigma = \mathbf{B}_\sigma{}^T$. Note that the matrices $\mathbf{A}_\varepsilon, \mathbf{D}_\varepsilon, \mathbf{A}_\sigma, \mathbf{D}_\sigma$ are symmetric.

3.2.1 Plane-Strain

The approximation known as "Plane-Strain" is founded on the assumption that the elastic body deformation is confined to the plane deformation components, while all strain components that include indices of the direction normal to the plane vanish. This simulates well the state in the inner sections of "long" elastic bodies, such as prismatic bodies that undergo uniform loading along their outer surface, see Figs. 3.7(a),(b). In such a case, each cross-section may be analyzed under the assumption that the displacements u, v are functions of x, y (only) and $w = 0$. Therefore, the associated strain components are

$$\varepsilon_x = u_{,x}, \qquad \varepsilon_y = v_{,y}, \qquad \varepsilon_z = \gamma_{yz} = \gamma_{xz} = 0, \qquad \gamma_{xy} = u_{,y} + v_{,x}, \tag{3.23}$$

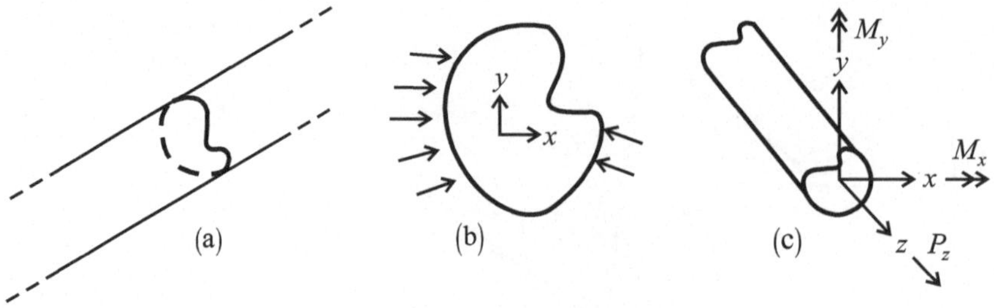

Figure 3.7: A slender cylindrical body in a state of Plane-Strain. The arrows in (b) represent the surface loading components.

or, according to (3.19) and (3.21),

$$\bar{\bar{\varepsilon}} = 0, \qquad \bar{\bar{\sigma}} = -\mathbf{D}_\varepsilon^{-1} \cdot \mathbf{C}_\varepsilon \cdot \bar{\sigma}, \qquad \bar{\varepsilon} = (\mathbf{A}_\varepsilon - \mathbf{B}_\varepsilon \cdot \mathbf{D}_\varepsilon^{-1} \cdot \mathbf{C}_\varepsilon)\bar{\sigma}, \tag{3.24a}$$

$$\bar{\bar{\sigma}} = \mathbf{C}_\sigma \bar{\varepsilon}, \qquad \bar{\varepsilon} = \mathbf{A}_\sigma^{-1} \cdot \bar{\sigma}. \tag{3.24b}$$

We shall now restrict the discussion to the MON13z-type material (see S.2.3), where $a_{14} = a_{15} = a_{24} = a_{25} = a_{34} = a_{35} = a_{46} = a_{56} = 0$, and similarly, $A_{14} = A_{15} = A_{24} = A_{25} = A_{34} = A_{35} = A_{46} = A_{56} = 0$. Then, the matrices $\mathbf{B}_\varepsilon$, $\mathbf{D}_\varepsilon$ and $\mathbf{B}_\sigma$, $\mathbf{D}_\sigma$ of (3.20) become

$$\mathbf{B}_\varepsilon = \begin{bmatrix} a_{13} & 0 & 0 \\ a_{23} & 0 & 0 \\ a_{36} & 0 & 0 \end{bmatrix}, \qquad \mathbf{D}_\varepsilon = \begin{bmatrix} a_{33} & 0 & 0 \\ 0 & a_{44} & a_{45} \\ 0 & a_{45} & a_{55} \end{bmatrix}, \tag{3.25a}$$

$$\mathbf{B}_\sigma = \begin{bmatrix} A_{13} & 0 & 0 \\ A_{23} & 0 & 0 \\ A_{36} & 0 & 0 \end{bmatrix}, \qquad \mathbf{D}_\sigma = \begin{bmatrix} A_{33} & 0 & 0 \\ 0 & A_{44} & A_{45} \\ 0 & A_{45} & A_{55} \end{bmatrix}. \tag{3.25b}$$

The stress component σ_z may be presented by two forms, using (3.24a) or (3.24b), respectively,

$$\sigma_z = -\frac{1}{a_{33}} \left(a_{13}\sigma_x + a_{23}\sigma_y + a_{36}\tau_{xy} \right), \tag{3.26a}$$

$$\sigma_z = A_{13}\varepsilon_x + A_{23}\varepsilon_y + A_{36}\gamma_{xy}. \tag{3.26b}$$

In both cases, $\tau_{yz} = \tau_{xz} = 0$. Equations (3.24a,b) also show that

$$\bar{\varepsilon} = \mathbf{b} \cdot \bar{\sigma} \quad \Leftrightarrow \quad \begin{Bmatrix} \varepsilon_x \\ \varepsilon_y \\ \gamma_{xy} \end{Bmatrix} = \begin{bmatrix} b_{11} & b_{12} & b_{16} \\ b_{12} & b_{22} & b_{26} \\ b_{16} & b_{26} & b_{66} \end{bmatrix} \cdot \begin{Bmatrix} \sigma_x \\ \sigma_y \\ \tau_{xy} \end{Bmatrix}. \tag{3.27}$$

Here the symmetric matrix $\mathbf{b} = \{b_{ij}\}$ $(i, j = 1, 2, 6)$ is defined by its coefficients

$$b_{ij} = a_{ij} - \frac{a_{i3}a_{j3}}{a_{33}}, \qquad i, j = 1, 2, 6 \tag{3.28}$$

referred to as the *reduced compliance elastic coefficients*. The inverse relation may be written as $\bar{\sigma} = \mathbf{A}_\sigma \cdot \bar{\varepsilon}$, see (3.21), and hence, $\mathbf{A}_\sigma = \mathbf{b}^{-1}$.

At this stage it is convenient to introduce the classical *Airy's stress function*, $\Phi(x, y)$, by its derivatives as

$$\sigma_x = \Phi_{,yy}, \qquad \sigma_y = \Phi_{,xx}, \qquad \tau_{xy} = -\Phi_{,xy}. \tag{3.29}$$

Reviewing the equilibrium equations (1.82a–c), and the compatibility equations (1.45a–f), shows that the definitions and assumptions described above identically satisfy all the above two sets of equations (for vanishing body forces the influence of which will be discussed later on), with the exception of the compatibility equation (1.45a). In terms of the stress function Φ and using (3.27), (3.29), the above compatibility equation turns to be the following *homogeneous biharmonic equation*:

$$\nabla_1^{(4)}\Phi = 0, \tag{3.30}$$

where the *generalized biharmonic operator* $\nabla_1^{(4)}$ is given by

$$\nabla_1^{(4)} = b_{22}\frac{\partial^4}{\partial x^4} - 2b_{26}\frac{\partial^4}{\partial x^3\partial y} + (2b_{12}+b_{66})\frac{\partial^4}{\partial x^2\partial y^2} - 2b_{16}\frac{\partial^4}{\partial x\partial y^3} + b_{11}\frac{\partial^4}{\partial y^4}. \tag{3.31}$$

In this operator notation, the superscript stands for the order of derivatives in the operator, and the subscript stands for its version number, which depend on the material and assumptions involved (see also Example 1.5). Throughout the derivation, we shall make use of additional versions of the biharmonic operator $\nabla_n^{(4)}$ documented in S.3.6.2. Correspondingly, Φ, is said to be *a generalized biharmonic function*, see e.g. (Lekhnitskii, 1981), (Lurie and Vasiliev, 1995).

For orthotropic materials, $b_{16} = b_{26} = 0$, since $a_{16} = a_{26} = a_{36} = 0$, and (3.31) is simplified to $\nabla_2^{(4)}$ given by (3.203). For this case, (3.26a) yields

$$\sigma_z = -\frac{1}{a_{33}}\left(a_{13}\sigma_x + a_{23}\sigma_y\right). \tag{3.32}$$

In the isotropic case (where E stands for Young's modulus and ν for Poisson's ratio, see S.2.8), one may substitute in (3.28) the values

$$a_{11} = a_{22} = a_{33} = \frac{1}{E}, \qquad a_{12} = a_{13} = a_{23} = -\frac{\nu}{E}, \qquad a_{66} = \frac{2(1+\nu)}{E}, \tag{3.33}$$

to obtain

$$b_{11} = b_{22} = \frac{1-\nu^2}{E}, \qquad b_{12} = -\nu\frac{1+\nu}{E}, \qquad b_{66} = \frac{2(1+\nu)}{E} \tag{3.34}$$

(also note that in this case, $2a_{12} + a_{66} = 2a_{11}$ and $2b_{12} + b_{66} = 2b_{11}$). Thus, the operator (3.31) is further streamlined to $\frac{1-\nu^2}{E}\nabla^{(4)}$, where we use the classical biharmonic operator of (3.205), while σ_z of (3.32) becomes

$$\sigma_z = \nu\left(\sigma_x + \sigma_y\right). \tag{3.35}$$

Remark 3.3 As shown by (3.26a), (3.32), (3.35), to create a state of plane-strain in a "long" cylindrical body as shown in Fig. 3.7(a) (where cross-sections are parallel to the x,y-plane), a specific distribution of σ_z must be maintained throughout the body and over the end cross-sections. Equivalently to this requirement one may supply an axial force, P_z, and two transverse moments M_x and M_y at the cylinder ends, see Fig. 3.7(c), which are given by

$$\{P_z, M_x, M_y\} = \iint_\Omega \sigma_z\{1, y, -x\}. \tag{3.36}$$

According to St. Venant's Principle (see S.5.1.3.1), the exact distribution of σ_z over the ends is immaterial, since the difference between the applied distribution and the one described by (3.26a) (for the general case), will constitute a self-equilibrated system the influence of which will diminish with the distance from the ends.

To eliminate the above need to supply tip resultants, the literature (see e.g. (Lekhnitskii, 1981)) offers an additional solution, which is superimposed on the discussed one. In this additional solution, we introduce a stress component $\sigma_z = \Delta\sigma(x,y)$ only. Clearly, such state of stress satisfies both the equilibrium and the compatibility equations. When the origin of the coordinate system is placed at the cross-section centroid, $\Delta\sigma(x,y)$ may be selected based on (3.36) as $\Delta\sigma = -\frac{P_z}{S_\Omega} - \frac{M_x}{I_x}y + \frac{M_y}{I_y}x$, where

$$\{S_\Omega, I_x, I_y\} = \iint_\Omega \{1, y^2, x^2\} \tag{3.37}$$

are the cross-section area and the corresponding moments (of inertia) about the x- and y- axes.

Indeed, a superposition of the above solution on the previously mentioned solution for plane-strain, eliminates the required end loads since

$$\iint_\Omega (\sigma_z + \Delta\sigma)\{1, y, -x\} = \{0,0,0\}, \tag{3.38}$$

and again, by invoking the St. Venant's principle, one may argue that the effect of the $\sigma_z + \Delta\sigma$ distribution and the exact one ($\sigma_z = 0$ at the tip) will decay far from the end cross-sections. However, one should note that according to (3.19), (3.25a), the above additional solution induces for the MON13z material under discussion the additional strain components

$$\Delta\varepsilon_x = a_{13}\Delta\sigma, \qquad \Delta\varepsilon_y = a_{23}\Delta\sigma, \qquad \Delta\varepsilon_z = a_{33}\Delta\sigma, \qquad \Delta\gamma_{xy} = a_{36}\Delta\sigma. \tag{3.39}$$

These strain components will induce additional plane deformation (which may be easily obtained by the procedure described in S.1.2 and **P.1.5**). In particular, it should be noted that ε_z does not vanish anymore. Hence, the above superimposed solution replaces the residual axial stress with a residual axial strain, see also Example 3.5.

3.2.2 *Plane-Stress*

The state of "Plane-Stress" is suitable for analyzing thin plates that are subjected to in-plane loading, see Fig. 3.8, and will be further discussed within the laminated plates theory of S.3.5.1. The notion "thin" means that the (constant) thickness of the plate, h, is much smaller than its typical in-plane dimension — see Fig. 3.8, where the x,y-plane coincides with the plate's middle plane. In this kind of analysis, we assume that all stress components that include the z index vanish, and therefore, set the assumption $\overline{\overline{\sigma}} = 0$ (namely, $\sigma_z = \tau_{yz} = \tau_{xz} = 0$) in (3.19), (3.25a), and consider σ_x, σ_y, and τ_{xy} as functions of x and y only. We carry out an analysis that is analogous to the plane-strain case, while replacing σ_i with ε_i and a_{ij} with A_{ij}. Hence, for general MON13z material (see S.2.3), $\gamma_{yz} = \gamma_{xz} = 0$ and

$$\varepsilon_z = a_{13}\sigma_x + a_{23}\sigma_y + a_{36}\tau_{xy}, \tag{3.40}$$

or $\varepsilon_z = -\frac{1}{A_{33}}(A_{13}\varepsilon_x + A_{23}\varepsilon_y + A_{36}\gamma_{xy})$. According to the above described analogy,

$$\overline{\sigma} = \mathbf{B}\cdot\overline{\varepsilon} \quad \Leftrightarrow \quad \begin{Bmatrix} \sigma_x \\ \sigma_y \\ \tau_{xy} \end{Bmatrix} = \begin{bmatrix} B_{11} & B_{12} & B_{16} \\ B_{12} & B_{22} & B_{26} \\ B_{16} & B_{26} & B_{66} \end{bmatrix} \cdot \begin{Bmatrix} \varepsilon_x \\ \varepsilon_y \\ \gamma_{xy} \end{Bmatrix}, \tag{3.41}$$

where the symmetric matrix $\mathbf{B} = \{B_{ij}\},\ i,j = 1,2,3$ is defined by its components

$$B_{ij} = A_{ij} - \frac{A_{i3}A_{j3}}{A_{33}}, \tag{3.42}$$

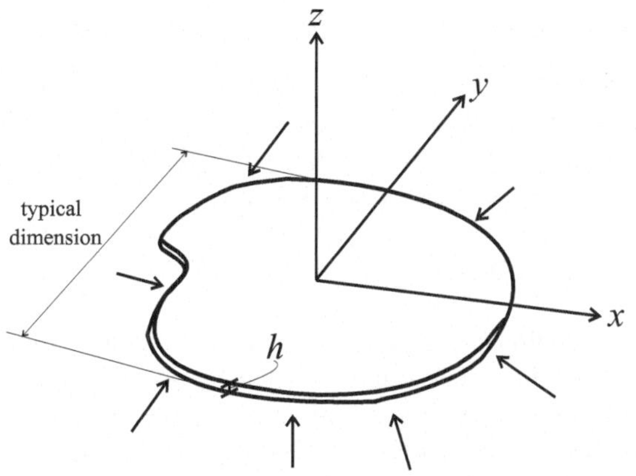

Figure 3.8: A thin plate in a state of plane-stress. The arrows represent the in-plane loading components.

referred as the *reduced stiffness elastic coefficients*. Since $\mathbf{B}^{-1} = \mathbf{A}_\varepsilon$, the inverse relation is given by

$$\overline{\varepsilon} = \mathbf{A}_\varepsilon \cdot \overline{\sigma}. \tag{3.43}$$

Employing Airy 's stress function for the present case, see (3.29), shows that the equilibrium equations are satisfied, and the compatibility equation (1.45a) is fulfilled provided that

$$\nabla_3^{(4)} \Phi = 0, \tag{3.44}$$

which constitute the general biharmonic operator definition for plane-stress

$$\nabla_3^{(4)} = a_{22}\frac{\partial^4}{\partial x^4} - 2a_{26}\frac{\partial^4}{\partial x^3 \partial y} + (2a_{12} + a_{66})\frac{\partial^4}{\partial x^2 \partial y^2} - 2a_{16}\frac{\partial^4}{\partial x \partial y^3} + a_{11}\frac{\partial^4}{\partial y^4}. \tag{3.45}$$

As already indicated, the state of plane-stress is only an approximate reduction of the general three-dimensional formulation. To show that, one should notice that the above derivation leaves three unsatisfied compatibility equations (1.45b,c,f), namely, $\varepsilon_{z,yy} = \varepsilon_{z,xx} = \varepsilon_{z,xy} = 0$. Hence, unless ε_z (see (3.40)) is linear, i.e. may be written as $\varepsilon_z = \widetilde{\kappa}_1 x + \widetilde{\kappa}_2 y + \widetilde{\varepsilon}_0$, the above plane-stress analysis is inaccurate.

For orthotropic materials, $a_{16} = a_{26} = a_{36} = 0$, and (3.40) becomes

$$\varepsilon_z = a_{13}\sigma_x + a_{23}\sigma_y, \tag{3.46}$$

and the biharmonic operator of (3.45) is simplified to $\nabla_4^{(4)}$ given by (3.207).

For the isotropic case the elastic moduli are reduced to

$$A_{11} = A_{22} = A_{33} = \frac{(1-\nu)E}{(1+\nu)(1-2\nu)}, \quad A_{12} = A_{13} = A_{23} = \frac{\nu E}{(1+\nu)(1-2\nu)}, \quad A_{66} = \frac{E}{2(1+\nu)}, \tag{3.47}$$

and therefore

$$B_{11} = B_{22} = \frac{E}{1-\nu^2}, \qquad B_{12} = \frac{\nu E}{1-\nu^2}, \qquad B_{66} = \frac{E}{2(1+\nu)}. \tag{3.48}$$

In addition

$$\varepsilon_z = -\frac{\nu}{E}(\sigma_x + \sigma_y), \tag{3.49}$$

and the operator of (3.45) is reduced to $\frac{1}{E}\nabla^{(4)}$, where $\nabla^{(4)}$ is the classical biharmonic operator (3.205).

3.2.3 Illustrative Examples of Prescribed Airy's Function

As a first study of a homogeneous body plane deformation, one may use a technique in which Airy 's stress function, Φ, is prescribed. This enables an initial and simple examination of the stress function and the resulting stress distribution properties. For that purpose, we expand Φ as a *homogeneous polynomial* of a given order $k_\Phi \geq 2$, as

$$\Phi^{(k_\Phi)} = \sum_{i=0}^{k_\Phi} B_{i,k_\Phi-i} x^i y^{k_\Phi-i}, \tag{3.50}$$

which yields an expression with $k_\Phi + 1$ undetermined coefficients. Since we are dealing with the biharmonic equation where stresses are obtained by second-order derivatives of $\Phi^{(k_\Phi)}$, values of $k_\Phi < 2$ are meaningless. Superposition of homogeneous polynomials of different k_Φ degree is clearly possible as shown further on in S.4.3.

Substitution of (3.50) in one of the above discussed versions of the biharmonic equation, $\nabla_n^{(4)} \Phi^{(k_\Phi)} = 0$, yields an equation where a homogeneous polynomial of order $\bar{k} = \max(0, k_\Phi - 4)$ must vanish, namely $\sum_{i=0}^{\bar{k}} D_{i,\bar{k}-i} x^i y^{\bar{k}-i} = 0$, where the coefficients $D_{i,\bar{k}-i}$ are linear functions of $B_{i,k_\Phi-i}$ and the elastic moduli. According to a known algebraic lemma, to satisfy the above equation identically for all x and y, one has to set $D_{i,\bar{k}-i} = 0$ $(0 \leq i \leq \bar{k})$. For $k_\Phi > 3$ this constitutes a system of $k_\Phi - 3$ linear equations that reflects dependencies between the coefficients $B_{i,k_\Phi-i}$ $(0 \leq i \leq k_\Phi)$. For $k_\Phi \leq 3$ no equations should be solved. Hence, for any value of $k_\Phi \geq 3$ there are four undetermined coefficients in (3.50), while for $k_\Phi = 2$ there are only three.

Example 3.4 *Prescribed Stress Function on a Rectangle.*

Illustrative application of a prescribed stress function may be created by activating **P.3.5**, which for $k_\Phi = 4$ (i.e., $\bar{k} = 0$) yields $\Phi^{(4)} = B_{04}y^4 + B_{13}xy^3 + B_{22}x^2y^2 + B_{31}x^3y + B_{40}x^4$. Using, for example, the biharmonic operator for plane-strain, (3.31), the only condition in this case turns to be

$$D_{00} = 24\,b_{22}B_{40} - 12\,b_{26}B_{31} + 4\,(2\,b_{12} + b_{66})\,B_{22} - 12\,b_{16}B_{13} + 24\,b_{11}B_{04} = 0, \tag{3.51}$$

which is a single equation with five unknown coefficients. Thus, one may arbitrarily choose the four coefficients B_{31}, B_{22}, B_{13} and B_{04}. By eliminating B_{40}, the biharmonic function may be written as

$$\Phi^{(4)} = (y^4 - \frac{b_{11}}{b_{22}}x^4)B_{04} + (\frac{b_{16}}{2b_{22}}x^4 + xy^3)B_{13} + (x^2y^2 - \frac{2\,b_{12} + b_{66}}{6b_{22}}x^4)B_{22} + (\frac{b_{26}}{2b_{22}}x^4 + x^3y)B_{31}. \tag{3.52}$$

The corresponding stress components (3.29) are homogeneous polynomials of order 2:

$$\begin{aligned}
\sigma_x &= 12\,B_{04}y^2 + 6\,B_{13}xy + 2\,B_{22}x^2, \\
\sigma_y &= (-\frac{12b_{11}}{b_{22}}B_{04} + \frac{6b_{16}}{b_{22}}B_{13} - 2\frac{2\,b_{12} + b_{66}}{b_{22}}B_{22} + \frac{6b_{26}}{b_{22}}B_{31})x^2 + 6\,B_{31}xy + 2\,B_{22}y^2, \\
\tau_{xy} &= -3\,B_{13}y^2 - 4\,B_{22}xy - 3\,B_{31}x^2.
\end{aligned} \tag{3.53}$$

According to (3.43), the strain components are also homogeneous polynomials of order 2. The displacement components $u(x,y)$, $v(x,y)$ in this case may be obtained by the integration procedure of S.1.2 and **P.1.6** as third-order homogeneous polynomials.

To further illustrate this example we activate **P.3.6** for $k_\Phi = 4$, and examine the rectangular domain $|x| \leq 1$, $|y| \leq 1$. We select all B_{ij} to vanish except for, $B_{22} = 1$. Figure 3.9 presents the above stress function and the resulting deformation field for the MON13z material obtained by a typical Graphite/Epoxy orthotropic material, which is rotated by 30° about the z-axis.

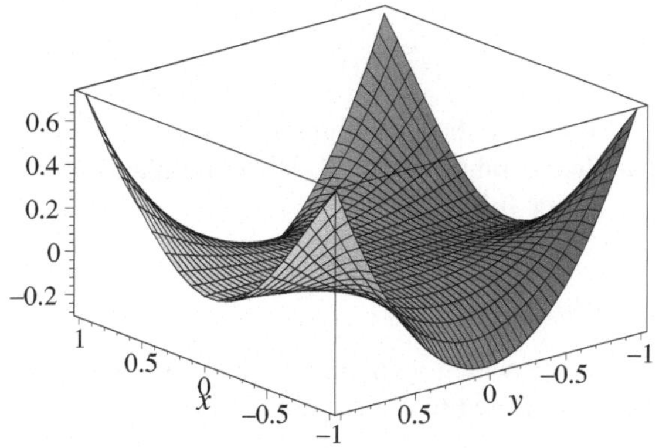

(a) $\Phi(x,y)$.

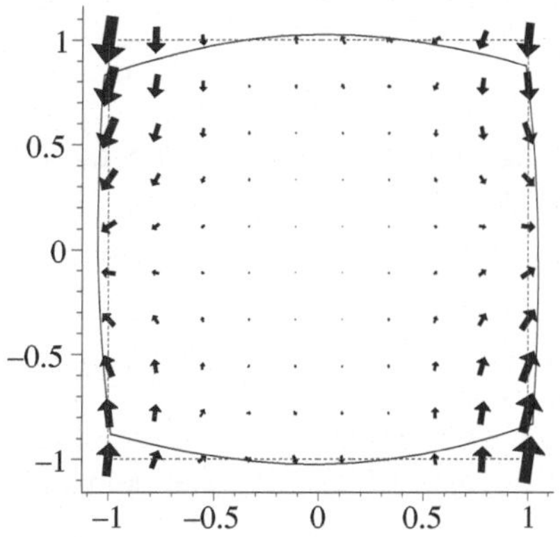

(b) Displacement field.

Figure 3.9: Stress function and displacement vector field over a rectangle (Example 3.4).

3.2.4 The Influence of Body Forces

As shown in previous sections, the plane-strain and plane-stress analyses yield different kinds of homogeneous biharmonic field equations. These results will be extended in this section to account for a generic distribution of body forces, $X_b(x,y)$, $Y_b(x,y)$. For that purpose, we replace the stresses of (3.29) by

$$\sigma_x = \Phi_{,yy} + \overline{U}_1, \qquad \sigma_y = \Phi_{,xx} + \overline{U}_2, \qquad \tau_{xy} = -\Phi_{,xy}, \tag{3.54}$$

which contains two separate potential functions $\overline{U}_1(x,y)$ and $\overline{U}_2(x,y)$ so that

$$X_b = -\overline{U}_{1,x}, \qquad Y_b = -\overline{U}_{2,y}. \tag{3.55}$$

It is a simple task to verify that the equilibrium equations are satisfied in this case as well. However, the compatibility equation (1.45a) (with the new strain expressions obtained by substituting (3.54) in (3.27)) may be written in terms of the stress function, Φ, as the non-homogeneous biharmonic equation

$$\nabla_1^{(4)}\Phi = F_0(x,y), \tag{3.56}$$

where

$$F_0 = -b_{12}\overline{U}_{1,xx} + b_{16}\overline{U}_{1,xy} - b_{11}\overline{U}_{1,yy} - b_{22}\overline{U}_{2,xx} + b_{26}\overline{U}_{2,xy} - b_{12}\overline{U}_{2,yy}. \tag{3.57}$$

Equation (3.56) replaces (3.30) in the case of plane-strain. Similarly, the above may apply to (3.44) for the plane-stress, which yields

$$\nabla_3^{(4)}\Phi = F_0(x,y), \tag{3.58}$$

where F_0 is identical to the one given by (3.57) except for b_{ij} that are replaced by a_{ij}. In the isotropic case, (3.56) takes the form (see (3.205))

$$\nabla^{(4)}\Phi = \frac{\nu}{1-\nu}(\overline{U}_{1,xx} + \overline{U}_{2,yy}) - (\overline{U}_{1,yy} + \overline{U}_{2,xx}), \tag{3.59}$$

while (3.58) becomes

$$\nabla^{(4)}\Phi = \nu(\overline{U}_{1,xx} + \overline{U}_{2,yy}) - (\overline{U}_{1,yy} + \overline{U}_{2,xx}). \tag{3.60}$$

When both X_b, Y_b emerge from the same potential, $\overline{U}$ (e.g. the case of gravity loads), the r.h.s. of (3.59), (3.60) are simplified to $\frac{2\nu-1}{1-\nu}\nabla^{(2)}\overline{U}$ for plane-strain and $(\nu-1)\nabla^{(2)}\overline{U}$ for plane-stress, where $\nabla^{(2)}$ is the simplest Laplace's operator, see (3.198).

3.2.5 Boundary and Single-Value Conditions

Let Ω be a plane domain with boundary $\partial\Omega$, along which a circumferential coordinate s is defined, see Fig. 3.10. The domain boundary is subjected to a distribution of forces per unit

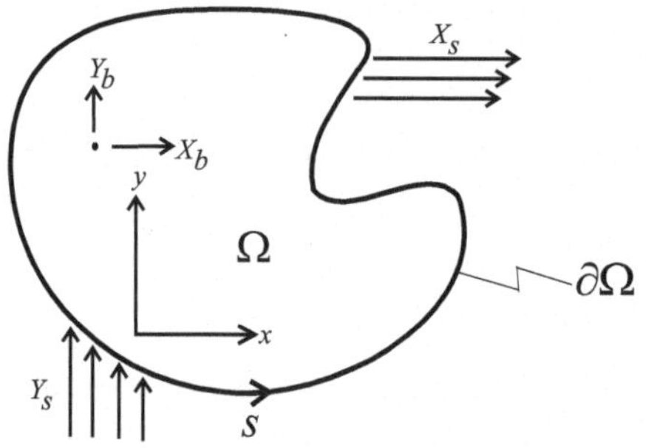

Figure 3.10: Notation for a simply connected domain.

length X_s and Y_s in the x- and y- directions, respectively. Hence $\cos(\bar{\mathbf{n}},z) = 0$, and the boundary conditions (1.85a,b) may be written as

$$\sigma_x \cos(\bar{\mathbf{n}},x) + \tau_{xy}\cos(\bar{\mathbf{n}},y) = X_s, \qquad \tau_{xy}\cos(\bar{\mathbf{n}},x) + \sigma_y \cos(\bar{\mathbf{n}},y) = Y_s, \tag{3.61}$$

which with the aid of (3.4) become

$$\sigma_x \frac{dy}{ds} - \tau_{xy}\frac{dx}{ds} = X_s, \qquad \tau_{xy}\frac{dy}{ds} - \sigma_y \frac{dx}{ds} = Y_s. \tag{3.62}$$

To enable further handling of the above boundary conditions, we shall also use the following general identities for C^2-differentiable functions:

$$\frac{d}{ds}\Phi_{,y} = \Phi_{,yx}\frac{dx}{ds} + \Phi_{,yy}\frac{dy}{ds} = \Phi_{,yy}\cos(\bar{\mathbf{n}},x) - \Phi_{,yx}\cos(\bar{\mathbf{n}},y), \tag{3.63a}$$

$$\frac{d}{ds}\Phi_{,x} = \Phi_{,xx}\frac{dx}{ds} + \Phi_{,xy}\frac{dy}{ds} = \Phi_{,xy}\cos(\bar{\mathbf{n}},x) - \Phi_{,xx}\cos(\bar{\mathbf{n}},y). \tag{3.63b}$$

By utilizing the above, one may express the boundary conditions (3.61) using (3.54) as

$$\frac{d}{ds}\{\Phi_{,x}, \Phi_{,y}\} = \{-F_1, F_2\} \qquad \text{on} \quad \partial\Omega, \tag{3.64}$$

where the boundary functions, $F_1(x,y)$ and $F_2(x,y)$ are defined on $\partial\Omega$ as

$$F_1 = Y_s + \overline{U}_2\frac{dx}{ds} = Y_s - \overline{U}_2\cos(\bar{\mathbf{n}},y), \qquad F_2 = X_s - \overline{U}_1\frac{dy}{ds} = X_s - \overline{U}_1\cos(\bar{\mathbf{n}},x). \tag{3.65}$$

There are certain conditions that the functions F_1 and F_2 should satisfy. In what follows these conditions will be derived independently from a pure mathematical point of view while physical argumentation will follow. The mathematical approach is founded on the requirement that Φ and its first partial derivatives will be single-valued, namely,

$$\oint_{\partial\Omega} \frac{d}{ds}\{\Phi_{,x}, \Phi_{,y}, \Phi\} = \{0,0,0\}. \tag{3.66}$$

Using (3.64), the first two integral conditions of (3.66) may be written as

$$\oint_{\partial\Omega} \{-F_1, F_2\} = \{0,0\}. \tag{3.67}$$

In addition, the single-valued property for Φ (the third integral condition of (3.66)) may be transformed to

$$\oint_{\partial\Omega} (\frac{d}{ds}\Phi) = \oint_{\partial\Omega} (\Phi_{,x}\frac{dx}{ds} + \Phi_{,y}\frac{dy}{ds}) = 0. \tag{3.68}$$

The above equations may be further derived via integration by parts as:

$$\oint_{\partial\Omega} \Phi_{,x}\frac{dx}{ds} = \underline{[x\Phi_{,x}]_{s=0}^{s=L}} - \oint_{\partial\Omega} x\frac{d}{ds}\Phi_{,x} = \underline{\underline{x\oint_{\partial\Omega}\frac{d}{ds}\Phi_{,x}}} - \oint_{\partial\Omega} x\frac{d}{ds}\Phi_{,x} \tag{3.69a}$$

$$\oint_{\partial\Omega} \Phi_{,y}\frac{dy}{ds} = \underline{[y\Phi_{,y}]_{s=0}^{s=L}} - \oint_{\partial\Omega} y\frac{d}{ds}\Phi_{,y} = \underline{\underline{y\oint_{\partial\Omega}\frac{d}{ds}\Phi_{,y}}} - \oint_{\partial\Omega} y\frac{d}{ds}\Phi_{,y}. \tag{3.69b}$$

The equality of the once underlined terms and the twice underlined terms in each of the above equations deserves a careful explanation. We first note that the x value is identical in both boundary points as $s = L$ (where L is the length of $\partial\Omega$) and $s = 0$ represent the same point. In addition, the differences in $\Phi_{,x}$ and $\Phi_{,y}$ is independent of the point that we select as the starting/end of the closed integral. Hence, (3.68) turns to be

$$x\oint_{\partial\Omega}\frac{d}{ds}\Phi_{,x} + y\oint_{\partial\Omega}\frac{d}{ds}\Phi_{,y} - \oint_{\partial\Omega}(x\frac{d}{ds}\Phi_{,x} + y\frac{d}{ds}\Phi_{,y}) = 0, \tag{3.70}$$

which may also be written as

$$-x\oint_{\partial\Omega} F_1 + y\oint_{\partial\Omega} F_2 - \underline{\oint_{\partial\Omega}(F_2y - F_1x) = 0}. \tag{3.71}$$

The above should be valid for each x and y and therefore takes the form

$$\oint_{\partial\Omega}\{F_1, F_2, F_2y - F_1x\} = \{0, 0, 0\}. \tag{3.72}$$

The conditions of (3.72) may also be examined from a physical point of view. To show that and following Green's Theorem, we derive the first two conditions as

$$\oint_{\partial\Omega}(Y_s + \overline{U}_2\frac{dx}{ds}) = \oint_{\partial\Omega} Y_s - \iint_\Omega \overline{U}_{2,y} = \underline{\underline{\oint_{\partial\Omega} Y_s + \iint_\Omega Y_b = 0}}, \tag{3.73a}$$

$$\oint_{\partial\Omega}(X_s - \overline{U}_1\frac{dy}{ds}) = \oint_{\partial\Omega} X_s - \iint_\Omega \overline{U}_{1,x} = \underline{\underline{\oint_{\partial\Omega} X_s + \iint_\Omega X_b = 0}}, \tag{3.73b}$$

while the underlined part of (3.71) becomes

$$\oint_{\partial\Omega}[(Y_s+\overline{U}_2\frac{dx}{ds})x-(X_s-\overline{U}_1\frac{dy}{ds})y]=\oint_{\partial\Omega}(Y_sx-X_sy)-\iint_{\Omega}(\overline{U}_{2,y}x-U_{1,x}y)=$$
$$\underline{\underline{\oint_{\partial\Omega}(Y_sx-X_sy)+\iint_{\Omega}(Y_bx-X_by)=0.}} \tag{3.74}$$

Hence, physical considerations show that (3.73a,b) are essentially the equilibrium equations for forces in the x- and y- directions, respectively. Similarly, (3.74) sets the condition for "zero net moment" (in the z-direction).

Solution methodologies and symbolic algorithms for solving the above discussed BVPs for general geometries, body-force and surface loads distributions are presented in Chapter 4.

To summarize the above, we should now distinguish between *three* cases. In the *first* one, both Φ and its derivatives $\Phi_{,x}$, $\Phi_{,y}$ are single-valued. The conditions for that are (3.72), which is the only meaningful case for a simply connected domain, as they are essential for the existence and uniqueness of the solution, see Example 3.7.

In a multiply connected domain, the stress function Φ is said to be single-valued with single-valued derivatives, if the surface loading acting on the contours $\partial\Omega_j$ satisfies (3.72) *for each boundary component separately*, namely,

$$\oint_{\partial\Omega_j}\{F_1^{(j)},F_2^{(j)},xF_1^{(j)}-yF_2^{(j)}\}=\{0,0,0\}. \tag{3.75}$$

Yet, the fact that Φ and its derivatives are not single-valued on some of the boundary components does not disqualify it as a solution as long as the sum of the above integral vanishes,

$$\sum_j\oint_{\partial\Omega_j}\{F_1^{(j)},F_2^{(j)},xF_1^{(j)}-yF_2^{(j)}\}=\{0,0,0\}. \tag{3.76}$$

We therefore define a *second* case, which may occur in multiply connected domains, where *only* the derivatives $\Phi_{,x}$, $\Phi_{,y}$ are single-valued functions. The conditions for that are (3.67), see also Example 3.8. In the *third* case, both Φ and its derivatives $\Phi_{,x}$, $\Phi_{,y}$ are *not* single-valued, see Example 3.9. In both cases, all conditions of (3.76) are satisfied.

Remark 3.4 In a state of plane-stress, where X_s and Y_s are also functions of z, one may replace them with their average (over the thickness, h) values $\overline{X}_s$ and $\overline{Y}_s$ depending on x, y, namely,

$$\overline{X}_s=\frac{1}{h}\int_{-h/2}^{h/2}X_s\,dz,\qquad \overline{Y}_s=\frac{1}{h}\int_{-h/2}^{h/2}Y_s\,dz. \tag{3.77}$$

The resulting stress and strain distribution will obviously be independent of z, and in thin plates, this operation, which "smears" all variations in the z-direction yields acceptable approximation. Further derivation of laminated (non-homogeneous) plates appears in S.3.5.1.

Example 3.5 *Prismatic Body Under Hydrostatic Load.*

Consider a "long" prismatic body of generic cross-section that undergoes a pure hydrostatic pressure, $\widetilde{P}$. The surface loads are therefore $X_s=-\widetilde{P}\cos(\bar{\mathbf{n}},x)$ and $Y_s=-\widetilde{P}\cos(\bar{\mathbf{n}},y)$, see Fig. 3.11(a) (the coordinate system origin is assumed to coincide with the cross-section area centroid). The stress function in this simple *plane-strain* case is independent of the material properties or domain shape and is given by

$$\Phi=-\frac{\widetilde{P}}{2}(x^2+y^2). \tag{3.78}$$

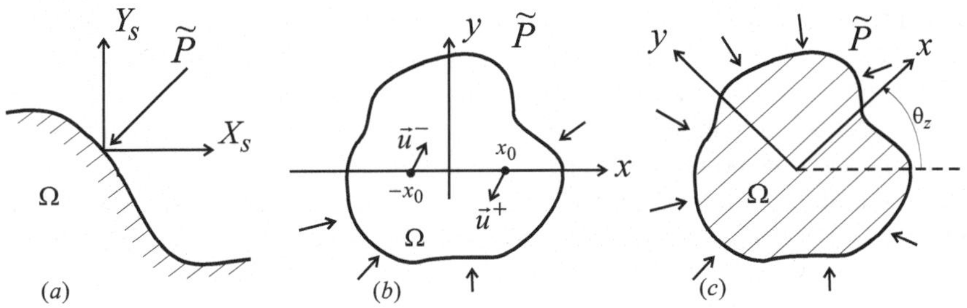

Figure 3.11: A generic domain under hydrostatic pressure in Example 3.5.

One may verify that both (3.56) and (3.64) are satisfied in this case (where no body forces are included). The resulting nonzero stress components are $\sigma_x = \sigma_y = -\widetilde{P}$ and $\sigma_z = \widetilde{P}(a_{13} + a_{23})/a_{33}$. The nonzero strain components are (see (3.27)) $\varepsilon_x = -\widetilde{P}(b_{11} + b_{12})$, $\varepsilon_y = -\widetilde{P}(b_{22} + b_{12})$ and $\gamma_{xy} = -\widetilde{P}(b_{16} + b_{26})$, and their integration by the procedure of S.1.2 yields

$$u = -\widetilde{P}[(b_{11} + b_{12})x + \frac{b_{16} + b_{26}}{2}y], \quad v = -\widetilde{P}[(b_{22} + b_{12})y + \frac{b_{16} + b_{26}}{2}x], \quad w = 0. \quad (3.79)$$

To create "free tips" of this long prismatic body, according to Remark 3.3, one needs to add to these tips a uniform stress distribution in the amount of $\Delta\sigma_z = -\widetilde{P}(a_{13} + a_{23})/a_{33}$. According to the analysis that will be derived within Chapter 5 for a material of generic anisotropy (see (5.29a–c) and note that for the MON13z material under discussion we set $a_{34} = a_{35} = 0$), the displacements of such an additional solution would be

$$\Delta u = \Delta\sigma_z(a_{13}x + \frac{1}{2}a_{36}y), \qquad \Delta v = \Delta\sigma_z(\frac{1}{2}a_{36}x + a_{23}y), \qquad \Delta w = \Delta\sigma_z a_{33} z. \qquad (3.80)$$

Superposition of Δu, Δv, Δw and u, v, w, respectively, yields

$$u_f = -\widetilde{P}[(a_{11} + a_{12})x + \underline{\frac{1}{2}(a_{16} + a_{26})y}], \quad v_f = -\widetilde{P}[(a_{12} + a_{22})y + \underline{\frac{1}{2}(a_{16} + a_{26})x}], \quad (3.81a)$$

$$w_f = -\widetilde{P}(a_{13} + a_{23})z. \qquad (3.81b)$$

Clearly, (3.81a,b) create a state of plane-stress, see S.3.2.2. If we further wish to simulate the effect of "all-around uniform pressure" on the prismatic body we need to superimpose on u_f, v_f, w_f an additional solution of (3.80) with $\Delta\sigma_z = -\widetilde{P}$, which yields

$$u_p = -\widetilde{P}[(a_{11} + a_{12} + a_{13})x + \underline{\frac{1}{2}(a_{16} + a_{26} + a_{36})y}], \qquad (3.82a)$$

$$v_p = -\widetilde{P}[(a_{12} + a_{22} + a_{23})y + \underline{\frac{1}{2}(a_{16} + a_{26} + a_{36})x}], \qquad (3.82b)$$

$$w_p = -\widetilde{P}(a_{13} + a_{23} + a_{33})z, \qquad (3.82c)$$

that obviously would emerge from a simple solution for which $\sigma_x = \sigma_y = \sigma_z = -\widetilde{P}$. Note that the deformation described above was derived for zero displacements and rotations at $x = y = z = 0$.

For a generic MON13z material it is clear that unlike the isotropic case, the in-plane deformation of both solutions of (3.81a,b) and (3.82a–c) is not "radial", namely, the displacements can not be written as $u = \widetilde{\alpha}x$ and $v = \widetilde{\alpha}y$, where $\widetilde{\alpha}$ is a constant. Thus, even for symmetric cross-section geometry, the deformation field will not exhibit symmetry. As an example, in the case of free tips, the displacements in the x direction of the two points $(x_0, 0)$ and $(-x_0, 0)$ will be symmetric with respect to the y-axis, i.e., $u^+ = -u^- = -\widetilde{P}(a_{11} + a_{12})x_0$, but the displacements in the y direction of the same points will be opposite as well, i.e., $v^+ = -v^- = -\frac{1}{2}\widetilde{P}(a_{16} + a_{26})x_0$, which results in a deformation as shown in Fig. 3.11(b). In the special case where a MON13z has been created by rotating an orthotropic material about the z-axis by an angle θ_z as shown in Fig. 3.11(c) (see also S.2.3, S.2.4), **P.2.7** shows that

$$a_{16} + a_{26} = 2\frac{E_{11} - E_{22}}{E_{11}E_{22}}\cos\theta_z \sin\theta_z, \tag{3.83a}$$

$$a_{16} + a_{26} + a_{36} = 2\frac{E_{11} - E_{22} + \nu_{13}E_{11} - \nu_{23}E_{22}}{E_{11}E_{22}}\cos\theta_z \sin\theta_z. \tag{3.83b}$$

Hence, by setting the system of coordinates to coincide with the orthotropic material principal directions (of anisotropy) as shown in Fig. 3.11(c), $\theta_z = 0$ should be taken, and the underlined terms in (3.81a), (3.82a,b) vanish. The deformation of cross-sections in such a case will be radial only over the axes (i.e. along the $x = 0$, $y = 0$ lines).

Example 3.6 *Rectangular Domain Under Low-Order body-force and Surface Loads.*
To illustrate a few elements of the above derivation, consider a rectangular domain (of dimensions $2a$, d), which is acted upon by a given uniform body force X_b, see Fig. 3.12. The stress solution is derived under the assumption $Y_s = \overline{U}_2 = \sigma_y = 0$ and $\overline{U}_1 = \sigma_x = \widetilde{A}x + \widetilde{B}$, where $\widetilde{A}$ and $\widetilde{B}$ are arbitrary constants. Equations (3.55) show that $Y_b = 0$, and $\widetilde{A} = -X_b$. In addition, according to (3.57), $F_0 = 0$, and subsequently the particular solution $\Phi_p = 0$ as well. To satisfy the boundary conditions (3.62) on the edges $x = -a$ (where $\frac{dy}{ds} = -1$) and $x = a$ (where $\frac{dy}{ds} = 1$), one should set $X_s{}^+ = \widetilde{A}a + \widetilde{B}$, $X_s{}^- = \widetilde{A}a - \widetilde{B}$. Thus, (3.65) show that both F_1, and F_2 vanish along all edges as well, and therefore $\Phi = \Phi_0 = 0$. This solution includes the case of symmetric loading where $\widetilde{B} = 0$ and $X_s{}^+ = X_s{}^- = \widetilde{A}a$ and $\sigma_x = \widetilde{A}x$, and the case of $\widetilde{B} = \widetilde{A}a$ where $X_s{}^+ = 2\widetilde{A}a$, $X_s{}^- = 0$ and $\sigma_x = \widetilde{A}(x + a)$. In both cases the body-force resultant $(-2\widetilde{A}ad)$ is cancelled out by the surface load resultant.

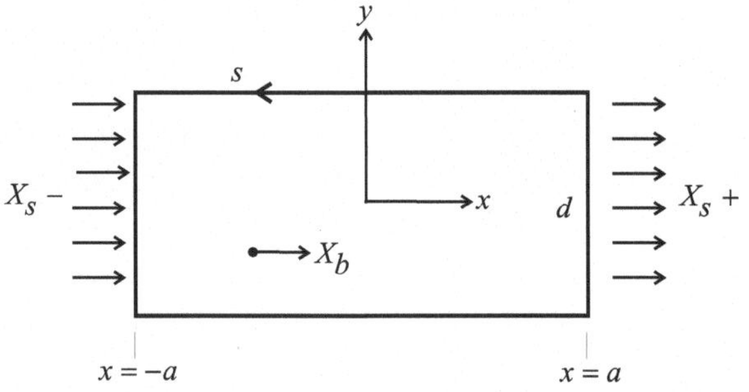

Figure 3.12: A rectangle subjected to a constant body force in the x-direction.

Example 3.7 *Single-Valued Stress Function with Single-Valued Derivatives.*
Consider, via polar coordinates (r, θ) notation, an isotropic circular ring with radii $R_2 > R_1 > 0$ loaded by normal stress $\sigma_{rr} = \frac{1}{R_1^2}$ on the inner boundary and $\sigma_{rr} = \frac{1}{R_2^2}$ on the outer boundary (i.e. the stresses are not functions of θ). The well-known solution in this case is $\Phi = \ln r$, and

it yields the stresses $\sigma_{rr} = -\sigma_{\theta\theta} = \frac{1}{r^2}$ and $\sigma_{r\theta} = 0$ or in Cartesian coordinates $\sigma_x = -\sigma_y = \frac{\cos(2\theta)}{r^2}$, $\tau_{xy} = \frac{\sin(2\theta)}{r^2}$, see **P.1.11**. The reader may verify that all single-value conditions of (3.75) are satisfied for each of the two boundary components of the circular ring.

Example 3.8 *Non Single-Valued Stress Function with Single-Valued Derivatives.*
Consider the ring of Example 3.7 loaded by tangential shear stress $\tau_{r\theta} = \frac{1}{R_1^2}$ over the inner boundary and $\tau_{r\theta} = \frac{1}{R_2^2}$ over the outer boundary. In polar coordinates, the solution of this simple problem is $\Phi = \theta = \arctan(y/x)$, which satisfies the biharmonic equation for isotropic materials, $\nabla^{(4)}\Phi = 0$ (see (3.213)), and all boundary conditions. Note that derivatives, $\Phi_{,x} = \frac{y}{x^2+y^2}$, $\Phi_{,y} = \frac{x}{x^2+y^2}$, are single-valued in the ring.

This solution yields the stresses $\sigma_{rr} = \sigma_{\theta\theta} = 0$, $\tau_{r\theta} = \frac{1}{r^2}$, see Fig. 3.13(a) for $R_1 = 1, R_2 = 2$. For Cartesian coordinates, the above is translated into $\sigma_x = -\sigma_y = -\frac{\sin(2\theta)}{r^2}$, $\tau_{xy} = \frac{\cos(2\theta)}{r^2}$. For the inner boundary $\cos(\bar{\mathbf{n}},x) = -\cos\theta$ and $\cos(\bar{\mathbf{n}},y) = -\sin\theta$, and therefore (3.61) shows that $X_s = \frac{\sin\theta}{R_1^2}$, $Y_s = -\frac{\cos\theta}{R_1^2}$. For the outer boundary $\cos(\bar{\mathbf{n}},x) = \cos\theta$, $\cos(\bar{\mathbf{n}},y) = \sin\theta$, and therefore (3.61) shows that $X_s = -\frac{\sin\theta}{R_2^2}$, $Y_s = \frac{\cos\theta}{R_2^2}$. Hence, (3.73a,b) are satisfied for each boundary component, namely,

$$\oint_{\partial\Omega_i} \{X_s, Y_s\} = \{0, 0\}, \qquad i = 1,2. \tag{3.84}$$

However, (3.74) is not satisfied for each boundary component separately, but for the entire domain only as required by (3.76), since

$$\oint_{\partial\Omega_1} (Y_s x - X_s y) = 2\pi, \qquad \oint_{\partial\Omega_2} (Y_s x - X_s y) = -2\pi, \tag{3.85}$$

which indicates again that Φ is not a single-valued function.

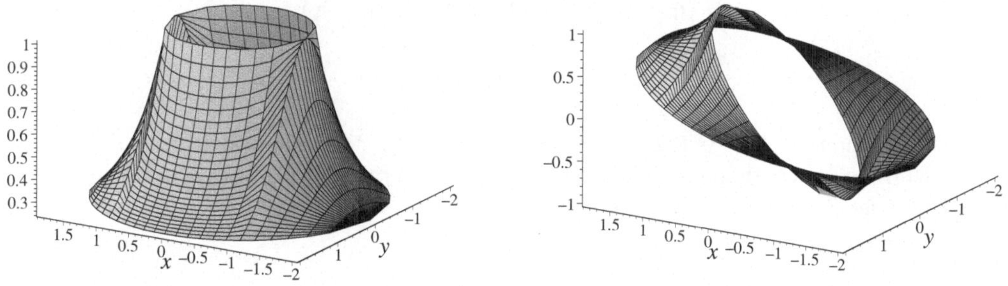

(a) $\sigma_{r\theta}$ of Example 3.8. Φ is not single-valued but its derivatives are.

(b) σ_{rr} of Example 3.9. Both Φ and its derivatives are not single-valued.

Figure 3.13: Examples of non single-valued stress functions.

Example 3.9 *Non Single-Valued Stress Function with non Single-Valued Derivatives.*
In this example, we re-employ the circular ring of Example 3.7 while it is loaded by a normal force $\sigma_{rr} = (\widetilde{c}_1 \cos\theta + \widetilde{c}_2 \sin\theta)/R_1$ on the inner boundary and a normal force $\sigma_{rr} = (\widetilde{c}_1 \cos\theta + \widetilde{c}_2 \sin\theta)/R_2$ on the outer boundary, where $\widetilde{c}_1$ and $\widetilde{c}_2$ are constants. The solution of

this problem is $\Phi = \frac{1}{2} r\theta(\widetilde{c}_1 \sin\theta - \widetilde{c}_2 \cos\theta)$. It yields the stresses $\sigma_{rr} = (\widetilde{c}_1 \cos\theta + \widetilde{c}_2 \sin\theta)/r$, $\sigma_{\theta\theta} = \tau_{r\theta} = 0$, see Fig. 3.13(b) (for $\widetilde{c}_1 = 1$, $\widetilde{c}_2 = 0$, $R_1 = 1$, $R_2 = 2$), or in Cartesian coordinates

$$\sigma_x = \frac{1}{r}(\widetilde{c}_1 \cos\theta + \widetilde{c}_2 \sin\theta)\cos^2\theta, \tag{3.86a}$$

$$\sigma_y = \frac{1}{r}(\widetilde{c}_1 \cos\theta + \widetilde{c}_2 \sin\theta)\sin^2\theta, \tag{3.86b}$$

$$\tau_{xy} = \frac{1}{r}(\widetilde{c}_1 \cos\theta + \widetilde{c}_2 \sin\theta)\cos\theta\sin\theta. \tag{3.86c}$$

It is easy to show that (3.73a,b) become

$$\oint_{\partial\Omega_1} \{X_s, Y_s\} = -\pi\{\widetilde{c}_1, \widetilde{c}_2\}, \qquad \oint_{\partial\Omega_2} \{X_s, Y_s\} = \pi\{\widetilde{c}_1, \widetilde{c}_2\}, \tag{3.87}$$

and thus Φ's derivatives are not single-valued. In addition, despite the fact that the condition of (3.74) is satisfied for each contour, namely, $\oint_{\partial\Omega_i} (Y_s x - X_s y) = 0$, $i = 1,2$, Φ is not single-valued on each boundary component since as shown by (3.71), for Φ to be single-valued, *both* (3.73a,b) *and* (3.74) should be satisfied. Yet, all conditions of (3.76) are satisfied in this case.

3.2.6 Plane Stress/Strain Analysis in a Non-Homogeneous Domain

Let Φ be an unknown stress function on a non-homogeneous plane domain, Ω, so that $\Phi^{[j]} = \Phi_{|\Omega_{[j]}}$ are C^4-continuous functions. We employ the basic relations of (3.54) for the stress components and require that the in-plane resultants, $F_1 = -\frac{d}{ds}\Phi_{,x}$, $F_2 = \frac{d}{ds}\Phi_{,y}$, and the displacements u, v will remain continuous over dividing surfaces. Subsequently, using the notation of (3.1), we formulate the biharmonic BVP for plane-stress/strain in a non-homogeneous domain, see Fig. 3.1(b,c), as

$$\nabla_n^{(4)}\Phi = F_0 \quad \text{over} \quad \Omega, \qquad n \in \{1,3\}, \tag{3.88a}$$

$$\frac{d}{ds}\{\Phi_{,x}, \Phi_{,y}\} = \{-F_1, F_2\} \quad \text{on} \quad \partial\Omega, \tag{3.88b}$$

$$\{\frac{d}{ds}\Phi_{,x}, \frac{d}{ds}\Phi_{,y}, u, v\}_{[i]}^{[j]} = \{0,0,0,0\} \quad \text{on} \quad \partial\Omega_{ij} \tag{3.88c}$$

while the functions F_i $(i = 0,1,2)$, are taken from (3.57), (3.65). Note that for $n = 1$, F_0 is written with the b_{ij} coefficients, while for $n = 3$, F_0 is written with the a_{ij} coefficients. For more details see S.8.1.3.

3.3 Plane-Shear

This section is focused on a two-dimensional problem, which differs from the problems discussed in S.3.2. As done previously, the x,y-plane contains the domain Ω, however, we assume that the only stress components considered are $\tau_{yz}(x,y)$ and $\tau_{xz}(x,y)$, see Fig. 3.14(a), and all other stress components are assumed to be known or to vanish. This type of analysis is classified as a "plane-shear" problem, since in MON13z material it includes the strain components $\gamma_{yz}(x,y)$ and $\gamma_{xz}(x,y)$ only. In other words, the stresses $\tau_{yz}(x,y)$ and $\tau_{xz}(x,y)$ and the strains $\gamma_{yz}(x,y)$ and $\gamma_{xz}(x,y)$ do not depend on the other stress or strain components. The deformation consists mainly of the axial component $w(x,y)$, with typically small participation of $u(x,y)$ and $v(x,y)$.

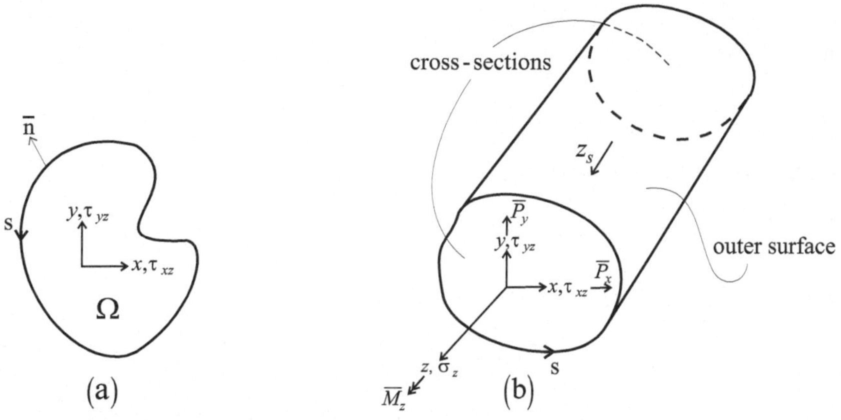

Figure 3.14: Plane-shear problem notation: (a) Two-dimensional domain. (b) A prismatic body.

The above kind of two-dimensional loading is useful in slender bodies analysis, when the loading over the end cross-sections consists of the $\tau_{yz}(x,y)$ and $\tau_{xz}(x,y)$ components only, in addition to the outer ("axial") surface loading component, $Z_s(s)$, see Fig. 3.14(b), e.g. torsion problems discussed further on within Chapters 4 and S.7.2.1.

We will separately consider two different analyses of the same problem — by stress function, and by warping function (both functions are conjugate in a certain sense).

3.3.1 *Analysis by Stress Function*

To carry out the analysis we examine the differential equilibrium equation in the z-direction only (as the remaining two equations are satisfied identically). To extend the scope and applicability of the present analysis, we shall write this equation as

$$\tau_{yz,y} + \tau_{xz,x} = Z(x,y), \tag{3.89}$$

where Z is a function that may include a generic body-force distribution, i.e., $Z = -Z_b$, or any other given function. The boundary condition for this case, see (1.85c) with $\cos(\bar{\mathbf{n}},z) = 0$, is

$$\tau_{xz}\cos(\bar{\mathbf{n}},x) + \tau_{yz}\cos(\bar{\mathbf{n}},y) = Z_s. \tag{3.90}$$

For reasons that will become clearer later on, we divide the compatibility equations (1.45a–f) into two sets: Set "A" that have a direct relation to the present two-dimensional analysis includes the compatibility equations (1.45d, e). Set "B" consists of the remaining four compatibility equations, (1.45a–c, f).

Suppose now that an auxiliary *stress solution* $\tau_{yz}^p(x,y)$ and $\tau_{xz}^p(x,y)$ of (3.89) with $Z = 0$, together with other components τ_{xy}^p, σ_x^p, σ_y^p, σ_z^p, has been found. The notion "stress solution" means that its components may not be legitimate ones for the theory of elasticity, since the corresponding strains $\varepsilon_i^p(x,y)$ $(i = 1,\ldots,6)$ do not necessarily satisfy the compatibility equations. Yet, the above stress solution has to satisfy the first two equilibrium equations, (1.82a,b) (with X_b, Y_b), the compatibility equations in set B, and the other two boundary conditions, see (1.85a,b),

$$\sigma_x^p\cos(\bar{\mathbf{n}},x) + \tau_{xy}^p\cos(\bar{\mathbf{n}},y) = X_s, \qquad \tau_{xy}^p\cos(\bar{\mathbf{n}},x) + \sigma_y^p\cos(\bar{\mathbf{n}},y) = Y_s. \tag{3.91}$$

When no stress solution is available, the following derivation remains valid with $\sigma_i^p = 0$ ($i = 1,\ldots,6$) as well. We now introduce the stress function $\psi(x,y)$ by assuming that

$$\tau_{yz} = \underline{-\psi_{,x}} + \tau_{yz}^p + \overline{U}_4, \qquad \tau_{xz} = \underline{\psi_{,y}} + \tau_{xz}^p + \overline{U}_3, \tag{3.92}$$

while the remaining stress components are

$$\tau_{xy} = \tau_{xy}^p, \qquad \sigma_x = \sigma_x^p, \qquad \sigma_y = \sigma_y^p, \qquad \sigma_z = \sigma_z^p. \tag{3.93}$$

In the above, $\overline{U}_3$, $\overline{U}_4$ are the body-force potentials, namely,

$$\overline{U}_{3,x} + \overline{U}_{4,y} = Z. \tag{3.94}$$

Note that (3.92) identically satisfy (3.89). Our task is therefore to determine a BVP for $\psi(x,y)$ that will make τ_{yz} and τ_{xz}, together with other stress components, a valid solution.

For the MON13z material (2.5) under discussion, from (1.82b) follow

$$\gamma_{yz} = \underline{-a_{44}\psi_{,x} + a_{45}\psi_{,y}} + a_{44}(\tau_{yz}^p + \overline{U}_4) + a_{45}(\tau_{xz}^p + \overline{U}_3), \tag{3.95a}$$

$$\gamma_{xz} = \underline{-a_{45}\psi_{,x} + a_{55}\psi_{,y}} + a_{45}(\tau_{yz}^p + \overline{U}_4) + a_{55}(\tau_{xz}^p + \overline{U}_3), \tag{3.95b}$$

while the other strain components are given as $\gamma_{xy} = \gamma_{xy}^p$, $\varepsilon_x = \varepsilon_x^p$, $\varepsilon_y = \varepsilon_y^p$, $\varepsilon_z = \varepsilon_z^p$.

Since according to (1.52) with $i = j = 3$ and f_{33} from (1.53c), $\gamma_{xz,y} - \gamma_{yz,x} = -2\,\omega_{z,z}$, where ω_z is the rotation about the z-axis, one may use (3.95a,b) and write

$$\nabla_3^{(2)}\psi = [a_{44}(\tau_{yz}^p + \overline{U}_4) + a_{45}(\tau_{xz}^p + \overline{U}_3)]_{,x} - [a_{45}(\tau_{yz}^p + \overline{U}_4) + a_{55}(\tau_{xz}^p + \overline{U}_3)]_{,y} - 2\,\omega_{z,z}. \tag{3.96}$$

To remain within a two-dimensional analysis, ω_z should be (at most) a linear function of z. Equation (1.9a) constitutes the governing equation for the stress function, ψ. The operator $\nabla_3^{(2)}$ is referred to as the *generalized Laplace's operator* (see also Example 1.4) and is given as

$$\nabla_3^{(2)} = a_{44}\frac{\partial^2}{\partial x^2} - 2a_{45}\frac{\partial^2}{\partial x \partial y} + a_{55}\frac{\partial^2}{\partial y^2}. \tag{3.97}$$

For orthotropic materials, $a_{45} = 0$, and the r.h.s. of (3.97) is simplified to another version of the Laplace's operator, given in (3.196). For isotropic materials, $a_{44} = a_{55} = \frac{2(1+\nu)}{E}$, and the operator of (3.97) becomes $\frac{2(1+\nu)}{E}\nabla^{(2)}$, i.e. proportional to the simplest version of the Laplace's operator $\nabla^{(2)} = \frac{\partial^2}{\partial x^2} + \frac{\partial^2}{\partial y^2}$ (see also Remark 3.5). Note that for MON13z material $a_{34} = a_{35} = 0$, and one may replace a_{44}, a_{45} and a_{55} with b_{44}, b_{45} and b_{55}, respectively, see (3.28) and S.3.6.1.

To determine $\omega_{z,z}$ of (3.96), we now write the compatibility equations of Set A as

$$-2\left(\omega_{z,z}\right)_{,x} = \left(\gamma_{xz,y} - \gamma_{yz,x}\right)_{,x} = 2\varepsilon_{x,yz}^p - \gamma_{xy,xz}^p, \tag{3.98a}$$

$$-2\left(\omega_{z,z}\right)_{,y} = \left(\gamma_{xz,y} - \gamma_{yz,x}\right)_{,y} = -2\varepsilon_{y,xz}^p + \gamma_{xy,yz}^p, \tag{3.98b}$$

which may be viewed as a set of two partial differential equations that determine $\omega_{z,z}$ of x, y. Suitable solutions for this system should be adapted for each specific case.

To formulate the boundary condition for (3.96) we use the surface loading $Z_s = Z_s^x \cos(\bar{\mathbf{n}},x) + Z_s^y \cos(\bar{\mathbf{n}},y)$ and employ the identity (see (3.4))

$$\frac{d}{ds}\psi = \psi_{,y}\cos(\bar{\mathbf{n}},x) - \psi_{,x}\cos(\bar{\mathbf{n}},y). \tag{3.99}$$

This enables us to write the boundary conditions for stress function ψ as

$$\frac{d}{ds}\psi = \left(Z_s^x - \tau_{xz}^p - \overline{U}_3\right)\cos(\bar{\mathbf{n}},x) + \left(Z_s^y - \tau_{yz}^p - \overline{U}_4\right)\cos(\bar{\mathbf{n}},y) \qquad \text{on} \quad \partial\Omega. \tag{3.100}$$

Therefore, (3.96), (3.100) constitute the Dirichlet BVP. The single-valued requirement for ψ is written as

$$\oint_{\partial\Omega} \left(Z_s^x - \tau_{xz}^p - \overline{U}_3\right)\cos(\bar{\mathbf{n}},x) + \left(Z_s^y - \tau_{yz}^p - \overline{U}_4\right)\cos(\bar{\mathbf{n}},y) = 0. \tag{3.101}$$

Hence, ψ is given on $\partial\Omega$ by $\psi(s) = \psi_0 + \int_0^s F_3\, ds$, where ψ_0 is an arbitrary constant. The resultant loads over Ω may be determined by the following integration scheme, Fig. 3.14(b):

$$\{\overline{P}_x, \overline{P}_y, \overline{M}_z\} = \iint_\Omega \{\tau_{xz}, \tau_{yz}, x\tau_{yz} - y\tau_{xz}\}, \tag{3.102}$$

in which (3.92) should be substituted. In some circumstances, the Green's Theorem version of (3.7) may be helpful in evaluating the integrals of (3.102).

For the simplest assumption $2\varepsilon_{x,yz}^p - \gamma_{xy,xz}^p = \widetilde{h}_1$ and $-2\varepsilon_{y,xz}^p + \gamma_{xy,yz}^p = \widetilde{h}_2$, where $\widetilde{h}_1$, $\widetilde{h}_2$ are constants, one may find from (3.98a, b)

$$-2\,\omega_{z,z} = \widetilde{h}_1 x + \widetilde{h}_2 y - 2\theta, \tag{3.103}$$

where the *twist* angle (per unit length)

$$\theta = \omega_{z0,z}, \tag{3.104}$$

may be interpreted as the rotation of the origin $x = y = 0$ about the z-axis per unit length (note that as already indicated, $\omega_{z0,z}$ is independent of z). The actual value of θ is out of the scope of the present two-dimensional discussion, see Remark 3.6.

Remark 3.5 Note that any arbitrary isotropic biharmonic function $\Phi(x,y)$ may be expressed in terms of three harmonic functions (λ_i, $i = 1,3$), as

$$\Phi(x,y) = x\lambda_1(x,y) + y\lambda_2(x,y) + \underline{\lambda_3(x,y)}. \tag{3.105}$$

The validity of λ_3 is trivial in view of $\nabla^{(4)} = \nabla^{(2)} \cdot \nabla^{(2)}$, while for λ_1, λ_2, it is easy to see that

$$\frac{\partial^4\Phi}{\partial x^4} + 2\frac{\partial^4\Phi}{\partial x^2\partial y^2} + \frac{\partial^4\Phi}{\partial y^4} = \tag{3.106}$$
$$x\,[(\nabla^{(2)}\lambda_1)_{xx} + (\nabla^{(2)}\lambda_1)_{yy}] + y\,[(\nabla^{(2)}\lambda_2)_{xx} + (\nabla^{(2)}\lambda_2)_{yy}] + 4[(\nabla^{(2)}\lambda_1)_x + (\nabla^{(2)}\lambda_2)_y].$$

Remark 3.6 The *torsion problem*, see e.g. S.7.2.1.2, is a special case of the present two-dimensional analysis, when $Z_s = Z_b = 0$, no body-force loads or auxiliary solution are included (i.e., $\sigma_{xx}^p = \sigma_{yy}^p = \sigma_{zz}^p = \tau_{xy}^p = \tau_{xz}^p = \tau_{yz}^p = \overline{U}_3 = \overline{U}_4 = \widetilde{h}_1 = \widetilde{h}_2 = 0$). So, (3.96), (3.100) become

$$\nabla_3^{(2)}\psi = -2\theta \quad \text{over} \quad \Omega, \qquad \frac{d}{ds}\psi = 0 \quad \text{on} \quad \partial\Omega. \tag{3.107}$$

For a simply connected cross-section, ψ is constant over the contour, say ψ_0, while for convenience, $\psi_0 = 0$ may always be selected.

The integration results of Remark 3.2 with $\Lambda = \psi$ show that the resultant loads are

$$\{\overline{P}_x, \overline{P}_y, \overline{M}_z\} = \{0, 0, -2S_\Omega\psi_0 + 2\iint_\Omega \psi\}, \tag{3.108}$$

which also clarifies the fact that the parts of torsional moment contributed by τ_{yz} and τ_{xz} are equal. The BVP under discussion may be derived for $\theta = 1$ and considering the $\overline{M}_z$ moment as the "resultant moment per unit twist", which essentially defines the torsional rigidity.

Example 3.10 *Isotropic Beam Under Tip Bending.*

This example presents a stress solution that satisfies the equilibrium equations and the compatibility equations of Set B (but not those of Set A).

Consider an isotropic "clamped-free" beam of a rectangular cross-section (that occupies the domain $|x| < \widetilde{a}/2$, $|y| < \widetilde{b}/2$, and has a cross-sectional moment of inertia $I_x = \frac{1}{12}\widetilde{a}\widetilde{b}^3$) subjected to a tip load P_y. The "stress solution" of this case will be

$$\sigma^p_{zz} = -\frac{yP_y}{I_x}(l-z), \quad \tau^p_{yz} = -\frac{P_y}{2I_x}[y^2 - \left(\frac{b}{2}\right)^2], \quad \sigma^p_{xx} = \sigma^p_{yy} = \tau^p_{xy} = \tau^p_{xz} = 0, \tag{3.109}$$

while the corresponding strain components are

$$\varepsilon^p_{zz} = \frac{1}{E}\sigma^p_{zz}, \qquad \varepsilon^p_{xx} = \varepsilon^p_{yy} = -\frac{\nu}{E}\sigma^p_{zz}, \qquad \gamma^p_{yz} = \frac{2(1+\nu)}{E}\tau^p_{yz}, \qquad \gamma^p_{xz} = \gamma^p_{xy} = 0. \tag{3.110}$$

One may verify that all equilibrium equations and the compatibility equations in Set B are identically satisfied. However, those of Set A are violated, and hence, this solution is not "exact". Yet, by employing the analysis in this section, one may determine the $\tau_{yz}(x,y)$ and $\tau_{xz}(x,y)$ stress distributions that will correct the above deficiency by setting

$$Z = -\sigma^p_{zz,z}, \; \widetilde{h}_1 = -\frac{2\nu P_y}{EI_x}, \; a_{44} = a_{55} = \frac{2(1+\nu)}{E}, \quad \widetilde{h}_2 = a_{45} = \overline{U}_3 = \overline{U}_4 = \theta = Z_s = 0. \tag{3.111}$$

Substituting these values in (3.96) and dividing by $\frac{2(1+\nu)}{E}$ leads to the governing equation

$$\nabla^{(2)}\psi = -\frac{\nu}{1+\nu}\cdot\frac{P_y}{I_x}x, \tag{3.112}$$

while (3.100) shows that $\frac{d}{ds}\psi = 0$ on $\partial\Omega$. Hence, by solving this BVP one obtains the desired correction to the above stress solution that will satisfy all requirements.

3.3.2 *Analysis by Warping Function*

In order to repeat the analysis of S.3.3.1 from the warping point of view, (i.e. using *out-of-plane warping*, $w(x,y)$, in the z-direction), we assume that the rotation of the domain about the z-axis does not depend on x, y, i.e., $\omega_z \equiv \omega_{z0}(z)$, and we write the displacements as

$$u = -y\omega_{z0}, \qquad v = x\omega_{z0}, \qquad w = w(x,y). \tag{3.113}$$

For simplicity, rigid body components of u, v are not placed here. This deformation field creates zero-strain components except for

$$\gamma_{xz} = w_{,x} - y\theta, \qquad \gamma_{yz} = w_{,y} + x\theta, \tag{3.114}$$

where the notation (3.104) for θ is used again. The above equations show that in order to remain within a two-dimensional analysis, $\omega_{z0}(z)$ should be (at most) a *linear function of z*, so that the twist, θ, is constant. This set of strains (that satisfies compatibility equation by definition) should now be substituted in the equilibrium equations, for which we first write the corresponding non-vanishing stress components in MON13z material as

$$\tau_{yz} = \frac{a_{55}}{a_0}(w_{,y} + x\theta) - \frac{a_{45}}{a_0}(w_{,x} - y\theta), \quad \tau_{xz} = -\frac{a_{45}}{a_0}(w_{,y} + x\theta) + \frac{a_{44}}{a_0}(w_{,x} - y\theta), \tag{3.115}$$

where $a_0 = a_{44}a_{55} - a_{45}^2$. Similar to S.3.3.1, we may extend the scope of the present derivation by superimposing the present solution with a given auxiliary two-dimensional deformation system that satisfy all compatibility equations (for example, a deformation system that has originated from a continuous set of displacements), the differential equilibrium equation in the x- and y- directions and the boundary conditions, but does not necessarily satisfy the equilibrium equation in the z-direction, (1.82c), and/or the boundary conditions associated with it, (1.85c). All quantities related to the above auxiliary solution will be denoted by a superscript "p". Subsequently, we reach the following Neumann BVP for w:

$$\nabla_3^{(2)} w = F_0^w \quad \text{over} \quad \Omega, \qquad D_1^n w = F_3^w \quad \text{on} \quad \partial\Omega, \tag{3.116}$$

where

$$F_0^w = -Z_b - \left(\sigma_{zz,z}^p + \tau_{yz,y}^p + \tau_{xz,x}^p\right), \tag{3.117a}$$

$$F_3^w = \left[a_0\left(Z_s^x - \tau_{xz}^p\right) + \underline{\theta(a_{45}x + a_{44}y)}\right]\cos(\bar{\mathbf{n}}, x) + \left[a_0\left(Z_s^y - \tau_{yz}^p\right) - \underline{\theta(a_{55}x + a_{45}y)}\right]\cos(\bar{\mathbf{n}}, y). \tag{3.117b}$$

In (3.116) we have used the Laplace's operator (3.97), and introduced the Neumann-type boundary operator, D_1^n, as

$$D_1^n = (a_{44}\frac{\partial}{\partial x} - a_{45}\frac{\partial}{\partial y})\cos(\bar{\mathbf{n}}, x) + (-a_{45}\frac{\partial}{\partial x} + a_{55}\frac{\partial}{\partial y})\cos(\bar{\mathbf{n}}, y). \tag{3.118}$$

For the sake of convenience, the solution of (3.116) is carried out under the integral condition $\iint_\Omega w = 0$ (i.e. zero average out-of-plane deformation) or, alternatively, $w(0,0) = 0$.

As will be shown within S.3.3.3, the existence and uniqueness of solution of the BVP (3.116) requires

$$\oint_{\partial\Omega} F_3^w = \iint_\Omega F_0^w. \tag{3.119}$$

In evaluating the above integrals, the terms containing θ in F_3^w may be neglected by virtue of Green's Theorem, see (3.8). This reduces the existence and uniqueness condition to

$$\oint_{\partial\Omega} Z_s - \oint_{\partial\Omega} \left[\tau_{xz}^p \cos(\bar{\mathbf{n}}, x) + \tau_{yz}^p \cos(\bar{\mathbf{n}}, y)\right] = -\frac{1}{a_0} \iint_\Omega \left[Z_b + \left(\sigma_{zz,z}^p + \tau_{yz,y}^p + \tau_{xz,x}^p\right)\right]. \tag{3.120}$$

The resultant loads may be derived using (3.102) and (3.7) in this case as well.

3.3.3 Generic Dirichlet/Neumann BVPs on a Homogeneous Domain

This section summarizes some basic properties of the Dirichlet/Neumann BVPs. Note that **P.3.7, P.3.8** illustrate prescribed polynomial solutions of Laplace's equation for MON13z and isotropic materials (i.e. prescribed solution of the field equation without the boundary conditions, analogously to **P.3.5, P.3.6** of Example 3.4).

3.3.3.1 The Neumann BVP

We shall keep the previous notation of the differential operators $\nabla_3^{(2)}$ of (3.97), and the Neumann-type boundary operator, D_1^n, (3.118), while for the moment, a_{ij} will be considered as known coefficients, the physical meaning of which is immaterial.

Let F_0^Λ and F_3^Λ be given functions on the homogeneous domain Ω and its boundary $\partial\Omega$, as shown in Fig. 3.1(a). We also define the function Λ, which is said to be a *generalized (Neumann) harmonic function* if it satisfies the following field equation and the Neumann-type boundary condition:

$$\nabla_3^{(2)} \Lambda = F_0^\Lambda \quad \text{over} \quad \Omega, \qquad D_1^n \Lambda = F_3^\Lambda \quad \text{on} \quad \partial\Omega. \tag{3.121}$$

The solution of the BVP (3.121) is carried out under the condition $\Lambda(0,0) = 0$.

We shall now document a few useful identities for generalized harmonic functions. First, substituting (3.7) in the expression of D_1^n in (3.118) yields

$$\oint_{\partial\Omega} yD_1^n\Lambda = \iint_\Omega [y(a_{44}\Lambda_{,xx} - 2a_{45}\Lambda_{,xy} + a_{55}\Lambda_{,yy}) + (-a_{45}\Lambda_{,x} + a_{55}\Lambda_{,y})], \qquad (3.122)$$

which, with the definition of $\nabla_3^{(2)}$ in (3.97), yields the first of the following equations:

$$\iint_\Omega (-a_{45}\Lambda_{,x} + a_{55}\Lambda_{,y}) = \oint_{\partial\Omega} yD_1^n\Lambda - \iint_\Omega y\nabla_3^{(2)}\Lambda, \qquad (3.123a)$$

$$\iint_\Omega (a_{44}\Lambda_{,x} - a_{45}\Lambda_{,y}) = \oint_{\partial\Omega} xD_1^n\Lambda - \iint_\Omega x\nabla_3^{(2)}\Lambda. \qquad (3.123b)$$

Analogously, the evaluation of $\oint_{\partial\Omega}(xD_1^n\Lambda)$, yields the second equation.

Moreover, for any generalized harmonic function, Λ, see (3.4), (3.118),

$$D_1^n\Lambda = (a_{44}\Lambda_{,x} - a_{45}\Lambda_{,y})\frac{dy}{ds} - (-a_{45}\Lambda_{,x} + a_{55}\Lambda_{,y})\frac{dx}{ds} \quad \text{on} \quad \partial\Omega, \qquad (3.124)$$

which, by virtue of Green's Theorem, becomes

$$\begin{aligned}\oint_{\partial\Omega} D_1^n\Lambda &= \oint_{\partial\Omega} [(a_{44}\Lambda_{,x} - a_{45}\Lambda_{,y})\,dy - (-a_{45}\Lambda_{,x} + a_{55}\Lambda_{,y})\,dx] \\ &= \iint_\Omega (a_{44}\Lambda_{,xx} - 2a_{45}\Lambda_{,xy} + a_{55}\Lambda_{,yy}) = \iint_\Omega \nabla_3^{(2)}\Lambda, \end{aligned} \qquad (3.125)$$

or in terms of the given functions in (3.121)

$$\oint_{\partial\Omega} F_3^\Lambda = \iint_\Omega F_0^\Lambda. \qquad (3.126)$$

Equation (3.126) is further used and denoted as *the solution existence condition* of the Neumann BVP, (3.121), in a simply connected domain. It is also customary to employ the notation $F_3^\Lambda = P^\Lambda \cos(\bar{\mathbf{n}},x) + Q^\Lambda \cos(\bar{\mathbf{n}},y)$ which yields

$$\iint_\Omega (-a_{45}\Lambda_{,x} + a_{55}\Lambda_{,y}) = \iint_\Omega [Q^\Lambda + y(P^\Lambda_{,x} + Q^\Lambda_{,y} - F_0^\Lambda)], \qquad (3.127a)$$

$$\iint_\Omega (a_{44}\Lambda_{,x} - a_{45}\Lambda_{,y}) = \iint_\Omega [P^\Lambda + x(P^\Lambda_{,x} + Q^\Lambda_{,y} - F_0^\Lambda)]. \qquad (3.127b)$$

3.3.3.2 The Dirichlet BVP

We shall present here the Dirichlet BVP

$$\nabla_3^{(2)}\Lambda = F_0^\Lambda \quad \text{over} \quad \Omega, \qquad \frac{d}{ds}\Lambda = F_3 \quad \text{on} \quad \partial\Omega. \qquad (3.128)$$

The following condition is necessary for Λ to be single-valued, and thus for the existence and uniqueness of the BVP (3.128) in a homogeneous simply connected domain:

$$\oint_{\partial\Omega} F_3 = 0. \qquad (3.129)$$

Remark 3.7 For the Dirichlet/Neumann BVPs in multiply connected homogeneous domain, Ω, with boundary $\partial\Omega = \bigcup_j \partial\Omega_j$, conditions (3.126), (3.129) are transformed using the orientation of the boundary components shown in Fig. 3.15, to the more complicated form

$$\sum_j \oint_{\partial\Omega_j} F_3^{\Lambda(j)} = \iint_\Omega F_0^\Lambda, \qquad \sum_j \oint_{\partial\Omega_j} F_3^{(j)} = 0. \qquad (3.130)$$

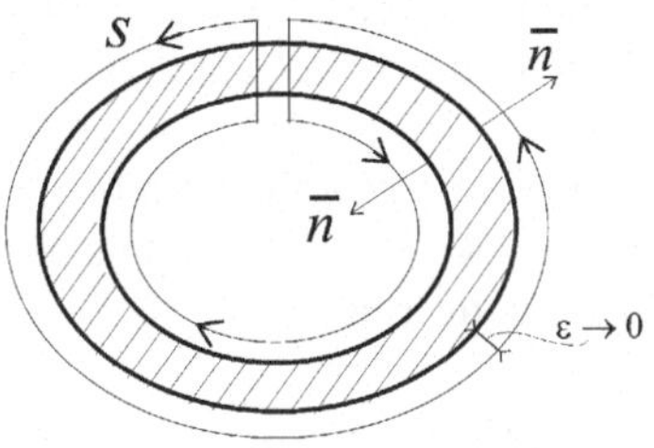

Figure 3.15: A multiply connected domain boundary.

3.3.4 Simplification of Generalized Laplace's and Boundary Operators

Employing the affine transformation

$$x' = \widetilde{a}x, \qquad y' = y + \widetilde{b}x, \tag{3.131}$$

where $\widetilde{a}$ and $\widetilde{b}$ are nonzero constants, the domain contour in the x, y-plane is transformed into a different contour in the $x'y'$-plane as schematically shown in Fig. 3.16.

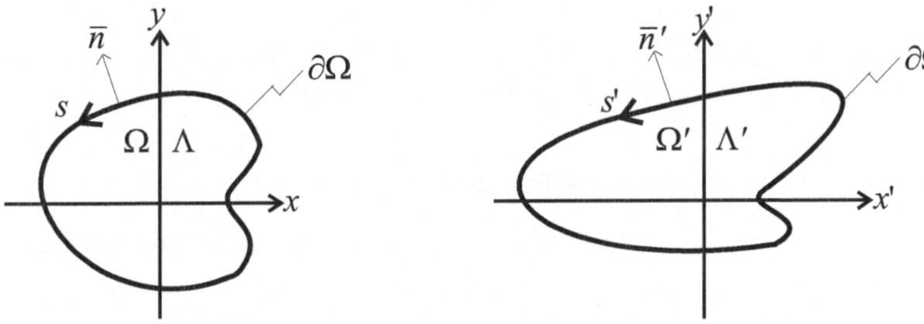

Figure 3.16: The "original" x, y-plane and the "transformed" $x'y'$-plane.

We parameterize the contour by two functions $x(t)$ and $y(t)$, and therefore the transformed parametrization functions $x'(t) = \widetilde{a}x(t)$ and $y'(t) = y(t) + \widetilde{b}x(t)$ of S.3.1.2, while a specific value of t relates the same point on the contours in both planes.

Suppose that a generalized harmonic function, $\Lambda(x,y)$, is defined on Ω. Using the inverse transformation $x = \frac{1}{\widetilde{a}}x'$, $y = y' - \frac{\widetilde{b}}{\widetilde{a}}x'$, one may present the following derivatives of $\Lambda(x,y)$:

$$\Lambda_{,x} = \widetilde{a}\Lambda_{,x'} + \widetilde{b}\Lambda_{,y'}, \qquad \Lambda_{,y} = \Lambda_{,y'}, \tag{3.132a}$$

$$\Lambda_{,xx} = \widetilde{a}^2\Lambda_{,x'x'} + 2\widetilde{a}\widetilde{b}\Lambda_{,x'y'} + \widetilde{b}^2\Lambda_{,y'y'}, \quad \Lambda_{,xy} = \widetilde{a}\Lambda_{,x'y'} + \widetilde{b}\Lambda_{,y'y'}, \quad \Lambda_{,yy} = \Lambda_{,y'y'}. \tag{3.132b}$$

One may reduce the operation $\nabla_3^{(2)}$ (see (3.97)) to a simpler one using derivatives with respect to x' and y', by requiring that the coefficients of $\frac{\partial^2}{(\partial x')^2}$ and $\frac{\partial^2}{(\partial y')^2}$ will be equal while the coefficient of $\frac{\partial^2}{\partial x' \partial y'}$ vanishes. This procedure yields

$$\widetilde{a} = \frac{1}{a_{44}}\sqrt{a_{44}a_{55} - a_{45}^2}, \qquad \widetilde{b} = \frac{a_{45}}{a_{44}}. \tag{3.133}$$

Therefore, the generalized Poisson's equation (3.128 :a) may be simplified to

$$\Lambda_{,x'x'} + \Lambda_{,y'y'} = \frac{1}{a_{44}\widetilde{a}^2}F_0^{\Lambda}(x',y') \qquad \text{on} \quad \partial\Omega'. \tag{3.134}$$

Hence, by applying the transformation (3.131) to suitable boundary conditions, one may reduce the Dirichlet/Neumann BVPs to ones of the same type in Ω' with the classical Laplace's operator.

The boundary condition $\Lambda_{|\partial\Omega} = G_3(x,y)$ (of the Dirichlet BVP) is obviously transformed to $\Lambda_{|\partial\Omega'} = G_3(x',y')$. By substituting (3.4) in (3.118) and using (3.132a) the boundary condition (3.121:b) (of the Neumann BVP) is converted into

$$\frac{d}{dn}\Lambda = \frac{1}{a_{44}\tilde{\tilde{a}}}[P(x',y')\cos(\bar{\mathbf{n}}',x') + \frac{1}{\tilde{\tilde{a}}}(\tilde{b}P(x',y') + Q(x',y'))\cos(\bar{\mathbf{n}}',y')], \tag{3.135}$$

where $\frac{d}{dn}$ is the *(geometrical) normal derivative*, see (3.219), and $\cos(\bar{\mathbf{n}}',x')$, $\cos(\bar{\mathbf{n}}',y')$ are the cosines of the angles between the normal $\bar{\mathbf{n}}$ to $\partial\Omega'$ and the respective axes. Hence, one may solve various Dirichlet/Neumann BVPs in the transformed plane as "equivalent isotropic cases", see Examples in Remarks 7.2, 7.4. **P.3.9** presents affine transformations for given $\tilde{a}$ and $\tilde{b}$.

3.3.5 *Plane-Shear Analysis of Non-Homogeneous Domain*

We shall extend here the governing equations for Plane-Shear analysis in non-homogeneous domain by considering the case where no auxiliary solutions exist, no surface loading (Z_s) is applied, and no body forces are involved, see notation in S.3.1.1.

3.3.5.1 The Dirichlet BVP for the Stress Function

The Dirichlet BVP for plane-shear in non-homogeneous domain Ω becomes (see also S.8.1)

$$\nabla_3^{(2)}\psi = F_0^{\psi} \quad \text{over} \quad \Omega, \tag{3.136a}$$

$$\frac{d}{ds}\psi = 0, \quad \text{on} \quad \partial\Omega, \tag{3.136b}$$

$$\{\frac{d}{ds}\psi,\, w\}_{[i]}^{[j]} = \{0,0\} \quad \text{on} \quad \partial\Omega_{ij}. \tag{3.136c}$$

In this case, $\psi^{[j]} = \psi_{|\Omega_{[j]}}$ are C^2-continuous functions inside each $\Omega_{[j]}$. Equations (3.136c) stand for force and displacement continuity (both in the z-direction). See also Remark 8.1.

3.3.5.2 The Neumann BVP for the Warping Function

The warping function w in non-homogeneous domain Ω is derived under the assumption that $w^{[j]} = w_{|\Omega_{[j]}}$ are C^2-continuous functions inside each $\Omega_{[j]}$. The analysis of S.3.3.2 yields

$$\nabla_3^{(2)} w = F_0^{w} \quad \text{over} \quad \Omega, \tag{3.137a}$$

$$D_1^n w = \theta[(a_{45}x + a_{44}y)\cos(\bar{\mathbf{n}},x) - (a_{55}x + a_{45}y)\cos(\bar{\mathbf{n}},y)] \quad \text{on} \quad \partial\Omega, \tag{3.137b}$$

$$[D_1^n w]_{[i]}^{[j]} = \theta\,[(a_{45}x + a_{44}y)\cos(\bar{\mathbf{n}},x) - (a_{55}x + a_{45}y)\cos(\bar{\mathbf{n}},y)]_{[i]}^{[j]} \quad \text{on} \quad \partial\Omega_{ij}, \tag{3.137c}$$

$$w_{[i]}^{[j]} = 0 \quad \text{on} \quad \partial\Omega_{ij}. \tag{3.137d}$$

Here θ is constant and a_{44}, a_{45}, a_{55} are piecewise constants in Ω. Equations (3.137c,d) stand for force and displacement continuity (both in the z-direction).

3.4 Coupled-Plane BVP

In this section we shall combine the problems of plane-strain and shear-stress and apply them for a generic anisotropic material. The result will therefore be a two-dimensional analysis where the only assumption is $\varepsilon_z = 0$, see Table 3.1. One may therefore eliminate σ_z as

$$\sigma_z = -\frac{1}{a_{33}}\left(a_{13}\sigma_x + a_{23}\sigma_y + a_{34}\tau_{yz} + a_{53}\tau_{xz} + a_{63}\tau_{xy}\right), \tag{3.138}$$

and express the constitutive relations as

$$\begin{Bmatrix} \varepsilon_x \\ \varepsilon_y \\ \gamma_{yz} \\ \gamma_{xz} \\ \gamma_{xy} \end{Bmatrix} = \begin{bmatrix} b_{11} & b_{12} & b_{14} & b_{15} & b_{16} \\ & b_{22} & b_{24} & b_{25} & b_{26} \\ & & b_{44} & b_{45} & b_{46} \\ & Sym. & & b_{55} & b_{56} \\ & & & & b_{66} \end{bmatrix} \begin{Bmatrix} \sigma_x \\ \sigma_y \\ \tau_{yz} \\ \tau_{xz} \\ \tau_{xy} \end{Bmatrix}, \tag{3.139}$$

where $b_{ij} = a_{ij} - \frac{a_{i3}a_{j3}}{a_{33}}$, $i,j \in \{1,2,4,5,6\}$, are referred to as the *reduced compliance elastic coefficients*, see also (3.28). We further define the stress components σ_x, σ_y, τ_{xy} by (3.54) and τ_{yz}, τ_{xz} by (3.92) with $\tau_{yz}^p = \tau_{xz}^p = 0$. Then, the compatibility equation (1.45a) yields

$$\begin{aligned} \nabla_1^{(4)}\Phi + \nabla_1^{(3)}\psi = &-b_{12}\overline{U}_{1,xx} + b_{16}\overline{U}_{1,xy} - b_{11}\overline{U}_{1,yy} - b_{22}\overline{U}_{2,xx} + b_{26}\overline{U}_{2,xy} - b_{12}\overline{U}_{2,yy} \\ &- b_{14}\overline{U}_{4,yy} - b_{24}\overline{U}_{4,xx} + b_{46}\overline{U}_{4,xy} - b_{15}\overline{U}_{3,yy} - b_{25}\overline{U}_{3,xx} + b_{56}\overline{U}_{3,xy}, \end{aligned} \tag{3.140}$$

where $\nabla_1^{(3)}$ is a third-order operator given by (3.214). In addition, the compatibility equations (1.45d, e) are written similar to (3.96) (with no auxiliary solution) as

$$\begin{aligned} \nabla_1^{(3)}\Phi + \nabla_1^{(2)}\psi = &-2\,\omega_{z,z} - b_{15}\overline{U}_{1,y} - b_{25}\overline{U}_{2,y} - b_{45}\overline{U}_{4,y} + b_{14}\overline{U}_{1,x} \\ &+ b_{24}\overline{U}_{2,x} + b_{44}\overline{U}_{4,x} + b_{45}\overline{U}_{3,x} - b_{55}\overline{U}_{3,y} \end{aligned} \tag{3.141}$$

where $\nabla_1^{(2)}$ is a version of Laplace's operator given by (3.194). The above (3.140), (3.141) constitute the field equations for the *Coupled-Plane problem.* In (Lekhnitskii, 1981), this set of equations is defined as *Generalized Plane-Strain*. Note that in MON13z material, b_{44}, b_{45}, b_{55} are identical to a_{44}, a_{45}, a_{55}, respectively, $\nabla_1^{(3)} = 0$, $\nabla_1^{(2)} = \nabla_3^{(2)}$, and the above coupled formulation is broken back to two uncoupled, Plane-Strain and Plane-Shear, problems.

To present the Coupled-Plane problem in a general form, we let g and f be the r.h.s. of (3.140), (3.141), respectively. In view of the definitions (3.65), (3.100) for the functions F_j ($j = 1,2,3$), one arrives at the following generic Coupled-Plane BVP:

$$\nabla_1^{(4)}\Phi + \nabla_1^{(3)}\psi = g, \qquad \nabla_1^{(3)}\Phi + \nabla_1^{(2)}\psi = f \qquad \text{over} \quad \Omega, \tag{3.142a}$$

$$\frac{d}{ds}\{\Phi_{,x}, \Phi_{,y}, \psi\} = \{-F_1, F_2, F_3\} \qquad \text{on} \quad \partial\Omega. \tag{3.142b}$$

In some cases (see Chapter 6), we shall define F_3 in a way that is broader then its definition in (3.100), using an additional potential, $\overline{U}_3$, as

$$F_3 = Z_s - \overline{U}_3\cos(\bar{\mathbf{n}}, x) - \overline{U}_4\cos(\bar{\mathbf{n}}, y), \tag{3.143}$$

so that $\overline{U}_{3,x} + \overline{U}_{4,y} = -Z_b$.

The single-valued conditions for the stress functions and their derivatives Φ, $\Phi_{,x}$, $\Phi_{,y}$, ψ in a homogeneous simply connected domain are those of (3.72), (3.126), namely,

$$\oint_{\partial\Omega}\{F_1, F_2, F_3, F_2y - F_1x\} = \{0, 0, 0, 0\}. \tag{3.144}$$

P.3.10, P.3.11 illustrate prescribed polynomial solutions of Coupled-Plane equations (3.142a).

For a non-homogeneous domain we put together with (3.142a,b) an additional set of conditions as appear in (3.88c), (3.136c)

$$\{\frac{d}{ds}\Phi_{,x},\ \frac{d}{ds}\Phi_{,y},\ \frac{d}{ds}\psi,\ u,\ v,\ w\}_{[i]}^{[j]} = \{0,0,0,0,0,0\} \quad \text{on} \quad \partial\Omega_{ij}. \tag{3.145}$$

Generic formulation of the Coupled-Plane BVP for a non-homogeneous domain appears in S.8.1.4.

Example 3.11 *Prismatic Generally Anisotropic Body Under Hydrostatic Load.*

Example 3.5 may be extended to the case of generic anisotropic by taking Φ as in (3.78) and $\psi = 0$, which may be proved to be a valid solution of the Coupled-Plane BVP in the case of hydrostatic load. Yet, in the present case, according to the analysis that will be derived within Chapter 5 (see (5.29a–c)) the additional displacements of (3.80) would be

$$\Delta u = \Delta\sigma_z(a_{13}x + \frac{1}{2}a_{36}y + \frac{1}{2}a_{35}z), \quad \Delta v = \Delta\sigma_z(\frac{1}{2}a_{36}x + a_{23}y + \frac{1}{2}a_{34}z),$$
$$\Delta w = \Delta\sigma_z[\frac{1}{2}(xa_{35} + ya_{34}) + za_{33}]. \tag{3.146}$$

Substituting $\Delta\sigma_z = -\widetilde{P}(a_{13} + a_{23})/a_{33}$ in the above and superposition of Δu, Δv, Δw and u, v, w of (3.79) yields the deformation of a prismatic body that is generally anisotropic under hydrostatic load and has free tips. As in Example 3.5, the deformation described above was derived for zero displacements and rotations at $x = y = z = 0$.

3.5 Analysis of Plates

3.5.1 *The Classical Laminated Plate Theory*

Plate structures are very important and common components of both classical and modern applications. The analysis of anisotropic plates is of special interest due to the tremendous potential in weight saving that may be gained by employing modern composite materials to fulfill the role of the traditional isotropic (mainly metal) materials in such structures.

The study of both isotropic and laminated (composite) plates is well established and the literature contains a vast range of analysis methodologies, including numerical and experimental studies with various levels of underlying assumptions. This section reviews the fundamental aspects of a well-known set of simplifying assumptions and equations that are collectively referred to as *"The Classical Laminated Plate Theory" (CLPT)* and were derived to simplify the analysis of a non-homogeneous laminated domain such as the one shown in Fig. 3.17. Without such a set of assumptions, the solution in a non-homogeneous laminated domain becomes complex as shown in Chapter 8. The reader should therefore bear in mind that in what follows, an approximate solution is under discussion.

The primary purpose of this section is to present the CLPT and to facilitate its common reduction into a two-dimensional bending analysis of plates (derived further on within S.3.5.2) and Beam-Plate Models (derived further on within S.5.3.1).

For additional reading see (Lekhnitskii, 1968), (Ambartsumyan, 1970), (Ambartsumyan, 1974), (Brush and Almroth, 1975), (Dudchenko *et al.*, 1984), (Whitney, 1987), (Ochoa and Reddy, 1992), (Ciarlet, 1997), (Vashakmadze, 1999), (Reddy, 1999), (Cheung and Zhou, 1999).

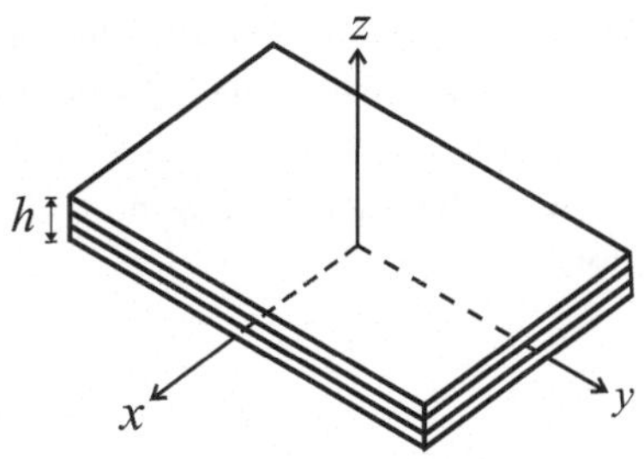

Figure 3.17: Scheme of a thin stack of laminates treated within the CLPT.

3.5.1.1 Underlying Assumptions

Consider an anisotropic plate that occupies a simply connected plane domain Ω, see Fig. 3.18.

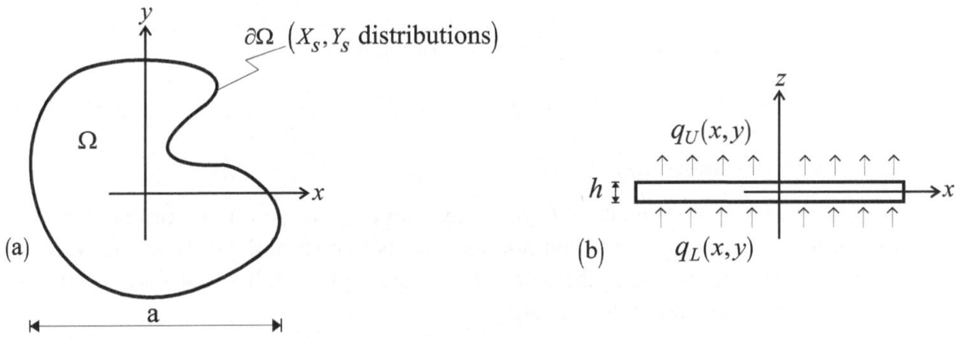

Figure 3.18: Notation for MON13z thin plate.

During the derivation we shall assume:

(a) The plate's middle plane coincides with the x,y-plane, and therefore, the z-axis is perpendicular to this plane.

(b) The plate's thickness is constant and denoted by h, while $h \ll a$, where a is a typical dimension of the Ω, see Fig. 3.18. For laminated plates, the thickness of each layer is constant as well (i.e. not a function of x and y).

(c) The plate is homogeneous, linear elastic and obeys the generalized Hook's law for MON13z materials, see S.2.3 and (2.5). Deformation is small and therefore linear analysis is applicable.

(d) The plate is loaded by a force per unit area $q_L(x,y)$ at its lower surface and $q_U(x,y)$ at its upper surface. Hence, the net vertical (upwards) loading is $q(x,y) = q_L(x,y) + q_U(x,y)$. In addition, an in-plane loading X_s, Y_s is applied and described by the distribution of the stress components σ_x, σ_y and τ_{xy}, along its contour, $\partial\Omega$, see (3.61). Similar to the argument in Remark 3.4, these stress components are constant across the thickness (i.e. equal to their average values over the thickness).

(e) The body forces are neglected.

(f) We shall adopt the well-known *Kirchhoff-Love hypothesis* that a plane that is perpendicular to the mid-plane before deformation will remain plane, undeformed and normal to the mid-plane after deformation.

(g) The normal stress, σ_z, is much smaller than σ_x, σ_y, and τ_{xy}. This assumption is physically demonstrated by Remark 3.8 below.

3.5.1.2 Deformation

To express the deformation for the laminate shown in Fig. 3.17, it is assumed that all layers are of uniform thickness, and are perfectly bonded. Based on the Kirchhoff-Love hypothesis (assumption (f)), the displacement components are expressed in terms of three functions, as

$$u(x,y,z) = u_0(x,y) - zw_{0,x}, \quad v(x,y,z) = v_0(x,y) - zw_{0,y}, \quad w(x,y,z) = w_0(x,y), \tag{3.147}$$

where u_0, v_0 and w_0 are the deformation of the laminate middle plane (i.e., $z = 0$).

Introducing the strain quantities

$$\varepsilon_x^0 = u_{0,x}, \qquad \varepsilon_y^0 = v_{0,y}, \qquad \gamma_{xy}^0 = u_{0,y} + v_{0,x}, \tag{3.148a}$$

$$\kappa_x = -w_{0,xx}, \qquad \kappa_y = -w_{0,yy}, \qquad \kappa_{xy} = -2w_{0,xy}, \tag{3.148b}$$

we arrive at a linear variation of the in-plane strain components

$$\varepsilon_x = \varepsilon_x^0 + z\kappa_x, \qquad \varepsilon_y = \varepsilon_y^0 + z\kappa_y, \qquad \gamma_{xy} = \gamma_{xy}^0 + z\kappa_{xy}. \tag{3.149}$$

3.5.1.3 Laminate Constitutive Relations

For the MON13z material with the $\sigma_z = 0$, one may adopt the assumptions for plane-stress of S.3.2.2, despite the fact that τ_{xz} and τ_{yz} do not necessarily vanish in the present case (as these components are uncoupled with the other). We therefore employ (3.20), (3.41), (3.43) to show that the compliance and stiffness matrices are

$$\begin{Bmatrix} \varepsilon_x \\ \varepsilon_y \\ \gamma_{xy} \end{Bmatrix} = \begin{bmatrix} a_{11} & a_{12} & a_{16} \\ a_{12} & a_{22} & a_{26} \\ a_{16} & a_{26} & a_{66} \end{bmatrix} \cdot \begin{Bmatrix} \sigma_x \\ \sigma_y \\ \tau_{xy} \end{Bmatrix}, \quad \begin{Bmatrix} \sigma_x \\ \sigma_y \\ \tau_{xy} \end{Bmatrix} = \begin{bmatrix} B_{11} & B_{12} & B_{16} \\ B_{12} & B_{22} & B_{26} \\ B_{16} & B_{25} & B_{66} \end{bmatrix} \cdot \begin{Bmatrix} \varepsilon_x \\ \varepsilon_y \\ \gamma_{xy} \end{Bmatrix}. \tag{3.150}$$

One may define resultant loads as shown in Fig. 3.19(a), which are essentially loads per unit length along the laminate edge as

$$[N_x, N_y, N_{xy}, M_x, M_y, M_{xy}, Q_x, Q_y] = \sum_{n=1}^{N} \int_{z_n}^{z_{n+1}} [\sigma_x, \sigma_y, \tau_{xy}, z\sigma_x, z\sigma_y, z\tau_{xy}, \tau_{xz}, \tau_{yz}]\, dz. \tag{3.151}$$

The above is a sum of integrals derived for each laminae, see the scheme in Fig. 3.19(b). Defining the quantities

$$[\mathbf{A}_{ij}, \mathbf{B}_{ij}, \mathbf{D}_{ij}] = \sum_{n=1}^{N} B_{ij}^{[n]} [(z_{n+1} - z_n), \frac{1}{2}(z_{n+1}^2 - z_n^2), \frac{1}{3}(z_{n+1}^3 - z_n^3)], \tag{3.152}$$

where $B_{ij}^{[n]}$ are the coefficients of the $n^{\underline{th}}$ laminae, we arrive at the *laminate constitutive relations*

$$\begin{bmatrix} \mathbf{A} & \mathbf{B} \\ \mathbf{B} & \mathbf{D} \end{bmatrix} \cdot [\varepsilon_x^0, \varepsilon_y^0, \gamma_{xy}^0, \kappa_x, \kappa_y, \kappa_{xy}]^T = [N_x, N_y, N_{xy}, M_x, M_y, M_{xy}]^T. \tag{3.153}$$

Since all layers are of uniform thickness (assumption (b)), the above $\mathbf{A}_{ij}$, $\mathbf{B}_{ij}$ and $\mathbf{D}_{ij}$ coefficients are piecewise constants of z (i.e. not functions of x or y).

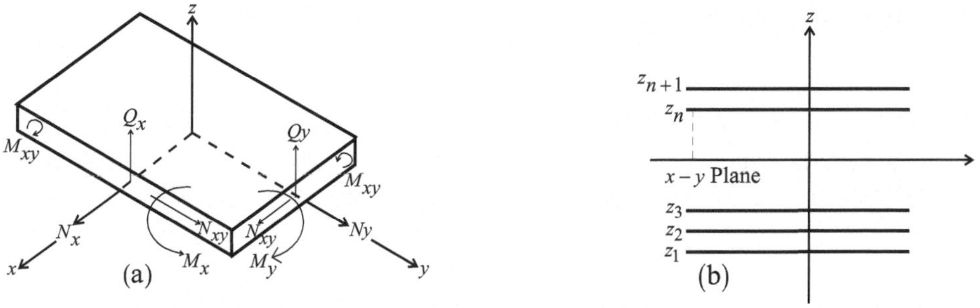

Figure 3.19: General notation for the CLPT.

Note that in some applications, such as homogeneous or "single-layer" laminates, $\mathbf{B}_{ij} = 0$, see (3.152). These $\mathbf{B}_{ij}$ terms are frequently referred to as the "Extensional-Bending" coupling terms as they relate κ_x, κ_y, κ_{xy} with normal (and shear loads), and ε_x^0, ε_y^0 and γ_{xy}^0 with bending (and twist) moments, and see (3.153). These coupling effects will be extensively dealt with in S.5.3.1 and further on in Chapters 6, 9, 10.

Example 3.12 *Strain Energy in a Homogeneous Plate.*
We shall elaborate here on the determination of the strain energy in a homogeneous plate. Substituting (3.150) in (2.27) yields the *volume density of the strain energy* as

$$W = \frac{1}{2}\left(B_{11}\varepsilon_x^2 + B_{22}\varepsilon_y^2 + B_{66}\gamma_{xy}^2 + 2B_{12}\varepsilon_x\varepsilon_y + 2B_{16}\varepsilon_x\gamma_{xy} + 2B_{26}\varepsilon_y\gamma_{xy}\right). \tag{3.154}$$

Substituting (3.149) and assuming that the plate is homogeneous enable us to carry out the integration with respect to z, which yields

$$U = \frac{h^3}{24}\iint_\Omega \left(B_{11}\kappa_x^2 + B_{22}\kappa_y^2 + B_{66}\kappa_{xy}^2 + 2B_{12}\kappa_x\kappa_y + 2B_{16}\kappa_x\kappa_{xy} + 2B_{26}\kappa_y\kappa_{xy}\right). \tag{3.155}$$

In the isotropic case, see (3.48), one obtains

$$U = \frac{Eh^3}{24(1-\nu^2)}\iint_\Omega \left[\kappa_x^2 + \kappa_y^2 + 2(1-\nu)\kappa_{xy}^2 + 2\nu\kappa_x\kappa_y\right]. \tag{3.156}$$

3.5.1.4 Integral Equilibrium Equations

To derive the laminate equilibrium equations, we integrate over the thickness the first two differential equilibrium equations (1.82a, b) to get

$$N_{x,x} + N_{xy,y} = 0, \qquad N_{y,y} + N_{xy,x} = 0. \tag{3.157}$$

In the above we exploited the identity

$$\int_{-\frac{h}{2}}^{\frac{h}{2}} \tau_{xz,z}dz = \tau_{xz}\Big|_{-\frac{h}{2}}^{\frac{h}{2}} = 0, \tag{3.158}$$

since $\tau_{xz} = 0$ for $z = \pm\frac{h}{2}$ as the laminate surfaces are loaded in the normal direction only. τ_{yz} is handled analogously. We now multiply the first two differential equilibrium equations (1.82a, b) by z and integrate over the thickness to get

$$Q_x = M_{x,x} + M_{xy,y}, \qquad Q_y = M_{y,y} + M_{xy,x}. \tag{3.159}$$

In the above we have substituted

$$\int_{-\frac{h}{2}}^{\frac{h}{2}} z\tau_{xz,z}dz = \underline{[z\tau_{xz}]|_{-\frac{h}{2}}^{\frac{h}{2}}} - \int_{-\frac{h}{2}}^{\frac{h}{2}} \tau_{xz}dz, \tag{3.160}$$

in which the underlined term vanishes similar to (3.158). Next we integrate over the thickness the third differential equilibrium equation, (1.82c), using $\sigma_z|_{z=\frac{h}{2}} = q_U$ and $\sigma_z|_{z=-\frac{h}{2}} = -q_L$, namely,

$$Q_{x,x} + Q_{y,y} + q = 0. \tag{3.161}$$

In addition, by substituting (3.159) in (3.161), the following equilibrium equation is obtained:

$$M_{x,xx} + 2M_{xy,xy} + M_{y,yy} + q = 0. \tag{3.162}$$

Further on, substituting (3.153) in (3.157), (3.162) constitutes three differential equations for $u_0(x,y)$ $v_0(x,y)$ and $w_0(x,y)$, that may be written as

$$\begin{aligned}
&[\mathbf{A}_{11}u_{0,x} + \mathbf{A}_{12}v_{0,y} + \mathbf{A}_{16}(u_{0,y} + v_{0,x})]_{,x} + [\mathbf{A}_{16}u_{0,x} + \mathbf{A}_{26}v_{0,y} + \mathbf{A}_{66}(u_{0,y} + v_{0,x})]_{,y} \\
&-[\mathbf{B}_{11}w_{0,xx} + \mathbf{B}_{12}w_{0,yy} + 2\mathbf{B}_{16}w_{0,xy}]_{,x} - [\mathbf{B}_{16}w_{0,xx} + \mathbf{B}_{26}w_{0,yy} + 2\mathbf{B}_{66}w_{0,xy}]_{,y} = 0, \qquad (3.163a)\\
&[\mathbf{A}_{16}u_{0,x} + \mathbf{A}_{26}v_{0,y} + \mathbf{A}_{66}(u_{0,y} + v_{0,x})]_{,x} + [\mathbf{A}_{12}u_{0,x} + \mathbf{A}_{22}v_{0,y} + \mathbf{A}_{26}(u_{0,y} + v_{0,x})]_{,y} \\
&-[\mathbf{B}_{16}w_{0,xx} + \mathbf{B}_{26}w_{0,yy} + 2\mathbf{B}_{66}w_{0,xy}]_{,x} - [\mathbf{B}_{12}w_{0,xx} + \mathbf{B}_{22}w_{0,yy} + 2\mathbf{B}_{26}w_{0,xy}]_{,y} = 0, \qquad (3.163b)\\
&[\mathbf{B}_{11}u_{0,x} + \mathbf{B}_{12}v_{0,y} + \mathbf{B}_{16}(u_{0,y} + v_{0,x})]_{,xx} + [\mathbf{B}_{12}u_{0,x} + \mathbf{B}_{22}v_{0,y} + \mathbf{B}_{26}(u_{0,y} + v_{0,x})]_{,yy} \\
&+2[\mathbf{B}_{16}u_{0,x} + \mathbf{B}_{26}v_{0,y} + \mathbf{B}_{66}(u_{0,y} + v_{0,x})]_{,xy} - [\mathbf{D}_{11}w_{0,xx} + \mathbf{D}_{12}w_{0,yy} + 2\mathbf{D}_{16}w_{0,xy}]_{,xx} \qquad (3.163c)\\
&-[\mathbf{D}_{12}w_{0,xx} + \mathbf{D}_{22}w_{0,yy} + 2\mathbf{D}_{26}w_{0,xy}]_{,yy} - 2[\mathbf{D}_{16}w_{0,xx} + \mathbf{D}_{26}w_{0,yy} + 2\mathbf{D}_{66}w_{0,xy}]_{,xy} + q = 0.
\end{aligned}$$

In cases where $\mathbf{B}_{ij} = 0$, the system of (3.163a–c) may be considerably simplified as the plane deformation (i.e., $u_0(x,y)$, $v_0(x,y)$) and the out-of-plane deformation (i.e., $w_0(x,y)$) become uncoupled. The first two equations describe a standard plane-stress problem (see S.3.2.2), while (3.163c) should be solved for $w_0(x,y)$, namely

$$\nabla_7^{(4)} w_0 = q, \tag{3.164}$$

where $\nabla_7^{(4)} = \mathbf{D}_{11}\frac{\partial^4}{\partial x^4} + 4\mathbf{D}_{16}\frac{\partial^4}{\partial x^3 \partial y} + 2(\mathbf{D}_{12} + 2\mathbf{D}_{66})\frac{\partial^4}{\partial x^2 \partial y^2} + 4\mathbf{D}_{26}\frac{\partial^4}{\partial x \partial y^3} + \mathbf{D}_{22}\frac{\partial^4}{\partial y^4}$.

An approximate way to couple the above equations with the in-plane solution while still working with $\mathbf{B}_{ij} = 0$, is founded on the incorporation of the in-plane solution via its resultants, see also Remark 3.9, in addition to the introduction of the out-of-plane solution into the in-plane one as will be described in what follows.

The out-of-plane solution in this approximation is written as

$$\nabla_7^{(4)} w_0 = q + \underline{N_x w_{0,xx} + 2N_{xy} w_{0,xy} + N_y w_{0,yy}}. \tag{3.165}$$

The above underlined additional terms are given by (3.172) of Remark 3.9, which shows that in the general case, N_x, N_{xy} and N_y are not constants (i.e. depend on (x,y)).

For the in-plane problem, we invoke the nonlinear terms for the strain expressions shown in (1.15b) to write

$$\varepsilon_x = u_{0,x} + \frac{1}{2}(w_{0,x})^2, \quad \varepsilon_y = v_{0,y} + \frac{1}{2}(w_{0,y})^2, \quad \gamma_{xy} = u_{0,y} + v_{0,x} + w_{0,x}w_{0,y}. \tag{3.166}$$

Then, the compatibility equation, (1.45a), that serves as the basis for the governing equation of the plane-stress problem, (3.44), becomes

$$\varepsilon_{x,yy} + \varepsilon_{y,xx} - \gamma_{xy,xy} = (w_{0,xy})^2 - w_{0,xx} w_{0,yy}. \tag{3.167}$$

We therefore arrive at a modified (by the out-of-plane solution) plane-stress biharmonic equation

$$\nabla_3^{(4)}\Phi = \underline{(w_{0,xy})^2 - w_{0,xx}w_{0,yy}}. \tag{3.168}$$

Hence, the above discussed approximation for the case of $\mathbf{B}_{ij} \neq 0$ enables separate solutions of the in-plane and out-of-plane problems, but still, (3.165), (3.168) remain coupled by the underlined terms in these equations. In both cases a biharmonic problem has to be solved. Two versions of an iterative solution scheme for these problems are shown in Fig. 3.20. Typically, such a process converges rapidly as the coupling terms are relatively small. See Remark 3.10 for stability analysis of (3.165), (3.168).

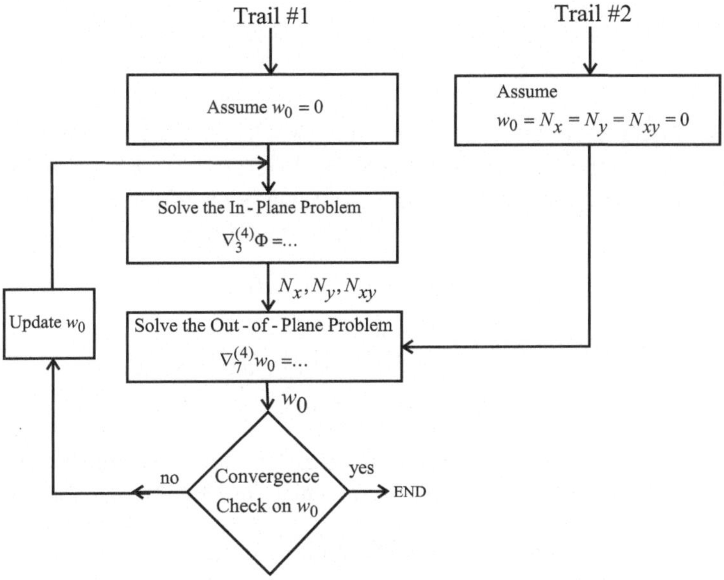

Figure 3.20: An iterative solution scheme for approximate coupled in-plane and out-of-plane behavior of a laminated plate.

Remark 3.8 To show that σ_z in a plate that undergoes bending is relatively small without solving the involved biharmonic problem, the clamped-free infinite (in the y-direction) strip shown in Fig. 3.21, is examined by a unit length of it. The strip is uniformly loaded by q (force per unit area), and the moment per unit length at the root becomes $M_q = \frac{1}{2}qd^2$. For linear variation of $\sigma_x = \sigma_0 z$, the internal moment at the root becomes

$$M_R = -\int_{-h/2}^{h/2} z\sigma_x\,dz = -\frac{1}{12}\sigma_0 h^3. \tag{3.169}$$

Equating M_R and M_q shows that

$$\sigma_x = -6q\frac{z}{h}\left(\frac{d}{h}\right)^2. \tag{3.170}$$

Since σ_z is on the order of q, for $d \gg h$, one may clearly assume $|\sigma_x| \gg |\sigma_z|$.

Remark 3.9 To determine the influence of in-plane loading, we solve a *plane-stress* problem, in which we specify $X_s(s)$ and $Y_s(s)$ over the domain contour (see assumption (d)), and find

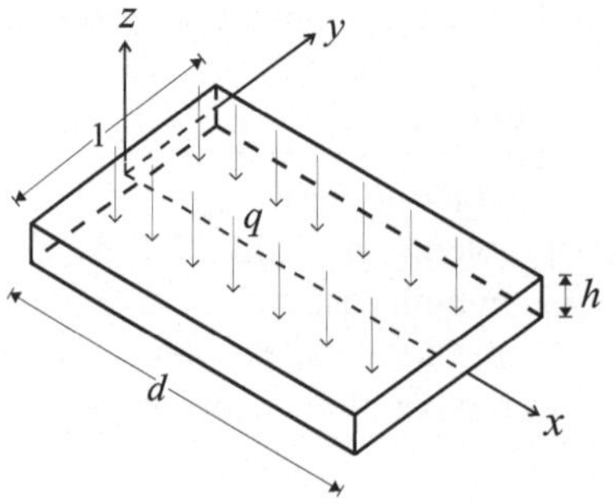

Figure 3.21: A unit length of an infinite strip (Remark 3.8).

the distributions of $\sigma_x(x,y)$, $\sigma_y(x,y)$ and $\tau_{xy}(x,y)$ as in S.3.2.2, S.3.2.5 (with the modification shown in (3.168)). The corresponding resultants are therefore $N_x = h\sigma_x(x,y)$, $N_y = h\sigma_y(x,y)$ and $N_{xy} = h\tau_{xy}(x,y)$. Finally, we construct an "equivalent perpendicular loading" q_e, using the projection of the in-plane stress components on the normal direction as

$$\begin{aligned} q_e\,dx\,dy = & [(N_x + N_{x,x}\,dx)\,(w_{0,x} + w_{0,xx}\,dx) - N_x\,w_{0,x}]\,dy \\ & + [(N_y + N_{y,y}\,dy)\,(w_{0,y} + w_{0,yy}\,dy) - N_y\,w_{0,y}]\,dx \\ & + [(N_{xy} + N_{xy,x}\,dx)\,(w_{0,x} + w_{0,xx}\,dx) - N_{xy}\,w_{0,x}]\,dy \\ & + [(N_{xy} + N_{x,y}\,dy)\,(w_{0,y} + w_{0,yy}\,dy) - N_{xy}w_{0,y}]\,dx. \end{aligned} \tag{3.171}$$

By neglecting high-order infinitesimal products (e.g., $(dx)^2dy$) and noting that the in-plane stress components satisfy the first two equilibrium equations (3.157), one obtains

$$q_e \approx N_x\,w_{0,xx} + 2N_{xy}\,w_{0,xy} + N_y\,w_{0,yy}. \tag{3.172}$$

Remark 3.10 Equations (3.165), (3.168) are also the source for stability (buckling) analysis of plates. The common way to look at that problem is to suppose that (3.168) has been solved (with $w_0 = 0$) and the in-plane resultants $\widetilde{N}_x(x,y)$, $\widetilde{N}_y(x,y)$ and $\widetilde{N}_{xy}(x,y)$ are known, see Remark 3.9. We then examine the laterally unloaded plate (i.e., $q = 0$) and write (3.165) as the following eigenvalue problem:

$$\nabla_7^{(4)} w_0 = \lambda(\widetilde{N}_x\,w_{0,xx} + 2\widetilde{N}_{xy}\,w_{0,xy} + \widetilde{N}_y\,w_{0,yy}). \tag{3.173}$$

In the above equation, we have assumed that the loads remain proportional to the initial ones (that were determined by the stress distributions over the boundaries), and ask the question: *what are the discrete values λ can take so that nonzero solution of* (3.173) *(i.e. $w_0 \neq 0$) will be obtained.* The lowest value of λ that supplies a nonzero solution of (3.173) is called the *critical value* as it yields the *critical loads* for which out-of-plane deformation is created by the in-plane loading. As an example, examination of the rectangular domain of Example 3.6 in the absence of body forces (i.e., $\widetilde{A} = 0$, $X_s{}^+ = -X_s{}^- = \widetilde{B}$) shows that $\widetilde{B} < 0$ creates a uniform compression load on both sides, and therefore, $\widetilde{N}_x = h\widetilde{B}$. Hence, we may set the eigenvalue problem as

$$\nabla_7^{(4)} w_0 + \lambda\,h\,w_{0,xx} = 0.$$

The resulting critical (compression) load will therefore be $\lambda_{\min}$ (i.e., $X_{s\,\mathrm{cr}}^{+} = -X_{s\,\mathrm{cr}}^{-} = -\lambda_{\min}$). Many other examples appear in (Lekhnitskii, 1968).

3.5.2 Bending of Anisotropic Plates

This section is focused on the "isolated" bending of anisotropic plate. Although the analysis may be derived as a special case of S.3.5.1 with $\mathbf{B}_{ij} = 0$ (i.e. (3.165), where the underlined terms represent prescribed in-plane loads), because of its wide applications, it will be briefly presented (based on the assumptions of S.3.5.1.1) in an almost independent way. We shall derive the governing equations to determine the deformation and the stress distribution in an anisotropic MON13z plate that is loaded normally over its outer surface, see Fig. 3.17. This case may be classified as an additional class of two-dimensional analysis of MON13z material, see Table 3.1.

3.5.2.1 Deformation

Based on assumption (f) of S.3.5.1.1, we consider the deflection in the z-direction of the mid-plane, $w_0(x,y)$, as the only deformation component (as $u_0 = v_0 = 0$). This enables us to write the u and v components as

$$u = -zw_{0,x}, \qquad v = -zw_{0,y}. \tag{3.174}$$

Following the definition of (3.149), we derive $\varepsilon_x^0 = \varepsilon_y^0 = \gamma_{xy}^0 = 0$. In view of (3.150), the constitutive relations become

$$\begin{Bmatrix} \sigma_x \\ \sigma_y \\ \tau_{xy} \end{Bmatrix} = -z \begin{bmatrix} B_{11} & B_{12} & 2B_{16} \\ B_{12} & B_{22} & 2B_{26} \\ B_{16} & B_{26} & 2B_{66} \end{bmatrix} \cdot \begin{Bmatrix} w_{0,xx} \\ w_{0,yy} \\ w_{0,xy} \end{Bmatrix}. \tag{3.175}$$

Unlike the case of plane-stress where $\tau_{yz} = \tau_{xz} = 0$, we shall now derive τ_{yz} and τ_{xz} by extracting their derivatives with respect to z from the equilibrium equations (1.82a, b), namely, $\tau_{xz,z} = -\sigma_{x,x} - \tau_{xy,y}$ and $\tau_{yz,z} = -\sigma_{y,y} - \tau_{xy,x}$. Therefore, $\tau_{xz,z}$ and $\tau_{yz,z}$ become linear functions of z as well. Their integration with respect to z, in view of their vanishing over the free surfaces $z = \pm\frac{h}{2}$, shows that

$$\begin{Bmatrix} \tau_{yz} \\ \tau_{xz} \end{Bmatrix} = \frac{1}{2}(z^2 - \frac{h^2}{4}) \begin{bmatrix} B_{16} & B_{12}+2B_{66} & 3B_{26} & B_{22} \\ B_{11} & 3B_{16} & B_{12}+2B_{66} & B_{26} \end{bmatrix} \cdot \begin{Bmatrix} w_{0,xxx} \\ w_{0,xxy} \\ w_{0,xyy} \\ w_{0,yyy} \end{Bmatrix}. \tag{3.176}$$

The above stress components yield the corresponding shear-strain components

$$\gamma_{yz} = \frac{1}{2}(z^2 - \frac{h^2}{4})a_v(x,y), \qquad \gamma_{xz} = \frac{1}{2}(z^2 - \frac{h^2}{4})a_u(x,y), \tag{3.177}$$

where in view of (2.5), (3.176), the functions $a_u(x,y)$ and $a_v(x,y)$ become

$$\begin{Bmatrix} a_v \\ a_u \end{Bmatrix} = \begin{bmatrix} a_{44} & a_{45} \\ a_{45} & a_{55} \end{bmatrix} \begin{bmatrix} B_{16} & B_{12}+2B_{66} & 3B_{26} & B_{22} \\ B_{11} & 3B_{16} & B_{12}+2B_{66} & B_{26} \end{bmatrix} \cdot \begin{Bmatrix} w_{0,xxx} \\ w_{0,xxy} \\ w_{0,xyy} \\ w_{0,yyy} \end{Bmatrix}. \tag{3.178}$$

These "additional" strain components, that did not originally emerge from the deformation assumption, violate compatibility and are kept as an approximation only, see also Remark 3.12.

3.5.2.2 Governing Equations

In contrast with (3.151), only resultant moments and shear loads are included in the present case. The moments are

$$\begin{Bmatrix} M_x \\ M_y \\ M_{xy} \end{Bmatrix} = - \begin{bmatrix} \mathbf{D}_{11} & \mathbf{D}_{12} & 2\mathbf{D}_{16} \\ \mathbf{D}_{12} & \mathbf{D}_{22} & 2\mathbf{D}_{26} \\ \mathbf{D}_{16} & \mathbf{D}_{26} & 2\mathbf{D}_{66} \end{bmatrix} \cdot \begin{Bmatrix} w_{0,xx} \\ w_{0,yy} \\ w_{0,xy} \end{Bmatrix}, \tag{3.179}$$

where $\mathbf{D}_{ij}$ were defined in (3.152). The shear loads are derived using (3.151), (3.176) as

$$\begin{Bmatrix} Q_y \\ Q_x \end{Bmatrix} = -\begin{bmatrix} \mathbf{D}_{16} & \mathbf{D}_{12}+2\mathbf{D}_{66} & 3\mathbf{D}_{26} & \mathbf{D}_{22} \\ \mathbf{D}_{11} & 3\mathbf{D}_{16} & \mathbf{D}_{12}+2\mathbf{D}_{66} & \mathbf{D}_{26} \end{bmatrix} \cdot \begin{Bmatrix} w_{0,xxx} \\ w_{0,xxy} \\ w_{0,xyy} \\ w_{0,yyy} \end{Bmatrix}. \tag{3.180}$$

We therefore employ (3.162) only and arrive at one integral equilibrium equation identical to (3.164) (or to (3.165) if prescribed in-plane loads are considered). On our way to derive the boundary conditions for this equation, we use the transformation technique for stresses of S.1.3.3 to get the three stress components over the contour, see Fig. 3.22(a), as

$$\sigma_n = \sigma_x \cos^2\psi + 2\tau_{xy}\sin\psi\cos\psi + \sigma_y \sin^2\psi, \tag{3.181a}$$
$$\tau_{ns} = (\sigma_y - \sigma_x)\sin\psi\cos\psi + \tau_{xy}(\cos^2\psi - \sin^2\psi), \tag{3.181b}$$
$$\tau_{nz} = \tau_{xz}\cos\psi + \tau_{yz}\sin\psi. \tag{3.181c}$$

We subsequently substitute $\cos\psi = \cos(\bar{\mathbf{n}},x)$, $\sin\psi = \cos(\bar{\mathbf{n}},y)$ (see (3.4)), and integrate over the thickness (see (3.151)) to evaluate the resultant loads along an arbitrarily oriented edge, see Fig. 3.22(b), as

$$\{M_n, M_{ns}, Q_n\} = \int_{-\frac{h}{2}}^{\frac{h}{2}} \{z\sigma_n, z\tau_{ns}, \tau_{nz}\}\, dz. \tag{3.182}$$

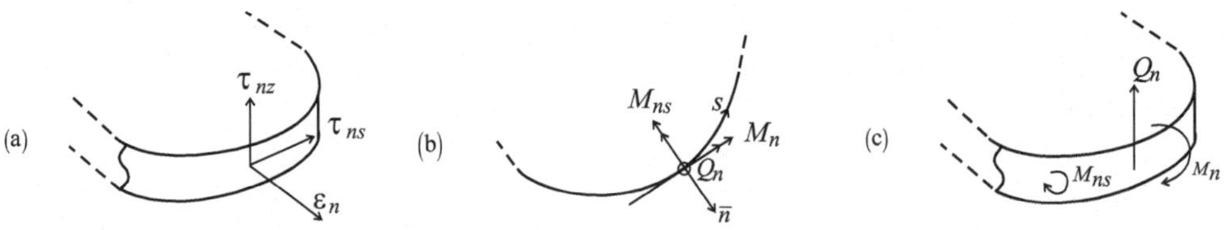

Figure 3.22: Load resultant over an arbitrarily oriented edge.

This process yields

$$M_n = M_x\cos^2(\bar{\mathbf{n}},x) + 2M_{xy}\cos(\bar{\mathbf{n}},y)\cos(\bar{\mathbf{n}},x) + M_y\cos^2(\bar{\mathbf{n}},y), \tag{3.183a}$$
$$M_{ns} = (M_y - M_x)\cos(\bar{\mathbf{n}},y)\cos(\bar{\mathbf{n}},x) + M_{xy}\left[\cos^2(\bar{\mathbf{n}},x) - \cos^2(\bar{\mathbf{n}},y)\right], \tag{3.183b}$$
$$Q_n = Q_x\cos(\bar{\mathbf{n}},x) + Q_y\cos(\bar{\mathbf{n}},y). \tag{3.183c}$$

Various boundary conditions may be applied here. We shall only mention the following three:

A "clamped" edge:

$$w_0 = 0, \qquad \frac{d}{dn}w_0 = 0 \quad \text{on} \quad \partial\Omega. \tag{3.184}$$

This type of boundary condition will be further on related to the biDirichlet BVP (see S.4.1).

A "simply supported" edge:

$$w_0 = 0, \qquad M_n = 0 \quad \text{on} \quad \partial\Omega. \tag{3.185}$$

A "free" edge:

$$Q_n + \frac{d}{ds} M_{ns} = 0, \qquad M_n = 0 \qquad \text{on} \quad \partial\Omega. \tag{3.186}$$

See Remarks 3.11, 3.12 for additional details.

A simple analytical solution of Example 3.13 demonstrates the above derivation.

Remark 3.11 By employing the third equilibrium equation, (1.82c), in its differential form, one obtains σ_z from $\sigma_{z,z} = -\tau_{xz,x} - \tau_{yz,y}$. In view of (3.176), σ_z turns to be a third-order polynomial of z. By accounting for the boundary conditions $\sigma_{z|z=\frac{h}{2}} = q_U(x,y)$ and $\sigma_{z|z=-\frac{h}{2}} = -q_L(x,y)$, we write

$$\sigma_z = -q_L + (q_U + q_L)[\frac{1}{2} + \frac{3}{2}\left(\frac{z}{h}\right) - 2\left(\frac{z}{h}\right)^3]. \tag{3.187}$$

Hence, one may determine the additional strain components $\varepsilon_x = a_{13}\sigma_z$, $\varepsilon_y = a_{23}\sigma_z$, $\gamma_{xy} = a_{36}\sigma_z$, and subsequently the additional stress components $\sigma_x, \sigma_y, \tau_{xy}$, see (3.150), to obtain a refined stress estimation as

$$\begin{Bmatrix} \sigma_x \\ \sigma_y \\ \tau_{xy} \end{Bmatrix} = -z \begin{bmatrix} B_{11} & B_{12} & 2B_{16} \\ B_{12} & B_{22} & 2B_{26} \\ B_{16} & B_{26} & 2B_{66} \end{bmatrix} \cdot \begin{Bmatrix} w_{0,xx} \\ w_{0,yy} \\ w_{0,xy} \end{Bmatrix} - \underline{\begin{bmatrix} B_{11} & B_{12} & B_{16} \\ B_{12} & B_{22} & B_{26} \\ B_{16} & B_{26} & B_{66} \end{bmatrix} \cdot \begin{Bmatrix} a_{13} \\ a_{23} \\ a_{36} \end{Bmatrix} \sigma_z}. \tag{3.188}$$

As already indicated, usually, σ_z is negligible, and the above underlined term represents small (approximate) correction only.

Remark 3.12 Similar to (Ambartsumyan, 1970), the theory may be "extended" by further approximation, where by integrating (3.177) we evaluate the additional displacement components

$$(\Delta u)_{,z} = \frac{1}{2}(z^2 - \frac{h^2}{4})\, a_u, \qquad (\Delta v)_{,z} = \frac{1}{2}(z^2 - \frac{h^2}{4})\, a_v. \tag{3.189}$$

Consequently, u and v may be refined as

$$u = -z w_{,x} + \frac{z}{2}(\frac{z^2}{3} - \frac{h^2}{4})\, a_u, \qquad v = -z w_{,y} + \frac{z}{2}(\frac{z^2}{3} - \frac{h^2}{4})\, a_v, \tag{3.190}$$

while $a_u(x,y)$ and $a_v(x,y)$ are considered as known functions. The "new" strain expressions become

$$\varepsilon_x = -z w_{,xx} + \underline{\frac{z}{2}(\frac{z^2}{3} - \frac{h^2}{4})}\, a_{u,x}, \qquad \varepsilon_y = -z w_{,yy} + \underline{\frac{z}{2}(\frac{z^2}{3} - \frac{h^2}{4})}\, a_{v,y},$$
$$\gamma_{xy} = -2z w_{,xy} + \underline{\frac{z}{2}(\frac{z^2}{3} - \frac{h^2}{4})}(a_{u,y} + a_{v,x}). \tag{3.191}$$

One may now proceed by repeating the analysis described in S.3.5.2 with the above strain expression, namely, by adding the underlined terms in (3.191), which are considered to be known. Then, based on (3.162), one may reach a refined governing equation similar to (3.164), where the r.h.s. includes an additional known function of x and y.

Example 3.13 *Uniformly Loaded Clamped Elliptical Domain.*

For a uniformly loaded clamped elliptical domain of semi-axes $\widetilde{a}$ and $\widetilde{b}$, the out-of-plane deflection is given by

$$w = \frac{q\widetilde{a}^4}{8\widetilde{D}}(1 - \frac{x^2}{\widetilde{a}^2} - \frac{y^2}{\widetilde{b}^2})^2, \tag{3.192}$$

where $\widetilde{D} = 3\mathbf{D}_{11} + 2\alpha^2(\mathbf{D}_{12} + 2\mathbf{D}_{66}) + 3\alpha^4\mathbf{D}_{22}$ and $\alpha = \frac{\widetilde{a}}{b}$, see (Lekhnitskii, 1968). The reader may verify that this solution satisfies (3.164), and the boundary conditions (3.184), and should note that $\mathbf{D}_{16}$ and $\mathbf{D}_{26}$ do not participate since w_{xxxy} and w_{xyyy} vanish in this case. Then, σ_x may be written as

$$\sigma_x = -6q\frac{z}{h}\left(\frac{\widetilde{a}}{h}\right)^2 \frac{B_{11}\left(3\xi^2+\eta^2-1\right)+B_{12}\alpha^2\left(3\eta^2+\xi^2-1\right)+4B_{16}\alpha\xi\eta}{3B_{11}+2\alpha^2\left(B_{12}+2B_{66}\right)+3\alpha^4 B_{22}}, \tag{3.193}$$

where $\xi = \frac{x}{\widetilde{a}}$, $\eta = \frac{y}{b}$. For σ_y and τ_{xy} one should replace the triad B_{11}, B_{12}, B_{16} in the above numerator only with the triads B_{12}, B_{22}, B_{26} and B_{16}, B_{26}, B_{66}, respectively.

As indicated in Remark 3.8, σ_z is on the order of q, which shows that for thin plates (where $\widetilde{a} \gg h$), $|\sigma_x| \gg |\sigma_z|$. Note that the leading term in (3.193) is identical to that of (3.170).

3.6 Appendix: Differential Operators

The differential operators that appear throughout the various derivations in this book are summarized in this appendix, which also examines the characteristics of some of these operators. Note that although additional versions of the operators exist, the following is confined to those that are relevant to the discussions in this book. Due to the desire to put the operators in a consistent sequence, their order of appearance here does not always match the one used throughout the derivation (mainly in Chapter 3).

In this section we denote the polar coordinates (ρ, θ) as (r, θ), respectively.

3.6.1 *Generalized Laplace's Operators*

Two versions of generalized Laplace's operators are of interest.

The generalized Laplace's operator in the xy-plane for *Cartesian anisotropy* is defined by

$$\nabla_1^{(2)} = b_{44}\frac{\partial^2}{\partial x^2} - 2b_{45}\frac{\partial^2}{\partial x \partial y} + b_{55}\frac{\partial^2}{\partial y^2}. \tag{3.194}$$

The following Laplace's operators have been developed in S.3.3.1 (see (3.97)), and may be obtained from (3.194) by replacing b_{ij} with a_{ij}, namely,

$$\nabla_3^{(2)} = a_{44}\frac{\partial^2}{\partial x^2} - 2a_{45}\frac{\partial^2}{\partial x \partial y} + a_{55}\frac{\partial^2}{\partial y^2}, \tag{3.195}$$

and its "orthotropic" version form (i.e., $a_{45} = 0$)

$$\nabla_4^{(2)} = a_{44}\frac{\partial^2}{\partial x^2} + a_{55}\frac{\partial^2}{\partial y^2}. \tag{3.196}$$

Recall that for MON13z material, and clearly for orthotropic material, the b_{44}, b_{45}, b_{55} set of coefficients is identical to the a_{44}, a_{45}, a_{55} set, respectively, see (3.28). Therefore, in such cases, $\nabla_3^{(2)} = \nabla_1^{(2)}$ and their orthotropic versions coincide.

The isotropic version of the operator (3.194) is given by

$$\nabla_0^{(2)} = \frac{2(1+\nu)}{E}\nabla^{(2)}, \tag{3.197}$$

where $\nabla^{(2)}$ is the simplest *Laplace's operator*, namely,

$$\nabla^{(2)} = \frac{\partial^2}{\partial x^2} + \frac{\partial^2}{\partial y^2}. \tag{3.198}$$

The isotropic version of operator (3.196) is identical to the one presented in (3.197).

In polar coordinates (r, θ) and for *cylindrical anisotropy*, see S.7.5.2, (7.152), Laplace's operator becomes

$$\nabla_5^{(2)} = a_{44}(\frac{\partial^2}{\partial r^2} + \frac{1}{r}\frac{\partial}{\partial r}) - \frac{2}{r}a_{45}\frac{\partial^2}{\partial r \partial \theta} + \frac{1}{r^2}a_{55}\frac{\partial^2}{\partial \theta^2}, \tag{3.199}$$

which for *cylindrical orthotropic* material takes the form, see S.7.5.3.2,

$$\nabla_6^{(2)} = a_{44}(\frac{\partial^2}{\partial r^2} + \frac{1}{r}\frac{\partial}{\partial r}) + \frac{1}{r^2}a_{55}\frac{\partial^2}{\partial \theta^2}. \tag{3.200}$$

In the isotropic case, Laplace's operator of (3.199) may be written as (3.197) while $\nabla^{(2)}$, the simplest Laplace's operator, is written in polar coordinates as

$$\nabla^{(2)} = \frac{\partial^2}{\partial r^2} + \frac{1}{r}\frac{\partial}{\partial r} + \frac{1}{r^2}\frac{\partial^2}{\partial \theta^2}. \tag{3.201}$$

The operators documented in (3.194–3.196) are all written in *Cartesian coordinates* and for *Cartesian anisotropy*. On the other hand, the operators documented in (3.199–3.201) are all written in *polar coordinates* and for *cylindrical anisotropy*. Therefore, employing the coordinate transformations of S.7.5.1 can not be used to obtain $\nabla_5^{(2)}$ from $\nabla_1^{(2)}$ or $\nabla_6^{(2)}$ from $\nabla_2^{(2)}$ and vice versa. Such a transformation is allowed only in the isotropic case. In other words, by simple coordinate transformation, (3.198) can be transformed into (3.201) and vice versa, and therefore, these operators are denoted identically as $\nabla^{(2)}$.

3.6.2 Biharmonic Operators

We shall first list the various biharmonic operators used in the book. As shown by (3.31), for *Cartesian anisotropy*, the biharmonic operator for plane-strain analysis is

$$\nabla_1^{(4)} = b_{22}\frac{\partial^4}{\partial x^4} - 2b_{26}\frac{\partial^4}{\partial x^3 \partial y} + (2b_{12} + b_{66})\frac{\partial^4}{\partial x^2 \partial y^2} - 2b_{16}\frac{\partial^4}{\partial x \partial y^3} + b_{11}\frac{\partial^4}{\partial y^4}. \tag{3.202}$$

The orthotropic version of this operator is given as

$$\nabla_2^{(4)} = b_{22}\frac{\partial^4}{\partial x^4} + (2b_{12} + b_{66})\frac{\partial^4}{\partial x^2 \partial y^2} + b_{11}\frac{\partial^4}{\partial y^4}. \tag{3.203}$$

The isotropic version of the biharmonic operator is given in (3.204) as

$$\nabla_{01}^{(4)} = \frac{1 - \nu^2}{E}\nabla^{(4)}, \tag{3.204}$$

where $\nabla^{(4)}$ is the simplest *biharmonic operator*,

$$\nabla^{(4)} = \nabla^{(2)} \cdot \nabla^{(2)} = \frac{\partial^4}{\partial x^4} + 2\frac{\partial^4}{\partial x^2 \partial y^2} + \frac{\partial^4}{\partial y^4}. \tag{3.205}$$

The biharmonic differential operator for xy-plane-stress analysis given in (3.45) is obtained from (3.202) by replacing b_{ij} with a_{ij}, namely,

$$\nabla_3^{(4)} = a_{22}\frac{\partial^4}{\partial x^4} - 2a_{26}\frac{\partial^4}{\partial x^3 \partial y} + (2a_{12}+a_{66})\frac{\partial^4}{\partial x^2 \partial y^2} - 2a_{16}\frac{\partial^4}{\partial x \partial y^3} + a_{11}\frac{\partial^4}{\partial y^4}, \quad (3.206)$$

and its orthotropic version is given by

$$\nabla_4^{(4)} = a_{22}\frac{\partial^4}{\partial x^4} + (2a_{12}+a_{66})\frac{\partial^4}{\partial x^2 \partial y^2} + a_{11}\frac{\partial^4}{\partial y^4}. \quad (3.207)$$

The isotropic version of (3.206) is

$$\nabla_{02}^{(4)} = \frac{1}{E}\nabla^{(4)}, \quad (3.208)$$

where $\nabla^{(4)}$ appears in (3.205).

Finally, for bending of a thin plate, the following operator has been found, see S.3.5.1.2 and (3.164):

$$\nabla_7^{(4)} = \mathbf{D}_{11}\frac{\partial^4}{\partial x^4} + 4\mathbf{D}_{16}\frac{\partial^4}{\partial x^3 \partial y} + 2(\mathbf{D}_{12}+2\mathbf{D}_{66})\frac{\partial^4}{\partial x^2 \partial y^2} + 4\mathbf{D}_{26}\frac{\partial^4}{\partial x \partial y^3} + \mathbf{D}_{22}\frac{\partial^4}{\partial y^4}. \quad (3.209)$$

For *cylindrical anisotropy*, the biharmonic differential operator for plane-strain in polar coordinates (r,θ) is derived in S.7.5.3.1, as

$$\begin{aligned}\nabla_5^{(4)} &= b_{22}\frac{\partial^4}{\partial r^4} - \frac{2}{r}b_{26}\frac{\partial^4}{\partial r^3 \partial \theta} + \frac{1}{r^2}(2b_{12}+b_{66})\frac{\partial^4}{\partial r^2 \partial \theta^2} - \frac{2}{r^3}b_{16}\frac{\partial^4}{\partial r \partial \theta^3} + \frac{1}{r^4}b_{11}\frac{\partial^4}{\partial \theta^4} \\ &+ \frac{2}{r}b_{22}\frac{\partial^3}{\partial r^3} - \frac{1}{r^3}(2b_{12}+b_{66})\frac{\partial^3}{\partial r \partial \theta^2} + \frac{2}{r^4}b_{16}\frac{\partial^3}{\partial \theta^3} - \frac{1}{r^2}b_{11}\frac{\partial^2}{\partial r^2} - \frac{2}{r^3}(2b_{16}+b_{26})\frac{\partial^2}{\partial r \partial \theta} \\ &+ \frac{1}{r^4}(2b_{11}+2b_{12}+b_{66})\frac{\partial^2}{\partial \theta^2} + \frac{1}{r^3}b_{11}\frac{\partial}{\partial r} + \frac{2}{r^4}(b_{16}+b_{26})\frac{\partial}{\partial \theta}. \end{aligned} \quad (3.210)$$

The orthotropic version of this operator is

$$\begin{aligned}\nabla_6^{(4)} &= b_{22}\frac{\partial^4}{\partial r^4} + \frac{1}{r^2}(2b_{12}+b_{66})\frac{\partial^4}{\partial r^2 \partial \theta^2} + \frac{1}{r^4}b_{11}\frac{\partial^4}{\partial \theta^4} + \frac{2}{r}b_{22}\frac{\partial^3}{\partial r^3} \\ &- \frac{1}{r^3}(2b_{12}+b_{66})\frac{\partial^3}{\partial r \partial \theta^2} - \frac{1}{r^2}b_{11}\frac{\partial^2}{\partial r^2} + \frac{1}{r^4}(2b_{11}+2b_{12}+b_{66})\frac{\partial^2}{\partial \theta^2} + \frac{1}{r^3}b_{11}\frac{\partial}{\partial r}. \end{aligned} \quad (3.211)$$

The isotropic version of the biharmonic operator of (3.210) is given by (3.204), where the simplest *biharmonic operator* $\nabla^{(4)}$ in polar coordinates is written as

$$\nabla^{(4)} = \nabla^{(2)} \cdot \nabla^{(2)} = (\frac{\partial^2}{\partial r^2} + \frac{1}{r}\frac{\partial}{\partial r} + \frac{1}{r^2}\frac{\partial^2}{\partial \theta^2})^2, \quad (3.212)$$

or, by opening brackets and differentiating, as

$$\nabla^{(4)} = \frac{\partial^4}{\partial r^4} + \frac{2}{r^2}\frac{\partial^4}{\partial r^2 \partial \theta^2} + \frac{1}{r^4}\frac{\partial^4}{\partial \theta^4} + \frac{2}{r}\frac{\partial^3}{\partial r^3} - \frac{2}{r^3}\frac{\partial^3}{\partial r \partial \theta^2} - \frac{1}{r^2}\frac{\partial^2}{\partial r^2} + \frac{4}{r^4}\frac{\partial^2}{\partial \theta^2} + \frac{1}{r^3}\frac{\partial}{\partial r}. \quad (3.213)$$

Analogously to the discussion in S.3.6.1, some clarification is required here. The operators documented in (3.202–3.208) are all written in *Cartesian coordinates* and for *Cartesian anisotropy*. On the other hand, the operators documented in (3.210–3.213) are all written in *polar coordinates* and for *cylindrical anisotropy*. Therefore, employing the coordinate transformations of S.7.5.1 can not be used to obtain $\nabla_5^{(4)}$ from $\nabla_1^{(4)}$ or $\nabla_6^{(4)}$ from $\nabla_2^{(4)}$ and vice versa. Such a transformation is allowed only in the isotropic case, namely, (3.205) can be transformed into (3.212) and vice versa, and therefore, these operators are denoted identically as $\nabla^{(4)}$.

3.6.3 Third-Order and Sixth-Order Differential Operators

The third-order differential operator $\nabla_1^{(3)}$ that appears for GEN21 materials but vanishes for MON13z (and simpler) materials, is given in S.3.4 for Cartesian anisotropy as

$$\nabla_1^{(3)} = -b_{24}\frac{\partial^3}{\partial x^3} + (b_{25}+b_{46})\frac{\partial^3}{\partial x^2 \partial y} - (b_{14}+b_{56})\frac{\partial^3}{\partial x \partial y^2} + b_{15}\frac{\partial^3}{\partial y^3}. \tag{3.214}$$

We also define an additional sixth-order operator $\nabla_1^{(6)}$ that may be written for Cartesian anisotropy using the $\nabla_1^{(2)}$, $\nabla_1^{(3)}$ and $\nabla_1^{(4)}$ operators as

$$\nabla_1^{(6)} = \nabla_1^{(4)} \cdot \nabla_1^{(2)} - \nabla_1^{(3)} \cdot \nabla_1^{(3)}. \tag{3.215}$$

No use of these operators with the a_{ij} coefficients appears in this book. The above operators are presented in (Lekhnitskii, 1981) for polar coordinates.

3.6.4 Generalized Normal Derivative Operators

The generalized normal derivative operators presented here are typically used along with the generalized Laplace's operators presented in S.3.6.1. For *Cartesian anisotropy*, the first operator is

$$D_1^n = (a_{44}\frac{\partial}{\partial x} - a_{45}\frac{\partial}{\partial y})\cos(\bar{\mathbf{n}},x) + (-a_{45}\frac{\partial}{\partial x} + a_{55}\frac{\partial}{\partial y})\cos(\bar{\mathbf{n}},y), \tag{3.216}$$

which in the orthotropic case becomes

$$D_2^n = a_{44}\frac{\partial}{\partial x}\cos(\bar{\mathbf{n}},x) + a_{55}\frac{\partial}{\partial y}\cos(\bar{\mathbf{n}},y). \tag{3.217}$$

In the isotropic case the operator of (3.216) is simplified to

$$D_0^n = \frac{2(1+\nu)}{E}\frac{d}{dn}, \tag{3.218}$$

where $\frac{d}{dn}$ is the *(geometrical) normal derivative*, namely,

$$\frac{d}{dn} \equiv \frac{\partial}{\partial x}\cos(\bar{\mathbf{n}},x) + \frac{\partial}{\partial y}\cos(\bar{\mathbf{n}},y). \tag{3.219}$$

No use of these operators with the b_{ij} coefficients appears in this book.

In polar coordinates (r,θ) and for *cylindrical anisotropy*, see S.7.5.2, (7.152), the generalized normal derivative operator becomes

$$D_3^n = (a_{44}\frac{\partial}{\partial r} - a_{45}\frac{1}{r}\frac{\partial}{\partial \theta})\cos(\bar{\mathbf{n}},r) + (-a_{45}\frac{\partial}{\partial r} + a_{55}\frac{1}{r}\frac{\partial}{\partial \theta})\cos(\bar{\mathbf{n}},\theta), \tag{3.220}$$

while for orthotropic *cylindrical anisotropy* we obtain, see S.7.5.3.2,

$$D_4^n = a_{44}\frac{\partial}{\partial r}\cos(\bar{\mathbf{n}},r) + a_{55}\frac{1}{r}\frac{\partial}{\partial \theta}\cos(\bar{\mathbf{n}},\theta). \tag{3.221}$$

In the isotropic case, we use (3.218) where $\frac{d}{dn}$ becomes again the (geometrical) normal derivative in the r, θ-plane and takes the form

$$\frac{d}{dn} \equiv \frac{\partial}{\partial r}\cos(\bar{\mathbf{n}},r) + \frac{1}{r}\frac{\partial}{\partial \theta}\cos(\bar{\mathbf{n}},\theta). \tag{3.222}$$

By employing the coordinate transformations of S.7.5.1 one may obtain D_3^n from D_1^n and D_4^n from D_2^n and vice versa only by substituting $\theta = 0$. Yet, for the isotropic case, (3.219) can be directly transformed into (3.222).

3.6.5 *Ellipticity of the Differential Operators*

In this section, we shall discuss the characteristic polynomials $p_{i\alpha}$, $\alpha \in \{a,b\}$, $i = 2,3,4,6$ corresponding to a_{ij} and b_{ij} versions of the operators defined by $i = 2$ in (3.194), (3.195), $i = 4$ in (3.202), (3.206), $i = 3$ in (3.214) and $i = 6$ in (3.215), namely

$$p_{2\alpha} = \alpha_{55}\mu^2 - 2\alpha_{45}\mu + \alpha_{44}, \tag{3.223a}$$

$$p_{3\alpha} = \alpha_{15}\mu^3 - (\alpha_{14} + \alpha_{56})\mu^2 + (\alpha_{25} + \alpha_{46})\mu - \alpha_{24}, \tag{3.223b}$$

$$p_{4\alpha} = \alpha_{11}\mu^4 - 2\alpha_{16}\mu^3 + (2\alpha_{12} + \alpha_{66})\mu^2 - 2\alpha_{26}\mu + \alpha_{22}, \tag{3.223c}$$

$$p_{6\alpha} = p_{4\alpha}p_{2\alpha} - p_{3\alpha}^2. \tag{3.223d}$$

Note that for the sake of completeness, we have documented here the polynomial p_{3a} and p_{6a} although no corresponding operator has been used.

To show that the differential operators (3.223a,c,d) have *elliptical properties*, i.e. their characteristic polynomials cannot have a real root, we write the strain energy, W, see (2.27), as

$$W = \frac{1}{2}\sum\nolimits_{i=1}^{6} \sigma_i \varepsilon_i = \frac{1}{2}\sum\nolimits_{i,j=1}^{6} a_{ij}\sigma_i\sigma_j, \tag{3.224}$$

and exploit the fact that W attains only positive values, see S.2.10. We treat each characteristic polynomial separately.

As an example, for the polynomial p_{2a}, by assigning the following values to the stress components: $\sigma_x = 0,\ \sigma_y = 0,\ \sigma_z = 0,\ \tau_{yz} = -1,\ \tau_{xz} = \mu,\ \tau_{xy} = 0,$ one obtains an expression for the energy that coincides with (3.223a) since $2W = p_{2a}(\mu)$, and therefore, one may conclude that $p_{2a}(\mu) > 0,\ \mu \in \mathbb{R}$.

Similarly, for the remaining operators of *even degree* (as odd-degree polynomials will always have a real root) we select the following five sets, where only the non-vanishing components are given:

$$\text{For } p_{2b}: \sigma_z = \frac{1}{a_{33}}(a_{34} - a_{35}\mu), \quad \tau_{yz} = -1, \quad \tau_{xz} = \mu. \tag{3.225a}$$

$$\text{For } p_{4a}: \sigma_x = \mu^2, \quad \sigma_y = 1, \quad \tau_{xy} = -\mu. \tag{3.225b}$$

$$\text{For } p_{4b}: \sigma_x = \mu^2, \ \sigma_y = 1, \ \sigma_z = -\frac{1}{a_{33}}(a_{13}\mu^2 + a_{23} - a_{36}\mu), \ \tau_{xy} = -\mu. \tag{3.225c}$$

$$\text{For } p_{6a}: \sigma_x = \mu^2, \quad \sigma_y = 1, \quad \tau_{yz} = -\lambda_a, \quad \tau_{xz} = \lambda_a\mu, \quad \tau_{xy} = -\mu. \tag{3.225d}$$

$$\begin{aligned} \text{For } p_{6b}: \sigma_x &= \mu^2, \quad \sigma_y = 1, \quad \sigma_z = -\frac{1}{a_{33}}(a_{13}\mu^2 + a_{23} - a_{36}\mu - a_{34}\lambda_b + a_{35}\lambda_b\mu), \\ \tau_{yz} &= -\lambda_b, \quad \tau_{xz} = \lambda_b\mu, \quad \tau_{xy} = -\mu. \end{aligned} \tag{3.225e}$$

In the above, we have used the notation $\lambda_a = -\frac{p_{3a}}{p_{2a}}$, $\lambda_b = -\frac{p_{3b}}{p_{2b}}$. Thus, it has been shown that $p_{2b}(\mu) > 0,\ p_{4a}(\mu) > 0,\ p_{4b}(\mu) > 0,\ p_{6a}(\mu) > 0,\ p_{6b}(\mu) > 0$ for $\mu \in \mathbb{R}$. Note that the polynomials in (3.223a,c,d) that contain b_{ij} were rewritten with a_{ij}. **P.3.12** illustrates the positiveness of polynomials as described in this section.

4

Solution Methodologies

This chapter discusses various solution methodologies for the two-dimensional BVPs introduced in Chapter 3 that may be adopted for diversified applications. In particular, these methods will be useful for the anisotropic beam analyses presented in subsequent chapters. Nevertheless, the same methods are beneficial for many other applications such as the analysis of plates, see S.3.5, or the analysis of anisotropic foundations discussed, for example, by (Muravskii, 2001).

In order to keep the discussion generic enough and capable of handling many and different domain shapes, the solutions are oriented to global polynomial analysis. Since this class of polynomial solutions does not always supply exact solutions (i.e. some combination of geometries and loading modes require other specific types of analytical solutions), some approximations become inevitable.

The proposed approach keeps its analytical nature in many senses. This is due to the fact that we shall use an *analytical elimination* process that ensures *exact* fulfillment of the field equations, while the entire approximation, if required, migrates to the boundary conditions. In addition, all approximation procedures that will be derived in this chapter are realized by a *least-squares* process, where all operations are carried out in a rational manner that loses no accuracy, and yields an "exact-approximate" solution. In other words, the resulting solution may be considered as exact since it can be analytically proved to be the "closest" (or the "most accurate") polynomial solution according to the least-squares criteria.

The above technique has several advantages. First, the exactness of fulfilling the field equations ensures integrability of the deformation components (since as shown in Chapter 3, the field equations emerge from compatibility requirements). Secondly, it enables a simple tracking of the error over the boundary line. Hence, the error interpretation becomes more intuitive (e.g., for a plane-strain problem, the error may be interpreted as "additional loading over the contour"). The influence of such an error may be estimated more easily than the consequence of an error in the fulfillment of the field equations. It also keeps the solution accuracy in areas (over the domain) that are far from the boundary.

To clarify the derivation, the general solutions of the field equations are separated into two parts: the particular solutions, and the general (or "prescribed") solutions of the homogeneous field equations. Although such a separation into particular and homogeneous solutions is not always essential, it greatly contributes to an insight into the solution ingredients.

The last part of this chapter is devoted to the complex potentials method which exhibits some analytical superiority. The method is derived and then implemented via Fourier series analysis.

While only a few, representative examples appear in the text, the solution schemes in this chapter are documented as programs as well, see S.P.4. The reader may rerun all examples and many others as required. In addition, these programs are extensively used in further chapters of this book.

4.1 Unified Formulation of Two-Dimensional BVPs

Prior to the solution methodologies presentation, we shall review here the general formulation of BVPs in a simply connected x,y-plane domain Ω, and their solutions' separation into particular and homogeneous ones. The BVPs will be expressed using harmonic (Λ) or biharmonic (Φ) functions (in some contexts, the above functions are also accompanied by the adjective "generalized"). In the equations documented below we assume $(x,y) \in \Omega$ for the field equations, and $(x,y) \in \partial\Omega$ for the boundary conditions. In addition, all given boundary functions are assumed to be single-valued. The $\frac{d}{ds}$-case of boundary conditions leaves the values of the solution functions Λ, Φ and the derivatives $\Phi_{,x}$, $\Phi_{,y}$ undetermined by a constant, and for the sake of simplicity one may put $\{\Lambda, \Phi, \Phi_{,x}, \Phi_{,y}\}(0,0) = \{0,0,0,0\}$.

The *Dirichlet/Neumann* BVPs are, see (3.121), (3.128), (3.195), (3.216),

$$\nabla_3^{(2)}\Lambda = F_0^\Lambda \qquad \text{over} \quad \Omega, \tag{4.1a}$$

$$\text{Dirichlet:} \qquad \Lambda = G_3 \quad \text{or} \quad \frac{d}{ds}\Lambda = F_3 \qquad \text{on} \quad \partial\Omega, \tag{4.1b}$$

$$\text{Neumann:} \qquad D_1^n\Lambda = F_3^\Lambda \qquad \text{on} \quad \partial\Omega. \tag{4.1c}$$

The above BVPs should be solved under the conditions (3.126), (3.129),

$$\oint_{\partial\Omega} F_3 = 0 \qquad \text{or} \qquad \oint_{\partial\Omega} F_3^\Lambda = \iint_\Omega F_0^\Lambda. \tag{4.2}$$

The *biharmonic* BVP for plane-strain is

$$\nabla_1^{(4)}\Phi = F_0 \qquad \text{over} \quad \Omega, \tag{4.3a}$$

$$\frac{d}{ds}\{\Phi_{,x}, \Phi_{,y}\} = \{-F_1, F_2\} \quad \text{or} \quad \{\Phi_{,x}, \Phi_{,y}\} = \{-G_1, G_2\} \qquad \text{on} \quad \partial\Omega \tag{4.3b}$$

which should be solved under the single-value conditions (3.72),

$$\oint_{\partial\Omega} \{F_1, F_2, F_2 y - F_1 x\} = \{0, 0, 0\}, \tag{4.4}$$

or, by integrating (by parts) the third equation of (4.4) and using $F_i = \frac{d}{ds}G_i$,

$$\oint_{\partial\Omega} \{G_2 \cos(\bar{\mathbf{n}}, x) + G_1 \cos(\bar{\mathbf{n}}, y)\} = 0. \tag{4.5}$$

For plane-stress analysis, $\nabla_1^{(4)}$ should be replaced by $\nabla_3^{(4)}$, see (3.58), (3.202), (3.206), (alternatively, b_{ij} may be replaced by a_{ij} in the resulting expressions).

The *biDirichlet* (biharmonic type) BVP, see also S.4.6.6, deals with (4.3a) under the following (mixed Dirichlet-Neumann type) boundary conditions:

$$\{\Phi, \frac{d}{dn}\Phi\} = \{G_1^D, G_2^N\} \qquad \text{on} \quad \partial\Omega \tag{4.6}$$

where the normal derivative may be expanded as $\frac{d}{dn}\Phi = \Phi_{,x}\cos(\bar{\mathbf{n}},x) + \Phi_{,y}\cos(\bar{\mathbf{n}},y)$.

To convert the necessary condition for solution existence given by (4.5) to the case of the biDirichlet BVP, it is more convenient to consider the single-value conditions as shown by (3.66). In view of the identity

$$\begin{Bmatrix} \Phi_{,x} \\ \Phi_{,y} \end{Bmatrix} = \begin{bmatrix} \cos(\bar{\mathbf{n}},x) & -\cos(\bar{\mathbf{n}},y) \\ \cos(\bar{\mathbf{n}},y) & \cos(\bar{\mathbf{n}},x) \end{bmatrix} \begin{Bmatrix} \frac{d}{dn}\Phi \\ \frac{d}{ds}\Phi \end{Bmatrix}, \tag{4.7}$$

we obtain

$$G_1 = -G_2^N\cos(\bar{\mathbf{n}},x) + (\frac{d}{ds}G_1^D)\cos(\bar{\mathbf{n}},y), \qquad G_2 = G_2^N\cos(\bar{\mathbf{n}},y) + (\frac{d}{ds}G_1^D)\cos(\bar{\mathbf{n}},x). \tag{4.8}$$

The first two conditions of (4.4) are clearly satisfied since $\frac{d}{ds}G_1^D$ and G_2^N are single-valued functions. The condition of (4.5) is identically satisfied since (see also (3.5))

$$\oint_{\partial\Omega} \{ [G_2^N\cos(\bar{\mathbf{n}},y) + (\frac{d}{ds}G_1^D)\cos(\bar{\mathbf{n}},x)]\cos(\bar{\mathbf{n}},x) \\ + [-G_2^N\cos(\bar{\mathbf{n}},x) + (\frac{d}{ds}G_1^D)\cos(\bar{\mathbf{n}},y)]\cos(\bar{\mathbf{n}},y)\} = \oint_{\partial\Omega} \frac{d}{ds}G_1^D = 0. \tag{4.9}$$

The *Coupled-Plane* BVP shown in (3.142a,b) is written as

$$\nabla_1^{(4)}\Phi + \nabla_1^{(3)}\Lambda = g, \qquad \nabla_1^{(3)}\Phi + \nabla_1^{(2)}\Lambda = f \qquad \text{over} \quad \Omega, \tag{4.10a}$$

$$\frac{d}{ds}\{\Phi_{,x}, \Phi_{,y}, \Lambda\} = \{-F_1, F_2, F_3\} \quad \text{or} \quad \{\Phi_{,x}, \Phi_{,y}, \Lambda\} = \{-G_1, G_2, G_3\} \quad \text{on} \quad \partial\Omega, \tag{4.10b}$$

where the corresponding single-value conditions are

$$\oint_{\partial\Omega}\{F_1, F_2, F_3, F_2y - F_1x\} = \{0, 0, 0, 0\}. \tag{4.11}$$

Table 4.1 presents generic terminology of the BVPs involved (note that the non-homogeneous Laplace's equation is also referred to as *Poisson's equation*). In some contexts the functions $F_j,\ j = 1,2,3$ of (4.10b) (and similarly F_3^Λ of (4.1c)) are written as

$$F_j(x,y) = P_j(x,y)\cos(\bar{\mathbf{n}},x) + Q_j(x,y)\cos(\bar{\mathbf{n}},y), \tag{4.12}$$

where $P_j,\ Q_j$ are continuous functions.

As already indicated we split the solution as $\Phi = \Phi_0 + \Phi_p,\ \Lambda = \Lambda_0 + \Lambda_p,$ where Φ_p,Λ_p are the "particular solutions", and $\Phi_0,\ \Lambda_0$ are the "homogeneous solutions". The first step is therefore the determination of particular solutions of the corresponding equations (4.1a), (4.3a), (4.6), (4.10a) that hold for $(x,y) \in \Omega$. Once such solutions are obtained, the equations become homogeneous, and Φ_0 and Λ_0 should satisfy the following BVPs.

The *Dirichlet/Neumann* BVP

$$\nabla_3^{(2)}\Lambda = 0 \qquad \text{over} \quad \Omega, \tag{4.13a}$$

$$\frac{d}{ds}\Lambda = F_3 - \underline{\frac{d}{ds}\Lambda_p} \quad \text{or} \quad \Lambda = G_3 - \underline{\Lambda_p} \quad \text{or} \quad D_1^n\Lambda = F_3^\Lambda - \underline{D_1^n\Lambda_p} \qquad \text{on} \quad \partial\Omega. \tag{4.13b}$$

BVP	Operator(s)	Function(s)	Equation(s)	BVP type
$\nabla_3^{(2)}\Lambda = \ldots,\ \frac{d}{ds}\Lambda_{\vert\partial\Omega} = \ldots$	Laplace	harmonic	Laplace	Dirichlet
$\nabla_3^{(2)}\Lambda = \ldots,\ \Lambda_{\vert\partial\Omega} = \ldots$	Laplace	harmonic	Laplace	Dirichlet
$\nabla_3^{(2)}\Lambda = \ldots,\ D_1^n\Lambda = \ldots$	Laplace	harmonic	Laplace	Neumann
$\nabla_j^{(4)}\Phi = \ldots,\ (j = 1 \text{ or } 3)$ $\frac{d}{ds}\Phi_{,x\vert\partial\Omega} = \ldots,\ \frac{d}{ds}\Phi_{,y\vert\partial\Omega} = \ldots$	biharmonic	biharmonic	biharmonic	biharmonic
$\nabla_j^{(4)}\Phi = \ldots,\ (j = 1 \text{ or } 3)$ $\frac{d}{dn}\Phi_{\vert\partial\Omega} = \ldots,\ \Phi_{\vert\partial\Omega} = \ldots$	biharmonic	biharmonic	biharmonic	biDirichlet
$\nabla_1^{(4)}\Phi + \nabla_1^{(3)}\Lambda = \ldots$ $\nabla_1^{(3)}\Phi + \nabla_1^{(2)}\Lambda = \ldots$ $\frac{d}{ds}\Phi_{,x\vert\partial\Omega} = \ldots,\ \frac{d}{ds}\Phi_{,y\vert\partial\Omega} = \ldots$ $\frac{d}{ds}\Lambda_{\vert\partial\Omega} = \ldots$	Laplace, biharmonic and third-order	coupled "harmonic and biharmonic"	Coupled-Plane	Coupled-Plane

Table 4.1: BVPs terminology.

For the *biharmonic* BVP

$$\nabla_1^{(4)}\Phi = 0 \quad \text{over} \quad \Omega, \tag{4.14a}$$

$$\frac{d}{ds}\{\Phi_{,x},\ \Phi_{,y}\} = \{-F_1 - \underline{\frac{d}{ds}\Phi_{p,x}},\ F_2 - \underline{\frac{d}{ds}\Phi_{p,y}}\} \quad \text{on} \quad \partial\Omega \quad \text{or} \tag{4.14b}$$

$$\{\Phi_{,x},\ \Phi_{,y}\} = \{-G_1 - \underline{\Phi_{p,x}},\ G_2 - \underline{\Phi_{p,y}}\} \quad \text{on} \quad \partial\Omega, \tag{4.14c}$$

while the reduced boundary conditions for the homogeneous biDirichlet BVP are

$$\{\Phi,\ \frac{d}{dn}\Phi\} = \{G_1^D - \underline{\Phi_p},\ G_2^N - \underline{\frac{d}{dn}\Phi_p}\} \quad \text{on} \quad \partial\Omega. \tag{4.15}$$

For the *Coupled-Plane* BVP

$$\nabla_1^{(4)}\Phi + \nabla_1^{(3)}\Lambda = 0, \qquad \nabla_1^{(3)}\Phi + \nabla_1^{(2)}\Lambda = 0 \quad \text{over} \quad \Omega, \tag{4.16a}$$

$$\frac{d}{ds}\{\Phi_{,x},\ \Phi_{,y},\ \Lambda\} = \{-F_1 - \underline{\frac{d}{ds}\Phi_{p,x}},\ F_2 - \underline{\frac{d}{ds}\Phi_{p,y}},\ F_3 - \underline{\frac{d}{ds}\Lambda_p}\} \quad \text{on} \quad \partial\Omega \quad \text{or}$$

$$\{\Phi_{,x},\ \Phi_{,y},\ \Lambda\} = \{-G_1 - \underline{\Phi_{p,x}},\ G_2 - \underline{\Phi_{p,y}},\ G_3 - \underline{\Lambda_p}\} \quad \text{on} \quad \partial\Omega. \tag{4.16b}$$

As shown, the underlined terms represent the influence of the selected particular solutions on the homogeneous solutions. In what follows we deal with the particular and homogeneous solutions one at a time.

4.2 Particular Polynomial Solutions

We shall discuss here some generic ways to derive particular solutions for the BVPs presented in S.4.1. Particular solutions satisfy only the field equations, and as a rule (except for an ellipse), the discussion in this book will not be related to a specific domain shape. Note that since the particular solutions are included in the compatibility requirements, their exactness is essential for displacement integration. Clearly, for each case, it is sufficient to find *one* solution out of the infinite number of valid particular solutions.

As a first step, we present the r.h.s. of the field equations as sums of *homogeneous polynomials*, namely

$$F_0^{\Lambda}=\sum_{k=0}^{P_\Lambda} F_0^{\Lambda(k)}, \qquad F_0^{\Lambda(k)}=\sum_{i=0}^{k} \overline{F}_{i,k-i}^{\Lambda} x^i y^{k-i}, \tag{4.17a}$$

$$F_0=\sum_{k=0}^{P_\Phi} F_0^{(k)}, \qquad F_0^{(k)}=\sum_{i=0}^{k} \overline{F}_{i,k-i} x^i y^{k-i}, \tag{4.17b}$$

$$g=\sum_{k=0}^{P_g} g^{(k)}, \qquad g^{(k)}=\sum_{i=0}^{k} \overline{g}_{i,k-i} x^i y^{k-i}, \tag{4.17c}$$

$$f=\sum_{k=0}^{P_f} f^{(k)}, \qquad f^{(k)}=\sum_{i=0}^{k} \overline{f}_{i,k-i} x^i y^{k-i}. \tag{4.17d}$$

In the above equations, P_Λ, P_Φ, P_g and P_f are the highest degrees of polynomials required to describe the respective field equations. At this stage, we shall consider each case of the BVPs discussed in S.4.1, in a separate way. A summary of particular solution programs appears in Table P.2.

Remark 4.1 When a plane problem is under discussion, the body forces X_b and Y_b in the x- and y- directions, respectively, see S.3.2.4, S.3.2.5, are also expanded into a sum of homogeneous polynomials as

$$X_b=\sum_{k=0}^{P_X} X^{(k)}, \qquad X^{(k)}=\sum_{i=0}^{k} \overline{X}_{i,k-i} x^i y^{k-i}, \tag{4.18a}$$

$$Y_b=\sum_{k=0}^{P_Y} Y^{(k)}, \qquad Y^{(k)}=\sum_{i=0}^{k} \overline{Y}_{i,k-i} x^i y^{k-i}. \tag{4.18b}$$

The body-force potentials may be found by integrating the above functions according to their definitions in (3.55), namely $\overline{U}_1=-\int_0^x X_b\,dx$, $\overline{U}_2=-\int_0^y Y_b\,dy$, and subsequently may be expanded as sums of homogeneous (but not complete) polynomials

$$\overline{U}_1=\sum_{k=1}^{P_X+1} U_1{}^{(k)}, \qquad U_1{}^{(k)}=\sum_{i=1}^{k} \frac{1}{i}\overline{X}_{i-1,k-i} x^i y^{k-i}, \tag{4.19a}$$

$$\overline{U}_2=\sum_{k=1}^{P_Y+1} U_2{}^{(k)}, \qquad U_2{}^{(k)}=\sum_{i=0}^{k-1} \frac{1}{k-i}\overline{Y}_{i,k-i-1} x^i y^{k-i}. \tag{4.19b}$$

These operations enable us to define the coefficients in (4.17b–d).

4.2.1 The Biharmonic BVP

The particular solution of (4.3a) with (4.17b) may be chosen as a $(P_\Phi+4)$-degree polynomial Φ_p, presented by a sum of homogeneous polynomials $\Phi_p^{(k)}$ as

$$\Phi_p=\sum_{k=0}^{P_\Phi} \Phi_p^{(k+4)}, \qquad \Phi_p^{(k)}=\sum_{i=0}^{k} \overline{B}_{i,k-i} x^i y^{k-i}. \tag{4.20}$$

When the biharmonic operator of (4.3a) is applied to $\Phi_p^{(k+4)}$ of (4.20), a homogeneous polynomial of degree k is obtained. It is therefore possible to set a linear system of $k+1$ equations ($k+1$ is the number of $F_0^{(k)}$ coefficients, see (4.17b)). However, the number of unknowns is $k+5$ (which is the number of $\Phi_p^{(k+4)}$ coefficients, see (4.20)). Thus, four undetermined coefficients are left and the system extracted from $\nabla_1^{(4)}\Phi_p^{(k)}=F_0^{(k)}$ may be written as $\mathbf{M}\cdot\overline{\mathbf{B}}=\overline{\mathbf{F}}$, where $\mathbf{M}$ is a $(k+1)\times(k+5)$ matrix, and

$$\overline{\mathbf{F}}=[\overline{F}_{0,k},\overline{F}_{1,k-1},\ldots,\overline{F}_{k,0}]^T, \qquad \overline{\mathbf{B}}=[\overline{B}_{0,k+4},\overline{B}_{1,k+3},\ldots,\overline{B}_{k+4,0}]^T \tag{4.21}$$

are vectors. As indicated, in the general case $rank(M) = k+1$, and it is possible to select arbitrarily four of the $\overline{B}_{ij}$ coefficients (e.g. as zeros). The resulting complete polynomial Φ_p of (4.20) contains $4P_\Phi + 4$ of such coefficients, see Example 4.1.

A special particular solution that has no influence along the boundary contour may be drawn for an elliptical domain (of semi-axes $\tilde{a} > \tilde{b} > 0$). In such a case, we replace (4.20:b) with

$$\Phi_p^{(k)} = (\sum_{i=0}^{k} \overline{B}_{i,k-i} x^i y^{k-i} + \sum_{i=0}^{k-2} \overline{B}_{i,k-2-i} x^i y^{k-2-i} + \cdots)(\frac{x^2}{\tilde{a}^2} + \frac{y^2}{\tilde{b}^2} - 1)^2, \tag{4.22}$$

where the summation process in the brackets is terminated by either $\overline{B}_{01}y + \overline{B}_{10}x$ or $\overline{B}_{00}$. In general, the above $\Phi_p^{(k)}$ is a non-homogeneous polynomial (except for the case of $k = 0, 1$). With this alternative definition, we obtain $\Phi_p^{(k)}$ (and not $\Phi_p^{(k+4)}$) as the particular solution for $F_0^{(k)}$. When the definition (4.22) is used, a consistent linear system (of rank $k+1$) for the $\overline{B}_{ij}$ coefficients is obtained and solved. Since the above particular solution vanishes over the ellipse contour, the underlined terms in (4.14b,c), (4.15) vanish, and hence the particular and the homogeneous solutions are uncoupled. This way of the particular solution evaluation has, therefore, a clear advantage. **P.4.1** constructs particular solutions for an elliptical domain according to the above described methodology.

Example 4.1 *Particular Polynomial Solutions of the Biharmonic Equation (Low Order).*
Let F_0 be a homogeneous polynomial of degree $k = 2$. We subsequently express $F_0^{(2)}$ and the corresponding particular solution $\Phi_p^{(6)}$ as

$$F_0^{(2)} = \overline{F}_{02}y^2 + \overline{F}_{11}xy + \overline{F}_{20}x^2, \tag{4.23a}$$

$$\Phi_p^{(6)} = \overline{B}_{06}y^6 + \overline{B}_{15}xy^5 + \overline{B}_{24}x^2y^4 + \underline{\overline{B}_{33}x^3y^3 + \overline{B}_{42}x^4y^2 + \overline{B}_{51}x^5y + \overline{B}_{60}x^6}, \tag{4.23b}$$

where we agree to set to zero the four underlined terms (with the highest power of x) as shown by (4.23b). **P.4.2** enables the construction of particular polynomial solutions according to the above described methodology. The corresponding reduced 3×3-order matrix **M** becomes

$$\mathbf{M} = \begin{bmatrix} 0 & 120b_{11} & -96b_{16} \\ 0 & 0 & 24b_{11} \\ 360b_{11} & -120b_{16} & 48b_{12} + 24b_{66} \end{bmatrix}. \tag{4.24}$$

The system $\mathbf{M} \cdot \overline{\mathbf{B}} = \overline{\mathbf{F}}$ solution yields

$$\overline{B}_{06} = \frac{(4b_{16}{}^2 - 2b_{12}b_{11} - b_{66}b_{11})\overline{F}_{20} + b_{16}b_{11}\overline{F}_{11} + b_{11}{}^2\overline{F}_{02}}{360b_{11}{}^3},$$

$$\overline{B}_{15} = \frac{4b_{16}\overline{F}_{20} + b_{11}\overline{F}_{11}}{120b_{11}{}^2}, \qquad \overline{B}_{24} = \frac{\overline{F}_{20}}{24b_{11}}, \tag{4.25}$$

which may be substituted in (4.23b) to establish the particular solution $\Phi_p^{(6)}$.

4.2.2 *The Dirichlet/Neumann BVPs*

Analogously to (4.20), the particular solution Λ_p of (4.1a), (4.17a) may be chosen as $(P_\Lambda + 2)$-degree polynomial, and presented by a sum of homogeneous polynomials $\Lambda_p^{(k)}$ as

$$\Lambda_p = \sum_{k=0}^{P_\Lambda} \Lambda_p^{(k+2)}, \qquad \Lambda_p^{(k)} = \sum_{i=0}^{k} \overline{H}_{i,k-i} x^i y^{k-i}. \tag{4.26}$$

The procedure here is analogous to one in S.4.2.1. For each $F_0^{\Lambda(k)}$ we expand a homogeneous polynomial $\Lambda_p^{(k+2)}$. The resulting **M** matrix in this case is of $(k+1)\times(k+3)$-order, and we therefore may arbitrarily select two of the $\overline{H}_{ij}$ coefficients. The resulting complete polynomial Λ_p of (4.26) therefore contains $2P_\Lambda+2$ of such coefficients. **P.4.3** derives particular solutions according to the above described methodology.

Similar to (4.22), a special particular solution that vanishes on the boundary contour may be drawn for an elliptical domain. Here, for each $F_0^{\Lambda(k)}$, the solution is assumed as

$$\Lambda_p^{(k)} = (\sum_{i=0}^{k}\overline{H}_{i,k-i}x^i y^{k-i} + \sum_{i=0}^{k-2}\overline{H}_{i,k-2-i}x^i y^{k-2-i} + \cdots)(\frac{x^2}{\tilde{a}^2} + \frac{y^2}{\tilde{b}^2} - 1), \tag{4.27}$$

where the summation process in the brackets is terminated by either $\overline{H}_{01}y+\overline{H}_{10}x$ or $\overline{H}_{00}$. Here, $\Lambda_p^{(k)}$ is a non-homogeneous polynomial (except for the case of $k=0,1$), and with this alternative definition, we obtain $\Lambda_p^{(k)}$ (and <u>not</u> $\Lambda_p^{(k+2)}$) as the particular solution for $F_0^{\Lambda(k)}$. Since $\Lambda_p^{(k)}$ of (4.27) vanishes over the ellipse boundary, the particular and the homogeneous solutions become uncoupled as the underlined terms in (4.13b: a,b) vanish (note that this may not be true for the Neumann BVP, i.e. for (4.13b: c)). **P.4.4** enables the construction of particular solutions for an elliptical domain according to the above described methodology.

4.2.3 The Coupled-Plane BVP

To derive a particular solution in this case, we use (4.20), (4.26), and for balancing powers of x and y, employ simultaneously the homogeneous polynomials $g^{(k)}$ and $f^{(k+1)}$ of (4.17c,d) on the r.h.s. of (4.10a). Hence we expand $\Phi_p^{(k+4)}$ as shown in (4.20) and select (for simplicity as zeros) arbitrarily four of the $k+5$ coefficients involved (leaving $k+1$ coefficients to be determined). Similarly, we expand $\Lambda_p^{(k+3)}$ as shown in (4.26) and select two out of the $k+4$ coefficients involved (leaving $k+2$ coefficients to be determined). Hence, overall, the resulting $\Phi_p^{(k+4)}$ and $\Lambda_p^{(k+3)}$ are polynomials of degrees $k+4$ and $k+3$, respectively, that contain a total of $2k+3$ coefficients to be determined selected coefficients.

P.4.5 enables the construction of particular polynomial solutions according to the above described methodology.

In the special case of an elliptical domain, for each set of the homogeneous polynomials $g^{(k)}$ and $f^{(k+1)}$ we assume (4.22), (4.27) and substitute them in (4.10a), to obtain a consistent linear system for the $\overline{B}_{ij}$, $\overline{H}_{ij}$ coefficients. Since, the underlined terms in (4.16b) vanish, the particular and the homogeneous solutions become uncoupled. **P.4.6** enables the particular solutions for an elliptical domain according to the above described methodology. Example 4.2 presents the cases of $P_g=0,1$.

Example 4.2 *Particular Polynomial Solutions of Coupled-Plane BVP (Low Order).*

1. Suppose we wish to determine a particular solution for $g=\overline{g}_{00}$, $f=\overline{f}_{00}+\overline{f}_{10}x+\overline{f}_{01}y$. We first activate **P.4.5** with $k_g=-1$ and $k_f=0$ to obtain

$$\Lambda_p^{(2)} = \frac{\overline{f}_{00}}{2b_{55}}y^2, \qquad \Phi_p^{(3)} = 0. \tag{4.28}$$

We then activate **P.4.5** with $k_g=0$ and $k_f=1$ to obtain

$$\Lambda_p^{(3)} = \overline{H}_{03}y^3 + \overline{H}_{12}xy^2, \qquad \Phi_p^{(4)} = \overline{B}_{04}y^4, \tag{4.29}$$

where

$$\overline{H}_{03} = -\frac{b_{15}b_{55}\overline{g}_{00} + b_{15}b_{14}\overline{f}_{10} + b_{15}b_{56}\overline{f}_{10} - b_{11}b_{55}\overline{f}_{01} - 2b_{11}b_{45}\overline{f}_{10}}{6b_{55}(b_{55}b_{11} - b_{15}{}^2)}, \qquad \overline{H}_{12} = \frac{\overline{f}_{10}}{2b_{55}},$$

$$\overline{B}_{04} = \frac{b_{55}^2\overline{g}_{00} + b_{55}b_{14}\overline{f}_{10} + b_{55}b_{56}\overline{f}_{10} - b_{15}b_{55}\overline{f}_{01} - 2b_{15}b_{45}\overline{f}_{10}}{24b_{55}(b_{55}b_{11} - b_{15}{}^2)}. \tag{4.30}$$

The required solution will be the sum of the above two cases (4.28), (4.29), namely $\Phi_p = \Phi_p^{(4)}$, $\Lambda_p = \Lambda_p^{(2)} + \Lambda_p^{(3)}$.

2. We shall now modify the previous calculation for the special case of an ellipse with semi-axes $\widetilde{a} > \widetilde{b} > 0$. We first activate **P.4.6** symbolically with $k_g = -1$ and get

$$\Phi_p^{(-1)} = 0, \qquad \Lambda_p^{(0)} = \overline{H}_{00}(\frac{x^2}{\widetilde{a}^2} + \frac{y^2}{\widetilde{b}^2} - 1), \tag{4.31}$$

where $\overline{H}_{00} = \frac{\overline{f}_{00}}{2}(\frac{b_{55}}{\widetilde{b}^2} + \frac{b_{44}}{\widetilde{a}^2})^{-1}$. We next activate **P.4.6** symbolically with $k_g = 0$ and get

$$\Phi_p^{(0)} = \overline{B}_{00}(\frac{x^2}{\widetilde{a}^2} + \frac{y^2}{\widetilde{b}^2} - 1)^2, \qquad \Lambda_p^{(1)} = (\overline{H}_{10}x + \overline{H}_{01}y)(\frac{x^2}{\widetilde{a}^2} + \frac{y^2}{\widetilde{b}^2} - 1), \tag{4.32}$$

where the unknown coefficients $\overline{B}_{00}, \overline{H}_{10}, \overline{H}_{01}$ are determined by the linear system

$$\begin{aligned}
8(3\frac{b_{11}}{\widetilde{b}^4} + \frac{2b_{12}+b_{66}}{\widetilde{b}^2\widetilde{a}^2} + 3\frac{b_{22}}{\widetilde{a}^4})\overline{B}_{00} - 2(\frac{b_{14}+b_{56}}{\widetilde{b}^2} + 3\frac{b_{24}}{\widetilde{a}^2})\overline{H}_{10} + 2(3\frac{b_{15}}{\widetilde{b}^2} + \frac{b_{25}+b_{46}}{\widetilde{a}^2})\overline{H}_{01} = \overline{g}_{00},\\
-8(\frac{b_{14}+b_{56}}{\widetilde{b}^2\widetilde{a}^2} + 3\frac{b_{24}}{\widetilde{a}^4})\overline{B}_{00} + 2(\frac{b_{55}}{\widetilde{b}^2} + 3\frac{b_{44}}{\widetilde{a}^2})\overline{H}_{10} - 4\frac{b_{45}}{\widetilde{a}^2}\overline{H}_{01} = \overline{f}_{10},\\
8(3\frac{b_{15}}{\widetilde{b}^4} + \frac{b_{25}+b_{46}}{\widetilde{b}^2\widetilde{a}^2})\overline{B}_{00} - 4\frac{b_{45}}{\widetilde{b}^2}\overline{H}_{10} + 2(3\frac{b_{55}}{\widetilde{b}^2} + \frac{b_{44}}{\widetilde{a}^2})\overline{H}_{01} = \overline{f}_{02}.
\end{aligned} \tag{4.33}$$

The resulting expressions are quite lengthy, see **P.4.6**. Again, the required solution will be the sum of the above two cases: $\Phi_p = \Phi_p^{(0)}$, $\Lambda_p = \Lambda_p^{(0)} + \Lambda_p^{(1)}$.

It is interesting to note that by assuming $\overline{f}_{00} = -2\theta$, and $\Lambda_p^{(0)} = \psi$, the first stage of the above result (i.e. the expression for $\Lambda_p^{(0)}$) serves as the solution of the *torsion problem* for an elliptical rod made of MON13z material, see Remark 3.6. Accordingly, the applied torsion moment is given by $M_z = 2\iint_\Omega \psi$, or

$$M_z = -2\theta(\frac{b_{55}}{\widetilde{b}^2} + \frac{b_{44}}{\widetilde{a}^2})^{-1}\iint_\Omega(\frac{x^2}{\widetilde{a}^2} + \frac{y^2}{\widetilde{b}^2} - 1) = \frac{\pi\widetilde{a}^3\widetilde{b}^3\theta}{a_{55}\widetilde{a}^2 + a_{44}\widetilde{b}^2}, \tag{4.34}$$

where the last equality has been based on the properties $a_{44} = b_{44}$ and $a_{55} = b_{55}$ that hold for MON13z material.

4.3 Homogeneous BVPs Polynomial Solutions

We shall be focused here on solving the homogeneous BVPs, i.e. (4.13a,b), (4.14a,b), (4.16a,b). For the sake of convenience, we shall make two technical agreements. First, we shall omit the subscripts "0" in the function notation, and secondly, we shall assume that the underlined terms in (4.13b), (4.14b,c), (4.16b), (4.15) vanish, or, in other words, the particular solution influence is already embedded in the F_1, F_2, F_3^{Λ}, F_3 and G_1, G_2, G_3, G_1^D, G_2^N functions. Further on we shall also write F_3 also for F_3^{Λ} keeping in mind that its version should be carefully selected depending on the specific BVP. Table P.2 summarizes the relevant programs.

As shown in (4.12), with the aid of (3.4), the boundary functions F_1, F_2, F_3 may be described by the auxiliary functions P_i, Q_i as $F_i = P_i \frac{dy}{ds} - Q_i \frac{dx}{ds}$, where s is a natural parameter of a contour. Also, using (3.4) we write the operator of (3.216) as

$$D_1^n = (a_{44}\frac{\partial}{\partial x} - a_{45}\frac{\partial}{\partial y})\frac{dy}{ds} - (-a_{45}\frac{\partial}{\partial x} + a_{55}\frac{\partial}{\partial y})\frac{dx}{ds}. \tag{4.35}$$

In view of identities

$$\frac{d}{ds} = \frac{dt}{ds}\cdot\frac{d}{dt}, \qquad \cos(\bar{\mathbf{n}},x) = \frac{dy}{dt}\cdot\frac{dt}{ds}, \qquad \cos(\bar{\mathbf{n}},y) = -\frac{dx}{dt}\cdot\frac{dt}{ds}, \tag{4.36}$$

it is possible to replace $\frac{dx}{ds}$ and $\frac{dy}{ds}$ by $\frac{dx}{dt}$ and $\frac{dy}{dt}$, respectively, on both sides of (4.13b), (4.14b), (4.16b) (which mathematically is equivalent to a multiplication of both sides by $\frac{ds}{dt}$). In what follows we shall also employ the following identities, see (3.63a,b), (3.99):

$$\frac{d}{ds}\Phi_{,y} = \Phi_{,yx}\frac{dx}{ds} + \Phi_{,yy}\frac{dy}{ds}, \quad \frac{d}{ds}\Phi_{,x} = \Phi_{,xx}\frac{dx}{ds} + \Phi_{,xy}\frac{dy}{ds}, \quad \frac{d}{ds}\Lambda = \Lambda_{,x}\frac{dx}{ds} + \Lambda_{,y}\frac{dy}{ds}. \tag{4.37}$$

To this end, we may modify the boundary conditions of the associated BVPs as follows:

(a) For homogeneous *Dirichlet/Neumann-type* boundary conditions, (4.13b: a,c) become

$$\Lambda_{,x}\frac{dx}{dt} + \Lambda_{,y}\frac{dy}{dt} = P_3\frac{dy}{dt} - Q_3\frac{dx}{dt}, \tag{4.38a}$$

$$(a_{44}\Lambda_{,x} - a_{45}\Lambda_{,y})\frac{dy}{dt} - (-a_{45}\Lambda_{,x} + a_{55}\Lambda_{,y})\frac{dx}{dt} = P_3\frac{dy}{dt} - Q_3\frac{dx}{dt} \tag{4.38b}$$

while (4.2) should be written as

$$\oint_{\partial\Omega} P_3\,dy - Q_3\,dx = 0. \tag{4.39}$$

(b) For homogeneous *biharmonic* BVP boundary conditions, (4.14b) becomes

$$\Phi_{,xx}\frac{dx}{dt} + \Phi_{,xy}\frac{dy}{dt} = Q_1\frac{dx}{dt} - P_1\frac{dy}{dt}, \qquad \Phi_{,yx}\frac{dx}{dt} + \Phi_{,yy}\frac{dy}{dt} = -Q_2\frac{dx}{dt} + P_2\frac{dy}{dt}. \tag{4.40}$$

This problem should be solved under the conditions, see (4.4),

$$\oint_{\partial\Omega}\{Q_1\,dx - P_1\,dy,\ P_2\,dy - Q_2\,dx,\ (xQ_1 - yQ_2)\,dx + (yP_2 - xP_1)\,dy\} = \{0,0,0\}. \tag{4.41}$$

(c) For homogeneous *Coupled-Plane* BVP, boundary conditions are those of (4.38a), (4.40). This problem should be solved under the combined conditions of (4.39), (4.41), see (4.11).

4.3.1 *Prescribing the Boundary Functions*

To create exact or conditional polynomial solutions for the above homogeneous BVPs, the following assumptions are made:

(1) The auxiliary boundary functions P_k, Q_k, $k = 1,2,3$ are polynomials of x, y of degree k_Q, namely,

$$P_j = \sum_{s=0}^{k_Q} \sum_{i=0}^{s} p_{j,i,s-i} x^i y^{s-i}, \qquad Q_j = \sum_{s=0}^{k_Q} \sum_{i=0}^{s} q_{j,i,s-i} x^i y^{s-i}. \tag{4.42}$$

(2) The contour parametrization functions $x(t)$, $y(t)$ are trigonometric polynomials of level k_p, see S.3.1. (Obviously, the derivatives $\frac{dx}{dt}$, $\frac{dy}{dt}$ are of the same level).

Substituting P_k, Q_k of (4.42) in the r.h.s. of (4.38a,b), (4.40) and employing trigonometric identities enable us to present F_j along the contour as a Fourier series of level $(k_Q+1)k_p$, namely,

$$F_j(t) = Q_i \frac{dx}{dt} - P_i \frac{dy}{dt} = F_j^{c0} + \sum_{m=1}^{(k_Q+1)k_p} F_j^{sm} \sin(mt) + F_j^{cm} \cos(mt). \tag{4.43}$$

Before carrying out the solution itself, and as a preliminary solution step, one may wish to verify the fulfillment of the four single-value conditions given by (4.39), (4.41). Since the integrals $\oint_{\partial\Omega} \cos(mt)\, dt = \oint_{\partial\Omega} \sin(mt)\, dt = 0$ for any $m > 0$, the first three conditions simply become $F_1^{c0} = F_2^{c0} = F_3^{c0} = 0$. Analogous trigonometric expansion may be derived for the fourth condition using the parametrization functions, namely,

$$\widetilde{R}(t) = (xQ_1 - yQ_2)\frac{dx}{dt} + (yP_2 - xP_1)\frac{dy}{dt} = \widetilde{R}^{c0} + \sum_{m=1}^{(k_Q+2)k_p} \widetilde{R}^{sm} \sin(mt) + \widetilde{R}^{cm} \cos(mt), \tag{4.44}$$

and the fourth condition becomes $\widetilde{R}^{c0} = 0$.

P.4.7 illustrates prescribed polynomials that satisfy the conditions (4.39), (4.41).

4.3.2 Prescribing the Field Equations

We now invoke the analysis in S.3.2 and present the solution, Φ, of the homogeneous biharmonic equation (4.14a) as a generic polynomial of degree k_Φ, which according to (4.42) should be not less than $k_Q + 2$ (as implied, for example, from the terms $\Phi_{,xx}\frac{dx}{dt}$, $Q_1\frac{dx}{dt}$, in (4.40)), namely,

$$\Phi = \sum_{k=2}^{k_\Phi} \Phi^{(k)}, \qquad \Phi^{(k)} = \sum_{i=0}^{k} B_{i,k-i} x^i y^{k-i}. \tag{4.45}$$

The lowest level (2) of the above polynomial is due to its appearance in the boundary condition (4.40), and hence, the conditions $\Phi(0,0) = \Phi_{,x}(0,0) = \Phi_{,y}(0,0) = 0$ are satisfied. As shown in S.3.2.3 and by **P.3.5**, each of the above homogeneous polynomials $\Phi^{(k)}$ fulfills the biharmonic equation (4.14a) in an exact manner, and contributes four free (i.e. to be determined) coefficients B_{ij}, except of $\Phi^{(2)}$ that contributes three free coefficients. Therefore, the resulting prescribed biharmonic polynomial (4.45) depends on $4k_\Phi - 5$ free coefficients.

Analogously, **P.3.7** illustrates prescribed harmonic polynomials Λ,

$$\Lambda = \sum_{k=1}^{k_\Lambda} \Lambda^{(k)}, \qquad \Lambda^{(k)} = \sum_{i=0}^{k} H_{i,k-i} x^i y^{k-i} \tag{4.46}$$

of degree k_Λ, which according to (4.38a) and (4.42) should be not less than $k_Q + 1$. Note that condition $\Lambda(0,0) = 0$ is satisfied. Each of the above homogeneous polynomials $\Lambda^{(k)}$ fulfills the Laplace's equation (4.13a) in an exact manner and contributes two free coefficients H_{ij}. Hence, the prescribed harmonic polynomial (4.46) contains $2k_\Lambda$ free coefficients.

P.3.10 illustrates prescribed polynomial solutions of Coupled-Plane equations (4.16a), where Φ and Λ are expressed as (4.45), (4.46) with $k_\Lambda = k_\Phi - 1$ (while the conditions $k_\Lambda \geq k_Q + 1$, $k_\Phi \geq k_Q + 2$ must be maintained). In general, Φ depends on $4k_\Phi - 5$ free coefficients B_{ij}, and Λ depends on $2k_\Phi - 2$ free coefficients H_{ij}. The total number of coefficients is $6k_\Phi - 7$.

4.3.3 *Exact and Conditional Polynomial Solutions*

4.3.3.1 The Dirichlet/Neumann BVPs

For Dirichlet/Neumann problems, using the contour parametrization functions $x(t), y(t)$ are trigonometric identities, we write l.h.s. of (4.38a) or (4.38b) as

$$J_\Lambda(t) = \sum_{m=1}^{k_\Lambda k_p} J_\Lambda^{sm} \sin(mt) + J_\Lambda^{cm} \cos(mt) \tag{4.47}$$

where $k_\Lambda = k_Q + 1$. Clearly, the functions J_Λ^{sm}, J_Λ^{cm} contain $2k_\Lambda$ free coefficients H_{ij} of Λ. Therefore, in order to ensure that $J_\Lambda(t) - F_3(t) \equiv 0$ one should enforce the equations

$$J_\Lambda^{cm} - F_3^{cm} = 0, \qquad J_\Lambda^{sm} - F_3^{sm} = 0, \qquad m = 1, \ldots, k_\Lambda k_p, \tag{4.48}$$

namely, a total of $2k_\Lambda k_p$ equations. In an elliptical domain where $k_p = 1$ (see Example 3.1), the number of equations and coefficients is $2k_\Lambda$, and the above linear system is solvable in an exact manner. When $k_p > 1$, *conditional solutions* may be obtained by producing a set of *"data relations"* between the coefficients of the boundary polynomials, $p_{j,i,s-i}, q_{j,i,s-i}$, that must be satisfied in order to enable polynomial solutions. **P.4.8** presents exact and conditional polynomial solutions for homogeneous Dirichlet/Neumann BVPs, see Examples 4.3, 4.4.

See also application of polynomial solutions involved for the generalized Neumann problem derived in (Kezerashvili, 1986).

Example 4.3 *Exact Polynomial Solution of the Neumann BVP in an Ellipse (Low Order).*

In order to examine the homogeneous Neumann BVP in an ellipse, see (4.13a), (4.38b), we activate **P.4.8** with BVP$=-1$, $iso = 0$, $k_3 (= k_Q) = 1$, $k_O = 0$. In addition, we work with symbolic a_{ij} and set $x(t) = \tilde{a} \cos t$, $y(t) = \tilde{b} \sin t$, $k_p = 1$. Accordingly, the given boundary polynomials are $P_3 = p_{00} + p_{10}x + p_{01}y$, $Q_3 = q_{00} + q_{10}x + q_{01}y$. Therefore, the r.h.s. of (4.38b) is written as

$$\begin{aligned} F_3(t) = & -\underline{\frac{1}{2}\tilde{a}\tilde{b}(q_{01} + p_{10})} - \tilde{a} q_{00} \sin t - \tilde{b} p_{00} \cos t \\ & - \frac{1}{2}(\tilde{b}^2 p_{01} + \tilde{a}^2 q_{10}) \sin(2t) - \frac{1}{2}\tilde{a}\tilde{b}(p_{10} - q_{01}) \cos(2t). \end{aligned} \tag{4.49}$$

The condition $F_3^{c0} = 0$ (for the BVP solution existence and uniqueness) is given by the above underlined term as

$$q_{01} + p_{10} = 0. \tag{4.50}$$

Hence, by replacing p_{10} by $-q_{01}$, we obtain a consistent version of the boundary function

$$F_3(t) = -\tilde{a} q_{00} \sin t - \tilde{b} p_{00} \cos t - \frac{1}{2}(\tilde{b}^2 p_{01} + \tilde{a}^2 q_{10}) \sin(2t) + \tilde{a}\tilde{b} q_{01} \cos(2t). \tag{4.51}$$

P.4.8 shows that the prescribed harmonic polynomial becomes (with $k_\Lambda = k_Q + 1 = 2$)

$$\Lambda = H_{10}x + H_{01}y + H_{20}x^2 + H_{11}xy + \frac{a_{45}H_{11} - a_{44}H_{20}}{a_{55}} y^2. \tag{4.52}$$

By balancing the trigonometric functions, we arrive at the following system of $2k_\Lambda = 4$ equations:

$$\begin{aligned} & a_{45}H_{01} - a_{44}H_{10} = -p_{00}, \quad a_{45}H_{10} - a_{55}H_{01} = -q_{00}, \quad a_{45}H_{11} - 2a_{44}H_{20} = -p_{10}, \\ & H_{11}(2a_{45}^2 - a_{44}a_{55} - 4a_{55}^2) - 2H_{20}a_{45}(a_{44} + 4a_{55}) = -a_{55}(4q_{10} + p_{01}). \end{aligned} \tag{4.53}$$

The solution of this system is

$$H_{10} = -\frac{1}{a_0}(p_{00}a_{55} + q_{00}a_{45}), \qquad H_{01} = -\frac{1}{a_0}(q_{00}a_{44} + p_{00}a_{45}),$$
$$H_{20} = \frac{1}{2\widetilde{\delta}_1 a_0}[\widetilde{a}^2 a_{55}(a_{45}q_{10} - q_{01}a_{55}) + \widetilde{b}^2(a_{45}a_{55}p_{01} + 2q_{01}a_{45}{}^2 - q_{01}a_{44}a_{55})],$$
$$H_{11} = -\frac{1}{\widetilde{\delta}_1 a_0}[\widetilde{a}^2 a_{55}(a_{45}q_{01} - q_{10}a_{44}) - \widetilde{b}^2 a_{44}(a_{55}p_{01} + a_{45}q_{01})], \tag{4.54}$$

where $a_0 = a_{44}a_{55} - a_{45}^2$ and $\widetilde{\delta}_1 = a_{55}\widetilde{a}^2 + a_{44}\widetilde{b}^2$.

Example 4.4 *Exact Polynomial Solution of the Neumann BVP in an Ellipse (Higher Order).*
In view of the superposition principle, we shall now restrict our attention to the second-order homogeneous polynomials $P_3 = p_{20}x^2 + p_{11}xy + p_{02}y^2$, $Q_3 = q_{20}x^2 + q_{11}xy + q_{02}y^2$. Hence, this example is identical to Example 4.3 except for the fact that we now activate **P.4.8** with $k_3(= k_Q) = 2$ and $k_0 = 2$. **P.4.8** shows that (4.39) is satisfied when Λ is expressed as

$$\Lambda = H_{10}x + H_{01}y + H_{30}x^3 + H_{21}x^2y + H_{12}xy^2 + H_{03}y^3, \tag{4.55}$$

where we have chosen to eliminate the coefficients

$$H_{12} = \frac{2a_{45}H_{21} - 3a_{44}H_{30}}{a_{55}}, \qquad H_{03} = -\frac{(a_{44}a_{55} - 4a_{45}^2)H_{21} + 6a_{45}a_{44}H_{30}}{3a_{55}^2}. \tag{4.56}$$

The resulting expressions for the remaining four coefficients are quite lengthy and require some "careful organization". **P.4.9** shows that the expressions documented below are identical to those obtained by **P.4.8**:

$$\begin{aligned} H_{01}\widetilde{\delta}_2 = {} & \widetilde{a}^2 a_{45}a_{55}(a_{55}\widetilde{a}^2 + a_{44}\widetilde{b}^2)(b^2 p_{02} + \widetilde{a}^2 q_{11}) \\ & + \widetilde{b}^2 a_{44}[(3a_{55}a_{44} - 2a_{45}{}^2)\widetilde{a}^2 + a_{44}{}^2\widetilde{b}^2](\widetilde{b}^2 p_{11} + \widetilde{a}^2 q_{20}) \\ & + [2a_{55}{}^2\widetilde{a}^4 + (9a_{55}a_{44} - 4a_{45}{}^2)\widetilde{b}^2\widetilde{a}^2 + 3a_{44}{}^2\widetilde{b}^4]\widetilde{a}^2 a_{45}p_{20} \\ & + [3a_{55}{}^2\widetilde{a}^4 + (7a_{55}a_{44} - 2a_{45}{}^2)\widetilde{b}^2\widetilde{a}^2 + 2a_{44}{}^2\widetilde{b}^4]\widetilde{b}^2 a_{44}q_{02}, \end{aligned} \tag{4.57a}$$

$$\begin{aligned} H_{10}\widetilde{\delta}_2 = {} & \widetilde{a}^2 a_{55}[a_{55}{}^2\widetilde{a}^2 + (3a_{55}a_{44} - 2a_{45}{}^2)\widetilde{b}^2](\widetilde{b}^2 p_{02} + \widetilde{a}^2 q_{11}) \\ & + \widetilde{b}^2 a_{44}a_{45}(a_{55}\widetilde{a}^2 + \widetilde{b}^2 a_{44})(\widetilde{b}^2 p_{11} + \widetilde{a}^2 q_{20}) \\ & + [2a_{55}{}^2\widetilde{a}^4 + (7a_{55}a_{44} - 2a_{45}{}^2)\widetilde{b}^2\widetilde{a}^2 + 3a_{44}{}^2\widetilde{b}^4]\widetilde{a}^2 a_{55}p_{20} \\ & + [3a_{55}{}^2\widetilde{a}^4 + (9a_{55}a_{44} - 4a_{45}{}^2)\widetilde{b}^2\widetilde{a}^2 + 2a_{44}{}^2\widetilde{b}^4]\widetilde{b}^2 a_{45}q_{02} \end{aligned} \tag{4.57b}$$

$$\begin{aligned} H_{30}\widetilde{\delta}_2 = {} & a_{55}[(2a_{45}{}^2 - a_{55}a_{44})\widetilde{b}^2 - \frac{1}{3}a_{55}^2\widetilde{a}^2](\widetilde{b}^2 p_{02} - \widetilde{a}^2 p_{20} + \widetilde{a}^2 q_{11}) \\ & + a_{45}[a_{55}{}^2\widetilde{a}^2 + (a_{55}a_{44} - \frac{4}{3}a_{45}{}^2)\widetilde{b}^2](\widetilde{b}^2 p_{11} - \widetilde{b}^2 q_{02} + \widetilde{a}^2 q_{20}), \end{aligned} \tag{4.57c}$$

$$\begin{aligned} H_{21}\widetilde{\delta}_2 = {} & a_{45}a_{55}(3a_{44}\widetilde{b}^2 - a_{55}\widetilde{a}^2)(\widetilde{b}^2 p_{02} - \widetilde{a}^2 p_{20} + \widetilde{a}^2 q_{11}) \\ & + a_{44}[3a_{55}{}^2\widetilde{a}^2 + (a_{55}a_{44} - 2a_{45}{}^2)\widetilde{b}^2](\widetilde{b}^2 p_{11} - \widetilde{b}^2 q_{02} + \widetilde{a}^2 q_{20}), \end{aligned} \tag{4.57d}$$

where $a_0 = a_{44}a_{55} - a_{45}^2$, $\widetilde{\delta}_1 = a_{55}\widetilde{a}^2 + a_{44}\widetilde{b}^2$ and $\widetilde{\delta}_2 = a_0(3\widetilde{\delta}_1^2 + 4a_0\widetilde{a}^2\widetilde{b}^2)$.

4.3.3.2 The Biharmonic BVP

Employing the parametrization functions $x(t)$, $y(t)$ and employing trigonometric identities, enable us to write the l.h.s. of (4.40) as

$$J_\Phi^x(t) = \sum_{m=1}^{(k_\Phi - 1)k_p} J_\Phi^{xsm}\sin(mt) + J_\Phi^{xcm}\cos(mt),$$

$$J^y_\Phi(t) = \sum_{m=1}^{(k_\Phi-1)k_p} J^{ysm}_\Phi \sin(mt) + J^{ycm}_\Phi \cos(mt). \tag{4.58}$$

Similar to S.4.3.3.1, in order to ensure that $J^x_\Phi(t) + F_1(t) \equiv 0$ and $J^y_\Phi(t) - F_2(t) \equiv 0$ one should enforce the following equations for $m = 1, \ldots, (k_\Phi - 1)k_p$:

$$J^{xcm}_\Phi + F^{cm}_1 = 0, \qquad J^{ycm}_\Phi - F^{cm}_2 = 0, \qquad J^{xsm}_\Phi + F^{sm}_1 = 0, \qquad J^{ysm}_\Phi - F^{sm}_2 = 0, \tag{4.59}$$

which yields a total of $4(k_\Phi - 1)k_p$ equations. For $k_p = 1$ there are $4k_\Phi - 4$ equations. Yet, in view of the single-value condition for Φ, the rank of this linear system is $4k_\Phi - 5$, see Remark 4.2. Hence, for $k_p = 1$ the number of equations and coefficients B_{ij} is the same (see discussion of S.4.3.2) and the above system is solvable in an exact manner. Again, when $k_p > 1$, conditional polynomial solutions may be obtained under assumed relations between the boundary polynomials coefficients $p_{j,i,s-i}, q_{j,i,s-i}$. **P.4.10** presents such exact and conditional polynomial solutions for homogeneous biharmonic BVPs.

Remark 4.2 The rank of the linear system (4.59) is lower by one than the number of equations due to the single-value conditions discussed in S.4.3.3. The conditions $F^{c0}_1 = F^{c0}_2 = 0$ are identically fulfilled since we earlier assumed $J^{xc0}_\Phi = 0$ and $J^{yc0}_\Phi = 0$ by taking the summation in (4.58) from $m = 1$. Hence only one condition should be imposed, i.e., $\widetilde{R}^{c0} = 0$, see (4.44). This additional condition lowers the system rank from $4k_\Phi - 4$ to $4k_\Phi - 5$. For higher k_p values (depending on the geometry, and mainly on the amount of "symmetry" it possesses), larger reduction in the system's rank is possible.

Example 4.5 *Exact Polynomial Solution of the Biharmonic BVP in an Ellipse (Low Order).*
For solving the homogeneous biharmonic problem for an ellipse, we activate **P.4.10** with $case = case_b$, $iso = 0$, $k_1 (= k_Q) = 1$, $k_0 = 0$. In addition, we work with symbolic b_{ij}, set $x(t) = \widetilde{a}\cos t$, $y(t) = \widetilde{b}\sin t$ and $k_p = 1$, and close "Ex=rot" command. We therefore assume the following third-degree polynomial solution (with $k_\Phi = k_Q + 2 = 3$)

$$\Phi = B_{20}x^2 + B_{11}xy + B_{02}y^2 + B_{30}x^3 + B_{21}x^2y + B_{12}xy^2 + B_{03}y^3, \tag{4.60}$$

while $P_j = p_{j00} + p_{j10}x + p_{j01}y$, $Q_j = q_{j00} + q_{j10}x + q_{j01}y$. The single-valued conditions in this case become

$$p_{100} = q_{200}, \qquad p_{110} = -q_{101}, \qquad p_{210} = -q_{201}. \tag{4.61}$$

We therefore arrive at $4k_\Phi - 4 = 8$ equations, but the system rank is only $4k_\Phi - 5 = 7$, which is the number of unknowns in (4.60). The system solution yields

$$B_{20} = \frac{1}{2}q_{100}, \quad B_{02} = \frac{1}{2}p_{200}, \quad B_{11} = -q_{200}, \quad B_{12} = -\frac{1}{2}q_{201}, \quad B_{21} = \frac{1}{2}q_{101},$$
$$B_{30} = \frac{1}{6\widetilde{a}^2}(\widetilde{b}^2 p_{101} + \widetilde{a}^2 q_{110} - \widetilde{b}^2 q_{201}), \quad B_{03} = \frac{1}{6\widetilde{b}^2}(\widetilde{a}^2 q_{101} + \widetilde{b}^2 p_{201} + \widetilde{a}^2 q_{210}). \tag{4.62}$$

As shown, the above result is not material-dependent since it deals with a low-order polynomial that satisfies the biharmonic equation identically.

Example 4.6 *Conditional Polynomial Solution of the Biharmonic BVP in a Triangle.*
To examine the conditional solution of the homogeneous biharmonic problem in a triangle, we activate **P.4.10** with $case = case_b$, $iso = 0$, $k_1 (= k_Q) = 1$, $k_0 = 1$. In addition, we work with symbolic b_{ij} and set $x(t) = -\sin t + \frac{1}{3}\sin(2t)$, $y(t) = \cos t + \frac{1}{3}\cos(2t)$ and $k_p = 2$ (which puts the system origin at the center of the triangle). The solution is therefore limited

to the loading $P_i = p_{i10}x + p_{i01}y$, $Q_i = q_{i10}x + q_{i01}y$ $(i = 1,2)$. **P.4.10** shows that the three single-valued conditions of (4.41) impose the data relations

$$p_{110} = -q_{101}, \qquad q_{201} = -p_{210} \tag{4.63}$$

and that this solution is a conditional one. In other words, an exact polynomial solution is possible only for specific loading distributions. In the present case, in order to form a valid loading distribution, the following additional relations are required:

$$q_{210} = -q_{101}, \qquad p_{210} = -p_{101}. \tag{4.64}$$

The stress function therefore becomes $\Phi = \frac{1}{6}p_{201}y^3 - \frac{1}{2}p_{101}xy^2 + \frac{1}{2}q_{101}x^2y + \frac{1}{6}q_{110}x^3$. As a special case, suppose that $q_{110} = 1$ and all three other coefficients vanish. Thus, we obtain $\Phi = \frac{1}{6}x^3$. This yields $F_2 = 0$ and $F_1(t) = -\frac{1}{6}\sin t - \frac{1}{2}\sin(2t) + \frac{1}{2}\sin(3t) - \frac{1}{9}\sin(4t)$. According to (3.65), this case may be generated by zero body-force and a surface loading of $X_s(t) = 0$, $Y_s(t) = F_1(t)$, as schematically shown in Fig. 4.1.

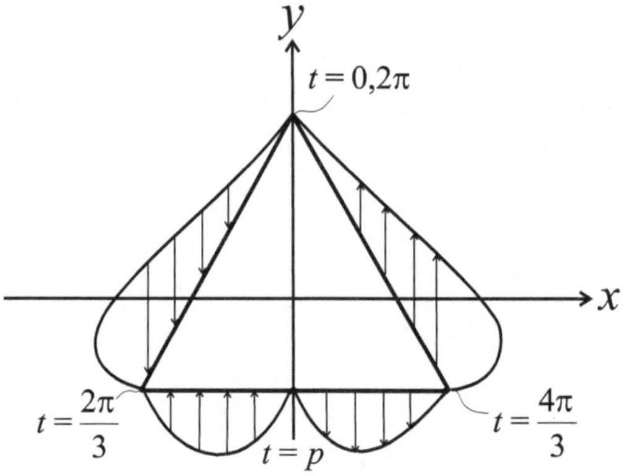

Figure 4.1: A conditional solution for the triangular domain of Example 4.6.

Example 4.7 *Exact Polynomial Solution of the Biharmonic BVP in an Ellipse (Higher Order).* We shall present here a symbolic solution of the homogeneous biharmonic problem in an ellipse, when the boundary functions, $F_j = P_j\cos(\bar{\mathbf{n}},x) + Q_j\cos(\bar{\mathbf{n}},y)$, $j = 1,2$, are described by (4.42) with $k_Q = 3$. Such a case encompasses most of the applications encountered in beam analysis with low-order loading, see Chapter 7. Activating **P.4.10** as in Example 4.5 but with $k_1(=k_Q) = 3$ yields the fifth-degree prescribed biharmonic polynomial

$$\begin{aligned}\Phi = {} & \underline{B_{50}x^5 + B_{41}x^4y + B_{32}x^3y^2 + B_{23}x^2y^3 + B_{14}xy^4 + B_{05}y^5} + B_{40}x^4 + B_{31}x^3y + B_{22}x^2y^2 \\ & + B_{13}xy^3 + B_{04}y^4 + B_{30}x^3 + B_{21}x^2y + B_{12}xy^2 + B_{03}y^3 + B_{20}x^2 + B_{11}xy + B_{02}y^2,\end{aligned} \tag{4.65}$$

while the three single-value conditions are

$$\begin{aligned} 4p_{210} &= -4q_{201}\underline{-\widetilde{a}^2(q_{221}+3p_{230}) - \widetilde{b}^2(p_{212}+3q_{203})}, \\ 4p_{110} &= -4q_{101}\underline{-\widetilde{a}^2(q_{121}+3p_{130}) - \widetilde{b}^2(p_{112}+3q_{103})}, \\ 4p_{100} &= 4q_{200} + \widetilde{a}^2(q_{220} - 3p_{120} - q_{111}) + \widetilde{b}^2(p_{211} + 3q_{202} - p_{102}). \end{aligned} \tag{4.66}$$

The underlined terms in the above and the following expressions should be ignored if one uses $k_1(=k_Q) = 2$ only. Since the detailed formulas for the involved B_{ij} terms are quite lengthy, we have organized them as shown below by suitable algebraic simplification. **P.4.11** proves that the following presentations and those obtained directly from **P.4.10** are identical:

$$
\begin{aligned}
\underline{B_{50}} &= -\frac{1}{80\widetilde{a}^4}[(-40B_{32}+4q_{112}-4p_{121}+q_{221}-p_{230})\widetilde{a}^2\widetilde{b}^2+(4p_{103}+p_{212}-q_{203})\widetilde{b}^4-4q_{130}\widetilde{a}^4],\\
\underline{B_{41}} &= -\frac{1}{16\widetilde{a}^2\widetilde{\delta}_3}[2\widetilde{b}^2(b_{16}\widetilde{a}^2+b_{26}\widetilde{b}^2)\widetilde{\delta}_4+\widetilde{a}^2(5b_{22}\widetilde{b}^4+\widetilde{a}^2\widetilde{b}^2(2b_{12}+b_{66})+b_{11}\widetilde{a}^4)\widetilde{\delta}_5],\\
\underline{B_{32}} &= -\frac{1}{8\widetilde{a}^2\widetilde{\delta}_3}[2\widetilde{a}^4(b_{16}\widetilde{a}^2+b_{26}\widetilde{b}^2)\widetilde{\delta}_5+(5b_{11}\widetilde{a}^4+\widetilde{a}^2\widetilde{b}^2(2b_{12}+b_{66})+b_{22}\widetilde{b}^4)\widetilde{\delta}_4],\\
\underline{B_{23}} &= -\frac{1}{8\widetilde{b}^2}[(q_{121}-p_{130}-16B_{41})\widetilde{a}^2+(p_{112}-q_{103})\widetilde{b}^2],\\
\underline{B_{14}} &= -\frac{1}{16\widetilde{b}^2}[(p_{230}-q_{221}-8B_{32})\widetilde{a}^2+(q_{203}-p_{212})\widetilde{b}^2],\\
\underline{B_{05}} &= -\frac{1}{40\widetilde{b}^4}[(-40B_{41}+3q_{121}+2q_{230}-3p_{130})\widetilde{a}^4+(3p_{112}+2p_{221}-3q_{103}-2q_{212})\widetilde{a}^2\widetilde{b}^2-2p_{203}\widetilde{b}^4],\\
B_{40} &= -\frac{1}{12\widetilde{a}^2}[(-6B_{22}+q_{102}-p_{111})\widetilde{b}^2-q_{120}\widetilde{a}^2],\\
B_{31} &= -\frac{1}{24\widetilde{a}^2}[(3p_{120}-3q_{111}-q_{220})\widetilde{a}^2+(q_{202}-p_{211}-3p_{102})\widetilde{b}^2],\\
B_{22} &= -\frac{1}{2(3b_{11}\widetilde{a}^4+\widetilde{a}^2\widetilde{b}^2(2b_{12}+b_{66})+3b_{22}\widetilde{b}^4)}[b_{11}(q_{211}-p_{220})\widetilde{a}^4\\
&\quad +b_{22}(p_{111}-q_{102})\widetilde{b}^4-(6b_{26}B_{31}+6b_{16}B_{13}-b_{22}q_{120}-b_{11}p_{202})\widetilde{a}^2\widetilde{b}^2],\\
B_{13} &= -\frac{1}{24\widetilde{b}^2}[(p_{120}-q_{111}-3q_{220})\widetilde{a}^2+(3q_{202}-3p_{211}-p_{102})\widetilde{b}^2],\\
B_{04} &= -\frac{1}{12\widetilde{b}^2}[(-6B_{22}-q_{211}+p_{220})\widetilde{a}^2-p_{202}\widetilde{b}^2],\\
B_{30} &= \frac{1}{6\widetilde{a}^2}[\underline{(-6B_{32}+q_{112})\widetilde{a}^2\widetilde{b}^2+(p_{103}-q_{203})\widetilde{b}^4}+(p_{101}-q_{201})\widetilde{b}^2+q_{110}\widetilde{a}^2],\\
B_{21} &= -\frac{1}{2}[\underline{(p_{130}+4B_{41})\widetilde{a}^2}+p_{110}], \quad B_{12}=\frac{1}{2}[\underline{(-2B_{32}+p_{230})\widetilde{a}^2}+p_{210}],\\
B_{03} &= \frac{1}{6\widetilde{b}^2}[\underline{(-12B_{41}+q_{230}-p_{130}+q_{121})\widetilde{a}^4+(p_{112}+p_{221})\widetilde{a}^2\widetilde{b}^2}+(q_{210}+q_{101})\widetilde{a}^2+p_{201}\widetilde{b}^2],\\
B_{11} &= -\frac{1}{8}[(5p_{120}+3q_{111}+q_{220})\widetilde{a}^2+(3p_{102}+p_{211}-q_{202})\widetilde{b}^2]-p_{100},\\
B_{20} &= \frac{1}{2}[(-2B_{22}+q_{102})\widetilde{b}^2+q_{100}], \quad B_{02}=\frac{1}{2}[(-2B_{22}+p_{220})\widetilde{a}^2+p_{200}],
\end{aligned}
\tag{4.67}
$$

where

$$
\begin{aligned}
\widetilde{\delta}_3 &= [5b_{22}\widetilde{b}^4+\widetilde{a}^2\widetilde{b}^2(2b_{12}+b_{66})+b_{11}\widetilde{a}^4][5b_{11}\widetilde{a}^4+\widetilde{a}^2\widetilde{b}^2(2b_{12}+b_{66})+b_{22}\widetilde{b}^4]\\
&\quad -4\widetilde{a}^2\widetilde{b}^2(b_{16}\widetilde{a}^2+b_{26}\widetilde{b}^2)^2,\\
\underline{\widetilde{\delta}_4} &= [2(q_{121}-p_{130})b_{16}+(q_{221}-p_{230})b_{11}]\widetilde{a}^6+(p_{230}-q_{221}+4p_{121}-4q_{112})b_{22}\widetilde{a}^2\widetilde{b}^4\\
&\quad +[2b_{16}(p_{112}-q_{103})+(p_{212}-q_{203})b_{11}+4b_{22}q_{130}]\widetilde{a}^4\widetilde{b}^2+(q_{203}-p_{212}-4p_{103})b_{22}\widetilde{b}^6,\\
\underline{\widetilde{\delta}_5} &= 2b_{11}(3p_{130}-2q_{230}-3q_{121})\widetilde{a}^4+[(2b_{12}+b_{66})(p_{130}-q_{121})+2b_{16}(p_{230}-q_{221})\\
&\quad +2b_{11}(3q_{103}+2q_{212}-2p_{221}-3p_{112})]\widetilde{a}^2\widetilde{b}^2\\
&\quad +[(2b_{12}+b_{66})(q_{103}-p_{112})+2(q_{203}-p_{212})b_{16}+4b_{11}p_{203}]\widetilde{b}^4.
\end{aligned}
\tag{4.68}
$$

Example 4.8 *Exact Polynomial Solution of Non-Homogeneous Biharmonic BVP in an Ellipse.*
Suppose now that the biharmonic problems discussed in Examples 4.5, 4.7 are non-homogeneous, and we wish to transform them into their homogeneous form by expressing the polynomial $\Phi(x,y)$ as the sum $\Phi = \Phi_p + \Phi_0$ of a particular and a homogeneous solution. By keeping the $k_Q = 3$ and $k_\Phi = 5$ selections (and again ignoring the underlined terms for $k_Q = 2$ and $k_\Phi = 4$), the maximal degree level of F_0 is $k_\Phi - 4 = 1$ (see S.4.2.1). Hence, $\nabla_1^{(4)}\Phi_p = \overline{F}_{00} + \underline{\overline{F}_{10}x + \overline{F}_{01}y}$, and we reach a particular solution of the form

$$\Phi_p = \underline{\frac{\overline{F}_{10}}{120b_{22}}x^5 + \frac{\overline{F}_{01}}{120b_{11}}y^5} + \frac{\overline{F}_{00}}{24b_{11}}y^4. \tag{4.69}$$

According to (4.14c), we may use the expressions derived in Example 4.7 to determine Φ_0, by modifying the coefficients of P_j, Q_j as

$$p_{203} \Rightarrow p_{203} - \frac{\overline{F}_{01}}{6b_{11}}, \quad p_{202} \Rightarrow p_{202} - \frac{\overline{F}_{00}}{2b_{11}}, \quad q_{130} \Rightarrow q_{130} - \frac{\overline{F}_{10}}{6b_{22}}. \tag{4.70}$$

Substituting the above p_{ijk}, q_{ijk} coefficients in the B_{ij} expressions derived in Example 4.7, we reach the following fifth-order polynomial $\Phi(x,y)$:

$$\begin{aligned}\Phi = &\underline{(B_{50} + \frac{\overline{F}_{10}}{120b_{22}})x^5 + B_{41}x^4y + B_{32}x^3y^2 + B_{23}x^2y^3 + B_{14}xy^4 + (B_{05} + \frac{\overline{F}_{01}}{120b_{11}})y^5} + B_{40}x^4 \\ &+ B_{31}x^3y + B_{22}x^2y^2 + B_{13}xy^3 + (B_{04} + \frac{\overline{F}_{00}}{24b_{11}})y^4 + B_{30}x^3 + B_{21}x^2y + B_{12}xy^2 + B_{03}y^3 \\ &+ B_{20}x^2 + B_{11}xy + B_{02}y^2.\end{aligned} \tag{4.71}$$

4.3.3.3 The Coupled-Plane BVP

Combining the definitions of S.4.3.3.1, S.4.3.3.2, we write the boundary conditions of the homogeneous Coupled-Plane BVP as the system (4.48), (4.59), where $k_\Lambda = k_\Phi - 1$.

For $k_p = 1$ the total number of equations and coefficients is the same, $6k_\Phi - 7$, and the above linear system is solvable in an exact manner. Conditional solutions may be obtained for $k_p > 1$ in this case as well. **P.4.12** presents exact and conditional polynomial solutions for the above Coupled-Plane BVP.

Example 4.9 *Exact Polynomial Solution of Coupled-Plane BVP in an Ellipse (Low Order).*
To examine the homogeneous problem presented by (4.16a), (4.38a), (4.40), we activate **P.4.12** with $k_1 (= k_Q) = 1$, $k_0 = 0$ and hence with $k_\Phi = 3$, $k_\Lambda = 2$. The program yields

$$\begin{aligned}\Phi &= B_{20}x^2 + B_{11}xy + B_{02}y^2 + \underline{B_{30}x^3 + B_{21}x^2y + B_{12}xy^2 + B_{03}y^3}, \\ \Lambda &= H_{10}x + H_{01}y + \underline{H_{20}x^2 + H_{11}xy + H_{02}y^2},\end{aligned} \tag{4.72}$$

where $H_{02} = \frac{1}{b_{55}}(b_{45}H_{11} - b_{44}H_{20} - 3b_{15}B_{03} + B_{12}b_{14} + B_{12}b_{56} - B_{21}b_{25} - B_{21}b_{46} + 3b_{24}B_{30})$. Here and below, the underlined terms are not used for the case of $k_1 (= k_Q) = 0$ (i.e., $k_\Phi = 2$, $k_\Lambda = 1$). According to the above selection,

$$\begin{aligned}&P_j = p_{j00} + \underline{p_{j10}x + p_{j01}y}, \qquad Q_j = q_{j00} + \underline{q_{j10}x + q_{j01}y}, \qquad j = 1,2, \\ &P_3 = p_{300}, \qquad Q_3 = q_{300}.\end{aligned} \tag{4.73}$$

P.4.12 also shows that the four conditions of (4.39), (4.41) impose (only) the following three relations:

$$\underline{p_{110} + q_{101} = 0}, \qquad \underline{p_{210} + q_{201} = 0}, \qquad q_{200} - p_{100} = 0. \tag{4.74}$$

Replacing q_{101} by $-p_{110}$, q_{201} by $-p_{210}$ and q_{200} by p_{100}, we obtain a consistent version of the boundary functions $F_1(t)$, $F_2(t)$,

$$F_1(t) = \tilde{a}q_{100}\sin t + \tilde{b}p_{100}\cos t \underline{- \frac{1}{2}(\tilde{b}^2 p_{101} + \tilde{a}^2 q_{110})\sin(2t) - \tilde{a}\tilde{b}p_{110}\cos(2t)},$$

$$F_2(t) = -\tilde{a}p_{100}\sin t + \tilde{b}p_{200}\cos t \underline{- \frac{1}{2}(\tilde{b}^2 p_{201} + \tilde{a}^2 q_{210})\sin(2t) - \tilde{a}\tilde{b}p_{210}\cos(2t)}. \tag{4.75}$$

Equations (4.38a), (4.40) on $\partial\Omega$ yield 12 linear equations of 11 variables (i.e. a system with a rank of 11), whose solution is

$$H_{10} = q_{300}, \qquad H_{01} = -p_{300}, \qquad \underline{H_{11}} = 0,$$

$$\underline{H_{20}} = \frac{1}{\tilde{a}^2(\tilde{a}^2 b_{55} + \tilde{b}^2 b_{44})}[b_{15}\tilde{b}^2\tilde{a}^2 p_{201} - \tilde{a}^2(b_{15}\tilde{a}^2 + b_{25}\tilde{b}^2 + b_{46}\tilde{b}^2)p_{110}$$
$$+ b_{15}\tilde{a}^4 q_{210} - \tilde{b}^2(b_{14}\tilde{a}^2 + b_{56}\tilde{a}^2 + b_{24}\tilde{b}^2)p_{210} - b_{24}\tilde{b}^4 p_{101} - b_{24}\tilde{b}^2\tilde{a}^2 q_{110}],$$

$$\underline{B_{03}} = -\frac{1}{6\tilde{b}^2}(\tilde{b}^2 p_{201} - \tilde{a}^2 p_{110} + \tilde{a}^2 q_{210}), \qquad \underline{B_{30}} = -\frac{1}{6\tilde{a}^2}(\tilde{b}^2 p_{101} + \tilde{b}^2 p_{210} + \tilde{a}^2 q_{110}),$$

$$\underline{B_{12}} = -\frac{1}{2}p_{210}, \quad \underline{B_{21}} = \frac{1}{2}p_{110}, \quad B_{20} = -\frac{1}{2}q_{100}, \quad B_{11} = p_{100}, \quad B_{02} = -\frac{1}{2}p_{200}. \tag{4.76}$$

4.3.4 *Approximate Polynomial Solutions*

As already indicated, for complex boundary contours and generic boundary functions that fulfill all required single-value conditions, the exact solution, if exists, may not be a polynomial. Hence, in all cases where an exact (global) polynomial solution may not be fitted, introducing an approximate polynomial is inevitable, see also Remark 4.2. As shown in S.4.3, once the field equations are fulfilled, we are left with $2k_\Lambda$ and $4k_\Phi - 5$ free coefficients of a prescribed polynomial solution. On the other hand, to fulfill the boundary conditions, we need to satisfy $2k_\Lambda k_p$ and $4(k_\Phi - 1)k_p$ equations, respectively. Hence, in general, we always reach an *"over-determined" system* and have *more equations to fulfill than coefficients to determine*.

For the Dirichlet/Neumann BVPs we generally write $k_\Lambda = k_Q + 1 + k_a$ where $k_a > 0$ is the solution augmentation degree, since more degrees of freedom must be incorporated when one wishes to raise solution accuracy. For the same reason we employ $k_\Phi = k_Q + 2 + k_a$.

The entire approximation for the prescribed homogeneous solution, if required, is shifted to the boundary conditions, and the exact fulfillment of the field equations (which have emerged from the compatibility equations) ensures consistent integrability of the strain components, see discussion in the introduction of this chapter. In what follows, we shall describe two methods to derive such an approximate solution that will minimize the overall inevitable error of the involved boundary equations.

The entire approximation process is carried out symbolically by rational numbers, see **P.4.13**, **P.4.14**, **P.4.15**. The resulting polynomials may be therefore denoted as "exact-approximate" solution, by which we mean that it is an *exact outcome of the least-squares process*, and can be proved to be the closest to the existing solution according to the least-squares criteria.

4.3.4.1 System Residual and Error Functionals

By harmonic balancing (i.e. by equating the coefficients of identical harmonics on both sides of the trigonometric equations), we express the "error functional" as a function of the residual harmonic coefficients. For the Dirichlet/Neumann BVPs we defined the error functional as

$$J_{DN}(H_{ij}) = \sum_{m=1}^{k_\Lambda k_p}[(J_\Lambda^{sm} - F_3^{sm})^2 + (J_\Lambda^{cm} - F_3^{cm})^2], \tag{4.77}$$

where H_{ij} are the free coefficients in the prescribed solution of Λ. Similarly, for the biharmonic BVP,

$$J_{BH}(B_{ij})=\sum_{m=1}^{(k_\Phi-1)k_p}[(J_\Phi^{xsm}+F_1^{sm})^2+(J_\Phi^{xcm}+F_1^{cm})^2+(J_\Phi^{ysm}-F_2^{sm})^2+(J_\Phi^{ycm}-F_2^{cm})^2], \quad (4.78)$$

where B_{ij} are the free coefficients in Φ. For the Coupled-Plane BVP, the error functional becomes

$$J_{CP}(H_{ij},B_{ij})=J_{DN}+J_{BH}. \quad (4.79)$$

We subsequently minimize the error functional by a least-square procedure as shown in Remark 4.3. This method is denoted in what follows as "Minimization of the System Residual".

4.3.4.2 Contour Error Functional

An alternative way to find an approximate polynomial solution of a homogeneous BVP is to integrate over the contour the square of the resulting error, which yields a positive-definite quadratic form as well. The minimization process leads again to a linear system of equations (differing from the one described in S.4.3.4.1), and again a least-square procedure may be applied.

For the Dirichlet/Neumann BVPs we define the error functional as

$$J_{DN}(H_{ij})=\oint_{\partial\Omega}\left[(\Lambda_{,x}+Q_3)\frac{dx}{dt}+(\Lambda_{,y}-P_3)\frac{dy}{dt}\right]^2 dt, \quad (4.80)$$

while for the biharmonic BVP we write

$$J_{BH}(B_{ij})=\oint_{\partial\Omega}\left\{\left[(\Phi_{,xx}-Q_1)\frac{dx}{dt}+(\Phi_{,xy}+P_1)\frac{dy}{dt}\right]^2+\left[(\Phi_{,yx}+Q_2)\frac{dx}{dt}+(\Phi_{,yy}-P_2)\frac{dy}{dt}\right]^2\right\}dt.$$

For the Coupled-Plane BVP, (4.79) holds in this case as well. In all of the above functionals, we replace x and y by the parametrization expressions described in S.3.1.2 and apply suitable trigonometric identities to express the integrands as Fourier series. In such a case, the integration procedure is simply replaced by considering the free coefficient only.

Note that when a polygon is under discussion, as described in S.3.1.4, one may directly use the piecewise-linear parametrization of (3.17) to derive the contour error functional.

This method is denoted in what follows as "Minimization of the Contour Error".

4.3.4.3 Error Index

As already indicated, when the homogeneous solution is founded on a prescribed polynomial, all errors are shifted to the boundary conditions. We therefore define a *relative error index*, which yields a quantitative index for the solution quality as shown by (see (4.38a,b), (4.40))

$$\varepsilon_{rel}=\frac{J_{min}}{\oint_{\partial\Omega}\sum_{i=i_0}^{k}\left(P_i\frac{dy}{dt}-Q_i\frac{dx}{dt}\right)^2 dt}, \quad (4.81)$$

where $i_0=3,\ k=3$ for the Dirichlet/Neumann BVPs, $i_0=1,\ k=2$ for the biharmonic BVP, and $i_0=1,\ k=3$ for the Coupled-Plane BVP. J_{min} is the calculated (minimal) value of the error functional. This definition holds for all error functional types described in S.4.3.4.1, S.4.3.4.2. Hence, per definition, $\varepsilon_{rel}\ll 1$ is expected for an acceptable approximate solution.

Remark 4.3 Consider an over-determined finite linear system $\mathbf{A} \cdot \mathbf{x} = \mathbf{b}$, where matrix $\mathbf{A}$ and vector $\mathbf{b}$ are known. To clarify the least-squares minimization of the error functionals, we build a non-negative definite quadratic form $J(\mathbf{x})$ (by ignoring the specific functional type) depending on a vector $\mathbf{x}$ that contains the participating coefficients H_{ij} and/or B_{ij}. To carry out the unconstrained minimization of $J(\mathbf{x})$,

$$J(\mathbf{x}) = ||\mathbf{A} \cdot \mathbf{x} - \mathbf{b}||^2 \to \min, \tag{4.82}$$

we wish to zero out the differential $dJ(\mathbf{x})$ by requiring $\mathbf{A}^T \cdot (\mathbf{A} \cdot \mathbf{x} - \mathbf{b})$=0. Since $\mathbf{A}$ is assumed to be of maximal rank, we find $\mathbf{x}_{\text{opt}} = (\mathbf{A}^T \cdot \mathbf{A})^{-1} \cdot \mathbf{A}^T \cdot \mathbf{b}$ where $\mathbf{A}^T \cdot \mathbf{A}$ is a symmetric and positive-definite matrix. The resulting $\mathbf{x}_{\text{opt}}$ coordinates give the optimal solution.

Example 4.10 *Torsion of a MON13z Rod with a Triangular Cross-Section.*
As discussed in S.3.3.3, the derivation of the generalized harmonic (warping) function in *torsion*, $\varphi(x,y)$, may be reduced to solution of the Neumann BVP of (3.121), where $\Lambda = \varphi$, $F_0^{\varphi} = 0$ and $F_3^{\varphi} = (a_{44}y + a_{45}x)\cos(\bar{\mathbf{n}}, x) + (-a_{45}y - a_{55}x)\cos(\bar{\mathbf{n}}, y)$.

To demonstrate the polynomial solution approach, we consider a simple example of isotropic *triangular* domain with the approximate parametrization $x = \sin t - \frac{1}{3}\sin(2t)$, $y = \cos t + \frac{1}{3}\cos(2t)$ of $k_p = 2$, and compare the obtained polynomial with the classical result.

Since $F_0^{\varphi} = 0$, no particular solution is required, and we activate **P.4.13** selecting BVP$= -1$, iso$= 1$, method_ $=$residual, $n = 3$, $ii = 2$, $R_0 = 1$, $k_3 = 1$, $kk = 1$, $dd = 0$. The program shows that $F_3(t) = \widetilde{A}\sin(3t)$, where $\widetilde{A}$ is a constant, and therefore, the first equation of (4.2) is satisfied. In the present case, $k_{\Lambda} = 3$, and the following prescribed polynomial (that satisfies (4.13a)) with six free coefficients is applied:

$$\Lambda_a = H_{10}x + H_{01}y - H_{02}x^2 + H_{11}xy + H_{02}y^2 + H_{30}x^3 - 3H_{03}x^2y - 3H_{30}xy^2 + H_{03}y^3. \tag{4.83}$$

The linear system to be solved contains ten (independent) equations with six unknowns. Employing the *least-squares method* yields the single nonzero term $H_{30} = -\frac{81}{325}$. So the approximate solution becomes

$$\Lambda_a = \widetilde{c}(y^2 - \frac{x^2}{3})x \tag{4.84}$$

where $\widetilde{c} = \frac{243}{325}$. As shown by (Novozhilov, 1961), the exact classical solution in an equilateral triangle of the height h is in the polynomial format of (4.84) with $\widetilde{c} = \frac{3}{2h}$. For the current value of $R_0 = 1$, one obtains $h = 2.053$, and therefore, the exact value for $\widetilde{c}$ is 0.7306. The value $\widetilde{c} = \frac{243}{325}$ therefore creates a relative error of $1 - \Lambda/\Lambda_a = 0.023$.

Additional execution of **P.4.13** with $ii = 3$ yields $\widetilde{c} = \frac{241280775}{330036809}$ and $1 - \Lambda/\Lambda_a = 0.001$. Note that in this case, since the exact solution is indeed polynomial, and accuracy may not be improved by increasing kk, since $kk = 1$ already supplies the required polynomial level for an exact solution. Hence, the only reason for error in this case emerges from the above approximate parametrization of the contour by truncated Fourier series.

We next activate **P.4.13** as described above but with iso$= 0$, $ii = 3$ and $t_z = 30°$. Figure 4.2 presents the differences in the warping function characteristics between the monoclinic and the isotropic cases. As shown, the level lines topologies in the triangle central area are different.

Example 4.11 *Torsion of a MON13z Rod with a Polygonal Cross-Section.*
Similar to Example 4.10, we activate **P.4.13** for a rectangle with BVP$= -1$, iso$= 0$ or 1, method_ $=$residual, $n = 4$, $ii = 3$, $R_0 = 1$, $k_3 = 1$, $kk = 2$, $t_z = 30°$ (for iso$= 0$). Figure 4.3 presents the differences in the warping function level lines topology in these two cases.

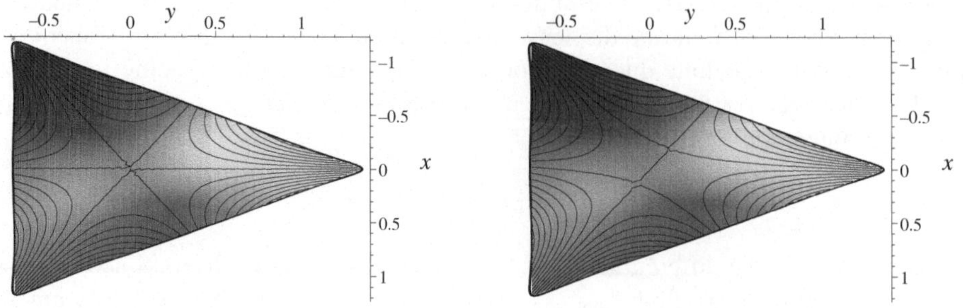

(a) Isotropic material. (b) MON13z material.

Figure 4.2: Level lines of the warping function in torsion for isotropic and monoclinic triangles.

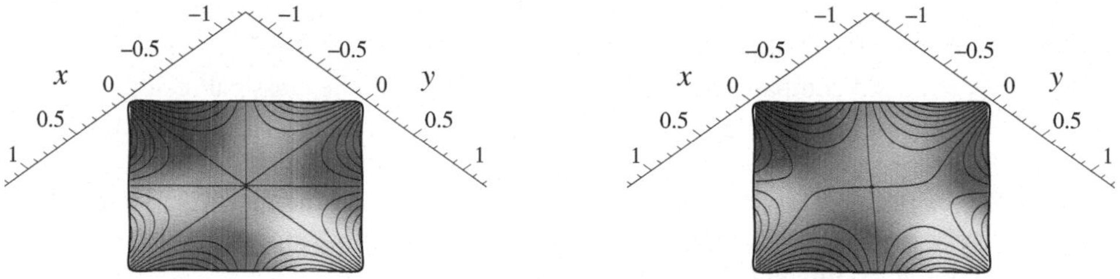

(a) Isotropic material. (b) Monoclinic material.

Figure 4.3: Level lines of the stress function in torsion for isotropic and monoclinic rectangles.

We next activate **P.4.13** with the modification of iso$=0$ or 1, $ii=1$, $m_0=2$, $kk=0$, then with iso=0, $n=5$, $ii=2$, $kk=4$, and again by replacing only $n=6$. We subsequently generate the three isotropic results presented in Fig. 4.4 for an ellipse, a pentagon and a hexagon.

The polynomial solution for an ellipse is exact, while for the pentagon the relative error is $\varepsilon_{rel}=0.017$, and for the hexagon $\varepsilon_{rel}=0.037$ has been obtained.

Example 4.12 *Ellipse and Polygons Under Centrifugal Loading.*
A domain that undergoes centrifugal loading is examined here. We adopt the plane-strain formulation discussed in S.3.2.5, and suppose no surface loads and general linear variation of body forces

$$X_b = X_{01}y + X_{10}x, \qquad Y_b = Y_{01}y + Y_{10}x. \tag{4.85}$$

The above selection is somewhat more general than required for centrifugal loading and shown at this stage for the sake of completeness only. Hence, the body-force potentials turn out to be, see Remark 4.1, $\overline{U}_1 = -X_{01}yx - \frac{1}{2}X_{10}x^2$, $\overline{U}_2 = -\frac{1}{2}Y_{01}y^2 - Y_{10}xy$. The boundary functions F_1 and F_2 are given by (3.65,b) (while $X_s = Y_s = 0$ in this case) as

$$F_1 = (\frac{1}{2}Y_{01}y^2 + Y_{10}xy)\cos(\bar{\mathbf{n}}, y), \qquad F_2 = (X_{01}yx + \frac{1}{2}X_{10}x^2)\cos(\bar{\mathbf{n}}, x). \tag{4.86}$$

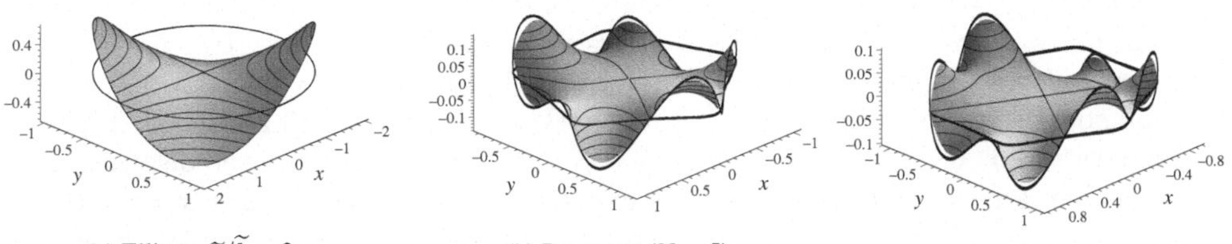

(a) Ellipse, $\widetilde{a}/\widetilde{b}=2$.　(b) Pentagon ($N=5$).　(c) Hexagon ($N=6$).

Figure 4.4: Level lines of a warping function in torsion.

For an ellipse of semi-axes $\widetilde{a}>\widetilde{b}>0$, the three single-valued integral conditions of (4.4) yield

$$X_{01}=\frac{\widetilde{a}^2}{\widetilde{b}^2}Y_{10}. \tag{4.87}$$

The above shows that a well-posed problem of this kind depends on three parameters only, X_{10}, Y_{01} and Y_{10}, as $X_b=\frac{\widetilde{a}^2}{\widetilde{b}^2}Y_{10}y+X_{10}x$, $Y_b=Y_{01}y+Y_{10}x$.

For the specific case of a beam that rotates with a constant angular velocity $\widetilde{\omega}$ about its axis (that passes through the $x=y=0$ point and is normal to the plane of the disc, see also (Lekhnitskii, 1968)), we select $X_{10}=Y_{01}=\gamma$, $Y_{10}=0$, and the body forces become

$$X_b=\gamma x, \qquad Y_b=\gamma y. \tag{4.88}$$

In the above equation, $\gamma=\rho_0\Omega_0^2$, where ρ_0 is the constant mass density of the material, and Ω_0 is the angular velocity. We therefore set $\overline{U}_1=-\frac{\gamma}{2}x^2$ and $\overline{U}_2=-\frac{\gamma}{2}y^2$ (despite the fact that in this specific case, a common potential for both force components of magnitude $\overline{U}=-\frac{\gamma}{2}(x^2+y^2)$ could be utilized), and the function F_0 of (3.57) becomes a constant: $2\gamma b_{12}$. Overall, using the notation of (4.42), p_{220} and q_{102} are the only nonzero coefficients,

The problem of a rotating beam may be examined from both the plane-strain or the plane-stress points of view. In the former case, we look at an (infinitely) long prism that rotates about an axis that coincides with the z-axis. In the latter case, we look at a thin plate that rotates about the same axis. We shall work with the first case, while the latter may be obtained by replacing the b_{ij} with the corresponding a_{ij} elastic coefficients.

We activate **P.4.10** with case=$case_b$, iso$=0$, $k_2(=k_Q)=2$, $k_0=2$. In addition, we work with symbolic b_{ij}, set $x(t)=\widetilde{a}\cos t$, $y(t)=\widetilde{b}\sin t$ and $k_p=1$, and open the "Ex=rot" command, which simulates the case of $\gamma=1$. This yields $\Phi=B_{40}x^4+B_{22}x^2y^2+B_{04}y^4+B_{20}x^2+B_{02}y^2$, where

$$B_{02}=\frac{6\widetilde{a}^2}{\widetilde{c}}(2\widetilde{b}^4b_{22}+\widetilde{a}^2\widetilde{b}^2b_{66}+2\widetilde{a}^4b_{11}), \quad B_{20}=\frac{6\widetilde{b}^2}{\widetilde{c}}(2\widetilde{b}^4b_{22}+\widetilde{b}^2\widetilde{a}^2b_{66}+2\widetilde{a}^4b_{11}),$$

$$B_{40}=\frac{\widetilde{b}^4}{\widetilde{c}}(4b_{12}-b_{66}), \quad B_{22}=\frac{6}{\widetilde{c}}(\widetilde{b}^4b_{22}+2b_{12}\widetilde{b}^2\widetilde{a}^2+b_{11}\widetilde{a}^4),$$

$$B_{04}=\frac{1}{\widetilde{c}b_{11}}[(\widetilde{a}^4b_{11}+2\widetilde{a}^2\widetilde{b}^2b_{12})(2b_{12}+b_{66})+6\widetilde{b}^4b_{12}b_{22}]+\underline{\frac{b_{12}}{12b_{11}}}, \tag{4.89}$$

and $\widetilde{c}=24(3\widetilde{b}^4b_{22}+2b_{12}\widetilde{b}^2\widetilde{a}^2+\widetilde{b}^2b_{66}\widetilde{a}^2+3\widetilde{a}^4b_{11})$. Note that the underlined term in B_{04} stands for the particular solution, and that b_{16} and b_{26} do not appear in the solution since in this case $\Phi_{,xxxy} = \Phi_{,yyyx} = 0$ (and hence, an "orthotropic-type" of solution is obtained). The corresponding analytical expression for stresses $\sigma_{xx} = \Phi_{,yy} + \overline{U}_1$, $\sigma_{yy} = \Phi_{,xx} + \overline{U}_2$, $\tau_{xy} = -\Phi_{,xy}$ then become identical to those of the second-degree polynomials of (Lekhnitskii, 1968). The strain components are obtained using the above stress components and (3.27). Then, displacements are derived by the integration scheme of S.1.2 and **P.1.6** (in this case, no displacements and no rotation at the system origin should be imposed, i.e., $u = v = \omega_z = 0$ at $x = y = 0$).

This example is unique as it clearly demonstrates that for the same loading and boundary conditions, the stress distribution is a function of the material properties. To illustrate that, note that for circular (i.e., $\widetilde{a} = \widetilde{b} = \widetilde{R}$) isotropic case, along the $x = 0$ symmetry line, $\sigma_{yy} = [(3+\nu)\widetilde{R}^2 - (1+3\nu)r^2]/8$ the distribution of which is clearly a function of ν, although its integral is not (i.e., $\int_0^{\widetilde{R}} \sigma_{yy} = \widetilde{R}^3/3$).

On our way to demonstrate the polynomial approximate solutions approach for the biharmonic BVP, we shall start with an elliptical domain, for which an exact solution has been derived above. We work here again with the plane-stress version (i.e. with the a_{ij} coefficients), assuming the semi-axes $\widetilde{a} = 2$, $\widetilde{b} = 1$, and body-force loading $X_b = x$, $Y_b = y$, which simulates a rotation (with $\gamma = 1$).

We activate **P.4.16**, which is an application of **P.4.14** with iso= 0, method_ =residual, $n > 2$, $ii = 1$, $R_0 = 1$, $m_0 = 2$, $kk = 0$ and $t_z = 30°$. In this special case, the code employs the exact solution, and the resulting error index is clearly $\varepsilon_{rel} = 0$. The resulting stress component σ_x and the deformation field are shown in Fig. 4.5, where for clarification purposes, the deformation has been magnified. The ellipses oblateness is clearly demonstrated.

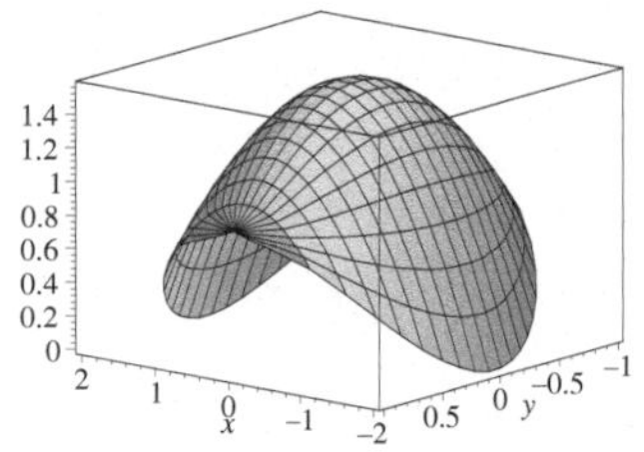

(a) σ_x (N/m^2).

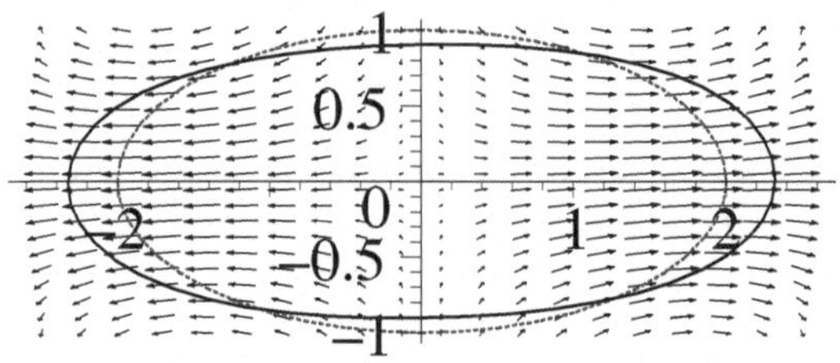

(b) Deformation field (not to scale).

Figure 4.5: Stress and deformation of a rotating ellipse, $\rho_0\Omega_0^2 = 1$.

We next activate **P.4.16** with the modifications $n = 3$, $ii = 2$, $kk = 2$ to obtain the case of a rotating triangle, see Fig. 4.6(a). The resulting relative error index in this case is $\varepsilon_{rel} = 0.030$. As shown, in contrast with the isotropic case, the deformation of an anisotropic triangle exhibits no perfect "triple-symmetry". This is due to the $\nabla_3^{(4)}$ biharmonic operator (see (3.206)) that reflects the nature of the MON13z Cartesian anisotropy.

The solution for a rectangle is obtained by the modification of $n = 4$ and yields $\varepsilon_{rel} = 0.019$, see Fig. 4.6(b), and the case of a pentagon is obtained by the modifications $n = 5$, $ii = 3$ and $kk = 3$, and yields $\varepsilon_{rel} = 0.026$, see Fig. 4.6(c).

To demonstrate the solution quality, we present in Fig. 4.7 the quantities that should be balanced over the boundary of the triangular domain, see (4.14b) and (3.65). Note that two lines are drawn in each case and that in this example, $X_s = Y_s = 0$ and $\overline{U}_1 = -\frac{1}{2}x^2, \overline{U}_2 = -\frac{1}{2}y^2$. The stress function Φ in this case is shown in Fig. 4.8(a).

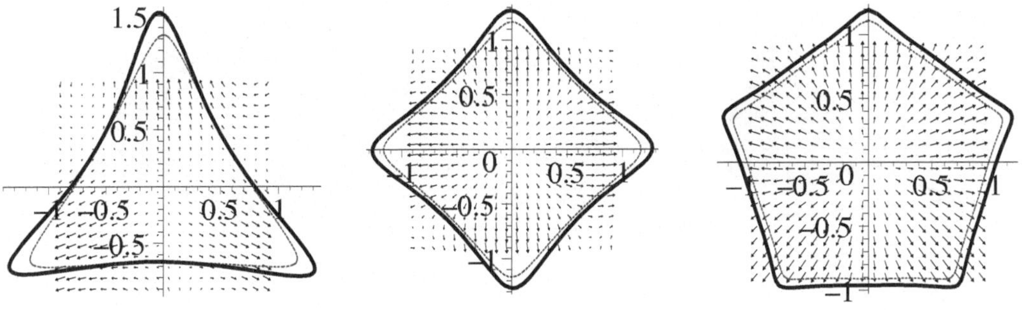

(a) Triangle. (b) Rectangle. (c) Pentagon.

Figure 4.6: Deformation field (magnified for clarification purposes) for a polygon.

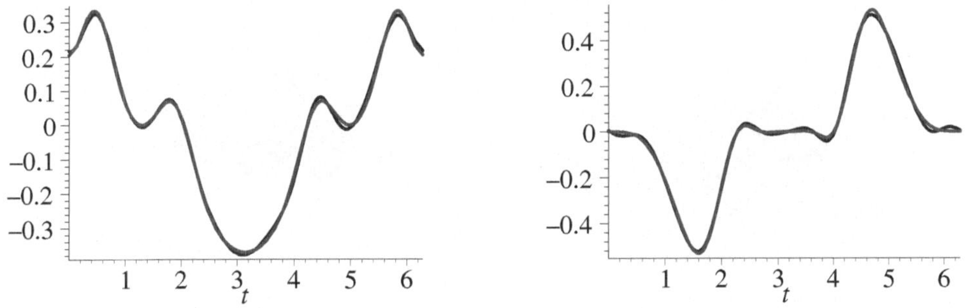

(a) $\frac{d}{dt}\Phi_{,x}$ and $-\overline{U}_2\frac{dx}{dt}$. (b) $\frac{d}{dt}(\Phi_{,y})$ and $-\overline{U}_1\frac{dy}{dt}$.

Figure 4.7: Comparison of boundary functions of rotating triangles.

Example 4.13 *Bending of Thin Plates of MON13z Material.*
To present a solution for the thin plates bending discussed in S.3.5.2, we activate **P.4.17** with iso$=0$, $n=4$, $ii=2$, $R_0=1$, $h=1$, $kk=12$, $t_z=30^\circ$, $k_0=k_{min}=0$, $S_1(0,0)=1$, $bc=1$. Hence, we are dealing with a clamped, uniformly loaded rectangular plate of unit thickness, made of orthotropic material, which is rotated by 30° about the z-axis. Figure 4.9(a) presents the $w(x,y)$ displacement, while Fig. 4.9(b) presents the resulting σ_x component, see (3.175). Both $w(x,y)$ and the $\sigma_x(x,y)/z$ are normalized by their values at $x=y=0$. To demonstrate the way in which the "approximate" technique deals with the problem under discussion, Fig. 4.8(b) presents the deflection of a clamped uniformly loaded triangular plate, normalized by the value at $x=y=0$. As shown, due to the boundary requirements, large "walls" are built outside the triangular area. This phenomena is the outcome of the requirement of (3.184) for "clamped" edge boundary conditions along the triangular contour.

Example 4.14 *Torsion of a MON13y Rod with a Polygonal Cross-Section.*
This is a relatively complex problem as the torsion is accompanied by deformation of the cross-section, and thus, both Φ and Λ are nonzero and one needs to solve a Coupled-Plane BVP. We assume no body-force loading, and according to (3.140), (3.141) solve the system (4.10a,b) with $g=0$, $f=1$ and $F_1=F_2=F_3=0$. For simplicity, we take here $2\theta=-1$.

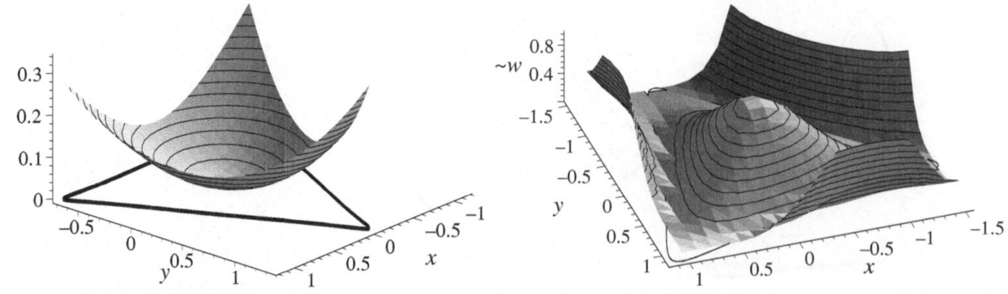

(a) The stress function Φ for a rotating triangle ($\gamma = 1$).

(b) The deflection $w(x,y)$ of a clamped uniformly loaded monoclinic triangular plate.

Figure 4.8: Results from Examples 4.12, 4.13.

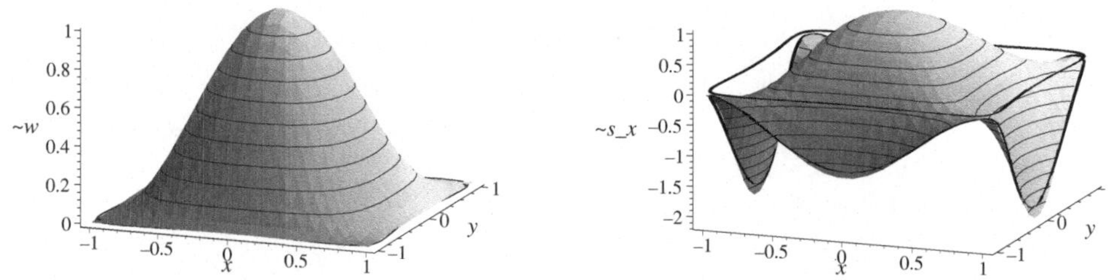

(a) $w(x,y)$.

(b) $\sigma_x(x,y)/z$.

Figure 4.9: The deflection $w(x,y)$ and the $\sigma_x(x,y)/z$ stress component of a clamped uniformly loaded monoclinic rectangular plate.

Therefore, both Φ and Λ will be proportional to θ, which should be determined by the applied moment, see Remark 3.6.

We first activate **P.4.5** with $k_g = -1$ and obtain the particular solution $\Lambda_p = \frac{y^2}{2b_{55}}$. To solve (4.16a), (4.38a), (4.40) in an anisotropic triangle with $P_1 = Q_1 = P_2 = Q_2 = Q_3 = 0$ and $P_3 = -\Lambda_{p,y} = -\frac{y}{b_{55}}$, we activate **P.4.15** with method_ = residual, $n = 3$, $ii = 2$, $R_0 = 1$, P_j, Q_j, $j = 1,3$ as indicated above, $k_1, k_2 = -1$, $k_3 = 1$, $kk_L = 9$, $kk_\Phi = 10$, $w_\Phi = 40$, $\{t_x, t_y, t_z\} = \{0, 30, 0\}$, $dd = 0$ and "corr =true". The relative error index obtained in this case is $\varepsilon_{rel} = 0.005$.

The resulting stress functions Φ and Λ are shown in Fig. 4.10, and the boundary line in Fig. 4.10(b) is determined by an integration of the $\frac{d}{dt}\Lambda_{|\partial\Omega} = F_3$, and therefore supplies an indication regarding the solution quality. As shown, Λ is much larger than Φ as the stress components τ_{xz} and τ_{yz} dominate the torsion mechanism. Yet, the in-plane stress components induced by Φ do not vanish in this anisotropic case. Figure 4.11 presents the boundary functions along the contour. As shown by Fig. 4.11(a), $\frac{d}{dt}\Phi_{,x|\partial\Omega}$ does not vanish as it should along the entire contour (since $F_1 = 0$ in this case). However, the "noisy" error is relatively small as will be shown later on. Figure 4.11(b) presents the values of $\frac{d}{dt}\Lambda_{|\partial\Omega}$ and F_3 (two lines). These two lines coherence represents excellent fulfillment of the requirement over the boundary. Figures 4.12(a),(b) present the distributions of the derivatives $\Phi_{,x}$ and $\Phi_{,y}$, respectively.

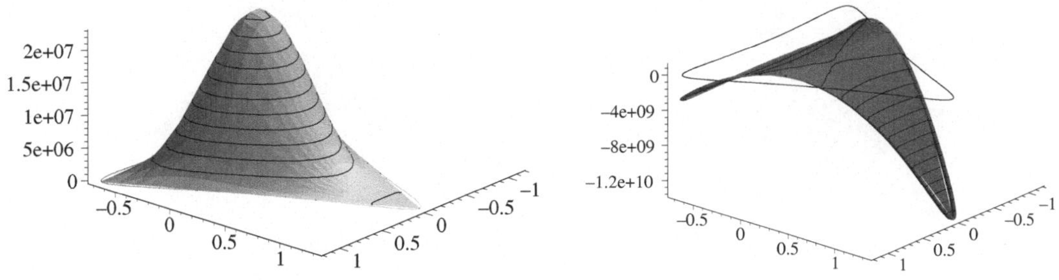

(a) $\Phi(x,y)$. (b) $\Lambda(x,y)$.

Figure 4.10: The stress functions in Example 4.14.

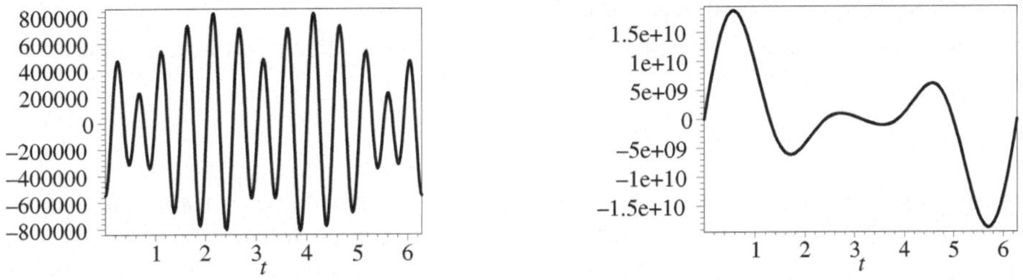

(a) $\frac{d}{dt}\Phi_{,x}$ vs. t (one line). (b) $\frac{d}{dt}\Lambda$ and F_3 vs. t (two lines).

Figure 4.11: Comparison of boundary functions in Example 4.14.

The solution quality may be examined by boundary lines, which are determined by $-F_1$ and F_2 integration, and their comparison with $\frac{d}{dt}\Phi_{,x|\partial\Omega}$ and $\frac{d}{dt}\Phi_{,y|\partial\Omega}$, respectively. In the present case, the conditions $\frac{d}{dt}\Phi_{,x|\partial\Omega} = \frac{d}{dt}\Phi_{,y|\partial\Omega} = 0$ are fairly fulfilled, and the error in Fig. 4.11(a) is shown to be relatively small.

4.4 The Method of Complex Potentials

An alternative solution methodology that may be employed for the solution of the BVPs described in this book is founded on the so-called "Functions of Composed Complex Variables". The term "composed" replaces here the term "complicated" that is used in (Lekhnitskii, 1968), (Lekhnitskii, 1981) which are the sources of the present method, see also (Lu, 1995).

To show the superiority of the method, we note that, as discussed in Chapter 1, the theory of elasticity requires the fulfillment of the differential equilibrium, the compatibility equations and the boundary conditions. By formulating the problem using stress functions, the differential equilibrium is fulfilled by definition, and one is left with the compatibility equations and the boundary conditions. However, as will be shown in what follows, by employing the method of complex potentials, compatibility is also inherently fulfilled. Hence, one is left with the boundary conditions only.

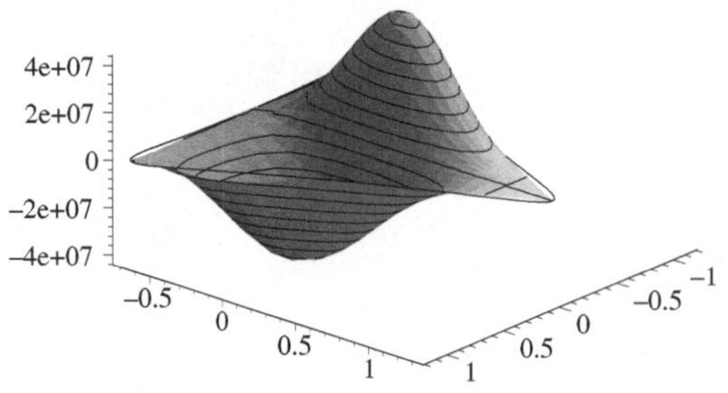

(a) $\Phi_{,x}(x,y)$.

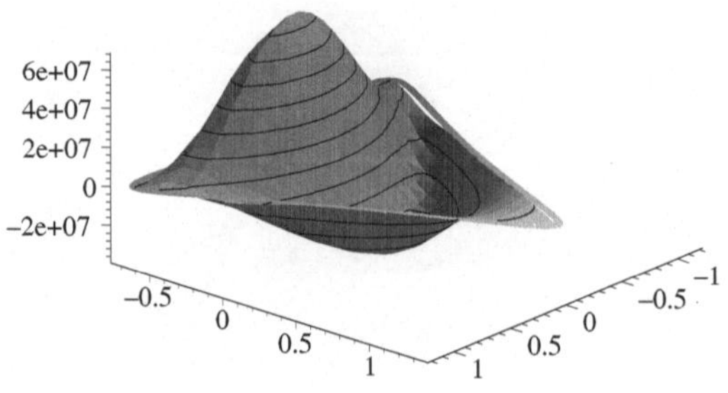

(b) $\Phi_{,y}(x,y)$.

Figure 4.12: Distribution of Φ derivatives in Example 4.14.

We shall first exploit complex potentials to formulate the *n-Coupled Dirichlet BVP*, and then show that it may be successfully used for the various BVPs discussed in this book.

4.4.1 n-Coupled Dirichlet BVP

We define the *complex potential* $\mathbf{H}_k(z_k) = \widetilde{\Psi}_{1k}(z_k) + \mathbf{i}\widetilde{\Psi}_{2k}(z_k)$ as an analytical function of complex variable $z_k = x_k + \mathbf{i}y_k$ in a plane domain Ω_k. In what follows, the indices k, m attain the values $1,\dots,n$, where $n > 0$. Since $\widetilde{\Psi}_{1k}(z_k)$, $\widetilde{\Psi}_{2k}(z_k)$ are harmonic conjugate functions in the z_k-plane, they satisfy Cauchy-Riemann conditions, namely,

$$\widetilde{\Psi}_{1k,x_k} = \widetilde{\Psi}_{2k,y_k}, \qquad \widetilde{\Psi}_{1k,y_k} = -\widetilde{\Psi}_{2k,x_k}, \tag{4.90}$$

and $(\frac{\partial^2}{\partial x_k^2} + \frac{\partial^2}{\partial y_k^2})\widetilde{\Psi}_{ik} = 0$ $(i = 1,2)$. A linear combination of the above functions, $\nu_1^{km}\widetilde{\Psi}_{1k} - \nu_2^{km}\widetilde{\Psi}_{2k}$, can be expressed by complex multiplication with complex potential, as $\Re[\nu^{km}\mathbf{H}_k]$, where $\nu^{km} = \nu_1^{km} + \mathbf{i}\nu_2^{km}$. Hence, linear mixtures of the above harmonic functions, $\widetilde{\Lambda}_m$, depending on $z_1,\dots,z_n$, may be considered as

$$\widetilde{\Lambda}_m(z_1,\dots,z_n) = \Re[\textstyle\sum_k \nu^{km}\mathbf{H}_k(z_k)]. \tag{4.91}$$

In order to derive the boundary conditions for $\widetilde{\Lambda}_m$ on the domain Ω, we introduce the *complex parameters* $\mu_k = \alpha_k + \mathbf{i}\beta_k$ $(\beta_k > 0)$, and define affine transformations $T_{\mu_k} : z \in \Omega \to z_k \in \Omega_k$, namely,

$$T_{\mu_k}(z) = x + \mu_k y, \qquad x_k = x + \alpha_k y, \quad y_k = \beta_k y. \tag{4.92}$$

One may imagine that T_{μ_k} is used to parameterize Ω_k by a given Ω. As an example, Fig. 4.13 presents a rectangular domain Ω, which is transformed to a parallelogram-shaped domain Ω_k.

From now on, the functions $\Psi(x,y)$ on the domain Ω and $\widetilde{\Psi}(x_k,y_k)$ on the domain Ω_k correspond (by the substitution $z_k = x + \mu_k y$), i.e., $\Psi = \widetilde{\Psi}(x + \mu_k y)$. Hence, the real functions $\Lambda_m(x,y) = \widetilde{\Lambda}_m(x + \mu_1 y,\dots,x + \mu_n y)$ on Ω are also defined.

At this stage, one may differentiate C^2-regular functions $\Psi(x,y) = \widetilde{\Psi}(x + \alpha_k y,\, \beta_k y)$ as

$$\begin{aligned}&\Psi_{,x} = \widetilde{\Psi}_{,x_k}, \qquad \Psi_{,y} = \alpha_k\widetilde{\Psi}_{,x_k} + \beta_k\widetilde{\Psi}_{,y_k}, \\ &\Psi_{,xx} = \widetilde{\Psi}_{,x_kx_k}, \;\; \Psi_{,xy} = \alpha_k\widetilde{\Psi}_{,x_kx_k} + \beta_k\widetilde{\Psi}_{,x_ky_k}, \;\; \Psi_{,yy} = \alpha_k^2\widetilde{\Psi}_{,x_kx_k} + 2\alpha_k\beta_k\widetilde{\Psi}_{,x_ky_k} + \beta_k^2\widetilde{\Psi}_{,y_ky_k}.\end{aligned} \tag{4.93}$$

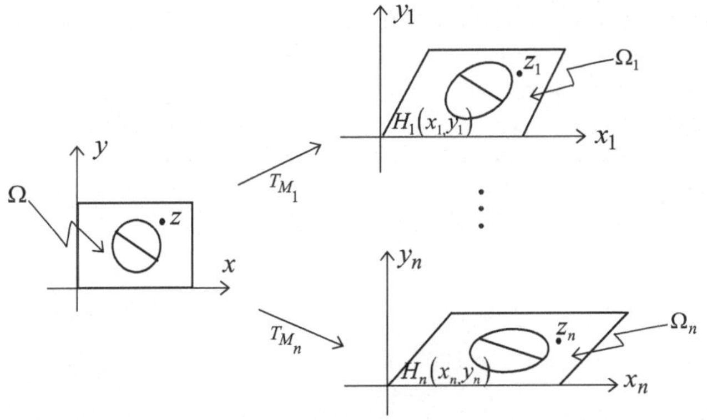

Figure 4.13: Transformation T_{μ_k} between the Ω and the Ω_k planes.

In view of (4.93), the following equality holds:

$$(\alpha_k^2+\beta_k^2)\Psi_{,xx}-2\alpha_k\Psi_{,xy}+\Psi_{,yy}=\beta_k^2(\frac{\partial^2}{\partial x_k^2}+\frac{\partial^2}{\partial y_k^2})\widetilde{\Psi}. \tag{4.94}$$

As shown, harmonic functions in Ω_k correspond to generalized harmonic functions in Ω.

Hence, the above represents a simplification of a generalized Laplace's equation similar to the one discussed in S.3.3.4. Note that the transformation in the present case may be put as $x_k=x+\widetilde{a}y,\ y_k=\widetilde{b}y$, while in the one used in (3.131), $x_k=\widetilde{a}x,\ y_k=y+\widetilde{b}x$.

We shall now define the *n-Coupled Dirichlet BVP* in terms of linear mixtures, Λ_m, of complex potentials, see (4.91), as follows: *For given plane domain Ω, functions G_m or F_m over $\partial\Omega$, sets of complex parameters μ_k with $\Im(\mu_k)>0$ and complex coefficients $\{\nu^{km}\}$, find complex potentials $\{\mathbf{H}_k(z_k)\}$, whose linear mixtures $\Lambda_m(x,y)=\Re[\sum_k\nu^{km}\mathbf{H}_k(x+\mu_k y)]$ satisfy the boundary conditions written in one of the following forms:*

$$\Lambda_m=G_m \qquad \text{or} \qquad \frac{d}{ds}\Lambda_m=F_m \qquad \text{on} \quad \partial\Omega. \tag{4.95}$$

In the case of (4.95:b) the expressions of Λ_m contain free constants of integration c_m, and, in addition, the single-value conditions along a contour

$$\oint_{\partial\Omega}F_m=0, \tag{4.96}$$

are essential for the BVP solution's existence and uniqueness on a simply connected domain, Ω.

Hence, we have established a coupled system of complex potentials $\mathbf{H}_k(z_k)$ that produce harmonic functions $\widetilde{\Psi}_{ik}$ in Ω_k, $i=1,2$, and generalized harmonic functions Ψ_{ik} in Ω. In the "diagonal" case, where $\nu^{km}=0,\ k\neq m,$ (4.95) can be solved separately as n independent Dirichlet BVPs, as will be shown in S.4.4.2.

When polynomial solutions are under discussion, one may write $\mathbf{H}_k(z_k)=\sum_{j=1}^N\widetilde{\alpha}_{jk}z_k^j$, where $\widetilde{\alpha}_{jk}=\widetilde{\alpha}_{1jk}+\mathbf{i}\widetilde{\alpha}_{2jk}$ are complex coefficients. The conjugate polynomials $\Re(z_k^j),\Im(z_k^j)$ form a base of harmonic (x_k,y_k)-polynomials of degree j. Simple examples of low degree pairs are $\{x_k^2,2x_ky_k\}$, $\{x_k^3-3x_ky_k^2,3x_k^2y_k-y_k^3\}$, etc. Hence the vector space of harmonic polynomials in x_k,y_k of degree $\leq n$ is $(2n+1)$-dimensional. The approximate polynomial solution of the n-Coupled Dirichlet problem for various simply connected domains is illustrated by **P.4.18**. The example in Fig. 4.14 has been produced for the illustrative values of BVP$=1$, $NF=3$, method_ $=$ residual, $n=3$, $ii=2$, $R_0=2$, $k_3=1$, $kk=5$, $\mu_1=2+\mathbf{i}$, $\mu_2=-1+2\mathbf{i}$, $\mu_3=3\mathbf{i}$, $P_1=y-4x$, $Q_1=4y+x$, $P_2=40y-60x$, $Q_2=60y+20x$, $Q_3=P_3=0$, $\nu^{11}=2+\mathbf{i}$, $\nu^{12}=3-5\mathbf{i}$, $\nu^{13}=8+5\mathbf{i}$, $\nu^{21}=-1$, $\nu^{22}=3-4\mathbf{i}$, $\nu^{23}=-5-2\mathbf{i}$, $\nu^{31}=-1-9\mathbf{i}$, $\nu^{32}=5+8\mathbf{i}$, $\nu^{33}=4-3\mathbf{i}$.

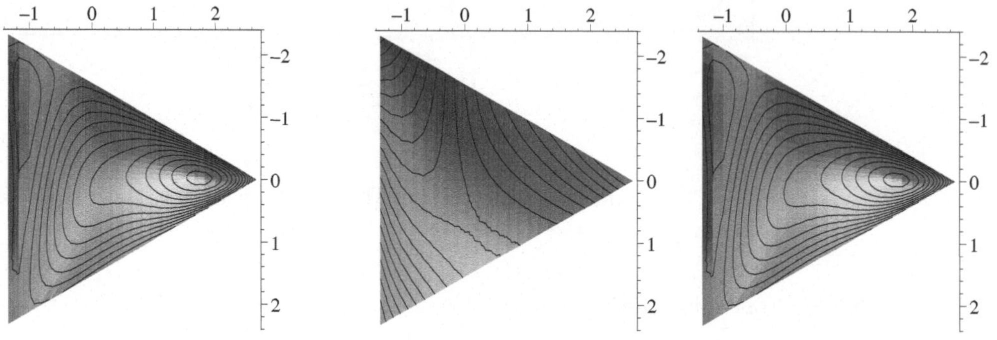

(a) $\Lambda_1(x,y)$. (b) $\Lambda_2(x,y)$. (c) $\Lambda_3(x,y)$.

Figure 4.14: Polynomial approximate solution for homogeneous 3-Coupled Dirichlet BVP in a triangle by complex potentials.

For the applications discussed in this book, the complex parameter μ_k appears as a root of the characteristic polynomial that may be typically written as $l_2^k \equiv c_k\mu^2 - 2b_k\mu + a_k$, associated with the differential operator $\nabla_3^{(2)k} = a_k\frac{\partial^2}{\partial x^2} - 2b_k\frac{\partial^2}{\partial x \partial y} + c_k\frac{\partial^2}{\partial y^2}$, see e.g., (3.195), where a_k, b_k, c_k are functions of elastic parameters, and are associated with the a_{44}, a_{45}, a_{55} elastic coefficients, respectively. In view of S.3.6.5, $a_k c_k - (b_k)^2 > 0$, and the roots of l_2^k are pairs of conjugate complex numbers $\mu_k = \alpha_k + \mathbf{i}\beta_k$, $\bar{\mu}_k = \alpha_k - \mathbf{i}\beta_k$, where $\beta_k > 0$. By Viet's Theorem, α_k and β_k may be determined by

$$\frac{b_k}{c_k} = \frac{1}{2}(\mu_k + \bar{\mu}_k) = \alpha_k, \qquad \frac{a_k}{c_k} = \mu_k\bar{\mu}_k = \alpha_k^2 + \beta_k^2. \tag{4.97}$$

By representing the characteristic polynomials equivalently as $l_2^k = c_k[(\alpha_k^2 - \beta_k^2) + 2\alpha_k\mu + \mu^2]$ and using (4.94), (4.97), one obtains

$$\nabla_3^{(2)k}\Psi_{ik} = c_k\beta_k^2\left(\frac{\partial^2}{\partial x_k^2} + \frac{\partial^2}{\partial y_k^2}\right)\widetilde{\Psi}_{ik}, \qquad i = 1,2, \tag{4.98}$$

which demonstrates again that $\Psi_{ik}(x,y)$ are generalized harmonic functions in Ω when the functions $\widetilde{\Psi}_{ik}(x_k, y_k)$ are harmonic in Ω_k.

4.4.2 *Application of Complex Potentials to the Dirichlet BVP*

We shall relate the homogeneous Dirichlet BVP, see S.4.1,

$$\nabla_3^{(2)}\Psi_{13} = 0 \qquad \text{over} \quad \Omega, \tag{4.99a}$$

$$\Psi_{13} = G_3 \qquad or \qquad \frac{d}{ds}\Psi_{13} = F_3 \qquad \text{on} \quad \partial\Omega, \tag{4.99b}$$

with a special case of S.4.4.1 for $n = 1$. To comply with previous notation, and without loss of generality, we use the index $k = 3$ (instead of $k = 1$). For the $\frac{d}{ds}$-case of (4.99b), the following single-valued condition along a contour should also be fulfilled by the "input function", see (4.96):

$$\oint_{\partial\Omega} F_3 = 0. \tag{4.100}$$

Let $\mu_3 = \alpha_3 + \mathbf{i}\beta_3$, $\bar{\mu}_3 = \alpha_3 - \mathbf{i}\beta_3$, where $\beta_3 > 0$, be the conjugate roots of the characteristic polynomial $l_2(\mu) = a_{55}\mu^2 - 2a_{45}\mu + a_{44}$, associated with the Laplacian $\nabla_3^{(2)}$. Hence affine transformation $T_{\mu_3} : \Omega \to \Omega_3$ is defined, namely, $T_{\mu_3}(z) = x + \mu_3 y$, i.e., $x_3 = x + \alpha_3 y$, $y_3 = \beta_3 y$. We also select $\nu^{11} = 2$, instead of the more natural one of $\nu^{11} = 1$ for "traditional" reasons only, see (Lekhnitskii, 1968), (Lekhnitskii, 1981). Hence, only one complex potential $\mathbf{H}_3(z_3)$ is in use in this case, and (4.91) becomes

$$\Lambda(x,y) = \Re[2\mathbf{H}_3(x + \mu_3 y)] = 2\Psi_3. \tag{4.101}$$

Therefore, the Dirichlet BVP under discussion is equivalently formulated, using the boundary conditions (4.99b), namely,

$$\Lambda = 2G_3 \qquad \text{or} \qquad \frac{d}{ds}\Lambda = 2F_3 \qquad \text{on} \quad \partial\Omega. \tag{4.102}$$

Using (3.92) and Remark 4.4, the stress components $\tau_{yz} = -\Lambda_{,x}$, $\tau_{xz} = \Lambda_{,y}$ are expressed as

$$\tau_{yz} = -2\Re[\mathbf{H}_3'(x + \mu_3 y)], \qquad \tau_{xz} = 2\Re[\mu_3 \mathbf{H}_3'(x + \mu_3 y)]. \tag{4.103}$$

Remark 4.4 We shall present a statement that facilitates the derivation in this section.
Let $\mu_k = \alpha_k + \mathbf{i}\beta_k$ be a complex number, and $\mathbf{H}_k(z_k) = \widetilde{\Psi}_{1k} + \mathbf{i}\widetilde{\Psi}_{2k}$ be an analytical function of the complex variable $z_k = x + \mu_k y = x_k + \mathbf{i}y_k$. Then,

$$\frac{\partial}{\partial x}\Re[\mathbf{H}_k(x + \mu_k y)] = \Re[\mathbf{H}_k'(x + \mu_k y)], \qquad \frac{\partial}{\partial y}\Re[\mathbf{H}_k(x + \mu_k y)] = \Re[\mu_k \mathbf{H}_k'(x + \mu_k y)]. \tag{4.104}$$

Proof. By definition, a complex derivative may be written as

$$\mathbf{H}_k'(x + \mu_k y) = \widetilde{\Psi}_{1k,x_k} + \mathbf{i}\widetilde{\Psi}_{2k,x_k} = \widetilde{\Psi}_{2k,y_k} - \mathbf{i}\widetilde{\Psi}_{1k,y_k}, \tag{4.105}$$

and $\Psi_{1k,x} = \widetilde{\Psi}_{1k,x_k}$, $\Psi_{1k,y} = \alpha_k \widetilde{\Psi}_{1k,x_k} + \beta_k \widetilde{\Psi}_{1k,y_k}$, where $\Psi_{ik}(x,y) = \widetilde{\Psi}_{ik}(x + \mu_k y)$. We therefore first calculate

$$\frac{\partial}{\partial y}\Re(\Psi_{1k} + \mathbf{i}\Psi_{2k}) = \frac{\partial}{\partial y}\Psi_{1k} = \alpha_k \widetilde{\Psi}_{1k,x_k} + \beta_k \widetilde{\Psi}_{1k,y_k}. \tag{4.106}$$

On the other hand, using (4.105), we have

$$\Re[\mu_k \mathbf{H}_k'(x + \mu_k y)] = \Re[(\alpha_k + \mathbf{i}\beta_k)(\widetilde{\Psi}_{1k,x_k} + \mathbf{i}\widetilde{\Psi}_{2k,x_k})] = \alpha_k \widetilde{\Psi}_{1k,x_k} - \beta_k \widetilde{\Psi}_{2k,x_k}. \tag{4.107}$$

From Cauchy-Riemann conditions of (4.90) one obtains the second condition of (4.104). The proof of the first one is analogous. □

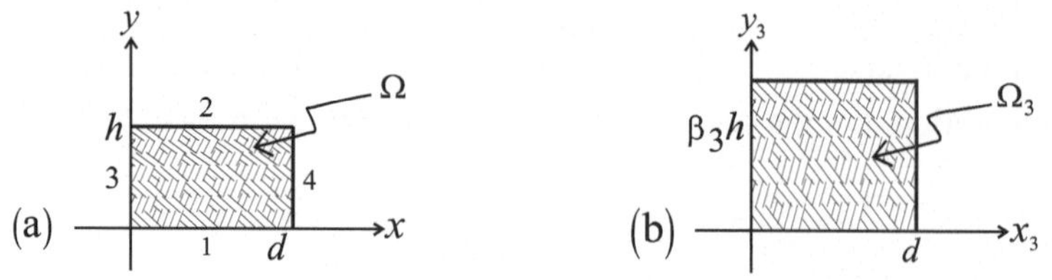

Figure 4.15: Transformation T_{μ_3} between Ω and Ω_3 planes.

4.4.2.1 Solution of the Dirichlet BVP by Complex Potentials and Fourier Series

Consider the Dirichlet BVP in the rectangle $\Omega = \{0 \le x \le d, 0 \le y \le h\}$ shown in Fig. 4.15(a). For orthotropic material, where $a_{45} = 0$, Laplace's operator is written as $\nabla_4^{(2)}$, see (3.196). The two roots of the characteristic polynomial $l_2(\mu) = a_{55}\mu^2 + a_{44}$ are purely imaginary and given by $\mu_3 = \mathbf{i}\beta_3, \bar{\mu}_3 = -\mathbf{i}\beta_3$, where $\beta_3 = \sqrt{\frac{a_{44}}{a_{55}}}$. Since $x_3 = x$, $y_3 = \beta_3 y$, the rectangle Ω is T_{μ_3}-transformed to the rectangle $\Omega_3 = \{0 \le x_3 \le d, 0 \le y_3 \le h\beta_3\}$ in the z_3-plane.

The discussion in what follows is founded on the well-known fact that a harmonic function (say, $\widetilde{\Psi}$, that satisfies $\widetilde{\Psi}_{,xx} + \widetilde{\Psi}_{,yy} = 0$) in a rectangle can be explicitly reconstructed by four real functions expanded in Fourier series, each of which satisfy the boundary conditions over one edge while leaving no influence on the remaining three edges.

Let the boundary function G_3 of (4.99b) along the rectangle Ω edges be represented in the following Fourier series form:

$$G_3(x,h) = \sum_{j=1}^{\infty} \widetilde{A}_{3j} \sin(\frac{\pi jx}{d}), \qquad G_3(x,0) = \sum_{j=1}^{\infty} \widetilde{B}_{3j} \sin(\frac{\pi jx}{d}),$$
$$G_3(d,y) = \sum_{j=1}^{\infty} \widetilde{C}_{3j} \sin(\frac{\pi jy}{h}), \qquad G_3(0,y) = \sum_{j=1}^{\infty} \widetilde{D}_{3j} \sin(\frac{\pi jy}{h}), \tag{4.108}$$

where the coefficients $\widetilde{A}_{3j}, \widetilde{B}_{3j}, \widetilde{C}_{3j}, \widetilde{D}_{3j}$ are assumed to be given. We now present the harmonic function $\widetilde{\Lambda}_3(x_3, y_3)$ as $\widetilde{\Lambda}_3 = 2\widetilde{\Psi}_{13}$, where

$$\widetilde{\Psi}_{13}(x_3,y_3) = \sum_{j=1}^{\infty} \{(A_{3j} \sinh(\frac{\pi j y_3}{d}) - B_{3j} \sinh(\frac{\pi j(y_3 - h\beta_3)}{d})) \sin(\frac{\pi j x_3}{d})$$
$$+ (C_{3j} \sinh(\frac{\pi j x_3}{h\beta_3}) - D_{3j} \sinh(\frac{\pi j (x_3 - d)}{h\beta_3})) \sin(\frac{\pi j y_3}{h\beta_3})\}, \tag{4.109}$$

while the harmonic conjugate function to $\widetilde{\Psi}_{13}(x_3, y_3)$ is

$$\widetilde{\Psi}_{23}(x_3,y_3) = \sum_{j=1}^{\infty} \{(A_{3j} \cosh(\frac{\pi j y_3}{d}) - B_{3j} \cosh(\frac{\pi j (y_3 - h\beta_3)}{d})) \cos(\frac{\pi j x_3}{d})$$
$$+ (-C_{3j} \cosh(\frac{\pi j x_3}{h\beta_3}) + D_{3j} \cosh(\frac{\pi j(x_3 - d)}{h\beta_3})) \cos(\frac{\pi j y_3}{h\beta_3})\}. \tag{4.110}$$

The reader may verify that $\widetilde{\Psi}_{i3}$, $i = 1,2$ are harmonic conjugate functions that satisfy (4.90), which in view of (4.94) one may write as

$$(\frac{\partial^2}{\partial x_3^2} + \frac{\partial^2}{\partial y_3^2})\widetilde{\Psi}_{i3}(x_3, y_3) = 0, \qquad \Leftrightarrow \qquad \nabla_4^{(2)} \Psi_{i3}(x,y) = 0. \tag{4.111}$$

We <u>first</u> wish to solve the Dirichlet BVP

$$\nabla_4^{(2)} \Psi_{13} = 0 \quad \text{over } \Omega, \qquad \Psi_{13} = G_3 \quad \text{on } \partial\Omega. \tag{4.112}$$

Substituting $x_3 = x,\ y_3 = \beta_3 y$ in (4.109), (4.110) and applying (4.102) to each edge, we find the coefficients

$$\{A_{3j}, B_{3j}\} = \frac{1}{\sinh(\frac{\pi j h \beta_3}{d})}\{\widetilde{A}_{3j}, \widetilde{B}_{3j}\}, \qquad \{C_{3j}, D_{3j}\} = \frac{1}{\sinh(\frac{\pi j d}{h\beta_3})}\{\widetilde{C}_{3j}, \widetilde{D}_{3j}\}. \quad (4.113)$$

This solution is illustrated by **P.4.19**, see also Example 4.15 and Fig. 4.16.

We next look at another version of the Dirichlet BVP

$$\nabla_4^{(2)}\Psi_{13} = 0 \qquad \text{over} \quad \Omega, \qquad \frac{d}{ds}\Psi_{13} = F_3 \qquad \text{on} \quad \partial\Omega. \quad (4.114)$$

One way to handle this problem is founded on the integration of $F_3(x,y)$ along $\partial\Omega$. In such a case we proceed as shown above by taking $G_3 = \int_0^{\widetilde{p}} F_3 ds + G_0$ (where $\widetilde{p}$ is the contour circumference and G_0 is the value of $G_3(x,y)$ at $s = 0, \widetilde{p}$).

Alternatively, we express the boundary function as $F_3 = P_3\cos(\bar{\mathbf{n}},x) + Q_3\cos(\bar{\mathbf{n}},y)$, see S.4.3, and let the boundary values of functions P_3, Q_3 in the rectangle Ω edges be represented in the following Fourier series form:

$$\begin{aligned} Q_3(x,h) &= \widetilde{A}_{30} + \sum\nolimits_{j=1}^{\infty}\widetilde{A}_{3j}\cos(\frac{\pi jx}{d}), \qquad & Q_3(x,0) &= \widetilde{B}_{30} + \sum\nolimits_{j=1}^{\infty}\widetilde{B}_{3j}\cos(\frac{\pi jx}{d}), \\ P_3(d,y) &= \widetilde{C}_{30} + \sum\nolimits_{j=1}^{\infty}\widetilde{C}_{3j}\cos(\frac{\pi jy}{h}), \qquad & P_3(0,y) &= \widetilde{D}_{30} + \sum\nolimits_{j=1}^{\infty}\widetilde{D}_{3j}\cos(\frac{\pi jy}{h}), \end{aligned} \quad (4.115)$$

where the coefficients $\widetilde{A}_{3j}, \widetilde{B}_{3j}, \widetilde{C}_{3j}, \widetilde{D}_{3j}$ are assumed to be given. Note that the single-value condition (4.100)

$$\oint_{\partial\Omega} F_3 = \int_0^d [Q_3(x,h) - Q_3(x,0)]\,dx + \int_0^h [P_3(d,y) - P_3(0,y)]\,dy, \quad (4.116)$$

yields in this case the requirement

$$d(\widetilde{A}_{30} - \widetilde{B}_{30}) + h(\widetilde{C}_{30} - \widetilde{D}_{30}) = 0. \quad (4.117)$$

To better cope with the above constants, we augment the harmonic function $\widetilde{\Psi}_{13}(x_3,y_3)$ of (4.109) by the harmonic polynomial $\alpha_{10}x + \alpha_{01}y + \alpha_{11}xy$ and present the derivatives of the resulting generalized harmonic function $\Psi_{13}(x,y)$ as

$$\begin{aligned} \Psi_{13,x} = \sum\nolimits_{j=1}^{\infty}\{&(A_{3j}\sinh(\frac{\pi jy\beta_3}{d}) - B_{3j}\sinh(\frac{\pi j(y-h)\beta_3}{d}))\frac{\pi j}{d}\cos(\frac{\pi jx}{d}) \\ &+ (C_{3j}\cosh(\frac{\pi jx}{h\beta_3}) - D_{3j}\cosh(\frac{\pi j(x-d)}{h\beta_3}))\frac{\pi j}{h\beta_3}\sin(\frac{\pi jy}{h})\} + \alpha_{10} + \alpha_{11}y, \end{aligned} \quad (4.118a)$$

$$\begin{aligned} \Psi_{13,y} = \sum\nolimits_{j=1}^{\infty}\{&(A_{3j}\cosh(\frac{\pi jy\beta_3}{d}) - B_{3j}\cosh(\frac{\pi j(y-h)\beta_3}{d}))\frac{\pi j\beta_3}{d}\sin(\frac{\pi jx}{d}) \\ &+ (C_{3j}\sinh(\frac{\pi jx}{h\beta_3}) - D_{3j}\sinh(\frac{\pi j(x-d)}{h\beta_3}))\frac{\pi j}{h}\cos(\frac{\pi jy}{h})\} + \alpha_{01} + \alpha_{11}x. \end{aligned} \quad (4.118b)$$

For a rectangle, $\cos(\bar{\mathbf{n}},x)$ and $\cos(\bar{\mathbf{n}},y)$ take the values 0 or ± 1, and therefore, the boundary conditions are specified as (see Fig. 4.15)

$$\begin{aligned} \Psi_{13,x}(x,0) &= -Q_3(x,0), \qquad & \Psi_{13,x}(x,h) &= -Q_3(x,h), \\ \Psi_{13,y}(0,y) &= P_3(0,y), \qquad & \Psi_{13,x}(d,y) &= P_3(d,y). \end{aligned} \quad (4.119)$$

Assuming

$$\alpha_{10} = -\widetilde{B}_{30}, \qquad \alpha_{01} = -\widetilde{D}_{30}, \qquad \alpha_{11} = \frac{1}{d}(\widetilde{C}_{30} - \widetilde{D}_{30}) = \frac{1}{h}(\widetilde{B}_{30} - \widetilde{A}_{30}), \tag{4.120}$$

we find the desired coefficients as

$$\{A_{3j}, B_{3j}\} = -\frac{d}{\pi j \sinh(\frac{\pi j h \beta_3}{d})}\{\widetilde{A}_{3j}, \widetilde{B}_{3j}\}, \quad \{C_{3j}, D_{3j}\} = \frac{h}{\pi j \sinh(\frac{\pi j d}{h\beta_3})}\{\widetilde{C}_{3j}, \widetilde{D}_{3j}\}. \tag{4.121}$$

Example 4.15 contains some applications of this solution.

Example 4.15 *Non-Homogeneous Laplace's Equation in an Orthotropic Rectangle.*
There are many practical cases when the Dirichlet BVP (4.1a, b) should be solved with a linear r.h.s. , for example

$$\nabla_4^{(2)}\Psi = f_0 + f_1 x + f_2 y \qquad \text{over} \quad \Omega, \qquad \Psi = 0 \qquad \text{on} \quad \partial\Omega. \tag{4.122}$$

Let the solution be presented as a sum of homogeneous and particular solutions, namely, $\Psi = \Psi_{13} + \Psi_p$. We first adopt a particular solution of the above field equation in the form

$$\Psi_p = \frac{f_0}{2a_{44}}x(x-d) + \frac{f_1}{6a_{44}}x(x^2-d^2) + \frac{f_2}{6a_{55}}y(y^2-h^2). \tag{4.123}$$

Since Ψ should vanish over the boundary $\partial\Omega$, the homogeneous solution Ψ_{13} should satisfy the boundary condition

$$\Psi_{13} = -\frac{f_0}{2a_{44}}x(x-d) - \frac{f_1}{6a_{44}}x(x^2-d^2) - \frac{f_2}{6a_{55}}y(y^2-h^2) \qquad \text{on} \quad \partial\Omega. \tag{4.124}$$

Using standard Fourier series we write

$$x(x-d) = \sum_{j=1}^{\infty} \frac{4d^2((-1)^j - 1)}{\pi^3 j^3}\sin(\frac{\pi j x}{d}), \qquad x(x^2-d^2) = \sum_{j=1}^{\infty} \frac{12d^3(-1)^j}{\pi^3 j^3}\sin(\frac{\pi j x}{d}), \tag{4.125}$$

while $y(y^2-h^2)$ is expanded in a similar way. We again expand Ψ_{13} as in (4.109). Then, employing (4.113) shows that $C_{3j} = D_{3j} = 0$ and

$$A_{3j} = B_{3j} = -\frac{2d^2 f_0((-1)^j - 1)}{\pi^3 j^3 a_{44}\sinh(\frac{\pi j h\beta_3}{d})} - \frac{2f_1 d^3(-1)^j}{\pi^3 j^3 a_{44}\sinh(\frac{\pi j h\beta_3}{d})} - \frac{2f_2 h^3(-1)^j}{\pi^3 j^3 a_{55}\sinh(\frac{\pi j d}{h\beta_3})}. \tag{4.126}$$

Hence Ψ_{13} becomes

$$\Psi_{13} = -\frac{f_0}{2}\Psi_{f_0} + f_1\Psi_{f_1} + f_2\Psi_{f_2}, \tag{4.127}$$

where, see (4.109),

$$\Psi_{f_0} = \frac{4d^2}{\pi^3 a_{44}}\sum_{j=1}^{\infty}\frac{(-1)^j - 1}{j^3}\cdot\frac{\cosh(\frac{\pi j\beta_3(y-\frac{h}{2})}{d})}{\cosh(\frac{\pi j h\beta_3}{2d})}\sin(\frac{\pi j x}{d}), \tag{4.128a}$$

$$\Psi_{f_1} = -\frac{2d^3}{\pi^3 a_{44}}\sum_{j=1}^{\infty}\frac{(-1)^j}{j^3}\cdot\frac{\cosh(\frac{\pi j\beta_3(y-\frac{h}{2})}{d})}{\cosh(\frac{\pi j h\beta_3}{2d})}\sin(\frac{\pi j x}{d}), \tag{4.128b}$$

$$\Psi_{f_2} = -\frac{2h^3}{\pi^3 a_{55}}\sum_{j=1}^{\infty}\frac{(-1)^j}{j^3}\cdot\frac{\cosh(\frac{\pi j(x-\frac{d}{2})}{h\beta_3})}{\cosh(\frac{\pi j d}{2h\beta_3})}\sin(\frac{\pi j y}{h}). \tag{4.128c}$$

As shown in S.3.3.1, in torsion related problems, one encounters the case of $f_0 = -2\theta$, $f_1 = f_2 = 0$, where θ is the twist per unit length. In such a case, Ψ_{f_0} is given by (4.128a) and $\Psi = \theta(\Psi_{f_0} - \frac{x(x-d)}{a_{44}})$. Note that as an alternative, one could select the particular solution

$$\Psi_p = -\frac{\theta}{a_{55}} y(y-h), \tag{4.129}$$

to obtain

$$\Psi = \theta(\Psi_{f_0} - \frac{y(y-h)}{a_{55}}), \qquad \Psi_{f_0} = \frac{4h^2}{\pi^3 a_{55}} \sum_{j=1}^{\infty} \frac{(-1)^j - 1}{j^3} \cdot \frac{\cosh(\frac{\pi j(x-\frac{d}{2})}{h\beta_3})}{\cosh(\frac{\pi j d}{2h\beta_3})} \sin(\frac{\pi j y}{h}). \tag{4.130}$$

The above expressions are equivalent to those that appear in (Lekhnitskii, 1981). The solution is illustrated by **P.4.19** from which Fig. 4.16 is extracted.

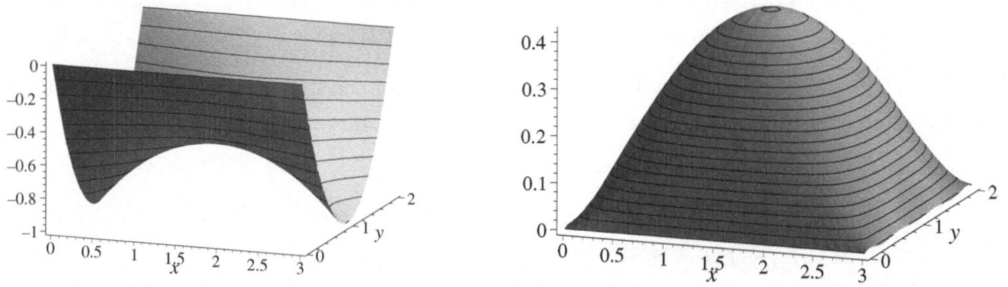

(a) Ψ_{f_0}. (b) $\Psi(\theta = 1)$.

Figure 4.16: Illustration of Example 4.15 for torsion, see (4.130).

4.4.3 *Application of Complex Potentials to the Biharmonic BVP*

In this section we shall present the homogeneous biharmonic BVP of (4.3a,b), with $F_0 = 0$, as a special case of 2-Coupled Dirichlet BVP derived in S.4.4.1. The characteristic polynomial of the differential operator $\nabla_1^{(4)}$ is $l_4(\mu) \equiv b_{22} - 2b_{26}\mu + (2b_{12} + b_{66})\mu^2 - 2b_{16}\mu^3 + b_{11}\mu^4$, and its roots cannot be real, see S.3.6.5. We denote these roots as $\mu_k = \alpha_k + \mathbf{i}\beta_k$, $\bar{\mu}_k = \alpha_k - \mathbf{i}\beta_k$, where $\beta_k > 0$, and so $l_4(\mu) = b_{11}[\mu^2 + 2\alpha_1\mu + (\alpha_1^2 + \beta_1^2)][\mu^2 + 2\alpha_2\mu + (\alpha_2^2 + \beta_2^2)]$.

In general, two cases should be considered: the case where the *complex parameters* are distinct, namely, $\mu_1 \neq \mu_2$, and the case where the complex parameters are pairwise equal, namely, $\mu_1 = \mu_2 = \alpha + \mathbf{i}\beta$ (for example, in the isotropic case $\mu_1 = \mu_2 = \mathbf{i}$). In what follows, we shall be focused on the first case. More details regarding the second one may be found in (Lekhnitskii, 1968). For the biharmonic function Φ, we employ two complex functions $\mathbf{F}_1(z_1)$ and $\mathbf{F}_2(z_2)$ and write $\Phi = 2\Re[\mathbf{F}_1(x + \mu_1 y) + \mathbf{F}_2(x + \mu_2 y)]$. We subsequently introduce two complex potentials $\mathbf{H}_k(z_k) = \widetilde{\Psi}_{1k}(z_k) + \mathbf{i}\widetilde{\Psi}_{2k}(z_k)$, which are the derivatives of the above functions $\mathbf{H}_k(z_k) = \mathbf{F}'_k$. Using Remark 4.4 we write,

$$\Phi_{,x} = 2\Re[\mathbf{H}_1(x + \mu_1 y) + \mathbf{H}_2(x + \mu_2 y)], \quad \Phi_{,y} = 2\Re[\mu_1 \mathbf{H}_1(x + \mu_1 y) + \mu_2 \mathbf{H}_2(x + \mu_2 y)]. \tag{4.131}$$

It is therefore shown that the biharmonic BVP may be equivalently formulated as a 2-Coupled Dirichlet BVP with the same set of boundary functions F_m or G_m, and $\Lambda_1 = -\Phi_{,x}$, $\Lambda_2 = \Phi_{,y}$, where the following complex constants are used:

$$\nu^{1k} = -2, \qquad \nu^{2k} = 2\mu_k. \tag{4.132}$$

Additional use of Remark 4.4 and (3.54) (without body-force influence since we are dealing here with the homogeneous BVP) shows that the general expressions for the stress components $\sigma_x = \Phi_{,yy}$, $\sigma_y = \Phi_{,xx}$, $\tau_{xy} = -\Phi_{,xy}$ are

$$\begin{aligned}
\sigma_x &= 2\Re[\mu_1^2\mathbf{H}_1'(x+\mu_1 y)+\mu_2^2\mathbf{H}_2'(x+\mu_2 y)],\\
\sigma_y &= 2\Re[\mathbf{H}_1'(x+\mu_1 y)+\mathbf{H}_2'(x+\mu_2 y)],\\
\tau_{xy} &= -2\Re[\mu_1\mathbf{H}_1'(x+\mu_1 y)+\mu_2\mathbf{H}_2'(x+\mu_2 y)].
\end{aligned} \tag{4.133}$$

Prior to the determination of the complex potentials, $\mathbf{H}_k(z_k)$, by applying the boundary conditions, we shall derive the expressions for the displacement components. Following (Lekhnitskii, 1968) we initially assume

$$\begin{aligned}
u &= 2\Re[p_1\mathbf{H}_1(x+\mu_1 y)+p_2\mathbf{H}_2(x+\mu_2 y)] - \omega_3^0 y + u^0,\\
v &= 2\Re[q_1\mathbf{H}_1(x+\mu_1 y)+q_2\mathbf{H}_2(x+\mu_2 y)] + \omega_3^0 x + v^0,
\end{aligned} \tag{4.134}$$

where ω_3^0, u^0, v^0 are the rigid body displacements' parameters and p_k, q_k are complex constants to be determined. We then derive the displacement derivatives using Remark 4.4 as

$$\begin{aligned}
u_{,x} &= 2\Re[p_1\mathbf{H}_1'(x+\mu_1 y)+p_2\mathbf{H}_2'(x+\mu_2 y)],\\
u_{,y} &= 2\Re[p_1\mu_1\mathbf{H}_1'(x+\mu_1 y)+p_2\mu_2\mathbf{H}_2'(x+\mu_2 y)] - \omega_3^0,\\
v_{,x} &= 2\Re[q_1\mathbf{H}_1'(x+\mu_1 y)+q_2\mathbf{H}_2'(x+\mu_2 y)] + \omega_3^0,\\
v_{,y} &= 2\Re[q_1\mu_1\mathbf{H}_1'(x+\mu_2 y)+q_2\mu_2\mathbf{H}_2'(x+\mu_2 y)].
\end{aligned} \tag{4.135}$$

Expressing the strain components $\varepsilon_x, \varepsilon_y$ by the above identities (see (1.9a,b)) and substituting these and (4.133) in (3.27), enables us by equating coefficients of $\mathbf{H}_k'(z_k)$, to deduce

$$p_k = b_{11}\mu_k^2 + b_{12} - b_{16}\mu_k, \qquad q_k = b_{12}\mu_k + \frac{b_{22}}{\mu_k} - b_{26}. \tag{4.136}$$

Since $l_4(\mu_k) = 0$, with the above p_k, q_k values, the third equation of (3.27) (i.e. $\gamma_{xy} = \cdots$) is satisfied as well.

We shall now present the solution for the first two fundamental problems of elasticity, see S.1.6.1. For the *first problem* when the boundary loads X_s, Y_s are given, the conditions on the contour take the form

$$\begin{aligned}
\frac{d}{ds}2\Re[\mathbf{H}_1(x+\mu_1 y)+\mathbf{H}_2(x+\mu_2 y)] &= -Y_s \qquad \text{on} \quad \partial\Omega,\\
\frac{d}{ds}2\Re[\mu_1\mathbf{H}_1'(x+\mu_1 y)+\mu_2\mathbf{H}_2'(x+\mu_2 y)] &= X_s \qquad \text{on} \quad \partial\Omega.
\end{aligned} \tag{4.137}$$

It should be noted that all three single-value conditions (4.4) should be applied here, and not just the two that emerge from (4.96). This is due to the fact that Λ_1, Λ_2 are derivatives of a common stress function, Φ, see (4.131).

For the *second problem*, when the displacements over the contour are given, the boundary conditions are, see (4.134),

$$\{u, v\} = \{u^*, v^*\} \qquad \text{on} \quad \partial\Omega, \tag{4.138}$$

where u^* and v^* are considered as known.

The above derivation may be realized for polynomial solutions by proper use of **P.4.18**, see also Example 4.16.

Example 4.16 *Polynomial Description of Complex Potentials.*

An effective way to apply solutions via complex potentials is the use of polynomial expansion. In this case, we express $\mathbf{H}_k(z_k)$ as

$$\mathbf{H}_k(z_k) = \sum_{l=2}^{L} \widetilde{\gamma}_{kl} z_k^l, \tag{4.139}$$

where $\widetilde{\gamma}_{kl}$ are unknown complex constants to be determined.

Consider, for example, the 2-Coupled Dirichlet BVP polynomial solution with $\mu_k = \alpha_k + \mathbf{i}\beta_k$, and let $\mathbf{H}_k(z_k) = \widetilde{\gamma}_{k2} z_k^2$ (i.e. $L = 2$), where $\widetilde{\gamma}_{k2} = A_k + \mathbf{i}B_k$. Since $\Lambda_1 = -\Phi_{,x}$, $\Lambda_2 = \Phi_{,y}$, we find using (4.131)

$$\begin{aligned}
\Phi_{,x} &= 4(A_1+A_2)x^2 + 8(-B_1\beta_1 - B_2\beta_2 + A_1\alpha_1 + A_2\alpha_2)yx \\
&\quad + 4(A_2{\alpha_2}^2 - 2B_1\alpha_1\beta_1 + A_1{\alpha_1}^2 - A_1{\beta_1}^2 - A_2{\beta_2}^2 - 2B_2\alpha_2\beta_2)y^2, \\
\Phi_{,y} &= 8(A_1{\alpha_1}^2 + A_2{\alpha_2}^2 - 2B_1\alpha_1\beta_1 - 2B_2\alpha_2\beta_2 - A_1{\beta_1}^2 - A_2{\beta_2}^2)yx \\
&\quad + 4(A_1\alpha_1 - B_2\beta_2 - B_1\beta_1 + A_2\alpha_2)x^2 + 4({\beta_2}^3 B_2 - 3B_1{\alpha_1}^2\beta_1 \\
&\quad - 3A_1\alpha_1{\beta_1}^2 + B_1{\beta_1}^3 - 3B_2\beta_2{\alpha_2}^2 - 3A_2\alpha_2{\beta_2}^2 + A_2{\alpha_2}^3 + A_1{\alpha_1}^3)y^2.
\end{aligned} \tag{4.140}$$

Integrating the system of (4.140) yields the stress function

$$\begin{aligned}
\Phi &= \frac{4}{3}(A_1+A_2)x^3 + 4(A_1\alpha_1 - B_2\beta_2 - B_1\beta_1 + A_2\alpha_2)yx^2 + 4[A_1\alpha_1(\frac{1}{3}{\alpha_1}^2 - {\beta_1}^2) \\
&\quad - B_1{\alpha_1}^2\beta_1 + \frac{1}{3}B_1{\beta_1}^3 - B_2{\alpha_2}^2\beta_2 + A_2\alpha_2(\frac{1}{3}{\alpha_2}^2 - {\beta_2}^2) + \frac{1}{3}B_2{\beta_2}^3]y^3 \\
&\quad + 4[-2B_1\alpha_1\beta_1 - 2B_2\alpha_2\beta_2 + A_1({\alpha_1}^2 - {\beta_1}^2) + A_2({\alpha_2}^2 - {\beta_2}^2)]y^2x.
\end{aligned} \tag{4.141}$$

Hence, we have reached a state where Φ is a prescribed biharmonic polynomial. Obviously, employing more terms in the series (4.139) yields more free coefficients and enables a higher accuracy solution. The “free coefficients” A_k, B_k of this function have to be determined so that the boundary conditions would be satisfied. Using continuous Fourier series based or piecewise-linear parametrization, i.e. substituting $x(t)$ and $y(t)$ in the boundary conditions (see S.3.1.4) enables us to write a contour error functional similar to S.4.3.4.2.

4.4.4 Application of Complex Potentials to a Coupled-Plane BVP

Here we shall present the Coupled-Plane BVP as a special case of the 3-Coupled Dirichlet BVP derived in S.4.4.1 with $f = g = 0$. In this case, the system's (six-order) characteristic polynomial $l_6 \equiv l_2 l_4 - l_3^2$ may be written as

$$\begin{aligned}
l_6(\mu) &= (b_{44} - 2b_{45}\mu + b_{55}\mu^2)(b_{22} - 2b_{26}\mu + (2b_{12}+b_{66})\mu^2 - 2b_{16}\mu^3 + b_{11}\mu^4) \\
&\quad - (-b_{24} + (b_{25}+b_{46})\mu - (b_{14}+b_{56})\mu^2 + b_{15}\mu^3)^2 = (b_{11}b_{55} - b_{15}^2)\mu^6 + \cdots.
\end{aligned} \tag{4.142}$$

The roots of the above polynomial can not be real, see Section 3.6.5, and we therefore have $\mu_k = \alpha_k + \mathbf{i}\beta_k$, $\bar{\mu}_k = \alpha_k - \mathbf{i}\beta_k$, where $\beta_k > 0$. Thus, $l_6(\mu)$ may also be written as

$$l_6(\mu) = (b_{11}b_{55} - b_{15}^2)\prod_{k=1}^{3}[\mu^2 + 2\alpha_k\mu + (\alpha_k^2 + \beta_k^2)]. \tag{4.143}$$

In addition, we assume that the *complex parameters* are distinct, namely, $\mu_1 \neq \mu_2 \neq \mu_3$.

To this end, the stress functions Φ and Λ are presented by the real parts of linear combinations of the three complex functions $\mathbf{F}_k(z_k)$, $k = 1,2,3$, see (Lekhnitskii, 1981), as

$$\begin{aligned}
\Phi &= 2\Re[\mathbf{F}_1(x+\mu_1 y) + \mathbf{F}_2(x+\mu_2 y) + \mathbf{F}_3(x+\mu_3 y)], \\
\Lambda &= 2\Re[\lambda_1\mathbf{F}_1'(x+\mu_1 y) + \lambda_2\mathbf{F}_2'(x+\mu_2 y) + \frac{1}{\lambda_3}\mathbf{F}_3'(x+\mu_3 y)],
\end{aligned} \tag{4.144}$$

where λ_i are the three complex numbers

$$\lambda_1 = -\frac{l_3(\mu_1)}{l_2(\mu_1)}, \qquad \lambda_2 = -\frac{l_3(\mu_2)}{l_2(\mu_2)}, \qquad \lambda_3 = -\frac{l_3(\mu_3)}{l_4(\mu_3)}. \tag{4.145}$$

Introducing the *complex potentials* $\mathbf{H}_k = \widetilde{\Psi}_{1k} + \mathbf{i}\widetilde{\Psi}_{2k}$, via differentiation of $\mathbf{F}_k$ as

$$\mathbf{H}_i(z_i) = \mathbf{F}_i'(z_i), \quad i = 1,2, \qquad \mathbf{H}_3(z_3) = \frac{1}{\lambda_3}\mathbf{F}_3'(z_i), \tag{4.146}$$

one may write using Remark 4.4

$$\begin{aligned}
\Phi_{,x} &= 2\Re[\mathbf{H}_1(x+\mu_1 y) + \mathbf{H}_2(x+\mu_2 y) + \lambda_3\mathbf{H}_3(x+\mu_3 y)],\\
\Phi_{,y} &= 2\Re[\mu_1\mathbf{H}_1(x+\mu_1 y) + \mu_2\mathbf{H}_2(x+\mu_2 y) + \mu_3\lambda_3\mathbf{H}_3(x+\mu_3 y)],\\
\Lambda &= 2\Re[\lambda_1\mathbf{H}_1(x+\mu_1 y) + \lambda_2\mathbf{H}_2(x+\mu_2 y) + \mathbf{H}_3(x+\mu_3 y)].
\end{aligned} \tag{4.147}$$

Hence the Coupled-Plane BVP may be formulated as a 3-coupled Dirichlet BVP with the same set of boundary functions F_m or G_m and $\Lambda_1 = -\Phi_{,x}$, $\Lambda_2 = \Phi_{,y}$, $\Lambda_3 = \Lambda$, where the following complex constants ν^{km} are used:

$$\begin{aligned}
\nu^{11} &= -2, & \nu^{12} &= -2, & \nu^{13} &= -2\lambda_3,\\
\nu^{21} &= 2\mu_1, & \nu^{22} &= 2\mu_2, & \nu^{23} &= 2\mu_3\lambda_3,\\
\nu^{31} &= 2\lambda_1, & \nu^{32} &= 2\lambda_2, & \nu^{33} &= 2.
\end{aligned} \tag{4.148}$$

Note that similar to the biharmonic equation in S.4.4.3, all three single-value conditions of (4.4) should be applied here since Λ_1, Λ_2 are derivatives of a common stress function, Φ.

Using (3.54), (3.92), (4.147) and Remark 4.4, one may express the stress components $\sigma_x = \Phi_{,yy}$, $\sigma_y = \Phi_{,xx}$, $\tau_{xy} = -\Phi_{,xy}$, $\tau_{yz} = -\Lambda_{,x}$, $\tau_{xz} = \Lambda_{,y}$ as

$$\begin{aligned}
\sigma_x &= 2\Re[\mu_1^2\mathbf{H}_1'(x+\mu_1 y) + \mu_2^2\mathbf{H}_2'(x+\mu_2 y) + \mu_3^2\lambda_3\mathbf{H}_3'(x+\mu_3 y)],\\
\sigma_y &= 2\Re[\mathbf{H}_1'(x+\mu_1 y) + \mathbf{H}_2'(x+\mu_2 y) + \lambda_3\mathbf{H}_3'(x+\mu_3 y)],\\
\tau_{xy} &= -2\Re[\mu_1\mathbf{H}_1'(x+\mu_1 y) + \mu_2\mathbf{H}_2'(x+\mu_2 y) + \mu_3\lambda_3\mathbf{H}_3'(x+\mu_3 y)],\\
\tau_{yz} &= -2\Re[\lambda_1\mathbf{H}_1'(x+\mu_1 y) + \lambda_2\mathbf{H}_2'(x+\mu_2 y) + \mathbf{H}_3'(x+\mu_3 y)],\\
\tau_{xz} &= 2\Re[\mu_1\lambda_1\mathbf{H}_1'(x+\mu_1 y) + \mu_2\lambda_2\mathbf{H}_2'(x+\mu_2 y) + \mu_3\mathbf{H}_3'(x+\mu_3 y)].
\end{aligned} \tag{4.149}$$

By assuming analogously to (4.134) the form of displacements

$$\begin{aligned}
u &= 2\Re[\textstyle\sum_{k=1}^3 p_k\mathbf{H}_k(x+\mu_k y)] - \omega_3^0 y + u^0,\\
v &= 2\Re[\textstyle\sum_{k=1}^3 q_k\mathbf{H}_k(x+\mu_k y)] + \omega_3^0 x + v^0,\\
w &= 2\Re[\textstyle\sum_{k=1}^3 r_k\mathbf{H}_k(x+\mu_k y)] + w^0,
\end{aligned} \tag{4.150}$$

where $\omega_3^0, u^0, v^0, w^0$ are the rigid displacement parameters, one may write

$$\begin{aligned}
u_{,x} &= 2\Re[\textstyle\sum_{k=1}^3 p_k\mathbf{H}_k'(x+\mu_k y)], & u_{,y} &= 2\Re[\textstyle\sum_{k=1}^3 p_k\mu_k\mathbf{H}_k'(x+\mu_k y)] - \omega_3^0,\\
v_{,x} &= 2\Re[\textstyle\sum_{k=1}^3 q_k\mathbf{H}_k'(x+\mu_k y)] + \omega_3^0, & v_{,y} &= 2\Re[\textstyle\sum_{k=1}^3 q_k\mu_k\mathbf{H}_k'(z_k)],\\
w_{,x} &= 2\Re[\textstyle\sum_{k=1}^3 r_k\mathbf{H}_k'(x+\mu_k y)], & w_{,y} &= 2\Re[\textstyle\sum_{k=1}^3 r_k\mu_k\mathbf{H}_k'(x+\mu_k y)].
\end{aligned} \tag{4.151}$$

Substituting (4.149), (4.150) in the expressions for the strain components $\varepsilon_x, \varepsilon_y, \gamma_{yz}$ of (3.139), we obtain for $i = 1,2$,

$$p_i = b_{11}\mu_i^2 + b_{12} - b_{16}\mu_i + \lambda_i(b_{15}\mu_i - b_{14}), \quad p_3 = \lambda_3(b_{11}\mu_3^2 + b_{12} - b_{16}\mu_3) + b_{15}\mu_3 - b_{14},$$
$$q_i = b_{12}\mu_i + \frac{b_{22}}{\mu_i} - b_{26} + \lambda_i(b_{25} - \frac{b_{24}}{\mu_i}), \quad q_3 = \lambda_3(b_{12}\mu_3 + \frac{b_{22}}{\mu_3} - b_{26}) + b_{25} - \frac{b_{24}}{\mu_3},$$
$$r_i = b_{14}\mu_i + \frac{b_{24}}{\mu_i} - b_{46} + \lambda_i(b_{45} - \frac{b_{44}}{\mu_i}), \quad r_3 = \lambda_3(b_{14}\mu_3 + \frac{b_{24}}{\mu_3} - b_{46}) + b_{45} - \frac{b_{44}}{\mu_3}. \qquad (4.152)$$

One may verify that the equations for the strain components γ_{xz}, γ_{xy} of (3.139) are satisfied in view of the definitions of (4.145) and the identities $l_6(\mu_k) = 0$.

Analogously to the biharmonic BVP case in S.4.4.3, we shall schematically discuss the solution for the first two fundamental problems of elasticity. For the *first problem* when the boundary loads X_s, Y_s, Z_s are given, the conditions on the contour take the form

$$\frac{d}{ds} 2\Re[\mathbf{H}_1(x+\mu_1 y) + \mathbf{H}_2(x+\mu_2 y) + \lambda_3 \mathbf{H}_3(x+\mu_3 y)] = -Y_s \quad \text{on} \quad \partial\Omega,$$
$$\frac{d}{ds} 2\Re[\mu_1 \mathbf{H}_1'(x+\mu_1 y) + \mu_2 \mathbf{H}_2'(x+\mu_2 y) + \mu_3\lambda_3 \mathbf{H}_3(x+\mu_3 y)] = X_s \quad \text{on} \quad \partial\Omega,$$
$$\frac{d}{ds} 2\Re[\lambda_1 \mathbf{H}_1(x+\mu_1 y) + \lambda_2 \mathbf{H}_2(x+\mu_2 y) + \mathbf{H}_3(x+\mu_3 y)] = Z_s \quad \text{on} \quad \partial\Omega. \qquad (4.153)$$

For the *second problem*, when the displacements u, v, w over the contour are given, the boundary conditions are, see (4.150),

$$\{u, v, w\} = \{u^*, v^*, w^*\} \quad \text{on} \quad \partial\Omega, \qquad (4.154)$$

where u^*, v^* and w^* are considered as known.

The above derivation may be realized for polynomial solution by **P.4.18**. An alternative, Fourier series based approach is presented in what follows.

4.4.5 Fourier Series Based Solution of a Coupled-Plane BVP

In this section, the homogeneous Coupled-Plane BVP (4.10a,b) in rectangle $\Omega = \{0 \le x \le d, 0 \le y \le h\}$, see Fig. 4.15(a), (with $f = g = 0$ and the first boundary condition version) is discussed. We assume MON13y material where the roots $\mu_k, \bar{\mu}_k$ of the characteristic polynomial $l_6 = l_4 l_2 - l_3^2$ (where $l_3 = (b_{25} + b_{46})\mu + b_{15}\mu^3$) are purely imaginary: $\mu_k = \mathbf{i}\beta_k, \bar{\mu}_k = -\mathbf{i}\beta_k$, while $\beta_k > 0$.

To support the above assumption of pure imaginary roots, we examine Figs. 4.17(a),(b) that describe the l_4 and l_6 characteristic polynomials as functions of μ^2 for a special MON13y material that has been created by rotating a typical orthotropic material by the angle $\theta_y = 30°$ about its y-axis. As shown, all roots are pure imaginary (since all μ_k^2 are pure negative). By activating **P.3.12**, the reader may verify that a similar behavior is obtained for any θ_y angle. Hence, the present analysis is confined to the above MON13y-type material. Therefore, in such a case, $x_k = x$, $y_k = \beta_k y$, and

$$v_i = \frac{b_{15}\beta_i^3 - \beta_i(b_{25} + b_{46})}{b_{44} - b_{55}\beta_i^2} \quad (i = 1,2), \qquad v_3 = \frac{b_{15}\beta_3^3 - \beta_3(b_{25} + b_{46})}{b_{22} - (2b_{12} + b_{66})\beta_3^2 + b_{11}\beta_3^4}. \qquad (4.155)$$

Note that for orthotropic (and simpler) materials, $v_2 = v_3 = 0$, $l_3 = 0$, and μ_1, μ_2 are the roots of l_4, while μ_3 is the root of l_2.

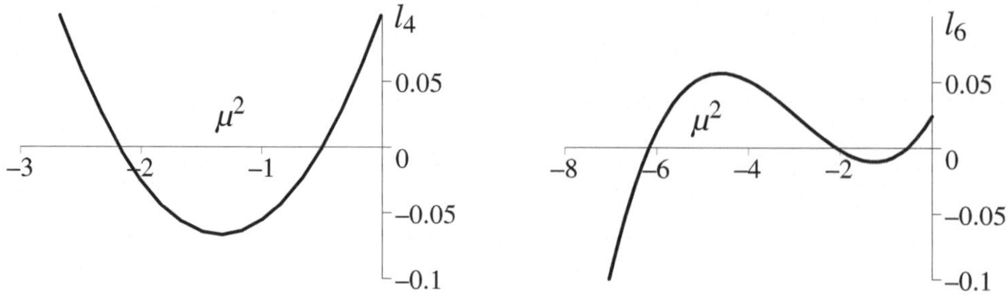

(a) l_4. (b) l_6.

Figure 4.17: The (normalized) operators for a special MON13y material.

The three *complex potentials* are presented as $\mathbf{H}_k(z_k) = \widetilde{\Psi}_{1k}(z_k) + \mathbf{i}\widetilde{\Psi}_{2k}(z_k)$. Three pairs of unknown harmonic conjugate functions $\widetilde{\Psi}_{1k}(x_k, y_k)$, $\widetilde{\Psi}_{2k}(x_k, y_k)$ should be defined over the rectangles $\Omega_k = \{0 \leq x \leq d, 0 \leq y \leq h\beta_k\}$. To this end, similar to S.4.4.2, one may augment the function Φ by the polynomial $\gamma_{11}xy + \gamma_{02}y^2 + \gamma_{12}xy^2$ and the function Λ by the polynomial $\alpha_{10}x + \alpha_{01}y + \alpha_{11}xy$, which do not violate the field equations for MON13y material, (4.10a), with $f = g = 0$. As in S.4.4.4, $\Lambda_1 = -\Phi_{,x}$, $\Lambda_2 = \Phi_{,y}$, $\Lambda_3 = -\Lambda$, and (4.147) show that

$$\begin{aligned}
\Phi_{,x} &= 2(\Psi_{11} + \Psi_{12} - \nu_3\Psi_{23}) + \gamma_{11}y + \gamma_{12}y^2, \\
\Phi_{,y} &= -2(\beta_1\Psi_{21} + \beta_2\Psi_{22} + \beta_3\nu_3\Psi_{13}) + \gamma_{11}x + 2\gamma_{02}y + 2\gamma_{12}xy, \\
\Lambda &= 2(-\nu_1\Psi_{21} - \nu_2\Psi_{22} + \Psi_{13}) + \alpha_{10}x + \alpha_{01}y + \alpha_{11}xy.
\end{aligned} \tag{4.156}$$

The following expressions participate in the boundary conditions under discussion:

$$\begin{aligned}
&\frac{d}{ds}\Phi_{,x} = \Phi_{,xy}\cos(\bar{\mathbf{n}}, x) - \Phi_{,xx}\cos(\bar{\mathbf{n}}, y), \qquad \frac{d}{ds}\Phi_{,y} = \Phi_{,yy}\cos(\bar{\mathbf{n}}, x) - \Phi_{,xy}\cos(\bar{\mathbf{n}}, y), \\
&\frac{d}{ds}\Lambda = \Lambda_{,y}\cos(\bar{\mathbf{n}}, x) - \Lambda_{,x}\cos(\bar{\mathbf{n}}, y).
\end{aligned} \tag{4.157}$$

For that purpose, Remark 4.4 enables us to write

$$\begin{aligned}
\Phi_{,yy} &= -2(\beta_1^2\Psi_{11,x} + \beta_2^2\Psi_{12,x} + \beta_3^2\nu_3\Psi_{23,x}) + 2\gamma_{02} + 2\gamma_{12}x, \\
\Phi_{,xx} &= 2(\Psi_{11,x} + \Psi_{12,x} - \nu_3\Psi_{23,x}), \\
-\Phi_{,xy} &= 2(\beta_1\Psi_{21,x} + \beta_2\Psi_{22,x} + \beta_3\nu_3\Psi_{13,x}) - \gamma_{11} - 2\gamma_{12}y, \\
\Lambda_{,y} &= -2(\nu_1\beta_1\Psi_{11,x} + \nu_2\beta_2\Psi_{12,x} - \beta_3\Psi_{23,x}) + \alpha_{01} + \alpha_{11}x, \\
-\Lambda_{,x} &= 2(\nu_1\Psi_{21,x} + \nu_2\Psi_{22,x} - \Psi_{13,x}) - \alpha_{10} - \alpha_{11}y,
\end{aligned} \tag{4.158}$$

where for the sake of convenience, we have replaced derivatives with respect to y by derivatives with respect to x according to Cauchy-Riemann conditions (4.90).

The conjugate harmonic functions $\widetilde{\Psi}_{1i}$, $\widetilde{\Psi}_{2i}$ for $i = 1, 2$ are presented in the *most generic* Fourier series form in the rectangle Ω_i, namely,

$$\begin{aligned}
\widetilde{\Psi}_{1i}(x, y_i) = \textstyle\sum_{j=1}^{\infty}\{&(-A_{ij}\cosh(\frac{\pi j y_i}{d}) + B_{ij}\cosh(\frac{\pi j(y_i - h\beta_i)}{d}))\cos(\frac{\pi j x}{d}) \\
&+ (C_{ij}\cosh(\frac{\pi j x}{h\beta_i}) - D_{ij}\cosh(\frac{\pi j(x-d)}{h\beta_i}))\cos(\frac{\pi j y_i}{h\beta_i})\},
\end{aligned} \tag{4.159a}$$

$$\widetilde{\Psi}_{2i}(x,y_i)=\sum_{j=1}^{\infty}\{(A_{ij}\sinh(\frac{\pi j y_i}{d})-B_{ij}\sinh(\frac{\pi j(y_i-h\beta_i)}{d}))\sin(\frac{\pi j x}{d})$$
$$+(C_{ij}\sinh(\frac{\pi j x}{h\beta_i})-D_{ij}\sinh(\frac{\pi j(x-d)}{h\beta_i}))\sin(\frac{\pi j y_i}{h\beta_i})\}. \quad (4.159b)$$

For $i=3$ (4.109), (4.110) are adopted. The functions Φ and Λ and their derivatives therefore become infinite series of *twelve terms* in each level (where the coefficients are A_{km}, B_{km}, C_{km}, D_{km}). Their derivatives are

$$\Phi_{,yy}=2\pi\sum_{m=1}^{\infty}\{[-A_{1m}\cosh(\frac{\pi m y\beta_1}{d})\beta_1^2+B_{1m}\cosh(\frac{\pi m(y-h)\beta_1}{d})\beta_1^2$$
$$-A_{2m}\cosh(\frac{\pi m y\beta_2}{d})\beta_2^2+B_{2m}\cosh(\frac{\pi m(y-h)\beta_2}{d})\beta_2^2$$
$$-A_{3m}\cosh(\frac{\pi m y\beta_3}{d})v_3\beta_3^2+B_{3m}\cosh(\frac{\pi m(y-h)\beta_3}{d})v_3\beta_3^2]\frac{m}{d}\sin(\frac{\pi m x}{d})$$
$$+[-C_{1m}\sinh(\frac{\pi m x}{h\beta_1})\beta_1+D_{1m}\sinh(\frac{\pi m(x-d)}{h\beta_1})\beta_1$$
$$-C_{2m}\sinh(\frac{\pi m x}{h\beta_2})\beta_2+D_{2m}\sinh(\frac{\pi m(x-d)}{h\beta_2})\beta_2 \quad (4.160a)$$
$$-C_{3m}\sinh(\frac{\pi m x}{h\beta_3})\beta_3 v_3+D_{3m}\sinh(\frac{\pi m(x-d)}{h\beta_3})\beta_3 v_3]\frac{m}{h}\cos(\frac{\pi m y}{h})\}+2\gamma_{02}+2\gamma_{12}x,$$

$$\Phi_{,xx}=2\pi\sum_{m=1}^{\infty}\{[A_{1m}\cosh(\frac{\pi m y\beta_1}{d})-B_{1m}\cosh(\frac{\pi m(y-h)\beta_1}{d})$$
$$+A_{2m}\cosh(\frac{\pi m y\beta_2}{d})-B_{2m}\cosh(\frac{\pi m(y-h)\beta_2}{d})$$
$$+A_{3m}\cosh(\frac{\pi m y\beta_3}{d})v_3-B_{3m}\cosh(\frac{\pi m(y-h)\beta_3}{d})v_3]\frac{m}{d}\sin(\frac{\pi m x}{d})$$
$$+[C_{1m}\sinh(\frac{\pi m x}{h\beta_1})\frac{1}{\beta_1}-D_{1m}\sinh(\frac{\pi m(x-d)}{h\beta_1})\frac{1}{\beta_1}$$
$$+C_{2m}\sinh(\frac{\pi m x}{h\beta_2})\frac{1}{\beta_2}-D_{2m}\sinh(\frac{\pi m(x-d)}{h\beta_2})\frac{1}{\beta_2}$$
$$+C_{3m}\sinh(\frac{\pi m x}{h\beta_3})\frac{v_3}{\beta_3}-D_{3m}\sinh(\frac{\pi m(x-d)}{h\beta_3})\frac{v_3}{\beta_3}]\frac{m}{h}\cos(\frac{\pi m y}{h})\}, \quad (4.160b)$$

$$\Phi_{,xy}=-2\pi\sum_{m=1}^{\infty}\{[A_{1m}\sinh(\frac{\pi m y\beta_1}{d})\beta_1-B_{1m}\sinh(\frac{\pi m(y-h)\beta_1}{d})\beta_1$$
$$+A_{2m}\sinh(\frac{\pi m y\beta_2}{d})\beta_2-B_{2m}\sinh(\frac{\pi m(y-h)\beta_2}{d})\beta_2$$
$$+A_{3m}\sinh(\frac{\pi m y\beta_3}{d})v_3\beta_3-B_{3m}\sinh(\frac{\pi m(y-h)\beta_3}{d})v_3\beta_3]\frac{m}{d}\cos(\frac{\pi m x}{d})$$
$$+[C_{1m}\cosh(\frac{\pi m x}{h\beta_1})-D_{1m}\cosh(\frac{\pi m(x-d)}{h\beta_1})$$
$$+C_{2m}\cosh(\frac{\pi m x}{h\beta_2})-D_{2m}\cosh(\frac{\pi m(x-d)}{h\beta_2})$$
$$+C_{3m}\cosh(\frac{\pi m x}{h\beta_3})v_3-D_{3m}\cosh(\frac{\pi m(x-d)}{h\beta_3})v_3]\frac{m}{h}\sin(\frac{\pi m y}{h})\}+\gamma_{11}+2\gamma_{12}y, \quad (4.160c)$$

$$\Lambda_{,x}=-2\pi\sum_{m=1}^{\infty}\{[A_{1m}\sinh(\frac{\pi m y\beta_1}{d})v_1-B_{1m}\sinh(\frac{\pi m(y-h)\beta_1}{d})v_1$$

$$+A_{2m}\sinh(\frac{\pi m y\beta_2}{d})\nu_2 - B_{2m}\sinh(\frac{\pi m(y-h)\beta_2}{d})\nu_2$$
$$-A_{3m}\sinh(\frac{\pi m y\beta_3}{d}) + B_{3m}\sinh(\frac{\pi m(y-h)\beta_3}{d})]\frac{m}{d}\cos(\frac{\pi m x}{d})$$
$$+[C_{1m}\cosh(\frac{\pi m x}{h\beta_1})\frac{\nu_1}{\beta_1} - D_{1m}\cosh(\frac{\pi m(x-d)}{h\beta_1})\frac{\nu_1}{\beta_1}$$
$$+C_{2m}\cosh(\frac{\pi m x}{h\beta_2})\frac{\nu_2}{\beta_2} - D_{2m}\cosh(\frac{\pi m(x-d)}{h\beta_2})\frac{\nu_2}{\beta_2}$$
$$-C_{3m}\cosh(\frac{\pi m x}{h\beta_3})\frac{1}{\beta_3} + D_{3m}\cosh(\frac{\pi m(x-d)}{h\beta_3})\frac{1}{\beta_3}]\frac{m}{h}\sin(\frac{\pi m y}{h})\} + \alpha_{10} + \alpha_{11}y, \quad (4.160d)$$

$$\Lambda_{,y} = 2\pi\sum_{m=1}^{\infty}\{[A_{3m}\cosh(\frac{\pi m y\beta_3}{d})\beta_3 - B_{3m}\cosh(\frac{\pi m(y-h)\beta_3}{d})\beta_3$$
$$-A_{1m}\cosh(\frac{\pi m y\beta_1}{d})\nu_1\beta_1 + B_{1m}\cosh(\frac{\pi m(y-h)\beta_1}{d})\nu_1\beta_1$$
$$-A_{2m}\cosh(\frac{\pi m y\beta_2}{d})\nu_2\beta_2 + B_{2m}\cosh(\frac{\pi m(y-h)\beta_2}{d})\nu_2\beta_2]\frac{m}{d}\sin(\frac{\pi m x}{d})$$
$$+[-C_{1m}\sinh(\frac{\pi m x}{h\beta_1})\nu_1 + D_{1m}\sinh(\frac{\pi m(x-d)}{h\beta_1})\nu_1$$
$$-C_{2m}\sinh(\frac{\pi m x}{h\beta_2})\nu_2 + D_{2m}\sinh(\frac{\pi m(x-d)}{h\beta_2})\nu_2$$
$$+C_{3m}\sinh(\frac{\pi m x}{h\beta_3}) - D_{3m}\sinh(\frac{\pi m(x-d)}{h\beta_3})]\frac{m}{h}\cos(\frac{\pi m y}{h})\} + \alpha_{01} + \alpha_{11}x. \quad (4.160e)$$

At this stage, we express boundary functions as $\widetilde{F}_k = \widetilde{P}_k\cos(\bar{\mathbf{n}},x) + \widetilde{Q}_k\cos(\bar{\mathbf{n}},y)$, see S.4.3, and let the functions $\widetilde{P}_k, \widetilde{Q}_k$ be represented in the Fourier series form

$$\widetilde{Q}_1(x,h) = \sum_{j=1}^{\infty}\widetilde{A}_{1j}\sin(\frac{\pi jx}{d}), \qquad \widetilde{Q}_1(x,0) = \sum_{j=1}^{\infty}\widetilde{B}_{1j}\sin(\frac{\pi jx}{d}),$$
$$\widetilde{P}_1(d,y) = \sum_{j=1}^{\infty}\widetilde{C}_{1j}\sin(\frac{\pi jy}{h}), \qquad \widetilde{P}_1(0,y) = \sum_{j=1}^{\infty}\widetilde{D}_{1j}\sin(\frac{\pi jy}{h}), \quad (4.161)$$
$$\widetilde{Q}_i(x,h) = \widetilde{A}_{i0} + \sum_{j=1}^{\infty}\widetilde{A}_{ij}\cos(\frac{\pi jx}{d}), \qquad \widetilde{Q}_i(x,0) = \widetilde{B}_{i0} + \sum_{j=1}^{\infty}\widetilde{B}_{ij}\cos(\frac{\pi jx}{d}), \quad i=2,3,$$
$$\widetilde{P}_i(d,y) = \widetilde{C}_{i0} + \sum_{j=1}^{\infty}\widetilde{C}_{ij}\cos(\frac{\pi jy}{h}), \qquad \widetilde{P}_i(0,y) = \widetilde{D}_{i0} + \sum_{j=1}^{\infty}\widetilde{D}_{ij}\cos(\frac{\pi jy}{h}), \quad i=2,3,$$

where the coefficients $\widetilde{A}_{ij}, \widetilde{B}_{ij}, \widetilde{C}_{ij}, \widetilde{D}_{ij}$ are assumed to be given.

We next exploit the identities (and analogous integral obtained by replacing $x \rightleftharpoons y$ and $h \rightleftharpoons d$)

$$\int_0^d \sin(\frac{\pi jx}{d})\,dx = -\frac{d((-1)^j-1)}{\pi j}, \qquad \int_0^d (x-\frac{d}{2})\sin(\frac{\pi jx}{d})\,dx = -\frac{d^2((-1)^j+1)}{2\pi j},$$
$$\int_0^h (y-\frac{h}{2})\cos(\frac{\pi jy}{h})\,dy = \frac{h^2((-1)^j-1)}{\pi^2 j^2}, \qquad \int_0^d \cos(\frac{\pi jx}{d})\,dx = 0, \quad (4.162)$$

to evaluate the single-value conditions of (4.11), which yield in this case a set of requirements, in which coefficients of (4.161) are involved, namely,

$$\oint_{\partial\Omega}\widetilde{F}_1 = \int_0^d [\widetilde{Q}_1(x,h) - \widetilde{Q}_1(x,0)]\,dx + \int_0^h [\widetilde{P}_1(d,y) - \widetilde{P}_1(0,y)]\,dy$$
$$= -\sum_{m=1}^{\infty}\frac{(-1)^m-1}{\pi m}[d(\widetilde{A}_{1m} - \widetilde{B}_{1m}) + h(\widetilde{C}_{1m} - \widetilde{D}_{1m})], \quad (4.163a)$$

$$\oint_{\partial\Omega} \widetilde{F}_i = \int_0^d [\widetilde{Q}_i(x,h) - \widetilde{Q}_i(x,0)]\,dx + \int_0^h [\widetilde{P}_i(d,y) - \widetilde{P}_i(0,y)]\,dy$$
$$= d(\widetilde{A}_{i0} - \widetilde{B}_{i0}) + h(\widetilde{C}_{i0} - \widetilde{D}_{i0}) = 0, \qquad i = 2,3, \qquad (4.163b)$$

$$\oint_{\partial\Omega} [(x - \frac{d}{2})\widetilde{F}_1 - (y - \frac{h}{2})\widetilde{F}_2] =$$
$$= \int_0^d (x - \frac{d}{2})[\widetilde{Q}_1(x,h) - \widetilde{Q}_1(x,0)]\,dx - \int_0^h (y - \frac{h}{2})[\widetilde{P}_2(d,y) - \widetilde{P}_2(0,y)]\,dy$$
$$+ \frac{d}{2}\int_0^h [\widetilde{P}_1(d,y) + \widetilde{P}_1(0,y)]\,dy - \frac{h}{2}\int_0^d [\widetilde{Q}_2(x,h) + \widetilde{Q}_2(x,0)]\,dx$$
$$= \sum_{m=1}^{\infty} (\widetilde{C}_{2m} - \widetilde{D}_{2m})\frac{h^2((-1)^m - 1)}{\pi^2 m^2} - \sum_{m=1}^{\infty} (\widetilde{A}_{1m} - \widetilde{B}_{1m})\frac{d^2((-1)^m + 1)}{2\pi m}$$
$$- \frac{d}{2}\sum_{m=1}^{\infty} (\widetilde{C}_{1m} + \widetilde{D}_{1m})\frac{h((-1)^m - 1)}{\pi m} - \frac{h}{2}(\widetilde{A}_{20} + \widetilde{B}_{20}). \qquad (4.163c)$$

We shall now construct the linear system of equations for the boundary conditions that will enable the determination of the coefficients A_{im}, B_{im}, C_{im}, D_{im}, $i = 1,3$. In view of Table 4.2,

	$y=0$	$x=d$	$y=h$	$x=0$
$\frac{dx}{ds}$	1	0	-1	0
$\frac{dy}{ds}$	0	1	0	-1
$\frac{d}{ds}\Phi_{,x}$	$\Phi_{,xx}$	$\Phi_{,xy}$	$-\Phi_{,xx}$	$-\Phi_{,xy}$
$\frac{d}{ds}\Phi_{,y}$	$\Phi_{,xy}$	$\Phi_{,yy}$	$-\Phi_{,xy}$	$-\Phi_{,yy}$
$\frac{d}{ds}\Lambda$	$\Lambda_{,x}$	$\Lambda_{,y}$	$-\Lambda_{,x}$	$-\Lambda_{,y}$
$\widetilde{F}_i$	$-\widetilde{Q}_i$	$\widetilde{P}_i$	$\widetilde{Q}_i$	$-\widetilde{P}_i$

Table 4.2: Boundary data for a Coupled-Plane BVP in a $h \times d$ rectangle.

Fig. 4.15(a) and (4.10b), (4.157) we write the twelve boundary conditions

$$\{\Phi_{,xy}, \Phi_{,xx}, \Lambda_{,x}\} = -\{\widetilde{Q}_2, -\widetilde{Q}_1, \widetilde{Q}_3\}, \qquad y \in \{0, h\},$$
$$\{\Phi_{,yy}, \Phi_{,xy}, \Lambda_{,y}\} = \{\widetilde{P}_2, -\widetilde{P}_1, \widetilde{P}_3\}, \qquad x \in \{0, d\}. \qquad (4.164)$$

We now deduce from (4.164) the desired system of linear algebraic equations.

The boundary conditions $\Phi_{,yy}(0,y) = \widetilde{P}_2(0,y)$ and $\Phi_{,yy}(d,y) = \widetilde{P}_2(d,y)$ yield

$$\beta_1 D_{1m}\sinh(\frac{\pi md}{h\beta_1}) + \beta_2 D_{2m}\sinh(\frac{\pi md}{h\beta_2}) + \beta_3 \nu_3 D_{3m}\sinh(\frac{\pi md}{h\beta_3}) = -\frac{h\widetilde{D}_{2m}}{2m\pi}, \qquad m > 0,$$
$$\beta_1 C_{1m}\sinh(\frac{\pi md}{h\beta_1}) + \beta_2 C_{2m}\sinh(\frac{\pi md}{h\beta_2}) + \beta_3 \nu_3 C_{3m}\sinh(\frac{\pi md}{h\beta_3}) = -\frac{h\widetilde{C}_{2m}}{2m\pi}, \qquad m > 0,$$
$$2\gamma_{02} = \widetilde{D}_{20}, \qquad 2\gamma_{02} + 2\gamma_{12}d = \widetilde{C}_{20}. \qquad (4.165)$$

The boundary conditions $\Lambda_{,y}(0,y) = \widetilde{P}_3(0,y)$ and $\Lambda_{,y}(d,y) = \widetilde{P}_3(d,y)$ yield

$$\nu_1 D_{1m}\sinh(\frac{\pi md}{h\beta_1}) + \nu_2 D_{2m}\sinh(\frac{\pi md}{h\beta_2}) - D_{3m}\sinh(\frac{\pi md}{h\beta_3}) = -\frac{h\widetilde{D}_{3m}}{2m\pi}, \qquad m > 0,$$

$$\nu_1 C_{1m} \sinh(\frac{\pi md}{h\beta_1}) + \nu_2 C_{2m} \sinh(\frac{\pi md}{h\beta_2}) - C_{3m} \sinh(\frac{\pi md}{h\beta_3}) = -\frac{h\widetilde{C}_{3m}}{2m\pi}, \quad m > 0,$$
$$\alpha_{01} = \widetilde{D}_{30}, \qquad \alpha_{01} + \alpha_{11} d = \widetilde{C}_{30}. \tag{4.166}$$

The boundary conditions $\Lambda_{,x}(x,0) = -\widetilde{Q}_3(x,0)$ and $\Lambda_{,x}(x,h) = -\widetilde{Q}_3(x,h)$ yield

$$\nu_1 B_{1m} \sinh(\frac{\pi mh\beta_1}{d}) + \nu_2 B_{2m} \sinh(\frac{\pi mh\beta_2}{d}) - B_{3m} \sinh(\frac{\pi mh\beta_3}{d}) = \frac{d\widetilde{B}_{3m}}{2m\pi}, \quad m > 0,$$
$$\nu_1 A_{1m} \sinh(\frac{\pi mh\beta_1}{d}) + \nu_2 A_{2m} \sinh(\frac{\pi mh\beta_2}{d}) - A_{3m} \sinh(\frac{\pi mh\beta_3}{d}) = \frac{d\widetilde{A}_{3m}}{2m\pi}, \quad m > 0,$$
$$\alpha_{10} = -\widetilde{B}_{30}, \qquad \alpha_{10} + \alpha_{11} h = -\widetilde{A}_{30}. \tag{4.167}$$

The boundary conditions $\Phi_{,xy}(x,0) = -\widetilde{Q}_2(x,0)$ and $\Phi_{,xy}(x,h) = -\widetilde{Q}_2(x,h)$ yield

$$\beta_1 B_{1m} \sinh(\frac{\pi mh\beta_1}{d}) + \beta_2 B_{2m} \sinh(\frac{\pi mh\beta_2}{d}) + \beta_3 \nu_3 B_{3m} \sinh(\frac{\pi mh\beta_3}{d}) = \frac{d\widetilde{B}_{2m}}{2m\pi}, \quad m > 0,$$
$$\beta_1 A_{1m} \sinh(\frac{\pi mh\beta_1}{d}) + \beta_2 A_{2m} \sinh(\frac{\pi mh\beta_2}{d}) + \beta_3 \nu_3 A_{3m} \sinh(\frac{\pi mh\beta_3}{d}) = \frac{d\widetilde{A}_{2m}}{2m\pi}, \quad m > 0,$$
$$\gamma_{11} = -\widetilde{B}_{20}, \qquad \gamma_{11} + 2\gamma_{12} h = -\widetilde{A}_{20}. \tag{4.168}$$

Hence, one may assume, see also (4.163b),

$$\alpha_{10} = -\widetilde{B}_{30}, \quad \alpha_{01} = \widetilde{D}_{30}, \quad \alpha_{11} = \frac{1}{d}(\widetilde{C}_{30} - \widetilde{D}_{30}) = \frac{1}{h}(\widetilde{B}_{30} - \widetilde{A}_{30}),$$
$$\gamma_{02} = \frac{1}{2}\widetilde{D}_{20}, \quad \gamma_{11} = -\widetilde{B}_{20}, \quad \gamma_{12} = \frac{1}{2d}(\widetilde{C}_{20} - \widetilde{D}_{20}) = \frac{1}{2h}(\widetilde{B}_{20} - \widetilde{A}_{20}). \tag{4.169}$$

So far, eight equations series have been derived. For the remaining four equations series we exploit the identities of Remark 4.5 to first write

$$\Phi_{,xy} = -2\pi \sum_{m=1}^{\infty} \{ \sum_{j=1}^{m} [\beta_1 (A_{1j} - (-1)^m B_{1j}) H^1_{jm} + \beta_2 (A_{2j} - (-1)^m B_{2j}) H^2_{jm}$$
$$+ \beta_3 \nu_3 (A_{3j} - (-1)^m B_{3j}) H^3_{jm}] \frac{j}{d} \cos(\frac{\pi jx}{d}) + [C_{1m} \cosh(\frac{\pi mx}{h\beta_1})$$
$$- D_{1m} \cosh(\frac{\pi m(x-d)}{h\beta_1}) + C_{2m} \cosh(\frac{\pi mx}{h\beta_2}) - D_{2m} \cosh(\frac{\pi m(x-d)}{h\beta_2})$$
$$+ C_{3m} \cosh(\frac{\pi mx}{h\beta_3}) \nu_3 - D_{3m} \cosh(\frac{\pi m(x-d)}{h\beta_3}) \nu_3] \frac{m}{h} \} \sin(\frac{\pi my}{h}) + \gamma_{11} + 2\gamma_{12} y. \tag{4.170}$$

Therefore, in view of the Fourier series expansion

$$\gamma_{11} + 2\gamma_{12} y = -\frac{2}{\pi} \sum_{m=1}^{\infty} \frac{((-1)^m - 1)\gamma_{11} + 2(-1)^m \gamma_{12}}{m} \sin(\frac{\pi my}{h}), \tag{4.171}$$

the boundary conditions $\Phi_{,xy}(0,y) = -\widetilde{P}_1(0,y)$ and $\Phi_{,xy}(d,y) = -\widetilde{P}_1(d,y)$ yield for $m > 0$

$$\frac{h}{d} \sum_{j=1}^{\infty} \frac{j}{m} [\beta_1 (A_{1j} - (-1)^m B_{1j}) H^1_{jm} + \beta_2 (A_{2j} - (-1)^m B_{2j}) H^2_{jm} + \beta_3 \nu_3 (A_{3j} - (-1)^m B_{3j}) H^3_{jm}]$$
$$+ C_{1m} + C_{2m} + \nu_3 C_{3m} - D_{1m} \cosh(\frac{\pi md}{h\beta_1}) - D_{2m} \cosh(\frac{\pi md}{h\beta_2}) - \nu_3 D_{3m} \cosh(\frac{\pi md}{h\beta_3})$$

$$= \frac{h}{2m\pi}[\widetilde{D}_{1m} + \frac{2}{\pi}\frac{((-1)^m - 1)\gamma_{11} + 2(-1)^m\gamma_{12}}{m}], \tag{4.172a}$$

$$\frac{h}{d}\sum_{j=1}^{\infty}\frac{j(-1)^j}{m}[\beta_1(A_{1j} - (-1)^m B_{1j})H^1_{jm} + \beta_2(A_{2j} - (-1)^m B_{2j})H^2_{jm} + \beta_3 v_3(A_{3j} - (-1)^m B_{3j})H^3_{jm}]$$

$$-D_{1m} - D_{2m} - v_3 D_{3m} + C_{1m}\cosh(\frac{\pi md}{h\beta_1}) + C_{2m}\cosh(\frac{\pi md}{h\beta_2}) + v_3 C_{3m}\cosh(\frac{\pi md}{h\beta_3})$$

$$= \frac{h}{2m\pi}[\widetilde{C}_{1m} + \frac{2}{\pi}\frac{((-1)^m - 1)\gamma_{11} + 2(-1)^m\gamma_{12}}{m}]. \tag{4.172b}$$

Analogously, for the boundary conditions $\Phi_{,xx}(x,0) = \widetilde{Q}_1(x,0)$ and $\Phi_{,xx}(x,h) = \widetilde{Q}_1(x,h)$ we write

$$\Phi_{,xx} = 2\pi\sum\nolimits_{m=1}^{\infty}\{[A_{1m}\cosh(\frac{\pi my\beta_1}{d}) - B_{1m}\cosh(\frac{\pi m(y-h)\beta_1}{d}) + A_{2m}\cosh(\frac{\pi my\beta_2}{d})$$

$$- B_{2m}\cosh(\frac{\pi m(y-h)\beta_2}{d}) + A_{3m}\cosh(\frac{\pi my\beta_3}{d})v_3 - B_{3m}\cosh(\frac{\pi m(y-h)\beta_3}{d})v_3]\frac{m}{d}$$

$$+\sum\nolimits_{j=1}^{m}[\frac{1}{\beta_1}(C_{1j} - (-1)^m D_{1j})G^1_{jm} + \frac{1}{\beta_2}(C_{2j} - (-1)^m D_{2j})G^2_{jm}$$

$$+ \frac{v_3}{\beta_3}(C_{3j} - (-1)^m D_{3j})G^3_{jm}]\frac{j}{h}\cos(\frac{\pi jy}{h})\}\sin(\frac{\pi mx}{d}), \tag{4.173}$$

and for $m > 0$

$$\frac{d}{h}\sum_{j=1}^{\infty}\frac{j}{m}[\frac{G^1_{jm}}{\beta_1}(C_{1j} - (-1)^m D_{1j}) + \frac{G^2_{jm}}{\beta_2}(C_{2j} - (-1)^m D_{2j}) + \frac{v_3}{\beta_3}G^3_{jm}(C_{3j} - (-1)^m D_{3j})] \tag{4.174}$$

$$+A_{1m} + A_{2m} + v_3 A_{3m} - B_{1m}\cosh(\frac{\pi nh\beta_1}{d}) - B_{2m}\cosh(\frac{\pi nh\beta_2}{d}) - v_3 B_{3m}\cosh(\frac{\pi nh\beta_3}{d}) = \frac{d\widetilde{B}_{1m}}{2m\pi},$$

$$\frac{d}{h}\sum_{j=1}^{\infty}\frac{j(-1)^j}{m}[\frac{G^1_{jm}}{\beta_1}(C_{1j} - (-1)^m D_{1j}) + \frac{G^2_{jm}}{\beta_2}(C_{2j} - (-1)^m D_{2j}) + \frac{v_3}{\beta_3}G^3_{jm}(C_{3j} - (-1)^m D_{3j})]$$

$$+A_{1m}\cosh(\frac{\pi mh\beta_1}{d}) + A_{2m}\cosh(\frac{\pi mh\beta_2}{d}) + v_3 A_{3m}\cosh(\frac{\pi mh\beta_3}{d}) - B_{1m} - B_{2m} - v_3 B_{3m} = \frac{d\widetilde{A}_{1m}}{2m\pi}.$$

The above methodology is implemented in **P.4.20**.

Remark 4.5 One may derive the functions H^k_{jm} and G^k_{jm} that satisfy the following identities of Fourier series:

$$\sinh(\frac{\pi jx}{h\beta_k}) = \sum\nolimits_{m=1}^{\infty} G^k_{jm}\sin(\frac{\pi mx}{d}), \qquad \sinh(\frac{\pi jy\beta_k}{d}) = \sum\nolimits_{m=1}^{\infty} H^k_{jm}\sin(\frac{\pi my}{h}), \tag{4.175}$$

where $k = 1,2,3,\ j > 0$. By definition, these functions are

$$H^k_{jm} = \frac{2}{h}\int_0^h \sinh(\frac{\pi jy\beta_k}{d})\sin(\frac{\pi my}{h})\,dy = -\frac{2d^2n(-1)^m\sinh(\frac{\pi hj\beta_k}{d})}{\pi(h^2j^2\beta_k^2 + d^2m^2)},$$

$$G^k_{jm} = \frac{2}{d}\int_0^d \sinh(\frac{\pi jx}{h\beta_k})\sin(\frac{\pi mx}{d})\,dx = -\frac{2nh^2\beta_k^2(-1)^m\sinh(\frac{\pi dj}{h\beta_k})}{\pi(d^2j^2 + h^2m^2\beta_k^2)}. \tag{4.176}$$

Note that using the identities $\sin(\frac{\pi m(y-h)}{h}) = (-1)^m\sin(\frac{\pi my}{h})$, $\sin(\frac{\pi m(x-d)}{d}) = (-1)^m\sin(\frac{\pi mx}{d})$, the following series are also obtained from (4.175):

$$\sinh(\frac{\pi j(x-d)}{h\beta_i}) = \sum\nolimits_{m=1}^{\infty} G^i_{jm}(-1)^m\sin(\frac{\pi mx}{d}),$$

$$\sinh(\frac{\pi j(y-h)\beta_i}{d}) = \sum_{m=1}^{\infty} H^i_{jm}(-1)^m \sin(\frac{\pi m y}{h}). \tag{4.177}$$

Example 4.17 *Non-Homogeneous Coupled-Plane BVP.*

There are many practical cases when the Coupled-Plane BVP (4.10a,b) with $g = 0, f = \overline{f}_{00} = const.$, and vanishing boundary conditions, i.e., $F_1 = F_2 = F_3 = 0$, should be solved, see Example 4.2. Let the solution be presented as a sum of homogeneous and particular solutions, namely, $\Phi = \Phi_0 + \Psi_p$, $\Lambda = \Lambda_0 + \Lambda_p$. We first seek a particular solution of the above field equations in the form $\Phi_p = 0$, $\Lambda_p = \frac{\overline{f}_{00}}{2b_{44}} x(x-d)$. Equation (4.16b) then shows that the homogeneous solutions Λ_0, Φ_0 should satisfy the following boundary conditions:

$$\frac{d}{ds}\{\Phi_{0,x}, \Phi_{0,y}\} = \{0, 0\}, \qquad \frac{d}{ds}\Lambda_0 = -\frac{d}{ds}\Lambda_p = -\Lambda_{p,x}\cos(\bar{\mathbf{n}}, y) \qquad \text{on} \quad \partial\Omega. \tag{4.178}$$

To determine the homogeneous solutions for MON13y material in a rectangular domain according to (4.16a,b), we activate **P.4.20** with $N = 12$, $h = d = 1$, vanishing boundary functions except for $Q_3 = \frac{d}{dx}\Lambda_{p|\partial\Omega}$, $t_y = 30^\circ$ and iso$= 0$. $\Lambda(x, y)$ and the particular solution $\Lambda_0(x, y)$ are presented in Figs. 4.18(a),(b), respectively, while Φ derivatives are presented in Figs. 4.19(a),(b). All functions are equally normalized. As shown, the derivatives of Φ are much smaller than Λ. As indicated in Example 4.2, when $\overline{f}_{00} = -2\theta$, the present solution represents a torsion problem, and therefore, this phenomena is similar to the one shown for orthotropic material in Example 4.15, and thus, it is shown again that Φ is relatively small (and frequently negligible) in pure torsion. This fact is exploited in the approximate analytical solutions presented in Chapters 9, 10.

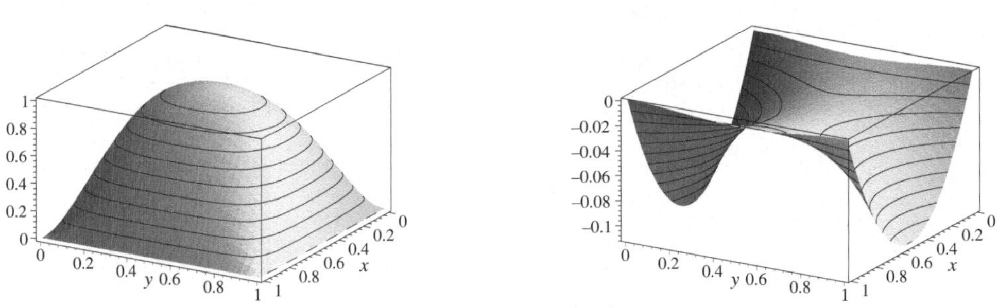

(a) $\Lambda(x, y)$. (b) $\Lambda_0(x, y)$.

Figure 4.18: Stress functions Λ, Λ_0 normalized by $\Lambda(\frac{d}{2}, \frac{h}{2})$ in Example 4.17.

In the orthotropic case, $\nabla_1^{(3)} = 0$ (and also $b_{44} = a_{44}$, $b_{55} = a_{55}$), and the above solution may be reduced accordingly. Since $l_3(\mu) \equiv 0$, (4.145) shows that $\lambda_i = 0$ and $v_i = 0$. Hence, the homogeneous part of Φ is set to zero ($\Phi_0 = 0$) and only Λ_0 has to be considered. In this case $\alpha_{ij} = \gamma_{ij} = 0$, $\widetilde{P}_j = \widetilde{Q}_j = 0$, $j = 1, 2$, and $\widetilde{P}_3 = 0$, $\widetilde{Q}_3 = \frac{\overline{f}_{00}}{a_{44}}(x - \frac{d}{2})$. We again employ the Fourier series (4.125). Hence

$$\begin{aligned} &\widetilde{C}_{3j} = \widetilde{D}_{3j} = \widetilde{A}_{ij} = \widetilde{B}_{ij} = \widetilde{C}_{ij} = \widetilde{D}_{ij} = 0, \qquad i = 1, 2, \quad j \geq 0, \\ &\widetilde{A}_{30} = \widetilde{B}_{30} = 0, \qquad \widetilde{A}_{3j} = \widetilde{B}_{3j} = \frac{\overline{f}_{00}}{a_{44}} \cdot \frac{d^2((-1)^j - 1)}{\pi^2 j^2}, \qquad j > 0. \end{aligned} \tag{4.179}$$

The above system of linear equations is decomposed into two systems, one of which has zero solution $A_{1j} = B_{1j} = A_{2j} = B_{2j} = C_{1j} = D_{1j} = C_{2j} = D_{2j} = C_{3j} = D_{3j} = 0$ (i.e. $\Phi_0 = 0$),

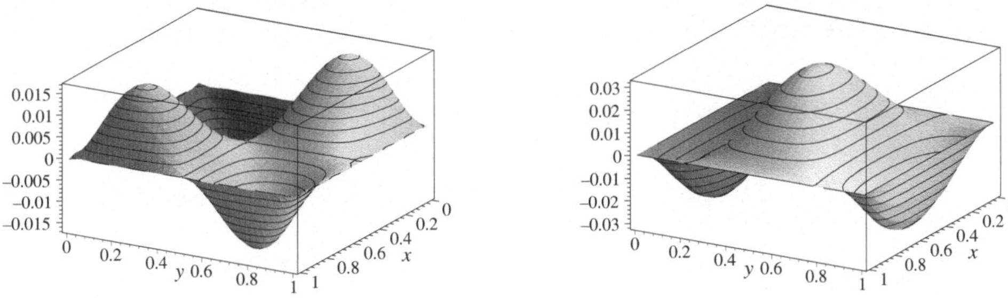

(a) $\Phi_{,x}$.

(b) $\Phi_{,y}$.

Figure 4.19: Stress function Φ derivatives normalized by $\Lambda(\frac{d}{2},\frac{h}{2})$ in Example 4.17.

while for the second system, (4.167) shows that

$$A_{3j} = B_{3j} = -\frac{d^3\overline{f}_{00}((-1)^j - 1)}{2\pi^3 j^3 a_{44}\sinh(\frac{\pi j h \beta_3}{d})}. \tag{4.180}$$

The stress function Λ obtained here is identical to the one presented in Example 4.15.

4.5 Three-Dimensional Prescribed Solutions

Not many analytical solutions for three-dimensional problems in elasticity are available for isotropic materials, and many fewer exist for anisotropic materials. In what follows we shall review and demonstrate some limited aspects of this problem.

4.5.1 Equilibrium Equations in Terms of Displacements

For general anisotropic material the stiffness matrix of (2.4) enables us to express the equilibrium equations (1.82a–c) in terms of displacements as

$$\sum_{j=1}^{3} L_{ji}u_j + F_{bi} = 0, \qquad i = 1,2,3 \tag{4.181}$$

where the $L_{ij} = L_{ji}$ are second-order differential operators, which take the following form for Cartesian anisotropy (see **P.4.21** which also produces analogous formulas for partial differential operators of various orthogonal coordinates):

$$\begin{aligned}
L_{11} &= A_{11}D_{11} + A_{66}D_{22} + A_{55}D_{33} + \underline{2A_{16}D_{12} + 2A_{56}D_{23} + 2A_{15}D_{13}},\\
L_{22} &= A_{66}D_{11} + A_{22}D_{22} + A_{44}D_{33} + \underline{2A_{26}D_{12} + 2A_{46}D_{13} + 2A_{24}D_{23}},\\
L_{33} &= A_{55}D_{11} + A_{44}D_{22} + A_{33}D_{33} + \underline{2A_{45}D_{12} + 2A_{34}D_{23} + 2A_{35}D_{13}},\\
L_{12} &= \underline{A_{16}D_{11} + A_{26}D_{22} + A_{45}D_{33} + (A_{25} + A_{46})D_{23} + (A_{14} + A_{56})D_{13}} + (A_{12} + A_{66})D_{12},\\
L_{13} &= \underline{A_{15}D_{11} + A_{46}D_{22} + A_{35}D_{33} + (A_{15} + A_{14})D_{12} + (A_{36} + A_{45})D_{23}} + (A_{13} + A_{55})D_{13},\\
L_{23} &= \underline{A_{56}D_{11} + A_{24}D_{22} + A_{34}D_{33} + (A_{24} + A_{25})D_{12} + (A_{36} + A_{45})D_{13}} + (A_{23} + A_{44})D_{23}.
\end{aligned} \tag{4.182}$$

Here, A_{ij} are constants, $D_{ij} = \frac{\partial^2}{\partial\alpha_i\partial\alpha_j}$ are partial derivative operators and $\alpha_1 = x$, $\alpha_2 = y$, $\alpha_3 = z$. The underlined terms vanish for orthotropic materials as described in S.2.4.

As indicated by (Baida, 1959), the general solution of the equilibrium equations shown by (4.181) may be presented by three potentials $\Phi_i^0(x,y,z)$ $(i=1,2,3)$ in the form

$$u_1 = \begin{vmatrix} L_{22} & L_{32} \\ L_{23} & L_{33} \end{vmatrix} \Phi_1^0 + \begin{vmatrix} L_{31} & L_{21} \\ L_{33} & L_{23} \end{vmatrix} \Phi_2^0 + \begin{vmatrix} L_{21} & L_{31} \\ L_{22} & L_{32} \end{vmatrix} \Phi_3^0 + u_{10},$$
$$u_2 = \begin{vmatrix} L_{32} & L_{12} \\ L_{33} & L_{13} \end{vmatrix} \Phi_1^0 + \begin{vmatrix} L_{11} & L_{31} \\ L_{13} & L_{33} \end{vmatrix} \Phi_2^0 + \begin{vmatrix} L_{31} & L_{11} \\ L_{32} & L_{12} \end{vmatrix} \Phi_3^0 + u_{20},$$
$$u_3 = \begin{vmatrix} L_{12} & L_{22} \\ L_{13} & L_{23} \end{vmatrix} \Phi_1^0 + \begin{vmatrix} L_{21} & L_{11} \\ L_{23} & L_{13} \end{vmatrix} \Phi_2^0 + \begin{vmatrix} L_{11} & L_{21} \\ L_{12} & L_{22} \end{vmatrix} \Phi_3^0 + u_{30}, \qquad (4.183)$$

where u_{i0}, $i=1,2,3$ are particular solutions of (4.181), and the second-order determinants create fourth-order differential operators. In the above formulation, the potential functions should satisfy the sixth-order partial differential equation

$$\begin{vmatrix} L_{11} & L_{12} & L_{13} \\ L_{21} & L_{22} & L_{23} \\ L_{31} & L_{32} & L_{33} \end{vmatrix} \Phi_i^0 = 0, \qquad i=1,2,3. \qquad (4.184)$$

Note that for the orthotropic material case, (4.184) contains derivatives (with respect to x, y, z) of even order only.

Example 4.18 *Orthotropic parallelepiped.*
As an example for the above discussion, consider the problem where all displacements are given on the six boundary faces of an orthotropic parallelepiped that occupies the domain $\Pi = \{|x| \leq d,\ |y| \leq h, |z| \leq l\}$, see Fig. 4.20. In order to satisfy boundary conditions the solution

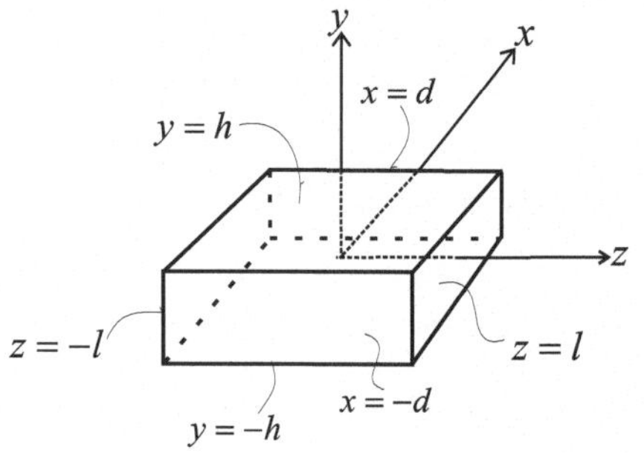

Figure 4.20: Parallelepiped notation.

of (4.184) should be expressed as the following double Fourier series:

$$\Phi_1^0 = \sum_{m,n=0}^{\infty} f_{nm}(x) \sin(\frac{n\pi y}{h}) \sin(\frac{m\pi z}{l}), \qquad (4.185a)$$
$$\Phi_2^0 = \sum_{k,m=0}^{\infty} f_{km}(y) \sin(\frac{k\pi x}{d}) \sin(\frac{m\pi z}{l}), \qquad (4.185b)$$
$$\Phi_3^0 = \sum_{k,n=0}^{\infty} f_{kn}(z) \sin(\frac{k\pi x}{d}) \sin(\frac{n\pi y}{h}). \qquad (4.185c)$$

By substituting (4.185a–c) in (4.184), one obtains three sixth-order ordinary differential equations for the functions $f_{nm}(x)$, $f_{km}(y)$, $f_{kn}(z)$, the characteristic equations of which are reduced to cubic algebraic equations that may be solved in an exact closed-form manner. One may present the above functions as

$$f_{nm} = \sum_{i=1}^{6} A_{nm}^{(i)} t_{1i}(x), \quad f_{km} = \sum_{i=1}^{6} B_{km}^{(i)} t_{2i}(y), \quad f_{kn} = \sum_{i=1}^{6} C_{kn}^{(i)} t_{3i}(z), \tag{4.186}$$

where $t_{1i}(x), t_{2i}(y), t_{3i}(z)$ are the linearly independent solutions of the corresponding ordinary differential equations (six for each equation). The arbitrary constants $A_{nm}^{(i)}, B_{km}^{(i)}, C_{kn}^{(i)}$ $(i = 1,\ldots,6)$ are sufficient to satisfy all given boundary conditions on Π. Finally, one may write the displacements (4.183), in view of (4.185a–c, 4.186), as

$$u_1 = \sum_{m,n=0}^{\infty} \Theta_{nm}^{(1)}(x)\sin\left(\frac{n\pi y}{h}\right)\sin\left(\frac{m\pi z}{l}\right) + \sum_{k,m=0}^{\infty} \Theta_{km}^{(2)}(y)\cos\left(\frac{k\pi x}{d}\right)\sin\left(\frac{m\pi z}{l}\right)$$
$$+ \sum_{k,n=0}^{\infty} \Theta_{kn}^{(3)}(z)\cos\left(\frac{k\pi x}{d}\right)\sin\left(\frac{n\pi y}{h}\right), \tag{4.187a}$$

$$u_2 = \sum_{n,m=0}^{\infty} \Theta_{nm}^{(4)}(x)\cos\left(\frac{n\pi y}{h}\right)\sin\left(\frac{m\pi z}{l}\right) + \sum_{k,m=0}^{\infty} \Theta_{km}^{(5)}(y)\sin\left(\frac{k\pi x}{d}\right)\sin\left(\frac{m\pi z}{l}\right)$$
$$+ \sum_{k,n=0}^{\infty} \Theta_{kn}^{(6)}(z)\sin\left(\frac{k\pi x}{d}\right)\cos\left(\frac{n\pi y}{h}\right), \tag{4.187b}$$

$$u_3 = \sum_{n,m=0}^{\infty} \Theta_{nm}^{(7)}(x)\sin\left(\frac{n\pi y}{h}\right)\cos\left(\frac{m\pi z}{l}\right) + \sum_{k,m=0}^{\infty} \Theta_{km}^{(8)}(y)\sin\left(\frac{k\pi x}{d}\right)\cos\left(\frac{m\pi z}{l}\right)$$
$$+ \sum_{k,n=0}^{\infty} \Theta_{kn}^{(9)}(z)\sin\left(\frac{k\pi x}{d}\right)\sin\left(\frac{n\pi y}{h}\right), \tag{4.187c}$$

where $\Theta_{ij}^{(s)}$ are nine series of functions of one variable which linearly depend on $A_{nm}^{(r)}, B_{km}^{(r)}, C_{kn}^{(r)}$.

Natural (stress) boundary conditions are treated analogously (i.e., by expressing them as derivatives of displacements via the material constitutive relations). Therefore, for generic surface load distributions that satisfy the integral conditions of (1.186),

$$\{\sigma_z, \tau_{xz}, \tau_{yz}\}_{|z=\pm l} = \{F_{11}^{\pm}, F_{12}^{\pm}, F_{13}^{\pm}\}(x,y), \tag{4.188a}$$
$$\{\sigma_x, \tau_{xy}, \tau_{xz}\}_{|x=\pm d} = \{F_{21}^{\pm}, F_{22}^{\pm}, F_{23}^{\pm}\}(y,z), \tag{4.188b}$$
$$\{\sigma_y, \tau_{yz}, \tau_{xy}\}_{|y=\pm h} = \{F_{31}^{\pm}, F_{32}^{\pm}, F_{33}^{\pm}\}(x,z), \tag{4.188c}$$

18 arbitrary functions $\{F_{ij}^{\pm}\}$, $i,j = 1,2,3$ may be presented as doubled Fourier series, and the stresses are expressed by displacements derivatives. One therefore obtains a system of 18 infinite groups of linear algebraic equations with respect to an infinite number of groups (each of 18 equations) of the variables $A_{nm}^{(s)}, B_{km}^{(s)}, C_{kn}^{(s)}$ $(m,n,k = 0,1,\ldots$ and $s = 1,\ldots,6)$. Numerical solution is possible by a truncated Fourier series, analogously to S.4.4.5.

4.5.2 *Fourier Series Based Solutions in an Isotropic Parallelepiped*

To present a Fourier series based closed form solution for a parallelepiped, we shall adopt the analytical approach presented in the previous section and follow the ideas of (Vishnyakov and Beresnev, 1964). By employing Lamé's constants for isotropic materials, see Remark 2.2, one may express the stress components shown by (2.21) as

$$\sigma_x = \lambda(u_{,x} + v_{,y} + w_{,z}) + 2\mu u_{,x}, \quad \sigma_y = \lambda(u_{,x} + v_{,y} + w_{,z}) + 2\mu v_{,y}, \quad \tau_{xy} = \mu(v_{,x} + u_{,y}),$$
$$\sigma_z = \lambda(u_{,x} + v_{,y} + w_{,z}) + 2\mu w_{,z}, \quad \tau_{yz} = \mu(w_{,y} + v_{,z}), \quad \tau_{xz} = \mu(w_{,x} + u_{,z}). \tag{4.189}$$

The reader may verify that in the isotropic case, the differential operators (4.182) are simplified as

$$L_{kk} = (\lambda + \mu)D_{kk} + \mu\Delta^{(2)}, \quad L_{kj} = (\lambda + \mu)D_{kj}, \qquad j,k = 1,2,3, \;\; j \neq k \tag{4.190}$$

where $\Delta^{(2)} = D_{11} + D_{22} + D_{33}$ is Laplace's operator in a three-dimensional space. The third-order determinant of (4.184) is reduced to $\mu^2(2\mu+\lambda)(\Delta^{(2)})^3$. Hence, when the stress components of (4.189) are substituted in the differential equilibrium equations, (1.82a–c), these become (see also 4.181)

$$\begin{aligned}(\lambda+\mu)(u_{,xx}+v_{,xy}+w_{,xz})+\mu(u_{,xx}+u_{,yy}+u_{,zz})+X_b=0,\\ (\lambda+\mu)(u_{,xy}+v_{,yy}+w_{,yz})+\mu(v_{,xx}+v_{,yy}+v_{,zz})+Y_b=0,\\ (\lambda+\mu)(u_{,xz}+v_{,yz}+w_{,zz})+\mu(w_{,xx}+w_{,yy}+w_{,zz})+Z_b=0.\end{aligned} \tag{4.191}$$

Further on, in the absence of body forces, and assuming $\Phi_i^0 = 0$, $i = 1,2$, one may prescribe the displacement components in the form

$$\begin{aligned}u_3 &= \sum\nolimits_{k,n} \Theta_{kn}^{(3)}(z)\cos(\frac{k\pi x}{d})\sin(\frac{n\pi y}{h}),\\ v_3 &= \sum\nolimits_{k,n} \Theta_{kn}^{(6)}(z)\sin(\frac{k\pi x}{d})\cos(\frac{n\pi y}{h}),\\ w_3 &= \sum\nolimits_{k,n} \Theta_{kn}^{(9)}(z)\sin(\frac{k\pi x}{d})\sin(\frac{n\pi y}{h}),\end{aligned} \tag{4.192}$$

where

$$\begin{aligned}\Theta_{kn}^{(3)}(z) &= -(A_{kn}^{(3)}+M_{kn}^{(3)})\cosh(\beta_{kn}^{(3)}z) - \frac{\pi k}{d}K_{kn}^{(3)}P_{kn}^{(3)}z\sinh(\beta_{kn}^{(3)}z),\\ \Theta_{kn}^{(6)}(z) &= -(A_{kn}^{(3)}+N_{kn}^{(3)})\cosh(\beta_{kn}^{(3)}z) - \frac{\pi n}{h}K_{kn}^{(3)}P_{kn}^{(3)}z\sinh(\beta_{kn}^{(3)}z),\\ \Theta_{kn}^{(9)}(z) &= -\beta_{kn}^{(3)}K_{kn}^{(3)}P_{kn}^{(3)}z\cosh(\beta_{kn}^{(3)}z) - \frac{1}{\beta_{kn}^{(3)}}(\frac{\pi k}{d}+\frac{\pi n}{h})A_{kn}^{(3)}\sinh(\beta_{kn}^{(3)}z).\end{aligned} \tag{4.193}$$

Here, $A_{kn}^{(3)}, M_{kn}^{(3)}, N_{kn}^{(3)}$ are three series of arbitrary constants, and

$$\beta_{kn}^{(3)} = \sqrt{(\frac{\pi k}{d})^2+(\frac{\pi n}{h})^2},\quad K_{kn}^{(3)} = \frac{\lambda+\mu}{(\lambda+3\mu)\beta_{kn}^{(3)}},\quad P_{kn}^{(3)} = \frac{\pi k}{d}M_{kn}^{(3)}+\frac{\pi n}{h}N_{kn}^{(3)}. \tag{4.194}$$

The subscript 3 stands for the fact that in the doubled Fourier series (4.192) for x and y, the coefficients are functions of z. Note that u_3, v_3 are even and w_3 is an odd function of z. As far as the boundary conditions for stresses are concerned, the following observations are important:

(a) The stress components $\sigma_x, \sigma_y, \sigma_z$ are of the form

$$\sum\nolimits_{k,n=0} \tilde{f}_{1kn}(z)\sin(\frac{k\pi x}{d})\sin(\frac{n\pi y}{h}). \tag{4.195}$$

(b) The stress component τ_{xz} is of the form

$$\sum\nolimits_{k,n=0} \tilde{f}_{2kn}(z)\cos(\frac{k\pi x}{d})\sin(\frac{n\pi y}{h}). \tag{4.196}$$

(c) The stress component τ_{yz} is of the form

$$\sum\nolimits_{k,n=0} \tilde{f}_{3kn}(z)\sin(\frac{k\pi x}{d})\cos(\frac{n\pi y}{h}). \tag{4.197}$$

(d) The stress component τ_{xy} is of the form

$$\sum\nolimits_{k,m=0} \tilde{f}_{4kn}(z)\cos(\frac{k\pi x}{d})\cos(\frac{n\pi y}{h}). \tag{4.198}$$

The first three functions that appear above are (to shorten the expressions we use $\beta=\beta_{kn}^{(3)}$, $K=K_{kn}^{(3)}$, $N=N_{kn}^{(3)}$, $M=M_{kn}^{(3)}$, $A=A_{kn}^{(3)}$):

$$\begin{aligned}\widetilde{f}_{1kn}=&[(\lambda((\frac{n\pi}{h})^2+(\frac{\pi k}{d})^2)-(\lambda+2\mu)\beta^2)\frac{n\pi}{h}zK\sinh(\beta z)+(\lambda-(\lambda+2\mu)\beta K)\frac{n\pi}{h}\cosh(\beta z)]N\\&+[(\lambda((\frac{n\pi}{h})^2+(\frac{\pi k}{d})^2)-(\lambda+2\mu)\beta^2)\frac{k\pi}{d}zK\sinh(\beta z)+(\lambda-(\lambda+2\mu)\beta K)\frac{k\pi}{d}\cosh(\beta z)]M\\&-2\mu(\frac{k\pi}{d}+\frac{n\pi}{h})\cosh(\beta z)A,\end{aligned}\tag{4.199a}$$

$$\begin{aligned}-\widetilde{f}_{2kn}=&((\frac{k\pi}{d})^2+\frac{n\pi}{h}\cdot\frac{k\pi}{d}+\beta^2)\mu\sinh(\beta z)A+\mu K\frac{k\pi}{d}\cdot\frac{n\pi}{h}(\sinh(\beta z)+2z\cosh(\beta z)\beta)\beta N\\&+((\beta+(\frac{k\pi}{d})^2K)\sinh(\beta z)+2(\frac{k\pi}{d})^2Kz\cosh(\beta z)\beta)\beta\mu M,\end{aligned}\tag{4.199b}$$

$$\begin{aligned}-\widetilde{f}_{3kn}=&(\frac{n\pi}{h}\frac{k\pi}{d}+(\frac{n\pi}{h})^2+\beta^2)\mu\sinh(\beta z)A+\mu K\frac{k\pi}{d}\cdot\frac{n\pi}{h}(2z\cosh(\beta z)\beta+\sinh(\beta z))\beta M\\&+((\beta+(\frac{n\pi}{h})^2K)\sinh(\beta z)+2(\frac{n\pi}{h})^2Kz\cosh(\beta z)\beta)\beta\mu N,\end{aligned}\tag{4.199c}$$

where the functions $\widetilde{f}_{1kn}$ and $\widetilde{f}_{4kn}$ are even and $\widetilde{f}_{2kn}, \widetilde{f}_{3kn}$ are odd with respect to z. To this end, one may select a set of five values k, n, $A_{kn}^{(3)}$, $M_{kn}^{(3)}$, $N_{kn}^{(3)}$, for which a prescribed three-dimensional consistent solution is obtained. As an example, we select $\widetilde{f}_{2kn}(\pm l)=\widetilde{f}_{3kn}(\pm l)=0$ and $\widetilde{f}_{1kn}(\pm l)=f_0$. By substituting the above in (4.199a–c), we obtain an algebraic system (which is not singular in a general case), the solution of which yields $A_{kn}^{(3)}$, $M_{kn}^{(3)}$, $N_{kn}^{(3)}$ per given k and n. The above selection yields the solution of following kind of the first fundamental problem (which is implemented in **P.4.22**), see S.1.6.1:

$$\{\sigma_z,\ \tau_{xz},\ \tau_{yz}\}_{z=\pm l}=\{f_0\sin(\frac{k\pi x}{d})\sin(\frac{n\pi y}{h}),0,0\},\tag{4.200a}$$

$$\{\sigma_x,\ \tau_{xy},\ \tau_{xz}\}_{x=\pm d}=\{\widetilde{f}_{1kn}(z)(-1)^{k+1}\sin(\frac{n\pi y}{h}),0,0\},\tag{4.200b}$$

$$\{\sigma_y,\ \tau_{yz},\ \tau_{xy}\}_{y=\pm h}=\{\widetilde{f}_{1kn}(z)(-1)^{n+1}\sin(\frac{k\pi x}{d})0,0\}.\tag{4.200c}$$

One interesting interpretation of this problem is shown in Fig. 4.21, where the three-dimensional parallelepiped is viewed as a beam where the dimension in the z direction is much larger than the other two, i.e., $l\gg h,d$. For $f_0(1,1)=1$, $\lambda=11/20$, $\mu=2/5$, $h=d=1$, $l=10$, Fig. 4.21 presents σ_z at $z=10$ (the fluctuated distribution) and $z=9.1$ (the "≈ 0" distribution). The above solution shows that while on the $z=\pm l$ edge, a double cosine distribution of σ_z is applied, and on the outer surface, single cosine distributions of σ_x and σ_y are applied at the tip, these stress components decay (very rapidly) with the distance from the tip according to St. Venant's Principle (see S.5.1.3.1).

4.5.3 Direct Solution in Terms of Displacements for Three-Dimensional Bodies

As an alternative course to the one presented above, one may obtain a prescribed polynomial form of displacement directly from (4.181), and then calculate the polynomial form of stress that satisfies the boundary conditions. In what follows, we shall review some applications for parallelepiped and three-dimensional bodies, the boundary surface S of which may be parameterized by Fourier series.

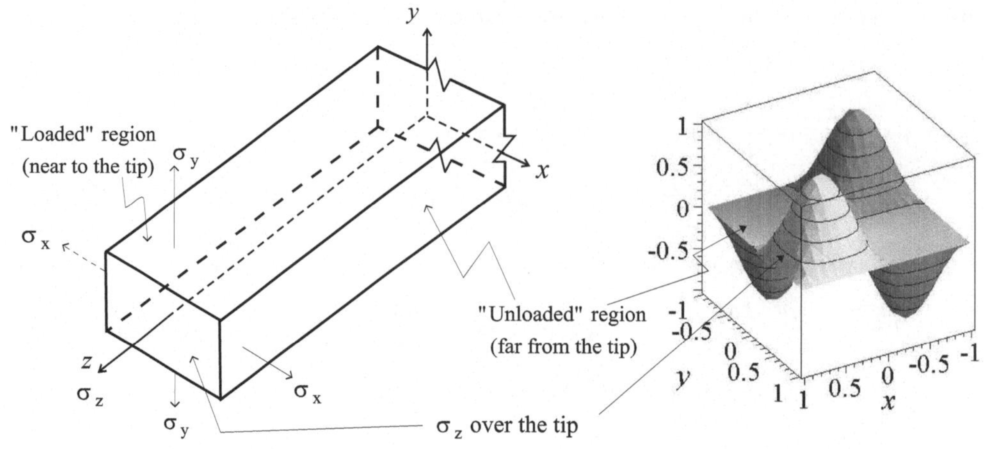

Figure 4.21: A prescribed solution in an isotropic parallelepiped for $k = n = 1$.

A regular surface S in $\mathbb{R}^3$ may be given in two analytical ways. First, one may implicitly define a single function of the three coordinates, i.e., $F(x,y,z) = 0$, that should be satisfied for each point over the surface. Instead, the surface may be defined by parametric equations as

$$\mathbf{r}(t_1,t_2) = [x(t_1,t_2),\ y(t_1,t_2),\ z(t_1,t_2)], \quad (t_1,t_2) \in G, \tag{4.201}$$

where the above three coordinate functions depend on two inner surface parameters t_1, t_2 which are properly bounded in G. Here we require that all functions are differentiable and that the normal vector $\mathbf{n}_0$ has no zeros on S. In the first method, $\mathbf{n}_0$ is defined as $\mathbf{n}_0 = \operatorname{grad} F = [F_{,x},\ F_{,y},\ F_{,z}]$, while in the second one, $\mathbf{n}_0 = \mathbf{r}_1 \times \mathbf{r}_2$ (vector product of the partial derivative vectors $\mathbf{r}_1 = [x_{,t_1}\, y_{,t_1}\, z_{,t_1}]$ and $\mathbf{r}_2 = [x_{,t_2}\, y_{,t_2}\, z_{,t_2}]$). The unit normal vector field, $\bar{\mathbf{n}}$, is therefore parallel to $\mathbf{n}_0$, see Example 4.19.

For demonstration purposes, we shall consider two types of loads: hydrostatic surface forces and centrifugal body forces induced by the body rotation about the z-axis. Figure 4.22 describes a generic homogeneous body of mass density, ρ_0, where the system of coordinates is located at its gravity center and its products of inertia vanish,

$$\iiint_V \{x,y,z,xy,yz,xz\} = \{0,0,0,0,0,0\}. \tag{4.202}$$

The hydrostatic problem is solved with zero body forces (except of Z_b required for static equilibrium, as the body and surface loads are assumed to satisfy the integral conditions of (1.186)), while the rotation problem is solved for zero surface loads, see Examples 4.20, 4.21.

We shall express the surface loads as

$$X_s = P_1 \cos(\bar{\mathbf{n}},x) + Q_1 \cos(\bar{\mathbf{n}},y) + R_1 \cos(\bar{\mathbf{n}},z), \tag{4.203a}$$
$$Y_s = P_2 \cos(\bar{\mathbf{n}},x) + Q_2 \cos(\bar{\mathbf{n}},y) + R_2 \cos(\bar{\mathbf{n}},z), \tag{4.203b}$$
$$Z_s = P_3 \cos(\bar{\mathbf{n}},x) + Q_3 \cos(\bar{\mathbf{n}},y) + R_3 \cos(\bar{\mathbf{n}},z) \tag{4.203c}$$

where P_i, Q_i, R_i are functions of x, y, z, and are defined on the boundary surface only.

To create exact/conditional or approximate polynomial solutions we shall assume that the functions P_i, Q_i, R_i are polynomials of degree k_Q. For the three-dimensional case under discussion, we shall implement the methodology of prescribed general and particular solutions, developed in S.4.2 — S.4.3.4 for plane problems.

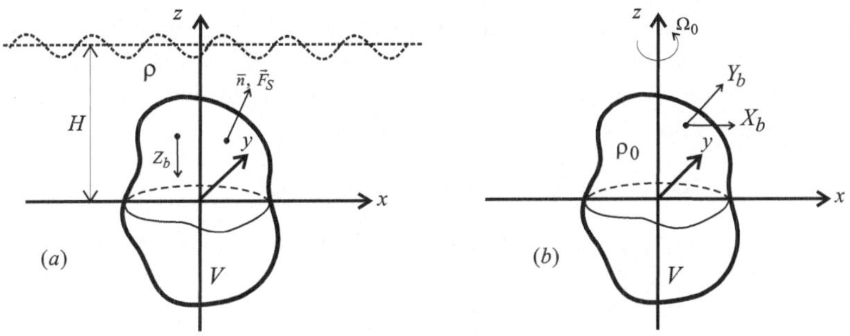

Figure 4.22: Hydrostatic and centrifugal loads over a three-dimensional body.

First, for the given body forces, the problem is reduced to a homogeneous one by introducing an elementary particular solution. Then, for the homogeneous solution, the displacement components are assumed to be polynomials of degree $k_u \geq k_Q + 1$, written as

$$u_i = \sum_{l+m+n \leq k_u} H_{i,l,m,n} x^l y^m z^n, \qquad i = 1,\ldots,3. \tag{4.204}$$

In the general case, the number of $H_{i,l,m,n}$ coefficients in (4.204) is larger than the number of equations obtained when (4.204) are substituted in (4.181). Therefore, some of the $H_{i,l,m,n}$ coefficients may be selected arbitrarily. This enables us to derive a prescribed solution that satisfies equilibrium in an exact manner, while having undetermined ("free") coefficients. The resulting prescribed solution is subsequently substituted in the surface boundary conditions, (1.82a–c), and the above free coefficients are determined by a least-square procedure.

Example 4.19 *Surface Definition of an Ellipsoid.*

The definition of an ellipsoid may be presented by the first method mentioned above, where F is a polynomial in x, y, z, as

$$F = \frac{x^2}{a^2} + \frac{y^2}{b^2} + \frac{z^2}{c^2} - 1, \tag{4.205}$$

or by the second method, where the components of $\mathbf{r}(t_1,t_2)$ are double Fourier polynomials in t_1, t_2, as

$$\mathbf{r} = [a\sin(t_1)\cos(t_2),\ b\sin(t_1)\sin(t_2),\ -c\cos(t_1)], \qquad 0 \leq t_1 \leq \pi,\ 0 \leq t_2 \leq 2\pi. \tag{4.206}$$

Subsequently, the corresponding directional cosines become

$$\widetilde{\lambda}\cos(\bar{\mathbf{n}},x) = \frac{x}{a^2} = \frac{1}{a}\sin(t_1)\cos(t_2), \tag{4.207a}$$

$$\widetilde{\lambda}\cos(\bar{\mathbf{n}},y) = \frac{y}{b^2} = \frac{1}{b}\sin(t_1)\sin(t_2), \tag{4.207b}$$

$$\widetilde{\lambda}\cos(\bar{\mathbf{n}},z) = \frac{z}{c^2} = -\frac{1}{c}\cos(t_1), \tag{4.207c}$$

where $\widetilde{\lambda} = \sqrt{\frac{x^2}{a^4} + \frac{y^2}{b^4} + \frac{z^2}{c^4}}$.

Example 4.20 *Three-Dimensional Body under Hydrostatic Loads*

Referring to Fig. 4.22(a), the hydrostatic surface forces over a *fully immersed* body in a liquid of mass density ρ are described by only three nonzero coefficients, namely: $P_1 = Q_2 = R_3 =$

$\gamma_1 z + \gamma_0$ and hence, $k_Q = 1$. Here, $\gamma_0 = -\rho g H$ and $\gamma_1 = \rho g$, where g is the gravity constant. For force and moment equilibrium as required by (1.186), we introduce additional body force $Z_b = -\gamma_1$ (uniform within the body). As expected, the solution (exact in this case) shows that the only nonzero stresses are

$$\sigma_x = \sigma_y = \sigma_z = \gamma_1 z + \gamma_0 = \rho g (z - H), \tag{4.208}$$

regardless of the material employed. Hence, this case becomes simple as one may derive the corresponding strains according to the constitutive relations of the material under discussion, and subsequently employ the integration procedure of S.1.2 to find the following displacement components (second-degree polynomials as $k_u = 2$ in this case):

$$u = [(a_{11} + a_{12} + a_{13})x + \frac{1}{2}(a_{16} + a_{26} + a_{36})y + \frac{1}{2}(a_{15} + a_{25} + a_{35})z]\rho g(z - H) + u_r, \tag{4.209a}$$

$$v = [\frac{1}{2}(a_{16} + a_{26} + a_{36})x + (a_{12} + a_{22} + a_{23})y + \frac{1}{2}(a_{14} + a_{24} + a_{34})z]\rho g(z - H) + v_r, \tag{4.209b}$$

$$w = [\frac{1}{2}(a_{15} + a_{25} + a_{35})x + \frac{1}{2}(a_{14} + a_{24} + a_{34})y + (a_{13} + a_{23} + a_{33})z]\rho g(z - H) + w_r, \tag{4.209c}$$

where u_r, v_r, w_r are the rigid body displacement. Since the displacements do not depend on the shape of the body, they may be used as a "test case" for any solution procedure.

Example 4.21 *Rotating Ellipsoid and Cube.*

Referring to Fig. 4.22(b), the body forces induced by the angular velocity Ω_0 around the z-axis are $X_b = \gamma x$, $Y_b = \gamma y$, $Z_b = 0$, where $\gamma = \rho_0 \Omega_0^2$. We first introduce a particular solution $\{u_{i,0}\}$ of the form of (4.204) with $k_u = 3$. Therefore for the prescribed homogeneous solution, $k_u \geq 3$ should be selected.

Based on Remark 4.6, for the case of an anisotropic ellipsoid, an exact solution has been obtained with $k_u = 3$, see Fig. 4.23(a) where $a = b = 1$, $c = 2$ and the material is MON13y. It is also interesting to compare this solution with the one of Example 4.12 for an ellipse under centrifugal loading.

For a parallelepiped only approximate polynomial solution is reachable. Figure 4.23(b) presents a deformed isotropic $(2 \times 2 \times 2)$ cube under centrifugal loading that has been obtained by $k_u = 13$, which yields a sufficiently accurate approximation.

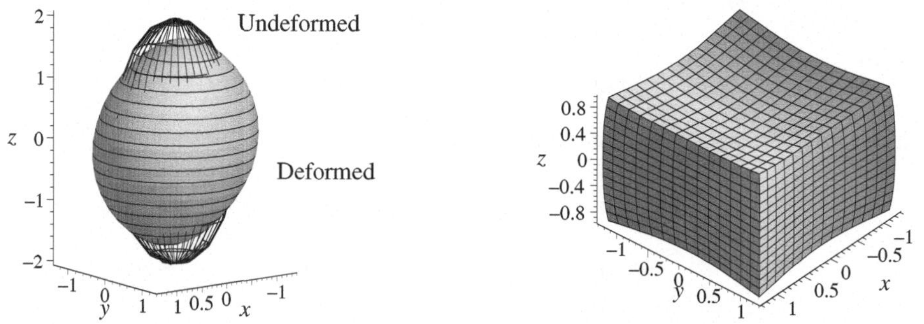

(a) MON13y ellipsoid.

(b) Isotropic cube.

Figure 4.23: Deformed three-dimensional bodies under centrifugal loading (deformation is not to scale).

Remark 4.6 Let the functions F_{bi} and P_i, Q_i, R_i be polynomials in x, y, z. Then, for the ellipsoid $\Omega : \frac{x^2}{a^2} + \frac{y^2}{b^2} + \frac{z^2}{c^2} = 1$, the solution of the BVP defined by (4.181), (1.82a–c), (1.186) results in displacements which are also polynomials of degree $k_u = \max\{k_b+2, k_s+1\}$, where $k_b = \max_i\{deg\,F_{bi}\}$ and $k_s = \max_i\{deg\,P_i, deg\,Q_i, deg\,R_i\}$,

In general, the three polynomials u_i of (4.204) contain $\frac{1}{2}(k_u+1)(k_u+2)(k_u+3)$ coefficients. By assuming $k_s = k_b+1$ (i.e. $k_u = k_b+2$), (4.181) yield $\frac{1}{2}(k_u-1)k_u(k_u+1)$ linear relations, and therefore leave $3(k_u+1)^2$ undetermined ("free") coefficients. It is easy to see that boundary conditions for an ellipsoid are transformed into double Fourier series of order k_u+1, and hence give the same number, $3(k_u+1)^2$, of linear relations. Thus, an exact solution is possible in the ellipsoid case.

4.6 Closed Form Solutions in Circular and Annular Isotropic Domains

We shall review here some classical solutions of the Dirichlet/Neumann and biDirichlet (biharmonic) BVPs in isotropic circular and annular domains. A summary of solution programs appears in Table P.4.

4.6.1 *Harmonic and Biharmonic Functions in Polar Coordinates*

We first define the *circular domain*, Ω, of radius $R > 0$, and its contour, $\partial\Omega$, as

$$\begin{aligned}\Omega &= \{(x,y):\ x^2+y^2 < R^2\} = \{(r,\theta):\ 0 \le \theta < 2\pi,\ 0 \le r < R\},\\ \partial\Omega &= \{x^2+y^2 = R^2\} = \{(R,\theta):\ 0 \le \theta < 2\pi\}.\end{aligned} \tag{4.210}$$

We then define the *circular ring*, Ω, of radii $R_2 > R_1 > 0$, and its boundary contours, $\partial\Omega_i$, as

$$\begin{aligned}\Omega &= \{(x,y):\ R_1^2 < x^2+y^2 < R_2^2\} = \{(r,\theta):\ 0 \le \theta < 2\pi,\ R_1 < r < R_2\},\\ \partial\Omega_i &= \{x^2+y^2 = R_i^2\} = \{(R_i,\theta):\ 0 \le \theta < 2\pi\},\quad i = 1,2.\end{aligned} \tag{4.211}$$

Employing polar coordinates (r,θ) (where $r^2 = x^2+y^2$, $\theta = \arctan\frac{y}{x}$), we express any boundary function, say $F(x,y)$ over a circular boundary of radius R, as $F(\theta) = F(R\cos\theta, R\sin\theta)$, and expand it into a Fourier series, namely,

$$F(\theta) = \frac{1}{2}A_0 + \sum\nolimits_{n=1}^{\infty}(A_n\cos(n\theta) + B_n\sin(n\theta)), \tag{4.212}$$

where coefficients are

$$A_n = \frac{1}{\pi}\int_{-\pi}^{\pi} F(\theta)\cos(n\theta)\,d\theta, \qquad B_n = \frac{1}{\pi}\int_{-\pi}^{\pi} F(\theta)\sin(n\theta)\,d\theta. \tag{4.213}$$

As a harmonic function in an isotropic domain with polar coordinates (r,θ), a generic single-valued harmonic function, $\Lambda(r,\theta)$ may be employed in one of the following two versions:

$$\Lambda^{(1)} = \frac{a_0}{2} + \sum\nolimits_{n=1}^{\infty} r^n(a_n\cos(n\theta) + b_n\sin(n\theta)), \tag{4.214a}$$

$$\Lambda^{(2)} = \frac{a_0}{2} + d_0\ln r + \sum_{n=1}^{\infty} r^n(a_n\cos(n\theta)+b_n\sin(n\theta)) + \sum_{n=1}^{\infty} r^{-n}(c_n\cos(n\theta)+d_n\sin(n\theta)), \tag{4.214b}$$

where a_0, a_n, b_n, d_0, c_n and d_n are coefficients to be determined. The functions $\Lambda^{(1)}$ and $\Lambda^{(2)}$ identically satisfy the field equation in the isotropic case (see (3.201)), namely,

$$\nabla^{(2)}\Lambda^{(i)} = \left(\frac{\partial^2}{\partial r^2} + \frac{1}{r}\frac{\partial}{\partial r} + \frac{1}{r^2}\frac{\partial^2}{\partial\theta^2}\right)\Lambda^{(i)} = 0, \quad i = 1,2. \tag{4.215}$$

Therefore, when one of the above $\Lambda^{(i)}$ versions is selected, it is sufficient to calculate the undetermined coefficients in order to satisfy the boundary conditions. Note that the c_n and d_n terms in (4.214b) are suitable only for cases where the point $r = 0$ is out of Ω, namely, the case of a circular ring. For a biharmonic function, Φ, we employ the prescribed format

$$\Phi(r,\theta) = (r^2 - r_0^2)\Lambda_1 + \Lambda_2 \tag{4.216}$$

where Λ_i, $i = 1,2$ is equal to one of the above $\Lambda^{(i)}$ functions that satisfy (4.215), and r_0 is an arbitrary constant. This prescribed solution satisfies the field equation in the isotropic case, $\nabla^{(4)}\Phi = 0$, see (3.212), (3.213). Therefore, when the above $\Phi(r,\theta)$ is selected, only the boundary conditions should be considered. A more general form of the biharmonic function is described in (Atanackovic and Guran, 2000).

4.6.2 *The Dirichlet BVP in a Circle*

Let Λ be a harmonic function in a circle, with the following boundary conditions:

$$\Lambda(R,\theta) = G_3(\theta) \qquad \text{or} \qquad \Lambda_{,\theta}(R,\theta) = F_3(\theta), \qquad 0 \le \theta \le 2\pi. \tag{4.217}$$

As previously discussed, see (4.39), $\int_0^{2\pi} F_3 d\theta = 0$ must be maintained for the second case in (4.217), and its Fourier series expansion shows that $A_0 = 0$. Adopting the form of (4.214a) for Λ, and by "harmonic balance" (also referred to as the "undetermined coefficients technique"), namely, by equating the coefficients of $\cos(n\theta)$, $\sin(n\theta)$ and the free terms, for the case of $\Lambda(R,\theta) = G_3$ boundary condition, we obtain

$$a_0 = A_0, \qquad a_n = \frac{A_n}{R^n}, \qquad b_n = \frac{B_n}{R^n}, \tag{4.218}$$

while for the case of $\frac{d}{d\theta}\Lambda(R,\theta) = F_3(\theta)$ boundary condition, a_0 is an arbitrary constant, and

$$a_n = -\frac{B_n}{nR^n}, \qquad b_n = \frac{A_n}{nR^n}, \qquad n > 0. \tag{4.219}$$

The above two versions of a solution are implemented in **P.4.23**.

4.6.3 *The Neumann BVP in a Circle*

Let Λ be a harmonic function in a circle, with the following boundary conditions:

$$\frac{d}{dn}\Lambda(R,\theta) = F_3(\theta), \qquad 0 \le \theta \le 2\pi, \tag{4.220}$$

where again, $\int_0^{2\pi} F_3\, d\theta = 0$ must be maintained, see (4.39). By using the same expansion as in S.4.6.2, it becomes clear that $A_0 = 0$. We adopt (4.214a) and by harmonic balance find that the coefficient a_0 is an arbitrary constant, and the other coefficients are

$$a_n = \frac{1}{nR^{n-1}}A_n, \qquad b_n = \frac{1}{nR^{n-1}}B_n, \qquad n > 0. \tag{4.221}$$

This solution is implemented in **P.4.23** by BVP$= -1$, and presented in Fig. 4.24(a) for $F_3 = 3\sin\theta$, $N = 8$, $R = 3$.

4.6.4 The Dirichlet BVP in a Circular Ring

Let Λ be a harmonic function in a circular ring, with the following two versions of boundary conditions:

$$\Lambda(R_i,\theta)=G_3^{(i)}(\theta) \quad \text{or} \quad \Lambda_{,\theta}(R_i,\theta)=F_3^{(i)}(\theta), \qquad 0\le\theta\le 2\pi, \quad i=1,2, \tag{4.222}$$

where the boundary functions $G_3^{(i)}(\theta)$ and $F_3^{(i)}(\theta)$ are expanded into Fourier series form as shown by (4.212), (4.213). In the second case of (4.222), two equations, $\int_0^{2\pi} F_3^{(i)}\,d\theta=0,\ i=1,2,$ (for each boundary component) should be satisfied. We adopt the solution format of (4.214b), apply (4.222) for both boundaries and apply harmonic balancing.

For the case of $\Lambda(R_i,\theta)=G_3^{(i)}(\theta)$, a_0, d_0 are derived from

$$\frac{a_0}{2}+d_0\ln(R_i)=\frac{A_{i0}}{2}, \qquad i=1,2, \tag{4.223}$$

while a_n, c_n, b_n, d_n are obtained separately for each level $n\ge 1$ from

$$R_i^n a_n+\frac{1}{R_i^n}c_n=A_{in}, \qquad R_i^n b_n+\frac{1}{R_i^n}d_n=B_{in}, \qquad i=1,2. \tag{4.224}$$

Solving the above systems in an explicit manner yields

$$d_0=\frac{A_{10}-A_{20}}{2\ln(R_1/R_2)}, \qquad a_0=\frac{A_{20}\ln(R_1)-A_{10}\ln(R_2)}{\ln(R_1/R_2)}, \tag{4.225a}$$

$$a_n=\frac{A_{2n}R_2^n-A_{1n}R_1^n}{R_2^{2n}-R_1^{2n}}, \qquad c_n=\frac{A_{1n}R_2^n-A_{2n}R_1^n}{R_2^{2n}-R_1^{2n}}(R_1R_2)^n, \tag{4.225b}$$

$$b_n=\frac{B_{2n}R_2^n-B_{1n}R_1^n}{R_2^{2n}-R_1^{2n}}, \qquad d_n=\frac{B_{1n}R_2^n-BA_{2n}R_1^n}{R_2^{2n}-R_1^{2n}}(R_1R_2)^n. \tag{4.225c}$$

For the case of $\Lambda_{,\theta}(R_i,\theta)=F_3^{(i)}(\theta)$, we find $A_{10}=A_{20}=0$ and a_0, d_0 are arbitrary constants, while a_n, c_n, b_n, d_n are obtained separately for each level $n\ge 1$ from

$$R_i^n a_n+\frac{1}{R_i^n}c_n=-\frac{R_i}{n}B_{in}, \qquad R_i^n b_n+\frac{1}{R_i^n}d_n=\frac{R_i}{n}A_{in}, \qquad i=1,2. \tag{4.226}$$

Solving the above two systems in an explicit form yields

$$a_n=\frac{B_{1n}R_1^n-B_{2n}R_2^n}{n(R_2^{2n}-R_1^{2n})}, \qquad c_n=\frac{B_{2n}R_1^n-B_{1n}R_2^n}{n(R_2^{2n}-R_1^{2n})}(R_1R_2)^n, \tag{4.227a}$$

$$b_n=\frac{A_{2n}R_2^n-A_{1n}R_1^n}{n(R_2^{2n}-R_1^{2n})}, \qquad d_n=\frac{A_{1n}R_2^n-A_{2n}R_1^n}{n(R_2^{2n}-R_1^{2n})}(R_1R_2)^n. \tag{4.227b}$$

This solution is implemented in **P.4.24**.

4.6.5 The Neumann BVP in a Circular Ring

Let Λ be a harmonic function in a circular ring, with the following boundary conditions:

$$\frac{d}{dn}\Lambda(R_i,\theta)=F_3^{(i)}(\theta), \qquad 0\le\theta\le 2\pi, \quad i=1,2. \tag{4.228}$$

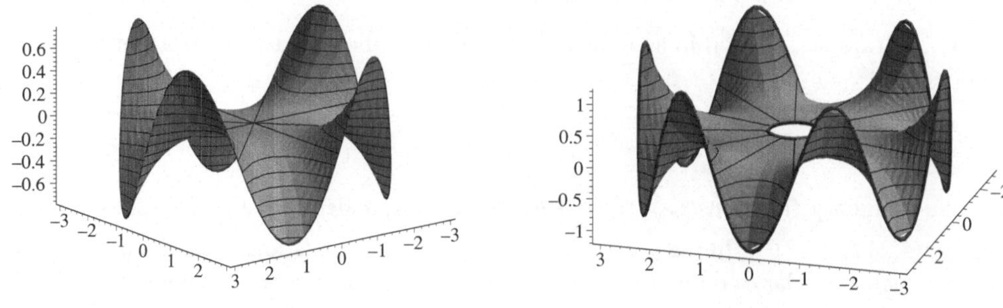

(a) Λ over a circle. (b) Λ over a circular ring.

Figure 4.24: Solutions of the Neumann BVP in circular domains.

As in S.4.6.4 we expand boundary functions $F_3^{(i)}(\theta)$ into a Fourier series. The condition of solution existence and uniqueness in the multiply connected domain under discussion is written as, see Remark 3.7,

$$\sum_{i=1}^{2} \int_0^{2\pi} F_3^{(i)}(\theta)\, d\theta = 0. \tag{4.229}$$

From (4.229) we obtain the following relation that should be satisfied by the data

$$A_{20}R_1 + A_{10}R_2 = 0. \tag{4.230}$$

The solution of (4.214b) is adopted again, and since $\frac{d}{dn}\Lambda(R_i,\theta) = (-1)^i \Lambda_{,r}(R_i,\theta)$, the boundary conditions (4.228) may be equivalently rewritten as $\Lambda_{,r}(R_i,\theta) = (-1)^i F_3^{(i)}(\theta)$, $i = 1,2$. By balancing the equations, we find that the coefficient a_0 is an arbitrary constant, and d_0 is obtained from (4.230) as $d_0 = -\frac{A_{10}R_2}{2} = \frac{A_{20}R_1}{2}$. We also deduce two linear systems for the unknown coefficients a_n, c_n, b_n, d_n separately for each level $n \geq 1$,

$$R_i^{n-1} a_n - (-1)^i \frac{c_n}{R_i^{n+1}} = \frac{A_{in}}{n}, \qquad R_i^{n-1} b_n - (-1)^i \frac{d_n}{R_i^{n+1}} = \frac{B_{in}}{n}, \qquad i = 1,2. \tag{4.231}$$

Solving the above system yields

$$a_n = \frac{A_{1n}R_2^2 {R_1}^{n+1} + A_{2n}R_2^{n+1}}{n\left({R_1}^{2n}R_2^2 + R_2^{2n}\right)}, \qquad c_n = \frac{A_{2n}R_2^{n+1}R_1^{2n} - A_{1n}R_1^{n+1}R_2^{2n}}{n\left(R_1^{2n}R_2^2 + R_2^{2n}\right)}, \tag{4.232a}$$

$$b_n = \frac{B_{1n}R_2^2 {R_1}^{n+1} + B_{2n}R_2^{n+1}}{n\left({R_1}^{2n}R_2^2 + R_2^{2n}\right)}, \qquad d_n = \frac{B_{2n}R_2^{n+1}R_1^{2n} - B_{1n}R_1^{n+1}R_2^{2n}}{n\left(R_1^{2n}R_2^2 + R_2^{2n}\right)}. \tag{4.232b}$$

This solution is implemented in **P.4.24** by BVP$= -1$, and presented in Fig. 4.24(b) for $F_3^{(1)} = 0$, $F_3^{(2)} = 10\cos(\sin(3\theta))$, $N = 20$, $R_1 = 0.5$, $R_2 = 3$.

4.6.6 The Biharmonic BVP in a Circle

Let Φ be a biharmonic function in a circle, with the following biDirichlet boundary conditions:

$$\{\Phi,\ \frac{d}{dn}\Phi\}(R,\theta) = \{G_1,\ G_2\}(\theta), \qquad \theta \in [0,\ 2\pi]. \tag{4.233}$$

To solve this problem we adopt Φ of (4.216) using $\Lambda^{(1)}$ format (see (4.214a)) for both Λ_1 and Λ_2, and hence, $\Phi = \left(r^2 - R^2\right)\Lambda_1 + \Lambda_2$, where $\Lambda_i = \frac{a_{i0}}{2} + \sum_{n=1}^{\infty} r^n (a_{in}\cos(nt) + b_{in}\sin(nt))$. Therefore, the undetermined coefficients are a_{i0}, a_{in} and b_{in}. We now expand the two boundary functions $G_i(\theta)$, $i = 1,2$ into Fourier series, with coefficients A_{i0}, A_{in} and B_{in}, $i = 1,2$, and in view of $\frac{d}{dn}\Lambda(R_i,\theta) = (-1)^i \Lambda_{,r}(R_i,\theta)$, rewrite the boundary conditions of (4.233) as the two coupled Dirichlet BVPs

$$\Lambda_2(R,\theta) = G_1(\theta), \tag{4.234a}$$
$$2R\Lambda_1(R,\theta) + \Lambda_{2,r}(R,\theta) = G_2(\theta). \tag{4.234b}$$

By comparing with the first case of (4.217), the above equations may be viewed as a sequence of solutions of two harmonic functions. Therefore, (4.234a) first shows that similar to (4.218),

$$a_{20} = A_{10}, \qquad a_{2n} = \frac{A_{1n}}{R^n}, \qquad b_{2n} = \frac{B_{1n}}{R^n}. \tag{4.235}$$

Then, Λ_1 is directly obtained by $\Lambda_1(R,\theta) = \frac{1}{2R}(G_2(\theta) - \Lambda_{2,r}(R,\theta))$, and one can write

$$a_{10} = \frac{A_{20}}{2R}, \qquad a_{1n} = \frac{1}{2R^{n+1}}(A_{2n} - \frac{n}{R}A_{1n}), \qquad b_{1n} = \frac{1}{2R^{n+1}}(B_{2n} - \frac{n}{R}B_{1n}). \tag{4.236}$$

This solution is implemented in **P.4.25** with $G_1 = 3 + \frac{1}{2}\sin(4\theta)$, $G_2 = 2$, $N = 10$, $R = 2$, see Fig. 4.25(a).

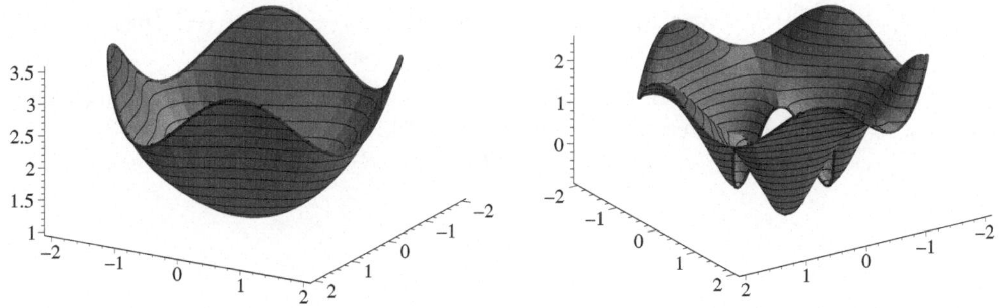

(a) Φ over a circle. (b) Φ over a circular ring.

Figure 4.25: Solutions of the biDirichlet BVP in circular domains.

4.6.7 The Biharmonic BVP in a Circular Ring

Let Φ be a biharmonic function in a circular ring, with the biDirichlet boundary conditions:

$$\{\Phi, \frac{d}{dn}\Phi\}(R_i,\theta) = \{G_{i1}, G_{i2}\}(\theta), \qquad \theta \in [0,2\pi], \qquad i = 1,2. \tag{4.237}$$

We may solve this problem by writing Φ of (4.216) using $\Lambda^{(2)}$ format (see (4.214b)) for both Λ_1 and Λ_2, therefore $\Phi = \left(r^2 - r_0^2\right)\Lambda_1 + \Lambda_2$, where r_0 is a selected parameter, and the undetermined coefficients are a_{i0}, a_{in}, b_{in} d_{i0}, c_{in} and d_{in}. We now expand the four boundary functions $G_{ik}(\theta)$, $i,k = 1,2$, into Fourier series form, with coefficients A_{ik0}, A_{ikn} and

B_{ikn}, $i,k=1,2$. For the boundary condition $\frac{d}{dn}\Phi(R_i,\theta)=(-1)^i\Phi_{,r}(R_i,\theta)$ we write $\Phi_{,r}=2r\Lambda_1+(r^2-r_0^2)\Lambda_{1,r}+\Lambda_{2,r}$, and the boundary conditions (4.237) for $i=1,2$ become

$$(R_i^2-r_0^2)\Lambda_1(R_i,\theta)+\Lambda_2(R_i,\theta)=G_{i1}(\theta), \tag{4.238a}$$

$$2R_i\Lambda_1(R_i,\theta)+(R_i^2-r_0^2)\Lambda_{1,r}(R_i,\theta)+\Lambda_{2,r}(R_i,\theta)=(-1)^iG_{i2}(\theta). \tag{4.238b}$$

By balancing the equations one obtains for a_{i0},d_{i0}:

$$(R_i^2-r_0^2)(\frac{a_{10}}{2}+d_{10}\ln(R_i))+\frac{a_{20}}{2}+d_{20}\ln(R_i)=\frac{A_{i1\,0}}{2}, \tag{4.239a}$$

$$2R_i(\frac{a_{10}}{2}+d_{10}\ln(R_i))+(R_i^2-r_0^2)\frac{d_{10}}{R_i}+\frac{d_{20}}{R_i}=(-1)^i\frac{A_{i2\,0}}{2}. \tag{4.239b}$$

For a_{in},c_{in}, $n\geq 1$ the equations are:

$$(R_i^2-r_0^2)(R_i^na_{1n}+\frac{c_{1n}}{R_i^n})+R_i^na_{2n}+\frac{c_{2n}}{R_i^n}=A_{i1\,n}, \tag{4.240a}$$

$$2R_i(R_i^na_{1n}+\frac{c_{1n}}{R_i^n})+(R_i^2-r_0^2)n(R_i^{n-1}a_{1n}-\frac{c_{1n}}{R_i^{n+1}})+n(R_i^{n-1}a_{2n}-\frac{c_{2n}}{R_i^{n+1}})=(-1)^iA_{i2\,n}, \tag{4.240b}$$

and finally, for b_{in},d_{in}, $n\geq 1$, the following set of equations is derived:

$$(R_i^2-r_0^2)(R_i^nb_{1n}+\frac{d_{1n}}{R_i^n})+R_i^nb_{2n}+\frac{d_{2n}}{R_i^n}=B_{i1\,n}, \tag{4.241a}$$

$$2R_i(R_i^nb_{1n}+\frac{d_{1n}}{R_i^n})+(R_i^2-r_0^2)n(R_i^{n-1}b_{1n}-\frac{d_{1n}}{R_i^{n+1}})+n(R_i^{n-1}b_{2n}-\frac{d_{2n}}{R_i^{n+1}})=(-1)^iB_{i2\,n}. \tag{4.241b}$$

By solving the above systems of four linear equations in an explicit form, we find coefficients $a_{in},b_{in},c_{in},d_{in}$. This solution is implemented in **P.4.26** with $G_{11}=\sin(\cos(3\theta))$, $G_{12}=0$, $G_{21}=2+\frac{1}{2}\cos(5\theta)$, $G_{22}=\sin(2\theta)$, $N=5$, $R_1=0.7$, $R_2=2$, see Fig. 4.25(b).

5

Foundations of Anisotropic Beam Analysis

The analysis of anisotropic beams has been the focal point of many research efforts during the last few decades, and the extensive literature in this area testifies to its potential in a vast range of engineering applications.

In this chapter we shall clarify the basic aspects of anisotropic beam analysis, review some exact and approximate analysis techniques, and discuss the associated coupling characteristics at both the material and the structural levels. In the following chapters, detailed analyses of various levels will be derived.

Traditionally, the term "beam" stands for a solid domain bounded by a cylindrical (prismatic) "outer surface" (frequently denoted as "free surface" as well) and two "ends" (or "end cross-sections"), see Fig. 5.1. Beams are assumed to be "slender", namely, their length is much larger than a typical size of their cross-section. In some cases the notion "beam" is replaced with "bar" or "rod" in various contexts, while the term selection is associated with the type of their degrees of freedom (DOF) or with their loading modes.

There are some main aspects that should be considered while selecting the proper analysis tool for an anisotropic beam. These aspects may have a tremendous effect on the complexity of their analysis:

(a) The cross-section geometry. One should first distinguish between homogeneous and non-homogeneous cross-sections. Here, the term "homogeneous" refers to constant material properties distribution over the cross-section (i.e. the a_{ij} coefficients are constants). Yet, in many modern applications, and in major components of common engineering applications, anisotropic beam analysis is mainly focused on *laminated* composite beams the cross-section of which is clearly non-homogeneous. See discussion of non-homogeneous two-dimensional domains in S.3.1.1 and Remark 3.1.

Another important classification of cross-section geometry deals with the complexity of its topology, and mainly, the level of its multiple-connectivity (see Chapter 3). This aspect includes the distinction between solid and ("open" and "closed") thin-wall cross-sections (see Chapter 10).

Typically, we shall be focused on longitudinally uniform beam structures from both geometric and material characteristics distribution points of view.

(b) Material/s type and orientation. These characteristics determine the number of elastic coefficients that should be taken into account to adequately model each material in the (generally non-homogeneous) cross-section. As shown in Chapter 2, for each material there are two (for the isotropic case) up to 21 (for generally anisotropic materials) such independent coefficients.

Physically, this aspect determines the amount of coupling at the material level, and thus in the overall beam behavior.

(c) The longitudinal uniformity of the loading. Within this respect, one should distinguish mainly between tip loads and distributed loads. In the case of distributed loads, there are many components that should be accounted for. Distributed loads may be introduced by detailed surface and body forces, or as integral quantities that represent the overall loading at each cross-section. In general, when linear analysis is under discussion, it is convenient to look at each component of the above loads separately.

(d) Level of solution approximation/refinement. Clearly, this selection plays a major role in the determination of the solution detailing level and analytical complexity. Beyond the basic solution that will be presented within this chapter, the solution methodologies for beams in this book appear as follows: Chapter 6 presents a complete model for homogeneous beams of general anisotropy. Chapter 7 deals with exact treatment of uncoupled homogeneous beams made of MON13z materials. For the same type of materials, Chapter 8 presents exact analysis of non-homogeneous beams under both tip and distributed loading. Chapters 9 and 10 derive approximate analytical solutions for coupled solid and thin-wall monoclinic beams, respectively.

For further reading, see literature citations in S.5.4.

5.1 Notation and Definitions

5.1.1 Geometrical Degrees of Freedom

A model of a slender, uniform cylindrical beam of a homogeneous simply connected cross-section is shown in Fig. 5.1. Here and henceforth, for all homogeneous cross-sections, we shall assume that the origin of the coordinate system is placed at the center of the cross-section area (centroid), and that the x- and y- axes are directed along the principal axes of inertia (i.e., $\iint_\Omega xy = 0$) of the cross-section. The z-axis is stretched perpendicular to the cross-sections (parallel to the cylindrical surface of the beam), and due to the beam's uniformity, passes at the same location at each cross-section. The above may be written together with the definition of the cross-section area, S_Ω, and the cross-sectional moments (of inertia) I_y and I_x about the x- and y- axes, respectively, as

$$\iint_\Omega \{1, x, y, x^2, xy, y^2\} = \{S_\Omega, 0, 0, I_y, 0, I_x\}. \tag{5.1}$$

Different placement of the coordinate system is required for non-homogeneous cross-sections, see Chapter 8. There are two ways to view and discuss the (geometrical) DOF of a beam. Clearly, like any other problem in elasticity, one may consider the displacements $u(x,y,z)$, $v(x,y,z)$ and $w(x,y,z)$ as the DOF, and seek a suitable solution for these three functions. This approach is certainly valid and exact (and will be used further on), but it somehow limits the understanding of the involved physical phenomena.

Hence, to simplify the present preliminary discussion, we apply a more global view, where we examine the deformation and rotation of the beam's "elastic axis" (the straight line that

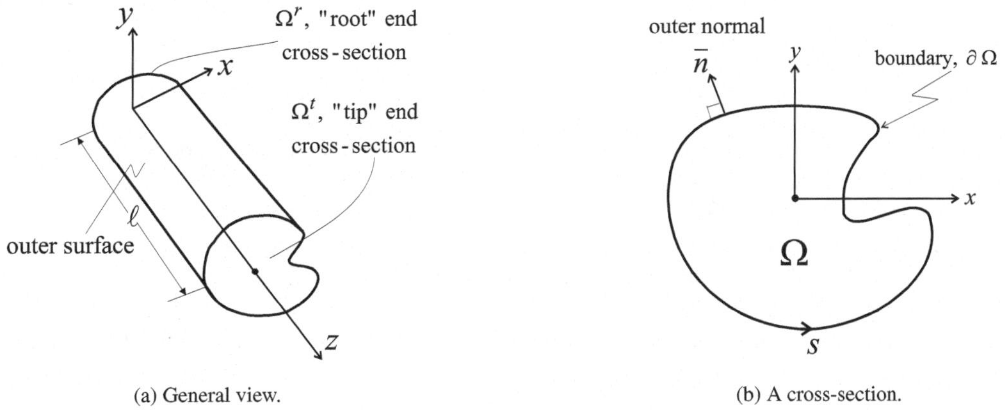

(a) General view. (b) A cross-section.

Figure 5.1: Notation for a slender beam.

coincides with the z-axis). This line will be also denoted in short as the "beam axis". The notion "elastic axis" has been widely discussed in the literature in connection with isotropic beams theories and as will become clearer later on and in further chapters, a large portion of these discussions becomes irrelevant when anisotropic beams are dealt with. Within the present context we just assume that the elastic axis is the line that passes through all $x = y = 0$ points of all cross-sections, which are selected so that (5.1) are fulfilled.

Subsequently, a global view of the beam deformation may be based on six longitudinal functions. The first three of them are the displacements along the beam axis $u_0(z) = u(0,0,z)$, $v_0(z) = v(0,0,z)$, $w_0(z) = w(0,0,z)$, while the other three are the rotations of the beam axis, namely, the two *bending rotations*

$$\omega_{x0}(z) = \frac{1}{2}(w_{,y} - v_{,z})\,|_{x=y=0}, \qquad \omega_{y0}(z) = \frac{1}{2}(u_{,z} - w_{,x})\,|_{x=y=0}, \tag{5.2}$$

and the *twist angle*

$$\omega_{z0}(z) = \phi(z) = \frac{1}{2}(v_{,x} - u_{,y})\,|_{x=y=0}. \tag{5.3}$$

The above six components constitute the global DOF. Referring to Fig. 5.2, we define the deformation $u_0(z)$ as the *Edgewise Bending*, the deformation $v_0(z)$ as the *Beamwise Bending* and the deformation $w_0(z)$ as the *Extension*.

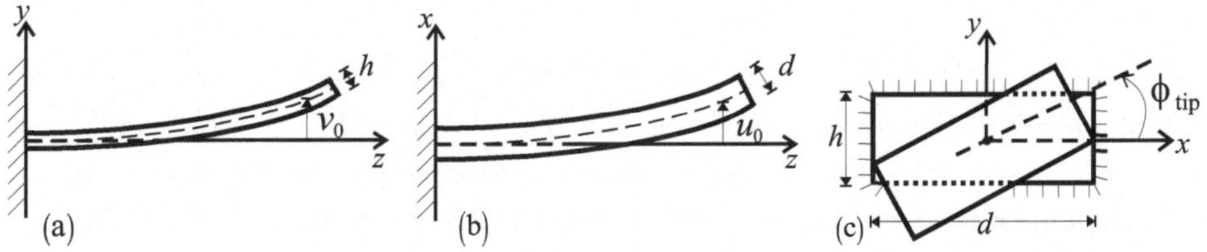

Figure 5.2: (a) The Beamwise Bending, (b) The Edgewise Bending, (c) The Twist.

Similarly, $\omega_{x0}(z)$ is denoted *Beamwise Bending Rotation*, $\omega_{y0}(z)$ — *Edgewise Bending Rotation* and $\omega_{z0}(z)$ — *Twist Angle*, which will occasionally be replaced by $\phi(z)$ as well. By substituting the strain expressions (see (1.9a,b)) in the above rotation definitions one may write

$$\omega_{y0} = u_{0,z} \underline{-\frac{1}{2}\gamma_{xz}(0,0,z)}, \qquad \omega_{x0} = -v_{0,z} \underline{+\frac{1}{2}\gamma_{yz}(0,0,z)}. \tag{5.4}$$

Typically, in slender beams, the underlined terms that represent the rotation due to shear strain, may be neglected (see Remark 5.1 and Fig. 5.3(a)), and thus, the rotations $\omega_{y0}(z)$, $\omega_{x0}(z)$ may be viewed as direct functions of $u_0(z)$ and $v_0(z)$, respectively. However, $\omega_z(z)$ is independent of the bending functions $u_0(z)$ and $v_0(z)$. Hence, as will be shown later on, many analyses are based on four *global deformation measures* only: $u_0(z)$, $v_0(z)$, $w_0(z)$ and $\phi(z)$.

To this end, one may define the *beam bending curvature* components. In general, for a given function, $\tilde{f}(z)$, the *curvature* of its graph is defined as *the change of the slope angle with respect to the arclength*, where in the present context, a *positive slope* means increasing in $\tilde{f}$ and vice versa, namely,

$$\tilde{\kappa}(z) = \frac{\frac{d^2\tilde{f}}{dz^2}}{\left[1+(\frac{d\tilde{f}}{dz})^2\right]^{3/2}} \cong \frac{d^2\tilde{f}}{dz^2} \qquad \text{for} \qquad \left(\frac{d\tilde{f}}{dz}\right)^2 \ll 1. \tag{5.5}$$

Hence, in the linear case we define the *edgewise curvature*, κ_u, and the *beamwise curvature*, κ_v, as

$$\kappa_u(z) = u_{0,zz}, \qquad \kappa_v(z) = v_{0,zz}. \tag{5.6}$$

In light of the above discussion, one may also write $\kappa_u(z) \cong \omega_{y0,z}$, $\kappa_v(z) \cong -\omega_{x0,z}$. In addition, this is a suitable point to introduce the *twist angle per unit length*, θ, as $\theta = \omega_{z0,z} = \phi_{,z}$, which will also be denoted in short as the "twist" or the "twist rate".

Regarding the cross-section deformation, one should distinguish between its two main components: the *in-plane warping* and the *out-of-plane warping*. The former is related to the deformation in the cross-section plane that might also cause a deformation in its shape. In the determination of the in-plane warping, only the components u and v participate. The latter refers to the deformation component w, and is essentially a reflection of the fact that after deformation, cross-sections that were perpendicular to the beam axis before deformation, are not planes anymore. The above does not preclude special cases, where the deformation is not accompanied with warping, or cases, where special warping shapes take place (e.g. a linear warping, with respect to x or y), that results in a plane cross-section, which is not perpendicular to the beam axis.

Remark 5.1 The reader may convince himself that $\omega_{y0} \cong u_{0,z}$ and $\omega_{x0} \cong -v_{0,z}$, by examining an isotropic beam (of length $\tilde{l}$, Young's modulus $\tilde{E}$, Poisson's ratio $\tilde{\nu}$, a rectangular cross-section $\Omega = \{|x| \le \frac{\tilde{d}}{2},\ |y| \le \frac{\tilde{h}}{2}\}$, area $S_\Omega = \tilde{d}\tilde{h}$, and a cross-sectional moment (of inertia) $I_x = \tilde{d}\tilde{h}^3/12$) about the x-axis with the elementary *Strength-of-Materials* theory, see S.5.3.4. Under a tip beamwise load, P_y, the maximal shear strain appears at $y=0$, namely, $\gamma_{yz}^{max} = \gamma_{yz}(x,0,z) = \frac{P_y}{\tilde{E}}\frac{3(1+\tilde{\nu})}{S_\Omega}$, while $v_{0,z}(z) = \frac{P_y}{2\tilde{E}I_x}z(2\tilde{l}-z)$, see Fig. 5.3(a). Hence, $\gamma_{yz}^{max}/v_{0,z}(z)$ is proportional to $(\tilde{h}/\tilde{l})^2$ (except for the root region, where $v_{0,z}$ approaches zero), and therefore, in general, $|\gamma_{yz}| \ll v_{0,z}$. In the case of a tip moment, M_x, where $\gamma_{yz}=0$ and $u_{0,z}(z) \cong -\frac{M_x}{\tilde{E}I_x}$, the assumption under discussion becomes exact.

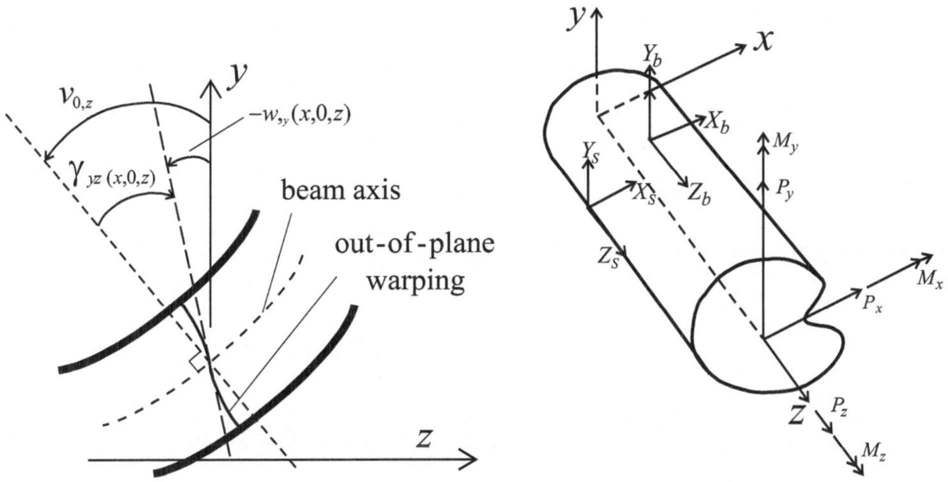

(a) The out-of-plane warping due to a tip beamwise load (γ_{yz} is unrealistically magnified for illustration purposes).

(b) Loading notation.

Figure 5.3: Beam notation.

5.1.2 *Tip and Distributed Loading*

Beam loading consists of two main categories: tip loading and distributed loading. These two categories induce different levels of complexity and therefore are treated separately in what follows. General notation is presented by Fig. 5.3(b).

Following the traditional terminology, we shall define the "St. Venant's Problem" as the case of "clamped-free" slender beam under generic tip loading. The notion "clamped-free" beam (or "a cantilever") simply means that the beam is fully clamped at its root end ($z = 0$), Ω^r and is free or loaded at its tip end ($z = l$), Ω^t, see Fig. 5.1. The prediction of the behavior of such a configuration is usually based on applying loads and boundary conditions over the end cross-sections in an integral manner as described in S.5.1.3. As shown in Fig. 5.3(b), the beam is assumed to be acted upon over the tip end cross-section, by three *tip forces* P_x, P_y, P_z and three *tip moments* M_x, M_y, M_z in the coordinate directions, where

$$\{P_z, M_x, -M_y, P_x, P_y, M_z\} = \iint_{\Omega^t} \{\sigma_z, \sigma_z y, \sigma_z x, \tau_{xz}, \tau_{yz}, x\tau_{yz} - y\tau_{xz}\}. \tag{5.7}$$

In this book, we shall usually deal with the "clamped-free" case as all other cases may be easily deduced from it, see S.5.1.3.2.

A different class of problems is associated with a distribution of loads over the outer surface, $\partial\Omega$, and a distribution of body forces that are applied on each material point in the cross-section, Ω. In the former case we generally write

$$\mathbf{F}_s = X_s(x,y,z)\,\mathbf{k}_1 + Y_s(x,y,z)\,\mathbf{k}_2 + Z_s(x,y,z)\,\mathbf{k}_3, \qquad \text{on} \quad \partial\Omega, \tag{5.8}$$

while for the latter case, the notation is

$$\mathbf{F}_b = X_b(x,y,z)\,\mathbf{k}_1 + Y_b(x,y,z)\,\mathbf{k}_2 + Z_b(x,y,z)\,\mathbf{k}_3 \qquad \text{over} \quad \Omega, \tag{5.9}$$

see S.1.3.2. In this notation, $\mathbf{k}_1$, $\mathbf{k}_2$ and $\mathbf{k}_3$ are unit vectors in the coordinate directions. Clearly, X_s, Y_s, Z_s are given as *forces per unit area*, and X_b, Y_b, Z_b are given as *forces per unit volume*.

For a given set of loads, by force and moment integration, one may determine the *local* resultants $\overline{P}_z$, $\overline{P}_y$, $\overline{P}_z$, $\overline{M}_x$, $\overline{M}_y$ and $\overline{M}_z$ at each cross-section. These are defined as the components of the equations

$$\mathbf{P} = \overline{P}_x \mathbf{k}_1 + \overline{P}_y \mathbf{k}_2 + \overline{P}_z \mathbf{k}_3, \qquad \mathbf{M} = \overline{M}_x \mathbf{k}_1 + \overline{M}_y \mathbf{k}_2 + \overline{M}_z \mathbf{k}_3. \tag{5.10}$$

Here, $\mathbf{P}$ and $\mathbf{M}$ are the local *force and moment resultant vectors.* For a clamped-free beam

$$\mathbf{P}(z) = P_x \mathbf{k}_1 + P_y \mathbf{k}_2 + P_z \mathbf{k}_3 + \int_z^l \Big(\oint_{\partial\Omega} \mathbf{F}_s(\tilde{z}) + \iint_{\Omega} \mathbf{F}_b(\tilde{z})\Big)\, d\tilde{z}, \tag{5.11a}$$

$$\begin{aligned}\mathbf{M}(z) &= M_x \mathbf{k}_1 + M_y \mathbf{k}_2 + M_z \mathbf{k}_3 - P_y(l-z)\,\mathbf{k}_1 + P_x(l-z)\,\mathbf{k}_2 \\ &\quad + \int_z^l \Big[\oint_{\partial\Omega} \mathbf{r}_s(z,\tilde{z}) \times \mathbf{F}_s(\tilde{z}) + \iint_{\Omega} \mathbf{r}_b(z,\tilde{z}) \times \mathbf{F}_b(\tilde{z})\Big]\, d\tilde{z}\end{aligned} \tag{5.11b}$$

and, obviously, $\mathbf{P}(l) = \{P_x, P_y, P_z\}$ and $\mathbf{M}(l) = \{M_x, M_y, M_z\}$. The vector $\mathbf{r}_s(z,\tilde{z})$ connects the points $(0,0,z)$ and $(x(s), y(s), \tilde{z})$, while the vector $\mathbf{r}_b(z,\tilde{z})$ connects the points $(0,0,z)$ and $(x, y, \tilde{z})$. Here, $s \in \partial\Omega$ and $(x,y) \in \Omega$.

5.1.3 Boundary Conditions

5.1.3.1 General Definitions

For a slender beam, one should distinguish between "end boundary conditions", which are applied at the ("root" and "tip") end cross-sections, and "outer surface boundary conditions", which are applied over the entire outer cylindrical surface, see Fig. 5.1(a). While the former type may include natural and/or geometrical conditions, the latter is usually confined to natural conditions only. One of the important differences between the solution we seek here for an anisotropic beam and the exact three-dimensional one is the manner in which both the geometrical and natural boundary conditions are applied at the beam's end cross-sections. To simplify the discussion, we shall be focused again on the case of a "clamped-free" beam.

First, the natural boundary conditions at the tip will be discussed. In this end cross-section, we satisfy the boundary conditions for the stresses in *an integral manner*, or in other words, we only make sure that the stress resultants will match the actual applied loads there, as opposed to *a differential manner*, where one should match the exact external loads distribution at each point over this cross-section. Justification for this approximation emerges from the St. Venant's Principle that states that *the difference between the actual applied stress distribution and the one that emerges from integral satisfaction of the natural boundary conditions will constitute a self-equilibrated system, the influence of which will diminish with the distance from the ends.*

While adopting the above principle, we bear in mind that it was originally proposed for the isotropic case and its applicability in the general anisotropic case should be carefully examined. An example for a study that questioned this point is (Horgan, 1972), where a plane problem of anisotropic MON13z domain was examined. It was concluded that the decay constants for the anisotropic case are smaller than those known for isotropic materials. The study has been focused on the lower bound of these decay constants and offered a more general conclusion: "the characteristic decay length associated with St. Venant's principle in plane elastostatics is smallest in the isotropic case".

To ensure that the distributions of the stress components over the tip end cross-section are equivalent in an integral manner to the tip forces and moments there, one should satisfy (5.7).

In a similar way, the geometrical boundary conditions (which are applied at the root of the clamped-free beam under discussion) are also satisfied in *an integral manner* or *at a specific point* of the root cross-section only. The most common version of these integral clamping conditions that will be denoted as "clamped-I", or "engineering" clamping conditions, is

$$u_0 = v_0 = w_0 = u_{0,z} = v_{0,z} = \omega_{z0} = 0, \qquad \text{at} \qquad z = 0. \tag{5.12}$$

Yet, another possible version for such an integral approximation that will be denoted as "clamped-II" conditions, is

$$u_0 = v_0 = w_0 = \omega_{x0} = \omega_{y0} = \omega_{z0} = 0, \qquad \text{at} \qquad z = 0. \tag{5.13}$$

Other versions that include integrals of displacements and rotations over the root cross-section are also possible.

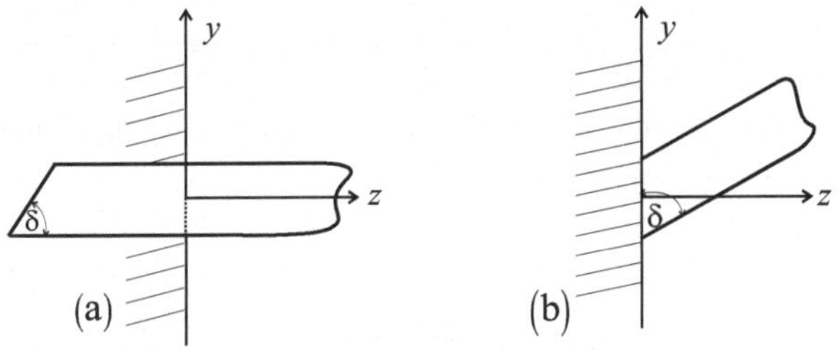

Figure 5.4: Root of a beam under two different clamping conditions. (a): "clamped-I", (b): "clamped-II".

In addition to Remark 5.1, Fig. 5.4 presents a scheme of the root of a beam that undergoes a constant shear strain, $\gamma_{yz} = w_{,y}$ (see (1.9b)) only, which is reflected by the fact that the angle δ is diminished (from $\delta = \pi/2$ in the undeformed case) to $\delta \cong \pi/2 - w_{,y}$ (see (1.39)). Thus, as shown by Fig. 5.4(a), $v_0 = v_{0,z} = 0$ would satisfy "clamped-I" conditions. However, to ensure $\omega_{x0} = 0$ as required by "clamped-II" conditions, $v_{0,z} = w_{,y}$ must be provided at the root. This would cause a *deflection due to shear strain* in the amount of $v_0 = zw_{,y}$ as shown by Fig. 5.4(b). Similar effects are induced by other components of shear strain. As shown below, the differences between "clamped-I" and "clamped-II" are relatively small, and create no significant modifications of the global response of the beam.

In most of the cases, we shall consider the "clamped-II" conditions as they present more canonical formulation. To convert displacements given for clamped-I conditions to those in clamped-II conditions one needs to determine the rotations at the origin, say $\widetilde{\omega}_x$, $\widetilde{\omega}_y$, and to superimpose the rigid body solution

$$\Delta u = -\widetilde{\omega}_y z, \qquad \Delta v = \widetilde{\omega}_x z, \qquad \Delta w = -\widetilde{\omega}_x y + \widetilde{\omega}_y x. \tag{5.14}$$

To convert displacements given for clamped-II conditions to those in clamped-I conditions, one needs to determine the derivatives $u_{0,z}(0)$ and $v_{0,z}(0)$ and to superimpose the rigid body solution

$$\Delta u = -u_{0,z}(0)\, z, \qquad \Delta v = -v_{0,z}(0)\, z, \qquad \Delta w = v_{0,z}(0)\, y + u_{0,z}(0)\, x. \tag{5.15}$$

The reader should note that from a "pure clamping conditions" point of view, the two versions of requirements at the root, (5.12) and (5.13), are both approximations. In both cases, the deformation at the root vanishes in the "averaged sense" only, while the root cross-section "warping" will be a strong function of the root-support stiffness, see Remark 5.2. Yet, the above integral application of the geometrical boundary conditions coincides with the global approach of viewing the beam deformation via its elastic axis, and the underlying assumptions of St. Venant's principle.

Regarding the natural conditions associated with the outer surface, these may be written similar to (1.85a–c) by setting $\cos(\bar{\mathbf{n}},z) = 0$ there (which is true for all points over the outer surface of a uniform beam), namely,

$$X_s = \sigma_x \cos(\bar{\mathbf{n}},x) + \tau_{xy} \cos(\bar{\mathbf{n}},y), \tag{5.16a}$$

$$Y_s = \tau_{xy} \cos(\bar{\mathbf{n}},x) + \sigma_y \cos(\bar{\mathbf{n}},y), \tag{5.16b}$$

$$Z_s = \tau_{xz} \cos(\bar{\mathbf{n}},x) + \tau_{yz} \cos(\bar{\mathbf{n}},y). \tag{5.16c}$$

Here, $\bar{\mathbf{n}}$ is the outward normal to the cross-section contour, $\partial\Omega$, as shown in Fig. 5.1(b).

Remark 5.2 Theoretically, for "infinitely" stiff support, one may superimpose an additional solution on the one obtained by applying the above integral conditions. This additional solution should consist of a self-equilibrated system of stress components, the influence of which will decay far from the root (see above discussion regarding the application of the natural tip conditions). Such a system has the potential to ensure the requirement of zero displacements and rotations over the entire root area.

The above should be read in light of the fact that in practice, a "pure" clamping conditions may not always be required and/or achieved.

5.1.3.2 Application of Various Boundary Conditions

Since the examples discussed in this book are primarily focused on the "clamped-free" boundary conditions, it is worth demonstrating the applicability of such analyses to other boundary conditions. In what follows, we shall modify the boundary conditions at the beam "root" and "tip" only to the case of a "simply-supported" beam. Other various types of boundary conditions, and cases where additional restrictions are applied at other points along the beam, should be considered analogously and are briefly discussed below.

Within the method proposed here, we employ a solution for a clamped-free beam (with Clamped-II boundary conditions at the root, see (5.13)), that has been derived for *zero tip loading* (i.e. for the distributed loads only). The derivation modifies the geometrical conditions at the root simultaneously with the tip loading to create the required new boundary conditions.

It is clear that for a "clamped-free" beam, the axis displacements and rotations at the root vanish, while in the general case the corresponding tip values are nonzero, and will be denoted as u^t, v^t, w^t, ω_x^t, ω_y^t and ω_z^t. In addition, we shall denote the resultant loads that are induced at the root as P_x^r, P_y^r, P_z^r, M_x^r, M_y^r and M_z^r.

We now introduce additional displacement and rotation components at the root, Δu, Δv, Δw, $\Delta\omega_x$, $\Delta\omega_y$ and $\Delta\omega_z$, and tip loads denoted as ΔP_x, ΔP_y, ΔP_z, ΔM_x, ΔM_y, ΔM_z. The additional displacements and rotations will be added uniformly along the beam, while the tip loads will induce a distribution of displacements and rotations along the beam. Hence, one may write the additional displacements and rotations at the tip as

$$\begin{Bmatrix} \Delta u^t \\ \Delta v^t \\ \Delta w^t \\ \Delta\omega_x^t \\ \Delta\omega_y^t \\ \Delta\omega_z^t \end{Bmatrix} = \begin{Bmatrix} \Delta u + l\Delta\omega_y \\ \Delta v - l\Delta\omega_x \\ \Delta w \\ \Delta\omega_x \\ \Delta\omega_y \\ \Delta\omega_z \end{Bmatrix} + \mathbf{U}^{tt} \cdot \begin{Bmatrix} \Delta P_x \\ \Delta P_y \\ \Delta P_z \\ \Delta M_x \\ \Delta M_y \\ \Delta M_z \end{Bmatrix}. \tag{5.17}$$

In the above equation, $\mathbf{U}^{tt}$ is the tip "load-displacement" sensitivity matrix that determines the amount of each displacement/rotation components that is induced at the tip by the tip loads (for a given clamped-free beam, once a solution methodology is adopted, this matrix may be extracted by applying the tip loads one at a time). Also, the additional tip loads will induce the following additional loads at the root

$$[\Delta P_x^r, \Delta P_y^r, \Delta P_z^r, \Delta M_x^r, \Delta M_y^r, \Delta M_z^r] = [\Delta P_x, \Delta P_y, \Delta P_z, \Delta M_x - \Delta P_y l, \Delta M_y + \Delta P_x l, \Delta M_z], \quad (5.18)$$

where l is the beam's length, and clearly, the additional tip loads will not induce any displacements components at the root since their corresponding deformation field is consistent with the clamped-free solution requirements.

We shall now deduce as an example, the correction required for a transformation of the solution of a clamped-free beam to the one that holds for the same beam when it is "simply-supported", see Fig. 5.5(a). By adopting the integral manner of fulfilling the boundary condi-

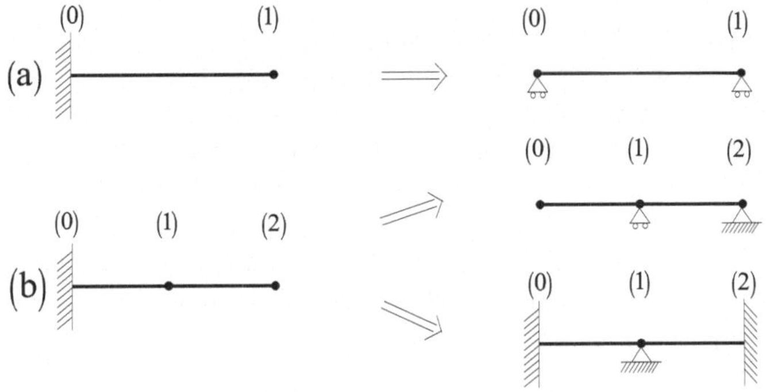

Figure 5.5: Modifying a "clamped-free" beam solution for other types boundary conditions.

tion described in S.5.1.3, for a "simply-supported" beam we require

$$u^r = v^r = w^r = M_x^r = M_y^r = \omega_z^r = 0, \quad \text{at the root,} \quad (5.19a)$$

$$u^t = v^t = w^t = M_x^t = M_y^t = \omega_z^t = 0, \quad \text{at the tip.} \quad (5.19b)$$

Putting these twelve conditions in a matrix form yields

$$\begin{bmatrix} 1 & 0 & 0 & 0 & 0 & 0 & 0 & 0 & 0 & 0 & 0 & 0 \\ 0 & 1 & 0 & 0 & 0 & 0 & 0 & 0 & 0 & 0 & 0 & 0 \\ 0 & 0 & 1 & 0 & 0 & 0 & 0 & 0 & 0 & 0 & 0 & 0 \\ 0 & 0 & 0 & 0 & 0 & 0 & 0 & -l & 0 & 1 & 0 & 0 \\ 0 & 0 & 0 & 0 & 0 & 0 & l & 0 & 0 & 0 & 1 & 0 \\ 0 & 0 & 0 & 0 & 0 & 1 & 0 & 0 & 0 & 0 & 0 & 0 \\ 1 & 0 & 0 & 0 & l & 0 & U_{11}^{tt} & U_{12}^{tt} & U_{13}^{tt} & U_{14}^{tt} & U_{15}^{tt} & U_{16}^{tt} \\ 0 & 1 & 0 & -l & 0 & 0 & U_{21}^{tt} & U_{22}^{tt} & U_{23}^{tt} & U_{24}^{tt} & U_{25}^{tt} & U_{26}^{tt} \\ 0 & 0 & 1 & 0 & 0 & 0 & U_{31}^{tt} & U_{32}^{tt} & U_{33}^{tt} & U_{34}^{tt} & U_{35}^{tt} & U_{36}^{tt} \\ 0 & 0 & 0 & 0 & 0 & 0 & 0 & 0 & 0 & 1 & 0 & 0 \\ 0 & 0 & 0 & 0 & 0 & 0 & 0 & 0 & 0 & 0 & 1 & 0 \\ 0 & 0 & 0 & 0 & 0 & 1 & U_{61}^{tt} & U_{62}^{tt} & U_{63}^{tt} & U_{64}^{tt} & U_{65}^{tt} & U_{66}^{tt} \end{bmatrix} \begin{Bmatrix} \Delta u \\ \Delta v \\ \Delta w \\ \Delta\omega_x \\ \Delta\omega_y \\ \Delta\omega_z \\ \Delta P_x \\ \Delta P_y \\ \Delta P_z \\ \Delta M_x \\ \Delta M_y \\ \Delta M_z \end{Bmatrix} = \begin{Bmatrix} 0 \\ 0 \\ 0 \\ -M_x^r \\ -M_y^r \\ 0 \\ -u^t \\ -v^t \\ -w^t \\ -M_x^t \\ -M_y^r \\ -\omega_z^t \end{Bmatrix}. \quad (5.20)$$

The above system solution yields 12 quantities required to modify the "clamped-free" boundary conditions to the "simply-supported" case. Any other version of (5.19a,b) could be handled in a similar way. To further illustrate the above procedure, it is worth mentioning that in the isotropic case, and per the analysis presented in S.5.3.4,

$$\mathbf{U}^{tt} = \begin{bmatrix} \frac{l^3}{3EI_y} & 0 & 0 & 0 & \frac{l^2}{2EI_y} & 0 \\ 0 & \frac{l^3}{3EI_x} & 0 & -\frac{l^2}{2EI_x} & 0 & 0 \\ 0 & 0 & \frac{l}{ES_\Omega} & 0 & 0 & 0 \\ 0 & -\frac{l^2}{2EI_x} & 0 & \frac{l}{EI_x} & 0 & 0 \\ \frac{l^2}{2EI_y} & 0 & 0 & 0 & \frac{l}{EI_y} & 0 \\ 0 & 0 & 0 & 0 & 0 & \frac{l}{D} \end{bmatrix}, \tag{5.21}$$

where E is Young's modulus, and D is the beam torsional rigidity.

The discussion in this section has been focused on modifying the boundary conditions at the beam root and tip as shown in Fig. 5.5(a). Yet, one may also introduce additional geometrical or natural conditions at other points along the beam. In such a case, the same approach of sensitivity matrices may be adopted. For example, let an additional condition be placed at the beam mid-point, see Fig. 5.5(b). Here, one needs to derive four sensitivity matrices that will reflect the influence of additional loads at points (1) and (2) on the displacement components in these locations (note that by reciprocity, two of these matrices will be identical, see S.1.4.3). Then, a system of eighteen equations should be set for the required boundary conditions at points (0), (1) and (2).

5.2 Elastic Coupling in General Anisotropic Beams

In recent years, composite beams have attracted considerable interest due to their unique elastic coupling characteristics, see S.5.4. This anisotropic capacity to adjust and "tailor" their design for specific needs substantially enlarges their spectrum of engineering applications.

At the material level, (anisotropic) composite materials usually offer better specific strength (i.e. strength per unit weight), better fatigue characteristics than metals, and their unique structure provides various kinds of multi-path-load mechanisms that augment their damage resistance. This kind of advantages will not be discussed here. On the other hand, the discussion will elaborate on the ability to construct anisotropic/composite beams, in which the global DOF are coupled.

Among other applications, elastic couplings may be exploited for "passive" improvement of both the static and the dynamic characteristics of composite beams. Improvement of the vibration characteristics and augmentation of the stability margins of aircraft wings and helicopter blades are typical examples of the potential applications in this area. For further reading see *Coupling Effects* of S.5.4.

5.2.1 Coupling at the Material Level

To provide some insight into the coupling effects in anisotropic beams, we first need to direct the discussion to the material level. For general anisotropic materials (GEN21, see S.2.2), we rewrite the strain-stress relationships in a matrix form as

$$\begin{Bmatrix} \varepsilon_x \\ \varepsilon_y \\ \varepsilon_z \\ \gamma_{yz} \\ \gamma_{xz} \\ \gamma_{xy} \end{Bmatrix} = \begin{bmatrix} a_{11} & a_{12} & a_{13} & a_{14} & a_{15} & a_{16} \\ & a_{22} & a_{23} & a_{24} & a_{25} & a_{26} \\ & & a_{33} & a_{34} & a_{35} & a_{36} \\ & & & a_{44} & a_{45} & a_{46} \\ & Sym. & & & a_{55} & a_{56} \\ & & & & & a_{66} \end{bmatrix} \begin{Bmatrix} \sigma_x \\ \sigma_y \\ \sigma_z \\ \tau_{yz} \\ \tau_{xz} \\ \tau_{xy} \end{Bmatrix}. \tag{5.22}$$

Examination of the above compliance matrix, $\mathbf{a} = \{a_{ij}\}$, shows that in this general case, all strain components are (linear) functions of all stress components, and vice versa. Hence, by applying any external loading over the outer surface, the end cross-sections, or any body-force loading that will induce some stress distribution in the elastic body, all strain components will appear. This phenomenon clearly stands in contrast with the isotropic case, where each stress component "activates" a limited number of strain components (e.g., uniform axial loading over the tip cross-section of a uniform isotropic beam will induce the σ_z component and subsequently the ε_x, ε_y and ε_z strain components only). The appearance of additional strain components in the anisotropic case, which subsequently activate additional deformation components, is generally termed as *coupling at the material level.*

The effect of elastic coupling on the beam behavior at the structural level may be clarified only by examination of the overall structural behavior, which, clearly, depend on other parameters such as body geometry, loading modes and boundary conditions.

In essence, coupling effects are of interest mainly from the global DOF point of view. Hence, the terms *coupled beam* and *uncoupled beams* stand for cases where the global DOF are coupled or uncoupled, respectively. Yet, it should be stressed that even in uncoupled beams, there may be many coupling effects between stress and strain components that do not affect the global DOF, but create diverse changes in the cross-section in-plane and out-of-plane deformation.

5.2.2 Coupling at the Structural Level

At the structural level of slender beams, the elastic coupling effects are those that couple the Beamwise Bending, the Edgewise Bending, the Extension and the Twist DOF. The resulting coupling effects will therefore be denoted by these global DOF. For example, a case where beamwise bending induces twist or torsional moment induces beamwise curvature, will be referred to as a *Beamwise Bending-Twist* or *Torsion-Beamwise Bending* Coupling, respectively, etc.

In this section, we shall review some cases for which simple and exact solutions may be drawn. These solutions are of great importance in providing physical insight into the structural coupling mechanisms. Subsequently, we shall only consider the influence of a tip axial load, P_z, and transverse bending moments, M_x and M_y, see Fig. 5.3(b). For these cases, a closed form exact solution for general homogeneous anisotropic beams with general cross-section geometry may be derived (including simply- and multiply- connected domains, thin-walled, etc.). For other tip loads and surface or body-force loads, and/or for non-homogeneous beams, much more involved analysis is required as will be shown in Chapters 6 – 10.

For the present analysis, we shall assume that σ_z is the only stress component that is induced by the tip loads P_z, M_x and M_y, and it is a linear function of x, y given by

$$\sigma_z = \widetilde{a} + \widetilde{b}x + \widetilde{c}y, \tag{5.23}$$

where we use the temporary substitution $\widetilde{a} = \frac{P_z}{S_\Omega}$, $\widetilde{b} = -\frac{M_y}{I_y}$ and $\widetilde{c} = \frac{M_x}{I_x}$. In view of (5.1), the above assumption clearly satisfies all tip conditions of (5.7), the lateral boundary conditions of

(5.16a,b) and the equilibrium equations (1.82a–c) as well. Further on, the strain components become

$$\varepsilon_i = a_{i3}\sigma_z, \qquad i = 1,\ldots,6, \tag{5.24}$$

which clearly satisfy the compatibility equations (1.45a–f), and therefore, the solution hypothesis of (5.23) constitutes a valid, unique and exact solution to the problem under discussion. For a "clamped-free" beam with "clamped-I" geometrical boundary conditions at its root, see (5.12), one obtains from the standard integration scheme presented in S.1.2 and **P.1.5**,

$$2u = \tilde{a}(\underline{2xa_{13} + ya_{36}}) + \tilde{b}(\underline{x^2a_{13} - y^2a_{23}} - \underline{\underline{zya_{34}}} - \overline{z^2a_{33}}) + \tilde{c}(\underline{2xya_{13} + y^2a_{36}} + \underline{\underline{zya_{35}}}), \tag{5.25a}$$

$$2v = \tilde{a}(\underline{2ya_{23} + xa_{36}}) + \tilde{b}(\underline{2yxa_{23} + x^2a_{36}} + \underline{\underline{xza_{34}}}) + \tilde{c}(\underline{y^2a_{23} - x^2a_{13}} - \overline{z^2a_{33}} - \underline{\underline{xza_{35}}}), \tag{5.25b}$$

$$2w = 2\tilde{a}(\overline{\overline{xa_{35} + ya_{34}}} + \overbrace{za_{33}}) + \tilde{b}(\overline{\overline{x^2a_{35} + yxa_{34}}} + \underbrace{2zxa_{33}}) + \tilde{c}(\overline{\overline{xya_{35} + y^2a_{34}}} + \underbrace{2zya_{33}}). \tag{5.25c}$$

These equations are exact for general anisotropic materials (GEN21), and may be now examined. The once underlined terms represent the in-plane warping. They appear in the equations for u and v only and are functions of x and y. The twice underlined terms represent the twist, namely, the rigid rotation of each cross-section about the z-axis. They appear only in the equations for u and v and are linear functions of x or y in each cross-section (i.e. for a fixed z). The once overlined terms represent the global bending, which is proportional to z^2 in this case. The twice overlined terms represent the out-of-plane warping. They appear in the equation for w only and are functions of x and y. The overbraced term represents the extension and the underbraced terms stand for the rotations of the cross-section due to the two bending components (for a constant z, they represent a uniform rotation of the cross-section). Keeping the same classification notation, the corresponding rotations are

$$2\omega_x = \overbrace{\tilde{a}a_{34}} + \tilde{c}(xa_{35} + ya_{34} + \overline{2za_{33}}), \tag{5.26a}$$

$$2\omega_y = \overbrace{-\tilde{a}a_{35}} - \tilde{b}(xa_{35} + ya_{34} + \overline{2za_{33}}), \tag{5.26b}$$

$$2\omega_z = \tilde{b}(2ya_{23} + xa_{36} + \underline{\underline{za_{34}}}) - \tilde{c}(2xa_{13} + \underline{\underline{za_{35}}} + ya_{36}), \tag{5.26c}$$

while the terms that are not marked represent the in-plane rotation (namely, rotation of each individual point that occurs on top of the rigid rotation).

For a "clamped-free" beam with "clamped-II" geometrical boundary conditions at its root (see (5.13)), one needs to make sure that all rotation components vanish at the root ($x = y = z = 0$), see S.5.1.3. Hence, the following rigid rotations should be added to those of (5.26a–c):

$$\Delta\omega_x = -\frac{\tilde{a}}{2}a_{34}, \qquad \Delta\omega_y = \frac{\tilde{a}}{2}a_{35}. \tag{5.27}$$

The above changes induce the following additional displacements, see (5.14):

$$\Delta u = \frac{\tilde{a}}{2}a_{35}z, \qquad \Delta v = \frac{\tilde{a}}{2}a_{34}z, \qquad \Delta w = -\frac{\tilde{a}}{2}(a_{35}x + a_{34}y), \tag{5.28}$$

which may be interpreted as additional transverse displacements u and v due to P_z, that as shown, induces a doubly-linear warping distribution at each cross-section. Although one may classify the above Δu and Δv as "coupling" between axial load and lateral bending, due to the fact that they introduce only a relatively small linear deflection distribution, they are not traditionally considered as such. Hence this phenomena should be kept within the context of the *deflection due to shear strain* as previously discussed and demonstrated by Fig. 5.4. Hence, for "clamped-II" boundary conditions, only the first terms of (5.25a–c) are modified as

$$2u = \tilde{a}(2xa_{13}+ya_{36}+za_{35})+\tilde{b}(\underline{x^2a_{13}-y^2a_{23}}-\underline{\underline{zya_{34}}}-\overline{z^2a_{33}})+\tilde{c}(\underline{2xya_{13}+y^2a_{36}}+\underline{\underline{zya_{35}}}), \quad (5.29a)$$

$$2v = \tilde{a}(2ya_{23}+xa_{36}+za_{34})+\tilde{b}(\underline{2yxa_{23}+x^2a_{36}}+\underline{\underline{xza_{34}}})+\tilde{c}(\underline{y^2a_{23}-x^2a_{13}}-\overline{z^2a_{33}}-\underline{\underline{xza_{35}}}), \quad (5.29b)$$

$$2w = 2\tilde{a}[\frac{1}{2}(xa_{35}+ya_{34})+za_{33}]+\tilde{b}(\overline{\overline{x^2a_{35}+yxa_{34}}}+\underbrace{2zxa_{33}})+\tilde{c}(\overline{\overline{xya_{35}+y^2a_{34}}}+\underbrace{2zya_{33}}). \quad (5.29c)$$

The corresponding rotations are therefore independent of P_z, namely

$$2\omega_x = \tilde{c}(xa_{35}+ya_{34}+\overline{2za_{33}}), \quad (5.30a)$$

$$2\omega_y = -\tilde{b}(xa_{35}+ya_{34}+\overline{2za_{33}}), \quad (5.30b)$$

$$2\omega_z = \tilde{b}(2ya_{23}+xa_{36}+\underline{\underline{za_{34}}})-\tilde{c}(2xa_{13}+\underline{\underline{za_{35}}}+ya_{36}). \quad (5.30c)$$

Finally, note that the beam axis deformation (see (5.25a–c), (5.29a–c)) may be written as

$$u_0 = \frac{1}{2}(\underline{\tilde{a}a_{35}z}-\tilde{b}a_{33}z^2), \quad v_0 = \frac{1}{2}(\underline{\tilde{a}a_{34}z}-\tilde{c}a_{33}z^2), \quad w_0 = \tilde{a}a_{33}z \quad (5.31)$$

where the underlined terms appear only for clamped-II boundary conditions.

Example 5.1 *Basic Structural Features of Anisotropic Beams.*
To ease the following examination, we shall first confine the discussion to MON13y-type materials, and consider the beam of rectangular cross-section shown in Fig. 2.2 of S.2.4 with "clamped-II" version of the boundary conditions. Hence, for the moment, we will assume $a_{34} = a_{36} = 0$ and consider each of the cases P_z, M_x, M_y separately.

(a) We shall first discuss the effect of the axial load, P_z. As shown by (5.29a,b), a tip axial load causes a z independent cross-section distortion, where u and v are linear functions of x and y, respectively. This would cause, for example, for a rectangular cross-section given by $\Omega = \{|x| \le \frac{d}{2}, |y| \le \frac{h}{2}\}$, to be deformed into the rectangle $\Omega^* = \{|x| \le \frac{d}{2}(1+\frac{P_z}{S\Omega}a_{13}), |y| \le \frac{h}{2}(1+\frac{P_z}{S\Omega}a_{23})\}$, which is essentially the well-known Poisson's effect in isotropic materials (note that a_{13}, a_{23} are typically negative).

Concerning the axial displacement w in (5.29c), the term $\frac{P_z}{S\Omega}a_{33}z$ is expected since, as in the isotropic case, it may be viewed as a linear (with respect to z) extension. However, the other terms reflect a linear out-of-plane distortion of the cross-section, which is not perpendicular to the longitudinal (z) axis anymore. This warping (that is a linear function of x in the present case, i.e., $\frac{P_z}{S\Omega}a_{35}x$) does not exist in isotropic or orthotropic cases, and is a clear outcome of the a_{35} "coupling module" that appears in MON13y beams.

To study the magnitude of the above coupling term, we may examine the resulting a_{35} term that is obtained by rotating an orthotropic material about the y-axis. In Fig. 2.2, the "lamination angle", θ_y, is the angle between the orthotropic material principal directions and the beam axis. Employing the definitions of S.2.4 and the transformation analysis of S.2.12, and by activating **P.2.7**, the amplitude of a_{35} turns to be

$$a_{35} = \tilde{A}\cos(2\theta_y)+\tilde{B}\cos(4\theta_y), \quad (5.32)$$

where

$$\tilde{A} = \frac{E_{33}G_{13}-E_{11}G_{13}}{2E_{11}G_{13}E_{33}}, \qquad \tilde{B} = \frac{-E_{33}G_{13}-2\nu_{13}G_{13}E_{33}+E_{11}E_{33}-E_{11}G_{13}}{4E_{11}G_{13}E_{33}}. \quad (5.33)$$

The above a_{35} term reaches its extreme value for $\theta_y^{\max} = \frac{1}{2}\arccos(\frac{-\widetilde{A}\pm\sqrt{\widetilde{A}^2+32\widetilde{B}^2}}{8\widetilde{B}})$, while the maximal value is obtained by the above plus sign solution, where

$$a_{35}^{\max} = \frac{\sqrt{2}}{32\widetilde{B}}\sqrt{16\widetilde{B}^2-\widetilde{A}^2+\widetilde{A}\sqrt{\widetilde{A}^2+32\widetilde{B}^2}\,(3\widetilde{A}+\sqrt{\widetilde{A}^2+32\widetilde{B}^2})}\,. \tag{5.34}$$

Variation of a_{35} as a function of the lamination angle θ_y for a typical Graphite/Epoxy orthotropic lamina is shown in Fig. 5.6(a).

(b) Regarding the effect of M_y, (5.29a) shows that for the beam axis, the displacement u is a quadratic function of z (just as in the isotropic case with Young's modulus of $1/a_{33}$). As far as the in-plane warping is concerned, it is somewhat more complicated than the axial loading case, as shown in Fig. 5.6(b) for rectangular cross-section deformation ($d=h=2,\ \widetilde{b}<0$). To

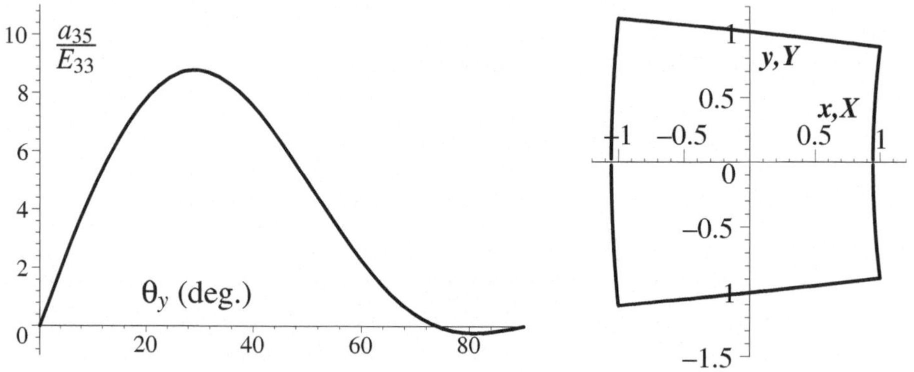

(a) a_{35} normalized by $1/E_{33}$ ($\theta_y^{\max} = 29°$, $a_{35}^{\max} = 8.77/E_{33}$).

(b) Deformation due to $M_y > 0$.

Figure 5.6: Results of Example 5.1.

derive the boundaries of the deformed cross-section, we ignore the global bending term $a_{33}z^2$ in (5.29a) since it represents a common displacement of the entire cross-section. For the upper and lower horizontal lines, $y=\pm\frac{h}{2}$, one may define their shape after deformation, $X=x+u$, $Y=y+v$, as

$$X = x+\frac{1}{2}\widetilde{b}(a_{13}x^2-a_{23}\frac{h^2}{4}), \qquad Y=\pm\frac{h}{2}(1+\widetilde{b}a_{23}x). \tag{5.35}$$

Eliminating x from the above equations yields

$$Y = \pm h(1-\frac{a_{23}}{a_{13}}+\frac{a_{23}}{a_{13}}\sqrt{1+a_{13}a_{23}\frac{h^2}{4}\widetilde{b}^2+2a_{13}X\widetilde{b}}\,). \tag{5.36}$$

For small bending moment, the above equation is close to a linear one, ($Y\cong\pm\frac{h}{2}(1+\widetilde{b}a_{23}X)$), see Fig. 5.6(b). For the vertical edges ($x=\pm\frac{d}{2}$), working with $X=x+u$, $Y=y+v$ yields

$$X=\pm\frac{d}{2}+\frac{1}{2}\widetilde{b}(a_{13}\frac{d^2}{4}-a_{23}y^2), \qquad Y=(1+\widetilde{b}\frac{d}{2}a_{23})y. \tag{5.37}$$

By eliminating y from the above equations, the vertical edges deformation becomes

$$Y=\pm\frac{1\pm\frac{d}{2}\widetilde{b}a_{23}}{\widetilde{b}a_{23}}\sqrt{\widetilde{b}a_{23}(\pm 2\frac{d}{2}+\widetilde{b}\frac{d^2}{4}a_{13}-2X\,)}\,. \tag{5.38}$$

As far as the out-of-plane warping due to M_y is concerned, (5.29c) shows a sum of a linear function of x with amplitude za_{33}, and a quadratic function (with respect to x) of amplitude $a_{35}/2$. The former appears in the isotropic case as well and is due to the rotation of the cross-section about the y-axis due to the applied bending moment (see also the ω_y rotation in (5.30b)). Hence this term represents a rotation with no distortion of the cross-section plane. However, the latter term, $w = \frac{1}{2}\widetilde{b}a_{35}x^2$, represents a quadratic out-of-plane warping, which is unique to anisotropic beams, and since it is proportional to a_{35}, Fig. 5.6(a) supplies information regarding the magnitude of this effect as well.

As already indicated, the above global bending stiffness is controlled by the a_{33} term. It is interesting to examine the variation of this term as a function of the lamination angle θ_y. Based on the analysis of S.2.12 and **P.2.7**, a_{33} may be expressed by the $\widetilde{A}$ and $\widetilde{B}$ coefficients of (5.33) as

$$a_{33} = \frac{1}{2}(\widetilde{B} + \frac{1}{E_{11}} + \frac{1}{E_{33}}) - \widetilde{A}\cos(2\theta_y) - \frac{1}{2}\widetilde{B}\cos(4\theta_y). \tag{5.39}$$

As shown in Fig. 5.7(a), a_{33} is reduced from a value of $1/E_{33}$ for $\theta_y = 0$ to a value of $1/E_{11}$ for $\theta_y = 90^\circ$. This dramatic decrease is in fact a reflection of the high moduli ratio $E_{33}/E_{11} \cong 13.7$ in this case.

(c) When a transverse tip moment, M_x, is applied ($\widetilde{c} \neq 0$), the beam behavior becomes more complex since this bending moment creates a twist angle in addition to bending curvature. First note that the bending curvature in this case is obtained by substituting $x = y = 0$ in (5.29b) and differentiation twice with respect to z, which yields $\kappa_v = v_{0,zz} = -\widetilde{c}a_{33}$. Hence, $1/a_{33}$ plays the role of the bending elastic modulus in this case as well. The cross-section in-plane warping is similar (but rotated by 90°) to the one discussed for M_y due to the similarities of the in-plane terms in (5.29a, b).

Further on, (5.30c) shows that M_x creates a longitudinally linear rigid twist angle, $\phi \equiv \omega_z|_{x=y=0}$, which is given by $\phi = -\frac{\widetilde{c}}{2}a_{35}z$. Hence, the above *Beamwise Bending-Twist* coupling may be defined as the amount of twist per unit bending curvature, and this non-dimensional value is given by

$$\frac{\phi_{,z}}{\kappa_v} = \frac{a_{35}}{2a_{33}}. \tag{5.40}$$

Using (5.32), (5.39) this coupling may be described as a function of θ_y as shown in Fig. 5.7(b). If we now remove the $a_{34} = a_{36} = 0$ assumption, two differences will take place. First, globally, a_{34} will create an *Edgewise Bending-Twist* coupling analogously to the *Beamwise Bending-Twist* discussed above. In addition, locally, the in-plane warping and the out-of-plane warping functions will be modified due to the presence of terms a_{36} and a_{34}, respectively.

To summarize this section, it should be emphasized again that the presented solutions are relatively simple and educating, but somehow limited since they are valid for a narrow spectrum of loading modes, and because of the *homogeneity* restriction on the cross-section geometry.

Example 5.2 *Strain Energy in a Beam Under Extension and Bending.*

We shall elaborate here on the determination of the strain energy in a beam that undergoes pure constant extension force and bending moments. Substituting (5.23), (5.24) in (2.27) yields the *volume density of the strain energy* which for the present case may be written as $W = \frac{1}{2}\sigma_z\varepsilon_z$. The total strain energy therefore becomes

$$U = \frac{1}{2}\iiint_V a_{33}\sigma_z^2 = \frac{1}{2}\int_0^l a_{33}\left(\widetilde{a}^2 S_\Omega + \widetilde{b}^2 I_y + \widetilde{c}^2 I_x\right)dz. \tag{5.41}$$

According to (5.31), one may write

$$[w_{0,z}, -u_{0,zz}, -v_{0,zz}] = a_{33}[\widetilde{a}, \widetilde{c}, \widetilde{b}], \tag{5.42}$$

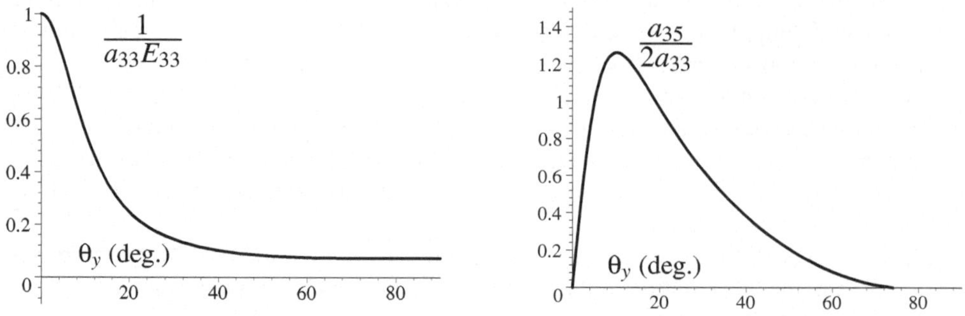

(a) Reduction in bending stiffness, $1/a_{33}$ (normalized by E_{33}).

(b) Bending-Twist coupling magnitude.

Figure 5.7: The lamination angle θ_y functions for a typical Graphite/Epoxy orthotropic lamina.

and therefore, (5.25a–c) show that the axial strain may be expressed in terms of the axis deformation as

$$\varepsilon_z = w_{,z} = w_{0,z} - u_{0,zz}x - v_{0,zz}y. \tag{5.43}$$

Thus, by using (5.1), U may be equivalently written as

$$U = \frac{1}{2}\iiint\limits_V \frac{1}{a_{33}}\varepsilon_z^2 = \frac{1}{2}\int_0^l \frac{1}{a_{33}}\left[(w_{0,z})^2 S_\Omega + (u_{0,zz})^2 I_y + (v_{0,zz})^2 I_x\right] dz. \tag{5.44}$$

It is also worth showing that differentiation of the *strain energy per unit length*, $\overline{U}$, and exploiting (5.41), (5.44), yield the following useful relations:

$$\frac{\partial \overline{U}}{\partial (w_{0,z})} = \frac{1}{a_{33}} w_{0,z} S_\Omega = \tilde{a} S_\Omega = P_z, \tag{5.45a}$$

$$\frac{\partial \overline{U}}{\partial (v_{0,zz})} = \frac{1}{a_{33}} v_{0,zz} I_x = -\tilde{c} I_y = -M_x, \tag{5.45b}$$

$$\frac{\partial \overline{U}}{\partial (u_{0,zz})} = \frac{1}{a_{33}} u_{0,zz} I_y = -\tilde{b} I_y = M_y. \tag{5.45c}$$

5.3 Simplified Beam Models

Prior to the presentation of detailed models for an anisotropic beam in the following chapters, in this section, we shall review three common types of simplified and approximate beam models that are frequently employed. Each of these models is restricted by some assumptions that on one hand, allow relatively simple solution, but on the other hand, pose a limitation on its applicability.

Note that Chapters 9, 10 present approximate models as well, yet they hold for a vast range of applications, much more involved, and therefore presented separately.

5.3.1 Beam-Plate Models

The derivation presented in what follows is largely founded on the Classical Laminated Plate Theory (CLPT) described in S.3.5.1.

We shall adopt here the beam of Fig. 5.8 where the laminae are laid parallel to the x,z-plane. This model is identical to Fig. 2.2, except for the fact that we take the beam thickness to be small, i.e., $h \ll d$. Hence, similar to S.3.5.1, we assume that the thin beam under discussion is

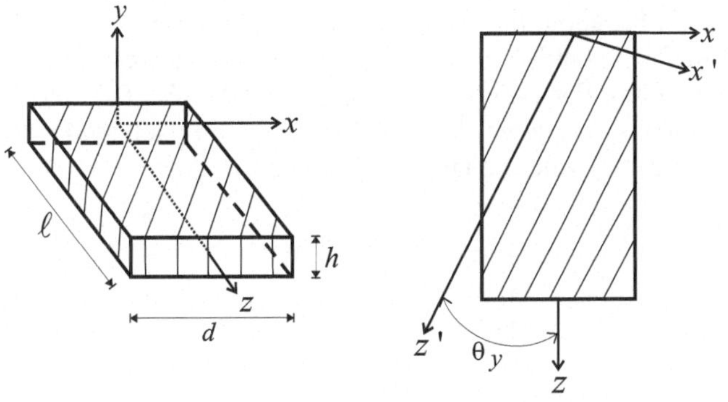

Figure 5.8: A thin MON13y homogeneous beam treated as a Beam-Plate model ($h \ll d$).

laminated, see schemes in Figs. 3.17, 3.19(b) by replacing $z \to y$, $y \to x$, $x \to z$.

The CLPT based Beam-Plate model presented in this section is restricted in various ways. First, as indicated, the analysis is confined to thin rectangular cross-sections. In addition, the edgewise bending is ignored. On the other hand, the analysis includes in-plane bending of the cross-section, which is not very important in slender beams. Moreover, as will be demonstrated in what follows, the torsion mechanism is approximate and may cause inconsistencies in some cases. Also, since Beam-Plate models are commonly employed for laminated (anisotropic, composite) beams, it should be noted that within such models, the interlaminar phenomena are treated in an approximate way only (see the exact approach to non-homogeneous beam derived in Chapter 8). Yet, keeping the above deficiencies in mind, for a considerable set of problems, Beam-Plate models may be adequately exploited as a replacement for a more complete and involved analysis.

As indicated in S.3.5.1, the CLPT models were improved over the years by the so-called "higher-order shear deformation theories", see e.g. (Ochoa and Reddy, 1992), which will not be discussed here. These theories substantially contribute to a better prediction of the shear stresses in Beam-Plate models, and therefore enable more accurate treatment of the interlaminar phenomena in laminated beams.

For further reading see *Laminated Beams, Delamination and Interlaminar Phenomena* of S.5.4. The reader is also referred to (Ip and Tse, 2001) where thick laminated composite beams are discussed.

5.3.1.1 Governing Equations

In what follows, we shall convert the derivation in S.3.5.1 to the case where the laminae are laid in the x,z-plane. We subsequently assume that stress components containing the y index vanish, namely, $\sigma_{yy} = \tau_{yz} = \tau_{xy} = 0$, and that σ_{xx}, σ_{zz}, and τ_{xz} are functions of x and z only. Hence, for MON13y material, $\varepsilon_{yz} = \varepsilon_{xy} = 0$ and $\varepsilon_{yy} = a_{12}\sigma_{xx} + a_{23}\sigma_{zz} + a_{25}\tau_{xz}$. The compliance and the stiffness matrices in this case are reduced to

$$\begin{Bmatrix} \varepsilon_{xx} \\ \varepsilon_{zz} \\ \gamma_{xz} \end{Bmatrix} = \begin{bmatrix} a_{11} & a_{13} & a_{15} \\ & a_{33} & a_{35} \\ Sym & & a_{55} \end{bmatrix} \cdot \begin{Bmatrix} \sigma_{xx} \\ \sigma_{zz} \\ \tau_{xz} \end{Bmatrix}, \quad \begin{Bmatrix} \sigma_{xx} \\ \sigma_{zz} \\ \tau_{xz} \end{Bmatrix} = \begin{bmatrix} \overline{B}_{11} & \overline{B}_{13} & \overline{B}_{15} \\ & \overline{B}_{33} & \overline{B}_{35} \\ Sym & & \overline{B}_{55} \end{bmatrix} \cdot \begin{Bmatrix} \varepsilon_{xx} \\ \varepsilon_{zz} \\ \gamma_{xz} \end{Bmatrix}, \quad (5.46)$$

where $\overline{B}_{ij} = A_{ij} - \frac{A_{i2}A_{j2}}{A_{22}}$ are the *reduced stiffness elastic coefficients for MON13y material.* Assuming that the beam is made of orthotropic material the principal axes of which are rotated by a "lamination angle", θ_y, about the y-axis, the procedure described in S.2.13 enables us to express the reduced stiffness coefficients $\overline{B}_{ij}$ as functions of θ_y as presented in Fig. 5.9 (for typical Graphite/Epoxy lamina). For symmetric terms $\overline{B}_{ij}(-\theta_y) = \overline{B}_{ij}(\theta_y)$, and for antisymmetric terms $\overline{B}_{ij}(-\theta_y) = -\overline{B}_{ij}(\theta_y)$. As shown, $\overline{B}_{15}$ and $\overline{B}_{35}$ are the only antisymmetric elastic moduli (i.e. they change sign together with the lamination angle θ_y). Another important feature of the

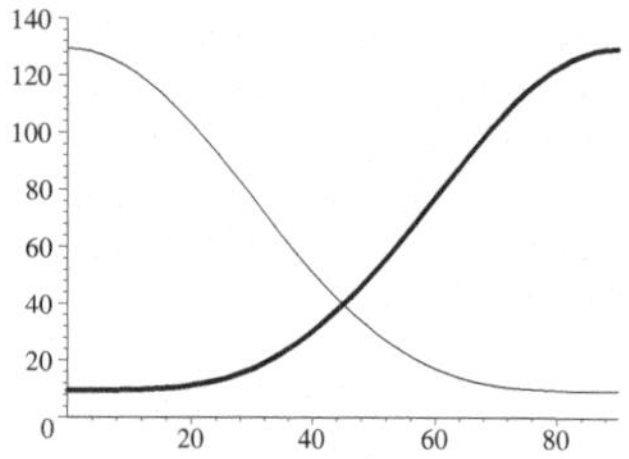

(a) $\overline{B}_{11}$ (Thick) and $\overline{B}_{33}$ (Thin), Symmetric.

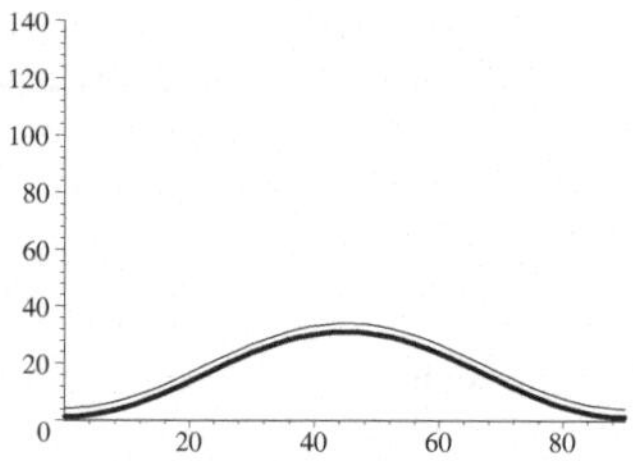

(b) $\overline{B}_{13}$ (Thick); $\overline{B}_{55}$ (Thin), Symmetric.

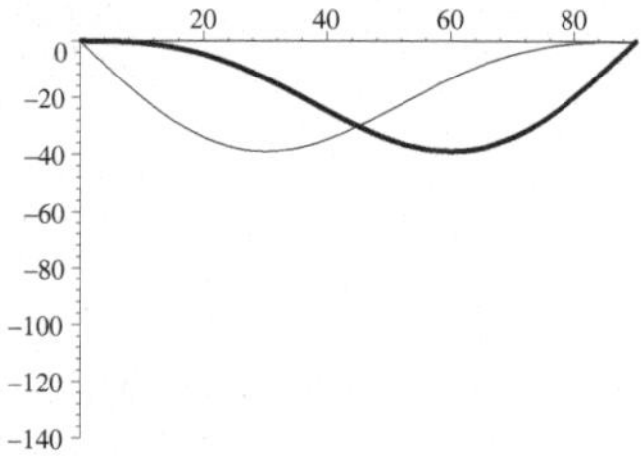

(c) $\overline{B}_{15}$ (Thick); $\overline{B}_{35}$ (Thin), Antisymmetric.

Figure 5.9: The reduced stiffness coefficients $\overline{B}_{ij}$ (GPa) vs. θ_y (deg.).

above characteristics is the dramatic decrease of $\overline{B}_{33}$ and the interchange between $\overline{B}_{33}$ and $\overline{B}_{11}$ as θ_y varies in the $[0, \pi/2]$ range.

We may now examine the "Beam-Plate Model" shown in Fig. 5.10. Following S.3.5.1 and

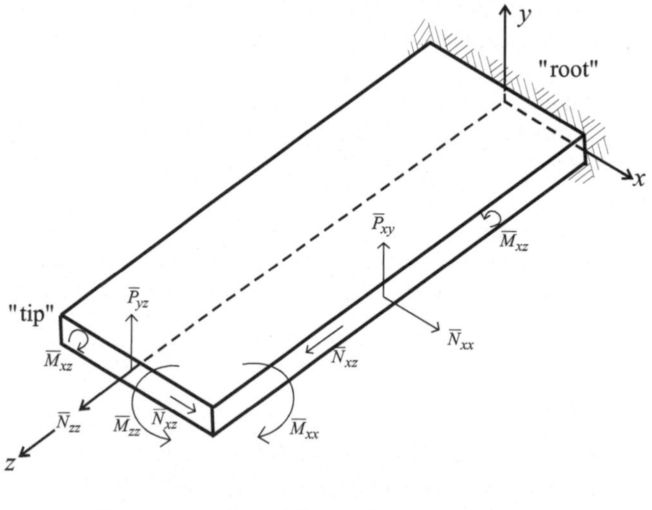

Figure 5.10: Notation for a Beam-Plate model.

similar to (3.147), the deformation components are assumed to be given by

$$u(x,y,z) = u_0(x,z) - y v_{0,x}, \quad v(x,y,z) = v_0(x,z), \quad w(x,y,z) = w_0(x,z) - y v_{0,z}. \tag{5.47}$$

We shall now write the corresponding strain components by the beam axis displacements as

$$\varepsilon_x = \varepsilon_x^0 - y\kappa_x, \qquad \varepsilon_z = \varepsilon_z^0 - y\kappa_z, \qquad \gamma_{xz} = \gamma_{xz}^0 - y\kappa_{xz}, \tag{5.48}$$

where

$$\varepsilon_x^0 = u_{0,x}, \qquad \varepsilon_z^0 = w_{0,z}, \qquad \gamma_{xz}^0 = u_{0,z} + w_{0,x}, \tag{5.49a}$$

$$\kappa_x = v_{0,xx}, \qquad \kappa_z = v_{0,zz}, \qquad \kappa_{xz} = 2v_{0,xz}. \tag{5.49b}$$

Note that the above κ_z curvature is identical to the previously defined κ_y curvature (see (5.6)), and that this notation is kept here for the sake of notation uniformity.

The governing equations of the beam under discussion are in fact the so-called *laminate constitutive relations* in S.3.5.1, and analogous to (3.151), yet with a slightly different notation (in order to avoid confusion with beam notation), we express the resultant loads *per unit edge length* as (see Fig. 5.10)

$$\{\overline{N}_{zz}, \overline{N}_{xx}, \overline{N}_{xz}, \overline{M}_{zz}, \overline{M}_{xx}, \overline{M}_{xz}, \overline{P}_{yz}, \overline{P}_{xy}\} = \sum_{n=1}^{N} \int_{y_n}^{y_{n+1}} \{\sigma_z, \sigma_x, \tau_{xz}, y\sigma_z, y\sigma_x, y\tau_{xz}, \tau_{yz}, \tau_{xy}\}\, dy. \tag{5.50}$$

The coefficients $\mathbf{A}_{ij}$, $\mathbf{B}_{ij}$ and $\mathbf{D}_{ij}$ in this case are

$$[\mathbf{A}_{ij}, \mathbf{B}_{ij}, \mathbf{D}_{ij}] = \sum_{n=1}^{N} \overline{B}_{ij}^{[n]} [(y_{n+1} - y_n), \frac{1}{2}(y_{n+1}^2 - y_n^2), \frac{1}{3}(y_{n+1}^3 - y_n^3)], \tag{5.51}$$

where $\overline{B}_{ij}^{[n]}$ are the elastic coefficients of the $n^{\underline{th}}$ laminae. Subsequently, the laminate constitutive relations that govern the beam response are given by

$$\begin{bmatrix} \mathbf{A} & \mathbf{B} \\ \mathbf{B} & \mathbf{D} \end{bmatrix} \cdot \begin{Bmatrix} \varepsilon_x^0 \\ \varepsilon_z^0 \\ \gamma_{xz}^0 \\ -\kappa_x \\ -\kappa_z \\ -\kappa_{xz} \end{Bmatrix} = \begin{Bmatrix} \overline{N}_{xx} \\ \overline{N}_{zz} \\ \overline{N}_{xz} \\ \overline{M}_{xx} \\ \overline{M}_{zz} \\ \overline{M}_{xz} \end{Bmatrix}. \tag{5.52}$$

5.3.1.2 Symmetric and Antisymmetric Configurations

To demonstrate the application of the Beam-Plate model, we shall first construct a simple model that consists of two layers as shown in Fig. 5.11. The laminae are assumed to be made of identical orthotropic material and have the same thickness ($h/2$), but they are rotated about the y-axis by different angles $\theta_y = \theta_1$ and $\theta_y = \theta_2$, see also Fig. 5.8.

We shall now define as "symmetric configuration" the case where $\theta_2 = \theta_1$. Then, $\mathbf{A}_{ij} = \overline{B}_{ij} h$, $\mathbf{B}_{ij} = 0$ and $\mathbf{D}_{ij} = \overline{B}_{ij} \frac{h^3}{12}$, and the laminate constitutive relations become

$$\begin{bmatrix} \mathbf{A}_{11} & \mathbf{A}_{13} & \mathbf{A}_{15} \\ & \mathbf{A}_{33} & \mathbf{A}_{35} \\ Sym & & \mathbf{A}_{55} \end{bmatrix} \cdot \begin{Bmatrix} \varepsilon_x^0 \\ \varepsilon_z^0 \\ \gamma_{xz}^0 \end{Bmatrix} = \begin{Bmatrix} \overline{N}_{xx} \\ \overline{N}_{zz} \\ \overline{N}_{xz} \end{Bmatrix}, \quad -\begin{bmatrix} \mathbf{D}_{11} & \mathbf{D}_{13} & \mathbf{D}_{15} \\ & \mathbf{D}_{33} & \mathbf{D}_{35} \\ Sym & & \mathbf{D}_{55} \end{bmatrix} \cdot \begin{Bmatrix} \kappa_x \\ \kappa_z \\ \kappa_{xz} \end{Bmatrix} = \begin{Bmatrix} \overline{M}_{xx} \\ \overline{M}_{zz} \\ \overline{M}_{xz} \end{Bmatrix}. \tag{5.53}$$

We shall also define as "antisymmetric configuration" the case where $\theta_2 = -\theta_1$, and denote by $\overline{B}_{ij}$ the elastic moduli of the lower lamina. It appears that for terms of the constitutive

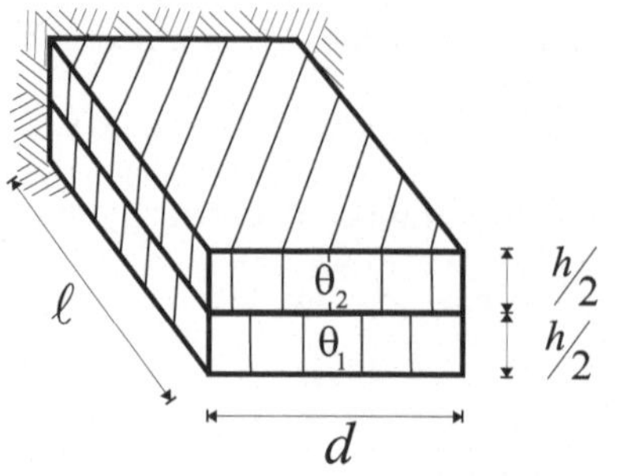

Figure 5.11: A Beam-Plate model made of two laminae.

relations matrix that are based on the symmetric elastic moduli (i.e., $\overline{B}_{11}, \overline{B}_{33}, \overline{B}_{13}$ and $\overline{B}_{55}$, see Figs. 5.9(a),(b)), $\mathbf{A}_{ij} = \overline{B}_{ij}h$, $\mathbf{B}_{ij} = 0$ and $\mathbf{D}_{ij} = \overline{B}_{ij}\frac{h^3}{12}$, while for the terms that consist of the antisymmetric moduli (i.e., $\overline{B}_{15}, \overline{B}_{35}$, see Fig. 5.9(c)), $\mathbf{A}_{ij} = 0$, $\mathbf{B}_{ij} = -\overline{B}_{ij}\frac{h^2}{4}$ and $\mathbf{D}_{ij} = 0$. The constitutive relations in this case are

$$\begin{bmatrix} \mathbf{A}_{11} & \mathbf{A}_{13} & 0 & 0 & 0 & \mathbf{B}_{15} \\ & \mathbf{A}_{33} & 0 & 0 & 0 & \mathbf{B}_{35} \\ & & \mathbf{A}_{55} & \mathbf{B}_{15} & \mathbf{B}_{35} & 0 \\ & & & \mathbf{D}_{11} & \mathbf{D}_{13} & 0 \\ & Sym & & & \mathbf{D}_{33} & 0 \\ & & & & & \mathbf{D}_{55} \end{bmatrix} \cdot \begin{Bmatrix} \varepsilon_x^0 \\ \varepsilon_z^0 \\ \gamma_{xz}^0 \\ -\kappa_x \\ -\kappa_z \\ -\kappa_{xz} \end{Bmatrix} = \begin{Bmatrix} \overline{N}_{xx} \\ \overline{N}_{zz} \\ \overline{N}_{xz} \\ \overline{M}_{xx} \\ \overline{M}_{zz} \\ \overline{M}_{xz} \end{Bmatrix}, \tag{5.54}$$

which also may be written as

$$\begin{bmatrix} \mathbf{A}_{11} & \mathbf{A}_{13} & \mathbf{B}_{15} \\ & \mathbf{A}_{33} & \mathbf{B}_{35} \\ Sym & & \mathbf{D}_{55} \end{bmatrix} \cdot \begin{Bmatrix} \varepsilon_x^0 \\ \varepsilon_z^0 \\ -\kappa_{xz} \end{Bmatrix} = \begin{Bmatrix} \overline{N}_{xx} \\ \overline{N}_{zz} \\ \overline{M}_{xz} \end{Bmatrix}, \quad \begin{bmatrix} \mathbf{A}_{55} & \mathbf{B}_{15} & \mathbf{B}_{35} \\ & \mathbf{D}_{11} & \mathbf{D}_{13} \\ Sym & & \mathbf{D}_{33} \end{bmatrix} \cdot \begin{Bmatrix} \gamma_{xz}^0 \\ -\kappa_x \\ -\kappa_z \end{Bmatrix} = \begin{Bmatrix} \overline{N}_{xz} \\ \overline{M}_{xx} \\ \overline{M}_{zz} \end{Bmatrix}. \tag{5.55}$$

For illustration purposes, we shall suppose that all deformation parameters, namely, ε_x^0, ε_z^0, γ_{xz}^0 and κ_x, κ_z, κ_{xz} are constants. Hence,

$$u_0(x,z) = x\varepsilon_x^0 + z\frac{\gamma_{xz}^0}{2}, \qquad v_0(x,z) = x^2\frac{\kappa_x}{2} + z^2\frac{\kappa_z}{2} + xz\frac{\kappa_{xz}}{2}, \qquad w_0(x,z) = x\frac{\gamma_{xz}^0}{2} + z\varepsilon_z^0. \tag{5.56}$$

Note that by substituting (5.56) in (5.47) one may restore (5.48). For clarification purposes we shall exploit simple kinematics considerations (see (1.19:c), (5.47), (5.49b:c)) and make the substitution $\kappa_{xz} = 2\phi_{,z}$ $(= 2\theta$, where $\phi_{,z} = \theta(z) = \omega_{z0,z}|_{x=y=0}$ is the beam axis twist angle per unit length).

Examples 5.3–5.5 present some solutions for Beam-Plate models. According to the general force and moment notation for beams shown by (5.7) and Fig. 5.3(b), and in order to compare the results of the present discussion (where all deformation parameters and resultant loads per unit edge are constants) the beam tip loads are defined as

$$P_x = d\overline{N}_{xz}, \quad P_y = d\overline{P}_{yz}, \quad P_z = d\overline{N}_{zz}, \quad M_x = d\overline{M}_{zz}, \quad M_y = 0, \quad M_z = -d\overline{M}_{xz}, \tag{5.57}$$

where d is the beam width (see Fig. 5.11).

Example 5.3 *Symmetric Beam Under Tip Bending and Torsional Moments.*

For the case of a symmetric homogeneous beam ($\theta_2 = \theta_1$) and $\overline{N}_{xx} = \overline{N}_{zz} = \overline{N}_{xz} = 0$, (5.53) shows that $\varepsilon_x^0 = \varepsilon_z^0 = \gamma_{xz}^0 = 0$, and

$$-[\mathbf{D}_{ij}] \cdot \begin{Bmatrix} \kappa_x \\ \kappa_z \\ \kappa_{xz} \end{Bmatrix} = \begin{Bmatrix} \overline{M}_{xx} \\ \overline{M}_{zz} \\ \overline{M}_{xz} \end{Bmatrix}, \qquad i,j = 1,3,5, \tag{5.58}$$

or by inverting, and exploiting (5.46, b), (5.51),

$$\begin{Bmatrix} \kappa_x \\ \kappa_z \\ \kappa_{xz} \end{Bmatrix} = -\frac{12}{h^3} [a_{ij}] \cdot \begin{Bmatrix} \overline{M}_{xx} \\ \overline{M}_{zz} \\ \overline{M}_{xz} \end{Bmatrix}, \qquad i,j = 1,3,5. \tag{5.59}$$

To simulate a beam subjected to a tip bending moment, $M_x (= d\overline{M}_{zz})$ (see (5.57)), we set $\overline{M}_{xx} = \overline{M}_{xz} = 0$ and obtain

$$\kappa_x = -\frac{12}{h^3} a_{13} \overline{M}_{zz}, \qquad \kappa_z = -\frac{12}{h^3} a_{33} \overline{M}_{zz}, \qquad \kappa_{xz} = -\frac{12}{h^3} a_{35} \overline{M}_{zz} \tag{5.60}$$

or

$$\frac{\kappa_{xz}}{2\kappa_z} = \frac{\phi_{,z}}{\kappa_z} = \frac{a_{35}}{2a_{33}}, \tag{5.61}$$

which is identical to the result presented in (5.40).

To simulate a beam subjected to a tip torsional moment $M_z (= -d\overline{M}_{xz})$ (see (5.57)), we set $\overline{M}_{xx} = \overline{M}_{zz} = 0$, which yields

$$\kappa_x = -\frac{12}{h^3} a_{15} \overline{M}_{xz}, \qquad \kappa_z = -\frac{12}{h^3} a_{35} \overline{M}_{xz}, \qquad \kappa_{xz} = -\frac{12}{h^3} a_{55} \overline{M}_{xz}, \tag{5.62}$$

or

$$\frac{\kappa_z}{\phi_{,z}} = \frac{2\kappa_z}{\kappa_{xz}} = 2\frac{a_{35}}{a_{55}}, \tag{5.63}$$

which is identical to the result obtained by the general solution derived further on within Chapter 9 when the latter is reduced to the case of thin homogeneous lamina, i.e., $h \ll d$, (see Remark 9.8). Employing the analysis of Chapter 10 allows us to clarify the amount of error that might be induced by the present model when one evaluates the torsional rigidity, D, for the beam as

$$D \equiv \frac{M_z}{\phi_{,z}} = -\frac{d\overline{M}_{xz}}{\kappa_{xz}/2} = \frac{dh^3}{6a_{55}}. \tag{5.64}$$

Comparison of the above result with the exact one developed for thin beams in S.10.3.2 (see (10.69)) shows that (5.64) presents only half of the exact value. To explain this phenomenon, Fig. 5.12 presents a scheme of the shear stress distribution over the beam's tip and its representation by the present model. As shown, in this case, the significant contribution of the τ_{yz} stress components is not included in the present analysis.

Example 5.4 *Symmetric Beam Under Tip Tensile Force.*

For the case of a symmetric homogeneous beam ($\theta_2 = \theta_1$) and $\overline{M}_{xx} = \overline{M}_{zz} = \overline{M}_{xz} = 0$, (5.53) shows that $\kappa_x = \kappa_z = \kappa_{xz} = 0$, and

$$[\mathbf{A}_{ij}] \cdot \begin{Bmatrix} \varepsilon_x^0 \\ \varepsilon_z^0 \\ \gamma_{xz}^0 \end{Bmatrix} = \begin{Bmatrix} \overline{N}_{xx} \\ \overline{N}_{zz} \\ \overline{N}_{xz} \end{Bmatrix}, \qquad i,j = 1,3,5, \tag{5.65}$$

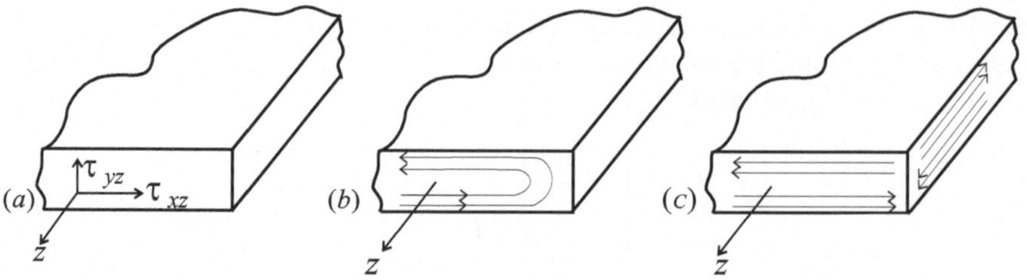

Figure 5.12: Shear stress over the beam's tip: (a) Notation, (b) Exact solution, (c) Beam-Plate model solution.

or by inverting, and exploiting (5.46, b), (5.51),

$$\begin{Bmatrix} \varepsilon_x^0 \\ \varepsilon_z^0 \\ \gamma_{xz}^0 \end{Bmatrix} = \frac{1}{h}[a_{ij}] \cdot \begin{Bmatrix} \overline{N}_{xx} \\ \overline{N}_{zz} \\ \overline{N}_{xz} \end{Bmatrix}, \qquad i, j = 1, 3, 5. \tag{5.66}$$

To examine the case of tensile force we set $\overline{N}_{xx} = \overline{N}_{xz} = 0$ and $P_z = d\overline{N}_{zz}$ (see (5.57)), and the extensional stiffness appears as $\frac{P_z}{hd\,\varepsilon_z^0} = \frac{\overline{N}_{zz}}{h\varepsilon_z^0} = \frac{1}{a_{33}}$, which is shown by the thin line in Fig. 5.13(a) for a two-laminae "Beam-Plate" beam made of typical Graphite/Epoxy orthotropic material as a function of the lamination angle of the lower lamina, and normalized (multiplied) by $a_{33}|_{\theta_y=0}$. This result also coincides with the exact one derived in S.9.2.3.1.

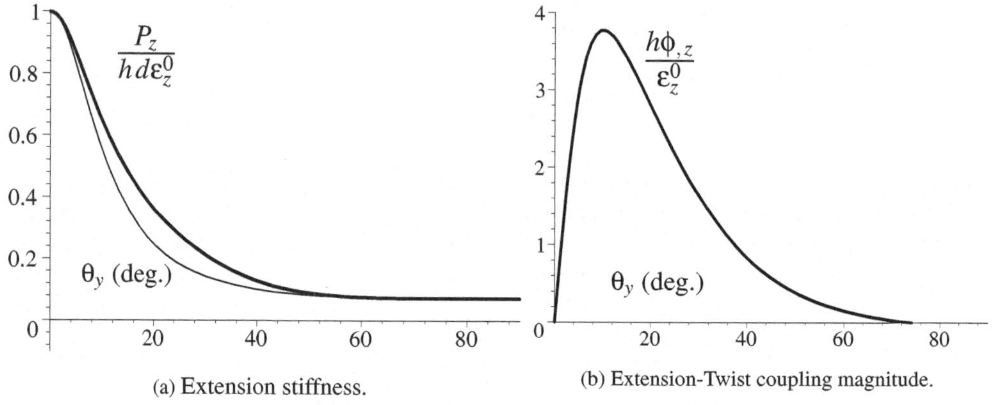

(a) Extension stiffness.

(b) Extension-Twist coupling magnitude.

Figure 5.13: Results from Examples 5.4, 5.5.

Example 5.5 *Antisymmetric Non-Homogeneous Beam Under Tip Tensile Force.*
To simulate the effect of tensile force on an antisymmetric beam ($\theta_2 = -\theta_1$), we set $\kappa_x = \kappa_z = 0$, and by inverting the first equation of (5.55) one may write

$$\begin{bmatrix} \widetilde{M}_{11} & \widetilde{M}_{13} & \widetilde{M}_{15} \\ & \widetilde{M}_{33} & \widetilde{M}_{35} \\ Sym & & \widetilde{M}_{55} \end{bmatrix} \cdot \begin{Bmatrix} \overline{N}_{xx} \\ \overline{N}_{zz} \\ \overline{M}_{xz} \end{Bmatrix} = \begin{Bmatrix} \varepsilon_x^0 \\ \varepsilon_z^0 \\ -\kappa_{xz} \end{Bmatrix}. \tag{5.67}$$

Assuming $\overline{N}_{xx} = \overline{M}_{xz} = 0$ and $\overline{N}_{zz} \neq 0$, the ratio $-\kappa_{xz}/\varepsilon_z^0 = \widetilde{M}_{35}/\widetilde{M}_{33}$, which stands for the Extension-Twist coupling in this case, may be written as

$$\left(-\frac{h}{2}\right)(-\kappa_{xz}/\varepsilon_z^0) = \frac{h\phi_{,z}}{\varepsilon_z^0} = \frac{h}{2} \cdot \frac{\mathbf{A}_{11}\mathbf{B}_{35} - \mathbf{A}_{13}\mathbf{B}_{15}}{\mathbf{A}_{11}\mathbf{D}_{55} - \mathbf{B}_{15}^2}. \tag{5.68}$$

The typical magnitude of this coupling is shown in Fig. 5.13(b) for a two-laminae antisymmetric "Beam-Plate" made of typical Graphite/Epoxy orthotropic material as a function of the lamination angle of the lower lamina. The extensional stiffness in this case is given by

$$\frac{P_z}{hd\varepsilon_z^0} = \frac{\overline{N}_{zz}}{h\varepsilon_z^0} = \frac{2\mathbf{A}_{13}\mathbf{B}_{15}\mathbf{B}_{35} + \mathbf{A}_{11}\mathbf{A}_{33}\mathbf{D}_{55} - \mathbf{A}_{11}\mathbf{B}_{35}^2 - \mathbf{D}_{55}\mathbf{A}_{13}^2 - \mathbf{A}_{33}\mathbf{B}_{15}^2}{h(\mathbf{A}_{11}\mathbf{D}_{55} - \mathbf{B}_{15}^2)}. \tag{5.69}$$

The above expression appears as the thick line in Fig. 5.13(a) and normalized (divided) by $\overline{B}_{33}|_{\theta_y=0}$. It should be noted that for zero lamination angle the extensional stiffness of the symmetric and the antisymmetric cases coincide as (5.69) becomes

$$\frac{P_z}{hd\varepsilon_z^0} = \frac{1}{h}(\mathbf{A}_{33} - \frac{\mathbf{A}_{13}^2}{\mathbf{A}_{11}}) = \overline{B}_{33} - \frac{\overline{B}_{13}^2}{\overline{B}_{11}} = \frac{1}{a_{33}}, \qquad (\theta_y = 0),$$

and thus, both the values at $\theta_y = 0$ and the normalization factors for the two curves in Fig. 5.13(a) are identical.

A similar problem for MON13z material is discussed in S.8.2.3.

5.3.2 Analysis by Cross-Section Stiffness Matrix

In this section, we shall review a common method that is applicable for both isotropic and anisotropic beams, in which the entire analysis is carried out by the so-called *cross-section stiffness matrix* (CSSM).

Generally speaking, this method is based on the combination of a two-dimensional analysis of each (or a representative) cross-section with a one-dimensional beam analysis. The outcomes of the two-dimensional analysis are the cross-section stiffness and coupling terms, which are fed into the one-dimensional beam analysis. Consistent separation of the overall three-dimensional problem into a (plane) two-dimensional, and an (axial) one-dimensional problems is quite involved as shown by the exact analysis of Chapter 6 and the multilevel approach described within S.9.3. Yet, the CSSM presented in what follows employs only a simplified version of the above separation.

To clarify the method, we shall first consider a *uniform* beam. As a preliminary stage, the CSSM method requires four distinct analyses for a clamped-free beam under the tip loads P_z, M_x, M_y and M_z. In these cases, the resultant loads are constants along the beam, namely, $\overline{P}_z(z) = P_z$, $\overline{M}_x(z) = M_x$, $\overline{M}_y(z) = M_y$ and $\overline{M}_z(z) = M_z$, and subsequently, constant distributions of global deformation measures $w_{0,z}(z)$, $u_{0,zz}(z)$, $v_{0,zz}(z)$, and $\phi_{,z}(z)$ are obtained. Applying the above tip loads one at a time enables us to fill the following 4×4 *"cross-section compliance matrix"*, $\widetilde{\mathbf{S}} = \{\widetilde{S}_{ij}\}$, as

$$\begin{Bmatrix} w_{0,z} \\ v_{0,zz} \\ u_{0,zz} \\ \phi_{,z} \end{Bmatrix} = \begin{bmatrix} \widetilde{S}_{11} & \widetilde{S}_{12} & \widetilde{S}_{13} & \widetilde{S}_{14} \\ \widetilde{S}_{21} & \widetilde{S}_{22} & \widetilde{S}_{23} & \widetilde{S}_{24} \\ \widetilde{S}_{31} & \widetilde{S}_{32} & \widetilde{S}_{33} & \widetilde{S}_{34} \\ \widetilde{S}_{41} & \widetilde{S}_{42} & \widetilde{S}_{43} & \widetilde{S}_{44} \end{bmatrix} \cdot \begin{Bmatrix} \overline{P}_z \\ \overline{M}_x \\ \overline{M}_y \\ \overline{M}_z \end{Bmatrix}. \tag{5.70}$$

Note that each of the four solutions supplies one column in the above matrix. The diagonal terms $\widetilde{S}_{ii}$ stand for the "direct" or "isotropic" relations, where P_z creates an extension $w_{0,z}$, M_x

creates a curvature $-v_{0,zz}$, M_y creates a curvature $u_{0,zz}$ and M_z creates a twist $\phi_{,z}$. Therefore, the off-diagonal terms represent global coupling of the beam's DOF. The terms $\widetilde{S}_{i1}$, $i = 2,3,4$ are the *Extension-Beamwise Bending*, the *Extension-Edgewise Bending* and the *Extension-Twist* coupling terms, respectively. Similarly, the terms $\widetilde{S}_{12}, \widetilde{S}_{32}, \widetilde{S}_{42}$ are the *Beamwise Bending-Extension*, the *Beamwise Bending-Edgewise Bending* and the *Beamwise Bending-Twist* coupling terms, respectively, etc. The relation of (5.70) may be rewritten by the *cross-section stiffness matrix*, $\widetilde{\mathbf{C}} = \{\widetilde{C}_{ij}\}$ (where $\widetilde{\mathbf{C}} = \widetilde{\mathbf{S}}^{-1}$), as

$$\begin{Bmatrix} \overline{P}_z \\ \overline{M}_x \\ \overline{M}_y \\ \overline{M}_z \end{Bmatrix} = \begin{bmatrix} \widetilde{C}_{11} & \widetilde{C}_{12} & \widetilde{C}_{13} & \widetilde{C}_{14} \\ \widetilde{C}_{21} & \widetilde{C}_{22} & \widetilde{C}_{23} & \widetilde{C}_{24} \\ \widetilde{C}_{31} & \widetilde{C}_{32} & \widetilde{C}_{33} & \widetilde{C}_{34} \\ \widetilde{C}_{41} & \widetilde{C}_{42} & \widetilde{C}_{43} & \widetilde{C}_{44} \end{bmatrix} \cdot \begin{Bmatrix} w_{0,z} \\ v_{0,zz} \\ u_{0,zz} \\ \phi_{,z} \end{Bmatrix}. \tag{5.71}$$

We shall now move to the application of the method in the general case.

For a given set of loads, by force and moment integration, one may determine the *local* resultants $\overline{P}_z$, $\overline{M}_x$, $\overline{M}_y$ and $\overline{M}_z$ at each cross-section as shown by (5.10), (5.11a) ($\overline{P}_x$ and $\overline{P}_y$, are not required here). Having these resultants in hand, (5.70) is then employed to determine the *local* values of $w_{0,z}(z)$, $u_{0,zz}(z)$, $v_{0,zz}(z)$ and $\phi_{,z}(z)$. In other words, it is assumed that *the four deformation measures, $w_{0,z}$, $u_{0,zz}$, $v_{0,zz}$, $\phi_{,z}$, are related (via (5.70)) to the four local resultants, $\overline{P}_z$, $\overline{M}_x$, $\overline{M}_y$, $\overline{M}_z$, as if they occur in an infinitely long uniformly loaded beam.*

To this end, the deformation measures may be integrated with respect to z. For a Clamped-I boundary conditions (see (5.12), where $\omega_{z0} \equiv \phi$)

$$u_{0,z}(z) = \int_0^z u_{0,zz}(\tilde{z})d\tilde{z}, \qquad u_0(z) = \int_0^z u_{0,z}(\tilde{z})d\tilde{z}, \tag{5.72a}$$

$$v_{0,z}(z) = \int_0^z v_{0,zz}(\tilde{z})d\tilde{z}, \qquad v_0(z) = \int_0^z v_{0,z}(\tilde{z})d\tilde{z}, \tag{5.72b}$$

$$w_0(z) = \int_0^z w_{0,z}(\tilde{z})d\tilde{z}, \qquad \phi(z) = \int_0^z \phi_{,z}(\tilde{z})d\tilde{z}, \tag{5.72c}$$

and the displacements become

$$u = u_0(z) - y\phi(z), \qquad v = v_0(z) + x\phi(z), \qquad w = w_0(z) - xu_{0,z}(z) - yv_{0,z}(z). \tag{5.73}$$

The main advantages of the CSSM method is its ability to handle non-uniform loading. It may also handle non-uniform beams where for each cross-section, a different cross-section compliance matrix is derived (based on a uniform beam analysis the cross-section of which is the one under discussion). In fact, many analyses for initially-twisted beams are founded on such an approach (where the beam's non-uniformity emerges from the initial twist). The method may also serve as a basis for geometrical nonlinear analysis, where (5.11a) are set to include the effect of large deformation, similar to the "Elastica" approach for isotropic beams — see Example 1.3 and (Stronge and Yu, 1993).

Yet, the above approach is clearly an approximate one for two main reasons: First, it yields an estimation of the global deformation only, i.e., $w_0(z)$, $u_0(z)$, $v_0(z)$ and $\phi(z)$, see also Example 5.6. Secondly, it ignores derivatives with respect to z in the case of a non-uniform loading/beam. For further reading of improved and advanced models, see e.g. (Cesnik and Hodges, 1997) and (Yu *et al.*, 2002).

Example 5.6 *An Isotropic Beam Under Tip Beamwise Load.*

To demonstrate the CSSM and to show that the above procedure is an approximate one, we shall adopt the exact solution derived in S.7.2.3 by reducing it to the case of a homogeneous isotropic beam under beamwise tip load P_y, which may be written for a clamped-I boundary conditions as (see (7.74a–c))

$$u^P = \nu \frac{P_y}{EI_x}(l-z)xy, \tag{5.74a}$$

$$v^P = \frac{P_y}{2EI_x}[(l-z)\nu(y^2-x^2) + \underline{(lz^2 - \frac{1}{3}z^3)}], \tag{5.74b}$$

$$w^P = -\frac{P_y}{EI_x}[\underline{(lz - \frac{z^2}{2})}y + x^2 y + \chi_2(x,y)], \tag{5.74c}$$

where E stands for Young's modulus, ν for Poisson's ratio and χ_2 is the *bending function*, which is the outcome of a solution of suitable Neumann BVP (see S.7.2.2.2). In order to employ the CSSM method for this case, we shall first write the exact solution for a tip moment, M_x, as shown by (7.74a–c)

$$u^M = -\nu \frac{M_x}{EI_x}xy, \qquad v^M = \frac{M_x}{2EI_x}[\nu(x^2-y^2) - z^2], \qquad w^M = \frac{M_x}{EI_x}yz. \tag{5.75}$$

The above equations show that along the beam axis ($x = y = 0$) $w_{0,z} = u_{0,zz} = \phi_{,z} = 0$ and $v_{0,zz} = -M_x/(EI_x)$. Hence, $\widetilde{S}_{22} = -1/(EI_x)$ and $\widetilde{C}_{22} = -EI_x$. We may now evaluate the influence of P_y by the CSSM method using (5.11a) as $\overline{P}_y = P_y$ and $\overline{M}_x = -P_y(l-z)$. Therefore, $v_{0,zz}(z) = P_y(l-z)/(EI_x)$, which by using the double integration of (5.72b) and (5.73) yields

$$u = 0, \qquad v = \frac{P_y}{2EI_x}(lz^2 - \frac{1}{3}z^3), \qquad w = -\frac{P_y}{EI_x}(lz - \frac{z^2}{2})y. \tag{5.76}$$

The above last two expressions are identical to the underlined terms in (5.74b,c) but all other terms in (5.74a–c) are lost by the current approximation. As far as the stress components are considered, (7.72a–d) show that the exact stress expressions are

$$\sigma_x^P = \sigma_y^P = \tau_{xy}^P = 0, \qquad \sigma_z^P = -\frac{P_y}{I_x}(l-z)y,$$
$$\tau_{yz}^P = -\frac{P_y}{4I_x}\cdot\frac{1}{1+\nu}[(2-\nu)x^2 + \nu y^2 + 2\chi_{2,y}], \qquad \tau_{xz}^P = -\frac{P_y}{2I_x}\cdot\frac{1}{1+\nu}[(2+\nu)xy + \chi_{2,x}]. \tag{5.77}$$

Equations (7.72a–d) also show that by analyzing the problem via $\overline{M}_x$ as suggested by the CSSM method yields the correct $\sigma_x = \sigma_y = \tau_{xy} = 0$ and correct σ_z, but erroneous $\tau_{yz} = \tau_{xz} = 0$.

There are cases where the CSSM approach yields erroneous results in predicting the axis deformation as well. As an example of such a deficiency, the reader is referred to the non-uniform axial load case, which is derived further on in Section 9.2.3.1, where lateral bending appears only when the axial resultant load $\overline{P}_z(z)$ is non-uniform.

5.3.3 The Influence of the In-Plane Deformation

The purpose of the derivation in this section is to identify and establish the relative importance of the in-plane warping components. Since the in-plane deformation is much more pronounced in beams of thin-walled cross-sections, an analytic description of the in-plane warping in beams of such single-cell circular and rectangular cross-sections will be derived (the reader may become familiar with thin-walled beam terminology by reading the introductory sections of Chapter 10). Along the same lines, the analysis may be extended to predict the effect of the in-plane warping on bending moment-curvature relationships of isotropic and anisotropic beams of generic cross-sections that undergo various loading distributions.

The "ovalization" of long isotropic "tubes" (i.e. thin-walled beams of circular cross-section) under pure bending moment has been extensively studied since the pioneer work of (Brazier, 1927). This problem is an example for a nonlinear phenomena that may be analyzed by *The Principle of Least Work* presented in S.1.4.2.

(Brazier, 1927) expressed the flattening of a circular thin-walled cross-section as a single cosine term of the radial component of the in-plane deformation. Following the above first study of Brazier, the ovalization of circular tubes is also referred to as *Brazier's effect*, see (Calladine, 1983). In general, it may be stated that the main conclusion that emerges from the above studies is that the in-plane deformation in elastic isotropic tubes induces nonlinear bending moment-curvature relationships, which should be accounted for when the bending moment reaches certain levels. Another important conclusion is that there is a maximal value for the moment that may be applied, even when the ultimate stress levels and the associated buckling phenomena are ignored.

We shall commence the discussion with a description of the classical *Brazier's effect* for a circular isotropic thin-walled beam and employ the results of S.5.2.2 for $M_x \neq 0$. The underlying assumption of Brazier's analysis was that a long tube under uniform bending moment will be deformed into an ellipse as shown in Fig. 5.14(a). To determine this effect we shall assume

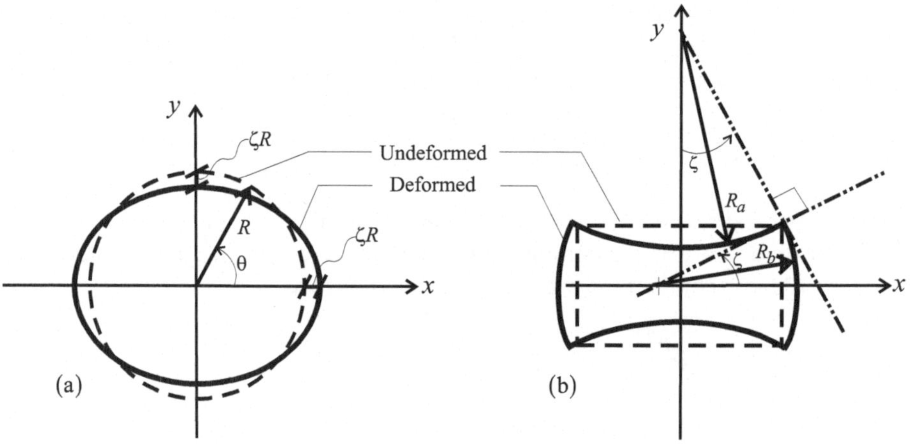

Figure 5.14: In-plane deformation of circular and rectangular thin-walled cross-sections.

that each cross-section deforms inextensionally in its own plane, and therefore for small deformation, one may assume that a circular cross-section will become an ellipse of semi-axes $R(1+\zeta)$ and $R(1-\zeta)$. We therefore write the resulting ellipse equation and its curvature as

$$y = R(1-\zeta)\sqrt{1-\frac{x^2}{R^2(1+\zeta)^2}}, \qquad \kappa = \frac{\frac{d^2y}{dx^2}}{\left[1+(\frac{dy}{dx})^2\right]^{3/2}}. \tag{5.78}$$

By substituting $x = R(1+\zeta)\cos\theta$ (see Fig. 5.14(a)) one may determine the first-order approximation for the cross-sectional moment (of inertia) about the x-axis, $I_x = \iint_\Omega y^2$, and the above curvature (note that in this case we set $d\Omega = t ds$, $ds = dx\sqrt{1+(\frac{dy}{dx})^2}$, $dx = -R(1+\zeta)\sin\theta d\theta$), as

$$I_x = \pi R^3 t(1-\frac{3}{2}\zeta) + O(\zeta^2), \qquad \kappa = \frac{1}{R} - \frac{3\cos(2\theta)}{R}\zeta + O(\zeta^2). \tag{5.79}$$

The first term in the expression for κ stands for the curvature before deformation, and thus, the second term is the linear approximation of the curvature due to the deformation.

We now write the *strain energy per unit length* that is induced by a given bending curvature $v_{0,zz}$. This energy consists of two parts: The bending energy, U_l, due to the axial stress as given by (5.44), and the bending energy, U_c, stored in the thin wall as expressed by (3.156) which we choose to exploit by setting $\kappa_x = -\frac{3\cos(2\theta)}{R}\zeta$, and $\kappa_y = \kappa_{xy} = 0$. Since we look for the energy per unit length, we integrate only along the circumference, namely

$$\overline{U}(\zeta) = \underbrace{\frac{1}{2}EI_x\left(v_{0,zz}\right)^2}_{U_l} + \underbrace{\frac{Eh^3}{24(1-\nu^2)}\oint\left[\frac{3\cos(2\theta)}{R}\zeta\right]^2 ds}_{U_c}, \tag{5.80}$$

where it is easy to show that $\oint \cos^2(2\theta)ds = \pi R + O(\zeta^2)$. Minimization of the strain energy as required by *The Principle of Least Work*, see S.1.4.2, namely, requiring $\partial\overline{U}/\partial\zeta = 0$, yields

$$\zeta = \left(\frac{R}{h}\right)^2 \left(Rv_{0,zz}\right)^2\left(1-\nu^2\right). \tag{5.81}$$

The above minimization process is illustrated in Fig. 5.15(a), where it is shown that the de-

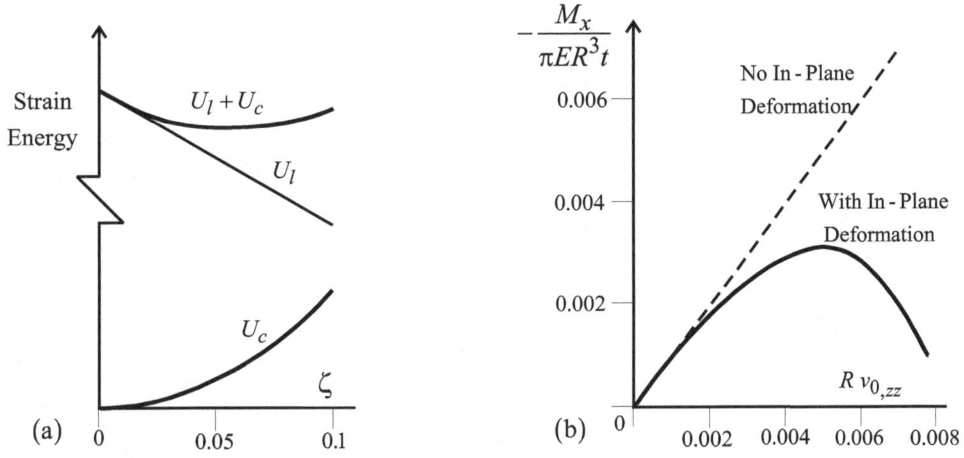

Figure 5.15: (a) Scheme of the minimization process; (b) Moment-curvature relationship for an isotropic circular thin-walled cross-section ($t/R = 0.01$, $\nu = 0.3$).

crease of the longitudinal bending energy and the increase of the thin-walled bending energy yield the desired state of stationary (minimum) energy solution.

The value of ζ in (5.81) represents the amount of ovalization induced by a given curvature. Substituting this value in the expression for the cross-sectional moment (of inertia) about the x-axis of (5.79:a) and equating the resulting I_x with its expression as obtained from (5.45b), namely, $-M_x/E/v_{0,zz}$, yields the beam moment-curvature relation

$$M_x = -\pi ER^3 h v_{0,zz}\left[1 - \frac{3}{2}\frac{R^4\left(v_{0,zz}\right)^2\left(1-\nu^2\right)}{h^2}\right]. \tag{5.82}$$

The above moment-curvature relationship is shown in Fig. 5.15(b) from which it is clear that unlike the case where the in-plane deformation is neglected and $M_x = -EI_x v_{0,zz}$, the introduction of this deformation creates an upper limit to the moment-carrying capacity of the cross-section which occurs at

$$v_{0,zz}|_{\max|M_x|} = \sqrt{\frac{2}{9}\frac{h}{R^2}}\frac{1}{\sqrt{1-\nu^2}}. \tag{5.83}$$

The above approach may be extended to other cross-section configurations as well. Consider for example, the thin-walled rectangular cross-section of dimensions a and b and a constant wall thickness, h, shown in Fig. 5.14(b). Based on the experimental results of (Vaze and Corona, 1997), and the numerical results of (Mazor and Rand, 2000), one may assume that the in-plane deformation causes the flanges and the webs (i.e. the horizontal and vertical walls, respectively) to deform as curved panels of constant curvature (i.e. to deform into circular panels), as shown in Fig. 5.14(b). Geometric considerations show that in order to maintain a right angle between the flange and the web, the angles created by the flange and the web arcs have to be identical and are denoted ζ in Fig. 5.14(b). It is further assumed that the cross-section deforms inextensionally (similar to Brazier's analysis) and therefore the flange and the web lengths should remain constants, and thus, $R_a = a/(2\zeta)$ and $R_b = b/(2\zeta)$ (which also shows that $R_a/R_b = a/b$). Hence, for the rectangular cross-section, the small deformation parameter is the angle ζ. One may therefore express the cross-sectional moment (of inertia) about the x-axis in this case as

$$I_x = a^3h\left(\frac{b}{a}\right)\left[\frac{1}{2}\frac{b}{a}+\frac{1}{6}\left(\frac{b}{a}\right)^2-\frac{1}{3}\zeta\right]+O(\zeta^2). \tag{5.84}$$

Based on the above, it is clear that the curvature of the flange and the web are $\kappa = 2\zeta/a$ and $\kappa = 2\zeta/b$, respectively, and the strain energy expression becomes similar to that of (5.80), namely

$$\overline{U}(\zeta) = \frac{1}{2}EI_x\left(v_{0,zz}\right)^2 + \frac{Eh^3}{3(1-\nu^2)}\left(\frac{1}{a}+\frac{1}{b}\right)\zeta^2. \tag{5.85}$$

The energy becomes stationary in this case for

$$\zeta = \frac{b}{4(a+b)}\left(\frac{a}{h}\right)^2\left(av_{0,zz}\right)^2\left(1-\nu^2\right), \tag{5.86}$$

and the resulting moment-curvature relationship may be expressed as

$$M_x = -\frac{1}{2}Ea^3hv_{0,zz}\left(\frac{b}{a}\right)\left[\frac{b}{a}+\frac{1}{3}\left(\frac{b}{a}\right)^2-\frac{1}{6}\frac{b}{a+b}\left(\frac{a}{h}\right)^2\left(av_{0,zz}\right)^2\left(1-\nu^2\right)\right]. \tag{5.87}$$

More details are given in (Rand, 2000b).

The above approach may be extended in various ways:

(a) By supplying enough in-plane deformation degrees of freedom, one may determine the deformation of various cross-section geometries. This applies to both closed and open thin-walled and to solid cross-sections as well, see also (Rand, 2001b).

(b) Since the above analysis is essentially two-dimensional (i.e. we consider the bending to be uniform along the beam), one may look at the corresponding effect in the local sense and replace M_x by the local resultant moment $\overline{M}_x$, see S.5.1.2. Similar to the CSSM approach of S.5.3.2, such a view enables us to estimate the in-plane warping effect in non-uniform beams under distributed loading.

(c) One may introduce the influence of anisotropy. This should be done for both the longitudinal bending strain energy and the in-plane warping strain energy.

For the longitudinal bending energy, the expressions of Example 5.2 may be used for the effect of the local resultants $\overline{P}_z$, $\overline{M}_x$ and $\overline{M}_y$. Expressions for the effect of the remaining $\overline{P}_x$, $\overline{P}_y$ and $\overline{M}_z$ resultants are more involved and should be carefully determined according to the specific anisotropy level as described in Chapters 6 – 10.

For the energy associated with the in-plane warping, (3.155) may be employed for thin-walled cross-sections (where a state of plane-stress is assumed). For a solid cross-section, one should write the *volume density of the strain energy* as in (3.154) but replace B_{ij} with A_{ij} to reflect a plane-strain analysis (which is consistent with the concept of viewing each cross-section as part of a long, uniformly loaded beam), see S.3.2.1 and (Rand, 2000c).

5.3.4 "Strength-of-Materials" Isotropic Beam Analysis

In many cases it is useful and educating to compare the results of detailed analyses with the results of the one-dimensional (elementary) "Strength-of-Materials" isotropic beam analysis. We shall summarize such results for a clamped-free (cantilever) beam. The loads are typically given as distributed body-force and surface loads as described in S.5.1.2, or as distributed force components per unit length $p_x(z),\ p_y(z),\ p_z(z)$ and distributed moment components per unit length $m_x(z),\ m_y(z),\ m_z(z)$. In both cases, there are additional tip force components $P_x,\ P_y,\ P_z$ and tip moment components $M_x,\ M_y,\ M_z$.

When the first way of introducing the loads is under discussion, the local resultants $\overline{P}_z$, $\overline{M}_x$, $\overline{M}_y$ and $\overline{M}_z$ at each cross-section are determined by (5.10), (5.11a). For the second way, the local resultants are determined as

$$\overline{M}_x(z) = M_x - P_y\,(l-z) + \int_z^l [-p_y(\tilde{z})\,(\tilde{z}-z) + m_x(\tilde{z})]\,d\tilde{z}, \tag{5.88a}$$

$$\overline{M}_y(z) = M_y + P_x\,(l-z) + \int_z^l [p_x(\tilde{z})\,(\tilde{z}-z) + m_y(\tilde{z})]\,d\tilde{z}, \tag{5.88b}$$

$$\overline{M}_z(z) = M_z + \int_z^l m_z(\tilde{z})\,d\tilde{z}, \qquad \overline{P}_z(z) = P_z + \int_z^l p_z(\tilde{z})\,d\tilde{z}. \tag{5.88c}$$

Subsequently, the curvature components, the extension and the twist are evaluated as

$$u_{0,zz} = \frac{\overline{M}_y}{EI_y}, \qquad v_{0,zz} = -\frac{\overline{M}_x}{EI_x}, \qquad w_{0,z} = \frac{\overline{P}_z}{ES_\Omega}, \qquad \phi_{,z} = \frac{\overline{M}_z}{D}, \tag{5.89}$$

where E is Young's modulus, and D is the torsional rigidity. Finally, the displacements are derived by the procedure described by (5.72a–c), (5.73) for a clamped-free beam with Clamped-I boundary conditions, see also Remark 5.3. **P.5.1** carries out the above elementary solution for an isotropic beam of elliptical cross-section.

Remark 5.3 By employing the identities

$$\int_z^l \widetilde{F}(\tilde{z})d\tilde{z} = \widetilde{F}(l)\,(l-z) - \int_z^l \frac{d\widetilde{F}}{dz}(\tilde{z})\,(\tilde{z}-z)\,d\tilde{z}, \tag{5.90a}$$

$$\frac{d}{dz}\int_z^l \widetilde{F}(\tilde{z})\,(\tilde{z}-z)\,d\tilde{z} = -\int_z^l \widetilde{F}(\tilde{z})d\tilde{z}, \qquad \frac{d}{dz}\int_z^l \widetilde{F}(\tilde{z})d\tilde{z} = -\widetilde{F}(z) \tag{5.90b}$$

the m_x term in (5.88a) can be replaced by an additional force distribution $\Delta p_y(z) = \frac{d}{dz}m_x(z)$, and a tip force $\Delta P_y = -m_x(l)$. Likewise, the m_y term in (5.88b) can be replaced by additional force distribution $\Delta p_x(z) = -\frac{d}{dz}m_y(z)$ and a tip force $\Delta P_x = m_y(l)$. These possible replacements are the basis for a common approximation, where moments distributions are represented by "equivalent force distributions". Strictly speaking, such replacements are not accurate as they misrepresent the shear force distribution along the beam. For example, in the former case of the m_x distribution, the above replacements will induce a shear force in the amount of

$$P_y(z) = \int_z^l \frac{d}{dz} m_x(z)(\tilde{z})\, d\tilde{z} - m_x(l) = -m_x(z),$$

which is erroneous and a pure outcome of the above approximation (more specifically, a constant m_x distribution will be replaced by a tip force, which, unlike constant moment distribution, induces a constant shear along the beam). Yet, this error has no influence on the resultant bending moments, $\overline{M}_x(z)$, and thus, it preserves the correct u_0, v_0, w_0, and ϕ_0 distributions.

In addition, in view of (5.90b), one may write (5.88a–d) in a differential form as

$$\overline{M}_{x,zz} = -p_y(z) - m_{x,z}(z), \quad \overline{M}_{y,zz} = p_x(z) - m_{y,z}(z), \quad \overline{M}_{z,z} = -m_z(z), \quad \overline{P}_{z,z} = -p_z(z), \tag{5.91}$$

which sometimes is more convenient for direct integration.

5.4 Literature

This section provides a representative body of recent literature in the anisotropic/composite beams analysis area. The following contains *only* books and articles that are directly related to this book's topics or has a direct influence on them. Many topics are *not* included (e.g. buckling, smart-structures and shape-memory-alloy beams, active control with piezoelectric actuators and sensors or magnetostrictive devices, plastic and elasto-plastic behavior, sandwich beams, active or passive damping and viscoelasticity effects, initially curved and/or initially twisted beams, impact and contact-stress phenomena, thermal loads, spinning and rotating beams, damage, crack and creep analyses, beams of multifunctional materials, manufacturing, fabrication and crashworthiness aspects, etc.).

Laminated Beams: (Whitney, 1987), (Sankar and Hu, 1991), (Singh *et al.*, 1991), (Wu and Sun, 1991), (Sankar, 1991), (Ochoa and Reddy, 1992), (Krishnaswamy *et al.*, 1992), (Harbert and Hogan, 1992), (Singh *et al.*, 1992), (Lee *et al.*, 1992), (Shi and Hull, 1992), (Liao *et al.*, 1992), (Qatu, 1992), (Savoia *et al.*, 1993), (Breivik *et al.*, 1993), (Pal and Ghosh, 1993), (Chandrashekhara and Bangera, 1993b), (Barbero *et al.*, 1993), (Liu *et al.*, 1993), (Lai *et al.*, 1993), (King and Chan, 1993), (Abramovich and Livshits, 1994), (Koo and Kwak, 1994), (Khdeir and Reddy, 1994), (Shi and Yee, 1994), (Yu, 1994), (Davalos *et al.*, 1994), (Tripathy *et al.*, 1994), (Bhaskar and Librescu, 1995), (Ackermann and Kozik, 1995), (Eisenberger *et al.*, 1995), (Whitney, 1995), (Krawczuk *et al.*, 1996), (Gadelrab, 1996), (Savoia, 1996), (Murakami and Yamakawa, 1996), (Soldatos and Watson, 1997), (Ratcliffe and Bagaria, 1998), (Mahajan, 1998), (Yildirim and Erhan, 1999), (Bassiouni *et al.*, 1999), (Shimpi and Ghugal, 1999), (Ray and Satsangi, 1999), (Davi and Milazzo, 1999), (Ozdil and Carlsson, 1999), (Mahapatra *et al.*, 2000), (Milazzo, 2000), (Qiao *et al.*, 2000), (Pinheiro and Sankar, 2000), (Tseng *et al.*, 2000), (Harrison and Butler, 2001), (Chen *et al.*, 2001), (Rand, 2001a), (Vlot and Gunnink, 2001), (Ghugal and Shimpi, 2001), (Nag *et al.*, 2002), (Cardoso *et al.*, 2002), (Ramtekkar *et al.*, 2002), (Mahapatra *et al.*, 2002), (Todoroki *et al.*, 2003), (Demakos, 2003).

Delamination and Interlaminar Phenomena: (Sankar and Hu, 1991), (Sankar, 1991), (Harbert and Hogan, 1992), (Lee *et al.*, 1992), (Shi and Hull, 1992), (Liao *et al.*, 1992), (Liu *et al.*, 1993), (King and Chan, 1993), (Shi and Yee, 1994), (Krawczuk *et al.*, 1996), (Gadelrab, 1996), (Wisnom and Jones, 1996), (Ratcliffe and Bagaria, 1998), (Rand, 1998b), (Milazzo, 2000), (Pinheiro and Sankar, 2000), (Icardi *et al.*, 2000), (Harrison and Butler, 2001), (Nag *et al.*, 2002), (Mahapatra *et al.*, 2002), (Todoroki *et al.*, 2003).

Thin-Walled Beams: (Timoshenko and Goodier, 1970), (Gjelsvik, 1981), (Bauld and Tzeng, 1984), (Murray, 1986), (Chandra *et al.*, 1990), (Bank, 1990), (Librescu and Song, 1991), (Atilgan and Hodges, 1991), (Wu and Sun, 1991), (Rand, 1991b), (Wu and Sun, 1992), (Librescu and Song, 1992), (Berdichevsky *et al.*, 1992), (Smith and Bank, 1992), (Chandra and Chopra, 1993), (Subrahmanyam, 1993), (Barbero *et al.*, 1993), (Kalfon and Rand, 1993), (Lai *et al.*, 1993), (Song and Librescu, 1993), (Bank and Cofie, 1993), (Rand, 1994a), (Bank and Cofie, 1994), (Armanios and Badir, 1995), (Bhaskar and Librescu, 1995), (Pollock *et al.*, 1995), (Rand, 1995), (Song and Librescu, 1995), (Suresh and Nagaraj, 1996), (Heredia *et al.*, 1996), (Song and Librescu, 1997b), (Floros and Smith, 1997), (Kim and White, 1997), (Song and Librescu, 1997a), (Suresh and Malhotra, 1997), (Loughlan and Ata, 1997), (Rand, 1998a), (Rand, 1998c), (Massa and Barbero, 1998), (Lentz and Armanios, 1998), (Omidvar, 1998), (Dancila *et al.*, 1999), (Volovoi and Hodges, 2000), (Qiao *et al.*, 2000), (Rappel and Rand, 2000), (Rand, 2000b), (Rand, 2000a), (Mazor and Rand, 2000), (Johnson *et al.*, 2001), (Qin and Librescu, 2001), (Rand, 2001b), (Volovoi and Hodges, 2002), (Kollár and Pluzsik, 2002), (Qin and Librescu, 2003), (Ko and Kim, 2003).

Optimization and Tailoring: (Librescu and Song, 1992), (Ascione and Fraternali, 1992), (Liao, 1993), (Song and Librescu, 1993), (Naik and Ganesh, 1993), (Li and Chen, 1993), (Miravete, 1996), (Taylor and Butler, 1997), (Lentz and Armanios, 1998), (Manne and Tsai, 1998), (Barkai and Rand, 1998), (Dancila *et al.*, 1999), (Evrard *et al.*, 2000), (Conceicao Antonio *et al.*, 2000), (Cardoso *et al.*, 2002), (Shokrieh and Rezaei, 2003).

Finite-Element Analyses: (Beltzer, 1990), (Al-Amery and Roberts, 1990), (Wu and Sun, 1991), (Sankar, 1991), (Ochoa and Reddy, 1992), (Singh *et al.*, 1992), (Gandhi and Lee, 1992), (Fish *et al.*, 1994), (Koo and Kwak, 1994), (Yu, 1994), (Mohamed Nabi and Ganesan, 1994), (Miller and Palazotto, 1995), (Ghosh *et al.*, 1995), (Xie and Adams, 1995), (Hodges *et al.*, 1996), (Krawczuk *et al.*, 1996), (Vinayak *et al.*, 1996), (Prathap *et al.*, 1996), (Floros and Smith, 1997), (Shi *et al.*, 1998), (Mahapatra *et al.*, 2000), (Pinheiro and Sankar, 2000), (Ramtekkar *et al.*, 2002).

Warping Effects: (Gandhi and Lee, 1992), (Lai *et al.*, 1993), (Floros and Smith, 1997), (Kim and White, 1997), (Rand, 1998a), (Rand, 2000b), (Rand, 2000c), (Mazor and Rand, 2000), (Tanghe Carrier, 2000), (Rand, 2001b), (Pluzsik and Kollár, 2002).

Higher-order Theories and Shear Deformation Effects: (Chandrashekhara *et al.*, 1990), (Segura, 1990), (Abramovich, 1992), (Lee *et al.*, 1992), (Savoia *et al.*, 1993), (Puspita *et al.*, 1993), (Liao, 1993), (Song and Librescu, 1993), (Nouri and Gay, 1994), (Koo and Kwak, 1994), (Yu, 1994), (Davalos *et al.*, 1994), (Pollock *et al.*, 1995), (Eisenberger *et al.*, 1995), (Xie and Adams, 1995), (Suresh and Nagaraj, 1996), (Murakami *et al.*, 1996), (Ponte-Castaneda, 1996), (Wisnom and Jones, 1996), (Vinayak *et al.*, 1996), (Prathap *et al.*, 1996), (McCarthy and Chattopadhyay, 1998), (Shi *et al.*, 1998), (Rand, 1998b), (Omidvar, 1998), (Shimpi and Ghugal, 1999), (Milazzo, 2000), (Pardini *et al.*, 2000), (Ip and Tse, 2001), (Ghugal and Shimpi, 2001), (Chandiramani *et al.*, 2002), (Pluzsik and Kollár, 2002), (Todoroki *et al.*, 2003).

Nonlinear and Large Deformation Analyses: (Al-Amery and Roberts, 1990), (Atilgan and Hodges, 1991), (Singh *et al.*, 1991), (Singh *et al.*, 1992), (Chandrashekhara and Bangera, 1993a), (Kalfon and Rand, 1993), (Creaghan and Palazotto, 1994), (Rand, 1994b), (Bhaskar and Librescu, 1995), (Miller and Palazotto, 1995), (Jeon *et al.*, 1995), (Hodges *et al.*, 1996), (Cesnik *et al.*, 1996), (Ponte-Castaneda, 1996), (Rand and Barkai, 1997), (Rand, 2001b).

Coupling Effects: (Chandra and Chopra, 1991), (Atanasoff and Vizzini, 1992), (Chandra and Chopra, 1992b), (Abdelnaser and Singh, 1993), (Bank and Cofie, 1993), (Song and Librescu, 1997a), (Lentz and Armanios, 1998), (Dancila *et al.*, 1999), (Rappel and Rand, 2000), (Alkahe and Rand, 2000), (Rand, 2001b).

Properties and Strength Prediction: (Suzuki, 1990), (Kardomateas, 1991), (Sankar and Marrey, 1993), (Puspita *et al.*, 1993), (Giurgiutiu and Reifsnider, 1994), (Steif and Trojnacki, 1994), (Bogomolov and Borisenko, 1994), (Ponte-Castaneda, 1996), (Wisnom and Jones, 1996), (Chen and Gibson, 1998), (Kocks *et al.*, 1998), (Massa and Barbero, 1998), (Zhang and Sikar skie, 1999), (Pardini *et al.*, 2000), (Tanghe Carrier, 2000), (Ugural and Fenster, 2003).

Structural dynamics and Vibrations: (Suresh *et al.*, 1990), (Chandrashekhara *et al.*, 1990), (Sankar and Hu, 1991), (Singh *et al.*, 1991), (King, 1991), (Rand, 1991a), (Wu and Sun, 1991), (Krishnaswamy *et al.*, 1992), (Chandra and Chopra, 1992a), (Abramovich, 1992), (Lee *et al.*, 1992), (Liao *et al.*, 1992), (Qatu, 1992), (Chandra and Chopra, 1993), (Chandrashekhara and Bangera, 1993b), (Lai *et al.*, 1993), (Abdelnaser and Singh, 1993), (Liao, 1993), (Song and Librescu, 1993), (Abramovich and Livshits, 1994), (Rand, 1994a), (Khdeir and Reddy, 1994), (Mohamed Nabi and Ganesan, 1994), (Armanios and Badir, 1995), (Banerjee and Williams, 1995), (Jeon *et al.*, 1995), (Eisenberger *et al.*, 1995), (Song and Librescu, 1995), (Karczmarzyk, 1996), (Nagaraj, 1996), (Song and Librescu, 1997b), (Sierakowski and Chaturvedi, 1997), (Krylov and Sorokin, 1997), (Song and Librescu, 1997a), (Volovoi *et al.*, 1998), (Ratcliffe and Bagaria, 1998), (Chen and Gibson, 1998), (McCarthy and Chattopadhyay, 1998), (Yildirim and Erhan, 1999), (Bassiouni *et al.*, 1999), (Dancila *et al.*, 1999), (Tseng *et al.*, 2000), (Icardi *et al.*, 2000), (Qin and Librescu, 2001), (Cortinez and Machado, 2001), (Ip and Tse, 2001), (Jung *et al.*, 2001), (Ghugal and Shimpi, 2001), (Volovoi *et al.*, 2001), (Chandiramani *et al.*, 2002), (Ramtekkar *et al.*, 2002), (Hodges, 2003), (Qin and Librescu, 2003), (Ko and Kim, 2003).

6

Beams of General Anisotropy

This chapter is devoted to the analysis of beams of general Cartesian anisotropy under axially non-uniform loading distribution. It provides a solution methodology that is founded on a *recurrence scheme*, where in each level, a suitable *Coupled-Plane problem* (see S.3.4) is formulated and solved based on results of similar BVPs in previous levels.

The origin of the analysis in this chapter is the classical approach to anisotropic beams that has been presented by (Lekhnitskii, 1981), who expressed the stress tensor in terms of stress functions and complex potentials, developed a rigorous derivation of the associated governing equations and boundary conditions, and offered analytic solutions for some specific cases. Yet, the formulation of (Lekhnitskii, 1981) was confined to anisotropic beams subjected loading for which all stress components do not vary along the beam's longitudinal axis.

The chapter extends the methodology of (Lekhnitskii, 1981) to the case where the loading varies in a generic manner in all directions, including the axial one, and allows generic tip loads as well. Removing the restriction for axially uniform stress distribution induces a level-based solution procedure, and also enforces the introduction of many additional deformation parameters, see also (Rovenski and Rand, 2001), (Rovenski and Rand, 2003).

To clarify the derivation, the present analysis is applied to beams of homogeneous and simply connected cross-sections. No substantial restriction prevents the extension of this approach to the non-homogeneous and/or multiply connected cross-sections. Examples for such configurations are dealt with in Chapter 8 for monoclinic beams.

Unless stated differently, we shall adopt here the DOF notation and definitions of S.5.1 for a general anisotropic beam. We shall make no restriction on the population of the constitutive relations matrix, and assume a fully populated compliance matrix with 21 independent material properties, i.e. GEN21-type of material (see S.2.2). Hence, the strain-stress relations of (5.22) will be under discussion in what follows.

6.1 Stress and Strain

As already discussed, the nature of the loading has a profound influence on the analysis complexity. Analogously to the level approach discussed in S.1.2.3, we shall adopt here the most general loading form where the beam is assumed to undergo *distributed loads* (per unit area), $\mathbf{F}_s = \{X_s, Y_s, Z_s\}$, along the outer surface of the beam, and *distributed body forces* (per unit volume), $\mathbf{F}_b = \{X_b, Y_b, Z_b\}$, which are applied at each material point (see S.5.1.2). The surface and body forces are expressed as vector polynomials of degree $K \geq 0$, and according to (1.60) we write their components in the coordinate directions as

$$\mathbf{F}_s = \sum_{k=0}^{K} \{X_s^{(k)}, Y_s^{(k)}, Z_s^{(k)}\}(x,y)z^k, \qquad \mathbf{F}_b = \sum_{k=0}^{K} \{X_b^{(k)}, Y_b^{(k)}, Z_b^{(k)}\}(x,y)z^k. \tag{6.1}$$

Hence, the case of $K=0$ stands for *uniform* distributed loads in the z direction, $K=1$ stands for *linear* distribution, etc. However, as will become clearer later on, the value $K=-1$ is reserved for the case where no surface and body loads are applied and the beam undergoes tip loads only.

Similar to the above polynomial expansion of the loading components, we let each one of the six stress components and, in view of the generalized Hooke's low, the six strain components to be expanded as polynomials in z of degree $K+2$, namely

$$\{\sigma_x, \sigma_y, \sigma_z, \tau_{xy}, \tau_{xz}, \tau_{yz}\} = \sum_{k=0}^{K+2} \{\sigma_x^{(k)}, \sigma_y^{(k)}, \sigma_z^{(k)}, \tau_{xy}^{(k)}, \tau_{xz}^{(k)}, \tau_{yz}^{(k)}\} z^k, \tag{6.2a}$$

$$\{\varepsilon_x, \varepsilon_y, \varepsilon_z, \gamma_{xy}, \gamma_{xz}, \gamma_{yz}\} = \sum_{k=0}^{K+2} \{\varepsilon_x^{(k)}, \varepsilon_y^{(k)}, \varepsilon_z^{(k)}, \gamma_{xy}^{(k)}, \gamma_{xz}^{(k)}, \gamma_{yz}^{(k)}\} z^k, \tag{6.2b}$$

where $\sigma_i^{(k)}, \tau_{ij}^{(k)}$ and $\varepsilon_i^{(k)}, \gamma_{ij}^{(k)}$ are functions of x, y. The constitutive relations of (5.22) also show that the axial stress may be expressed by the axial strain as

$$\sigma_z = \frac{1}{a_{33}}(\varepsilon_z - a_{13}\sigma_x - a_{23}\sigma_y - a_{43}\tau_{yz} - a_{53}\tau_{xz} - a_{63}\tau_{xy}). \tag{6.3}$$

We shall now eliminate σ_z from the constitutive relations by substituting its expression of (6.3) into (5.22), which yields

$$\varepsilon_i = b_{i1}\sigma_x + b_{i2}\sigma_y + b_{i4}\tau_{yz} + b_{i5}\tau_{xz} + b_{i6}\tau_{xy} + \frac{a_{i3}}{a_{33}}\varepsilon_z, \qquad i \in \{1,2,4,5,6\}, \tag{6.4}$$

where the *reduced elastic constants* (see also (3.28)) are defined as

$$b_{ij} = a_{ij} - \frac{a_{i3}\, a_{3j}}{a_{33}}, \qquad i,j \in \{1,2,4,5,6\}. \tag{6.5}$$

We shall now assume (and justify later on) that the highest component of axial strain ε_z, is a linear function in x, y, namely,

$$\varepsilon_z^{(K+2)} = \frac{a_{33}}{K+2}\left[\frac{P_1^{(K+1)}}{I_y}x + \frac{P_2^{(K+1)}}{I_x}y\right] \tag{6.6}$$

while the coefficients $P_1^{(K+1)}, P_2^{(K+1)}$ are derived later on by (6.15a) for $k=K$. Also, we let the five stress components be polynomials in z of degree $K+1$ only, namely:

$$\sigma_x^{(K+2)} = \sigma_y^{(K+2)} = \tau_{yz}^{(K+2)} = \tau_{xz}^{(K+2)} = \tau_{xy}^{(K+2)} = 0. \tag{6.7}$$

From (6.3), (6.4), (6.6), (6.7) one obtains the expressions for the highest axial stress and strain components

$$\sigma_z^{(K+2)} = \frac{1}{K+2}\left[\frac{P_1^{(K+1)}}{I_y}x + \frac{P_2^{(K+1)}}{I_x}y\right], \tag{6.8a}$$

$$\varepsilon_i^{(K+2)} = \frac{a_{i3}}{K+2}\left[\frac{P_1^{(K+1)}}{I_y}x + \frac{P_2^{(K+1)}}{I_x}y\right]. \tag{6.8b}$$

Hence, $\varepsilon_i^{(K+2)}(0,0) = 0$ for $i = 1,\ldots,6$.

It should be noted that the solution presented in this chapter exhibits "symmetry" in the sense that some expressions may be directly obtained by the parameter interchange: $x \leftrightarrow y$, $X \leftrightarrow Y$, $1 \leftrightarrow 2$, $4 \leftrightarrow 5$. Corollaries of the above are for example $\Theta_2 = Sym(\Theta_1)$, $\kappa_2 = Sym(\kappa_1)$, etc. For quantities that will be defined later on one may write: $v = Sym(u)$, $V = Sym(U)$, $\omega_y = -Sym(\omega_x)$, $\omega_z = -Sym(\omega_z)$ and $M_3 = -Sym(M_3)$. This symmetry will also be written in short as $\sigma_y \leftrightarrow \sigma_x$, $v =\leftrightarrow u$, etc.

6.1.1 Stress Resultants

To maintain a simple numerical indexing, we formally define the external loading *force resultant vector* as $\mathbf{P} = \{P_1, P_2, P_3\}$ and the external *moment resultant vector* as $\mathbf{M} = \{M_1, -M_2, M_3\}$ via an integration of the stress components over a cross-section, Ω, which are functions of the spanwise location, z, as

$$\mathbf{P}(z) = \iint_\Omega \{\tau_{xz}, \tau_{yz}, \sigma_z\}, \qquad \mathbf{M}(z) = \iint_\Omega \{\sigma_z y, -\sigma_z x, x\tau_{yz} - y\tau_{xz}\}. \tag{6.9}$$

Note that the components of the force and moments resultant vectors of Chapter 5 (see e.g. (5.10)) are related to the above as

$$\{P_1, P_2, P_3, M_1, M_2, M_3\} = \{\overline{P}_x, \overline{P}_y, \overline{P}_z, \overline{M}_x, -\overline{M}_y, \overline{M}_z\}. \tag{6.10}$$

Differentiating (5.11a,b) one obtains

$$\frac{d}{dz}\mathbf{P} = -\oint_{\partial\Omega} \mathbf{F}_s - \iint_\Omega \mathbf{F}_b, \tag{6.11a}$$

$$\frac{d}{dz}\mathbf{M} = P_y \mathbf{k}_1 - P_x \mathbf{k}_2 + \oint_{\partial\Omega} \mathbf{r}_s(z,z) \times \mathbf{F}_s(z) + \iint_\Omega \mathbf{r}_b(z,z) \times \mathbf{F}_b(z). \tag{6.11b}$$

The integral terms of (6.11b) may be written as

$$\oint_{\partial\Omega} \mathbf{r}_s(z,z) \times \mathbf{F}_s(z) = -\oint_{\partial\Omega} \{yZ_s, xZ_s, xY_s - yX_s\},$$
$$\iint_\Omega \mathbf{r}_b(z,z) \times \mathbf{F}_b(z) = -\iint_\Omega \{yZ_b, xZ_b, xY_b - yX_b\}. \tag{6.12}$$

The above definitions enable us to develop the following fundamental relations between the resultant loads, the surface loads and the body forces, see (6.11a,b), (6.12):

$$\frac{d}{dz}\mathbf{P} = -\oint_{\partial\Omega} \mathbf{F}_s - \iint_\Omega \mathbf{F}_b, \tag{6.13a}$$

$$\frac{d}{dz}\{M_1, M_2, M_3\} = -\oint_{\partial\Omega} \{yZ_s, xZ_s, xY_s - yX_s\} - \iint_\Omega \{yZ_b, xZ_b, xY_b - yX_b\} + \{P_2, P_1, 0\}. \tag{6.13b}$$

The reader may verify the above identities (6.13a, b) by employing the results documented in S.6.5. More specifically, referring to lines 1–3 of Table 6.4, the first ("x-direction") component of (6.13a) is obtained by using $P = \sigma_x$, $Q = \tau_{xy}$, $R = \tau_{xz}$, the second one is obtained by $P = \tau_{xy}$, $Q = \sigma_y$, $R = \tau_{yz}$ while the third one is obtained by $P = \tau_{xz}$, $Q = \tau_{yz}$, $R = \sigma_z$. Further on, the identities of (6.13b) are obtained from lines 4–7 of Table 6.4.

We may now expand $P_i(z)$ and $M_i(z)$ in a polynomial form as well, namely,

$$P_i = \sum_{k=0}^{K+1} P_i^{(k)} z^k, \; i = 1,2,3, \quad M_i = \sum_{k=0}^{K+2} M_i^{(k)} z^k, \; i = 1,2, \quad M_3 = \sum_{k=0}^{K+1} M_3^{(k)} z^k, \tag{6.14}$$

while the exact expansion degree is based on considerations that will become clearer later on. These expressions and (6.1), (6.13a,b) indicate that

$$(k+1)\mathbf{P}^{(k+1)} = -\oint_{\partial\Omega} \mathbf{F}_s^{(k)} - \iint_{\Omega} \mathbf{F}_b^{(k)}, \tag{6.15a}$$

$$\begin{aligned}(k+1)\{M_1,M_2,M_3\}^{(k+1)} = &-\oint_{\partial\Omega} \{yZ_s^{(k)}, xZ_s^{(k)}, xY_s^{(k)} - yX_s^{(k)}\} \\ &- \iint_{\Omega} \{yZ_b^{(k)}, xZ_b^{(k)}, xY_b^{(k)} - yX_b^{(k)}\} + \{P_2^{(k)}, P_1^{(k)}, 0\}.\end{aligned} \tag{6.15b}$$

The above system is a simple recurrent procedure. Equations (6.1), (6.15a) show that by definition, $P_i^{(K+2)} = 0$ $(i = 1,2,3)$, and (6.15b) confirms that

$$M_1^{(K+2)} = \frac{P_2^{(K+1)}}{K+2}, \qquad M_2^{(K+2)} = \frac{P_1^{(K+1)}}{K+2}, \qquad M_3^{(K+2)} = 0. \tag{6.16}$$

We also assume that the *resultant loads* are known at the beam "root" end cross-section, Ω^r (i.e. at $z = 0$ — see Fig. 5.1), and thus, $P_i^{(0)}$ and $M_i^{(0)}$ are given by (6.9) as

$$\{P_1^{(0)}, P_2^{(0)}, P_3^{(0)}, M_1^{(0)}, M_2^{(0)}, M_3^{(0)}\} = \iint_{\Omega^r} \{\tau_{xz}, \tau_{yz}, \sigma_z, \sigma_z y, \sigma_z x, x\tau_{yz} - y\tau_{xz}\}. \tag{6.17}$$

Cases where the resultant loads are known at other cross-sections (for example, the common case where tip loads are given), are handled by using the loads together with suitable integrals of the distributed surface and body forces in order to determine the required root resultant loads, see also Remark 6.1.

Remark 6.1 In the general case, one may determine $P_i^{(0)}, M_i^{(0)}$ $(i = 1,2,3)$, by employing (5.10), (5.11a,b) for $z = 0$. Hence, in the simple case of a cantilever of length l with *tip loads*, $P_x, P_y, P_z, M_x, M_y, M_z$, we write,

$$\{P_1^{(0)}, P_2^{(0)}, P_3^{(0)}, M_1^{(0)}, M_2^{(0)}, M_3^{(0)}\} = \{P_x, P_y, P_z, M_x - P_y l, -M_y - P_x l, M_z\}. \tag{6.18}$$

Remark 6.2 One may wish to inquire about cases where the stress level does not exceed the loading level, K. Equation (6.16) shows that the requirements for $M_2^{(K+2)} = M_1^{(K+2)} = 0$ are equivalent to the conditions $P_1^{(K+1)} = P_2^{(K+1)} = 0$, respectively. Hence, one is left with the requirements

$$P_i^{(K+1)} = M_i^{(K+1)} = 0, \qquad i = 1,2,3. \tag{6.19}$$

Equation (6.15a) shows that $P_i^{(K+1)} = 0$ yield

$$\oint_{\partial\Omega} \{X_s^{(K)}, Y_s^{(K)}, Z_s^{(K)}\} + \iint_{\Omega} \{X_b^{(K)}, Y_b^{(K)}, Z_b^{(K)}\} = 0, \tag{6.20}$$

while (6.15b) shows that the requirements $M_i^{(K+1)} = 0$ yield

$$\{P_2^{(K)}, P_1^{(K)}, 0\} = \oint_{\partial\Omega} \{yZ_s^{(K)}, xZ_s^{(K)}, xY_s^{(K)} - yX_s^{(K)}\} + \iint_{\Omega} \{yZ_b^{(K)}, xZ_b^{(K)}, xY_b^{(K)} - yX_b^{(K)}\}. \tag{6.21}$$

However, for $K > 0$, one can replace $P_1^{(K)}, P_2^{(K)}$ in (6.21) by the following values obtained from (6.15a):

$$\{P_1^{(K)}, P_2^{(K)}\} = -\frac{1}{K+1}\left[\oint_{\partial\Omega} \{X_s^{(K-1)}, Y_s^{(K-1)}\} + \iint_{\Omega} \{X_b^{(K-1)}, Y_b^{(K-1)}\}\right]. \tag{6.22}$$

Thus, overall (6.20) constitute three conditions, while (6.21), (6.22) constitute an additional three conditions that together ensure the six requirements of (6.19).

For the special $K = 0$ case of axially uniform distributed loading, where $P_1^{(0)} = P_2^{(0)} = 0$ and $Z_b = Z_s = 0$ discussed by (Lekhnitskii, 1981), we arrive at the following three integral conditions:

$$\oint_{\partial\Omega} \{X_s, Y_s, xY_s - yX_s\} + \iint_{\Omega} \{X_b, Y_b, xY_b - yX_b\} = \{0, 0, 0\}. \tag{6.23}$$

6.1.2 *Compatibility Conditions*

The above definitions enable us to express the (six) *compatibility equations* (see (1.45a–f)) by explicit execution of the derivatives with respect to z. We shall split these equations into three groups. In the *first* one, we include (1.45a) only which is written in levels as

$$\varepsilon_{x,yy}^{(k)} + \varepsilon_{y,xx}^{(k)} - \gamma_{xy,xy}^{(k)} = 0. \tag{6.24}$$

In the *second* group, we include (1.45d, e) and write them in short as (see also (1.55))

$$f_{zz,x} = f_{zx,z}, \qquad f_{zz,y} = f_{zy,z}. \tag{6.25}$$

Since $f_{zx,y} = f_{zy,x}$ is equivalent to compatibility equation (6.30c), the compatibility equations (6.25) may be replaced by (1.52) with $i = j = 3$, namely:

$$\gamma_{xz,y} - \gamma_{yz,x} = -2\omega_{z,z} \tag{6.26}$$

(see also (1.52), (1.53c)), which may be written in levels as

$$\gamma_{xz,y}^{(k)} - \gamma_{yz,x}^{(k)} = -2(k+1)\omega_z^{(k+1)}. \tag{6.27}$$

Writing (6.24), (6.26) using (6.4) leads to the two conditions that represent three of the compatibility equations (1.45a, d, e) as

$$\begin{aligned}(b_{11}\sigma_x^{(k)} + \cdots + b_{16}\tau_{xy}^{(k)})_{,yy} + (b_{12}\sigma_x^{(k)} + \cdots + b_{26}\tau_{xy}^{(k)})_{,xx}\\ -(b_{16}\sigma_x^{(k)} + \cdots + b_{66}\tau_{xy}^{(k)})_{,xy} = \frac{a_{36}}{a_{33}}\varepsilon_{z,xy}^{(k)} - \frac{a_{13}}{a_{33}}\varepsilon_{z,yy}^{(k)} - \frac{a_{23}}{a_{33}}\varepsilon_{z,xx}^{(k)},\end{aligned} \tag{6.28a}$$

$$(b_{15}\sigma_x^{(k)} + \cdots + b_{56}\tau_{xy}^{(k)})_{,y} - (b_{14}\sigma_x^{(k)} + \cdots + b_{46}\tau_{xy}^{(k)})_{,x} = \frac{a_{34}}{a_{33}}\varepsilon_{z,x}^{(k)} - \frac{a_{35}}{a_{33}}\varepsilon_{z,y}^{(k)} - 2(k+1)\omega_z^{(k+1)}. \tag{6.28b}$$

Applying (1.58) of Remark 1.2 in levels, we obtain

$$\omega_z^{(k)} = \omega_z^{(k)}(0,0) - \int_0^x \varepsilon_{x,y}^{(k)}\,dx + \int_0^y \varepsilon_{y,x}^{(k)}(0,y)\,dy + \frac{1}{2}\gamma_{xy}^{(k)} - \gamma_{xy}^{(k)}(0,y) + \frac{1}{2}\gamma_{xy}^{(k)}(0,0). \tag{6.29}$$

The *third* group of equations contains the remaining set of three compatibility equations, (1.45b,c,f), which are written in levels as

$$\frac{1}{k+1}\varepsilon_{z,yy}^{(k)} = \gamma_{yz,y}^{(k+1)} - (k+2)\varepsilon_y^{(k+2)}, \tag{6.30a}$$

$$\frac{1}{k+1}\varepsilon_{z,xx}^{(k)} = \gamma_{xz,x}^{(k+1)} - (k+2)\varepsilon_x^{(k+2)}, \tag{6.30b}$$

$$\frac{2}{k+1}\varepsilon_{z,xy}^{(k)} = \gamma_{xz,y}^{(k+1)} + \gamma_{yz,x}^{(k+1)} - (k+2)\gamma_{xy}^{(k+2)}. \tag{6.30c}$$

As shown, the terms on the r.h.s. of (6.30a–c) belong to higher levels, and therefore, for the moment, we shall consider them here as known. These equations will be used to determine ε_z within the next section.

6.1.3 Axial Strain Integration

The derivation of ε_z is carried out by two integration steps.

In the *first step*, we determine $\varepsilon^{(k)}_{z,x}$ and $\varepsilon^{(k)}_{z,y}$ by integrating (6.30a,b) with respect to y and x, respectively. The integrability condition required for the above operations may be put as

$$(\varepsilon^{(k)}_{z,xx})_{,yy}+(\varepsilon^{(k)}_{z,yy})_{,xx}-2(\varepsilon^{(k)}_{z,xy})_{,xy}=0. \tag{6.31}$$

As a corollary of the fact that the complete set of the compatibility equations is both essential and sufficient to assure integrability of the above equations, the reader may verify that substituting (6.30a–c) in (6.31), results in (6.24) (with indices $k+2$ instead of k). Hence, formally, the compatibility condition (6.24) ensures the integrability of (6.30a–c).

In the *second step*, we integrate the system $\varepsilon^{(k)}_{z,x}$ and $\varepsilon^{(k)}_{z,y}$ to determine $\varepsilon^{(k)}_{z}$.

Thus, we first integrate (6.30b) with respect to x, and (6.30a) with respect to y under the conditions

$$\varepsilon^{(k)}_{z,x}(0,0)=-\kappa^{(k)}_1, \qquad \varepsilon^{(k)}_{z,y}(0,0)=-\kappa^{(k)}_2, \tag{6.32}$$

and obtain

$$\frac{\varepsilon^{(k)}_{z,x}}{k+1}=-\frac{\kappa^{(k)}_1}{k+1}+(k+2)\Theta^{(k+1)}_1(y)+\gamma^{(k+1)}_{xz}-\gamma^{(k+1)}_{xz}(0,0)-(k+2)\int_0^x \varepsilon^{(k+2)}_x\,dx, \tag{6.33a}$$

$$\frac{\varepsilon^{(k)}_{z,y}}{k+1}=Sym\Big(\frac{\varepsilon^{(k)}_{z,x}}{k+1}\Big). \tag{6.33b}$$

Subsequently, the above equations should be solved with

$$\varepsilon^{(k)}_{z}(0,0)=\varepsilon^{(k)}_{0}. \tag{6.34}$$

The constants of integration $\kappa^{(k)}_1$, $\kappa^{(k)}_2$, $\varepsilon^{(k)}_0$, $k\geq 0$, that appear in (6.32)–(6.34) will be shown to be the components of the *beam axis curvatures* and the *beam axis axial strain*. To this end, these functions of z are defined as

$$\kappa_i(z)=\sum\nolimits_{k=0}^{K+2}\kappa^{(k)}_i z^k, \quad i=1,2, \qquad \varepsilon_0(z)=\sum\nolimits_{k=0}^{K+1}\varepsilon^{(k)}_0 z^k. \tag{6.35}$$

In (6.33a,b) $\Theta^{(k+1)}_i$ $(i=1,2)$ are functions of one variable, the meaning of which will become clearer later on. Note that $\Theta^{(k+1)}_i$ are assumed to vanish for $y=0$, and $x=0$, respectively.

The integrability condition for (6.33a,b) may be written as $(\varepsilon^{(k)}_{z,x})_{,y}=(\varepsilon^{(k)}_{z,y})_{,x}$, or

$$\gamma^{(k)}_{xz,y}-\gamma^{(k)}_{yz,x}=(k+1)\Big(\frac{d}{dx}\Theta^{(k)}_2-\frac{d}{dy}\Theta^{(k)}_1+\int_0^x \varepsilon^{(k+1)}_{x,y}\,dx-\int_0^y \varepsilon^{(k+1)}_{y,x}\,dy\Big) \tag{6.36}$$

where $k>0$ ($k=0$ will be discussed later on within S.6.2.4).

Comparing the r.h.s. of (6.36) with that of (6.27) shows that

$$\frac{d}{dx}\Theta^{(k)}_2-\frac{d}{dy}\Theta^{(k)}_1=-2\omega^{(k+1)}_z+\int_0^y \varepsilon^{(k+1)}_{y,x}\,dy-\int_0^x \varepsilon^{(k+1)}_{x,y}\,dx. \tag{6.37}$$

On the other hand, calculating the quantity $(\varepsilon^{(k)}_{z,x})_{,y}+(\varepsilon^{(k)}_{z,y})_{,x}$ from the r.h.s. of (6.33a,b) and equating it to $2\varepsilon^{(k)}_{z,xy}$ of (6.30c), for $k>0$, leads to

$$\frac{d}{dy}\Theta^{(k)}_1+\frac{d}{dx}\Theta^{(k)}_2=\int_0^x \varepsilon^{(k+1)}_{x,y}\,dx+\int_0^y \varepsilon^{(k+1)}_{y,x}\,dy-\gamma^{(k+1)}_{xy}. \tag{6.38}$$

Hence, from (6.37), (6.38) we obtain

$$\frac{d}{dy}\Theta_1^{(k)} = \omega_z^{(k+1)} + \int_0^x \varepsilon_{x,y}^{(k+1)}\,dx - \frac{1}{2}\gamma_{xy}^{(k+1)}, \qquad \frac{d}{dx}\Theta_2^{(k)} = Sym(\frac{d}{dy}\Theta_1^{(k)}). \tag{6.39}$$

Integrating and assuming $x = 0$ and $y = 0$ in (6.39:a, 6.39:b), respectively, yields for $k > 0$,

$$\Theta_1^{(k)} = \int_0^y (\omega_z^{(k+1)} - \frac{1}{2}\gamma_{xy}^{(k+1)})_{|x=0}\,dy, \qquad \Theta_2^{(k)} = Sym(\Theta_1^{(k)}). \tag{6.40}$$

We now wish to present the axial strain component as

$$\varepsilon_z^{(k)} = -\kappa_1^{(k)}x - \kappa_2^{(k)}y + \varepsilon_0^{(k)} + \bar{\varepsilon}_z{}^{(k)}, \tag{6.41}$$

where the function $\bar{\varepsilon}_z^{(k)}$ is expressed from (6.33a,b), by selecting the polygonal integration trajectory $(0,0) \to (0,y) \to (x,y)$, as

$$\bar{\varepsilon}_z^{(k)} = (k+1)\{(k+2)[\Theta_1^{(k+1)}x - \int_0^x (\int_0^x \varepsilon_x^{(k+2)}dx)dx - \int_0^y (\int_0^y \varepsilon_y^{(k+2)}(0,y)dy)dy]$$
$$+ \int_0^x \gamma_{xz}^{(k+1)}\,dx - \gamma_{xz}^{(k+1)}(0,0)x + \int_0^y \gamma_{yz}^{(k+1)}(0,y)\,dy - \gamma_{yz}^{(k+1)}(0,0)y\}. \tag{6.42}$$

Note that $\bar{\varepsilon}_z{}^{(k)}(0,0) = \bar{\varepsilon}_{z,x}^{(k)}(0,0) = \bar{\varepsilon}_{z,y}^{(k)}(0,0) = 0$, and that the functions ε_z, $\bar{\varepsilon}_z$ are

$$\varepsilon_z = \sum\nolimits_{k=0}^{K+2} \varepsilon_z{}^{(k)}(x,y)z^k, \qquad \bar{\varepsilon}_z = \sum\nolimits_{k=0}^{K+1} \bar{\varepsilon}_z{}^{(k)}(x,y)z^k. \tag{6.43}$$

Remark 6.3 The auxiliary functions Θ_i discussed in (6.40) are written as

$$\Theta_1(y,z) = \sum\nolimits_{k=0}^{K+2} \Theta_1^{(k)}(y)z^k, \qquad \Theta_2(x,z) = \sum\nolimits_{k=0}^{K+2} \Theta_2^{(k)}(x)z^k. \tag{6.44}$$

From (6.40) we deduce $\Theta_1(0,z) = \Theta_2(0,z) = 0$ and

$$\Theta_1 = \frac{1}{z}\int_0^y (\omega_z - \frac{1}{2}\gamma_{xy})_{|x=0}\,dy, \qquad \Theta_2 = Sym(\Theta_1) = -\frac{1}{z}\int_0^x (\omega_z + \frac{1}{2}\gamma_{xy})_{|y=0}\,dx. \tag{6.45}$$

At this stage, we use the notion of the *twist angle*, $\theta(z) = \omega_{z0,z}(z)$, see Chapter 5, which may also be defined by its components as

$$\theta(z) = \sum\nolimits_{k=0}^{K+2} \theta^{(k)}z^k, \qquad \theta^{(k)} = (k+1)\omega_z^{(k+1)}(0,0). \tag{6.46}$$

In addition, from (6.37) with $x = y = 0$ one obtains the following level and continuous presentations of $\theta(z)$:

$$\frac{2}{k+1}\theta^{(k)} = \frac{d}{dy}\Theta_1^{(k)}(0) - \frac{d}{dx}\Theta_2^{(k)}(0) \;\Leftrightarrow\; 2\int_0^z \theta(z)\,dz = z[\Theta_{1,y}(0,z) - \Theta_{2,x}(0,z)]. \tag{6.47}$$

6.1.4 Stress Functions and the Coupled-Plane BVP

The $k^{\underline{th}}$ -order level of the differential equilibrium equations, (1.82a–c), may be presented by carrying out the differentiation with respect to z (see (1.61:b)) as

$$\sigma_{x,x}^{(k)} + \tau_{xy,y}^{(k)} + (k+1)\tau_{xz}^{(k+1)} + X_b^{(k)} = 0, \tag{6.48a}$$
$$\tau_{xy,x}^{(k)} + \sigma_{y,y}^{(k)} + (k+1)\tau_{yz}^{(k+1)} + Y_b^{(k)} = 0, \tag{6.48b}$$
$$\tau_{xz,x}^{(k)} + \tau_{yz,y}^{(k)} + (k+1)\sigma_z^{(k+1)} + Z_b^{(k)} = 0. \tag{6.48c}$$

In addition, the *generalized body-forces potential functions* are defined as

$$\overline{U}_1 = -\int_0^x X_b\,dx = \sum_{k=0}^{K} \overline{U}_1^{(k)}(x,y)z^k \qquad \overline{U}_1^{(k)} = -\int_0^x [(k+1)\tau_{xz}^{(k+1)} + X_b^{(k)}]\,dx, \tag{6.49a}$$

$$\overline{U}_2 = -\int_0^y Y_b\,dy = \sum_{k=0}^{K} \overline{U}_2^{(k)}(x,y)z^k, \qquad \overline{U}_2^{(k)} = -\int_0^y [(k+1)\tau_{yz}^{(k+1)} + Y_b^{(k)}]\,dy, \tag{6.49b}$$

and $\overline{U}_i = \sum_{k=0}^{K+1} \overline{U}_i^{(k)}(x,y)z^k$ $(i = 3,4)$, where

$$\overline{U}_{4,y}^{(k)} + \overline{U}_{3,x}^{(k)} = -(k+1)\sigma_z^{(k+1)} - Z_b^{(k)} \tag{6.50}$$

while for $k = K+1$, (6.8a) shows that

$$\overline{U}_{4,y}^{(K+1)} + \overline{U}_{3,x}^{(K+1)} = -\frac{P_1^{(K+1)}}{I_y}x - \frac{P_2^{(K+1)}}{I_x}y. \tag{6.51}$$

Actually, $U_3^{(k)}, U_4^{(k)}$ are particular solutions of (6.50), (6.51). The reader should note that many particular solutions may be selected here, and the exact choice is usually founded on convenience considerations that are related to the cross-section geometry. With the above definitions, (6.48a–c) become

$$(\sigma_x^{(k)} - \overline{U}_1^{(k)})_{,x} + \tau_{xy,y}^{(k)} = 0, \tag{6.52a}$$

$$\tau_{xy,x}^{(k)} + (\sigma_y^{(k)} - \overline{U}_2^{(k)})_{,y} = 0, \tag{6.52b}$$

$$(\tau_{xz}^{(k)} - \overline{U}_3^{(k)})_{,x} + (\tau_{yz}^{(k)} - \overline{U}_4^{(k)})_{,y} = 0. \tag{6.52c}$$

We also adopt the definition of the stress function Φ of (3.54) and the stress function ψ of (3.92) (with $\tau_{yz}^P = \tau_{xz}^P = 0$), namely,

$$\tau_{yz} = -\psi_{,x} + \overline{U}_4, \qquad \tau_{xz} = \psi_{,y} + \overline{U}_3, \tag{6.53a}$$

$$\sigma_x = \Phi_{,yy} + \overline{U}_1, \qquad \sigma_y = \Phi_{,xx} + \overline{U}_2, \qquad \tau_{xy} = -\Phi_{,xy}. \tag{6.53b}$$

We therefore express these functions in levels as

$$\Phi = \sum_{k=0}^{K+1} \Phi^{(k)}(x,y)z^k, \qquad \psi = \sum_{k=0}^{K+1} \psi^{(k)}(x,y)z^k, \tag{6.54}$$

and thus, the stress components become

$$\sigma_x^{(k)} = \Phi_{,yy}^{(k)} + \underline{\overline{U}_1^{(k)}}, \qquad \sigma_y^{(k)} = \Phi_{,xx}^{(k)} + \underline{\overline{U}_2^{(k)}}, \qquad \tau_{xy}^{(k)} = -\Phi_{,xy}^{(k)},$$

$$\tau_{yz}^{(k)} = -\psi_{,x}^{(k)} + \overline{U}_4^{(k)}, \qquad \tau_{xz}^{(k)} = \psi_{,y}^{(k)} + \overline{U}_3^{(k)}, \tag{6.55}$$

where the underlined terms vanish for $k = K+1$. With the above stress functions definition, the equilibrium equations (6.48a–c) are satisfied identically. Once the stress functions are determined, the stress components may also be calculated, and subsequently, (6.3) shows that the components of $\sigma_z^{(k)}$ take the form

$$\sigma_z^{(k)} = \frac{1}{a_{33}}(\varepsilon_z^{(k)} - a_{13}\sigma_x^{(k)} - a_{23}\sigma_y^{(k)} - a_{43}\tau_{yz}^{(k)} - a_{53}\tau_{xz}^{(k)} - a_{36}\tau_{xy}^{(k)}), \tag{6.56}$$

where $\varepsilon_z^{(k)}$ are given by (6.41). Using the relations of (6.55) and the differential operators definitions of S.3.6, one may write the compatibility conditions of (6.28a,b) for $k \le K+1$ in a compact form as

$$\nabla_1^{(4)}\Phi^{(k)} + \nabla_1^{(3)}\psi^{(k)} = g^{(k)}(x,y), \tag{6.57a}$$

$$\nabla_1^{(3)}\Phi^{(k)} + \nabla_1^{(2)}\psi^{(k)} = f_0^{(k)} + f^{(k)}(x,y). \tag{6.57b}$$

The above $f^{(k)}(x,y)$, $g^{(k)}(x,y)$ functions and the constant $f_0^{(k)}$ are

$$f_0^{(k)} = -2\theta^{(k)} - \underline{\frac{1}{a_{33}}(a_{34}\kappa_1^{(k)} - a_{35}\kappa_2^{(k)})}, \tag{6.58a}$$

$$f^{(k)} = \underline{\frac{a_{34}}{a_{33}}\bar{\varepsilon}_{z,x}^{(k)} - \frac{a_{35}}{a_{33}}\bar{\varepsilon}_{z,y}^{(k)} - 2(k+1)\bar{\omega}_z^{(k+1)}} + b_{14}\overline{U}_{1,x}^{(k)} + b_{24}\overline{U}_{2,x}^{(k)} - b_{15}\overline{U}_{1,y}^{(k)} - b_{25}\overline{U}_{2,y}^{(k)}$$
$$+ \underline{b_{45}\overline{U}_{3,x}^{(k)}} + \underline{b_{44}\overline{U}_{4,x}^{(k)}} - \underline{b_{55}\overline{U}_{3,y}^{(k)}} - \underline{b_{45}\overline{U}_{4,y}^{(k)}}, \tag{6.58b}$$

$$g^{(k)} = \underline{\frac{1}{a_{33}}(a_{36}\bar{\varepsilon}_{z,xy}^{(k)} - a_{13}\bar{\varepsilon}_{z,yy}^{(k)} - a_{23}\bar{\varepsilon}_{z,xx}^{(k)})} - b_{11}\overline{U}_{1,yy}^{(k)} - b_{12}\overline{U}_{2,yy}^{(k)} - b_{22}\overline{U}_{2,xx}^{(k)} + b_{16}\overline{U}_{1,xy}^{(k)} + b_{26}\overline{U}_{2,xy}^{(k)}$$
$$- b_{12}\overline{U}_{1,xx}^{(k)} - \underline{b_{15}\overline{U}_{3,yy}^{(k)} - b_{14}\overline{U}_{4,yy}^{(k)}} - \underline{b_{25}\overline{U}_{3,xx}^{(k)} - b_{24}\overline{U}_{4,xx}^{(k)}} + \underline{b_{56}\overline{U}_{3,xy}^{(k)} + b_{46}\overline{U}_{4,xy}^{(k)}}, \tag{6.58c}$$

where $\bar{\omega}_z = \omega_z - \omega_{0z}$. Note that *only* the underlined terms are used for $k = K+1$, while for $k = K+2$ the entire expressions vanish. Note that $f^{(k)}(0,0) = 0$ when $\mathbf{F}_b = 0$.

The general solution $\Phi^{(k)}$ and $\psi^{(k)}$ of (6.57a,b) should be founded on the following boundary conditions, see (3.64), (3.65), (3.100):

$$\frac{d}{ds}\{\Phi_{,y}^{(k)}, \Phi_{,x}^{(k)}, \psi^{(k)}\} = \{-F_1^{(k)}, F_2^{(k)}, F_3^{(k)}\} \quad \text{on} \quad \partial\Omega, \tag{6.59}$$

where

$$\begin{aligned} F_1^{(k)} &= -X_s^{(k)} + \overline{U}_1^{(k)}\cos(\bar{\mathbf{n}},x), \\ F_2^{(k)} &= -Y_s^{(k)} + \overline{U}_2^{(k)}\cos(\bar{\mathbf{n}},y), \\ F_3^{(k)} &= Z_s^{(k)} - \overline{U}_3^{(k)}\cos(\bar{\mathbf{n}},x). \end{aligned} \tag{6.60}$$

To show that the four single-valued conditions of (3.144) for the functions $\Phi^{(k)}$, $\Phi_{,x}^{(k)}$, $\Phi_{,y}^{(k)}$ and $\psi^{(k)}$, namely,

$$\oint_{\partial\Omega}\{F_1^{(k)}, F_2^{(k)}, F_3^{(k)}, F_2^{(k)}y - F_1^{(k)}x\} = \{0,0,0,0\}, \tag{6.61}$$

are always satisfied. we employ Green's Theorem and (6.9), (6.15a,b), (6.49a,b), and verify the above conditions as follows:

$$-\oint_{\partial\Omega}F_1^{(k)} = \oint_{\partial\Omega}[X_s^{(k)} - \overline{U}_1^{(k)}\cos(\bar{\mathbf{n}},x)] = \oint_{\partial\Omega}X_s^{(k)} - \iint_\Omega \overline{U}_{1,x}^{(k)}$$
$$= \oint_{\partial\Omega}X_s^{(k)} + \iint_\Omega X_b^{(k)} + (k+1)\iint_\Omega \tau_{xz}^{(k+1)} = (k+1)(-P_1^{(k+1)} + \iint_\Omega \tau_{xz}^{(k+1)}) = 0, \tag{6.62a}$$

$$\oint_{\partial\Omega}F_2^{(k)} = Sym(\oint_{\partial\Omega}F_1^{(k)}) = 0, \tag{6.62b}$$

$$\oint_{\partial\Omega}F_3^{(k)} = \oint_{\partial\Omega}[Z_s^{(k)} - \overline{U}_3^{(k)}\cos(\bar{\mathbf{n}},x) - \overline{U}_4^{(k)}\cos(\bar{\mathbf{n}},y)] = \oint_{\partial\Omega}Z_s^{(k)} - \iint_\Omega[\overline{U}_{3,x}^{(k)} + \overline{U}_{4,y}^{(k)}]$$
$$= \oint_{\partial\Omega}Z_s^{(k)} + \iint_\Omega Z_b^{(k)} + (k+1)\iint_\Omega \sigma_z^{(k+1)} = (k+1)(-P_3^{(k+1)} + \iint_\Omega \sigma_z^{(k+1)}) = 0, \tag{6.62c}$$

$$\oint_{\partial\Omega}F_2^{(k)}y - F_1^{(k)}x = \oint_{\partial\Omega}(yX_s^{(k)} - xY_s^{(k)}) + \iint_\Omega(-y\overline{U}_{1,x}^{(k)} + x\overline{U}_{2,y}^{(k)})$$
$$= \oint_{\partial\Omega}(yX_s^{(k)} - xY_s^{(k)}) + \iint_\Omega(yX_b^{(k)} - xY_b^{(k)}) + (k+1)\iint_\Omega(y\tau_{xz}^{(k+1)} - x\tau_{yz}^{(k+1)})$$
$$= (k+1)[M_3^{(k+1)} - \iint_\Omega(x\tau_{yz}^{(k+1)} - y\tau_{xz}^{(k+1)})] = 0. \tag{6.62d}$$

Thus, the systems of (6.57a,b), (6.59) constitute a consistent Coupled-Plane BVP for which all solution methodologies described in Chapter 4 may be employed. For the sake of simplicity, and due to the degrees of the operators in field and boundary conditions, one may assume

$$\{\Phi^{(k)}, \Phi^{(k)}_{,x}, \Phi^{(k)}_{,y}, \psi^{(k)}\}(0,0) = \{0,0,0,0\}. \tag{6.63}$$

As shown by (6.58a), $f_0^{(k)}$ is a linear function of $\theta^{(k)}$, which is in general, an unknown parameter. We therefore solve (6.57a,b) twice. In "case I" we set $\theta^{(k)} = 0$ and in "case II" we set $f_0^{(k)} = 1$ and $f^{(k)} = g^{(k)} = 0$. We also denote the resulting stress quantities (which are obtained from (6.55)) by upper indices I and II, respectively. The stress components therefore become a linear combination of the above two solutions, for example,

$$\tau_{yz}^{(k)} = \tau_{yz}^{I\,(k)} - 2\theta^{(k)}\tau_{yz}^{II\,(k)}, \qquad \tau_{xz}^{(k)} = \tau_{xz}^{I\,(k)} - 2\theta^{(k)}\tau_{xz}^{II\,(k)}. \tag{6.64}$$

Further on, (6.9) shows that

$$M_3^{(k)} = \iint_\Omega (x\tau_{yz}^{(k)} - y\tau_{xz}^{(k)}) = M_3^{I\,(k)} - 2\theta^{(k)} M_3^{II\,(k)}. \tag{6.65}$$

Hence, for each level we may first calculate the torsional moment components $M_3^{I\,(k)}$ and $M_3^{II\,(k)}$. Then, the value of $\theta^{(k)}$ is determined as

$$\theta^{(k)} = \frac{M_3^{I\,(k)} - M_3^{(k)}}{2M_3^{II\,(k)}}. \tag{6.66}$$

Remark 6.4 The general solution of (6.57a,b) may be expanded as

$$\Phi^{(k)} = \Phi_0^{(k)} + \Phi_p^{(k)}, \qquad \psi^{(k)} = \psi_0^{(k)} + \psi_p^{(k)}. \tag{6.67}$$

Here, $\Phi_0^{(k)}$, $\psi_0^{(k)}$ are the general solutions of the homogeneous system (6.57a,b) and $\Phi_p^{(k)}$, $\psi_p^{(k)}$ are particular solutions of the non-homogeneous system. In cases where the expressions on the r.h.s. of the above system are simple enough (e.g. polynomials of x, y, or a composition of other elementary functions), the determination of the particular solutions should not be a difficult task. Once suitable particular solutions are obtained, the boundary conditions (6.59) take the following equivalent form for $\Phi_0^{(k)}$ and $\psi_0^{(k)}$:

$$\frac{d}{ds}\Phi_{0,y}^{(k)} = X_s^{(k)} - \overline{U}_1^{(k)}\cos(\bar{\mathbf{n}},x) - \frac{d}{ds}\Phi_{p,y}^{(k)} \quad \text{on} \quad \partial\Omega, \tag{6.68a}$$

$$\frac{d}{ds}\Phi_{0,x}^{(k)} = -Y_s^{(k)} + \overline{U}_2^{(k)}\cos(\bar{\mathbf{n}},y) - \frac{d}{ds}\Phi_{p,x}^{(k)} \quad \text{on} \quad \partial\Omega, \tag{6.68b}$$

$$\frac{d}{ds}\psi_0^{(k)} = Z_s^{(k)} - \overline{U}_3^{(k)}\cos(\bar{\mathbf{n}},x) - \overline{U}_4^{(k)}\cos(\bar{\mathbf{n}},y) - \frac{d}{ds}\psi_p^{(k)} \quad \text{on} \quad \partial\Omega. \tag{6.68c}$$

6.2 Displacements and Rotations

At this stage we wish to determine the displacement components by integrating the strain components along the lines of the general approach of S.1.2 but with the special expressions obtained in this chapter.

6.2.1 Continuous Expressions

We first integrate the axial strain definition $\varepsilon_z = w_{,z}$, namely,

$$w = \int_0^z \varepsilon_z \, dz + w^{(0)}, \tag{6.69}$$

where $w^{(0)}(x,y)$ is a function to be determined that includes rigid body displacements and rotations. Differentiation of the above expression yields

$$w_{,x} = \int_0^z \varepsilon_{z,x} \, dz + w_{,x}^{(0)}, \qquad w_{,y} = \int_0^z \varepsilon_{z,y} \, dz + w_{,y}^{(0)}, \tag{6.70}$$

which are subsequently substituted in the strain definitions, $\gamma_{xz} = u_{,z} + w_{,x}$ and $\gamma_{yz} = v_{,z} + w_{,y}$, to produce the following derivatives:

$$u_{,z} = \gamma_{xz} - \int_0^z \varepsilon_{z,x} \, dz - w_{,x}^{(0)}, \qquad v_{,z} = \gamma_{yz} - \int_0^z \varepsilon_{z,y} \, dz - w_{,y}^{(0)}. \tag{6.71}$$

Integrating the above with respect to z enables us to derive u and v as

$$u = -\int_0^z \Big(\int_0^z \varepsilon_{z,x} \, dz\Big) dz + \int_0^z \gamma_{xz} \, dz - w_{,x}^{(0)} z + u^{(0)}, \qquad v = Sym(u) \tag{6.72}$$

where $u^{(0)}(x,y)$ and $v^{(0)}(x,y)$ are functions to be determined that also include rigid body displacements and rotations.

At this stage, it is convenient to introduce three *root warping functions* U, V and W of x, y. These functions represent the cross-section deformation up to a rigid body constant at the root (i.e., $z = 0$), and are defined by their relations with $u^{(0)}, v^{(0)}$ and $w^{(0)}$, respectively, as

$$u^{(0)} = U - \omega_z^0 y + u^0, \qquad v^{(0)} = V + \omega_z^0 x + v^0, \qquad w^{(0)} = W + \omega_x^0 y - \omega_y^0 x + w^0 \tag{6.73}$$

where $\omega_x^0, \omega_y^0, \omega_z^0$ are the components of the rigid rotation about the coordinate axes, x, y, z, respectively, and u^0, v^0, w^0 are the components of the rigid body displacements along the respective coordinate axes. Then, using (6.73), we rewrite (6.69), (6.72) as

$$u = \int_0^z \Big(\int_0^z \kappa_1 \, dz\Big) dz - \int_0^z \Big(\int_0^z \bar{\varepsilon}_{z,x} \, dz\Big) dz + \int_0^z \gamma_{xz} \, dz - W_{,x} z + U + u_r, \qquad v = Sym(u), \tag{6.74a}$$

$$w = \int_0^z (-\kappa_1 x - \kappa_2 y + \varepsilon_0 + \bar{\varepsilon}_z) \, dz + W + w_r. \tag{6.74b}$$

Here, u_r, v_r and w_r are rigid body displacements, see Chapter 1,

$$u_r = \omega_y^0 z - \omega_z^0 y + u^0, \qquad v_r = \omega_z^0 x - \omega_x^0 + v^0, \qquad w_r = \omega_x^0 y - \omega_y^0 x + w^0, \tag{6.75}$$

which are all determined by the geometrical boundary conditions at the root.

Analogously, one may present the rotations, see (1.19), (1.54) and (6.29), as

$$\omega_x = -\int_0^z \kappa_2 \, dz + \int_0^z \bar{\varepsilon}_{z,y} \, dz - \frac{1}{2}\gamma_{yz} + W_{,y} + \omega_x^0, \qquad \omega_y = -Sym(\omega_x), \tag{6.76a}$$

$$\omega_z = \int_0^z \theta \, dz - \int_0^x \varepsilon_{x,y} \, dx + \int_0^y \varepsilon_{y,x}(0,y) \, dy + \frac{1}{2}\gamma_{xy} - \gamma_{xy}(0,y) + \frac{1}{2}\gamma_{xy}(0,0) + \omega_z^0. \tag{6.76b}$$

6.2.2 Level Expressions

Based on (6.2b), (6.6), (6.69), (6.72), and (1.19), the displacements and rotations may be expanded as

$$u=\sum_{k=0}^{K+4} u^{(k)} z^k, \qquad v=\sum_{k=0}^{K+4} v^{(k)} z^k, \qquad w=\sum_{k=0}^{K+3} w^{(k)} z^k, \tag{6.77a}$$

$$\omega_i=\sum_{k=0}^{K+3} \omega_i^{(k)} z^k, \quad i\in\{x,y,z\}, \qquad \bar{\omega}_z=\sum_{k=0}^{K+2} \bar{\omega}_z^{(k)} z^k, \tag{6.77b}$$

where $u^{(k)}, v^{(k)}, w^{(k)}$ and $\omega_i^{(k)}$, $\bar{\omega}_z^{(k)}$ are functions of x, y.

6.2.2.1 General Level

The $k^{\underline{th}}$ component of the displacements u, v for $k>1$ and w for $k>0$ are extracted from (6.74a,b) as

$$u^{(k)}=\frac{1}{k(k-1)}(\kappa_1^{(k-2)}-\varepsilon_{z,x}^{(k-2)})+\frac{1}{k}\gamma_{xz}^{(k-1)}, \qquad v^{(k)}=Sym(u^{(k)}),$$
$$w^{(k)}=\frac{1}{k}(-\kappa_1^{(k-1)}x-\kappa_2^{(k-1)}y+\varepsilon_0^{(k-1)}+\bar{\varepsilon}_z^{(k-1)}). \tag{6.78}$$

The components of the rotations for $k>0$ are extracted from (6.76a,b) as

$$\omega_x^{(k)}=-\frac{1}{k}\kappa_2^{(k-1)}-\frac{1}{2}\gamma_{yz}^{(k)}+\frac{1}{k}\bar{\varepsilon}_{z,y}^{(k-1)}, \quad \omega_y^{(k)}=-Sym(\omega_x^{(k)}), \quad \omega_z^{(k)}=\frac{2\theta^{(k-1)}}{k+1}+\bar{\omega}_z^{(k)}. \tag{6.79}$$

6.2.2.2 High Levels

Using (6.6), (6.32), (6.78), the upper displacement and rotation components become

$$u^{(K+4)}=-\frac{a_{33}P_1^{(K+1)}}{(K+2)(K+3)(K+4)I_y}, \qquad v^{(K+4)}=Sym(u^{(K+4)}), \tag{6.80a}$$

$$w^{(K+3)}=\frac{a_{33}}{(K+2)(K+3)}(\frac{P_1^{(K+1)}}{I_y}x+\frac{P_2^{(K+1)}}{I_x}y), \tag{6.80b}$$

$$\omega_x^{(K+3)}=\frac{a_{33}P_2^{(K+1)}}{(K+2)(K+3)I_x}, \qquad \omega_y^{(K+3)}=-Sym(\omega_x^{(K+3)}), \tag{6.80c}$$

$$\omega_z^{(K+3)}=\frac{1}{2(K+2)(K+3)}(a_{34}\frac{P_1^{(K+1)}}{I_y}-a_{35}\frac{P_2^{(K+1)}}{I_x}). \tag{6.80d}$$

The remaining high levels of displacements are obtained directly from (6.78) as

$$u^{(K+3)}=\frac{1}{K+3}[\frac{1}{K+2}\kappa_1^{(K+1)}-y\theta^{(K+2)}], \qquad v^{(K+3)}=Sym(u^{(K+3)}), \tag{6.81a}$$

$$u^{(K+2)}=\frac{1}{K+2}[\frac{1}{K+1}\kappa_1^{(K)}-y\theta^{(K+1)}+\gamma_{xz}^{(K+1)}(0,0) \tag{6.81b}$$
$$+\frac{P_1^{(K+1)}}{2I_y}(a_{13}x^2-a_{23}y^2)+\frac{P_2^{(K+1)}}{2I_x}(a_{36}y^2+2a_{13}xy)],$$
$$v^{(K+2)}=Sym(u^{(K+2)}).$$

Note that $w^{(K+2)}$ is explicitly presented using expressions for $\kappa_1^{(K+1)}, \kappa_2^{(K+1)}, \varepsilon_0^{(K+1)}$ and $\bar{\varepsilon}_z^{(K+1)}$, given later on by (6.106), (6.107).

Similarly, the remaining high levels of rotations are obtained directly from (6.76a,b) as

$$\omega_x^{(K+2)} = -\frac{\kappa_2^{(K+1)}}{K+2} - \frac{x}{2}\theta^{(K+2)} + \frac{P_1^{(K+1)}}{2(K+2)I_y}a_{34}x + \frac{P_2^{(K+1)}}{2(K+2)I_x}(2a_{34}y + a_{35}x), \tag{6.82a}$$

$$\omega_y^{(K+2)} = -Sym(\omega_x^{(K+2)}), \tag{6.82b}$$

$$\omega_z^{(K+2)} = \frac{1}{K+2}[\theta^{(K+1)} + \frac{P_1^{(K+1)}}{2I_y}(2a_{23}y + a_{36}x) - \frac{P_2^{(K+1)}}{2I_x}(2a_{13}x + a_{36}y)]. \tag{6.82c}$$

6.2.2.3 Low Levels

For $k = 1$ we use (6.74a) to obtain

$$u^{(1)} = \gamma_{xz}^{(0)} - W_{,x} + \omega_y^0, \qquad v^{(1)} = Sym(u^{(1)}). \tag{6.83}$$

For $u^{(0)}, v^{(0)}, w^{(0)}$ see (6.73), while the rotations for $k = 0$ are obtained from (6.76a,b) as

$$\omega_x^{(0)} = -\frac{1}{2}\gamma_{yz}^{(0)} + W_{,y} + \omega_x^0, \qquad \omega_y^{(0)} = -Sym(\omega_x^{(0)}) \tag{6.84}$$

while $\omega_z^{(0)}$ is given by (6.29). The root warping functions, U, V, W, are derived within S.6.2.4.

6.2.3 Axis Deformation

We shall now derive some corollaries from the above results for the beam axis deformation.

6.2.3.1 Continuous Expressions

First, from (6.74a,b), by replacing the derivatives of W with those of (6.97c) that will be derived below, we determine the z-axis displacement components $u_0(z)$, $v_0(z)$, $w_0(z)$, as

$$u_0 = \int_0^z (\int_0^z \kappa_1(z)\,dz)\,dz + \int_0^z \gamma_{xz}(0,0,z)\,dz - \frac{1}{2}\gamma_{xz}(0,0,0)z + \omega_y^0 z + u^0, \tag{6.85a}$$

$$v_0 = Sym(u_0), \qquad w_0 = \int_0^z \varepsilon_0(z)\,dz + w^0. \tag{6.85b}$$

Also, in the expressions for the rotation components ω_{i0} along the beam axis, (6.76a,b), we replace the derivatives of W by those of (6.97c) and write:

$$\omega_{x0} = -\int_0^z \kappa_2(z)\,dz - \frac{1}{2}[\gamma_{yz}(0,0,z) - \gamma_{yz}(0,0,0)] + \omega_x^0, \tag{6.86a}$$

$$\omega_{y0} = -Sym(\omega_{x0}), \qquad \omega_{z0} = \int_0^z \theta(z)\,dz + \omega_z^0. \tag{6.86b}$$

6.2.3.2 Level Expressions

For the $k^{\underline{th}}$ level, where $k > 1$ for u and v, and $k > 0$ for w, from (6.78) one may write

$$u_0^{(k)} = \frac{\kappa_1^{(k-2)}}{k(k-1)} + \frac{1}{k}\gamma_{xz}^{(k-1)}(0,0), \qquad v_0^{(k)} = Sym(u_0^{(k)}), \qquad w_0^{(k)} = \frac{1}{k}\varepsilon_0^{(k-1)}, \tag{6.87}$$

while for $k = 1$ from (6.83) one may extract

$$u_0^{(1)} = \frac{1}{2}\gamma_{xz}^{(0)}(0,0) + \omega_y^0, \qquad v_0^{(1)} = Sym(u_0^{(1)}). \tag{6.88}$$

In view of (6.86a,b), the initial conditions for the above rotation components for $k > 0$ are

$$\omega_{x0}^{(k)} = -\frac{\kappa_2^{(k-1)}}{k} - \frac{1}{2}\gamma_{yz}^{(k)}(0,0), \qquad \omega_{y0}^{(k)} = -Sym(\omega_{x0}^{(k)}), \qquad \omega_{z0}^{(k)} = \frac{\theta^{(k-1)}}{k}. \tag{6.89}$$

For $k = 0$ the rigid body displacement components, u^0, v^0, w^0, and the rigid body rotation components, ω_x^0, ω_y^0, ω_z^0, are specified as

$$u_0^{(0)} = u^0, \qquad v_0^{(0)} = v^0, \qquad w_0^{(0)} = w^0, \tag{6.90a}$$

$$\omega_{x0}^{(0)} = \omega_x^0, \qquad \omega_{y0}^{(0)} = \omega_y^0, \qquad \omega_{z0}^{(0)} = \omega_z^0. \tag{6.90b}$$

6.2.3.3 Curvature and Extension

Differentiating u_0, v_0 form (6.85a,b) yields the following expressions for axis curvature $\kappa_1(z)$ and $\kappa_2(z)$:

$$\kappa_1 = u_{0,zz}(z) - \gamma_{xz,z}(0,0,z), \qquad \kappa_2 = Sym(\kappa_1). \tag{6.91}$$

Note that in the absence of γ_{yz} and γ_{xz} (e.g. the case of $K = 0$ treated by (Lekhnitskii, 1981)), κ_1 and κ_2 coincide with the edgewise curvature κ_u and the beamwise curvature κ_v of (5.6), respectively.

From (6.3), (6.9) and the identity $\varepsilon_z = \bar{\varepsilon}_z - \kappa_1 x - \kappa_2 y + \varepsilon_0$ (see (6.41)) one may deduce the following equations that relate the curvature and the axial strain functions $\kappa_1(z), \kappa_2(z)$, $\varepsilon_0(z)$ with the resultant loads and moments:

$$M_1 = \iint_\Omega \frac{y}{a_{33}}(\bar{\varepsilon}_z - \kappa_1 x - \kappa_2 y + \varepsilon_0 - a_{13}\sigma_x - a_{23}\sigma_y - a_{34}\tau_{yz} - a_{35}\tau_{xz} - a_{36}\tau_{xy}), \tag{6.92a}$$

$$M_2 = Sym(M_1), \tag{6.92b}$$

$$P_3 = \iint_\Omega \frac{1}{a_{33}}(\bar{\varepsilon}_z - \kappa_1 x - \kappa_2 y + \varepsilon_0 - a_{13}\sigma_x - a_{23}\sigma_y - a_{34}\tau_{yz} - a_{35}\tau_{xz} - a_{36}\tau_{xy}). \tag{6.92c}$$

Applying (5.1), one may equivalently write

$$-\kappa_1 I_y = a_{33}M_2 + \iint_\Omega (a_{13}\sigma_x + a_{23}\sigma_y + a_{34}\tau_{yz} + a_{35}\tau_{xz} + a_{36}\tau_{xy} - \bar{\varepsilon}_z)x, \tag{6.93a}$$

$$\kappa_2 I_x = Sym(\kappa_1 I_y), \tag{6.93b}$$

$$\varepsilon_0 S_\Omega = a_{33}P_3 + \iint_\Omega (a_{13}\sigma_x + a_{23}\sigma_y + a_{34}\tau_{yz} + a_{35}\tau_{xz} + a_{36}\tau_{xy} - \bar{\varepsilon}_z). \tag{6.93c}$$

Using the integration identities from S.6.5, (6.93a–c) may be expressed as

$$\begin{aligned} -\kappa_1 I_y = {} & a_{33}M_2 + \frac{a_{34}}{2}M_3 - \iint_\Omega \bar{\varepsilon}_z x + \iint_\Omega [\frac{1}{2}(a_{13}x^2 - a_{23}y^2)(X_b + \tau_{xz,z}) \\ & + (a_{23}xy + \frac{a_{36}}{2}x^2)(Y_b + \tau_{yz,z}) + \frac{1}{2}(a_{34}xy + a_{35}x^2)(Z_b + \sigma_{z,z})] \\ & + \oint_{\partial\Omega} [\frac{1}{2}(a_{13}x^2 - a_{23}y^2)X_s + (a_{23}xy + \frac{a_{36}}{2}x^2)Y_s + \frac{1}{2}(a_{34}xy + a_{35}x^2)Z_s], \end{aligned} \tag{6.94a}$$

$$\kappa_2 I_x = Sym(\kappa_1 I_y), \tag{6.94b}$$

$$\begin{aligned} \varepsilon_0 S_\Omega = {} & a_{33}P_3 + \oint_{\partial\Omega} [(a_{13}x + \frac{a_{36}}{2}y)X_s + (\frac{a_{36}}{2}x + a_{23}y)Y_s + (a_{35}x + a_{34}y)Z_s] - \iint_\Omega \bar{\varepsilon}_z \\ & + \iint_\Omega [(a_{13}x + \frac{a_{36}}{2}y)(X_b + \tau_{xz,z}) + (\frac{a_{36}}{2}x + a_{23}y)(Y_b + \tau_{yz,z}) + (a_{35}x + a_{34}y)(Z_b + \sigma_{z,z})], \end{aligned} \tag{6.94c}$$

or, equivalently, by their components, as

$$-\kappa_1^{(k)} I_y = a_{33} M_2^{(k)} + \frac{a_{34}}{2} M_3^{(k)} - \iint_\Omega \bar{\varepsilon}_z{}^{(k)} x + \iint_\Omega [\frac{1}{2}(a_{13}x^2 - a_{23}y^2)(X_b^{(k)} + (k+1)\tau_{xz}^{(k+1)})$$
$$+ (a_{23}xy + \frac{a_{36}}{2}x^2)(Y_b^{(k)} + (k+1)\tau_{yz}^{(k+1)}) + \frac{1}{2}(a_{34}xy + a_{35}x^2)(Z_b^{(k)} + (k+1)\sigma_z^{(k+1)})]$$
$$+ \oint_{\partial\Omega} [\frac{1}{2}(a_{13}x^2 - a_{23}y^2) X_s^{(k)} + (a_{23}xy + \frac{a_{36}}{2}x^2) Y_s^{(k)} + \frac{1}{2}(a_{34}xy + a_{35}x^2) Z_s^{(k)}], \quad (6.95a)$$

$$\kappa_2^{(k)} I_x = Sym(\kappa_1^{(k)} I_y), \quad (6.95b)$$

$$\varepsilon_0^{(k)} S_\Omega = a_{33} P_3^{(k)} - \iint_\Omega \bar{\varepsilon}_z{}^{(k)} + \iint_\Omega [(a_{13}x + \frac{a_{36}}{2}y)(X_b^{(k)} + (k+1)\tau_{xz}^{(k+1)})$$
$$+ (\frac{a_{36}}{2}x + a_{23}y)(Y_b^{(k)} + (k+1)\tau_{yz}^{(k+1)}) + (a_{35}x + a_{34}y)(Z_b^{(k)} + (k+1)\sigma_z^{(k+1)})]$$
$$+ \oint_{\partial\Omega} [(a_{13}x + \frac{a_{36}}{2}y) X_s^{(k)} + (\frac{a_{36}}{2}x + a_{23}y) Y_s^{(k)} + (a_{35}x + a_{34}y) Z_s^{(k)}]. \quad (6.95c)$$

6.2.4 Root Warping Integration

From (6.73), (6.90a) it is clear that

$$U(0,0) = 0, \qquad V(0,0) = 0, \qquad W(0,0) = 0, \quad (6.96)$$

while (6.73) and (1.67c) for $k = 0$ show that

$$U_{,x}(0,0) = \varepsilon_x^{(0)}(0,0), \qquad U_{,y}(0,0) = \frac{1}{2}\gamma_{xy}^{(0)}(0,0), \quad (6.97a)$$

$$V_{,x}(0,0) = \frac{1}{2}\gamma_{xy}^{(0)}(0,0), \qquad V_{,y}(0,0) = \varepsilon_y^{(0)}(0,0), \quad (6.97b)$$

$$W_{,x}(0,0) = \frac{1}{2}\gamma_{xz}^{(0)}(0,0), \qquad W_{,y}(0,0) = \frac{1}{2}\gamma_{yz}^{(0)}(0,0). \quad (6.97c)$$

Integrating two equations of (1.69a) for $k = 1$ with respect to x and y, respectively, yields

$$u^{(1)} = \int_0^x \varepsilon_x^{(1)}\, dx - \Theta_1^{(0)}(y) + u_0^{(1)}, \qquad v^{(1)} = Sym(u^{(1)}). \quad (6.98)$$

From (6.98), using (6.83), (6.88) the equations required for the derivation of W are obtained as

$$W_{,x} = \Theta_1^{(0)}(y) + \gamma_{xz}^{(0)} - \int_0^x \varepsilon_x^{(1)}\, dx - \frac{1}{2}\gamma_{xz}^{(0)}(0,0), \qquad W_{,y} = Sym(W_{,x}) \quad (6.99)$$

which should be solved under the initial boundary condition $W(0,0) = 0$, see (6.96). In this case, the requirement $(W_{,x})_{,y} = (W_{,y})_{,x}$ serves as the integrability condition, and it is easy to see that it becomes identical to (6.36) for $k = 0$. Hence, the definitions of the functions $\Theta_i^{(0)}$, $(i = 1,2)$, appear in (6.40) for $k = 0$ (previously they were defined only for $k > 0$).

Using (6.73), we rewrite (1.69a) for $k = 0$, in the form

$$U_{,x} = \varepsilon_x^{(0)}, \qquad V_{,y} = \varepsilon_y^{(0)}, \qquad U_{,y} + V_{,x} = \gamma_{xy}^{(0)}. \quad (6.100)$$

Equation (6.24) for $k = 0$ ensures integrability of (6.100). We subsequently find the functions $U(x,y)$ and $V(x,y)$ directly from the following two systems, see (1.67c), where the rotation component $\omega_z^{(0)}$ is given by (6.29) for $k = 0$:

$$U_{,x} = \varepsilon_x^{(0)}, \quad U_{,y} = \frac{1}{2}\gamma_{xy}^{(0)} - \omega_z^{(0)} + \omega_z^0, \qquad V_{,x} = Sym(U_{,y}), \quad V_{,y} = \varepsilon_y^{(0)}. \quad (6.101)$$

The above systems should be solved under the initial conditions $U(0,0) = 0$ and $V(0,0) = 0$, see (6.96).

6.3 Recurrence Solution Scheme

As already indicated, the overall solution for axially non-uniform loading of an anisotropic beam is based on a recurrence scheme. As shown in Fig. 6.1, the procedure starts at the highest order and handles all other levels in a decreasing order down to $k = 0$ where each solution depends on data obtained in higher solutions.

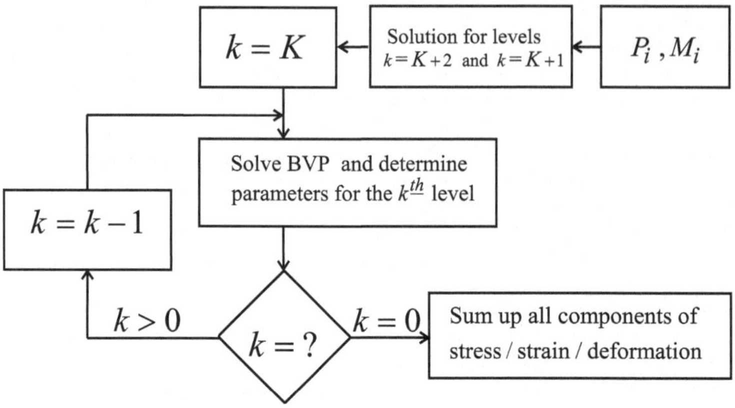

Figure 6.1: Recurrence solution scheme for axially non-uniform loading of anisotropic beam.

6.3.1 Solution Steps

The *solution scheme* may be described by the following steps, see Table 6.1, where (*) indicates that the equation may be broken down to its level components as is.

	$k=K+2$	$k=K+1$	$0 \le k \le K$
$\overline{\varepsilon}_z$	0	(6.106)	(6.42)
$\Theta_1 \leftrightarrow \Theta_2$	(6.103)	(6.105)	(6.40)
$\kappa_1 \leftrightarrow \kappa_2, \varepsilon_0$	(6.102)	(6.107)	(6.95a-c)
ε_z	(6.8b)	(6.41)	(6.41)
$\overline{U}_1 \leftrightarrow \overline{U}_2$	0	0	(6.49a,b)
$\overline{U}_3, \overline{U}_4$	0	(6.51)	(6.50)
f_0, f, g	0	(6.109a–c)	(6.58a–c)
Φ, ψ	0	(6.57a,b), (6.110a, b)	(6.57a, b), (6.59)
$\sigma_x \leftrightarrow \sigma_y$ $\tau_{yz} \leftrightarrow \tau_{xz}, \tau_{xy}$	0	(6.55)	(6.55)
σ_z	(6.8a)	(6.56)	(6.56)
$\varepsilon_x \leftrightarrow \varepsilon_y$ $\gamma_{yz} \leftrightarrow \gamma_{xz}, \gamma_{xy}$	(6.8b)	(6.4) (*)	(6.4) (*)
$\theta \leftrightarrow -\theta$	(6.104)	(6.66)	(6.66)

Table 6.1: Equations used to evaluate level components of various parameters.

(**1**) From the initial data, calculate the tip constants $M_i^{(k)}$, $P_i^{(k)}$ — see summary or relevant equations in Table 6.2.

	$k=K+2$	$k=K+1$	$1\leq k\leq K$	$k=0$
$X_s \leftrightarrow Y_s, Z_s$	0	0	Data	Data
$X_b \leftrightarrow Y_b, Z_b$	0	0	Data	Data
$P_1 \leftrightarrow P_2, P_3$	0	(6.15a)	(6.15a)	Data, see (6.17) and Remark 6.1
$M_1 \leftrightarrow M_2$	(6.16)	(6.15b)	(6.15b)	Data, see (6.17) and Remark 6.1
$M_3 \leftrightarrow -M_3$	0	(6.15b)	(6.15b)	Data, see (6.17) and Remark 6.1

Table 6.2: The loading components.

(**2**) Calculate all parameters of level $k=K+2$ as shown in S.6.3.2.1.

(**3**) Calculate all parameters of level $k=K+1$ as shown in S.6.3.2.2.

(**4**) For all levels $0\leq k\leq K$:
(**a**) Calculate $\bar{\varepsilon}_z{}^{(k)}$ from (6.42).
(**b**) Calculate $\kappa_i^{(k)}, \varepsilon_0^{(k)}$ from (6.95a-c).
(**c**) Calculate $\varepsilon_z^{(k)}$ from (6.41).
(**d**) Calculate $\overline{U}_1^{(k)}$, $\overline{U}_2^{(k)}$ from (6.49a,b).
(**e**) Calculate $\overline{U}_3^{(k)}$, $\overline{U}_4^{(k)}$ from (6.50).
(**f**) Calculate $f_0^{(k)}$, $f^{(k)}$, $g^{(k)}$, from (6.58a–c).
(**g**) Solve the Coupled-Plane BVP of (6.57a,b) under the boundary conditions (6.59) and initial conditions (6.63)
including the parameter $\theta^{(k)}$ of (6.66). This step yields $\Phi^{(k)}$ and $\psi^{(k)}$.
(**h**) Calculate the stress components $\sigma_x^{(k)}$, $\sigma_y^{(k)}$, $\tau_{yz}^{(k)}$, $\tau_{xz}^{(k)}$ and $\tau_{xy}^{(k)}$ by (6.55).
Calculate $\sigma_z^{(k)}$ by (6.56).
(**j**) Calculate the strain components $\varepsilon_x^{(k)}$, $\varepsilon_y^{(k)}$, $\gamma_{yz}^{(k)}$, $\gamma_{xz}^{(k)}$ and $\gamma_{xy}^{(k)}$ by (6.4).
(**k**) Calculate $\Theta_i^{(k)}$ from (6.40).

(**5**) Determine the root warping functions:
(**a**) Calculate $\omega_z^{(0)}$ from (6.29) with $k=0$.
(**b**) Solve (6.101) for U and V, and (6.99) for W.

(**6**) Evaluate all relevant functions from their level components.

	$k=K+4$	$k=K+3$	$k=K+2$	$K+1\geq k\geq 2$	1	0
$u \leftrightarrow v$	(6.80a)	(6.81a)	(6.81b)	(6.78)	(6.83)	(6.73)
w	0	(6.80b)	(6.78)	(6.78)	(6.78)	(6.73)
$\omega_x \leftrightarrow -\omega_y$	0	(6.80c)	(6.82a,b)	(6.79)	(6.79)	(6.84)
$\omega_z \leftrightarrow -\omega_z$	0	(6.80d)	(6.82c)	(6.29)	(6.29)	(6.29)

Table 6.3: Equations used to evaluate deformation level components.

(**7**) Determine displacements and rotations from (6.74a,b), (6.76a,b) or by their level components, see (6.77a) and Table 6.3. In this case, level contribution should be collected (summed up) for $k = 0,\dots,K+4$.

6.3.2 *Expressions for the High Solution Levels*

In this section, we shall document detailed expressions for the high solution levels $k = K+2$ and $k = K+1$.

6.3.2.1 Level $K+2$ Solution

The following equations summarize the quantities that should be determined in level $K+2$. Refer also to the "$K+2$" column of Table 6.1.

First, $\varepsilon_0^{(K+2)}$, $\kappa_1^{(K+2)}$ and $\kappa_2^{(K+2)}$ are derived from (6.95a-c) (for $k = K+2$) as

$$\varepsilon_0^{(K+2)} = 0, \qquad \kappa_1^{(K+2)} = -a_{33}\frac{P_1^{(K+1)}}{(K+2)I_y}, \qquad \kappa_2^{(K+2)} = -a_{33}\frac{P_2^{(K+1)}}{(K+2)I_x}. \tag{6.102}$$

Recall that $\sigma_z^{(K+2)}$, $\varepsilon_i^{(K+2)}$ are given by (6.6), (6.8a,b) and $\varepsilon_i^{(K+2)}(0,0) = 0 \quad (i = 1,\dots,6)$. Then, (6.40) with $k = K+2$ show that

$$\Theta_1^{(K+2)} = \frac{\theta^{(K+2)}}{K+3}y, \qquad \Theta_2^{(K+2)} = -\frac{\theta^{(K+2)}}{K+3}x. \tag{6.103}$$

Equations (6.58b,c) show that $f^{(K+2)} = g^{(K+2)} = 0$ since the involved components vanish for $k > K+1$. Since $\Phi^{(K+2)} = \psi^{(K+2)} = 0$, (6.57b) shows that $f_0^{(K+2)} = 0$, and hence

$$\theta^{(K+2)} = -\frac{a_{34}}{2a_{33}}\kappa_1^{(K+2)} + \frac{a_{35}}{2a_{33}}\kappa_2^{(K+2)} = \frac{1}{K+2}(a_{34}\frac{P_1^{(K+1)}}{2I_y} - a_{35}\frac{P_2^{(K+1)}}{2I_x}). \tag{6.104}$$

6.3.2.2 Level $K+1$ Solution

We document here the quantities required in level $K+1$. Refer also to the "$K+1$" column of Table 6.1. From (6.40) follow

$$\Theta_1^{(K+1)} = \frac{1}{K+2}[\theta^{(K+1)}y + (a_{23}\frac{P_1^{(K+1)}}{I_y} - a_{36}\frac{P_2^{(K+1)}}{I_x})\frac{y^2}{2}], \quad \Theta_2^{(K+1)} = Sym(\Theta_1^{(K+1)}). \tag{6.105}$$

Then, from (6.42) follows

$$\bar{\varepsilon}_z^{(K+1)} = (a_{35}x + a_{34}y)(\frac{P_1^{(K+1)}}{2I_y}x + \frac{P_2^{(K+1)}}{2I_x}y). \tag{6.106}$$

From (6.95a-c) using (6.106) we obtain

$$\kappa_1^{(K+1)} = -a_{33}\frac{M_2^{(K+1)}}{I_y} - a_{34}\frac{M_3^{(K+1)}}{2I_y} + (a_{35}I_{30} + a_{34}I_{21})\frac{P_1^{(K+1)}}{2(I_y)^2} + (a_{35}I_{21} + a_{34}I_{12})\frac{P_2^{(K+1)}}{2I_xI_y},$$

$$\kappa_2^{(K+1)} = Sym(\kappa_1^{(K+1)}), \qquad \varepsilon_0^{(K+1)} = a_{33}\frac{P_3^{(K+1)}}{S_\Omega} + a_{35}\frac{P_1^{(K+1)}}{2S_\Omega} + a_{34}\frac{P_2^{(K+1)}}{2S_\Omega}, \tag{6.107}$$

where the quantities I_{ij} are defined as

$$\{I_{30}, I_{21}, I_{12}, I_{03}\} = \iint_{\Omega} \{x^3, x^2y, xy^2, y^3\}. \tag{6.108}$$

Note that in cases where both x and y are the symmetry axes of the cross-section the above integrals vanish. Also, one may calculate $\varepsilon_z^{(K+1)}$ from (6.41).

Next, (6.57a,b) for $k = K+1$ should be solved with the r.h.s. functions, (see (6.58a–c))

$$\begin{aligned} f_0^{(K+1)} &= -2\theta^{(K+1)} + a_{34}\frac{M_2^{(K+1)}}{I_y} - a_{35}\frac{M_1^{(K+1)}}{I_x} + (\frac{a_{34}^2}{I_y} + \frac{a_{35}^2}{I_x})\frac{M_3^{(K+1)}}{2a_{33}} \\ &+ \frac{1}{a_{33}}\{a_{35}[(a_{35}I_{21} + a_{34}I_{12})\frac{P_1^{(K+1)}}{2I_xI_y} + (a_{35}I_{12} + a_{34}I_{03})\frac{P_2^{(K+1)}}{2(I_x)^2}] \\ &- a_{34}[(a_{35}I_{30} + a_{34}I_{21})\frac{P_1^{(K+1)}}{2(I_y)^2} + (a_{35}I_{21} + a_{34}I_{12})\frac{P_2^{(K+1)}}{2I_xI_y}]\}, \end{aligned} \tag{6.109a}$$

$$\begin{aligned} f^{(K+1)} &= [(\frac{a_{34}^2}{a_{33}} - 4a_{23})y + (\frac{a_{34}a_{35}}{a_{33}} - 2a_{36})x]\frac{P_1^{(K+1)}}{2I_y} \\ &+ [(2a_{36} - \frac{a_{34}a_{35}}{a_{33}})y + (4a_{13} - \frac{a_{35}^2}{a_{33}})x]\frac{P_2^{(K+1)}}{2I_x} \\ &+ (b_{45}\overline{U}_3^{(K+1)} + b_{44}\overline{U}_4^{(K+1)})_{,x} - (b_{55}\overline{U}_3^{(K+1)} + b_{45}\overline{U}_4^{(K+1)})_{,y}, \end{aligned} \tag{6.109b}$$

$$\begin{aligned} g^{(K+1)} &= \frac{a_{36}a_{34} - 2a_{23}a_{35}}{a_{33}} \cdot \frac{P_1^{(K+1)}}{2I_y} + \frac{a_{36}a_{35} - 2a_{13}a_{34}}{a_{33}} \cdot \frac{P_2^{(K+1)}}{2I_x} \\ &- (b_{15}\overline{U}_3^{(K+1)} + b_{14}\overline{U}_4^{(K+1)})_{,yy} - (b_{25}\overline{U}_3^{(K+1)} + b_{24}\overline{U}_4^{(K+1)})_{,xx} \\ &+ (b_{56}\overline{U}_3^{(K+1)} + b_{46}\overline{U}_4^{(K+1)})_{,xy}. \end{aligned} \tag{6.109c}$$

The boundary conditions associated with $\Phi^{(K+1)}, \psi^{(K+1)}$ are given by (6.59) as

$$\frac{d}{ds}\{\Phi_{,y}^{(K+1)}, \Phi_{,x}^{(K+1)}\} = \{0, 0\} \qquad \text{on} \quad \partial\Omega, \tag{6.110a}$$

$$\frac{d}{ds}\psi^{(K+1)} = -\overline{U}_3^{(K+1)}\cos(\bar{\mathbf{n}}, x) - \overline{U}_4^{(K+1)}\cos(\bar{\mathbf{n}}, y) \qquad \text{on} \quad \partial\Omega. \tag{6.110b}$$

Note that in the case of a simply connected domain, Ω, the potentials $\overline{U}_3^{(K+1)}, \overline{U}_4^{(K+1)}$ can be chosen so as to make the r.h.s. of (6.110b) zero on the contour. Also, for an elliptical cross-section, one may find closed form expressions for $\Phi^{(K+1)}$, $\psi^{(K+1)}$, see Examples 6.1, 6.2 below.

Once $\Phi^{(K+1)}$ and $\psi^{(K+1)}$ are determined, the stress components $\sigma_x^{(K+1)}$, $\sigma_y^{(K+1)}$, $\tau_{yz}^{(K+1)}$, $\tau_{xz}^{(K+1)}$ and $\tau_{xy}^{(K+1)}$ are derived by (6.55). This requires two solutions of the system, which also provides the value of the twist angle component $\theta^{(K+1)}$, see (6.66) and the related discussion. Subsequently, $\sigma_z^{(K+1)}$ is evaluated by (6.56), and based on (6.4) the strain components $\varepsilon_x^{(K+1)}$, $\varepsilon_y^{(K+1)}$, $\gamma_{yz}^{(K+1)}$, $\gamma_{xz}^{(K+1)}$ and $\gamma_{xy}^{(K+1)}$ are also determined.

Remark 6.5 In the case of MON13z material (3.214) shows that $\nabla_1^{(3)} = 0$ and (6.109a) yields

$$f_0^{(K+1)} = -2\theta^{(K+1)}, \qquad g^{(K+1)} = 0, \tag{6.111a}$$

$$\begin{aligned} f^{(K+1)} &= -(2a_{23}y + a_{36}x)\frac{P_1^{(K+1)}}{I_y} + (2a_{13}x + a_{36}y)\frac{P_2^{(K+1)}}{I_x} \\ &+ (b_{45}\overline{U}_3^{(K+1)} + b_{44}\overline{U}_4^{(K+1)})_{,x} - (b_{55}\overline{U}_3^{(K+1)} + b_{45}\overline{U}_4^{(K+1)})_{,y}. \end{aligned} \tag{6.111b}$$

In such a case, and for vanishing in-plane loading, the biharmonic function $\Phi^{(K+1)}$ is zero, and the problem is reduced to the determination of the generalized harmonic function $\psi^{(K+1)}$. Moreover, (6.111a,b) show that one may represent $\psi^{(K+1)}$ as a linear combination of three generalized harmonic functions: one function for the influence of $P_1^{(K+1)}$, a second function for the influence of $P_2^{(K+1)}$, and a third function for the influence of $\theta^{(K+1)}$ and $\overline{U}_3^{(K+1)}, \overline{U}_4^{(K+1)}$. These functions are "similar", but not equal, to the bending stress functions χ_1, χ_2 and the out-of-plane warping function φ, derived for MON13z material in S.7.2.1, S.7.2.2.1, S.7.2.2.2, respectively. The above term "similar" stands for the fact that the definition of relations between the stress functions χ_1, χ_2 and φ and the stress components in Chapter 7 is different from the one used in the present analysis, but essentially, the problems are equivalent.

6.4 Applications

P.6.1 demonstrates a complete analytical solution for a beam of general anisotropy and elliptical cross-sections. This program may therefore be used for many educating purposes as the involved Coupled-Plane BVPs have relatively simple solution in this case.

6.4.1 *Tip Moments and Axial Force*

For the problems in this class no body or surface loads are applied, namely, $X_b = Y_b = Z_b = 0$ and $X_s = Y_s = Z_s = 0$, and only the tip loads $M_1^{(0)}$, $M_2^{(0)}$, $M_3^{(0)}$ and $P_3^{(0)}$ are accounted for. Hence, these problems are formally characterized by $K = -1$, where all quantities of level $k = K+2 = 1$ vanish, and only the parameters of level $k = K+1 = 0$ are considered. For the sake of convenience, in this section, we shall omit the upper index (0). Also, in the cases under discussion, (6.51) shows that $\overline{U}_{4,y} = \overline{U}_{3,x} = 0$ may be conveniently assumed.

6.4.1.1 Bending Moments

Consider first the case of $M_i \neq 0,\ i = 1,2$ only. The derivation for $k = K+1 = 0$ in S.6.3.2.2 yields:

$$\bar{\varepsilon}_z = 0, \quad \Theta_1 = \theta y, \quad \Theta_2 = -\theta x, \quad \kappa_1 = -a_{33}\frac{M_2}{I_y}, \quad \kappa_2 = -a_{33}\frac{M_1}{I_x}, \quad \varepsilon_0 = 0,$$

$$\varepsilon_z = \Big(\frac{M_1}{I_x}y + \frac{M_2}{I_y}x\Big)a_{33}, \qquad \sigma_z = \frac{M_1}{I_x}y + \frac{M_2}{I_y}x. \tag{6.112}$$

In view of (6.109a–c) we have $f = g = 0$, and f_0 is presented as a sum of -2θ and a constant, see (6.109a). Since $\overline{U}_i = 0$, the boundary conditions for the stress functions vanish as well. Suppose now that $f_0 = 0$. Equation (6.109a) then gives

$$\theta = \frac{M_2}{2I_y}a_{34} - \frac{M_1}{2I_x}a_{35}. \tag{6.113}$$

Hence, the stress functions Φ, ψ vanish as well and subsequently $\sigma_x = \sigma_y = \tau_{xz} = \tau_{yz} = \tau_{xy} = 0$, and (6.4) shows that the strain components are

$$\varepsilon_i = \Big(\frac{M_1}{I_x}y + \frac{M_2}{I_y}x\Big)a_{i3}, \qquad i \in \{1,2,4,5,6\}. \tag{6.114}$$

Equations (6.99) show that the out-of-plane root warping derivatives are

$$W_{,x} = \theta y + (\frac{M_1}{I_x}y + \frac{M_2}{I_y}x)a_{35} = (\frac{M_2}{I_y}a_{34} + \frac{M_1}{I_x}a_{35})\frac{y}{2} + \frac{M_2}{I_y}a_{35}x, \quad W_{,y} = Sym(W_{,x}) \quad (6.115)$$

and their integration for the value of θ of (6.113) in view of $W(0,0) = 0$ yields

$$W = (\frac{M_2}{2I_y}a_{34} + \frac{M_1}{2I_x}a_{35})xy + \frac{M_2}{2I_y}a_{35}x^2 + \frac{M_1}{2I_x}a_{34}y^2. \quad (6.116)$$

Similarly, (6.101) show that the in-plane root warping derivatives are

$$U_{,x} = (\frac{M_1}{I_x}y + \frac{M_2}{I_y}x)a_{13}, \qquad U_{,y} = \frac{M_1}{I_x}a_{13}x + (\frac{M_1}{I_x}a_{36} - \frac{M_2}{I_y}a_{23})y, \quad (6.117)$$

and their integration in view of $U(0,0) = 0$ yields

$$U = \frac{M_1}{I_x}a_{13}xy + (\frac{M_1}{I_x}a_{36} - \frac{M_2}{I_y}a_{23})\frac{y^2}{2} + \frac{M_2}{2I_y}a_{13}x^2, \qquad V = Sym(U). \quad (6.118)$$

From the level expressions in S.6.2.2 or by the continuous expressions in S.6.2.1, the displacements become

$$u = \frac{M_1}{2I_x}(2a_{13}x + a_{36}y + a_{35}z)y + \frac{M_2}{2I_y}(a_{13}x^2 - a_{23}y^2 - a_{34}yz - a_{33}z^2) + u_r, \quad v = Sym(v),$$
$$w = (\frac{M_1}{2I_x}y + \frac{M_2}{2I_y}x)(a_{35}x + a_{34}y + 2a_{33}z) + w_r, \quad (6.119)$$

where u_r, v_r and w_r are rigid body displacements given in (6.75). The corresponding rotations are

$$\omega_x = \frac{M_1}{2I_x}(a_{35}x + a_{34}y + 2a_{33}z) + \omega_x^0, \qquad \omega_y = -Sym(\omega_x),$$
$$\omega_z = -\frac{M_1}{2I_x}(2a_{13}x + a_{36}y + a_{35}z) + \frac{M_2}{2I_y}(a_{36}x + 2a_{23}y + a_{34}z) + \omega_z^0. \quad (6.120)$$

The above results coincide with those of (Lekhnitskii, 1981).

6.4.1.2 Axial Force

In this case P_3 is the only nonzero loading component and the derivation in S.6.3.2.2 yields:

$$\kappa_1 = \kappa_2 = \bar{\varepsilon}_z = 0, \qquad \Theta_1 = \theta y, \qquad \Theta_2 = -\theta x,$$
$$\varepsilon_0 = \frac{P_3}{S_\Omega}a_{33}, \qquad \varepsilon_z = \frac{P_3}{S_\Omega}a_{33}, \qquad \sigma_z = \frac{P_3}{S_\Omega}. \quad (6.121)$$

Equations (6.58a–c) show that $f_0 = f = g = 0$, and subsequently, the stress functions $\Phi = \psi = 0$ and so the stress components σ_x, σ_y, τ_{xz}, τ_{yz} and τ_{xy} vanish. Equation (6.66) then shows that $\theta = 0$ as well and from (6.4) one obtains $\varepsilon_i = \frac{P_3}{S_\Omega}a_{i3}$ for $i = 1,2,4,5,6$. In addition, (6.99), (6.101) show that the root warping derivatives are

$$U_{,x} = \frac{P_3}{S_\Omega}a_{13}, \qquad U_{,y} = \frac{P_3}{2S_\Omega}a_{36}, \quad (6.122a)$$
$$W_{,x} = \frac{P_3}{S_\Omega}a_{35}, \qquad W_{,y} = \frac{P_3}{S_\Omega}a_{34}, \quad (6.122b)$$

and their integration in view of $U(0,0) = V(0,0) = W(0,0) = 0$ yields

$$U = \frac{P_3}{2S_\Omega}(2a_{13}x + a_{36}y), \qquad V = Sym(U), \qquad W = \frac{P_3}{S_\Omega}(a_{35}x + a_{34}y). \tag{6.123}$$

From the level expressions in S.6.2.2 or by the continuous expressions in S.6.2.1, the following linear expressions are obtained for the displacements:

$$u = \frac{P_3}{2S_\Omega}(2a_{13}x + a_{36}y + a_{35}z) + u_r, \;\; v = Sym(u), \;\; w = \frac{P_3}{S_\Omega}(a_{35}x + a_{34}y + a_{33}z) + w_r, \tag{6.124}$$

while the rotations are constants, i.e., $\omega_i = \omega_i^0$, $i = 1,2,3$. These results are in complete agreement with those presented in S.5.2.2 (for $\widetilde{a} \neq 0$), with the results presented in Remark 9.3 for a MON13y material, and also corresponds to the results of (Lekhnitskii, 1981).

6.4.1.3 Torsional Moment

In this case, only M_3 is applied, and the derivation in S.6.3.2.2 yields:

$$\begin{aligned} &\varepsilon_0 = \bar{\varepsilon}_z = 0, \qquad \Theta_1 = \theta y, \qquad \Theta_2 = -\theta x, \\ &\kappa_1 = -a_{34}\frac{M_3}{2I_y}, \qquad \kappa_2 = a_{35}\frac{M_3}{2I_x}, \qquad \varepsilon_z = a_{34}\frac{M_3}{2I_y}x - a_{35}\frac{M_3}{2I_x}y. \end{aligned} \tag{6.125}$$

For $k = 0$, (6.51) shows that $\overline{U}_3 = \overline{U}_4 = 0$. From (6.109a–c) with $k = K+1$ follow $f = g = 0$ and $f_0 = -2\theta + \frac{M_3}{2a_{33}}(\frac{a_{34}^2}{I_y} + \frac{a_{35}^2}{I_x})$. The boundary conditions, see (6.110a,b), take the form

$$\frac{d}{ds}\{\Phi_{,y}, \Phi_{,x}, \psi\} = \{0, 0, 0\} \qquad \text{on} \;\; \partial\Omega. \tag{6.126}$$

Once the above Coupled-Plane BVP is solved, the stress components may be derived from (6.55), (6.56). In this case, in view of the discussion related to (6.64), one may write

$$M_3^I = \left[(\frac{a_{34}^2}{I_y} + \frac{a_{35}^2}{I_x})\frac{M_3}{2a_{33}} \right] M_3^{II}. \tag{6.127}$$

Therefore, the twist angle, θ, is derived from (6.66) as

$$\theta = \frac{M_3}{2M_3^{II}} \left[(\frac{a_{34}^2}{I_y} + \frac{a_{35}^2}{I_x})\frac{1}{2a_{33}} M_3^{II} - 1 \right], \tag{6.128}$$

from which the definition of torsional rigidity $D = M_3/\theta$ becomes

$$D = \frac{2}{(\frac{a_{34}^2}{I_y} + \frac{a_{35}^2}{I_x})\frac{1}{2a_{33}} - \frac{1}{M_3^{II}}}. \tag{6.129}$$

The displacements u, v, w in the x, y, z directions, respectively, are obtained from the level expressions in S.6.2.2 or by the continuous expressions in S.6.2.1 as

$$\begin{aligned} &u = -\frac{M_3}{4I_y}a_{34}z^2 - \theta yz + \frac{1}{2}\gamma_{xz}(0,0)z + U + u_r, \qquad v = Sym(u), \\ &w = (\frac{M_3}{2I_y}a_{34}x + \frac{M_3}{2I_x}a_{35}y)z + W + w_r, \end{aligned} \tag{6.130}$$

while the rotation components are

$$\omega_x = \frac{M_3}{2I_x} a_{35} z + \frac{1}{2}(W_{,y} - \frac{1}{2}\gamma_{yz}(0,0)) - \frac{\theta}{2} x + \omega_x^0, \qquad \omega_y = -Sym(\omega_x),$$

$$\omega_z = \theta z + \frac{1}{2}(V_{,x} - U_{,y}) + \omega_z^0. \tag{6.131}$$

Example 6.1 *A Beam of Elliptical Cross-Section Under Tip Torsion.*

For the elliptical cross-section $\frac{x^2}{\widetilde{a}^2} + \frac{y^2}{\widetilde{b}^2} = 1$, where $I_x = \frac{1}{4}\pi \widetilde{a} \widetilde{b}^3$, $I_y = \frac{1}{4}\pi \widetilde{a}^3 \widetilde{b}$ are the cross-sectional principal moments (of inertia), we first note that $\widetilde{b} \leftrightarrow \widetilde{a}$ and

$$\kappa_1 = -a_{34} \frac{2M_3}{\pi \widetilde{a}^3 \widetilde{b}}, \qquad \kappa_2 = Sym(\kappa_1) = a_{35} \frac{2M_3}{\pi \widetilde{a} \widetilde{b}^3}. \tag{6.132}$$

Then, we establish a solution for case II (i.e. for $f_0 = 1$ and $f = g = 0$) in the form

$$\psi = \widetilde{A}(\frac{x^2}{\widetilde{a}^2} + \frac{y^2}{\widetilde{b}^2} - 1), \qquad \Phi = 0, \tag{6.133}$$

where $\widetilde{A}$ is a constant. Substituting in $\nabla_1^{(2)} \psi = 1$ ($\nabla_1^{(3)} \Phi = 0$ in this case) we obtain

$$\psi^{II} = \frac{1}{2(\frac{b_{44}}{\widetilde{a}^2} + \frac{b_{55}}{\widetilde{b}^2})}(\frac{x^2}{\widetilde{a}^2} + \frac{y^2}{\widetilde{b}^2} - 1), \tag{6.134}$$

which yields, see (7.29),

$$M_3^{II} = -\frac{\pi \widetilde{a}^3 \widetilde{b}^3}{2(b_{44} \widetilde{b}^2 + b_{55} \widetilde{a}^2)}. \tag{6.135}$$

Then, (6.128) shows that

$$\theta = \frac{M_3}{a_{33}} \frac{a_{35}^2 \widetilde{a}^2 + a_{34}^2 \widetilde{b}^2}{\pi \widetilde{a}^3 \widetilde{b}^3} + M_3 \underline{\frac{b_{44} \widetilde{b}^2 + b_{55} \widetilde{a}^2}{\pi \widetilde{a}^3 \widetilde{b}^3}}. \tag{6.136}$$

Comparing the above result with those of the models discussed in further chapters, we first note that for a MON13z material where $a_{34} = a_{35} = 0$, only the underlined term remains, and the value of torsional rigidity coincides with the one reported in (7.38), ($b_{44} = a_{44}$ and $b_{55} = a_{55}$ in such a case — see definition of b_{ij} in (6.5)). Moreover, using the notation of S.9.2.3.4 where $a_{34} = 0$ but $a_{35} \neq 0$ one may write

$$\frac{1}{D_{eq}} = \frac{1}{a_{33}} \frac{a_{35}^2 \widetilde{a}^2 + a_{34}^2 \widetilde{b}^2}{\pi \widetilde{a}^3 \widetilde{b}^3} + \frac{b_{44} \widetilde{b}^2 + b_{55} \widetilde{a}^2}{\pi \widetilde{a}^3 \widetilde{b}^3}, \qquad \frac{1}{D} = \frac{b_{44} \widetilde{b}^2 + b_{55} \widetilde{a}^2}{\pi \widetilde{a}^3 \widetilde{b}^3}, \tag{6.137}$$

which shows that (6.136) is in a complete agreement with (9.62) as well.

Since the relation between ψ^I and ψ^{II} is similar to the one written for the moments in (6.127), we finally obtain

$$\psi = -\frac{M_3}{\pi \widetilde{a} \widetilde{b}}(\frac{x^2}{\widetilde{a}^2} + \frac{y^2}{\widetilde{b}^2} - 1). \tag{6.138}$$

Hence, see (6.55),

$$\sigma_x = \sigma_y = \sigma_z = \tau_{xy} = 0, \qquad \tau_{xz} = -\frac{2M_3}{\pi \widetilde{a} \widetilde{b}^3} y, \qquad \tau_{yz} = \frac{2M_3}{\pi \widetilde{a}^3 \widetilde{b}} x, \tag{6.139}$$

and from (5.22), the strains are given by

$$\varepsilon_i = \frac{2M_3}{\pi\tilde{a}\tilde{b}}\left(\frac{a_{i4}}{\tilde{a}^2}x - \frac{a_{i5}}{\tilde{b}^2}y\right) \qquad (i=1,\ldots,6). \tag{6.140}$$

The out-of-plane warping derivatives of (6.99) for value of θ of (6.136) are

$$W_{,x} = \theta y + \gamma_{xz} = \frac{M_3}{\pi\tilde{a}\tilde{b}}\left(\frac{2a_{45}}{\tilde{a}^2}x + \frac{a_{44}}{\tilde{a}^2}y - \frac{a_{55}}{\tilde{b}^2}y\right), \qquad W_{,y} = Sym(W_{,x}), \tag{6.141}$$

which by integration with $W(0,0)=0$ yield

$$W = \frac{M_3}{\pi\tilde{a}\tilde{b}}\left[\left(\frac{a_{44}}{\tilde{a}^2} - \frac{a_{55}}{\tilde{b}^2}\right)xy - \frac{a_{45}}{\tilde{b}^2}y^2 + \frac{a_{45}}{\tilde{a}^2}x^2\right]. \tag{6.142}$$

In addition, the in-plane warping derivatives of (6.101) are

$$U_{,x} = \frac{2M_3}{\pi\tilde{a}\tilde{b}}\left(\frac{a_{14}}{\tilde{a}^2}x - \frac{a_{15}}{\tilde{b}^2}y\right), \qquad U_{,y} = -\frac{2M_3}{\pi\tilde{a}\tilde{b}}\left[\frac{a_{15}}{\tilde{b}^2}x + \left(\frac{a_{24}}{\tilde{a}^2} + \frac{a_{56}}{\tilde{b}^2}\right)y\right], \tag{6.143}$$

from which in view of $U(0,0)=0$ we find

$$U = \frac{M_3}{\pi\tilde{a}\tilde{b}}\left[\frac{a_{14}}{\tilde{a}^2}x^2 - \frac{2a_{15}}{\tilde{b}^2}xy - \left(\frac{a_{24}}{\tilde{a}^2} + \frac{a_{56}}{\tilde{b}^2}\right)y^2\right], \qquad V = Sym(U). \tag{6.144}$$

Hence, see S.6.2.1, S.6.2.2,

$$\begin{aligned} u &= \frac{M_3}{\pi\tilde{a}\tilde{b}}\left[-\frac{a_{34}}{\tilde{a}^2}z^2 - \left(\frac{a_{44}}{\tilde{a}^2} + \frac{a_{55}}{\tilde{b}^2}\right)yz + \frac{a_{14}}{\tilde{a}^2}x^2 - \frac{2a_{15}}{\tilde{b}^2}xy - \left(\frac{a_{24}}{\tilde{a}^2} + \frac{a_{56}}{\tilde{b}^2}\right)y^2\right] + u_r, \quad v = Sym(u), \\ w &= \frac{M_3}{\pi\tilde{a}\tilde{b}}\left[2\left(\frac{a_{34}}{\tilde{a}^2}x - \frac{a_{35}}{\tilde{b}^2}y\right)z + \left(\frac{a_{44}}{\tilde{a}^2} - \frac{a_{55}}{\tilde{b}^2}\right)xy - \frac{a_{45}}{\tilde{b}^2}y^2 + \frac{a_{45}}{\tilde{a}^2}x^2\right] + w_r, \end{aligned} \tag{6.145}$$

and

$$\begin{aligned} \omega_x &= -\frac{M_3}{\pi\tilde{a}\tilde{b}^3}(a_{55}x + a_{45}y + 2a_{35}z) + \omega_x^0, \qquad \omega_y = -Sym(\omega_x), \\ \omega_z &= \frac{M_3}{\pi\tilde{a}\tilde{b}}\left[\left(\frac{a_{46}}{\tilde{a}^2} + \frac{2a_{15}}{\tilde{b}^2}\right)x + \left(\frac{2a_{24}}{\tilde{a}^2} + \frac{a_{56}}{\tilde{b}^2}\right)y + \left(\frac{a_{44}}{\tilde{a}^2} + \frac{a_{55}}{\tilde{b}^2}\right)z\right] + \omega_z^0. \end{aligned} \tag{6.146}$$

The lateral deformation of the beam axis due to the applied torsional moment is therefore

$$u_0(z) = -\frac{M_3}{\pi\tilde{a}^3\tilde{b}}a_{34}z^2 + \omega_y^0 z + u^0, \qquad v_0(z) = Sym(u_0(z)). \tag{6.147}$$

Note that in this case, the edgewise and beamwise curvature functions (see (5.6)) coincide with κ_1, κ_2, respectively, and are in a complete agreement with the result of the simplified approach for a MON13y material, shown by (9.59).

6.4.2 Tip Bending Forces

Similar to the applications discussed in S.6.4.1, the problems in this class contain no body or surface loads, namely, $X_b = Y_b = Z_b = 0$ and $X_s = Y_s = Z_s = 0$ and only the tip loads $P_i = P_i^{(0)} \neq 0,\ i = 1,2$ are considered. Hence, these problems are formally characterized by $K=-1$ as well, however, not all components that are related to $K+2$ vanish. Thus, we have a two-steps solution for $k = K+1 = 0$ and for $k = K+2 = 1$. Note that $M_1^{(0)} = M_2^{(0)} = 0$, but as shown by Remark 6.1, for a cantilever of length l one should set $M_1^{(0)} = -P_2 l$ and $M_2^{(0)} = -P_1 l$.

6.4.2.1 Solution Levels

Level $K+2=1$: From (6.16) we first note that $M_1^{(1)}=P_2$, $M_2^{(1)}=P_1$. For this level, S.6.3.2.1 shows that

$$\varepsilon_0^{(1)}=0,\qquad \kappa_1^{(1)}=-a_{33}\frac{P_1}{I_2},\qquad \kappa_2^{(1)}=-a_{33}\frac{P_2}{I_1},$$
$$\sigma_z^{(1)}=\frac{P_1}{I_2}x+\frac{P_2}{I_1}y,\qquad \varepsilon_i^{(1)}=a_{i3}(\frac{P_1}{I_2}x+\frac{P_2}{I_1}y),\quad i=1,\dots,6,\tag{6.148}$$

and also

$$\theta^{(1)}=a_{34}\frac{P_1}{2I_2}-a_{35}\frac{P_2}{2I_1},$$
$$\Theta_1^{(1)}=\frac{1}{2}\theta^{(1)}y=(a_{34}\frac{P_1}{4I_2}-a_{35}\frac{P_2}{4I_1})y,\qquad \Theta_2^{(1)}=Sym(\Theta_1^{(1)}).\tag{6.149}$$

Level $K+1=0$: For this level (6.106) shows that

$$\overline{\varepsilon}_z^{(0)}=(a_{35}x+a_{34}y)(\frac{P_1}{2I_2}x+\frac{P_2}{2I_1}y)\tag{6.150}$$

and (6.105) yield

$$\Theta_1^{(0)}=\theta^{(0)}y+(a_{23}\frac{P_1}{2I_2}-a_{36}\frac{P_2}{2I_1})y^2,\qquad \Theta_2^{(0)}=Sym(\Theta_1^{(0)}).\tag{6.151}$$

From (6.107) we derive

$$\kappa_1^{(0)}=(a_{35}I_{30}+a_{34}I_{21})\frac{P_1}{2(I_y)^2}+(a_{35}I_{21}+a_{34}I_{12})\frac{P_2}{2I_xI_y},\qquad \kappa_2^{(0)}=Sym(\kappa_1^{(0)}),\tag{6.152a}$$
$$\varepsilon_0^{(0)}=a_{35}\frac{P_1}{2S_\Omega}+a_{34}\frac{P_2}{2S_\Omega}.\tag{6.152b}$$

At this stage, $\varepsilon_z^{(0)}$ may be determined from (6.41) as

$$\varepsilon_z^{(0)}=-\kappa_1^{(0)}x-\kappa_2^{(0)}y+a_{34}\frac{P_2}{2S_\Omega}+a_{35}\frac{P_1}{2S_\Omega}+(a_{35}x+a_{34}y)(\frac{P_1}{2I_y}x+\frac{P_2}{2I_x}y).\tag{6.153}$$

The Coupled-Plane system of (6.57a,b) for $k=K+1$ and $K=-1$ are then written as

$$\nabla_1^{(4)}\Phi+\nabla_1^{(3)}\psi=g,\qquad \nabla_1^{(3)}\Phi+\nabla_1^{(2)}\psi=f_0+f\qquad \text{over}\quad \Omega,\tag{6.154}$$

where $f_0=f_0^{(0)}$, $f=f^{(0)}$, $g=g^{(0)}$ are expressed using (6.109a–c) with $K=-1$. The boundary conditions, see (6.110a,b), take the form

$$\frac{d}{ds}\{\Phi_{,x},\Phi_{,y},\psi\}=\{0,0,-\overline{U}_3\cos(\bar{\mathbf{n}},x)-\overline{U}_4\cos(\bar{\mathbf{n}},y)\}\qquad \text{on}\quad \partial\Omega,\tag{6.155}$$

where the potentials $\overline{U}_3=\overline{U}_3^{(0)}$, $\overline{U}_4=\overline{U}_4^{(0)}$ are particular solutions of (6.51), namely,

$$\overline{U}_{4,y}+\overline{U}_{3,x}=-\frac{P_1}{I_y}x-\frac{P_2}{I_x}y.\tag{6.156}$$

In the general case one may select the potentials $\overline{U}_3, \overline{U}_4$ of (6.156) as quadratic functions of x and y. With this selection, $\underset{\sim}{g}$ becomes a constant and $f = \widetilde{f_1}x + \widetilde{f_2}y$ becomes a linear function with two constants $\widetilde{f_1}$ and $\widetilde{f_2}$. One may therefore adopt the particular solutions ψ_p and Φ_p of Example 4.2 for this case.

Once the Coupled-Plane BVP of (6.154), (6.155) is solved, the stress components are obtained from (6.55) for $k = K+1 = 0$. Likewise, $\sigma_z^{(0)}$ is derived from (6.56) for $k = 0$. The above solution also provides the twist angle $\theta^{(0)}$ derived from (6.66) with $k = K+1$ (i.e. by two sequential solutions of the BVP that yield $M_3^{I(0)}$ and $M_3^{II(0)}$, while $M_3^{(0)} = 0$). Overall, since level $k = K+2$ produced no stresses except for $\sigma_z^{(1)}$, one may write

$$\sigma_x = \sigma_z^{(0)}, \quad \sigma_y = \sigma_y^{(0)}, \quad \tau_{yz} = \tau_{yz}^{(0)}, \quad \tau_{xz} = \tau_{xz}^{(0)}, \quad \tau_{xy} = \tau_{xy}^{(0)}, \quad \sigma_z = \sigma_z^{(0)} + \sigma_z^{(1)} z. \tag{6.157}$$

The overall twist angle then becomes $\theta = \theta^{(0)} + \theta^{(1)} z$.

6.4.2.2 Displacements and Rotations

Using (6.74a,b), the displacements u, v, w are directly derived as

$$\begin{aligned}
u = & -\frac{\kappa_1^{(0)}}{2} z^2 - \theta^{(0)} yz + \frac{P_1}{2I_2}(-a_{33}\frac{z^2}{3} - \frac{1}{2}a_{34}yz + a_{13}x^2 - a_{23}y^2)z \\
& + \frac{P_2}{2I_1}(a_{36}y + \frac{1}{2}a_{35}z + 2a_{13}x)yz + \gamma_{xz}^{(0)}(0,0)\frac{z}{2} + U + u_r, && (6.158a) \\
v = & \, Sym(u), && (6.158b) \\
w = & \frac{P_1}{2I_2}(a_{33}z + a_{35}x + a_{34}y)xz + \frac{P_2}{2I_1}(a_{33}z + a_{35}x + a_{34}y)yz \\
& + (\varepsilon_0^{(0)} - \kappa_1^{(0)} x - \kappa_2^{(0)} y)z + W + w_r, && (6.158c)
\end{aligned}$$

where the root warping functions U, V, W should be determined as indicated by step **(5)(b)** of the solution scheme of S.6.3. The deflection of the beam axis then becomes

$$\begin{aligned}
u_0(z) &= -\frac{P_1}{2I_2}a_{33}\frac{z^3}{3} - \frac{\kappa_1^{(0)}}{2}z^2 + \gamma_{xz}^{(0)}(0,0)\frac{z}{2} + \omega_y^0 + u^0, \qquad v_0(z) = Sym(u_0(z)), \\
w_0(z) &= (a_{35}\frac{P_1}{2S_\Omega} + a_{34}\frac{P_2}{2S_\Omega})z + w^0. && (6.159)
\end{aligned}$$

In this case, the rotations are

$$\begin{aligned}
\omega_x = & -(a_{36}\frac{x^2}{2} + a_{23}xy)\frac{P_1}{2I_2} + [a_{33}z^2 + (a_{35}x + a_{34}y)z + a_{13}\frac{x^2}{2} - a_{23}\frac{y^2}{2}]\frac{P_2}{2I_1} \\
& - \kappa_2^{(0)} z - \frac{\theta^{(0)}}{2}x - \frac{1}{4}\gamma_{yz}^{(0)}(0,0) + \omega_x^0, \\
\omega_y = & -Sym(\omega_x), \\
\omega_z = & (a_{34}\frac{z}{4} + a_{23}y + a_{36}\frac{x}{2})z\frac{P_1}{I_2} - (a_{35}\frac{z}{4} + a_{13}x + a_{36}\frac{y}{2})z\frac{P_2}{I_1} + \theta^{(0)}z + \omega_z^0, && (6.160)
\end{aligned}$$

while the beam axis rotations are

$$\begin{aligned}
\omega_{x0} &= a_{33}z^2\frac{P_2}{2I_1} - \kappa_2^{(0)}z - \frac{1}{4}\gamma_{yz}^{(0)}(0,0) + \omega_x^0, \qquad \omega_{y0} = -Sym(\omega_{x0}), \\
\omega_{z0} &= (a_{34}\frac{P_1}{I_2} - a_{35}\frac{P_2}{I_1})\frac{z^2}{4} + \theta^{(0)}z + \omega_z^0. && (6.161)
\end{aligned}$$

As an example, one may examine a beam of doubly-symmetric cross-sections (where $\kappa_2^{(0)} = 0$), which is loaded by $P_2 \neq 0$ only. In this case, the ratio between the twist rate, $\theta^{(1)} z$, and the axis beamwise curvature $v_{0,zz}(z)$ is given by

$$\frac{\theta^{(1)} z}{v_{0,zz}(z)} = \frac{a_{35}}{2a_{33}}, \tag{6.162}$$

which is in a complete agreement with the simplified analysis reported in S.9.2.3.3, see (9.44). Likewise, one may also verify these results with formulas of (Lekhnitskii, 1981) for a cantilever.

Example 6.2 *A Beam of Elliptical Cross-Section Under Tip Bending.*
We employ here the elliptical cross-section described in Example 6.1.

Level $K+2$: Equations (6.148) show that:

$$\kappa_1^{(1)} = -a_{33}\frac{4P_1}{\pi\tilde{a}^3\tilde{b}}, \qquad \kappa_2^{(1)} = -a_{33}\frac{4P_2}{\pi\tilde{a}\tilde{b}^3}, \qquad \theta^{(1)} = a_{34}\frac{2P_1}{\pi\tilde{a}^3\tilde{b}} - a_{35}\frac{2P_2}{\pi\tilde{a}\tilde{b}^3}, \tag{6.163a}$$

$$\sigma_z^{(1)} = \frac{4P_1}{\pi\tilde{a}^3\tilde{b}}x + \frac{4P_2}{\pi\tilde{a}\tilde{b}^3}y, \qquad \varepsilon_i^{(1)} = 4a_{i3}\left(\frac{P_1}{\pi\tilde{a}^3\tilde{b}}x + \frac{P_2}{\pi\tilde{a}\tilde{b}^3}y\right) \qquad (i = 1,\ldots,6). \tag{6.163b}$$

Level $K+1$: Equations (6.150), (6.152a–c) show that:

$$\kappa_1^{(0)} = \kappa_2^{(0)} = 0, \qquad \varepsilon_0^{(0)} = \frac{1}{2\pi\tilde{a}\tilde{b}}(a_{35}P_1 + a_{34}P_2), \tag{6.164a}$$

$$\bar{\varepsilon}_z^{(0)} = \frac{2}{\pi\tilde{a}\tilde{b}}(a_{35}x + a_{34}y)\left(\frac{P_1}{\tilde{a}^2}x - \frac{P_2}{\tilde{b}^2}y\right). \tag{6.164b}$$

To satisfy (6.156) one may assume the following $\overline{U}_3$ and $\overline{U}_4$ potentials

$$\overline{U}_4 = -\frac{2P_2}{\pi\tilde{a}\tilde{b}}\left(\frac{y^2}{\tilde{b}^2} + \frac{x^2}{\tilde{a}^2} - 1\right), \qquad \overline{U}_3 = Sym(\overline{U}_4) = -\frac{2P_1}{\pi\tilde{a}\tilde{b}}\left(\frac{x^2}{\tilde{a}^2} + \frac{y^2}{\tilde{b}^2} - 1\right). \tag{6.165}$$

These expressions yield $\overline{U}_3\cos(\bar{\mathbf{n}},x) + \overline{U}_4\cos(\bar{\mathbf{n}},y) = 0$ over the cross-section contour. With this selection, for the system of field equations described in (6.57a,b), (6.109a–c), g and $f_0 = -2\theta^{(0)}$ are constants, and $f = f_1 x + f_2 y$, where

$$\begin{aligned} g = {} & \left[\left(a_{15} - \frac{a_{13}a_{35}}{a_{33}}\right)\frac{2}{\tilde{b}^2} + \left(2a_{25} + \frac{a_{36}a_{34} - 4a_{23}a_{35}}{a_{33}}\right)\frac{1}{\tilde{a}^2}\right]\frac{2P_1}{\pi\tilde{a}\tilde{b}} \\ & + \left[\left(2a_{14} + \frac{a_{36}a_{35} - 4a_{13}a_{34}}{a_{33}}\right)\frac{1}{\tilde{b}^2} + \left(a_{24} - \frac{a_{23}a_{34}}{a_{33}}\right)\frac{2}{\tilde{a}^2}\right]\frac{2P_2}{\pi\tilde{a}\tilde{b}}, \end{aligned} \tag{6.166}$$

$$f_1 = \left(\frac{3a_{34}a_{35}}{a_{33}} - 2a_{36} - 2a_{45}\right)\frac{2P_1}{\pi\tilde{a}^3\tilde{b}} + \left[\left(4a_{13} - \frac{a_{35}{}^2}{a_{33}}\right)\frac{1}{\tilde{b}^2} + \left(\frac{a_{34}^2}{a_{33}} - a_{44}\right)\frac{2}{\tilde{a}^2}\right]\frac{2P_2}{\pi\tilde{a}\tilde{b}},$$

$$f_2 = -Sym(f_1^{(0)}).$$

The boundary conditions of (6.110a,b) take the form

$$\frac{d}{ds}\{\Phi_{,y}, \Phi_{,x}, \psi\} = \{0,0,0\} \qquad \text{on} \quad \partial\Omega. \tag{6.167}$$

To determine $\theta^{(0)}$ according to (6.66), we define analogously to (6.64)

$$\Phi = \Phi^I - 2\theta^{(0)}\Phi^{II}, \qquad \psi = \psi^I - 2\theta^{(0)}\psi^{II}, \tag{6.168}$$

while a similar notation is used for stress components. To determine Φ^I and ψ^I we set $\theta^{(0)} = 0$. This yields the problem treated in Example 4.2 and should be solved accordingly. The above solution yields stress functions of the form

$$\Phi^I_P = \widetilde{\alpha}_0(\frac{x^2}{\widetilde{a}^2} + \frac{y^2}{\widetilde{b}^2} - 1)^2, \qquad \psi^I_P = (\widetilde{\alpha}_1 x + \widetilde{\alpha}_2 y)(\frac{x^2}{\widetilde{a}^2} + \frac{y^2}{\widetilde{b}^2} - 1), \tag{6.169}$$

where $\widetilde{\alpha}_0, \widetilde{\alpha}_1, \widetilde{\alpha}_2$ are constants. Thus, the corresponding stress components may be written as

$$\sigma_x^{I(0)} = \widetilde{\alpha}_0(\frac{x^2}{\widetilde{a}^2} + \frac{3y^2}{\widetilde{b}^2} - 1)\frac{4}{\widetilde{b}^2}, \quad \sigma_y^{I(0)} = \widetilde{\alpha}_0(\frac{3x^2}{\widetilde{a}^2} + \frac{y^2}{\widetilde{b}^2} - 1)\frac{4}{\widetilde{a}^2}, \quad \tau_{xy}^{I(0)} = -\widetilde{\alpha}_0\frac{8xy}{\widetilde{a}^2\widetilde{b}^2},$$
$$\tau_{xz}^{I(0)} = \widetilde{\alpha}_1\frac{2xy}{\widetilde{b}^2} + \widetilde{\alpha}_2(\frac{x^2}{\widetilde{a}^2} + \frac{3y^2}{\widetilde{b}^2} - 1) - \frac{2P_1}{\pi\widetilde{a}^3\widetilde{b}}(x^2 + \frac{\widetilde{a}^2}{\widetilde{b}^2}y^2 - \widetilde{a}^2),$$
$$\tau_{yz}^{I(0)} = \widetilde{\alpha}_1(\frac{3x^2}{\widetilde{a}^2} + \frac{y^2}{\widetilde{b}^2} - 1) + \widetilde{\alpha}_2\frac{2xy}{\widetilde{a}^2} - \frac{2P_2}{\pi\widetilde{a}\widetilde{b}^3}(y^2 + \frac{\widetilde{b}^2}{\widetilde{a}^2}x^2 - \widetilde{b}^2). \tag{6.170}$$

The above expressions show that $M_3^{I(0)} = \iint_\Omega (x\tau_{yz}^{I(0)} - y\tau_{xz}^{I(0)}) = 0$. Then, since $M_3^{(0)} = 0$, (6.66) shows that $\theta^{(0)} = 0$. It is therefore unnecessary to determine Φ^{II} and ψ^{II}.

One may activate **P.6.1** for a cantilever (i.e., by adding $M_1^{(0)} = -P_2 l$ to the above described formulation). Typical results for a beam of elliptical cross-section made of MON13y material (that has been obtained by orthotropic material rotation as shown in Fig. 2.2) are presented here. The distributions of the τ_{yz} stress component and the stress function ψ (for $P_1 = 0, P_2 = 1$) are shown in Figs. 6.2(a), (b), respectively. It turns out that in this case $\widetilde{\alpha}_0 = \widetilde{\alpha}_2 = 0$, and similar to

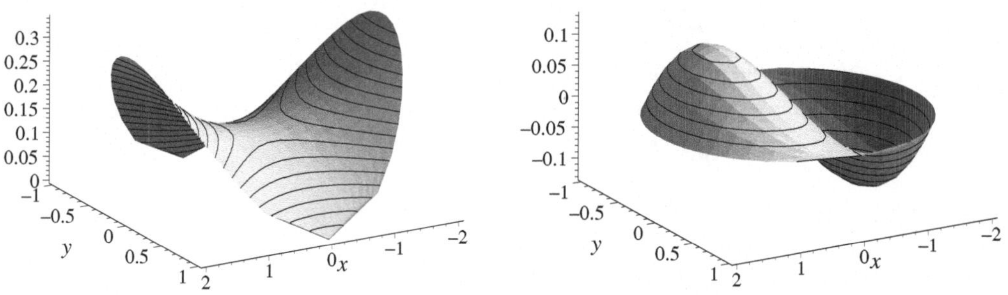

(a) τ_{yz}. (b) ψ.

Figure 6.2: Typical shear stress and stress function of a MON13y beam of elliptic cross-section.

(6.162), the coupling induced twist (rate) is given by $\theta/v_{0,zz}(z) = a_{35}/(2a_{33})$ (in this case the twist angle is a quadratic function and the axis bending function is a third-order polynomial).

6.4.3 *Axially Uniform Distributed Loading*

We shall now consider a rod that is uniformly loaded (with respect to the z-direction) while the body and surface forces, $\mathbf{F}_b$, $\mathbf{F}_s$, do not depend on the axis variable z, and therefore, analysis with $K = 0$ is required. In order not to include the previously discussed cases of tip loads, we shall also assume that adequate tip loads are applied so that $P_i^{(0)} = M_i^{(0)} = 0$, $i = 1,2,3$, and the only non-vanishing coefficients are $M_1^{(2)}$, $M_2^{(2)}$ and $P_i^{(1)}, M_i^{(1)}$, $i = 1,2,3$.

Level $k=K+2=2$: The following may be directly derived from S.6.3.2.1, while (6.16) show that $M_1^{(2)}=\frac{1}{2}P_2^{(1)}$, $M_2^{(2)}=\frac{1}{2}P_1^{(1)}$. The expressions are therefore

$$\kappa_1^{(2)}=-a_{33}\frac{P_1^{(1)}}{2I_2},\qquad \kappa_2^{(2)}=-a_{33}\frac{P_2^{(1)}}{2I_1},\qquad \theta^{(2)}=a_{34}\frac{P_1^{(1)}}{4I_2}-a_{35}\frac{P_2^{(1)}}{4I_1},$$

$$\sigma_z^{(2)}=\frac{P_1^{(1)}}{2I_2}x+\frac{P_2^{(1)}}{2I_1}y,\qquad \varepsilon_i^{(2)}=a_{i3}\Big(\frac{P_1^{(1)}}{2I_2}x+\frac{P_2^{(1)}}{2I_1}y\Big)\qquad (i=1,\ldots,6).\tag{6.171}$$

Also

$$\Theta_1^{(2)}=\Big(a_{34}\frac{P_1^{(1)}}{I_2}-a_{35}\frac{P_2^{(1)}}{I_1}\Big)\frac{y}{12},\qquad \Theta_2^{(2)}=Sym(\Theta_1^{(2)}).\tag{6.172}$$

Level $k=K+1=1$: The following may be directly derived from S.6.3.2.2. First, (6.106) shows that

$$\bar{\varepsilon}_z^{(1)}=(a_{35}x+a_{34}y)\Big(\frac{P_1^{(1)}}{2I_2}x+\frac{P_2^{(1)}}{2I_1}y\Big).\tag{6.173}$$

From (6.105) we write

$$\Theta_1^{(1)}=\theta^{(1)}\frac{y}{2}+\Big(a_{23}\frac{P_1^{(1)}}{I_2}-a_{36}\frac{P_2^{(1)}}{I_1}\Big)\frac{y^2}{4},\qquad \Theta_2^{(1)}=Sym(\Theta_1^{(1)})\tag{6.174}$$

while (6.107) for $K=0$ allow us to express $\kappa_1^{(1)},\kappa_2^{(1)}$ and $\varepsilon_0^{(1)}$. At this stage one may calculate $\varepsilon_z^{(1)}$ from (6.41) with $k=1$ as

$$\varepsilon_z^{(1)}=-\kappa_1^{(1)}x-\kappa_2^{(1)}y+a_{33}\frac{P_3^{(1)}}{S_\Omega}+a_{34}\frac{P_2^{(1)}}{2S_\Omega}+a_{35}\frac{P_1^{(1)}}{2S_\Omega}+(a_{35}x+a_{34}y)\Big(\frac{P_1^{(1)}}{2I_y}x+\frac{P_2^{(1)}}{2I_x}y\Big).\tag{6.175}$$

The system (6.57a,b) for $k=1$ takes the form of

$$\nabla_1^{(4)}\Phi^{(1)}+\nabla_1^{(3)}\psi^{(1)}=\tilde{g}^{(1)},\qquad \nabla_1^{(3)}\Phi^{(1)}+\nabla_1^{(2)}\psi^{(1)}=\tilde{f}_0^{(1)}+\tilde{f}^{(1)}\tag{6.176}$$

where $\tilde{f}_0^{(1)},\tilde{f}^{(1)},\tilde{g}^{(1)}$ are expressed from (6.109a–c) with $K=0$. The boundary conditions (6.110a,b) are expressed with $K=0$, and the potentials $\overline{U}_3^{(1)},\overline{U}_4^{(1)}$ which are particular solutions of (6.51) with $K=0$, are written as

$$\overline{U}_{3,x}^{(1)}+\overline{U}_{4,y}^{(1)}=-\frac{P_1^{(1)}}{I_y}x-\frac{P_2^{(1)}}{I_x}y.\tag{6.177}$$

As discussed in relation with (6.66), the twist angle parameter $\theta^{(1)}$ is also derived within the solution process of the above system of equations.

Level $k=K=0$: One should carry out the solution steps **(4)(a)** through **(4)(k)** that are described by the general scheme of S.6.3 for $k=K=0$.

Finally, solution steps **(5)** through **(7)** of the general scheme of S.6.3 should be executed.

Example 6.3 *The $K+1$ Level Solution for a Beam of Elliptical Cross-Section.*

It should be noted that for elliptical cross-section (the characteristics of which are described in Example 6.1) the solution of this $K+1$ level may be determined analogously to the solution described for level $K+1$ in Example 6.2. Hence, to satisfy (6.177) one may assume (similar to (6.165)) the following potentials $\overline{U}_3^{(1)}$ and $\overline{U}_4^{(1)}$:

$$\overline{U}_4^{(1)} = -\frac{2P_2^{(1)}}{\pi\widetilde{a}\widetilde{b}}\left(\frac{y^2}{\widetilde{b}^2}+\frac{x^2}{\widetilde{a}^2}-1\right), \qquad \overline{U}_3^{(1)} = Sym(\overline{U}_4^{(1)}) = -\frac{2P_1^{(1)}}{\pi\widetilde{a}\widetilde{b}}\left(\frac{x^2}{\widetilde{a}^2}+\frac{y^2}{\widetilde{b}^2}-1\right). \quad (6.178)$$

These expressions yield $\overline{U}_3^{(1)}\cos(\bar{\mathbf{n}},x)+\overline{U}_4^{(1)}\cos(\bar{\mathbf{n}},y)=0$ over the entire cross-section contour. Therefore, for the system (6.176) we obtain $g^{(1)}=\widetilde{g}^{(1)}$, $\widetilde{f}^{(1)}=\widetilde{f}_1^{(1)}x+\widetilde{f}_2^{(1)}y$, where the four constants are

$$\widetilde{f}_0^{(1)} = -2\theta^{(1)} - 4a_{35}\frac{M_1^{(1)}}{\pi\widetilde{b}^3\widetilde{a}} + 4a_{34}\frac{M_2^{(1)}}{\pi\widetilde{a}^3\widetilde{b}} + \left(\frac{2a_{35}^2}{a_{33}\widetilde{b}^2}+\frac{2a_{34}^2}{a_{33}\widetilde{a}^2}\right)\frac{M_3^{(1)}}{\pi\widetilde{a}\widetilde{b}},$$

$$\widetilde{g}^{(1)} = \left[\frac{2b_{15}}{\widetilde{b}^2}+\left(2b_{25}+\frac{a_{36}a_{34}-2a_{23}a_{35}}{a_{33}}\right)\frac{1}{\widetilde{a}^2}\right]\frac{2P_1^{(1)}}{\pi\widetilde{a}\widetilde{b}} + \left[\left(2b_{14}+\frac{a_{36}a_{35}-2a_{13}a_{34}}{a_{33}}\right)\frac{1}{\widetilde{b}^2}+\frac{4b_{24}}{\widetilde{a}^2}\right]\frac{2P_2^{(1)}}{\pi\widetilde{a}\widetilde{b}},$$

$$\widetilde{f}_1^{(1)} = \left(\frac{a_{34}a_{35}-2a_{36}a_{33}}{a_{33}}-2b_{45}\right)\frac{2P_1^{(1)}}{\pi\widetilde{a}^3\widetilde{b}} + \left(-\frac{2b_{44}}{\widetilde{a}^2}+\frac{4a_{13}a_{33}-a_{35}{}^2}{a_{33}\widetilde{b}^2}\right)\frac{2P_2^{(1)}}{\pi\widetilde{a}\widetilde{b}},$$

$$\widetilde{f}_2^{(1)} = -Sym(\widetilde{f}_1^{(1)}). \quad (6.179)$$

Hence, the solution becomes a superposition of the solutions presented in Examples 6.1, 6.2, namely,

$$\Phi_P^{I(1)} = \widetilde{\beta}_0\left(\frac{x^2}{\widetilde{a}^2}+\frac{y^2}{\widetilde{b}^2}-1\right)^2, \qquad \psi_P^{I(1)} = (\widetilde{\beta}_1+\widetilde{\beta}_2x+\widetilde{\beta}_3y)\left(\frac{x^2}{\widetilde{a}^2}+\frac{y^2}{\widetilde{b}^2}-1\right). \quad (6.180)$$

Similar to Example 6.2, one may construct a linear system of equations for the $\widetilde{\beta}_i$ $(i=0,\ldots,3)$ coefficients. Solving the above system for $\theta^{(1)}=0$ yields a set of $\widetilde{\beta}_i$ coefficient where only $\widetilde{\beta}_1$ induces a torsional moment, and therefore,

$$M_3^{I(1)} = -\widetilde{\beta}_1\pi\widetilde{a}\widetilde{b} \quad (6.181)$$

while (6.135) shows that

$$M_3^{II(1)} = -\frac{\pi\widetilde{a}^3\widetilde{b}^3}{2(b_{44}\widetilde{b}^2+b_{55}\widetilde{a}^2)}. \quad (6.182)$$

Substitution of (6.181), (6.182) in (6.66) enables us to determine $\theta^{(1)}$.

Example 6.4 *Lateral Deflection due to Linearly Distributed Axial Loading.*

Consider the case where the only non-vanishing loading parameter is $P_3^{(1)}$, which represents a linear (with respect to z) variation of axial loading. In level $K+2$ all components are zero. In level $K+1$, in view of (6.107), (6.175), the only non-vanishing parameters are $\varepsilon_0^{(1)}=\varepsilon_z^{(1)}=a_{33}\frac{P_3^{(1)}}{S_\Omega}$. Equations (6.176) show that $\Phi^{(1)}=\psi^{(1)}=0$, and thus, the only stress component is given by (6.56) as $\sigma_z^{(1)}=\varepsilon_z^{(1)}/a_{33}$. Equation (5.22) then shows that $\gamma_{yz}^{(1)}=a_{34}\sigma_z^{(1)}$ and $\gamma_{xz}^{(1)}=a_{35}\sigma_z^{(1)}$. In addition, according to (6.87), (6.88), (6.90a), the beam axis deformation is given by

$$u_0^{(2)} = \frac{\kappa_1^{(0)}}{2}+\frac{1}{2}\gamma_{xz}^{(1)}(0,0), \qquad v_0^{(2)} = Sym(u_0^{(2)}), \qquad w_0^{(2)} = \frac{1}{2}\varepsilon_0^{(1)}, \quad (6.183a)$$

$$u_0^{(1)} = \frac{1}{2}\gamma_{xz}^{(0)}(0,0)+\omega_y^0, \qquad v_0^{(1)} = Sym(u_0^{(1)}), \qquad w_0^{(1)} = \varepsilon_0^{(0)}, \quad (6.183b)$$

$$u_0^{(0)} = u^0, \qquad v_0^{(0)} = v^0, \qquad w_0^{(0)} = w^0. \quad (6.183c)$$

Hence, since $\kappa_1^{(0)} = \kappa_2^{(0)} = 0$, the curvature components of (5.6) are

$$\kappa_u = \kappa_1^{(0)} + \gamma_{xz}^{(1)}(0,0) = a_{35}\frac{P_3^{(1)}}{S_\Omega}, \qquad \kappa_v = Sym(\kappa_u) = a_{34}\frac{P_3^{(1)}}{S_\Omega}. \tag{6.184}$$

The first of (6.184) is in complete agreement with the approximate model derived in Chapter 9, see (9.34), and represents bending due to non-uniform extension.

6.4.4 Additional Examples

Only a few illustrative additional examples are given here, while the reader may activate **P.6.1** for many other cases. The program supplies all solution details including stress functions, stress and strain components, displacement components and all involved parameters.

In the following examples, we set $\widetilde{a} = 2,\ \widetilde{b} = 1,\ l = 10$ while parameters that are not mentioned are set to zero.

Example 6.5 *A Cantilever Isotropic Beam Under Tip Load.*

We activate **P.6.1** (using rational numbers in order to prove the symbolic exactness) with *material* $= 2$ and $E = 72999999983,\ G = 364999999915/13$ ($\nu = 3/10$). We also set $K = -1,\ P_x = 1$ and root loads of $P_1^{(0)} = P_x,\ M_2^{(0)} = -P_x l$. The resulting stress functions are

$$\psi = \frac{1544230768879}{4744999998895\pi}(y^2 + \frac{1}{4}x^2 - 1)y, \qquad \Phi = 0, \tag{6.185}$$

which are identical to the solution presented by (Timoshenko and Goodier, 1970). Hence this simple case of tip bending does not include in-plane bending at all.

Example 6.6 *A Cantilever Beam of General Anisotropy Under Tip Load.*

Consider the beam of Example 6.5 by replacing the material with orthotropic material that is rotated by three orthogonal angles and therefore create a fully populated constitutive relations matrix. We therefore set material$= 1, K = -1$ and $t_x = t_y = t_z = 30°$. The resulting stress functions are

$$\psi = (-0.135x + 0.00922y)(y + 0.250x - 1)x, \qquad \Phi = 0.00375(y^2 + 0.250x^2 - 1). \tag{6.186}$$

It is clear that in-plane stresses would be induced in this case as well.

Example 6.7 *A Cantilever Rotating Beam of General Anisotropy.*

To simulate the influence body force induced by the centrifugal loading when the beam is rotating around the y-axis (see discussion in Example 4.12), we assume $\rho_0\Omega_0^2 = 1$, and therefore select $K = 1,\ X_b^{(0)} = x,\ Z_b^{(1)} = 1$. In addition, according to (5.11a), the only nonzero root load is $P_3^{(0)} = \frac{1}{2}S_\Omega l^2$, where $S_\Omega = \pi a_0 b_0$. For beams of different materials, we activate **P.6.1** with the following selections: ISO2: material$= 2$ and arbitrary t_x, t_y, t_z, ORT9: material$= 1$, $t_x = t_y = t_z = 0$, MON13x: material$= 1, t_x = 30°, t_y = t_z = 0$, MON13y: material$= 1, t_x = t_z = 0, t_y = 30°$, MON13z: material$= 1, t_x = t_y = 0, t_z = 30°$. For clamped-II boundary conditions, the reader may verify that in the cases of ISO2, ORT9 and MON13z materials, the beam axis deformation is confined to cubic axial displacements, namely: $w_0 = \widetilde{w}_1 z + \widetilde{w}_3 z^3$. However, for MON13x material, one also obtains $u_0 = \widetilde{u}_1 z + \widetilde{u}_3 z^3$. Likewise, for MON13y material, in addition to the above w_0, one also obtains $v_0 = \widetilde{v}_1 z + \widetilde{v}_3 z^3$. In the general anisotropic case, all of the above u_0, v_0 and w_0 are created. This example demonstrates again the fact that ISO2, ORT9 and MON13z beams are fundamentally uncoupled.

Example 6.8 *A Cantilever Isotropic Beam Under Quadratic Distributed Load.*

Consider the beam of Example 6.5 by removing the tip load and applying a distributed body force $Y_b = z^2$. We therefore set $K = 2$, $Y_b^{(2)} = 1$, $P_2^{(0)} = \frac{\pi}{3}\tilde{a}\tilde{b}l^3$ and $M_1^{(0)} = -\frac{\pi}{4}\tilde{a}\tilde{b}l^4$. The resulting stress functions are presented in Fig. 6.3. As shown, unlike the case of tip loading discussed in Example 6.5, the in-plane stress components would not vanish in this case. Two of

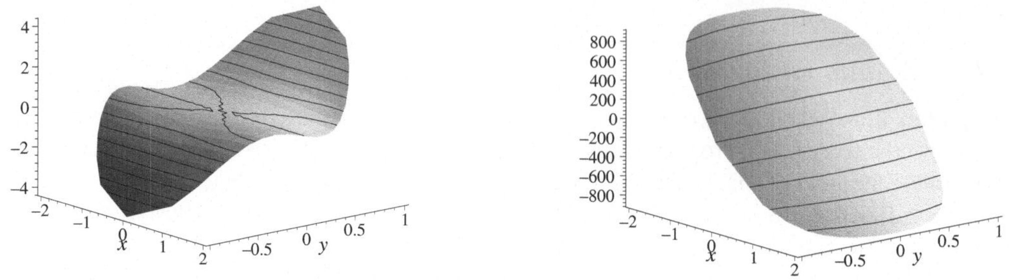

(a) Φ.

(b) ψ.

Figure 6.3: Stress functions in Example 6.8 for $z = l/2$.

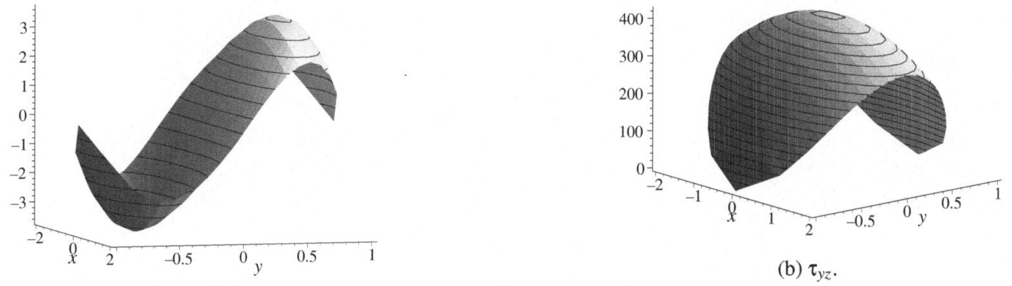

(a) σ_y.

(b) τ_{yz}.

Figure 6.4: Stress components in Example 6.8 for $z = l/2$.

the corresponding stress components are shown in Fig. 6.4. Also the axis lateral deformation in this case may be schematically written as $v_0 = \tilde{v}_2 z^2 + \tilde{v}_3 z^3 + \underline{\tilde{v}_4 z^4} + \tilde{v}_6 z^6$, while an elementary "Strength-of-Materials" isotropic beam analysis does not yield the underlined term (see S.5.3.4 and **P.5.1**), which may be predicted only by a complete and exact approach such as the present methodology.

6.5 Appendix: Integral Identities

As shown, throughout the derivation it became necessary to use the integral identities that will be derived in what follows. First, we modify Green's Theorem by adding the underlined terms in the following equation:

$$\iint_{\Omega}(P_{,x}+Q_{,y}+\underline{R_{,z}})=\underline{\iint_{\Omega}R_{,z}}+\oint_{\partial\Omega}[P\cos(\bar{\mathbf{n}},x)+Q\cos(\bar{\mathbf{n}},y)] \tag{6.187}$$

where $\bar{\mathbf{n}}$ is the outward normal to the contour of the cross-section.

By substituting appropriate functions in the above integrands, one may produce a variety of identities that, as stated before, appear to be useful throughout the analysis. As an example, we consider the case $P=\tau_{xz}$, $Q=\tau_{yz}$, $R=\sigma_z$. The l.h.s. of (6.187) in view of equilibrium equation (1.82b) becomes

$$\iint_{\Omega}P_{,x}+Q_{,y}+R_{,z}=-\iint_{\Omega}Z_b. \tag{6.188}$$

In addition, using the boundary condition of (5.16c), one obtains

$$\oint_{\partial\Omega}P\cos(\bar{\mathbf{n}},x)+Q\cos(\bar{\mathbf{n}},y)=\oint_{\partial\Omega}Z_s \tag{6.189}$$

while clearly, since $R=\sigma_z$,

$$\iint_{\Omega}R_{,z}=\iint_{\Omega}\sigma_{z,z}. \tag{6.190}$$

In Table 6.4, various selections of the functions P, Q and R are presented. In addition, the quantities $P_{,x}+Q_{,y}+R_{,z}$ and $P\cos(\bar{\mathbf{n}},x)+Q\cos(\bar{\mathbf{n}},y)$ as obtained by the relevant equilibrium equation (i.e. one of (1.82a−c)) or boundary condition (i.e. one of (5.16a−c)), respectively, are presented. Note that lines $3,5,\ldots,17$ of Table 6.4 may be obtained by the "*Sym*" operation, in addition to $P\leftrightarrow Q$, from lines $2,4,\ldots,16$, respectively.

As an example for the use of Table 6.4, the first three lines of it yield the three equations presented by (6.13a), lines 6, 7 yield the first two equations of (6.13b), and lines 4, 5 yield the third equation of (6.13b).

The remaining lines of Table 6.4 yield the following integral equalities:

$$\iint_{\Omega}\tau_{xy}=\oint_{\partial\Omega}xY_s+\iint_{\Omega}xY_b+\iint_{\Omega}\tau_{yz,z}x, \tag{6.191a}$$

$$\iint_{\Omega}\tau_{xy}=Sym(\iint_{\Omega}\tau_{xy})=\oint_{\partial\Omega}yX_s+\iint_{\Omega}yX_b+\iint_{\Omega}\tau_{xz,z}y, \tag{6.191b}$$

$$\iint_{\Omega}\tau_{xz}=\oint_{\partial\Omega}xZ_s+\iint_{\Omega}xZ_b+\iint_{\Omega}\sigma_{z,z}x, \qquad \iint_{\Omega}\tau_{yz}=Sym(\iint_{\Omega}\tau_{xz}), \tag{6.191c}$$

$$\iint_{\Omega}\sigma_y=\oint_{\partial\Omega}yY_s+\iint_{\Omega}yY_b+\iint_{\Omega}\tau_{yz,z}y, \qquad \iint_{\Omega}\sigma_x=Sym(\iint_{\Omega}\sigma_y), \tag{6.191d}$$

$$\iint_{\Omega}\sigma_x x=\oint_{\partial\Omega}\frac{x^2}{2}X_s+\iint_{\Omega}\frac{x^2}{2}X_b+\iint_{\Omega}\tau_{xz,z}\frac{x^2}{2}, \qquad \iint_{\Omega}\sigma_y y=Sym(\iint_{\Omega}\sigma_x x), \tag{6.191e}$$

$$\iint_{\Omega}\tau_{xy}x=\oint_{\partial\Omega}\frac{x^2}{2}Y_s+\iint_{\Omega}\frac{x^2}{2}Y_b+\iint_{\Omega}\tau_{yz,z}\frac{x^2}{2}, \qquad \iint_{\Omega}\tau_{xy}y=Sym(\iint_{\Omega}\tau_{xy}x), \tag{6.191f}$$

$$\iint_{\Omega}\tau_{yz}y=\oint_{\partial\Omega}\frac{y^2}{2}Z_s+\iint_{\Omega}\frac{y^2}{2}Z_b+\iint_{\Omega}\sigma_{z,z}\frac{y^2}{2}, \qquad \iint_{\Omega}\tau_{xz}x=Sym(\iint_{\Omega}\tau_{yz}y), \tag{6.191g}$$

$$\iint_{\Omega}\tau_{xz}y+\tau_{yz}x=\oint_{\partial\Omega}xyZ_s+\iint_{\Omega}xyZ_b+\iint_{\Omega}\sigma_{z,z}xy, \tag{6.191h}$$

$$\iint_{\Omega}\sigma_x y+\tau_{xy}x=\oint_{\partial\Omega}xyX_s+\iint_{\Omega}xyX_b+\iint_{\Omega}\tau_{xz,z}xy, \tag{6.191i}$$

$$\iint_{\Omega}\sigma_y x+\tau_{xy}y=Sym(\iint_{\Omega}\sigma_x y+\tau_{xy}x). \tag{6.191j}$$

	P	Q	R	$P_{,x}+Q_{,y}+R_{,z}$	$P\cos(\bar{\mathbf{n}},x)+Q\cos(\bar{\mathbf{n}},y)$
1	τ_{xz}	τ_{yz}	σ_z	$-Z_b$	Z_s
2	σ_x	τ_{xy}	τ_{xz}	$-X_b$	X_s
3	τ_{xy}	σ_y	τ_{yz}	$-Y_b$	Y_s
4	$x\tau_{xy}$	$x\sigma_y$	$x\tau_{yz}$	$\tau_{xy}-xY_b$	xY_s
5	$y\sigma_x$	$y\tau_{xy}$	$y\tau_{xz}$	$\tau_{xy}-yX_b$	yX_s
6	$x\tau_{xz}$	$x\tau_{yz}$	$x\sigma_z$	$\tau_{xz}-xZ_b$	xZ_s
7	$y\tau_{xz}$	$y\tau_{yz}$	$y\sigma_z$	$\tau_{yz}-yZ_b$	yZ_s
8	$y\tau_{xy}$	$y\sigma_y$	$y\tau_{yz}$	σ_y-yY_b	yY_s
9	$x\sigma_x$	$x\tau_{xy}$	$x\tau_{xz}$	σ_x-xX_b	xX_s
10	$x^2\sigma_x$	$x^2\tau_{xy}$	$x^2\tau_{xz}$	$2x\sigma_x-x^2X_b$	x^2X_s
11	$y^2\tau_{xy}$	$y^2\sigma_y$	$y^2\tau_{yz}$	$2y\sigma_y-y^2Y_b$	y^2Y_s
12	$x^2\tau_{xy}$	$x^2\sigma_y$	$x^2\tau_{yz}$	$2x\tau_{xy}-x^2Y_b$	x^2Y_s
13	$y^2\sigma_x$	$y^2\tau_{xy}$	$y^2\tau_{xz}$	$2y\tau_{xy}-y^2X_b$	y^2X_s
14	$y^2\tau_{xz}$	$y^2\tau_{yz}$	$y^2\sigma_z$	$2y\tau_{yz}-y^2Z_b$	y^2Z_s
15	$x^2\tau_{xz}$	$x^2\tau_{yz}$	$x^2\sigma_z$	$2x\tau_{xz}-x^2Z_b$	x^2Z_s
16	$xy\sigma_x$	$xy\tau_{xy}$	$xy\tau_{xz}$	$y\sigma_x+x\tau_{xy}-xyX_b$	xyX_s
17	$xy\tau_{xy}$	$xy\sigma_y$	$xy\tau_{yz}$	$y\tau_{xy}+x\sigma_y-xyY_b$	xyY_s
18	$xy\tau_{xz}$	$xy\tau_{yz}$	$xy\sigma_z$	$y\tau_{xz}+x\tau_{yz}-xyZ_b$	xyZ_s

Table 6.4: Summary of integral identities terms.

In addition, the following integrals may also be derived based on the above results:

$$\iint_\Omega \sigma_y x = \oint_{\partial\Omega}(xyY_s - \frac{y^2}{2}X_s) + \iint_\Omega (xyY_b - \frac{y^2}{2}X_b) + \iint_\Omega (\tau_{yz,z}xy - \tau_{xz,z}\frac{y^2}{2}), \tag{6.192a}$$

$$\iint_\Omega \tau_{yz}x = \frac{1}{2}M_3 + \oint_{\partial\Omega}\frac{xy}{2}Z_s + \iint_\Omega \frac{xy}{2}Z_b + \iint_\Omega \sigma_{z,z}\frac{xy}{2}, \tag{6.192b}$$

$$\iint_\Omega \sigma_x y = Sym(\iint_\Omega \sigma_y x), \qquad \iint_\Omega \tau_{xz}y = Sym(\iint_\Omega \tau_{yz}x). \tag{6.192c}$$

Equation (6.192a) is obtained using (6.191j) and the second integral of (6.191e). Equation (6.192b) is obtained using the integral (6.191h) and the definition of M_3 in (6.9). Particular cases of these integrals, derived for uniform stress distribution, appear in (Lekhnitskii, 1981).

7
Homogeneous, Uncoupled Monoclinic Beams

This chapter is focused on homogeneous beams made of MON13z (monoclinic) material type (see S.2.3), and thus, as discussed in S.5.2.2, an uncoupled behavior at the structural level is expected. Due to the simplification offered by MON13z materials, and unlike the exact solution for beams of general anisotropy presented in Chapter 6, a different approach is described in what follows. More specifically, the analysis restriction to MON13z material converts the fundamental *Coupled-Plane BVP* that plays a major role in each level solution (see Chapter 6), into an uncoupled pair of the biharmonic and the Dirichlet/Neumann BVPs. This simplifies the analysis, and allows explicit solution for various loading modes.

Unlike the derivation in Chapter 6, the entire reasoning of the analytical approach in this chapter is founded on the *St. Venant's semi-inverse method of solution* discussed in S.1.6.3. Subsequently, the class of solutions presented here is based on a set of initially assumed expressions for the stress components ("solution hypothesis") that contain series of unknown coefficients, and generalized harmonic and biharmonic stress functions. By employing the governing equations (equilibrium, compatibility, single-valued conditions, etc.), the field equations and the boundary conditions for the stress functions are established, and the involved parameters and functions are determined (and hence, the validity of the pre-assumed stress expressions is proved). In light of the above, the analysis presented in this chapter may be considered as "dual" or "conjugate" to the one presented in Chapter 6.

The origin of the present methodology seems to be (Michell, 1901), where a solution for a homogeneous *isotropic* beam under surface loads that do not vary along the generators was expressed by three harmonic and one biharmonic functions. In an independent way, (Almansi, 1901) presented an analytical (level-based) solution for the same problem when the surface loads are polynomials of the beam axis coordinate. Hence, in many occasions, the above works are generally referred to as the "Michell-Almansi (recursive) method".

Similar problems were also discussed by (Liebenson, 1940), (Love, 1944), (Muskhelishvili, 1953), (Kosmodamianskii, 1956), (Dzhanelidze, 1960), (Iesan, 1987) and others. The above solutions were further evolved by (Ruchadze, 1975), (Ruchadze and Berekashvili, 1980) and (Zivzivadze and Berekashvili, 1984) for MON13z beams undergoing both surface and body generic polynomial loading.

The formulation in this chapter is an *improved, unified and consistent* derivation of the various solutions found in the above described literature, and is based on an experience and know-how that has been acquired over a long period, see also (Kazar (Kezerashvili) *et al.*, 2004), (Grebshtein *et al.*, 2004). To prove the symbolic exactness of the expressions, and to eliminate errors and typographical mistakes, the entire methodology is documented, verified and proved to be exact by a suitable system of symbolic programs which are referenced in what follows.

In this chapter, we shall examine beams of simply connected homogeneous cross-sections under generic tip and distributed loads. When required, we shall exploit the limited number of exact solutions for tip loads derived in S.5.2.2. Multiply connected and non-homogeneous beams are separately handled in Chapter 8. Thus, the derivation presented in this chapter serves as an explicit, complete and exact model for a homogeneous MON13z beam, and is clearly valid for all simpler materials (i.e. orthotropic, tetragonal, transversely isotropic, cubic, and isotropic, see S.2.4 — S.2.8).

This chapter contains three main parts. First, solutions for the tip loads effect are presented in S.7.2. Subsequently, explicit formulation for the effect of generic distributions of surface and body loads appears in S.7.3. Both cases are accompanied in S.7.4 by illustrative examples of polynomial solutions for various fundamental problems. Finally, analysis and examples of beams of cylindrical anisotropy are dealt with in S.7.5.

7.1 Background

The analysis presented in this chapter deals with MON13z materials, the compliance matrix of which is written as (see S.2.3)

$$\begin{Bmatrix} \varepsilon_x \\ \varepsilon_y \\ \varepsilon_z \\ \gamma_{yz} \\ \gamma_{xz} \\ \gamma_{xy} \end{Bmatrix} = \begin{bmatrix} a_{11} & a_{12} & a_{13} & 0 & 0 & a_{16} \\ & a_{22} & a_{23} & 0 & 0 & a_{26} \\ & & a_{33} & 0 & 0 & a_{36} \\ & & & a_{44} & a_{45} & 0 \\ & Sym. & & & a_{55} & 0 \\ & & & & & a_{66} \end{bmatrix} \begin{Bmatrix} \sigma_x \\ \sigma_y \\ \sigma_z \\ \tau_{yz} \\ \tau_{xz} \\ \tau_{xy} \end{Bmatrix}. \quad (7.1)$$

Similar to the discussion in S.5.2.1, examination of the above compliance matrix shows that ε_z is not influenced by τ_{yz} or τ_{xz}, and similarly, σ_z is not influenced by γ_{yz} or γ_{xz}. Since on one hand, bending and extension activate the σ_z, ε_z components, and on the other hand, torsion activate the τ_{yz},γ_{yz} and τ_{xz},γ_{xz} components, the above material will not create coupling between global DOF such as the Bending-Twist coupling or Extension-Twist coupling (see also S.5.2.2). For that reason, MON13z beams are generally referred to as "uncoupled".

As discussed in S.2.4, an important subset of MON13z materials may be obtained by rotating orthotropic materials about the z-axis as demonstrated in Fig. 7.1. In such a case, only 10 independent parameters exist (9 material coefficients and a rotating angle). Yet, the analysis presented in this chapter treats *all 13 material properties as independent.*

7.2 Tip Loads

We shall first derive a separate analysis of a clamped-free beam for each of the tip loads that are not covered by S.5.2.2, and then summarize all tip effects in S.7.2.3. In classical terminology, the collection of these solutions is referred to as the solutions for the "St. Venant's Problem", see also S.5.1.2.

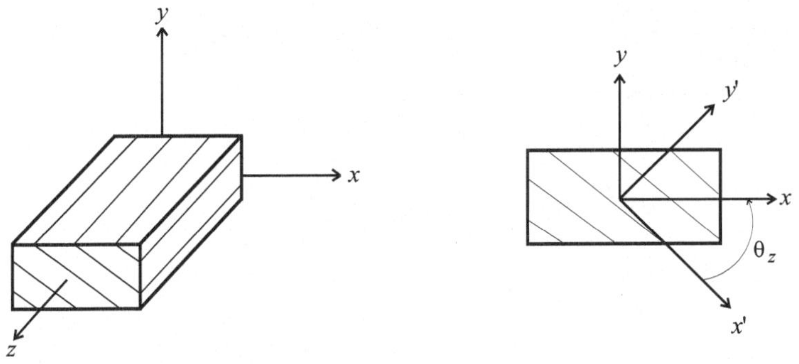

Figure 7.1: A homogeneous, MON13z beam made of orthotropic material which is rotated about the z-axis (x', y', z' are the orthotropic material principal axes).

It should be emphasized here that the tip loads are introduced in an integral manner as described in S.5.1.2, and likewise the root boundary conditions. In fact, this is the only approximation included in the derivation in this chapter. Note that unless stated differently, all relevant notation and definitions of S.5.1 will be adopted here. All tip load solutions are verified and documented in **P.7.1**.

7.2.1 Torsional Moment

The effect of torsional moment is analyzed by two complementary ways:

The first one is the *Out-of-Plane Warping* approach, in which equations are derived for the torsional related warping (see Solution Trail "A" or "Deformation Hypothesis" of S.1.6.5.1).

Within the second, *Stress Function* approach, the analysis derives the governing equations for the stress function of this problem (see Solution Trail "C" or "Stress Functions Hypothesis" of S.1.6.5.3).

7.2.1.1 Out-of-Plane Warping Approach

We shall commence our discussion assuming that the beam is acted upon by a tip (i.e. applied at the free end $z = l$) torsional moment, M_z, only. In this case the resultant moment at each cross-section is constant, i.e., $\overline{M}_z = M_z$. The analytical solution originates from the displacement hypothesis (3.113),

$$u = -\frac{M_z}{D} yz - \omega_z^0 y + \omega_y^0 z + u^0, \tag{7.2a}$$

$$v = \frac{M_z}{D} xz + \omega_z^0 x - \omega_x^0 z + v^0, \tag{7.2b}$$

$$w = \frac{M_z}{D} \varphi(x,y) - \omega_y^0 x + \omega_x^0 y + w^0, \tag{7.2c}$$

where u^0, v^0, and w^0, are rigid displacements, while ω_x^0, ω_y^0 and ω_z^0, are rigid rotations.

As was assumed in S.3.3.2, the rotation $\omega_z = \frac{M_z}{D} z + \omega_z^0$ is a linear function of z. The warping, $w(x,y)$, is a product of the beam twist $\theta = \frac{M_z}{D}$ (the warping amplitude), which is a constant, and an unknown function, φ, that physically represents the shape of out-of-plane warping. Due to its role in (7.2c), φ is usually termed as the *torsion function*. The constant D (not known as yet) will be shown to play the role of the beam *torsional rigidity*. Therefore, D may be defined as the ratio $\frac{M_z}{\theta}$, and unlike the isotropic case, it combines in an inseparable manner elastic moduli and geometry parameters.

The above displacement assumption yields a state where, see (3.114), all strain components vanish except for

$$\gamma_{yz} = \frac{M_z}{D}(\varphi_{,y} + x), \qquad \gamma_{xz} = \frac{M_z}{D}(\varphi_{,x} - y). \tag{7.3}$$

Hence the only nonzero stress components are, see (7.1) and (3.115),

$$\tau_{yz} = \frac{M_z}{Da_0}[-a_{45}(\varphi_{,x} - y) + a_{55}(\varphi_{,y} + x)], \qquad \tau_{xz} = \frac{M_z}{Da_0}[a_{44}(\varphi_{,x} - y) - a_{45}(\varphi_{,y} + x)], \tag{7.4}$$

where $a_0 = a_{44}a_{55} - a_{45}^2$. The above two stress components need to comply with the requirement of free outer surface of the beam. Examination of Fig. 7.2 shows that the normal stress τ_N

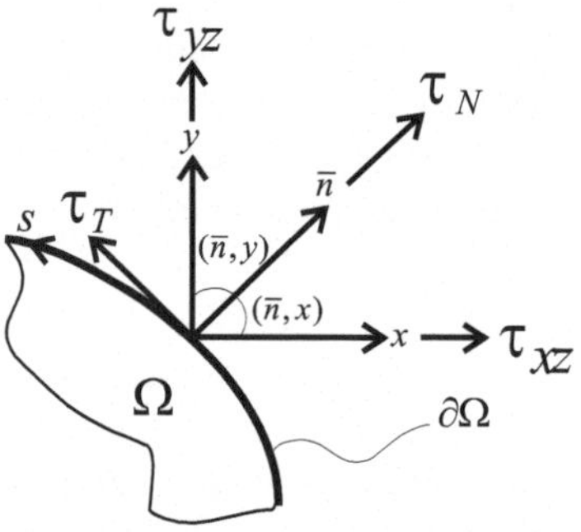

Figure 7.2: The normal and tangential stress resultants of τ_{xz} and τ_{yz} over the contour.

is essentially Z_s of (5.16c), and for future use we shall write it again together with the tangential component τ_T as

$$\tau_N = \tau_{xz}\cos(\bar{\mathbf{n}}, x) + \tau_{yz}\cos(\bar{\mathbf{n}}, y), \qquad \tau_T = -\tau_{xz}\cos(\bar{\mathbf{n}}, y) + \tau_{yz}\cos(\bar{\mathbf{n}}, x). \tag{7.5}$$

Hence, the boundary condition, or the requirement that the lateral (cylindrical) surface will remain "traction-free" (which is essential in the case of a pure tip torsional moment) is

$$\tau_N = 0 \quad \text{on} \quad \partial\Omega. \tag{7.6}$$

Examination of the compatibility equations (1.45a–f) shows that they are satisfied identically (due to the fact that they include second-order derivatives and that φ does not depend on z). The first two equilibrium equations, (1.82a, b), are satisfied identically as well, while the third one, (1.82c), together with the boundary condition (7.6), yield the following governing equation and boundary condition for φ:

$$\nabla_3^{(2)}\varphi = 0 \quad \text{over} \quad \Omega, \tag{7.7a}$$
$$D_1^n\varphi = P^\varphi\cos(\bar{\mathbf{n}}, x) + Q^\varphi\cos(\bar{\mathbf{n}}, y) \quad \text{on} \quad \partial\Omega, \tag{7.7b}$$

where the differential operators $\nabla_3^{(2)}$, D_1^n are defined in S.3.3.2, S.3.6, and

$$P^\varphi = a_{45}x + a_{44}y, \qquad Q^\varphi = -a_{55}x - a_{45}y. \tag{7.8}$$

Note that the above BVP follows from (3.116) when $\mathbf{F}_b = \mathbf{F}_s = 0$ and the auxiliary stress solution vanishes. Therefore, $F_0^{\varphi} = 0$ and only the underlined terms of (3.117b) are used in (7.8).

It has been therefore established that φ is a generalized harmonic function that should be determined as a solution of the Neumann BVP per the discussion in S.3.3.3. The existence condition (3.126) is clearly satisfied in view of Green's Theorem and the equality $P^{\varphi}_{,x} + Q^{\varphi}_{,y} \equiv 0$. Since only derivatives of φ are included in the formulation, one may select $\varphi(0,0) = 0$.

The tip conditions $P_z = M_x = M_y = 0$ at $z = l$ of (5.7) are fulfilled identically in view of $\sigma_z = 0$. To show that $P_x = P_y = 0$, one needs to integrate (7.4) over the cross-section area. By employing the system origin location definition (5.1), we first use (3.123a,b) and (7.7a) to show that

$$\{P_y, P_x\} = \oint_{\partial\Omega} \{y, x\} D_1^n \varphi. \tag{7.9}$$

Then, we use (7.7b), (7.8) to evaluate $D_1^n \varphi$, from which it becomes clear that indeed $P_x = P_y = 0$, and (7.8) enable us to rewrite (7.4) as

$$\tau_{yz} = -\frac{M_z}{D a_0}(Q^{\varphi} + a_{45}\varphi_{,x} - a_{55}\varphi_{,y}), \qquad \tau_{xz} = -\frac{M_z}{D a_0}(P^{\varphi} - a_{44}\varphi_{,x} + a_{45}\varphi_{,y}). \tag{7.10}$$

The (third) condition of (5.7) for M_z yields the *torsional rigidity*, see (Sokolnikoff, 1983), and Remark 7.1,

$$D = \iint_{\Omega} \frac{1}{a_0}[y P^{\varphi} - x Q^{\varphi} - (a_{45}x + a_{44}y)\varphi_{,x} + (a_{55}x + a_{45}y)\varphi_{,y}]. \tag{7.11}$$

For clamped-II conditions, see (5.13), one obtains by integrating the strain components (see also S.1.2)

$$u^0 = v^0 = \omega_z^0 = 0, \quad w^0 = \underline{-\frac{M_z}{D}\varphi(0,0)}, \quad \omega_y^0 = \frac{M_z}{2D}\varphi_{,x}(0,0), \quad \omega_x^0 = -\frac{M_z}{2D}\varphi_{,y}(0,0), \tag{7.12}$$

and thus, by substituting in (7.2a–c),

$$u = \frac{M_z}{D} z(-y + \frac{1}{2}\varphi_{,x}(0,0)), \qquad v = \frac{M_z}{D} z(x - \frac{1}{2}\varphi_{,y}(0,0)), \tag{7.13a}$$

$$w = \frac{M_z}{D}[\varphi - \underline{\varphi(0,0)} - \frac{1}{2}\varphi_{,x}(0,0)x - \frac{1}{2}\varphi_{,y}(0,0)y], \tag{7.13b}$$

where as indicated, the underlined term may be assumed to vanish.

To shift to clamped-I conditions as described in S.5.1.3.1, we first note that at the root

$$u_{,z} = \frac{M_z}{2D}\varphi_{,x}(0,0), \qquad v_{,z} = -\frac{M_z}{2D}\varphi_{,y}(0,0). \tag{7.14}$$

Therefore, one needs to supplement the displacement components u and v by $-\frac{M_z}{2D}\varphi_{,x}(0,0)\,z$ and $-\frac{M_z}{2D}\varphi_{,y}(0,0)\,z$, respectively, which yields,

$$u = -\frac{M_z}{D} yz, \qquad v = \frac{M_z}{D} xz, \qquad w = \frac{M_z}{D}[\varphi - \varphi_{,x}(0,0)x - \varphi_{,y}(0,0)y]. \tag{7.15}$$

Illustrative derivation of the torsion warping function $\varphi(x,y)$ for an elliptical cross-section are given in Example 7.1.

Remark 7.1 To prove that $D > 0$ and to establish an upper bound for the torsional rigidity, we write (7.11) as

$$D = \iint_\Omega \frac{1}{a_0}(yP^\varphi - xQ^\varphi) - \iint_\Omega \frac{1}{a_0}(P^\varphi \varphi_{,x} + Q^\varphi \varphi_{,y}). \tag{7.16}$$

Using Green's Theorem, in view of $P_{,x} + Q_{,y} = 0$, one may show that

$$\iint_\Omega \frac{1}{a_0}(P^\varphi \varphi_{,x} + Q^\varphi \varphi_{,y}) = \oint_{\partial\Omega} \frac{1}{a_0}\varphi D_1^n \varphi = \iint_\Omega \frac{1}{a_0}(a_{44}\varphi_{,x}^2 - 2a_{45}\varphi_{,x}\varphi_{,y} + a_{55}\varphi_{,y}^2). \tag{7.17}$$

Adding the first and the third integrals of (7.17) with plus and minus signs, respectively, to (7.16) yields

$$D = \iint_\Omega \frac{1}{a_0}\left[a_{44}(\varphi_{,x} - y)^2 - 2a_{45}(\varphi_{,x} - y)(\varphi_{,y} + x) + a_{55}(\varphi_{,y} + x)^2\right]. \tag{7.18}$$

In view of (3.223a) with $\alpha_{ij} = a_{ij}$ and $\mu = (\varphi_{,y} + x)/(\varphi_{,x} - y)$, it is clear that $D \geq 0$. For $D = 0$, the equalities $\varphi_{,x} - y = 0$, $\varphi_{,y} + x = 0$ should be satisfied, which conflicts with the identity $\varphi_{,xy} = \varphi_{,yx}$, and thus $D > 0$.

For a similar reason, the integral $\iint_\Omega(a_{44}\varphi_{,x}^2 - 2a_{45}\varphi_{,x}\varphi_{,y} + a_{55}\varphi_{,y}^2)$ is non-negative and vanishes if, and only if, φ is a constant (i.e. zero). We subsequently conclude that (see also (5.1))

$$D \leq \frac{1}{a_0}\iint_\Omega (a_{44}y^2 + 2a_{45}xy + a_{55}x^2) = \frac{1}{a_0}(a_{44}I_x + a_{55}I_y). \tag{7.19}$$

When the above is taken as equality, (7.7b) shows that $\frac{d}{ds}(a_{44}y^2 + 2a_{45}xy + a_{55}x^2) = 0$ over the cross-section contour. Hence, $\varphi = 0$ is possible only for a cross-section of elliptical boundaries of the type $a_{44}y^2 + 2a_{45}xy + a_{55}x^2 = const$. For the isotropic case, the torsional rigidity does not exceed GI_0, where $I_0 = \iint_\Omega(y^2 + x^2)$ is the polar moment of inertia of Ω, and equality is reached when Ω is a circle or a concentric circular ring. For further details see Example 7.1 below.

Remark 7.2 The above solution may be somewhat simplified if one employs the linear transformation of S.3.3.4 that is based on a new coordinate system $x' = \tilde{a}x$ and $y' = y + \tilde{b}x$, where $\tilde{a}$ and $\tilde{b}$ are given in (3.133). Subsequently, the cross-section contour in the xy-plane is transformed into a different contour in the $x'y'$-plane, see Fig. 3.16. Thus, it is sufficient to consider an equivalent isotropic case for the modified shape in the $x'y'$-plane. According to (3.135), equations (7.7a,b), (7.8) together with P^φ and Q^φ of (7.8) yield the following governing equation and boundary condition for $\varphi'(x', y') = \tilde{a}\varphi(x, y)$:

$$\varphi'_{,x'x'} + \varphi'_{,y'y'} = 0 \qquad \text{over} \quad \Omega', \tag{7.20a}$$

$$\frac{d}{dn'}\varphi' = y'\cos(\bar{\mathbf{n}}', x') - x'\cos(\bar{\mathbf{n}}', y') \qquad \text{on} \quad \partial\Omega'. \tag{7.20b}$$

Hence, we arrive at a "pure" isotropic case in the transformed plane Ω', in which the stresses are

$$\tau'_{yz} = G\theta(\varphi'_{,y'} + x'), \qquad \tau'_{xz} = G\theta(\varphi'_{,x'} - y'), \tag{7.21}$$

while G is the shear modulus. With the help of (7.4) and the definitions in S.3.3.4, one may show that once τ'_{yz} and τ'_{xz} (the stresses in the "isotropic" domain Ω') are found, the actual cases in the anisotropic case under discussion are given by

$$\tau_{yz} = \frac{1}{a_{44}\tilde{a}}\frac{\tau'_{yz}}{G} - \frac{a_{45}}{a_{44}^2\tilde{a}^2}\frac{\tau'_{xz}}{G}, \qquad \tau_{xz} = \frac{1}{a_{44}\tilde{a}^2}\frac{\tau'_{xz}}{G}. \tag{7.22}$$

For the "isotropic" domain, the torsional rigidity is given by (see (7.11) with $a_{44} = a_{55} = 1/G$, $a_{45} = 0$)

$$D' = G \iint_{\Omega'} (\varphi'_{,y'} x' - \varphi'_{,x'} y' + x'^2 + y'^2). \tag{7.23}$$

Substituting $\varphi = \varphi'/\tilde{a}$ and the definitions of S.3.3.4, one obtains

$$D = \frac{M_z}{\theta} = \frac{D'}{G a_{44} \tilde{a}^3}. \tag{7.24}$$

Hence, once the torsional rigidity of the "isotropic" cross-section is determined, the torsional rigidity of the original (anisotropic) cross-section is obtained from (7.24).

7.2.1.2 Stress Function Analysis

We shall now examine the same problem of a MON13z beam, which is acted upon by a tip torsional moment, using a suitable stress function. The discussion here is a reduced and torsion-oriented version of the Plane-Shear analysis derived in Remark 3.6 of S.3.3.1, namely,

$$\nabla_3^{(2)} \psi = -2\theta \qquad \text{over} \quad \Omega, \tag{7.25a}$$

$$\frac{d}{ds}\psi = 0 \qquad \text{on} \quad \partial\Omega. \tag{7.25b}$$

The non-vanishing stress components are defined by the underlined terms of (3.92), as

$$\tau_{yz} = -\psi_{,x} \qquad \tau_{xz} = \psi_{,y}. \tag{7.26}$$

Hence, the non-vanishing strain components are given by the underlined terms of (3.95a,b) as

$$\gamma_{yz} = a_{45}\psi_{,y} - a_{44}\psi_{,x}, \qquad \gamma_{xz} = -a_{45}\psi_{,x} + a_{55}\psi_{,y}. \tag{7.27}$$

The normal and tangential stress components of (7.5) may now be written as

$$\tau_N = \psi_{,x}\frac{dx}{ds} + \psi_{,y}\frac{dy}{ds} = \frac{d}{ds}\psi, \tag{7.28a}$$

$$\tau_T = -\psi_{,x}\cos(\bar{\mathbf{n}}, x) - \psi_{,y}\cos(\bar{\mathbf{n}}, y) = -\frac{d}{dn}\psi. \tag{7.28b}$$

Equation (7.28a) shows that $\tau_N = 0$ along level lines of ψ. This characteristic explains these lines' definition as *stress lines*. In other words, *the resultant stress at each point coincides with the direction of the tangent to the level line of* ψ *that passes through that point.*

In addition, (7.28b) shows that the magnitude of the resultant tangential stress at each point is given by the normal derivative of the stress function. Also, the resultant loads in the present case are, see (3.108),

$$\{P_x, P_y, M_z\} = \iint_{\Omega} \{\tau_{xz}, \tau_{yz}, x\tau_{yz} - y\tau_{xz}\} = \{0, 0, -2S_{\Omega}\psi_0 + 2\iint_{\Omega}\psi\}. \tag{7.29}$$

For further reading regarding solution of problems in this class see (Rand, 1992).

Remark 7.3 Equations (7.3), (7.27) and the identity $\theta = \frac{M_z}{D}$ yield the following relations between the torsion warping function and the torsion stress function derivatives:

$$\begin{Bmatrix} \varphi_{,x} \\ \varphi_{,y} \end{Bmatrix} = \frac{1}{\theta}\begin{bmatrix} a_{45} & a_{55} \\ a_{44} & a_{45} \end{bmatrix}\begin{Bmatrix} -\psi_{,x} \\ \psi_{,y} \end{Bmatrix} + \begin{Bmatrix} y \\ -x \end{Bmatrix}. \tag{7.30}$$

The above serves as a general conversion between φ and ψ. By defining a slightly different stress function, ψ_1, so that

$$\psi = \frac{\theta}{a_0}[\sqrt{a_0}\psi_1 - \frac{1}{2}(a_{55}x^2 + 2a_{45}xy + a_{44}y^2)], \tag{7.31}$$

one reaches the following field equation and boundary condition for ψ_1:

$$\nabla_3^{(2)}\psi_1 = 0 \qquad \text{over} \quad \Omega, \tag{7.32a}$$

$$\psi_1 = \frac{\sqrt{a_0}}{\theta}\psi_0 + \frac{1}{2\sqrt{a_0}}(a_{55}x^2 + 2a_{45}xy + a_{44}y^2) \qquad \text{on} \quad \partial\Omega. \tag{7.32b}$$

Then, $\psi_{1,x} = \frac{1}{\sqrt{a_0}}(a_{45}\varphi_{,x} - a_{55}\varphi_{,y})$, $\psi_{1,y} = \frac{1}{\sqrt{a_0}}(a_{44}\varphi_{,x} - a_{45}\varphi_{,y})$, which simplifies the mutual relations between the torsion warping and the stress functions.

In the isotropic case the torsion warping and the stress functions are conjugate, i.e., $\psi_{1,x} = -\varphi_{,y}$, $\psi_{1,y} = \varphi_{,x}$, and may be analytically defined as the real and imaginary parts of an analytical complex function. A rigorous derivation of this point is presented in S.4.4.

Remark 7.4 When the linear transformation of S.3.3.4 is applied to (7.25a) one obtains

$$\psi_{,x'x'} + \psi_{,y'y'} = -\frac{2\theta}{a_{44}\tilde{a}^2} \qquad \text{over} \quad \Omega', \tag{7.33}$$

while the condition (7.25b), $\frac{d}{ds}\psi(x,y)_{|\partial\Omega} = 0$, converts to $\frac{d}{ds'}\psi(x',y')_{|\partial\Omega'} = 0$. Further definition of $\psi' = (Ga_{44}\tilde{a}^2)\psi$ converts the formulation into an "isotropic" one that should be solved in the Ω' domain with a shear modulus G. It is therefore shown that similar to Remark 7.2, for a MON13z material, one may first solve the "isotropic" problem in Ω', and then convert $\psi(x',y')$ to $\psi(x,y)$ by using (3.131).

Employing (7.29) enables us to determine the moment in the "isotropic case" as

$$M_z' = -2S_{\Omega'}\psi_0' + 2\iint_{\Omega'}\psi' = Ga_{44}\tilde{a}^2(-2S_{\Omega}\psi_0 + 2\iint_{\Omega}\psi). \tag{7.34}$$

The above results may be written as $M_z' = (Ga_{44}\tilde{a}^3)M_z$. Recalling that the torsional rigidity is defined as $D = \frac{M_z}{\theta}$, it is shown again that $D' = (Ga_{44}\tilde{a}^3)D$, which is identical to the result of Remark 7.2.

Example 7.1 *Torsion of a Beam of an Elliptical Cross-Section.*

In this example, we shall present solutions for the torsion warping function $\varphi(x,y)$ and the torsion stress function $\psi(x,y)$ in an elliptical cross-section. By substituting $\Lambda = \varphi$ in Example 4.3, (7.8) show that the only non-vanishing coefficients are

$$P_{10}^{\varphi} = a_{45}, \qquad P_{01}^{\varphi} = a_{44}, \qquad Q_{10}^{\varphi} = -a_{55}, \qquad Q_{01}^{\varphi} = -a_{45}. \tag{7.35}$$

By activating **P.7.2** with $Ex_h = 2$, we arrive at the following second-order polynomial:

$$\varphi = H_{20}^{\varphi}x^2 + H_{11}^{\varphi}xy + H_{02}^{\varphi}y^2, \tag{7.36}$$

where

$$H_{20}^{\varphi} = \frac{a_{45}\tilde{b}^2}{a_{55}\tilde{a}^2 + a_{44}\tilde{b}^2}, \qquad H_{11}^{\varphi} = \frac{a_{44}\tilde{b}^2 - a_{55}\tilde{a}^2}{a_{55}\tilde{a}^2 + a_{44}\tilde{b}^2}, \qquad H_{02}^{\varphi} = -\frac{a_{45}\tilde{a}^2}{a_{55}\tilde{a}^2 + a_{44}\tilde{b}^2}. \tag{7.37}$$

It should also be noted that the warping function derivatives $\varphi_{,x}(0,0)$, $\varphi_{,y}(0,0)$ that appear in the deformation expressions (see e.g. (7.13b)) vanish in this case (since the cross-section exhibits axial symmetries). Substituting $\varphi(x,y)$ in (7.11) yields an expression for the torsional rigidity

$$D = \frac{\pi \widetilde{a}^3 \widetilde{b}^3}{a_{55}\widetilde{a}^2 + a_{44}\widetilde{b}^2}, \tag{7.38}$$

which appears to be independent of a_{45} (see also Example 4.2). However, the torsion warping function does depend on a_{45}, while for orthotropic materials, where $a_{45} = 0$, one obtains

$$\varphi_{\text{ort}} = \frac{a_{44}\widetilde{b}^2 - a_{55}\widetilde{a}^2}{a_{55}\widetilde{a}^2 + a_{44}\widetilde{b}^2} xy, \tag{7.39}$$

which clearly vanish for $a_{44}/a_{55} = (\widetilde{a}/\widetilde{b})^2$. Note that in the isotropic case $a_{44} = a_{55} = 1/G$ (where G stands for the shear modulus) and the above expression becomes material-independent. Using (7.30) it is easy to show that the torsion stress function derivatives are

$$\psi_{,x} = -\frac{2\theta \widetilde{b}^2}{a_{55}\widetilde{a}^2 + a_{44}\widetilde{b}^2} x, \qquad \psi_{,y} = -\frac{2\theta \widetilde{a}^2}{a_{55}\widetilde{a}^2 + a_{44}\widetilde{b}^2} y, \tag{7.40}$$

which by integration and selection of $\psi_0 = 0$ yields

$$\psi = -\frac{\theta \widetilde{a}^2 \widetilde{b}^2}{a_{55}\widetilde{a}^2 + a_{44}\widetilde{b}^2} \left(\frac{x^2}{\widetilde{a}^2} + \frac{y^2}{\widetilde{b}^2} - 1\right). \tag{7.41}$$

Equations (7.38), (7.40) also show that the stress components are

$$\tau_{yz} = \frac{2M_z}{\pi \widetilde{a}^3 \widetilde{b}} x, \qquad \tau_{xz} = -\frac{2M_z}{\pi \widetilde{a} \widetilde{b}^3} y. \tag{7.42}$$

The expression for D in (7.38) may also be confirmed by employing (7.29), (7.41).

Remark 7.5 We shall express here the strain energy in a MON13z beam that undergoes pure torsion. Substituting (7.3), (7.4) in (2.27) yields the *volume density of the strain energy* which for the present case may be written as $W = \frac{1}{2}(\tau_{yz}\gamma_{yz} + \tau_{yz}\gamma_{yz})$. In terms of the warping function derivatives, the total *strain energy per unit length*, $\overline{U}$, therefore becomes

$$2\overline{U} = \frac{M_z^2}{D} + \frac{M_z^2}{D^2 a_0} \iint_\Omega \left(a_{55}\varphi_{,y}^2 - 2a_{45}\varphi_{,x}\varphi_{,y} + a_{44}\varphi_{,x}^2 - P^\varphi \varphi_{,x} - Q^\varphi \varphi_{,y}\right). \tag{7.43}$$

Alternatively, substituting (7.26), (7.27) in (2.27) yields the total strain energy in terms of the stress function as

$$2\overline{U} = \iint_\Omega \left(a_{55}\psi_{,y}^2 - 2a_{45}\psi_{,x}\psi_{,y} + a_{44}\psi_{,x}^2\right). \tag{7.44}$$

For the case presented in Example 7.1, the integral term in (7.43) vanishes and the strain energy is reduced to $\overline{U} = \frac{M_z^2}{2D}$.

Example 7.2 *Application of the Minimum Complementary Energy Theorem to Torsion.*

In this example, we shall demonstrate an approximate solution of the torsion problem for a uniform beam of rectangular cross-section by employing the Theorem of Minimum Complementary Energy derived in S.1.4.2.

Referring to Fig. 7.3(a), we denote the outer surface as S_L, and the end and tip cross-sections as S_D. At the root, the displacements vanish, namely, $u = v = w = 0$, while at the tip, $u =$

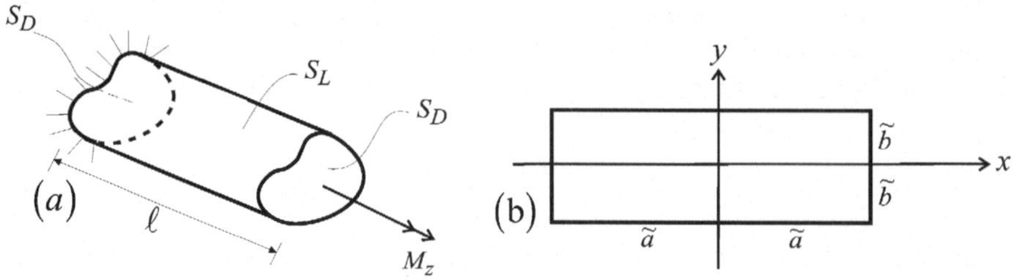

Figure 7.3: A uniform beam under torsion analyzed by the Minimum Complementary Energy Theorem.

$-\theta ly$, $v = \theta lx$, $w = 0$, where l is the beam's length and $\theta = \frac{M_z}{D}$ is the twist per unit length. According to (1.128), the complementary energy is given by

$$V^* = \frac{l}{2}\iint_{\Omega} (\gamma_{xz}\tau_{xz} + \gamma_{yz}\tau_{yz}) - \theta l \iint_{\Omega_{tip}} (x\tau_{yz} - y\tau_{xz}). \tag{7.45}$$

Since the admissible stresses must satisfy equilibrium, we use (7.26), (7.27), (7.29), and write V^* in terms of the stress function by assuming $\psi_0 = 0$, as

$$V^* = \frac{l}{2}\iint_{\Omega} \left[a_{44}\psi_{,x}{}^2 - 2a_{45}\psi_{,x}\psi_{,y} + a_{55}\psi_{,y}{}^2\right] - 2\theta l \iint_{\Omega_{tip}} \psi. \tag{7.46}$$

Following (Timoshenko and Goodier, 1970) (and although originally suggested for the isotropic case), we approximate ψ for the rectangular cross-section shown in Fig. 7.3(b) as the polynomial

$$\psi \approx (x^2 - \widetilde{a}^2)(y^2 - \widetilde{b}^2) \sum_{m=0,2,\ldots}^{M} \sum_{n=0,2,\ldots}^{N} c_{mn}x^m y^n. \tag{7.47}$$

Substituting (7.47) for $M = N = 0$ in (7.46) yields

$$\frac{9}{32l\widetilde{a}^3\widetilde{b}^3}V^* = \frac{2}{5}c_{00}^2(a_{44}\widetilde{b}^2 + a_{55}\widetilde{a}^2) - \theta c_{00}. \tag{7.48}$$

Looking for a stationary value of V^* by solving $\frac{\partial V^*}{\partial c_{00}} = 0$ shows that the minimum of V^* is reached at $c_{00} = \frac{5}{4}\frac{\theta}{a_{55}\widetilde{a}^2 + a_{44}\widetilde{b}^2}$. By evaluating the torsional moment as $M_z = 2\iint_{\Omega_{tip}} \psi = D\theta$ where $\psi = c_{00}(x^2 - \widetilde{a}^2)(y^2 - \widetilde{b}^2)$, we deduce the torsional rigidity as

$$D = \frac{40}{9} \cdot \frac{\widetilde{a}^3\widetilde{b}^3}{a_{55}\widetilde{a}^2 + a_{44}\widetilde{b}^2}. \tag{7.49}$$

The expression for the torsional rigidity in this lowest level approximation is similar to the exact solution for an ellipse given by (7.38) (while $\frac{40}{9} = 4.44$ replaces π). Note that for a thin cross-section where $\widetilde{a} \gg \widetilde{b}$, the present approximation yields a torsional rigidity of $D \cong \frac{5}{18}\frac{(2\widetilde{a})(2\widetilde{b})^3}{a_{55}}$. This result may be compared with that of (10.69) where the exact expression for the same case is derived and is shown to be identical when the value of the $\frac{5}{18}$ coefficient is replaced by $\frac{1}{3}$.

When a two-terms expression is employed, namely, $\psi = (x^2 - \widetilde{a}^2)(y^2 - \widetilde{b}^2)(c_{00} + c_{22}x^2y^2)$, a stationary value of V^* is sought in a similar way, by solving two coupled equations for c_{00}

and c_{22}. This process yields

$$D = \frac{1120\widetilde{a}^4}{9a_{44}}\widetilde{\beta}^3\widetilde{\alpha}\frac{1+\widetilde{\beta}^2(\widetilde{\beta}^4\widetilde{\alpha}+9\widetilde{\beta}^2+9\widetilde{\alpha})}{25+\widetilde{\beta}^2[25\widetilde{\beta}^6\widetilde{\alpha}^2+280\widetilde{\beta}^4\widetilde{\alpha}+\widetilde{\beta}^2(252\widetilde{\alpha}^2+252-6\widetilde{\alpha})+280\widetilde{\alpha}]}, \tag{7.50}$$

where $\widetilde{\alpha} = \frac{a_{44}}{a_{55}}$ and $\widetilde{\beta} = \frac{\widetilde{b}}{\widetilde{a}}$. For $\widetilde{a} \gg \widetilde{b}$ (i.e., $\widetilde{\beta} \to 0$) the above becomes

$$D = \frac{224\widetilde{a}^4\widetilde{\beta}^3\widetilde{\alpha}}{45\,a_{44}} = \frac{14}{45}\cdot\frac{(2\widetilde{a})\,(2\widetilde{b})^3}{a_{55}} \cong 0.311\,\frac{(2\widetilde{a})\,(2\widetilde{b})^3}{a_{55}}, \tag{7.51}$$

which is closer to the above mentioned exact value for a thin rectangular cross-section.

In the isotropic case (where $\widetilde{\alpha} = 1$), both the first and the second approximation yield a torsional rigidity that may be written as $D_i = \widetilde{k}_i G(2\widetilde{a})(2\widetilde{b})^3$ $(i = 1,2)$, where

$$\widetilde{k}_1 = \frac{5}{18}\cdot\frac{\widetilde{\gamma}^2}{1+\widetilde{\gamma}^2}, \qquad \widetilde{k}_2 = \frac{14}{45}\cdot\frac{\widetilde{\gamma}^2(1+9\widetilde{\gamma}^2+9\widetilde{\gamma}^4+\widetilde{\gamma}^6)}{1+\frac{56}{5}\widetilde{\gamma}^2+\frac{498}{25}\widetilde{\gamma}^4+\frac{56}{5}\widetilde{\gamma}^6+\widetilde{\gamma}^8}, \tag{7.52}$$

while $\widetilde{\gamma} = \frac{\widetilde{a}}{\widetilde{b}} = \frac{1}{\widetilde{\beta}}$. The variation of $\widetilde{k}_1, \widetilde{k}_2$ as functions of $\widetilde{\gamma}$ is presented in Fig. 7.4. Practically, the second approximation coincides with the exact solution.

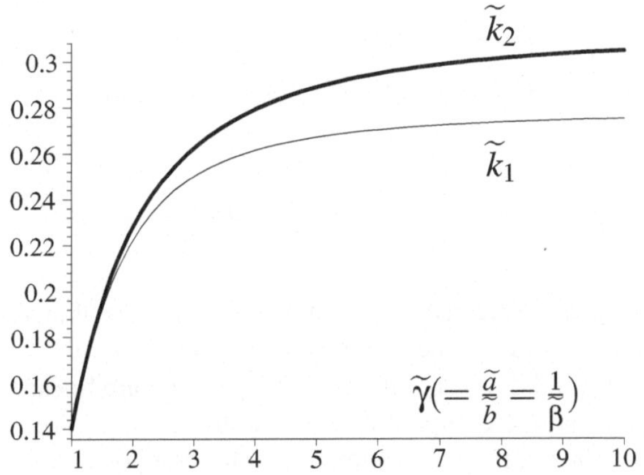

Figure 7.4: $\widetilde{k}_1$ and $\widetilde{k}_2$ vs. $\widetilde{\gamma}$ for the isotropic cross-section in Example 7.2.

7.2.2 Tip Bending Forces

We shall now seek a solution for the case where the beam is acted upon by tip bending forces P_x, P_y. Detailed derivation of the case of P_x will be provided, while as will be shown later on, the case of P_y may be easily derived in an analogous way. The solution methodology in what follows may be generally classified as "Stress/Strain Hypothesis" or Trail "B" in S.1.6.5.2.

7.2.2.1 Edgewise Bending Force

In the case of a tip edgewise bending force, P_x, we adopt a *solution hypothesis* that states that all stress components vanish except

$$\sigma_z = -\frac{P_x(l-z)}{I_y}x, \tag{7.53a}$$

$$\tau_{yz} = \frac{P_x a_{33}}{I_y a_0}(Q^{\chi_1} + a_{45}\chi_{1,x} - a_{55}\chi_{1,y}), \qquad \tau_{xz} = \frac{P_x a_{33}}{I_y a_0}(P^{\chi_1} - a_{44}\chi_{1,x} + a_{45}\chi_{1,y}) \quad (7.53b)$$

where

$$P^{\chi_1} = p_{20}^{\chi_1} x^2 + p_{11}^{\chi_1} xy + p_{02}^{\chi_1} y^2, \qquad Q^{\chi_1} = q_{20}^{\chi_1} x^2 + q_{11}^{\chi_1} xy + q_{02}^{\chi_1} y^2, \tag{7.54}$$

and

$$p_{20}^{\chi_1} = \frac{a_{44}a_{13} - a_{36}a_{45}}{2a_{33}}, \qquad p_{11}^{\chi_1} = \frac{a_{45}(2a_{33} - a_{23})}{a_{33}}, \qquad p_{02}^{\chi_1} = -\frac{a_{44}(a_{23} + 2a_{33})}{2a_{33}},$$
$$q_{20}^{\chi_1} = \frac{a_{55}a_{36} - a_{45}a_{13}}{2a_{33}}, \qquad q_{11}^{\chi_1} = \frac{a_{55}(a_{23} - 2a_{33})}{a_{33}}, \qquad q_{02}^{\chi_1} = \frac{a_{45}(a_{23} + 2a_{33})}{2a_{33}}. \tag{7.55}$$

Here $\chi_1(x,y)$ is an unknown function that will be referred to as the *edgewise bending function*, and is the only non-trivial function in the above expressions. One may show that the equilibrium equations and the boundary conditions are satisfied identically by the above stress expressions, except for the equilibrium equation (1.82c) and the boundary condition (5.16c) (i.e. the equations in the "z-direction") that establish the following BVP for $\chi_1(x,y)$ (see differential operators definition in S.3.6):

$$\nabla_3^{(2)}\chi_1 = F_0^{\chi_1} \quad \text{over} \quad \Omega, \tag{7.56a}$$
$$D_1^n\chi_1 = P^{\chi_1}\cos(\bar{\mathbf{n}},x) + Q^{\chi_1}\cos(\bar{\mathbf{n}},y) \quad \text{on} \quad \partial\Omega, \tag{7.56b}$$

where

$$F_0^{\chi_1} = \bar{F}_{10}^{\chi_1} x + \bar{F}_{01}^{\chi_1} y, \quad \bar{F}_{10}^{\chi_1} = \frac{a_0 + a_{13}a_{44} + a_{23}a_{55} - a_{36}a_{45} - 2a_{33}a_{55}}{a_{33}}, \quad \bar{F}_{01}^{\chi_1} = 4a_{45}. \tag{7.57}$$

Hence, χ_1 is shown to be a *generalized* harmonic function (note that for the isotropic case $F_0^{\chi_1} = 0$ and the function χ_1 is harmonic). The existence condition (3.126) is satisfied in view of Green's Theorem, (5.1), and the fact that $F_0^{\chi_1}$ and $P_{,x}^{\chi_1} + Q_{,y}^{\chi_1}$ are linear functions of x and y.

The reader may verify that the compatibility equations (1.45a–f) are satisfied and pose no additional condition, by substituting the following strain components (that emerge from the above stress expressions and (7.1))

$$\varepsilon_i = -a_{i3}\frac{P_x(l-z)}{I_y}x, \qquad i \in \{1,2,3,6\}, \tag{7.58a}$$
$$\gamma_{yz} = a_{45}\tau_{xz} + a_{44}\tau_{yz}, \qquad \gamma_{xz} = a_{55}\tau_{xz} + a_{45}\tau_{yz}. \tag{7.58b}$$

To examine the end conditions, (5.7), we first note that in view of (5.1), the natural tip conditions at $z = l$ for M_x, M_y and P_z are fulfilled identically. As far as P_x and P_y are concerned, we employ the identities of S.3.3.3 and (5.1), which show that

$$\iint_\Omega \{\tau_{yz}, \tau_{xz}\} = \{0, P_x\}. \tag{7.59}$$

The condition for the tip torsional moment in (5.7) yields $M_z = -a_{33}\frac{P_x}{I_y}D_x$, where

$$D_x = \iint_\Omega \frac{1}{a_0}[yP^{\chi_1} - xQ^{\chi_1} - (a_{44}y + a_{45}x)\chi_{1,x} + (a_{55}x + a_{45}y)\chi_{1,y}], \tag{7.60}$$

which is not zero unless the center of gravity coincides with the *shear center*, see Remark 7.6.

Hence, in the general case, in order to satisfy the natural boundary condition of zero tip torsional moment, one needs to superimpose a "torsion solution" such as the one derived in S.7.2.1, for a tip moment of $a_{33}\frac{P_x}{I_y}D_x$. This solution will clearly induce additional shear stresses and a twist angle.

To determine the displacements, we integrate the strain components of (7.58a,b) by the procedure of S.1.2, which yields

$$u = \frac{P_x}{2I_y}[(l-z)(-a_{13}x^2 + a_{23}y^2) + a_{33}(l - \frac{1}{3}z)z^2] + u^0 - \omega_z^0 y + \omega_y^0 z, \tag{7.61a}$$

$$v = -\frac{P_x}{2I_y}(a_{36}x^2 + 2a_{23}xy)(l-z) + v^0 - \omega_x^0 z + \omega_z^0 x, \tag{7.61b}$$

$$w = -\frac{P_x a_{33}}{I_y}[(l - \frac{1}{2}z)zx + xy^2 + \chi_1] + w^0 - \omega_y^0 x + \omega_x^0 y. \tag{7.61c}$$

For a beam with "clamped-II" conditions at its root (see (5.13)) one obtains

$$u^0 = v^0 = 0, \qquad w^0 = \frac{P_x}{I_y}a_{33}\chi_1(0,0),$$

$$\omega_z^0 = 0, \qquad \omega_y^0 = -\frac{P_x}{2I_y}a_{33}\chi_{1,x}(0,0), \qquad \omega_x^0 = \frac{P_x}{2I_y}a_{33}\chi_{1,y}(0,0). \tag{7.62}$$

Similar to the case discussed in S.7.2.1, for convenience, one may select $\chi_1(0,0) = 0$, which yields $w^0 = 0$. An illustrative derivation of the edgewise bending function $\chi_1(x,y)$ in an elliptical cross-section is discussed in Example 7.3.

7.2.2.2 Beamwise Bending Force

Analogously to the previous case, we may reach a solution for the case of tip load P_y by solving the system:

$$\nabla_3^{(2)}\chi_2 = F_0^{\chi_2} \quad \text{over} \quad \Omega, \tag{7.63a}$$

$$D_1^n\chi_2 = P^{\chi_2}\cos(\bar{\mathbf{n}},x) + Q^{\chi_2}\cos(\bar{\mathbf{n}},y) \quad \text{on} \quad \partial\Omega, \tag{7.63b}$$

where χ_2 will be referred to as the *beamwise bending function*, as it plays a role analogous to the one of χ_1 in edgewise bending. Here

$$F_0^{\chi_2} = \overline{F}_{10}^{\chi_2}x + \overline{F}_{01}^{\chi_2}y, \tag{7.64a}$$

$$P^{\chi_2} = p_{20}^{\chi_2}x^2 + p_{11}^{\chi_2}xy + p_{02}^{\chi_2}y^2, \qquad Q^{\chi_2} = q_{20}^{\chi_2}x^2 + q_{11}^{\chi_2}xy + q_{02}^{\chi_2}y^2, \tag{7.64b}$$

where

$$\overline{F}_{10}^{\chi_2} = 4a_{45}, \qquad \overline{F}_{01}^{\chi_2} = \frac{a_0 + a_{23}a_{55} + a_{13}a_{44} - a_{36}a_{45} - 2a_{33}a_{44}}{a_{33}}, \tag{7.65}$$

$$p_{20}^{\chi_2} = \frac{a_{45}(a_{13} + 2a_{33})}{2a_{33}}, \qquad p_{11}^{\chi_2} = \frac{a_{44}(a_{13} - 2a_{33})}{a_{33}}, \qquad p_{02}^{\chi_2} = \frac{a_{44}a_{36} - a_{45}a_{23}}{2a_{33}},$$

$$q_{20}^{\chi_2} = -\frac{a_{55}(a_{13} + 2a_{33})}{2a_{33}}, \qquad q_{11}^{\chi_2} = \frac{a_{45}(2a_{33} - a_{13})}{a_{33}}, \qquad q_{02}^{\chi_2} = \frac{a_{55}a_{23} - a_{36}a_{45}}{2a_{33}},$$

and one may select $\chi_2(0,0) = 0$. Comparison with the corresponding equations in S.7.2.2.1 shows that a solution for this case may be directly deduced from the solution for P_x by suitable coordinates and parameters interchange. An illustrative derivation of the beamwise bending function $\chi_2(x,y)$ in an elliptical cross-section is discussed in Example 7.3.

Remark 7.6 The fact that the resulting D_x does not vanish enables the y coordinate calculation of the shear center, which for the present purpose is defined as *the location where transverse bending induces no twist.* Practically, the y coordinate of the shear center is derived as $y_{sc} = a_{33}\frac{D_x}{I_y}$.

Similarly, the residual moment $M_z = -a_{33}\frac{P_y}{I_x}D_y$ obtained from S.7.2.2.2 for beamwise bending may be used to determine the x coordinate of the shear center as $x_{sc} = -a_{33}\frac{D_y}{I_x}$, where

$$D_y = \iint_\Omega \frac{1}{a_0}[yP^{\chi_2} - xQ^{\chi_2} - (a_{44}y + a_{45}x)\chi_{2,x} + (a_{55}x + a_{45}y)\chi_{2,y}]. \tag{7.66}$$

It should be emphasized that unlike the isotropic case (see Example 3.10), in the MON13z beam, the shear center location is a function of the material properties as well. Note that in other cases of anisotropy, such as the coupled MON13y beams shown in Remark 9.5, the entire concept of shear center collapses, as the twist $\phi_{,z}(z)$ is no longer a linear function, and thus, adding a tip moment by shifting the tip transverse force is not capable of cancelling the twist along the entire beam simultaneously.

Example 7.3 *The Bending Functions in an Elliptical Cross-Section.*

The bending functions $\chi_i(x,y)$ $(i = 1,2)$ for an elliptical cross-section are third-degree polynomials. To derive them, we shall first seek suitable particular solutions. We note that $F_0^{\chi_i}$ of (7.57), (7.64a) are linear functions of x and y. We therefore select particular solutions of (7.56a), (7.63a) in the form $\chi_i^p(x,y) = \widetilde{A}_i x^3 + \widetilde{B}_i y^3$ where

$$\widetilde{A}_i = \frac{F_{10}^{\chi_i}}{6a_{44}}, \qquad \widetilde{B}_i = \frac{F_{01}^{\chi_i}}{6a_{55}}. \tag{7.67}$$

For the homogeneous solution we shall adopt the results of Example 4.4. For that purpose, we may directly use the coefficients $p_{20}^{\chi_i}$, $p_{02}^{\chi_i}$, $p_{11}^{\chi_i}$, $q_{20}^{\chi_i}$, $q_{02}^{\chi_i}$ and $q_{11}^{\chi_i}$ from (7.55), (7.65).

Due to the above particular solution, and according to (3.216), the boundary conditions should be corrected by the additional terms

$$\Delta P^{\chi_i} = -a_{44}\chi_{i,x}^p + a_{45}\chi_{i,y}^p, \qquad \Delta Q^{\chi_i} = a_{45}\chi_{i,x}^p - a_{55}\chi_{i,y}^p. \tag{7.68}$$

Therefore, the following coefficients should be modified:

$$p_{20}^{\chi_i} \Rightarrow p_{20}^{\chi_i} - 3\widetilde{A}_i a_{44}, \qquad p_{02}^{\chi_i} \Rightarrow p_{02}^{\chi_i} + 3\widetilde{B}_i a_{45}, \tag{7.69a}$$
$$q_{20}^{\chi_i} \Rightarrow q_{20}^{\chi_i} + 3\widetilde{A}_i a_{45}, \qquad q_{02}^{\chi_i} \Rightarrow q_{02}^{\chi_i} - 3\widetilde{B}_i a_{55}. \tag{7.69b}$$

According to (4.55), the general harmonic (bending) functions, χ_i, are written as the following third-order polynomials:

$$\chi_i = (H_{30}^{\chi_i} + \widetilde{A}_i)x^3 + H_{21}^{\chi_i}x^2y + H_{12}^{\chi_i}xy^2 + (H_{03}^{\chi_i} + \widetilde{B}_i)y^3 + H_{10}^{\chi_i}x + H_{01}^{\chi_i}y. \tag{7.70}$$

One should therefore substitute the above $p_{ij}^{\chi_i}$, $q_{ij}^{\chi_i}$ coefficients in the general solution of Example 4.4 and calculate the $H_{ij}^{\chi_i}$ coefficients from (4.56), (4.57a–d). Note that the derivatives

$$\chi_{i,x}(0,0) = H_{10}^{\chi_i}, \qquad \chi_{i,y}(0,0) = H_{01}^{\chi_i} \tag{7.71}$$

(required for example in (7.62)) do not vanish.

By activating **P.7.2** with $Ex_h = 2$ and $Ex_h = 3$, one may create the detailed analytical solutions for χ_1 and χ_2, respectively.

7.2.3 Summarizing the Tip Loading Effects

For future purposes we shall summarize here the "St. Venant's Problem" solutions, namely, the effect of the three *tip forces* P_x, P_y, P_z, and three *tip moments* M_x, M_y, M_z, discussed so far for a MON13z cantilever beam. Note that the effects of P_z, M_x, M_y, have already appeared in S.5.2.2 for the general anisotropic case. The stress components are

$$\sigma_x = \sigma_y = \tau_{xy} = 0, \tag{7.72a}$$

$$\sigma_z = \frac{P_z}{S_\Omega} + \frac{M_x}{I_x} y - \frac{M_y}{I_y} x - \frac{P_x}{I_y}(l-z)x - \frac{P_y}{I_x}(l-z)y, \tag{7.72b}$$

$$\begin{aligned} \tau_{yz} = &\frac{P_x a_{33}}{I_y a_0}(Q^{\chi_1} + a_{45}\chi_{1,x} - a_{55}\chi_{1,y}) + \frac{P_y a_{33}}{I_x a_0}(Q^{\chi_2} + a_{45}\chi_{2,x} - a_{55}\chi_{2,y}) \\ &- \frac{M_z^*}{D a_0}(Q^\varphi + a_{45}\varphi_{,x} - a_{55}\varphi_{,y}), \end{aligned} \tag{7.72c}$$

$$\begin{aligned} \tau_{xz} = &\frac{P_x a_{33}}{I_y a_0}(P^{\chi_1} - a_{44}\chi_{1,x} + a_{45}\chi_{1,y}) + \frac{P_y a_{33}}{I_x a_0}(P^{\chi_2} - a_{44}\chi_{2,x} + a_{45}\chi_{2,y}) \\ &- \frac{M_z^*}{D a_0}(P^\varphi - a_{44}\varphi_{,x} + a_{45}\varphi_{,y}) \end{aligned} \tag{7.72d}$$

where $M_z^* = M_z + a_{33}(\frac{P_x}{I_y}D_x + \frac{P_y}{I_x}D_y)$ reflects the fact that P_x and P_y might cause a moment resultant at the tip.

The above expressions are based on three non-trivial functions: the torsion function $\varphi(x,y)$, and the bending functions $\chi_1(x,y)$, $\chi_2(x,y)$, that should be solved for each specific cross-section geometry. Note that for the MON13z material under discussion, the in-plane stress components $(\sigma_x, \sigma_y, \tau_{xy})$ vanish for all tip loads.

The strain components are obtained from the stress-strain relationships (7.1) as

$$\varepsilon_x = a_{13}[\frac{P_z}{S_\Omega} + \frac{M_x}{I_x} y - \frac{M_y}{I_y} x - \frac{P_x}{I_y} x(l-z) - \frac{P_y}{I_x} y(l-z)], \tag{7.73a}$$

$$\varepsilon_y = a_{23}[\frac{P_z}{S_\Omega} + \frac{M_x}{I_x} y - \frac{M_y}{I_y} x - \frac{P_x}{I_y} x(l-z) - \frac{P_y}{I_x} y(l-z)], \tag{7.73b}$$

$$\varepsilon_z = a_{33}[\frac{P_z}{S_\Omega} + \frac{M_x}{I_x} y - \frac{M_y}{I_y} x - \frac{P_x}{I_y} x(l-z) - \frac{P_y}{I_x} y(l-z)], \tag{7.73c}$$

$$\begin{aligned} \gamma_{yz} = &(\varphi_{,y} + x)\frac{M_z^*}{D} + [\frac{1}{2}a_{36}x^2 + (a_{23} - 2a_{33})yx - a_{33}\chi_{1,y}]\frac{P_x}{I_y} \\ &+ [-(a_{33} + \frac{1}{2}a_{13})x^2 + \frac{1}{2}a_{23}y^2 - a_{33}\chi_{2,y}]\frac{P_y}{I_x}, \end{aligned} \tag{7.73d}$$

$$\begin{aligned} \gamma_{xz} = &(\varphi_{,x} - y)\frac{M_z^*}{D} + [a_{13}\frac{x^2}{2} - (a_{33} + \frac{1}{2}a_{23})y^2 - a_{33}\chi_{1,x}]\frac{P_x}{I_y} \\ &+ [(a_{13} - 2a_{33})yx + \frac{1}{2}a_{36}y^2 - a_{33}\chi_{2,x}]\frac{P_y}{I_x}, \end{aligned} \tag{7.73e}$$

$$\gamma_{xy} = a_{36}[\frac{P_z}{S_\Omega} + \frac{M_x}{I_x} y - \frac{M_y}{I_y} x - \frac{P_x}{I_y} x(l-z) - \frac{P_y}{I_x} y(l-z)]. \tag{7.73f}$$

Integrating the above strain components yields the displacements

$$u = \frac{P_x}{2I_y}[(l-z)(a_{23}y^2 - a_{13}x^2) + a_{33}(l - \frac{1}{3}z)z^2] \tag{7.74a}$$

$$
\begin{aligned}
&-\frac{P_y}{2I_x}(l-z)(2a_{13}xy+a_{36}y^2)+\frac{P_z}{S_\Omega}(a_{13}x+\frac{1}{2}a_{36}y)-\frac{M_z^*}{D}yz\\
&+\frac{M_x}{2I_x}(2a_{13}xy+a_{36}y^2)+\frac{M_y}{2I_y}(a_{23}y^2-a_{13}x^2+a_{33}z^2)+u^0-\omega_z^0 y+\omega_y^0 z,\\
v=&-\frac{P_x}{2I_y}(l-z)(a_{36}x^2+2a_{23}xy)+\frac{P_y}{2I_x}[(l-z)(a_{13}x^2-a_{23}y^2)+a_{33}(l-\frac{1}{3}z)z^2]\\
&+\frac{P_z}{S_\Omega}(\frac{1}{2}a_{36}x+a_{23}y)+\frac{M_z^*}{D}xz\\
&-\frac{M_x}{2I_x}(a_{13}x^2-a_{23}y^2+a_{33}z^2)-\frac{M_y}{2I_y}(a_{36}x^2+2a_{23}xy)+v^0-\omega_x^0 z+\omega_z^0 x, \qquad (7.74b)\\
w=&-\frac{P_x}{I_y}a_{33}[(l-\frac{1}{2}z)zx+xy^2+\chi_1]-\frac{P_y}{I_x}a_{33}[(l-\frac{1}{2}z)zy+x^2y+\chi_2]\\
&+\frac{P_z}{S_\Omega}a_{33}z+\frac{M_x}{I_x}a_{33}yz-\frac{M_y}{I_y}a_{33}xz+\frac{M_z^*}{D}\varphi+w^0-\omega_y^0 x+\omega_x^0 y. \qquad (7.74c)
\end{aligned}
$$

For a clamped-free beam with clamped-II type of geometrical conditions at its root we apply

$$
\begin{aligned}
&u^0=v^0=0, \qquad \omega_z^0=0,\\
&w^0=\frac{P_x}{I_y}a_{33}\underline{\chi_1(0,0)}+\frac{P_y}{I_x}a_{33}\underline{\chi_2(0,0)}-\frac{M_z^*}{D}\underline{\varphi(0,0)},\\
&\omega_y^0=-\frac{P_x}{2I_y}a_{33}\chi_{1,x}(0,0)-\frac{P_y}{2I_x}a_{33}\chi_{2,x}(0,0)+\frac{M_z^*}{2D}\varphi_{,x}(0,0),\\
&\omega_x^0=\frac{P_x}{2I_y}a_{33}\chi_{1,y}(0,0)+\frac{P_y}{2I_x}a_{33}\chi_{2,y}(0,0)-\frac{M_z^*}{2D}\varphi_{,y}(0,0). \qquad (7.75)
\end{aligned}
$$

As discussed, $\chi_1(0,0)=\chi_2(0,0)=\varphi(0,0)=0$ may be assumed with no loss of generality for which $w^0=0$ is obtained. Under these assumptions, the beam axis deformation (i.e. the deformation of the $x=y=0$ line) becomes

$$
\begin{aligned}
&u_0=\frac{P_x}{2I_y}a_{33}(l-\frac{1}{3}z)z^2+\frac{M_y}{2I_y}a_{33}z^2\underline{+\omega_y^0 z}, \quad v_0=\frac{P_y}{2I_x}a_{33}(l-\frac{1}{3}z)z^2-\frac{M_x}{2I_x}a_{33}z^2\underline{-\omega_x^0 z},\\
&w_0=\frac{P_z}{S_\Omega}a_{33}z. \qquad (7.76)
\end{aligned}
$$

In addition, the beam axis rotation components turn out to be

$$
\omega_{x0}=-v_{0,z}\underline{-\omega_x^0}, \qquad \omega_{y0}=u_{0,z}\underline{-\omega_y^0}, \qquad \omega_{z0}=\frac{M_z^*}{D}z(\equiv\phi). \qquad (7.77)
$$

Hence, by ignoring the relatively small underlined terms, the above results become identical to those obtained by the elementary theory (see S.5.3.4) of isotropic beams (with $a_{33}=1/E$ where E stands for Young's modulus). This also supports the fact that MON13z beams are generally termed as "uncoupled".

As already indicated, all tip loads solutions are verified and documented in **P.7.1**.

Remark 7.7 To gain a global look at the above presented solutions for the M_z, P_x and P_y tip loads, one may adopt the stress expressions (7.72a–d) as the initial stress hypothesis and carry out a verification procedure that proves that these proposed expressions constitute a valid solution of the St. Venant's problem. This procedure may be summarized as follows:

(a) We first note that the equilibrium equations, (1.82a, b), are satisfied identically. From the third equilibrium equation, (1.82c), we deduce the field equations (in Ω) for the functions φ, χ_1 and χ_2. We do that by isolating products of $\frac{M_z^*}{D}$, $\frac{P_x}{I_y}a_{33}$ and $\frac{P_y}{I_x}a_{33}$, respectively.

(b) Similar to the above, the boundary conditions (5.16a, b) are satisfied identically. From the third boundary condition, (5.16c), we deduce the boundary conditions (on the contour $\partial\Omega$) for the harmonic functions φ, χ_1 and χ_2.

(c) By employing (7.73a–f), one may verify that all compatibility equations, (1.45a–f), are satisfied identically.

7.3 Axially Distributed Loads

Throughout the derivation in this section, we shall seek analytical formulation for the problem of a MON13z beam that is acted upon by a general distribution of body forces $X_b(x,y,z)$, $Y_b(x,y,z)$, $Z_b(x,y,z)$, $(x,y) \in \Omega$, that act over each material volume, and external surface loads $X_s(x,y,z)$, $Y_s(x,y,z)$, $Z_s(x,y,z)$, $(x,y) \in \partial\Omega$, that act over the beam surface. Unless stated differently, all notation and preliminaries considered in S.5.1 will be adopted here as well.

We shall adopt here the polynomial expansion (with respect to z) for the *distributed loads* (per unit area), $\vec{\mathbf{F}}_s = \{X_s, Y_s, Z_s\}$, and the *distributed body forces* (per unit volume), $\vec{\mathbf{F}}_b = \{X_b, Y_b, Z_b\}$ that appear in (6.1). For further use we recall the definition (6.49a) of the body-force potentials

$$\overline{U}_1 = -\int_0^x X_b\,dx = \sum\nolimits_{k=0}^{K} \overline{U}_1^{(k)}(x,y)z^k, \qquad \overline{U}_2 = -\int_0^y Y_b\,dy = \sum\nolimits_{k=0}^{K} \overline{U}_2^{(k)}(x,y)z^k. \quad (7.78)$$

Moreover, the functions $X_s^{(k)}(x,y)$, $Y_s^{(k)}(x,y)$, $Z_s^{(k)}(x,y)$ are expressed in the form

$$X_s^{(k)} = X_1^{(k)}(x,y)\cos(\bar{\mathbf{n}},x) + X_2^{(k)}(x,y)\cos(\bar{\mathbf{n}},y), \quad (7.79a)$$

$$Y_s^{(k)} = Y_1^{(k)}(x,y)\cos(\bar{\mathbf{n}},x) + Y_2^{(k)}(x,y)\cos(\bar{\mathbf{n}},y), \quad (7.79b)$$

$$Z_s^{(k)} = Z_1^{(k)}(x,y)\cos(\bar{\mathbf{n}},x) + Z_2^{(k)}(x,y)\cos(\bar{\mathbf{n}},y). \quad (7.79c)$$

7.3.1 *Solution Methodology*

A scheme of the level-based solution methodology is presented in Fig. 7.5. As already implied by (6.1), K is the highest power level of the (given) loading distribution. The process is initiated for $k = K$ and continues for lower levels down to $k = 0$. For each $K \geq k \geq 0$ level, a set of the biharmonic and the Neumann BVPs should be solved. As will be shown, each solution level is driven by the loading at its level and the quantities obtained in previous (higher) levels. Once all levels are solved, their contributions are summed up for the overall stress, strain and deformation. Finally tip corrections are produced as will be clarified later on within S.7.3.6.

In what follows we shall describe the solution hypothesis and a formal verification for a generic $k^{\underline{th}}$ level. All functions that are level-dependent are denoted by a superscript in brackets, while level-dependent constants are denoted by a subscript. For example, the biharmonic function that is calculated at the $k^{\underline{th}}$ level is denoted $\Phi^{(k)}$, and a constant obtained in the $k^{\underline{th}}$ level is denoted p_k. Each component has its maximal level of appearance. Hence, by definition, we shall assume that *all quantities with level index that is greater than their maximal level of appearance or less than zero*, vanish.

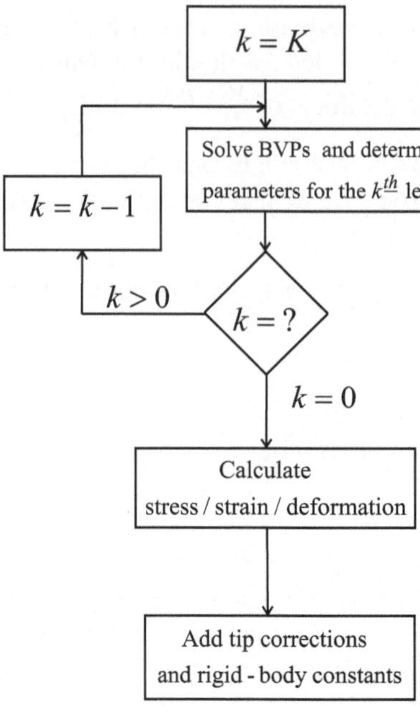

Figure 7.5: Solution procedure for axially non-uniform loading of a MON13z beam.

We shall first present all components participating in the solution hypothesis, and then describe the process that proves the validity of the above solution hypothesis (see discussion in the introduction to this chapter). The entire derivation presented in this section is verified *symbolically* by **P.7.3, P.7.4**.

7.3.2 Solution Hypothesis

The solution hypothesis is given in terms of the stress components followed by a description of the various ingredients of the stress expressions, and therefore, may be generally classified as "Stress/Strain Hypothesis" or Trail "B" of S.1.6.5.2. In general, we express each of the three stress components $\sigma_x, \sigma_y, \tau_{xy}$ as polynomials of degree K in z, each of the two stress components τ_{yz}, τ_{xz} as polynomials of degree $K+1$ in z, and σ_z as a polynomial of degree $K+2$ in z , as follows

$$\{\sigma_x, \sigma_y, \tau_{xy}\} = \sum_{k=0}^{K} \{\sigma_x^{(k)}, \sigma_y^{(k)}, \tau_{xy}^{(k)}\} z^k, \quad \{\tau_{xz}, \tau_{yz}\} = \sum_{k=0}^{K+1} \{\tau_{xz}^{(k)}, \tau_{yz}^{(k)}\} z^k, \quad \sigma_z = \sum_{k=0}^{K+2} \sigma_z^{(k)} z^k \tag{7.80}$$

where $\sigma_i^{(k)}, \tau_{ij}^{(k)}$ are functions of x, y. The detailed stress expressions are

$$\sigma_x = \sum_{k=0}^{K} z^k \{\underbrace{\Phi_{,yy}^{(k)} + \overline{U}_1^{(k)}} - \frac{k+1}{a_0}[a_{44} H^{(k+1)} + \frac{d_{k+1} a_0}{4 a_{33}}(x^2 - \frac{3a_{23} + a_{44}}{a_{13}} y^2) + S_x^{(k+1)}]\}, \tag{7.81a}$$

$$\sigma_y = \sum_{k=0}^{K} z^k \{\underbrace{\Phi_{,xx}^{(k)} + \overline{U}_2^{(k)}} - \frac{k+1}{a_0}[a_{55} H^{(k+1)} + \frac{d_{k+1} a_0}{4 a_{33}}(y^2 - \frac{3a_{13} + a_{55}}{a_{23}} x^2) + S_y^{(k+1)}]\}, \tag{7.81b}$$

$$\sigma_z = \sum_{k=0}^{K} \frac{z^k}{a_{33}} \{\underbrace{-a_{13}(\Phi_{,yy}^{(k)} + \overline{U}_1^{(k)}) - a_{23}(\Phi_{,xx}^{(k)} + \overline{U}_2^{(k)}) + a_{36}\Phi_{,xy}^{(k)}} - \frac{z^2(p_k x + q_k y)}{(k+1)(k+2)} \tag{7.81c}$$

$$-\frac{zd_k}{k+1}+(p_k y+q_k x)xy+\frac{k+1}{a_0}[(a_{13}a_{44}+a_{23}a_{55}-a_{36}a_{45}+a_0)H^{(k+1)}$$
$$+\frac{d_{k+1}a_0}{2a_{33}}(a_{45}+a_{36})xy+a_{13}S_x^{(k+1)}+a_{23}S_y^{(k+1)}]\},$$

$$\tau_{yz}=\sum_{k=0}^{K+1}\frac{z^k}{a_0}\{\underbrace{a_{55}H^{(k)}_{,y}-a_{45}H^{(k)}_{,x}}+\frac{d_k a_0}{2a_{33}}y+S^{(k)}_{y,y}\}, \quad (7.81d)$$

$$\tau_{xz}=\sum_{k=0}^{K+1}\frac{z^k}{a_0}\{\underbrace{a_{44}H^{(k)}_{,x}-a_{45}H^{(k)}_{,y}}+\frac{d_k a_0}{2a_{33}}x+S^{(k)}_{x,x}\}, \quad (7.81e)$$

$$\tau_{xy}=\sum_{k=0}^{K}z^k[\underbrace{-\Phi^{(k)}_{,xy}}+(k+1)\frac{a_{45}}{a_0}H^{(k+1)}]. \quad (7.81f)$$

From (7.81c) it is clear that the higher components of the axial stress are

$$\sigma_z^{(K+2)}=-\frac{p_K x+q_K y}{(K+1)(K+2)a_{33}},\qquad \sigma_z^{(K+1)}=\begin{cases}-\frac{p_{K-1}x+q_{K-1}y}{K(K+1)a_{33}}-\frac{d_K}{(K+1)a_{33}}, & K>0,\\ -\frac{d_0}{a_{33}}, & K=0.\end{cases} \quad (7.82)$$

As expected, using (7.99), $\sigma_z^{(K+2)}$ may be shown to be identical to that of (6.8a).

As a general perspective on the above hypothesis, note that by setting $K=0$ (a uniform loading case), the under-braced terms in (7.81a–c,f) are identical to those obtained for plane-strain, see (3.26a), (3.54). In addition, the under-braced terms in (7.81d,e) are identical for $K=0$ to (3.115) (with $\theta=0$) that were developed for plane shear while $H(x,y)$ takes the role of an axial deformation function.

The above stress expressions together with the constitutive relations of (7.1) yield the strain components given in (7.104a–f). Furthermore, the integration scheme of S.1.2 is used to determine the displacement and rotation components via integration of (7.104a–f) as documented in (7.107a–c), (7.109a–c).

The loading constants p_k, q_k, τ_k, d_k $(k=0,\ldots,K)$ that are included in the stress expressions are given below by (7.98). In addition, the expressions for $S_x^{(k)}$ and $S_y^{(k)}$ $(k=0,\ldots,K+1)$ are

$$S_x^{(k)}=\int_0^x[(k+1)(a_{44}L^{(k+1)}-a_{45}M^{(k+1)})-\widetilde{\delta}_{k0}(p_{k-1}P^{\chi_1}+q_{k-1}P^{\chi_2}-\tau_{k-1}P^{\varphi})]\,dx, \quad (7.83a)$$

$$S_y^{(k)}=\int_0^y[(k+1)(a_{55}M^{(k+1)}-a_{45}L^{(k+1)})-\widetilde{\delta}_{k0}(p_{k-1}Q^{\chi_1}+q_{k-1}Q^{\chi_2}-\tau_{k-1}Q^{\varphi})]\,dy, \quad (7.83b)$$

where $\widetilde{\delta}_{k0}=\frac{1}{k}$ for $k\neq 0$ and $\widetilde{\delta}_{00}=0$, while P^{χ_k}, Q^{χ_k}, P^{φ}, Q^{φ} are defined in S.7.2.

As shown by (7.81a–f), the present formulation also includes two series of auxiliary functions by $L^{(k)}(x,y)$, $M^{(k)}(x,y)$ $(k=0,\ldots,K)$. These functions are defined by (7.96a,b).

7.3.3 The Harmonic Stress Functions

The hypothesis (7.81a–f) contains a series of generalized harmonic stress functions $H^{(k)}(x,y)$ $(k=0,\ldots,K+1)$, which depend on the cross-section geometry, the material coefficients and the loading, and are governed by the Neumann BVPs

$$\nabla_3^{(2)}H^{(k)}=F_0^{H^{(k)}} \qquad \text{over} \quad \Omega, \quad (7.84a)$$

$$D_1^n H^{(k)}=P^{H^{(k)}}\cos(\bar{\mathbf{n}},x)+Q^{H^{(k)}}\cos(\bar{\mathbf{n}},y) \qquad \text{on} \quad \partial\Omega. \quad (7.84b)$$

$F_0^{H^{(k)}}$ is given below by (7.102a) while $P^{H^{(k)}}$, $Q^{H^{(k)}}$ are given in (7.102b). Again, for convenience, we shall assume $H^{(k)}(0,0)=0$.

7.3.4 The Biharmonic and Longitudinal Stress Functions

The present case of axially non-uniform distributed loads, also contains a series of generalized biharmonic stress functions $\Phi^{(k)}(x,y)$ $(k=0,\ldots,K)$, which are based on the biharmonic operator "for plane-strain" (3.202). The functions $\Phi^{(k)}$ depend on the cross-section geometry, the material coefficients and the loading. The field equations and the associated boundary conditions for the $\Phi^{(k)}$ functions are

$$\nabla_1^{(4)}\Phi^{(k)} = F_0^{(k)} \qquad \text{over} \quad \Omega, \tag{7.85a}$$

$$\frac{d}{ds}\{\Phi_{,x}^{(k)}, \Phi_{,y}^{(k)}\} = \{-F_1^{(k)}, F_2^{(k)}\} \qquad \text{on} \quad \partial\Omega, \tag{7.85b}$$

where $F_1^{(k)}$, $F_2^{(k)}$ are written as

$$F_i^{(k)} = P_i^{(k)}\cos(\bar{\mathbf{n}},x) + Q_i^{(k)}\cos(\bar{\mathbf{n}},y), \qquad i=1,2. \tag{7.86}$$

The expression for $F_0^{(k)}(x,y)$ is given below by (7.100), while those of $P_i^{(k)}$, $Q_i^{(k)}$ are given by (7.101). In general, for the sake of convenience we assume

$$\Phi^{(k)}(0,0) = \Phi_{,x}^{(k)}(0,0) = \Phi_{,y}^{(k)}(0,0) = 0. \tag{7.87}$$

For the single-value conditions for $\Phi^{(k)}$ and its derivatives $\Phi_x^{(k)}$, $\Phi_y^{(k)}$ that will be derived later on by item **(e)** in S.7.3.5, we assume (should be considered as part of the hypothesis)

$$H^{(k)} = \omega^{(k)} + \widetilde{\delta}_{k0}(p_{k-1}\chi_1 + q_{k-1}\chi_2 - \tau_{k-1}\varphi), \tag{7.88}$$

where $\omega^{(k)}(x,y)$ is an additional series of *longitudinal stress functions*. Employing the BVPs for $H^{(k)}$, χ_1, χ_2 and φ, we reach the following Neumann BVP for $\omega^{(k)}$:

$$\nabla_3^{(2)}\omega^{(k)} = F_0^{\omega^{(k)}} \qquad \text{over} \quad \Omega, \tag{7.89a}$$

$$D_1^n\omega^{(k)} = P^{\omega^{(k)}}\cos(\bar{\mathbf{n}},x) + Q^{\omega^{(k)}}\cos(\bar{\mathbf{n}},y) \qquad \text{on} \quad \partial\Omega, \tag{7.89b}$$

where $F_0^{\omega^{(k)}}$ is given below by (7.103a) and $P^{\omega^{(k)}}$, $Q^{\omega^{(k)}}$ are given in (7.103b,c). We also select $\omega^{(k)}(0,0)=0$. One may verify that by definition, $F_0^{\omega^{(K+1)}} = P^{\omega^{(K+1)}} = Q^{\omega^{(K+1)}} = 0$, and therefore, $\omega^{(k)}=0$ for $k>K$. Thus, in view of (7.88) it is clear that

$$H^{(K+1)} = p_K\chi_1 + q_K\chi_2 - \tau_K\varphi. \tag{7.90}$$

7.3.5 Verification of Solution Hypothesis

We shall now prove that the solution hypothesis of (7.81a–f) provides an exact solution that satisfies all requirements of the theory of elasticity, see **P.7.3**, **P.7.4**.

To carry out the above task, we employ the equilibrium equations, (1.82a–c), the compatibility equations (1.45a–f) and the outer surface boundary conditions (5.16a–c). The process is broken into subsequent steps, where in general, at each step, we use all relevant relations that were found in previous steps.

(a) Equilibrium Equations. Substituting the stress components of (7.81a–f) in the equilibrium equations while using the generic identity

$$\sum_{k=0}^{K-j}\widetilde{f}_{k+j} = \sum_{k=j}^{K}\widetilde{f}_k, \qquad j=1,2, \tag{7.91}$$

shows that (1.82a,b) are satisfied identically. From the third equilibrium equation, (1.82c), the governing equations for the generalized harmonic functions $H^{(k)}$ $(k = 0,\dots,K+1)$, namely, $\nabla_3^{(2)} H^{(k)} = F_0^{H^{(k)}}$, documented in (7.84a), are obtained as the coefficients of the terms z^{k+1}.

(b) Boundary Conditions. From the outer surface boundary condition (5.16a), one may deduce (as the coefficients of the terms z^{k+1}) the conditions for the derivatives $\frac{d}{ds}\Phi^{(k)}_{,y}$ of the biharmonic generalized functions $\Phi^{(k)}$ over the contour $\partial\Omega$, see (7.85b:a). Similarly, from the outer surface boundary condition (5.16b), the conditions for the derivatives $\frac{d}{ds}\Phi^{(k)}_{,x}$ are deduced, see (7.85b:b).
From the third boundary condition (5.16c) the normal derivatives $D_1^n H^{(k)}$ used in (7.84b) for the contour $\partial\Omega$ are obtained.

(c) Compatibility Equations. To examine the compatibility equations we employ the strain expressions of (7.104a–f). Satisfying the compatibility equation (1.45a), by means of balancing powers of z yields

$$\varepsilon^{(k)}_{M,xx} + \varepsilon^{(k)}_{L,yy} = \gamma^{(k)}_{LM,xy}. \tag{7.92}$$

Substituting $\varepsilon^{(k)}_M$ $\varepsilon^{(k)}_L$ and $\gamma^{(k)}_{LM}$ from (7.96a,b) in the above yields the sequence of governing equations for the generalized biharmonic functions $\Phi^{(k)}$ $(k = 0,\dots,K)$ in Ω, namely, $\nabla_1^{(4)}\Phi^{(k)} = F_0^{(k)}$ as documented in (7.85a). As far as (1.45b, c) are concerned, they are satisfied identically. Equations (1.45d–f) then yield

$$[M^{(k)}_{,x} + L^{(k)}_{,y} - \gamma^{(k)}_{LM}]_{,x} = 0, \quad [M^{(k)}_{,x} + L^{(k)}_{,y} - \gamma^{(k)}_{LM}]_{,y} = 0, \quad M^{(k)}_{,x} + L^{(k)}_{,y} - \gamma^{(k)}_{LM} = 0. \tag{7.93}$$

To show that the above equations are also fulfilled, the following equality is invoked:

$$\underline{\underline{\gamma^{(k)}_{LM} = M^{(k)}_{,x} + L^{(k)}_{,y}} \; \underline{- \int_0^y \Big(\int_0^x (\nabla_1^{(4)}\Phi^{(k)} - F_0^{(k)})\, dx \Big)\, dy}}. \tag{7.94}$$

When $\nabla_1^{(4)}\Phi^{(k)} = F_0^{(k)}$, the above once-underlined term vanishes, and the twice-underlined equality is clearly valid, which proves (7.93).

(d) Existence of the generalized harmonic stress function. Applying the existence condition of a generalized harmonic problem given in (3.125) to the functions $H^{(k)}(x,y)$ $(k = 0,\dots,K+1)$, namely, $\oint_{\partial\Omega} D_1^n H^{(k)} = \iint_\Omega \nabla_3^{(2)} H^{(k)}$, yields the definition of (7.100) for the constant d_k.

The existence condition for the longitudinal stress function $\omega^{(k)}$ is satisfied since the same condition is fulfilled for $H^{(k)}, \chi_1, \chi_2$ and φ.

(e) Single-Value Conditions. We impose the single-valued conditions for $\Phi^{(k)}(x,y)$ and their first derivatives as derived in S.3.2.5, namely,

$$\oint_{\partial\Omega} \Big\{ \frac{d}{ds}\Phi^{(k)}_{,x}, \frac{d}{ds}\Phi^{(k)}_{,y}, x\frac{d}{ds}\Phi^{(k)}_{,x} + y\frac{d}{ds}\Phi^{(k)}_{,y} \Big\} = \{0,0,0\}. \tag{7.95}$$

Using (7.85b), (7.86) and Green's Theorem (in the form of (3.7), (3.123a,b), (5.1), in addition to (5.1)), shows that the above conditions are satisfied provided that the constants τ_k and p_k, q_k are those documented by (7.98).

7.3.6 Solution Procedure

As already indicated in S.7.3.1, the analysis is initiated by the $K^{\underline{th}}$ level (where K is the maximal level of the surface loads and body forces), and continues down to lower levels. As shown by Table 7.1 the solution ingredients are gradually introduced according to their level of appearance.

Component	Level
$\sigma_x \leftrightarrow \sigma_y, \tau_{xy} \leftrightarrow \tau_{xy}$	K
$\tau_{yz} \leftrightarrow \tau_{xz}, \gamma_{yz} \leftrightarrow \gamma_{xz}$	$K+1$
$\sigma_z \leftrightarrow \sigma_z, \varepsilon_z \leftrightarrow \varepsilon_z, \varepsilon_x \leftrightarrow \varepsilon_y, \gamma_{xy} \leftrightarrow \gamma_{xy}$	$K+2$
$u \leftrightarrow v, u_0 \leftrightarrow v_0$	$K+4$
$w \leftrightarrow w, w_0 \leftrightarrow w_0$	$K+3$
$\omega_x \leftrightarrow -\omega_y, \omega_{x0} \leftrightarrow -\omega_{y0}$	$K+3$
$\omega_z \leftrightarrow -\omega_z, \omega_{z0} \leftrightarrow -\omega_{z0}$	$K+2$
$H, S_x \leftrightarrow S_y$	$K+1$
$p \leftrightarrow q, \tau \leftrightarrow -\tau, d \leftrightarrow d$	K
$L \leftrightarrow M$	K
Φ, ω	K
$X_b, Y_b, Z_b, X_s, Y_s, Z_s, \overline{U}_1 \leftrightarrow \overline{U}_2$	K
$Q^{\omega} \leftrightarrow P^{\omega}, Q_2 \leftrightarrow P_1, Q_1 \leftrightarrow P_2$	K

Table 7.1: Maximal level of appearance and symmetry of various solution components.

In practice, only the BVPs for $\Phi^{(k)}$ and $\omega^{(k)}$, and subsequently the constants p_k, q_k, τ_k and d_k and the functions $\overline{U}_1^{(k)}$, $\overline{U}_2^{(k)}$, $L^{(k)}$ and $M^{(k)}$ should be derived in each level. Hence, in all relevant expressions, one should replace $H^{(k)}$ by their equivalent form given by (7.88). Once all levels have been derived, the expressions for the overall strain, stress, displacement and rotation components may be used.

As shown, the determination of the $H^{(k)}$ functions is not necessary as they are given by (7.88). Yet, one may choose to solve the BVPs for $H^{(k)}$ instead of those for $\omega^{(k)}$. In such a case, the constants p_k, q_k, τ_k will not be determined (i.e. will remain arbitrary), and Φ will be valid except for the fact that it will not be single-valued. Also as shown by (7.90), no BVP solution is required for $H^{(K+1)}$.

The solution presented above does not ensure that the tip loads of the beam vanish, see stress integral of (5.7). Hence, one needs to superimpose on the above solution a series of solutions for tip loads (i.e. the St. Venant's problem solutions summarized in S.7.2.3) in order to cancel out the residual loads. These additional solutions will also include rigid body displacements and rotations.

It should be noted that similar to the symmetry notation of S.6.1, the present solution is "symmetric" under the following parameter interchange: $x \leftrightarrow y$, $X_b \leftrightarrow Y_b$, $X_s \leftrightarrow Y_s$, $1 \leftrightarrow 2$, $4 \leftrightarrow 5$, etc., see Table 7.1. For example, $\sigma_y \leftrightarrow \sigma_x$ means $\sigma_y = Sym(\sigma_x)$ and $\sigma_x = Sym(\sigma_y)$. Also, $Q^{\chi_1} = Sym(P^{\chi_2})$, $Q^{\chi_2} = Sym(P^{\chi_1})$, $Q^{\varphi} = -Sym(P^{\varphi})$, $D_y = -Sym(D_x)$ and $\sigma_z = Sym(\sigma_z)$, $D = -Sym(D)$.

7.3.7 Detailed Solution Expressions

This section provides detailed expressions for various components of the solution described in S.7.3. In these expressions, when necessary, $H^{(k)}$ may be replaced by their equivalent form shown by (7.88).

The Auxiliary Functions:

$$L^{(k)} = b_{12}\Phi^{(k)}_{,x} - b_{16}\Phi^{(k)}_{,y} + \int_0^x \{b_{11}(\Phi^{(k)}_{,yy} + \overline{U}^{(k)}_1) + b_{12}\overline{U}^{(k)}_2 \tag{7.96a}$$

$$- (k+1)[(\frac{b_{11}a_{44} + b_{12}a_{55} - b_{16}a_{45}}{a_0} - \frac{a_{13}}{a_{33}})H^{(k+1)} + \frac{b_{11}}{a_0}S^{(k+1)}_x + \frac{b_{12}}{a_0}S^{(k+1)}_y]\}\,dx$$

$$+ \frac{p_k}{a_{33}}\{a_{13}\frac{x^2y^2}{2} + [\frac{(b_{26}a_{45} + b_{12}a_{44})(a_{23} + 2a_{33}) + a_{55}b_{22}(2a_{33} - a_{23})}{a_0} - 2a_{23}]\frac{y^4}{24}\}$$

$$+ \frac{q_k}{a_{33}}\{a_{13}\frac{x^3y}{3} - \frac{a_{45}b_{22}(2a_{33}-a_{13})+b_{12}(a_{36}a_{44}-a_{23}a_{45})-b_{26}(a_{23}a_{55}-a_{36}a_{45})}{a_0}\frac{y^4}{24}\}$$

$$+\frac{\tau_k}{a_0}(b_{12}a_{44}-b_{22}a_{55}+b_{26}a_{45})\frac{y^3}{6}+(k+1)\frac{d_{k+1}}{4a_{33}}\{[\frac{a_{12}(3a_{13}+a_{55})}{a_{23}}-a_{11}]\frac{x^3}{3}+\frac{a_{13}(a_{45}+a_{36})}{a_{33}}x^2y$$

$$+[\frac{a_{11}(3a_{23}+a_{44})}{a_{13}}-a_{12}]xy^2-[a_{26}+\frac{a_{23}(a_{45}+a_{36})}{a_{33}}-\frac{a_{16}(3a_{23}+a_{44})}{a_{13}}]\frac{y^3}{3}\}-\int_0^y l^{(k)}(y)\,dy,$$

$$M^{(k)} = Sym(L^{(k)}) = b_{12}\Phi^{(k)}_{,y} - b_{26}\Phi^{(k)}_{,x} + \int_0^y \{b_{12}\overline{U}^{(k)}_1 + b_{22}(\Phi^{(k)}_{,xx} + \overline{U}^{(k)}_2) \tag{7.96b}$$

$$- (k+1)[(\frac{b_{12}a_{44} + b_{22}a_{55} - b_{26}a_{45}}{a_0} - \frac{a_{23}}{a_{33}})H^{(k+1)} + \frac{b_{12}}{a_0}S^{(k+1)}_x + \frac{b_{22}}{a_0}S^{(k+1)}_y]\}\,dy$$

$$- \frac{p_k}{a_{33}}\{\frac{a_{45}b_{11}(2a_{33} - a_{23})+b_{12}(a_{36}a_{55}-a_{13}a_{45})-b_{16}(a_{13}a_{44}-a_{36}a_{45})}{a_0}\frac{x^4}{24}-a_{23}\frac{xy^3}{3}\}$$

$$+ \frac{q_k}{a_{33}}\{[\frac{(b_{16}a_{45} + b_{12}a_{55})(a_{13} + 2a_{33}) + a_{44}b_{11}(2a_{33} - a_{13})}{a_0} - 2a_{13}]\frac{x^4}{12} + a_{23}\frac{x^2y^2}{2}\}$$

$$+ \frac{\tau_k}{a_0}(b_{11}a_{44}-b_{12}a_{55}-b_{16}a_{45})\frac{x^3}{6}-(k+1)\frac{d_{k+1}}{4a_{33}}\{[a_{16}+\frac{a_{13}(a_{45}+a_{36})}{a_{33}}-\frac{a_{26}(3a_{13}+a_{55})}{a_{23}}]\frac{x^3}{3}$$

$$+[a_{12}-\frac{a_{22}(3a_{13}+a_{55})}{a_{23}}]x^2y-\frac{a_{23}(a_{45}+a_{36})}{a_{33}}xy^2+[a_{22}-\frac{a_{12}(3a_{23}+a_{44})}{a_{13}}]\frac{y^3}{3}\}-\int_0^x m^{(k)}(x)\,dx,$$

where the functions $l^{(k)}(y)$, $m^{(k)}(x) = Sym(l^{(k)}(y))$ are

$$l^{(k)}(y) = \{\int_0^y [b_{22}(\Phi^{(k)}_{,xxx} + \overline{U}^{(k)}_{2,x}) + b_{12}\overline{U}^{(k)}_{1,x}]\,dy - (k+1)[(\frac{b_{12}a_{44} + b_{22}a_{55} - b_{26}a_{45}}{a_0} \tag{7.97a}$$

$$- \frac{a_{23}}{a_{33}})\int_0^y H^{(k+1)}_{,x}\,dy - (\frac{b_{16}a_{44} + b_{26}a_{55} - b_{66}a_{45}}{a_0} - \frac{a_{36}}{a_{33}})H^{(k+1)}]$$

$$+ b_{12}[2\Phi^{(k)}_{,xy} - \Phi^{(k)}_{,xy}(0,0)] - b_{16}[2\Phi^{(k)}_{,yy} - \Phi^{(k)}_{,yy}(0,0) + \overline{U}^{(k)}_1] - b_{26}[2\Phi^{(k)}_{,xx} - \Phi^{(k)}_{,xx}(0,0) + \overline{U}^{(k)}_2]$$

$$+ b_{66}[\Phi^{(k)}_{,xy} - \frac{1}{2}\Phi^{(k)}_{,xy}(0,0)] + (k+1)(k+2)[\frac{b_{12}}{a_0}\int_0^y (a_{44}L^{(k+2)} - a_{45}M^{(k+2)})\,dy$$

$$- \frac{b_{22}}{a_0}\int_0^y \left(\int_0^y (a_{55}M^{(k+2)} - a_{45}L^{(k+2)})_{,x}\,dy\right)dy + \frac{b_{26}}{a_0}\int_0^y (a_{55}M^{(k+2)} - a_{45}L^{(k+2)})\,dy]\}_{x=0},$$

$$m^{(k)}(x) = \{\int_0^x [b_{11}(\Phi^{(k)}_{,yyy} + \overline{U}^{(k)}_{1,y}) + b_{12}\overline{U}^{(k)}_{2,y}]\,dx - (k+1)[(\frac{b_{11}a_{44} + b_{12}a_{55} - b_{16}a_{45}}{a_0} \tag{7.97b}$$

$$- \frac{a_{13}}{a_{33}})\int_0^x H^{(k+1)}_{,y}\,dx - (\frac{b_{16}a_{44} + b_{26}a_{55} - b_{66}a_{45}}{a_0} - \frac{a_{36}}{a_{33}})H^{(k+1)}]$$

$$+ b_{12}[2\Phi^{(k)}_{,xy} - \Phi^{(k)}_{,xy}(0,0)] - b_{16}[2\Phi^{(k)}_{,yy} - \Phi^{(k)}_{,yy}(0,0) + \overline{U}^{(k)}_1] - b_{26}[2\Phi^{(k)}_{,xx} - \Phi^{(k)}_{,xx}(0,0) + \overline{U}^{(k)}_2]$$

$$+ b_{66}[\Phi^{(k)}_{,xy} - \frac{1}{2}\Phi^{(k)}_{,xy}(0,0)] - (k+1)(k+2)[\frac{b_{11}}{a_0}\int_0^x \left(\int_0^x (a_{44}L^{(k+2)} - a_{45}M^{(k+2)})_{,y}\,dx\right)dx$$

$$+\frac{b_{12}}{a_0}\int_0^x (a_{55}M^{(k+2)} - a_{45}L^{(k+2)})\,dx - \frac{b_{16}}{a_0}\int_0^x (a_{44}L^{(k+2)} - a_{45}M^{(k+2)})\,dx]\}_{y=0}\,.$$

Note that $L^{(k)}(0,0) = M^{(k)}(0,0) = 0$ for all k.

The Loading Constants:

$$p_k = \frac{a_{33}}{I_y}\oint_{\partial\Omega} X_s^{(k)} + \frac{a_{33}}{I_y}\iint_\Omega \{X_b^{(k)} + \frac{k+1}{a_0}[a_{44}\omega_{,x}^{(k+1)} - a_{45}\omega_{,y}^{(k+1)} - (k+2)(a_{44}L^{(k+2)} - a_{45}M^{(k+2)})]\},$$

$$q_k = \frac{a_{33}}{I_x}\oint_{\partial\Omega} Y_s^{(k)} + \frac{a_{33}}{I_x}\iint_\Omega \{Y_b^{(k)} + \frac{k+1}{a_0}[a_{55}\omega_{,y}^{(k+1)} - a_{45}\omega_{,x}^{(k+1)} - (k+2)(a_{55}M^{(k+2)} - a_{45}L^{(k+2)})]\},$$

$$\tau_k = p_k\frac{D_x}{D} + q_k\frac{D_y}{D} + \frac{1}{D}\oint_{\partial\Omega}(xY_s^{(k)} - yX_s^{(k)}) + \frac{1}{D}\iint_\Omega \{xY_b^{(k)} - yX_b^{(k)}$$

$$+\frac{k+1}{a_0}[(a_{55}x + a_{45}y)(\omega_{,y}^{(k+1)} + (k+2)M^{(k+2)}) - (a_{45}x + a_{44}y)(\omega_{,x}^{(k+1)} + (k+2)L^{(k+2)})]\},$$

$$d_k = \frac{a_{33}}{S_\Omega}\oint_{\partial\Omega} Z_s^{(k)} + \frac{a_{33}}{S_\Omega}\iint_\Omega \{Z_b^{(k)} + \frac{k+1}{a_{33}}[-a_{13}(\Phi_{,yy}^{(k+1)} + \overline{U}_1^{(k+1)}) - a_{23}(\Phi_{,xx}^{(k+1)} + \overline{U}_2^{(k+1)})$$

$$+a_{36}\Phi_{,xy}^{(k+1)} + (p_{k+1}y + q_{k+1}x)xy + (k+2)(\frac{a_{13}a_{44} + a_{23}a_{55} - a_{36}a_{45} + a_0}{a_0}H^{(k+2)}$$

$$+\frac{a_{13}}{a_0}S_x^{(k+2)} + \frac{a_{23}}{a_0}S_y^{(k+2)})]\}. \tag{7.98}$$

One may compare p_K, q_K, d_K, τ_K with the quantities $P_1^{(K+1)}, P_2^{(K+1)}, P_3^{(K+1)}, M_3^{(K+1)}$ of (6.15a,b):

$$p_K = -(K+1)\frac{a_{33}}{I_y}P_1^{(K+1)}, \qquad q_K = -(K+1)\frac{a_{33}}{I_x}P_2^{(K+1)},$$

$$d_K = -(K+1)\frac{a_{33}}{S_\Omega}P_3^{(K+1)}, \qquad \tau_K = -\frac{K+1}{D}M_3^{(K+1)} + \widetilde{\alpha}_1 P_1^{(K+1)} + \widetilde{\alpha}_2 P_2^{(K+1)}, \tag{7.99}$$

where $\widetilde{\alpha}_1, \widetilde{\alpha}_2$ are constants.

The Biharmonic Stress Functions:

$$F_0^{(k)} = -(b_{11}\overline{U}_1^{(k)} + b_{12}\overline{U}_2^{(k)})_{,yy} + (b_{16}\overline{U}_1^{(k)} + b_{26}\overline{U}_2^{(k)})_{,xy} - (b_{12}\overline{U}_1^{(k)} + b_{22}\overline{U}_2^{(k)})_{,xx} \tag{7.100}$$

$$-\frac{2p_k}{a_{33}}(a_{13}x - a_{36}y) + \frac{2q_k}{a_{33}}(a_{36}x - a_{23}y) + (k+1)\{(\frac{b_{11}a_{44} + b_{12}a_{55} - b_{16}a_{45}}{a_0} - \frac{a_{13}}{a_{33}})H_{,yy}^{(k+1)}$$

$$+(\frac{b_{12}a_{44} + b_{22}a_{55} - b_{26}a_{45}}{a_0} - \frac{a_{23}}{a_{33}})H_{,xx}^{(k+1)} - (\frac{b_{16}a_{44} + b_{26}a_{55} - b_{66}a_{45}}{a_0} - \frac{a_{36}}{a_{33}})H_{,xy}^{(k+1)}$$

$$+\frac{d_{k+1}}{2a_{33}}[2a_{12} + \frac{a_{36}}{a_{33}}(a_{36} + a_{45}) - \frac{a_{11}}{a_{13}}(3a_{23} + a_{44}) - \frac{a_{22}}{a_{23}}(3a_{13} + a_{55})]$$

$$+\frac{1}{a_0}[(b_{11}S_x^{(k+1)} + b_{12}S_y^{(k+1)})_{,yy} + (b_{12}S_x^{(k+1)} + b_{22}S_y^{(k+1)})_{,xx} + (b_{16}S_x^{(k+1)} + b_{26}S_y^{(k+1)})_{,xy}]\},$$

and

$$P_1^{(k)} = Y_1^{(k)} - (k+1)\frac{a_{45}}{a_0}H^{(k+1)},$$

$$Q_1^{(k)} = Y_2^{(k)} - \overline{U}_2^{(k)} + \frac{k+1}{a_0}[a_{55}H^{(k+1)} - \frac{d_{k+1}a_0}{4a_{33}}(\frac{3a_{13} + a_{55}}{a_{23}}x^2 - y^2) + S_y^{(k+1)}],$$

$$P_2^{(k)} = X_1^{(k)} - \overline{U}_1^{(k)} + \frac{k+1}{a_0}[a_{44}H^{(k+1)} + \frac{d_{k+1}a_0}{4a_{33}}(x^2 - \frac{3a_{23} + a_{44}}{a_{13}}y^2) + S_x^{(k+1)}],$$

$$Q_2^{(k)} = X_2^{(k)} - (k+1)\frac{a_{45}}{a_0}H^{(k+1)}. \tag{7.101}$$

The Harmonic Stress Functions:

$$F_0^{H(k)} = -Z_b^{(k)} a_0 - \frac{a_0}{a_{33}}(k+1)\{-a_{13}(\Phi_{,yy}^{(k+1)} + \overline{U}_1^{(k+1)}) - a_{23}(\Phi_{,xx}^{(k+1)} + \overline{U}_2^{(k+1)}) + a_{36}\Phi_{,xy}^{(k+1)}$$
$$+ (p_{k+1}y + q_{k+1}x)xy + (k+2)[\frac{a_{13}a_{44} + a_{23}a_{55} - a_{36}a_{45} + a_0}{a_0} H^{(k+2)} + \frac{d_{k+2}}{2a_{33}^2}(a_{45} + a_{36})xy$$
$$+ \frac{a_{13}}{a_0} S_x^{(k+2)} + \frac{a_{23}}{a_0} S_y^{(k+2)}]\} - S_{x,xx}^{(k)} - S_{y,yy}^{(k)} - \widetilde{\delta}_{k0} \cdot \frac{a_0}{a_{33}}(p_{k-1}x + q_{k-1}y), \qquad (7.102a)$$

$$P^{H(k)} = a_0(Z_1^{(k)} - \frac{d_k}{2a_{33}}x) - S_{x,x}^{(k)}, \qquad Q^{H(k)} = a_0(Z_2^{(k)} - \frac{d_k}{2a_{33}}y) - S_{y,y}^{(k)}. \qquad (7.102b)$$

The Longitudinal Stress Functions:

$$F_0^{\omega(k)} = -Z_b^{(k)} a_0 - \frac{a_0}{a_{33}}(k+1)\{-a_{13}(\Phi_{,yy}^{(k+1)} + \overline{U}_1^{(k+1)}) - a_{23}(\Phi_{,xx}^{(k+1)} + \overline{U}_2^{(k+1)}) + a_{36}\Phi_{,xy}^{(k+1)} \qquad (7.103a)$$
$$+ (p_{k+1}y + q_{k+1}x)xy + (k+2)[\frac{a_{13}a_{44} + a_{23}a_{55} - a_{36}a_{45} + a_0}{a_0} H^{(k+2)} + \frac{d_{k+2}}{2a_{33}^2}(a_{45} + a_{36})xy$$
$$+ \frac{a_{13}}{a_0} S_x^{(k+2)} + \frac{a_{23}}{a_0} S_y^{(k+2)}]\} - (k+1)(a_{44}L_{,x}^{(k+1)} - a_{45}M_{,x}^{(k+1)} + a_{55}M_{,y}^{(k+1)} - a_{45}L_{,y}^{(k+1)}),$$

$$P^{\omega(k)} = a_0(Z_1^{(k)} - \frac{d_k}{2a_{33}}x) - (k+1)(a_{44}L^{(k+1)} - a_{45}M^{(k+1)}), \qquad (7.103b)$$

$$Q^{\omega(k)} = a_0(Z_2^{(k)} - \frac{d_k}{2a_{33}}y) - (k+1)(a_{55}M^{(k+1)} - a_{45}L^{(k+1)}). \qquad (7.103c)$$

The Strain Components:

$$\varepsilon_x = \sum_{k=0}^{K} z^k \{\varepsilon_L^{(k)} - \frac{a_{13}}{a_{33}}[z\frac{d_k}{k+1} - z^2 \frac{p_k x + q_k y}{(k+1)(k+2)}]\}, \qquad (7.104a)$$

$$\varepsilon_y = \sum_{k=0}^{K} z^k \{\varepsilon_M^{(k)} - \frac{a_{23}}{a_{33}}[z\frac{d_k}{k+1} - z^2 \frac{p_k x + q_k y}{(k+1)(k+2)}]\}, \qquad (7.104b)$$

$$\varepsilon_z = \sum_{k=0}^{K} z^k \{-(p_k y + q_k x)xy - z\frac{d_k}{k+1} - z^2 \frac{p_k x + q_k y}{(k+1)(k+2)} \qquad (7.104c)$$
$$+ (k+1)[H^{(k+1)} + \frac{d_{k+1}}{4a_{33}}((2a_{13} + a_{55})x^2 + 2(a_{36} + a_{45})xy + (2a_{23} + a_{44})y^2)]\},$$

$$\gamma_{yz} = \sum_{k=0}^{K} z^k \{H_{,y}^{(k)} + \frac{d_k}{2a_{33}}(a_{45}x + a_{44}y) + (k+1)M^{(k+1)} \qquad (7.104d)$$
$$+ \frac{z}{k+1}[(a_{36}x^2 + 2(a_{23} - 2a_{33})xy)\frac{p_k}{2a_{33}} + (-(a_{13} + 2a_{33})x^2 + a_{23}y^2)\frac{q_k}{2a_{33}} - x\tau_k]\},$$

$$\gamma_{xz} = \sum_{k=0}^{K} z^k \{H_{,x}^{(k)} + \frac{d_k}{2a_{33}}(a_{55}x + a_{45}y) + (k+1)L^{(k+1)} \qquad (7.104e)$$
$$+ \frac{z}{k+1}[(a_{13}x^2 - (2a_{33} + a_{23})y^2)\frac{p_k}{2a_{33}} + (2(a_{13} - 2a_{33})xy + a_{36}y^2)\frac{q_k}{2a_{33}} + y\tau_k]\},$$

$$\gamma_{xy} = \sum_{k=0}^{K} z^k \{\gamma_{LM}^{(k)} - \frac{a_{36}}{a_{33}}[z\frac{d_k}{k+1} - z^2 \frac{p_k x + q_k y}{(k+1)(k+2)}]\}, \qquad (7.104f)$$

where, see (7.96a,b),

$$\varepsilon_L^{(k)} = L_{,x}^{(k)}, \qquad \varepsilon_M^{(k)} = M_{,y}^{(k)}, \qquad (7.105)$$

and

$$\gamma_{LM}^{(k)} = b_{16}(\Phi_{,yy}^{(k)} + \overline{U}_1^{(k)}) + b_{26}(\Phi_{,xx}^{(k)} + \overline{U}_2^{(k)}) - b_{66}\Phi_{,xy}^{(k)} - \frac{a_{36}}{a_{33}}(p_k y + q_k x)xy \tag{7.106}$$
$$- (k+1)[(\frac{b_{16}a_{44} + b_{26}a_{55} - b_{66}a_{45}}{a_0} - \frac{a_{36}}{a_{33}})H^{(k+1)} + \frac{b_{16}}{a_0}S_x^{(k+1)} + \frac{b_{26}}{a_0}S_y^{(k+1)}$$
$$+ \frac{d_{k+1}}{4a_{33}}\left((\frac{a_{26}(3a_{13}+a_{55})}{a_{23}} - a_{16})x^2 + 2\frac{a_{36}(a_{45}+a_{36})}{a_{33}}xy + (\frac{a_{16}(3a_{23}+a_{44})}{a_{13}} - a_{26})y^2\right)].$$

As expected, using (7.99), $\varepsilon_i^{(K+2)}$ may be shown to be identical to those of (6.8b).

The Displacement Components:

$$u = \sum_{k=0}^{K} z^k\{L^{(k)} - (2a_{13}x + a_{36}y)\frac{z\,d_k}{2a_{33}(k+1)} - \frac{z^4\,p_k}{(k+1)(k+2)(k+3)(k+4)} \tag{7.107a}$$
$$- \frac{z^2}{(k+1)(k+2)}[(a_{23}y^2 - a_{13}x^2)\frac{p_k}{2a_{33}} - (a_{36}y^2 + 2a_{13}xy)\frac{q_k}{2a_{33}} - y\tau_k]\},$$
$$v = \sum_{k=0}^{K} z^k\{M^{(k)} - (a_{36}x + 2a_{23}y)\frac{z\,d_k}{2a_{33}(k+1)} - \frac{z^4\,q_k}{(k+1)(k+2)(k+3)(k+4)}$$
$$- \frac{z^2}{(k+1)(k+2)}[-(2a_{23}xy + a_{36}x^2)\frac{p_k}{2a_{33}} + (a_{13}x^2 - a_{23}y^2)\frac{q_k}{2a_{33}} + x\tau_k]\}, \tag{7.107b}$$
$$w = \sum_{k=0}^{K+1} z^k\{H^{(k)} + [(2a_{13} + a_{55})x^2 + 2(a_{36} + a_{45})xy + (2a_{23} + a_{44})y^2]\frac{d_k}{4a_{33}}$$
$$- \frac{z}{k+1}(p_k y + q_k x)xy - \frac{z^2\,d_k}{(k+1)(k+2)} + z^3\frac{p_k x + q_k y}{(k+1)(k+2)(k+3)}\}. \tag{7.107c}$$

For the beam axis ($x = y = 0$), the following displacements are obtained:

$$u_0 = \sum_{k=0}^{K} \frac{z^{k+4}p_k}{(k+1)(k+2)(k+3)(k+4)}, \qquad v_0 = \sum_{k=0}^{K} \frac{z^{k+4}q_k}{(k+1)(k+2)(k+3)(k+4)},$$
$$w_0 = -\sum_{k=0}^{K} \frac{z^{k+2}d_k}{(k+1)(k+2)}. \tag{7.108}$$

Note that in the above expressions, no rigid body displacements are included as they are introduced by the tip loads correction discussed in S.7.3.1.

The Rotation Components:

$$\omega_x = \sum_{k=0}^{K+1} \frac{z^k}{2}\{H_{,y}^{(k)} + \frac{d_k}{a_{33}}[(a_{45}+2a_{36})x + (4a_{23}+a_{44})y] - (k+1)M^{(k+1)} - \frac{2z^3 q_k}{(k+1)(k+2)(k+3)}$$
$$+ \frac{z}{k+1}[(a_{36}x^2 + 2(2a_{33}+a_{23})xy)\frac{p_k}{2a_{33}} + ((2a_{33} - a_{13})x^2 + a_{23}y^2)\frac{q_k}{2a_{33}} + x\tau_k]\}, \tag{7.109a}$$
$$\omega_y = -\sum_{k=0}^{K+1} \frac{z^k}{2}\{H_{,x}^{(k)} + \frac{d_k}{a_{33}}[(4a_{13}+a_{55})x + (a_{45}+2a_{36})y] - (k+1)L^{(k+1)} - \frac{2z^3 p_k}{(k+1)(k+2)(k+3)}$$
$$+ \frac{z}{k+1}[(a_{13}x^2 + (2a_{33} - a_{23})y^2)\frac{p_k}{2a_{33}} + (2(2a_{33}+a_{13})xy + a_{36}y^2)\frac{q_k}{2a_{33}} - y\tau_k]\}, \tag{7.109b}$$
$$\omega_z = \sum_{k=0}^{K} \frac{z^k}{2}\{L_{,y}^{(k)} - M_{,x}^{(k)} - \frac{z^2}{(k+1)(k+2)}[(a_{36}x + 2a_{23}y)\frac{p_k}{a_{33}} - (2a_{13}x + a_{36}y)\frac{q_k}{a_{33}} + 2\tau_k]\}, \tag{7.109c}$$

while the corresponding rotations along the beam axis are

$$\omega_{x0} = \sum_{k=0}^{K+1} \frac{z^k}{2} \{ \frac{2z^3 q_k}{(k+1)(k+2)(k+3)} + \underline{H_{,y}^{(k)}(0,0)} \}, \tag{7.110}$$

$$\omega_{y0} = -\sum_{k=0}^{K+1} \frac{z^k}{2} \{ \frac{2z^3 p_k}{(k+1)(k+2)(k+3)} + \underline{H_{,x}^{(k)}(0,0)} \},$$

$$\omega_{z0} = -\sum_{k=0}^{K} \frac{z^k}{2} \{ \underline{L_{,y}^{(k)}(0,0) - M_{,x}^{(k)}(0,0)} \}.$$

In practice, the underlined terms are relatively small, and their physical origin is similar to the shear effects discussed in S.5.1.3.

7.4 Applications

Example 7.4 *The Longitudinal Stress Function for Constant Body Loads.*
We shall derive here the generalized harmonic stress function $\omega = \omega^{(0)}$ for a homogeneous MON13z beam acted upon by constant body forces. These solutions are implemented in **P.7.5**.
Hence, we set $X_b = X_0, Y_b = Y_0, Z_b = Z_0$, where X_0, Y_0, Z_0 are constants, while no other surface or tip loads are applied. The loading constants of (7.100) are

$$p_0 = \frac{S_\Omega}{I_y} a_{33} X_0, \qquad q_0 = \frac{S_\Omega}{I_x} a_{33} Y_0, \qquad \tau_0 = 0, \qquad d_0 = a_{33} Z_0. \tag{7.111}$$

We further employ (7.102a,b) to find

$$F_0^{\omega} = -Z_0 a_0, \qquad P^{\omega} = -\frac{1}{2} a_0 Z_0 x, \qquad Q^{\omega} = -\frac{1}{2} a_0 Z_0 y. \tag{7.112}$$

In view of $F_0^{\omega} = P_{,x}^{\omega} + Q_{,y}^{\omega}$ and Green's Theorem, the condition (4.2) for the existence of ω is clearly satisfied. The solution is independent of the domain shape:

$$\omega = -\frac{Z_0}{4} (a_{55} x^2 + 2a_{45} xy + a_{44} y^2). \tag{7.113}$$

P.7.2 (with $Ex_h = 6$) confirms (7.113) for elliptical domain Ω, by introducing a particular solution $-\frac{a_0}{2a_{55}} Z_0 y^2$, see also Example 4.3.

Example 7.5 *The Biharmonic Stress Function for Constant Body Loads.*
Consider a homogeneous MON13z beam that undergoes constant body forces. For the biharmonic stress function, $\Phi = \Phi^{(0)}$, (7.100), (7.101) show that:

$$F_0 = 2S_\Omega [\frac{X_0}{I_y} (a_{36} y - a_{13} x) + \frac{Y_0}{I_x} (a_{36} x - a_{23} y)], \tag{7.114a}$$

$$F_1 = [Y_0 y - \frac{a_{55}}{a_0} (p_0 \chi_1 + q_0 \chi_2)] \cos(\bar{\mathbf{n}}, y), \quad F_2 = [X_0 x - \frac{a_{44}}{a_0} (p_0 \chi_1 + q_0 \chi_2)] \cos(\bar{\mathbf{n}}, x). \tag{7.114b}$$

Here we used (7.88) for $H^{(1)}$ and (7.78) for $\overline{U}_1, \overline{U}_2$.

In the case of an elliptical cross-section, the expressions for $\cos(\bar{\mathbf{n}}, x)$ and $\cos(\bar{\mathbf{n}}, y)$ are given by (3.6), while the cross-sectional area and moments of inertia about its principal axes (namely, the x and y coordinate lines) are

$$S_\Omega = \pi \tilde{a} \tilde{b}, \qquad I_y = \frac{\pi}{4} \tilde{a}^3 \tilde{b}, \qquad I_x = \frac{\pi}{4} \tilde{a} \tilde{b}^3. \tag{7.115}$$

Subsequently, the loading constants p_0, q_0, τ_0, d_0 of (7.100) are

$$p_0 = \frac{4a_{33}}{\widetilde{a}^2} X_0, \qquad q_0 = \frac{4a_{33}}{\widetilde{b}^2} Y_0, \qquad \tau_0 = 0, \qquad d_0 = a_{33} Z_0. \tag{7.116}$$

Since the bending functions χ_1, χ_2, are polynomials of third-degree, see Example 7.3, P_2, Q_1 are also polynomials of third-degree. Note that φ does not participate since $\tau_0 = 0$ in this case. The solution of the biharmonic BVP in an ellipse for linear F_0, and third-degree polynomials P_j, Q_j is presented by Example 4.8.

Example 7.6 *A Beam Under Constant Distributed Body Force.*

By activating **P.7.6** for *material* $= 1$, $t_z = 30°$, $\widetilde{a} = 2$, $\widetilde{b} = 1$, $X_0 = 0$, $Y_0 = -1$, $Z_0 = 0$, $\gamma_1 = 0$ and $l = 10$, we simulate a beam made of typical Graphite/Epoxy orthotropic material oriented at $\psi = 30°$ (see Fig. 7.1), which is loaded by a constant body force in the y direction. All results presented in this example include the tip loads correction discussed in S.7.3.6.

First, Figs. 7.6(a), (b) present the beam axis deflection $v_0(z)/(a_{33}l^4)$ compared with the elementary solution derived in S.5.3.4 (with $E = 1/a_{33}$) for two values of beam slenderness. As shown, the two results coincide for relatively "long" beams, while considerable differences are obtained for "short" beams. These differences are mainly due to the beam slenderness and are not affected by the anisotropic characteristics. The corresponding out-of-plane warping,

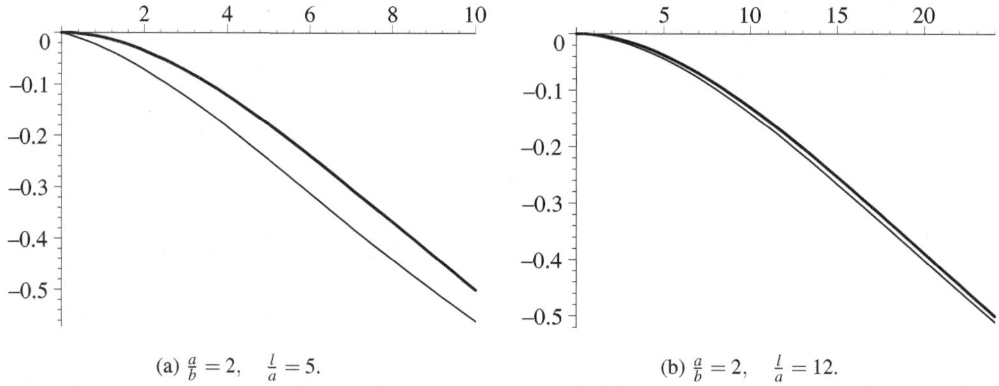

Figure 7.6: Axis deflection of a MON13z beam of elliptical cross-section (thin line) and similar isotropic beam (thick line).

$w(x,y)$ is presented in Figs. 7.7(a), (b). As shown, close to the root, the $w(x,y)$ distribution is relatively small and is influenced by the effective shear force accumulated over the beam and the relatively low influence of the bending (which is represented by the $yv_{0,z}$ term that is more easy to identify in the simplified analysis of S.5.3.4, see (5.73:c)). Far from the root, the $w(x,y)$ distribution is larger and almost linear with respect to y due to the lower accumulated shear force and the higher $yv_{0,z}$ values. For illustration purposes, Figs. 7.8(a), (b) present the deformation field at the above two cross-sections of the beam. Note that close to the root, the cross-section distortion is much more pronounced and the elliptical shape is more distorted due to the relatively high effective shear force accumulated over the beam. Finally, Fig. 7.9 presents the corresponding biharmonic function Φ. In addition, as shown in Example 7.4, ω is proportional to Z_0 and therefore vanishes for $Y_0 = -1$ loading.

Many additional symbolic results, including plotting options for all stress, strain and displacement components of the present solution, may be obtained by activating **P.7.6**.

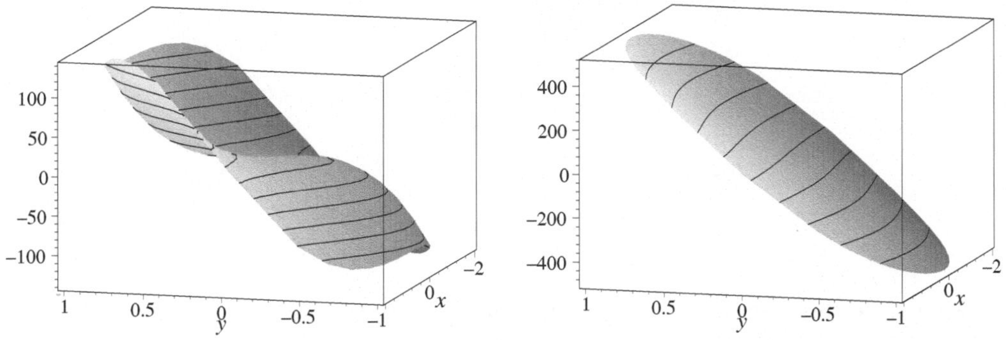

(a) $z/l = 0.1$. (b) $z/l = 0.5$.

Figure 7.7: Normalized out-of-plane warping, $w(x,y)/a_{33}$, for a MON13z beam of elliptical cross-section under a constant body force $Y_0 = -1$.

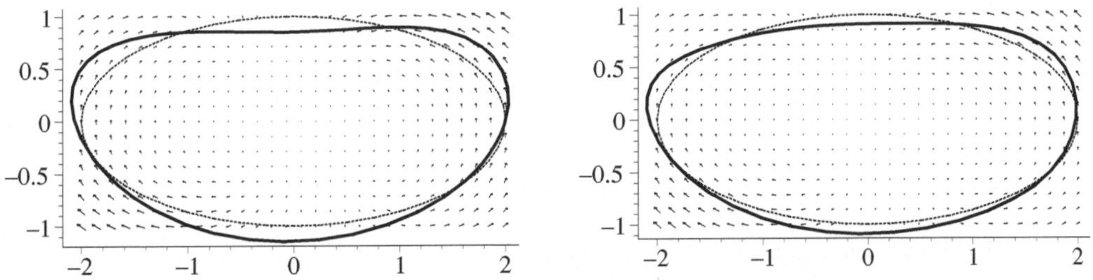

(a) $z/l = 0.1$, magnification factor $= 10^9$ (b) $z/l = 0.5$, magnification factor $= 10^{9.3}$

Figure 7.8: Normalized field deformation for an elliptical MON13z beam under a constant body force $Y_0 = -1$. Deformed (not to scale): full line. Undeformed: dotted line.

Example 7.7 *A Beam Under Hydrostatic Surface Loads.*
We shall now deal with a MON13z beam that undergoes two types of hydrostatic surface loads:

(a) The case of a horizontally-placed beam: Let a beam be positioned at a depth $H \geq \widetilde{b}$ in a liquid of mass density ρ. As shown in Fig. 7.10(a) the beam is loaded laterally in a non-uniform fashion, but this loading is not a function of z. For the sake of simplicity, we suppose that in this case, the beam cross-section has two orthogonal symmetry axes (for example, the beam discussed in Example 7.6). The static pressure at each point is therefore $\rho g(H-y)$ where g is the gravity constant. Employing the constants $\gamma_0 = -\rho g H$ and $\gamma_1 = \rho g$, one may write

$$X_s = (\gamma_0 + \gamma_1 y)\cos(\bar{\mathbf{n}}, x), \qquad Y_s = (\gamma_0 + \gamma_1 y)\cos(\bar{\mathbf{n}}, y), \qquad Z_s = 0. \tag{7.117}$$

It is therefore clear that the only nonzero load components are $X_1 = Y_2 = \gamma_1 y + \gamma_0$. Based on the fact that the integrals $\iint_\Omega \{x^3, x^2 y, xy^2, y^3\}$ vanish for an axes-symmetrical cross-section, and with the aid of (3.7), one may use (7.98) to show that for the present loading,

$$p_0 = \tau_0 = d_0 = 0, \qquad q_0 = \frac{S_\Omega}{I_x}\gamma_1 a_{33}. \tag{7.118}$$

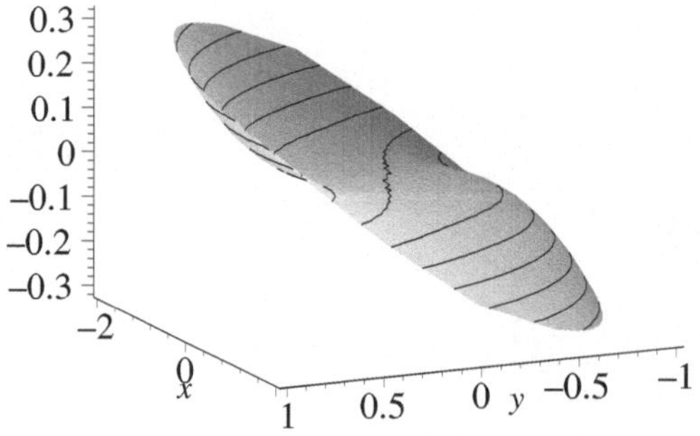

Figure 7.9: $\Phi(x,y)$ distribution for a MON13z beam of elliptical cross-section with $Y_0 = -1$.

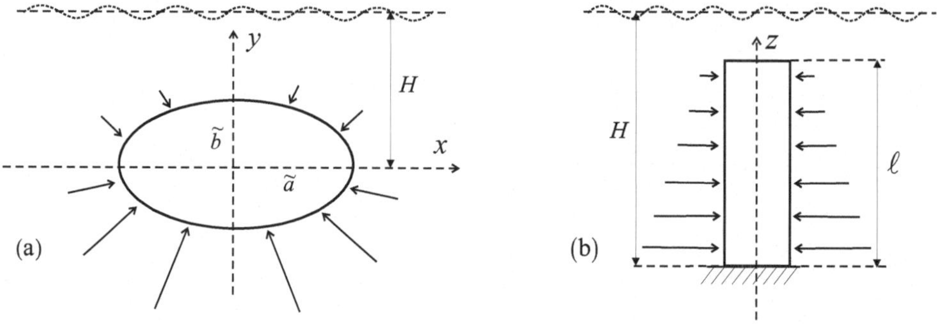

Figure 7.10: Physical interpretation of a beam under hydrostatic-type surface loads: (a) a cross-section of a horizontally-placed beam of elliptical cross-section, (b) a vertically-placed beam of generic cross-section.

Note that the longitudinal stress function vanishes, i.e., $\omega = 0$, where S_Ω and I_x are given in (7.115).

By activating **P.7.6** for an elliptical cross-section, made of typical Graphite/Epoxy orthotropic material oriented at $\psi = 30°$, as in Example 7.6 but with $Y_0 = 0, \gamma_1 = 1$ and $H = 1, 20$, we have drawn Figs. 7.11(a), (b) that show the cross-section deformation for low and high depth values. As expected, for $H = 20\tilde{b}$ the deformation obtained is characterized by relatively high contraction around the entire circumference, while for $H = \tilde{b}$ most of the deformation occurs at the lower part of the ellipse. Note that although the loading is symmetric about the y (vertical) axis and rigid body rotations are applied so that $\omega_z = 0$ at $x = y = z = 0$, both deformed shapes are non-symmetric (which is easier to see in Fig. 7.11(b)), see also discussion in Example 3.5.

(b) The case of a vertically-placed beam: Here we consider the case where a beam of generic cross-section is placed vertically as shown in Fig. 7.10(b) for at a depth $H \geq l$. Hence, the beam is loaded laterally in a uniform fashion, but this loading is a linear function of z. The static pressure at each point is therefore

$$X_s = (\gamma_0 + \gamma_1 z)\cos(\bar{\mathbf{n}}, x), \qquad Y_s = (\gamma_0 + \gamma_1 z)\cos(\bar{\mathbf{n}}, y), \qquad Z_s = 0. \tag{7.119}$$

In the present case we face a two-level solution ($K = 1$) where the only nonzero loading components are $X_1^{(1)} = Y_2^{(1)} = \gamma_1$ and $X_1^{(0)} = Y_2^{(0)} = \gamma_0$.

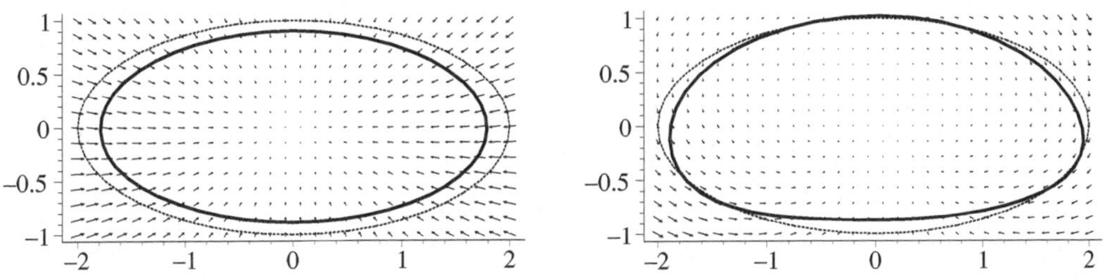

(a) $H = 20b$, magnification factor $= 10^8$.

(b) $H = b$, magnification factor $= 10^9$.

Figure 7.11: Normalized field deformation for a MON13z beam of elliptical cross-section under hydrostatic surface loads. $z/l = 0.5$. Deformed (not to scale): full line, Undeformed: dotted line.

Level $k = 1$: Equations (7.98) show that the loading constants in this level are $p_1 = q_1 = \tau_1 = d_1 = 0$. In such a case, the solution of the biharmonic problem for level $k = 1$ takes the form described in Example 3.5, namely

$$\Phi^{(1)} = \frac{\gamma_1}{2}(x^2 + y^2), \tag{7.120}$$

and the longitudinal stress function for this level vanishes, i.e., $\omega^{(1)} = 0$. At this stage, all functions and constants may be found, and in particular

$$L^{(1)} = \gamma_1[(b_{11} + b_{12})x + b_{26}y], \qquad M^{(1)} = \gamma_1[(b_{12} + b_{22})y + b_{16}x]. \tag{7.121}$$

Level $k = 0$: It is easy to show that $p_0 = q_0 = \tau_0 = 0$, $d_0 = -a_{33}(a_{13} + a_{23})\gamma_1$, and therefore, the biharmonic function takes again the form described in Example 3.5, namely

$$\Phi^{(0)} = \frac{\gamma_0}{2}(x^2 + y^2), \tag{7.122}$$

while the auxiliary functions are

$$L^{(0)} = \gamma_0[(b_{11} + b_{12})x + b_{26}y], \qquad M^{(0)} = \gamma_0[(b_{12} + b_{22})y + b_{16}x]. \tag{7.123}$$

One may also verify that

$$\begin{aligned} \omega^{(0)} = \gamma_1\{\frac{1}{2}(b_{16} - b_{26})\varphi + \frac{x^2}{2}[\frac{a_{13} + a_{23}}{2a_{33}}a_{55} - (b_{11} + b_{12})] \\ + xy[\frac{a_{13} + a_{23}}{2a_{33}}a_{45} - (b_{16} + b_{26})] + \frac{y^2}{2}[\frac{a_{13} + a_{23}}{2a_{33}}a_{44} - (b_{12} + b_{22})]\}, \end{aligned} \tag{7.124}$$

where φ is the torsional warping function, see S.7.2.1.1.

Summing all level contributions:

At this stage one may evaluate the stress components by (7.81a–f), which yield

$$\sigma_x = \gamma_0 + z\gamma_1, \quad \sigma_y = \gamma_0 + z\gamma_1, \tag{7.125}$$

and nonzero distributions of *constant* σ_z, in addition to τ_{yz}, and τ_{xz} that are *proportional* to those of (7.4). Hence, these stress components leave nonzero resultant force, P_z, and a resultant moment, M_z, over the tip cross-section. Superimposing of suitable tip loads solutions, see

S.7.3.6, shows that the only remaining nonzero stress components are σ_x and σ_y of (7.125). Note that due to the above unique distribution of the σ_z, τ_{yz}, and τ_{xz} stresses, the tip loads correction creates a state where the tip cross-section is free of stresses at each point of it (and not only in an integral manner). One may argue that this result was expected, yet, the entire solution process well demonstrates the offered solution methodology.

Example 7.8 *A Beam Under Linear body-force Distribution.*

We shall consider here a MON13z beam of generic cross-section when the body-force distribution is given by

$$X_b = \gamma_x x, \qquad Y_b = 0, \qquad Z_b = \gamma_z z, \tag{7.126}$$

while all other surface and tip loads vanish. Hence, in this case

$$\begin{aligned} &X_b^{(0)} = \gamma_x x, \quad Y_b^{(0)} = 0, \quad Z_b^{(0)} = 0, \quad X_b^{(1)} = 0, \quad Y_b^{(1)} = 0, \quad Z_b^{(1)} = \gamma_z, \\ &U_1^{(0)} = -\int_0^x X_b^{(0)}\,dx = -\frac{1}{2}\gamma_x x^2, \qquad U_2^{(0)} = 0, \quad U_1^{(1)} = 0, \quad U_2^{(1)} = 0. \end{aligned} \tag{7.127}$$

A physical example for such a loading is the rotating beam shown in Fig. 7.12.

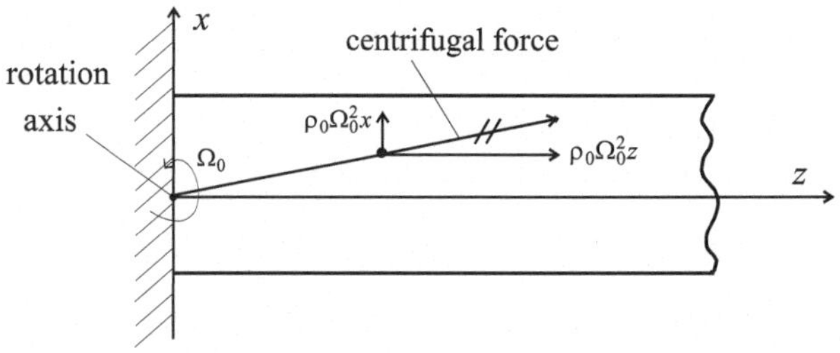

Figure 7.12: Notation for rotating beam in Example 7.8.

In such a case, $\gamma_x = \gamma_z = \rho_0\Omega_0^2$ where ρ_0 is the (uniform) mass density of the material, and Ω_0 is the angular velocity.

The level-independent generalized harmonic functions χ_1, χ_2, φ and the torsional rigidity D for the elliptical cross-section were already derived in Examples 7.1, 7.3 and are applicable here. We shall now discuss the two levels of this problem.

Level $k = 1$: Equations (7.98) show that

$$p_1 = q_1 = \tau_1 = 0, \qquad d_1 = \frac{a_{33}}{S_\Omega}\iint_\Omega Z_b^{(1)} = a_{33}\gamma_z. \tag{7.128}$$

According to the discussion in S.7.3, for the generalized harmonic function $\omega^{(1)}$ we find

$$F_0^{\omega^{(1)}} = -\gamma_z a_0, \qquad P^{\omega^{(1)}} = -\frac{1}{2}a_0\gamma_z x, \qquad Q^{\omega^{(1)}} = -\frac{1}{2}a_0\gamma_z y. \tag{7.129}$$

Analogously to the solution presented in Example 7.4, see (7.113), the expression for $\omega^{(1)}$ is independent of the domain shape, namely

$$\omega^{(1)} = -\frac{\gamma_z}{4}(a_{55}x^2 + 2a_{45}xy + a_{44}y^2). \tag{7.130}$$

Level $k=0$: Equations (7.98), (3.123a,b) and Green's Theorem show that

$$p_0 = \frac{a_{33}}{I_y}\iint_\Omega [X_b^{(0)} + \frac{1}{a_0}(a_{44}\omega_{,x}^{(1)} - a_{45}\omega_{,y}^{(1)})] = 0, \tag{7.131a}$$

$$q_0 = \frac{a_{33}}{a_0 I_y}\iint_\Omega (a_{55}\omega_{,y}^{(1)} - a_{45}\omega_{,x}^{(1)}) = 0, \tag{7.131b}$$

and that $d_0 = 0$, $\tau_0 \neq 0$. In addition, (7.103a–c) show that $\omega^{(0)} = 0$. At this stage one should solve the biharmonic BVP (7.85a,b) for $k=0$.

For an elliptical cross-section, $\Phi^{(0)}$ would be a fourth-degree polynomial, which may be determined using a particular solution (e.g., $\frac{F_0^{(0)}}{24b_{22}}x^4$) and the expressions of Example 4.7.

Summing all level contributions:

At this stage one may evaluate the stress components by (7.81a–f) the strain components by (7.104a–f), and the deformation expressions of (7.107a–c), (7.109a–c). Further on, by integrating the stress components according to (5.7), the tip loads resultants are obtained. This enables the determination of the additional tip loads solutions that should be superimposed on the above solution, see S.7.3.6, which also include the rigid body displacements and rotations.

7.5 Beams of Cylindrical Anisotropy

In this section we shall examine some aspects of the behavior of uncoupled, homogeneous, MON13z beams with cylindrical anisotropy. The derivation will therefore be founded on some of the results presented in S.2.17. Note that for the sake of convenience, *in this section we denote the cylindrical coordinates* (ρ, θ^c, z) *as* (r, θ, z)*, respectively.*

As shown in Fig. 7.13, in the general case, the beam is acted upon by external forces $R_s(r,\theta)$, $\Theta_s(r,\theta)$, $Z_s(r,\theta)$ and the body forces $R_b(r,\theta)$, $\Theta_b(r,\theta)$, $Z_b(r,\theta)$. We also define the cosines of the angles between the normal $\bar{\mathbf{n}}$ (directed outwards) and the $\hat{r}$ and $\hat{\theta}$ directions by the $\cos(\bar{\mathbf{n}}, r)$ and $\cos(\bar{\mathbf{n}}, \theta)$, respectively.

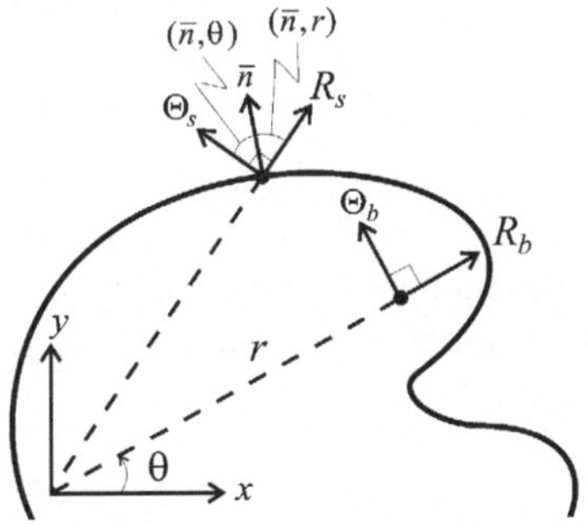

Figure 7.13: Body and surface loads acting over a beam cross-section in cylindrical coordinates.

Note that in order to ensure a *constant constitutive relations matrix*, the position vector of a point, r, is always measured from the origin of the cylindrical anisotropy, while θ is measured from the x coordinate line of a Cartesian coordinate system, the origin of which is placed at the same point, see domain "A" in Fig. 7.14. Note that in the general case which will not be dealt here, the cross-section centroid does not coincide with the origin of the cylindrical anisotropy as shown by domain "B" in Fig. 7.14. Hence, unlike the case described in S.5.1.1, we do not

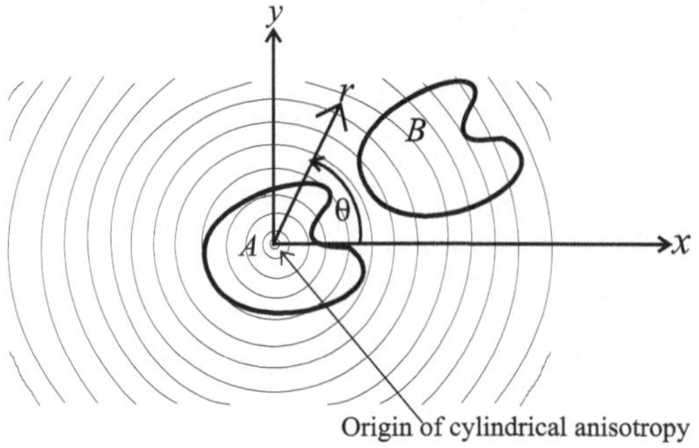

Figure 7.14: Relations between the domain and the origin of the cylindrical anisotropy. For domain "A" the origin of the cylindrical anisotropy coincides with the cross-section centroid, for "B" it does not.

have the freedom to place the origin of the coordinate system at the cross-section centroid, and to zero out the products of inertia. Subsequently, (5.1) are written as

$$\iint_{\Omega}\{1, r\cos\theta, r\sin\theta, r^2\cos^2\theta, r^2\cos\theta\sin\theta, r^2\sin^2\theta\}r\,dr\,d\theta = \{S_\Omega, x_c, y_c, I_y, I_{xy}, I_x\} \quad (7.132)$$

where $d\Omega = r\,dr\,d\theta$ has been employed. Yet, to clarify some of the examples presented in what follows, we shall discuss simple cases where the origin of the cylindrical anisotropy coincides with the cross-section centroid and the system axes coincide with the principal directions (i.e., $x_c = y_c = I_{xy} = 0$) as schematically shown by domain "A" in Fig. 7.14.

Based on the analysis presented in Chapter 1, and in particular in S.1.1.3, S.1.7.2, by activating **P.1.1** and making the following replacements:

$$\rho \Rightarrow r,\quad e_{11} \Rightarrow e_r,\quad e_{22} \Rightarrow e_\theta,\quad e_{33} \Rightarrow e_z,\quad 2e_{12} \Rightarrow \gamma_{r\theta},\quad 2e_{13} \Rightarrow \gamma_{rz},\quad 2e_{23} \Rightarrow \gamma_{\theta z},$$
$$U_1 \Rightarrow u,\quad U_2 \Rightarrow v,\quad U_3 \Rightarrow w, \quad (7.133)$$

the (linear) strains in cylindrical coordinates appear as

$$\varepsilon_r = u_{,r},\qquad \varepsilon_\theta = u_{,r} + \frac{1}{r}v_{,\theta},\qquad \gamma_{r\theta} = -\frac{v}{r} + \frac{1}{r}u_{,\theta} + v_{,r}, \quad (7.134a)$$

$$\varepsilon_z = w_{,z},\quad \gamma_{\theta z} = v_{,z} + \frac{1}{r}w_{,\theta},\quad \gamma_{rz} = w_{,r} + u_{,z}. \quad (7.134b)$$

With the same replacements and based on the analysis presented in S.1.2.1, by activating (the linear version of) **P.1.3**, the compatibility equations in cylindrical coordinates become

$$(r\varepsilon_{r\theta})_{,r\theta} = \varepsilon_{r,\theta\theta} + r(r\varepsilon_\theta)_{,rr} - r\varepsilon_{r,r}, \quad (7.135a)$$

$$(r\varepsilon_{\theta z})_{\theta z} = r^2\varepsilon_{\theta,zz} - r\varepsilon_{rz,z} + r\varepsilon_{z,r} + \varepsilon_{z,\theta\theta}, \quad (7.135b)$$

$$\varepsilon_{rz,rz} = \varepsilon_{z,rr} + \varepsilon_{r,zz}, \quad (7.135c)$$

$$(-\gamma_{r\theta,z} + \frac{1}{r}\gamma_{rz,\theta} + \gamma_{\theta z,r})_{,z} = 2(\frac{\varepsilon_z}{r})_{,r\theta} + \frac{1}{r}\gamma_{\theta z,z}, \quad (7.135d)$$

$$(\gamma_{r\theta,z} - \frac{1}{r}\gamma_{rz,\theta} + \gamma_{\theta z,r})_{,\theta} = 2(r\varepsilon_\theta)_{,rz} - 2\varepsilon_{r,z} - \frac{1}{r}\gamma_{\theta z,\theta}, \quad (7.135e)$$

$$(\gamma_{r\theta,z} + \frac{1}{r}\gamma_{rz,\theta} - \gamma_{\theta z,r})_{,r} = \frac{2}{r}\varepsilon_{r,\theta z} - \frac{2}{r}\varepsilon_{r\theta,z} + (\frac{\varepsilon_{\theta z}}{r})_{,r}. \quad (7.135f)$$

In addition, based on the analysis presented in S.1.3.2, by activating **P.1.7** and replacing the notation as

$$\rho \Rightarrow r, \quad \sigma_{11} \Rightarrow \sigma_r, \quad \sigma_{22} \Rightarrow \sigma_\theta, \quad \sigma_{33} \Rightarrow \sigma_z, \quad \sigma_{12} \Rightarrow \tau_{r\theta}, \quad \sigma_{13} \Rightarrow \tau_{rz}, \quad \sigma_{23} \Rightarrow \tau_{\theta z}, \tag{7.136}$$

the equilibrium equations in the present case become, see also (1.87a–c),

$$\sigma_{r,r} + \frac{1}{r}\tau_{r\theta,\theta} + \tau_{rz,z} + \frac{\sigma_r - \sigma_\theta}{r} + R_b = 0, \tag{7.137a}$$

$$\tau_{r\theta,r} + \frac{1}{r}\sigma_{\theta,\theta} + \tau_{\theta z,z} + \frac{2\tau_{r\theta}}{r} + \Theta_b = 0, \tag{7.137b}$$

$$\tau_{rz,r} + \frac{1}{r}\tau_{\theta z,\theta} + \sigma_{z,z} + \frac{\tau_{rz}}{r} + Z_b = 0. \tag{7.137c}$$

Also, the boundary conditions on the lateral (cylindrical) surface may be written for the case under discussion as

$$R_s = \sigma_r \cos(\bar{\mathbf{n}}, r) + \tau_{r\theta} \cos(\bar{\mathbf{n}}, \theta), \tag{7.138a}$$

$$\Theta_s = \tau_{r\theta} \cos(\bar{\mathbf{n}}, r) + \sigma_\theta \cos(\bar{\mathbf{n}}, \theta), \tag{7.138b}$$

$$Z_s = \tau_{rz} \cos(\bar{\mathbf{n}}, r) + \tau_{\theta z} \cos(\bar{\mathbf{n}}, \theta). \tag{7.138c}$$

To ensure that the distributions of the stress components over the tip ($z = l$) cross-section, Ω^t, are equivalent in an integral manner to the tip forces and moments there, one should require similar to (5.7)

$$\iint_{\Omega^t} \{\sigma_z, \sigma_z \sin\theta\, r, \sigma_z \cos\theta\, r, \tau_{rz}\cos\theta - \tau_{\theta z}\sin\theta, \tau_{rz}\sin\theta + \tau_{\theta z}\cos\theta, \tau_{\theta z} r\} r\, dr\, d\theta$$
$$= \{P_z, M_x, -M_y, P_x, P_y, M_z\}. \tag{7.139}$$

For "clamped-I" conditions at the root of a clamped-free beam, see (5.12), we write

$$u = v = w = u_{,z} = v_{,z} = \omega_z = 0 \qquad \text{at} \qquad r = z = 0, \tag{7.140}$$

where according to (1.30)

$$2\omega_z = v_{,r} + \frac{v}{r} - \frac{1}{r}u_{,\theta}. \tag{7.141}$$

7.5.1 Geometrical Aspects

We shall adopt here the relations of (1.218a) between Cartesian and cylindrical coordinates (with $\alpha_1 = r$, $\alpha_2 = \theta$) to express the following identities:

$$x = r\cos\theta, \quad y = r\sin\theta, \quad r^2 = x^2 + y^2, \quad \theta = arc\tan\frac{y}{x}, \tag{7.142a}$$

$$r_{,x} = \cos\theta, \qquad r_{,y} = \sin\theta, \qquad \theta_{,x} = -\frac{\sin\theta}{r}, \qquad \theta_{,y} = \frac{\cos\theta}{r}, \tag{7.142b}$$

$$x_{,r} = \cos\theta, \qquad y_{,r} = \sin\theta, \qquad x_{,\theta} = -r\sin\theta, \qquad y_{,\theta} = r\cos\theta, \tag{7.142c}$$

$$\frac{\partial}{\partial x} = \cos\theta\frac{\partial}{\partial r} - \frac{\sin\theta}{r}\cdot\frac{\partial}{\partial\theta}, \qquad \frac{\partial}{\partial y} = \sin\theta\frac{\partial}{\partial r} + \frac{\cos\theta}{r}\cdot\frac{\partial}{\partial\theta}. \tag{7.142d}$$

As shown in Fig. 7.15 and similar to (3.4), we write the cosines $\cos(\bar{\mathbf{n}}, r)$ and $\cos(\bar{\mathbf{n}}, \theta)$ as

$$\cos(\bar{\mathbf{n}}, r) = r\frac{d\theta}{ds} = \frac{dr}{dn}, \qquad \cos(\bar{\mathbf{n}}, \theta) = -\frac{dr}{ds} = r\frac{d\theta}{dn}. \tag{7.143}$$

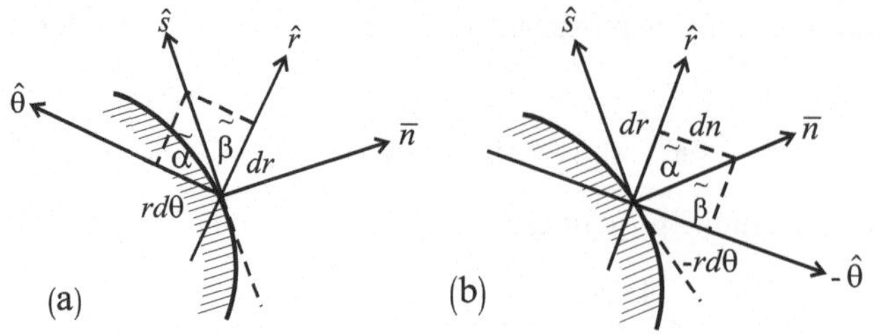

Figure 7.15: Directional cosines notation, $\tilde{\alpha} \equiv (\bar{\mathbf{n}}, r)$, $\tilde{\beta} \equiv -(\bar{\mathbf{n}}, \theta)$.

These cosines are related to $\cos(\bar{\mathbf{n}}, x)$ and $\cos(\bar{\mathbf{n}}, y)$ of (3.4) by

$$\begin{Bmatrix} \cos(\bar{\mathbf{n}}, r) \\ \cos(\bar{\mathbf{n}}, \theta) \end{Bmatrix} = \begin{bmatrix} \cos\theta & \sin\theta \\ -\sin\theta & \cos\theta \end{bmatrix} \begin{Bmatrix} \cos(\bar{\mathbf{n}}, x) \\ \cos(\bar{\mathbf{n}}, y) \end{Bmatrix}. \tag{7.144}$$

To facilitate the discussion in what follows, we also present here the relations between stress components in Cartesian and cylindrical coordinates (note that σ_z is invariant)

$$\begin{aligned} &\sigma_r = \sigma_x \cos^2\theta + \sigma_y \sin^2\theta + \tau_{xy} \sin 2\theta, \qquad \sigma_\theta = \sigma_x \sin^2\theta + \sigma_y \cos^2\theta - \tau_{xy} \sin 2\theta, \\ &\tau_{\theta z} = \tau_{yz} \cos\theta - \tau_{xz} \sin\theta, \quad \tau_{rz} = \tau_{yz} \sin\theta + \tau_{xz} \cos\theta, \quad \tau_{r\theta} = (\sigma_y - \sigma_x)\frac{\sin 2\theta}{2} + \tau_{xy} \cos 2\theta, \end{aligned} \tag{7.145}$$

and

$$\begin{aligned} &\sigma_x = \sigma_r \cos^2\theta + \sigma_\theta \sin^2\theta - \tau_{r\theta} \sin 2\theta, \qquad \sigma_y = \sigma_r \sin^2\theta + \sigma_\theta \cos^2\theta + \tau_{r\theta} \sin 2\theta, \\ &\tau_{yz} = \tau_{\theta z} \cos\theta + \tau_{rz} \sin\theta, \quad \tau_{xz} = \tau_{rz} \cos\theta - \tau_{\theta z} \sin\theta, \quad \tau_{xy} = (\sigma_r - \sigma_\theta)\frac{\sin 2\theta}{2} + \tau_{r\theta} \cos 2\theta. \end{aligned} \tag{7.146}$$

The corresponding strain relations are (while ε_z is invariant)

$$\begin{aligned} &\varepsilon_r = \varepsilon_x \cos^2\theta + \varepsilon_y \sin^2\theta + \frac{1}{2}\gamma_{xy} \sin 2\theta, \qquad \varepsilon_\theta = \varepsilon_x \sin^2\theta + \varepsilon_y \cos^2\theta - \frac{1}{2}\gamma_{xy} \sin 2\theta, \\ &\gamma_{\theta z} = \gamma_{yz} \cos\theta - \gamma_{xz} \sin\theta, \quad \gamma_{rz} = \gamma_{yz} \sin\theta + \gamma_{xz} \cos\theta, \quad \gamma_{r\theta} = (\varepsilon_y - \varepsilon_x) \sin 2\theta + \gamma_{xy} \cos 2\theta, \end{aligned} \tag{7.147}$$

and

$$\begin{aligned} &\varepsilon_x = \varepsilon_r \cos^2\theta + \varepsilon_\theta \sin^2\theta - \frac{1}{2}\gamma_{r\theta} \sin 2\theta, \qquad \varepsilon_y = \varepsilon_r \sin^2\theta + \varepsilon_\theta \cos^2\theta + \frac{1}{2}\gamma_{r\theta} \sin 2\theta, \\ &\gamma_{yz} = \gamma_{\theta z} \cos\theta + \gamma_{rz} \sin\theta, \quad \gamma_{xz} = \gamma_{rz} \cos\theta - \gamma_{\theta z} \sin\theta, \quad \gamma_{xy} = (\varepsilon_r - \varepsilon_\theta) \sin 2\theta + \gamma_{r\theta} \cos 2\theta. \end{aligned} \tag{7.148}$$

7.5.2 Torsion of a Beam of Cylindrical Anisotropy

In this case we consider a beam of a cylindrical MON13z anisotropy which is acted upon by a tip torsion moment, M_z, only. We subsequently assume cylindrical displacement components u, v, w (i.e., u_r, u_θ, u_z) of the form

$$u = 0, \qquad v = \frac{M_z}{D} rz, \qquad w = \frac{M_z}{D} \varphi(r, \theta), \tag{7.149}$$

where D is the torsional rigidity, and $\varphi(r, \theta)$ is a torsion warping function to be determined (recall that $\frac{M_z}{D}$ stands here for the twist per unit length). Then, (7.134a,b) show that

$$\varepsilon_r = \varepsilon_\theta = \gamma_{r\theta} = \varepsilon_z = 0, \qquad \gamma_{\theta z} = \frac{M_z}{D}\left(\frac{1}{r}\varphi_{,\theta} + r\right), \qquad \gamma_{rz} = \frac{M_z}{D}\varphi_{,r}. \tag{7.150}$$

Hence, by using the constitutive relations for cylindrical MON13z anisotropy of (2.72), the following stress components are obtained:

$$\sigma_r = \sigma_\theta = \sigma_z = \tau_{r\theta} = 0, \tag{7.151a}$$

$$\tau_{\theta z} = \frac{M_z}{Da_0}[-a_{45}\varphi,_r + a_{55}(\frac{1}{r}\varphi,_\theta + r)], \qquad \tau_{rz} = \frac{M_z}{Da_0}[a_{44}\varphi,_r - a_{45}(\frac{1}{r}\varphi,_\theta + r)]. \tag{7.151b}$$

Substituting (7.151a,b) in the equilibrium equations (7.137a–c), shows that the first two equations are satisfied identically, and likewise, the first two equations of the boundary conditions (7.138a–c). The third equilibrium equation and the third boundary condition constitute the following Neumann BVP for the torsion warping function $\varphi(r, \theta)$:

$$\nabla_5^{(2)}\varphi = 2a_{45} \quad \text{over} \quad \Omega, \qquad D_3^n\varphi = a_{45}r\cos(\bar{\mathbf{n}}, r) - a_{55}r\cos(\bar{\mathbf{n}}, \theta) \qquad \text{on} \quad \partial\Omega, \tag{7.152}$$

where the differential operators $\nabla_5^{(2)}$ and D_3^n are given by (3.199), (3.220), respectively.

For the specific case of a circular cross-section of radius R, $\cos(\bar{\mathbf{n}}, r) = 1$, $\cos(\bar{\mathbf{n}}, \theta) = 0$, and we assume a torsion warping function of the form $\varphi(r) = \tilde{c}r^2$, where $\tilde{c}$ is a constant. Employing (7.152) shows that $\varphi = \frac{a_{45}}{2a_{44}}r^2$ (which as expected, vanishes for $a_{45} = 0$, i.e. in the orthotropic or isotropic cases). The shear stress components then become, see (7.151a–c),

$$\tau_{\theta z} = \frac{M_z}{Da_{44}}r, \qquad \tau_{rz} = 0. \tag{7.153}$$

As far as the fulfillment of the tip conditions is concerned, the first five of (7.139) are satisfied identically, while the last one shows that the torsional rigidity is given by

$$D = \iint_\Omega \frac{1}{a_0}[-a_{45}\varphi,_r + a_{55}(\frac{1}{r}\varphi,_\theta + r)]r^2 dr\, d\theta = \frac{\pi R^4}{2a_{44}}. \tag{7.154}$$

It is interesting to note that (7.38) yields for a circular cross-section of Cartesian anisotropy a torsional rigidity of

$$D_{\text{Cartesian}} = \frac{\pi R^4}{a_{55} + a_{44}}. \tag{7.155}$$

Comparison of (7.154) and (7.155) shows that these two cases are indeed different, as illustrated by Fig. 2.11(c),(d) (see also Example 10.1). In addition, it is shown that a_{55} does not participate in the cylindrical anisotropy case (as it reflects shear stiffness in the radial direction) and its role is "fulfilled" by a_{44}. As another comparison, we may transform (7.153) into Cartesian direction using (7.146), which yields

$$\tau_{yz} = \frac{2M_z}{\pi R^4}x, \qquad \tau_{xz} = -\frac{2M_z}{\pi R^4}y. \tag{7.156}$$

This result is identical to the one obtained for a circular cross-section of Cartesian orthotropic material, see (7.42). An extension of this torsion problem to the case of a non-homogeneous beam with *continuously varying elastic moduli* appears in Remark 8.3.

7.5.3 Extension and Bending of a Beam of Cylindrical Anisotropy

Two different methods will be described in what follows for the analysis of a beam of cylindrical MON13z anisotropy under extension and bending induced by tip loads. The first "Stress Function Method" developed in S.7.5.3.1 (see Trail "C" of S.1.6.5.3) for tip extension load,

P_z, and tip bending moments M_x, M_y, is a detailed and comprehensive version of the one presented by (Lekhnitskii, 1981). The second "Stress Hypothesis Method" developed in S.7.5.3.2 (see Trail "B" of S.1.6.5.2) for the transverse tip load P_y is founded on the stress hypothesis methodology discussed in S.7.2.

Overall, together with the torsion solution of S.7.5.2, it will be shown that the tip loads P_x, P_y and M_z induce the shear stress components $\tau_{\theta z}$, τ_{rz} while the tip loads P_z, M_x and M_y induce the stress components σ_r, σ_θ, $\tau_{r\theta}$, while all loads, except for M_z, induce σ_z as well.

7.5.3.1 Stress Function Method

The derivation is founded on the vanishing shear stresses assumption, namely, $\tau_{\theta z} = \tau_{rz} = 0$, and therefore, the constitutive relations for cylindrical MON13z anisotropy of (2.72) show that $\gamma_{\theta z} = \gamma_{rz} = 0$ as well, and that ε_z may be further expressed as

$$\varepsilon_z(r,\theta) = a_{13}\sigma_r + a_{23}\sigma_\theta + a_{33}\sigma_z + a_{36}\tau_{r\theta}. \tag{7.157}$$

Integrating (7.134b) yields

$$u = -\frac{z^2}{2}\varepsilon_{z,r} - z\widetilde{W}_{,r} + \widetilde{U}, \qquad v = -\frac{z^2}{2r}\varepsilon_{z,\theta} - \frac{z}{r}\widetilde{W}_{,\theta} + \widetilde{V}, \qquad w = z\varepsilon_z(r,\theta) + \widetilde{W}, \tag{7.158}$$

where $\widetilde{U}(r,\theta)$, $\widetilde{V}(r,\theta)$, $\widetilde{W}(r,\theta)$ are functions to be determined. Substituting the above expressions in (7.134a) enables us to write

$$\varepsilon_r = -\frac{z^2}{2}\varepsilon_{z,rr} - z\widetilde{W}_{,rr} + \widetilde{U}_{,r}, \tag{7.159a}$$

$$\varepsilon_\theta = -\frac{z^2}{2r}(\frac{1}{r}\varepsilon_{z,\theta\theta} + \varepsilon_{z,r}) - \frac{z}{r}(\frac{1}{r}\widetilde{W}_{,\theta\theta} + \widetilde{W}_{,r}) + \frac{1}{r}\widetilde{V}_{,\theta} + \frac{\widetilde{U}}{r}, \tag{7.159b}$$

$$\gamma_{r\theta} = \frac{z^2}{r}(\frac{1}{r}\varepsilon_{z,\theta} - \varepsilon_{z,r\theta}) + \frac{2z}{r}(\frac{1}{r}\widetilde{W}_{,\theta} - \widetilde{W}_{,r\theta}) - \frac{\widetilde{V}}{r} + \frac{1}{r}\widetilde{U}_{,\theta} + \widetilde{V}_{,r}. \tag{7.159c}$$

We shall now "balance" coefficients of z^2, z and the free terms on both sides of (7.159a–c). Equating terms with z^2 yields the following differential equations for ε_z:

$$\varepsilon_{z,rr} = 0, \qquad \frac{1}{r}\varepsilon_{z,\theta\theta} + \varepsilon_{z,r} = 0, \qquad \frac{1}{r}\varepsilon_{z,\theta} - \varepsilon_{z,r\theta} = 0. \tag{7.160}$$

Equating terms with z yields the differential equations for $\widetilde{W}$ (identical to the above):

$$\widetilde{W}_{,rr} = 0, \qquad \frac{1}{r}\widetilde{W}_{,\theta\theta} + \widetilde{W}_{,r} = 0, \qquad \frac{1}{r}\widetilde{W}_{,\theta} - \widetilde{W}_{,r\theta} = 0. \tag{7.161}$$

Equating the free terms yields the following set of differential equations for $\widetilde{U}$ and $\widetilde{V}$:

$$\widetilde{U}_{,r} = a_{11}\sigma_r + a_{12}\sigma_\theta + a_{13}\sigma_z + a_{16}\tau_{r\theta}, \tag{7.162a}$$

$$\frac{1}{r}\widetilde{V}_{,\theta} + \frac{\widetilde{U}}{r} = a_{12}\sigma_r + a_{22}\sigma_\theta + a_{23}\sigma_z + a_{26}\tau_{r\theta}, \tag{7.162b}$$

$$-\frac{\widetilde{V}}{r} + \frac{1}{r}\widetilde{U}_{,\theta} + \widetilde{V}_{,r} = a_{16}\sigma_r + a_{26}\sigma_\theta + a_{36}\sigma_z + a_{66}\tau_{r\theta}. \tag{7.162c}$$

Integrating (7.160) and (7.161) yields

$$\varepsilon_z(r,\theta) = a_{33}(\widetilde{\alpha} r\cos\theta + \widetilde{\beta} r\sin\theta + \widetilde{\gamma}), \tag{7.163a}$$

$$\widetilde{W}(r,\theta) = \widetilde{A} r\cos\theta + \widetilde{B} r\sin\theta + \widetilde{C}, \tag{7.163b}$$

where $\widetilde{\alpha}, \widetilde{\beta}$, $\widetilde{\gamma}$ and $\widetilde{A}, \widetilde{B}$, $\widetilde{C}$ are constants to be determined (the slightly different form of the above equations is for convenience only). Substituting (7.163a) in (7.157) enables us to write

$$\sigma_z = -\frac{1}{a_{33}}(a_{13}\sigma_r + a_{23}\sigma_\theta + +a_{36}\tau_{r\theta}) + \widetilde{\alpha} r\cos\theta + \widetilde{\beta} r\sin\theta + \widetilde{\gamma}. \tag{7.164}$$

Further substitution of (7.164) in (7.162a–c) yields

$$\widetilde{U}_{,r} = b_{11}\sigma_r + b_{12}\sigma_\theta + b_{16}\tau_{r\theta} + a_{13}(\widetilde{\alpha} r\cos\theta + \widetilde{\beta} r\sin\theta + \widetilde{\gamma}), \tag{7.165a}$$

$$\frac{1}{r}\widetilde{V}_{,\theta} + \frac{\widetilde{U}}{r} = b_{12}\sigma_r + b_{22}\sigma_\theta + b_{26}\tau_{r\theta} + a_{23}(\widetilde{\alpha} r\cos\theta + \widetilde{\beta} r\sin\theta + \widetilde{\gamma}), \tag{7.165b}$$

$$-\frac{\widetilde{V}}{r} + \frac{1}{r}\widetilde{U}_{,\theta} + \widetilde{V}_{,r} = b_{16}\sigma_r + b_{26}\sigma_\theta + b_{66}\tau_{r\theta} + a_{36}(\widetilde{\alpha} r\cos\theta + \widetilde{\beta} r\sin\theta + \widetilde{\gamma}). \tag{7.165c}$$

We now assume that the stress components are the following partial derivatives of a stress function, $\Phi(r,\theta)$, in the classical manner (i.e., employed for the isotropic case as well):

$$\sigma_r = \frac{1}{r}\Phi_{,r} + \frac{1}{r^2}\Phi_{,\theta\theta}, \qquad \sigma_\theta = \Phi_{,rr}, \qquad \tau_{r\theta} = -(\frac{\Phi}{r})_{,r\theta} = \frac{1}{r^2}\Phi_{,\theta} - \frac{1}{r}\Phi_{,r\theta}. \tag{7.166}$$

Evidently, the r.h.s. of (7.165a–c) are ε_r, ε_θ and $\gamma_{r\theta}$, respectively (see 7.162a-c). Substituting these expressions in the first compatibility equation (7.135a) yields

$$\nabla_5^{(4)}\Phi = 2[(a_{13} - a_{23})\widetilde{\beta} - a_{36}\widetilde{\alpha}]\frac{\sin\theta}{r} + 2[a_{36}\widetilde{\beta} + (a_{13} - a_{23})\widetilde{\alpha}]\frac{\cos\theta}{r} \tag{7.167}$$

where the biharmonic operator $\nabla_5^{(4)}$ is defined by (3.210). All other compatibility equations of (7.135b,f) are satisfied identically.

Since $\tau_{\theta z} = \tau_{rz} = 0$, the stress components in this case must satisfy only the boundary conditions of (7.138a,b) over the lateral surface. For the present case of tip loading and zero surface loading we write

$$(\frac{1}{r}\Phi_{,r} + \frac{1}{r^2}\Phi_{,\theta\theta})\cos(\bar{\mathbf{n}},r) + (\frac{1}{r^2}\Phi_{,\theta} - \frac{1}{r}\Phi_{,r\theta})\cos(\bar{\mathbf{n}},\theta) = 0, \tag{7.168a}$$

$$(\frac{1}{r^2}\Phi_{,\theta} - \frac{1}{r}\Phi_{,r\theta})\cos(\bar{\mathbf{n}},r) + \Phi_{,rr}\cos(\bar{\mathbf{n}},\theta) = 0. \tag{7.168b}$$

Finally, the displacements are determined by (7.158) as (see also Remark 7.8)

$$u = -\frac{z^2}{2}a_{33}(\widetilde{\alpha}\cos\theta + \widetilde{\beta}\sin\theta) - z(\widetilde{A}\cos\theta + \widetilde{B}\sin\theta) + \widetilde{U}(r,\theta), \tag{7.169a}$$

$$v = -\frac{z^2}{2}a_{33}(-\widetilde{\alpha}\sin\theta + \widetilde{\beta}\cos\theta) - z(-\widetilde{A}\sin\theta + \widetilde{B}\cos\theta) + \widetilde{V}(r,\theta), \tag{7.169b}$$

$$w = za_{33}(\widetilde{\alpha} r\cos\theta + \widetilde{\beta} r\sin\theta + \widetilde{\gamma}) + \widetilde{A} r\cos\theta + \widetilde{B} r\sin\theta + \widetilde{C}. \tag{7.169c}$$

The following are the solution steps one should carry out to implement the method under discussion:

(1) Solve the biharmonic BVP (7.167), (7.168a,b). Note that in the general case, $\widetilde{\alpha}$ and $\widetilde{\beta}$ are not known at this stage. It is therefore useful to express the stress function as $\Phi = \Phi^0 + \widetilde{\alpha}\Phi^\alpha + \widetilde{\beta}\Phi^\beta$, where Φ^0 is the solution for $\widetilde{\alpha} = \widetilde{\beta} = 0$, Φ^α is the solution for $\widetilde{\alpha} = 1$, $\widetilde{\beta} = 0$, and Φ^β is the solution for $\widetilde{\beta} = 1$, $\widetilde{\alpha} = 0$.

(2) Calculate the stress components σ_r, σ_θ and $\tau_{r\theta}$ from (7.166). When (7.167) is not homogeneous, the resulting stress components should be expressed as $\sigma_r = \sigma_r^0 + \widetilde{\alpha}\sigma_r^\alpha + \widetilde{\beta}\sigma_r^\beta$, etc., similarly to the above discussion in (1).

(3) Substitute σ_z of (7.164) in the three first equations of (7.139), and calculate $\widetilde{\alpha}$, $\widetilde{\beta}$, $\widetilde{\gamma}$.

(4) Apply the root boundary conditions of (7.140) while these are written with (7.169a–c) components. This yields

$$\widetilde{A} = \widetilde{B} = \widetilde{C} = \widetilde{U} = \widetilde{V} = \widetilde{V}_{,r} + \frac{\widetilde{V}}{r} - \frac{1}{r}\widetilde{U}_{,\theta} = 0 \quad \text{at} \quad r = z = 0. \tag{7.170}$$

Integrate (7.165a–c) and find $\widetilde{U}$ and $\widetilde{V}$ under the above conditions. An immediate result of (7.140) and (7.163b) is that $\widetilde{W}(r,\theta) = 0$. The above solution is illustrated by Example 7.9.

Remark 7.8 The displacement triad in the present context is $u \equiv u_r$, $v \equiv u_\theta$ and $w \equiv u_z$, while for Cartesian analysis we use $u \equiv u_x$, $v \equiv u_y$ and $w \equiv u_z$. These components are related as $u_x = u_r \cos\theta - u_\theta \sin\theta$ and $u_y = u_r \sin\theta + u_\theta \cos\theta$. Therefore, (7.169a–c) may be written in Cartesian coordinates as

$$u = u_x = -a_{33}\frac{z^2}{2}\widetilde{\alpha} - z\widetilde{A} + \widetilde{U}\cos\theta - \widetilde{V}\sin\theta, \tag{7.171a}$$

$$v = u_y = -a_{33}\frac{z^2}{2}\widetilde{\beta} - z\widetilde{B} + \widetilde{U}\sin\theta + \widetilde{V}\cos\theta, \tag{7.171b}$$

$$w = u_z = a_{33}(\widetilde{\alpha}x + \widetilde{\beta}y + \widetilde{\gamma}) + \widetilde{A}x + \widetilde{B}y + \widetilde{C}. \tag{7.171c}$$

Employing the definition of the edgewise curvature, κ_u, and the beamwise curvature, κ_v, of (5.6) we find

$$\kappa_u = -a_{33}\widetilde{\alpha}, \qquad \kappa_v = -a_{33}\widetilde{\beta}, \qquad w_{,z} = a_{33}\widetilde{\gamma}. \tag{7.172}$$

Hence, $\widetilde{\alpha}$, $\widetilde{\beta}$ and $\widetilde{\gamma}$ are the curvature and extension measures in this case.

Example 7.9 *Annular Cross-Section Under Tip Axial Force.*
For an orthotropic beam of annular cross-section (of radii $R_2 > R_1 > 0$) under axial force P_z, we assume $\Phi = \Phi(r)$ and for axisymmetric behavior $\widetilde{\alpha} = \widetilde{\beta} = 0$ must also be assumed. Hence, (7.163a) shows that $\varepsilon_z(r,\theta) = \widetilde{\gamma}$ and (7.167) becomes homogeneous, i.e., $\nabla_5^{(4)}\Phi = 0$, or

$$b_{22}\Phi_{,rrrr} + \frac{2}{r}b_{22}\Phi_{,rrr} - \frac{1}{r^2}b_{11}\Phi_{,rr} + \frac{1}{r^3}b_{11}\Phi_{,r} = 0. \tag{7.173}$$

For the annular cross-section case ($\cos(\bar{\mathbf{n}}, r) = \pm 1$, $\cos(\bar{\mathbf{n}}, \theta) = 0$), (7.168a,b) yield the conditions $\Phi_{,r} = 0$ at $r = R_1, R_2$. We shall now proceed by considering two different cases:

Case I: $a_{13} = a_{23}$ and $b_{11} = b_{22}$. We assume $\Phi = \widetilde{c}r^2$, which satisfies (7.173) and yields the stress components

$$\sigma_r = 2\widetilde{c}, \qquad \sigma_\theta = 2\widetilde{c}, \qquad \tau_{r\theta} = 0, \qquad \sigma_z = \widetilde{\gamma} - \widetilde{c}\frac{a_{13} + a_{23}}{a_{33}}. \tag{7.174}$$

The boundary conditions require $\widetilde{c} = 0$, which leaves us with zero stress components except for $\sigma_z = \widetilde{\gamma}$. Equation (7.139:a) then shows that $\widetilde{\gamma} = \frac{P_z}{S_\Omega}$, and (7.158), (7.162a–c) yield $u = a_{13}\widetilde{\gamma}r$, $v = 0$ and $w = za_{33}\widetilde{\gamma}$. Note that the condition $a_{13} = a_{23}$ is necessary here for the single-valued property of the displacement v, which formally takes the form $v = (a_{23} - a_{13})\widetilde{\gamma} r\theta$. The above displacements are also identical to those of (7.74a–c) when the latter are written for the effect of P_z in orthotropic Cartesian material (see relations in Remark 7.8).

Case II: $b_{11} \neq b_{22}$. To solve (7.173) in this case we assume the following form of stress function Φ:

$$\Phi = \widetilde{c}_1 \frac{r^2}{2} + \widetilde{c}_2 \frac{r^{\kappa+1}}{\kappa+1} + \widetilde{c}_3 \frac{r^{1-\kappa}}{1-\kappa}, \tag{7.175}$$

where $\widetilde{c}_1$, $\widetilde{c}_2$ and $\widetilde{c}_3$ are arbitrary constants and $\kappa = \sqrt{\frac{b_{11}}{b_{22}}} \neq 1$. With this format, the stress components are

$$\sigma_r = \widetilde{c}_1 + \widetilde{c}_2 r^{\kappa-1} + \widetilde{c}_3 r^{-\kappa-1}, \qquad \sigma_\theta = \widetilde{c}_1 + \widetilde{c}_2 \kappa r^{\kappa-1} - \widetilde{c}_3 \kappa r^{-\kappa-1}, \qquad \tau_{r\theta} = 0, \tag{7.176a}$$

$$\sigma_z = \frac{\varepsilon_z}{a_{33}} - \widetilde{c}_1 \frac{a_{13}+a_{23}}{a_{33}} - \widetilde{c}_2 \frac{a_{13}+\kappa a_{23}}{a_{33}} r^{\kappa-1} - \widetilde{c}_3 \frac{a_{13}-\kappa a_{23}}{a_{33}} r^{-\kappa-1}. \tag{7.176b}$$

Equations (7.165a–c) then become

$$\widetilde{U}_{,r} = b_{11}(\widetilde{c}_1 + \widetilde{c}_2 r^{\kappa-1} + \widetilde{c}_3 r^{-\kappa-1}) + b_{12}(\widetilde{c}_1 + \widetilde{c}_2 k r^{\kappa-1} - \widetilde{c}_3 k r^{-\kappa-1}) + a_{13}\frac{\varepsilon_z}{a_{33}}, \tag{7.177a}$$

$$\frac{1}{r}\widetilde{V}_{,\theta} + \frac{\widetilde{U}}{r} = b_{12}(\widetilde{c}_1 + \widetilde{c}_2 r^{\kappa-1} + \widetilde{c}_3 r^{-\kappa-1}) + b_{22}(\widetilde{c}_1 + \widetilde{c}_2 k r^{\kappa-1} - \widetilde{c}_3 k r^{-\kappa-1}) + a_{23}\frac{\varepsilon_z}{a_{33}}, \tag{7.177b}$$

$$-\frac{\widetilde{V}}{r} + \frac{1}{r}\widetilde{U}_{,\theta} + \widetilde{V}_{,r} = 0. \tag{7.177c}$$

Integrating under the conditions of (7.170) yields

$$\widetilde{U} = b_{11}(\widetilde{c}_1 r + \widetilde{c}_2 \frac{r^\kappa}{\kappa} - \widetilde{c}_3 \frac{r^{-\kappa}}{\kappa}) + b_{12}(\widetilde{c}_1 r + \widetilde{c}_2 r^\kappa + \widetilde{c}_3 r^{-\kappa}) + a_{13}\frac{\varepsilon_z}{a_{33}} r, \tag{7.178a}$$

$$\widetilde{V} = [(a_{23} - a_{13})\frac{\varepsilon_z}{a_{33}} + (b_{22} - b_{11})\widetilde{c}_1] r\theta. \tag{7.178b}$$

As shown, $\widetilde{V}$ will be single-valued only for

$$\widetilde{c}_1 = \widetilde{h}\frac{\varepsilon_z}{a_{33}} \qquad \text{where} \qquad \widetilde{h} = \frac{a_{23} - a_{13}}{b_{11} - b_{22}}, \tag{7.179}$$

for which $\widetilde{V} = 0$. The boundary conditions on the inner and outer surfaces $r = R_1, R_2$ yield

$$\widetilde{c}_1 + \widetilde{c}_2 R_1^{\kappa-1} + \widetilde{c}_3 R_1^{-\kappa-1} = 0, \qquad \widetilde{c}_1 + \widetilde{c}_2 R_2^{\kappa-1} + \widetilde{c}_3 R_2^{-\kappa-1} = 0. \tag{7.180}$$

Substituting (7.179) in (7.180) and solving for $\widetilde{c}_2$ and $\widetilde{c}_3$ shows that

$$\widetilde{c}_2 = -\widetilde{h}\widetilde{\gamma}\frac{R_2^{\kappa+1} - R_1^{\kappa+1}}{R_2^{2\kappa} - R_1^{2\kappa}}, \qquad \widetilde{c}_3 = -\widetilde{h}\widetilde{\gamma} R_1^{\kappa+1} R_2^{\kappa+1} \frac{R_2^{\kappa-1} - R_1^{\kappa-1}}{R_2^{2\kappa} - R_1^{2\kappa}}. \tag{7.181}$$

Application of (7.139:a) leads to $\frac{\varepsilon_z}{a_{33}} = \frac{P_z}{S_e}$, where S_e may be viewed as the "effective" area given by

$$S_e = \pi(R_2^2 - R_1^2) - \frac{2\pi\widetilde{h}}{a_{33}}\Big[(a_{13}+a_{23})\frac{(R_2^2 - R_1^2)}{2} - \frac{a_{13}+\kappa a_{23}}{\kappa+1}\cdot\frac{(R_2^{\kappa+1} - R_1^{\kappa+1})^2}{R_2^{2\kappa} - R_1^{2\kappa}}$$
$$- \frac{a_{13}-\kappa a_{23}}{\kappa-1}\cdot\frac{R_1^2 R_2^2 (R_2^{\kappa-1} - R_1^{\kappa-1})^2}{R_2^{2\kappa} - R_1^{2\kappa}}\Big]. \tag{7.182}$$

The displacements in this case are, see (7.169a–c),

$$
\begin{aligned}
u\frac{S_e}{P_z} &= b_{11}\tilde{h}\Big(1-\frac{R_2^{\kappa+1}-R_1^{\kappa+1}}{R_2^{2\kappa}-R_1^{2\kappa}}\cdot\frac{r^{\kappa}}{\kappa}+R_1^{\kappa+1}R_2^{\kappa+1}\frac{R_2^{\kappa-1}-R_1^{\kappa-1}}{R_2^{2\kappa}-R_1^{2\kappa}}\cdot\frac{r^{-\kappa}}{\kappa}\Big)\\
&\quad + b_{12}\tilde{h}\Big(1-\frac{R_2^{\kappa+1}-R_1^{\kappa+1}}{R_2^{2\kappa}-R_1^{2\kappa}}r^{\kappa}-R_1^{\kappa+1}R_2^{\kappa+1}\frac{R_2^{\kappa-1}-R_1^{\kappa-1}}{R_2^{2\kappa}-R_1^{2\kappa}}r^{-\kappa}\Big)+a_{13}r,\\
v &= 0, \qquad w\frac{S_e}{P_z} = a_{33}z.
\end{aligned}
\tag{7.183}
$$

When $a_{23}=a_{13}$, $\tilde{h}$ vanishes, $S_e=S_\Omega$, and the above displacements become identical to those obtained in Case I of this example. The final stress expressions become

$$
\sigma_r\frac{S_e}{P_z} = \tilde{h}\Big(1-\frac{R_2^{\kappa+1}-R_1^{\kappa+1}}{R_2^{2\kappa}-R_1^{2\kappa}}r^{\kappa-1}-R_1^{\kappa+1}R_2^{\kappa+1}\frac{R_2^{\kappa-1}-R_1^{\kappa-1}}{R_2^{2\kappa}-R_1^{2\kappa}}r^{-\kappa-1}\Big), \tag{7.184a}
$$

$$
\sigma_\theta\frac{S_e}{P_z} = \tilde{h}\Big(1-\frac{R_2^{\kappa+1}-R_1^{\kappa+1}}{R_2^{2\kappa}-R_1^{2\kappa}}\kappa r^{\kappa-1}+R_1^{\kappa+1}R_2^{\kappa+1}\frac{R_2^{\kappa-1}-R_1^{\kappa-1}}{R_2^{2\kappa}-R_1^{2\kappa}}\kappa r^{-\kappa-1}\Big), \tag{7.184b}
$$

$$
\tau_{\theta z}=\tau_{r\theta}=\tau_{rz}=0, \tag{7.184c}
$$

$$
\begin{aligned}
\sigma_z\frac{S_e}{P_z} &= 1-\frac{\tilde{h}}{a_{33}}\Big[a_{13}+a_{23}-\frac{R_2^{\kappa+1}-R_1^{\kappa+1}}{R_2^{2\kappa}-R_1^{2\kappa}}(a_{13}+\kappa a_{23})r^{\kappa-1}\\
&\quad -R_1^{\kappa+1}R_2^{\kappa+1}\frac{R_2^{\kappa-1}-R_1^{\kappa-1}}{R_2^{2\kappa}-R_1^{2\kappa}}(a_{13}-\kappa a_{23})r^{-\kappa-1}\Big].
\end{aligned}
\tag{7.184d}
$$

For a (full) circular cross-section, $R_1=\tilde{c}_3=0$. In such a case, when $\kappa<1$ the stresses are singular at $r=0$ and only a solution of an annular cross-section is possible. For $\kappa>1$, the stresses are non-singular at $r=0$ and solutions of both annular and circular cross-sections are possible (the case of $\kappa=1$ is discussed within Case I of this example). Illustrative examples are given in Fig. 7.16. Note that as shown by (7.72b), for Cartesian anisotropy, a constant σ_z distribution is obtained by a tip axial load (in such a case for the annular cross-section shown in Fig. 7.16(a) $\frac{R_2^2}{P_z}\sigma_z\cong 0.349$, and for the circular cross-section shown in Fig. 7.16(b), $\frac{R_2^2}{P_z}\sigma_z\cong 0.318$).

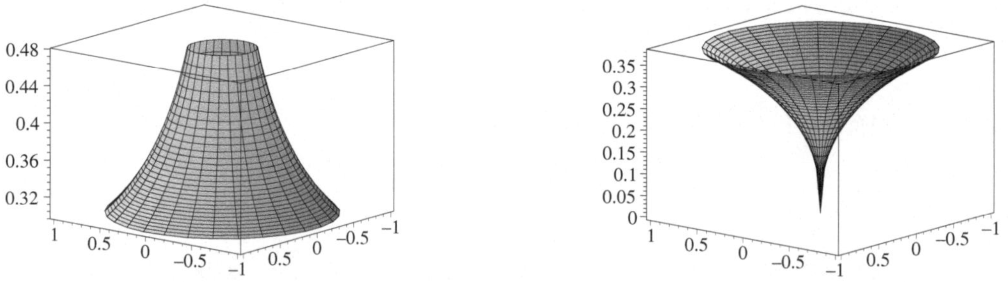

(a) $\kappa=0.61$, $\frac{R_1}{R_2}=0.3$. (b) $\kappa=1.38$, $\frac{R_1}{R_2}=0.$.

Figure 7.16: $\frac{\sigma_z R_2^2}{P_z}$ in an annular/circular cross-section of cylindrical anisotropy.

Example 7.10 *Annular Cross-Section Under Tip Bending Moment.*

We shall discuss here the behavior of the orthotropic beam of Example 7.9 under a tip bending moment M_x. Assuming $\widetilde{\alpha} = \widetilde{\gamma} = 0$, the function $\Phi(r, \theta)$ must satisfy the following equation:

$$\nabla_6^{(4)}\Phi = 2\widetilde{\beta}(a_{13} - a_{23})\frac{\sin\theta}{r}, \tag{7.185}$$

where $\nabla_6^{(4)}$ is the orthotropic version of the biharmonic operator, see (3.211). We shall seek a solution of the form

$$\Phi(r,\theta) = (\frac{\widetilde{\gamma}_1}{\lambda_1}r^{\lambda_1+1} + \frac{\widetilde{\gamma}_2}{\lambda_1}r^{1-\lambda_1} + \widetilde{\gamma}_3 r\ln r + \widetilde{\gamma}_4 r + \frac{\widetilde{\beta}\lambda_2}{2}r^3)\sin\theta, \tag{7.186}$$

where $\widetilde{\gamma}_1, \ldots, \widetilde{\gamma}_4$ are arbitrary constants, and

$$\lambda_1 = \sqrt{1 + \frac{b_{11} + 2b_{12} + b_{66}}{b_{22}}}, \qquad \lambda_2 = \frac{a_{23} - a_{13}}{b_{11} + 2b_{12} + b_{66} - 3b_{22}}. \tag{7.187}$$

The stress components then become

$$\sigma_r = (\widetilde{\gamma}_1 r^{\lambda_1 - 1} - \widetilde{\gamma}_2 r^{-\lambda_1 - 1} + \frac{\widetilde{\gamma}_3}{r} + \widetilde{\beta}\lambda_2 r)\sin\theta, \tag{7.188a}$$

$$\sigma_\theta = [\widetilde{\gamma}_1(\lambda_1 + 1)r^{\lambda_1 - 1} - \widetilde{\gamma}_2(1 - \lambda_1)r^{-\lambda_1 - 1} + \frac{\widetilde{\gamma}_3}{r} + 3\widetilde{\beta}\lambda_2 r]\sin\theta, \tag{7.188b}$$

$$\tau_{r\theta} = -(\widetilde{\gamma}_1 r^{\lambda_1 - 1} - \widetilde{\gamma}_2 r^{-\lambda_1 - 1} + \frac{\widetilde{\gamma}_3}{r} + \widetilde{\beta}\lambda_2 r)\cos\theta, \tag{7.188c}$$

$$\sigma_z = -\frac{1}{a_{33}}\{a_{13}(\widetilde{\gamma}_1 r^{\lambda_1 - 1} - \widetilde{\gamma}_2 r^{-\lambda_1 - 1} + \frac{\widetilde{\gamma}_3}{r} + \widetilde{\beta}\lambda_2 r)$$
$$+ a_{23}[\widetilde{\gamma}_1(\lambda_1 + 1)r^{\lambda_1 - 1} - \widetilde{\gamma}_2(1 - \lambda_1)r^{-\lambda_1 - 1} + \frac{\widetilde{\gamma}_3}{r} + 3\widetilde{\beta}\lambda_2 r] - a_{33}\widetilde{\beta}r\}\sin\theta. \tag{7.188d}$$

Equations (7.162a–c) are then written as

$$\widetilde{U}_{,r} = b_{11}\sigma_r + b_{12}\sigma_\theta + a_{13}\widetilde{\beta}r\sin\theta, \tag{7.189a}$$

$$\frac{1}{r}\widetilde{V}_{,\theta} + \frac{\widetilde{U}}{r} = b_{12}\sigma_r + b_{22}\sigma_\theta + a_{23}\widetilde{\beta}r\sin\theta, \tag{7.189b}$$

$$-\frac{\widetilde{V}}{r} + \frac{1}{r}\widetilde{U}_{,\theta} + \widetilde{V}_{,r} = b_{66}\tau_{r\theta}. \tag{7.189c}$$

By integrating (7.189a) one obtains

$$\widetilde{U}(r,\theta) = \{b_{11}(\widetilde{\gamma}_1\frac{r^{\lambda_1}}{\lambda_1} + \widetilde{\gamma}_2\frac{r^{-\lambda_1}}{\lambda_1} + \widetilde{\gamma}_3\ln r + \widetilde{\beta}\lambda_2\frac{r^2}{2})$$
$$+ b_{12}[\widetilde{\gamma}_1(\lambda_1 + 1)\frac{r^{\lambda_1}}{\lambda_1} + \widetilde{\gamma}_2(1 - \lambda_1)\frac{r^{-\lambda_1}}{\lambda_1} + \widetilde{\gamma}_3\ln r + 3\widetilde{\beta}\lambda_2\frac{r^2}{2}] + a_{13}\widetilde{\beta}\frac{r^2}{2}\}\sin\theta + \widetilde{\Gamma}_1(\theta), \tag{7.190}$$

where $\widetilde{\Gamma}_1(\theta)$ is a function to be determined. Substituting the above $\widetilde{U}$ in (7.189b) and integrating yields

$$\widetilde{V}(r,\theta) = \{b_{11}(\widetilde{\gamma}_1\frac{r^{\lambda_1}}{\lambda_1} + \widetilde{\gamma}_2\frac{r^{-\lambda_1}}{\lambda_1} + \widetilde{\gamma}_3\ln r + \widetilde{\beta}\lambda_2\frac{r^2}{2}) + b_{12}[\widetilde{\gamma}_1(\lambda_1 + 1)\frac{r^{\lambda_1}}{\lambda_1} + \widetilde{\gamma}_2(1 - \lambda_1)\frac{r^{-\lambda_1}}{\lambda_1}$$
$$+ \widetilde{\gamma}_3\ln r + 3\widetilde{\beta}\lambda_2\frac{r^2}{2}] + a_{13}\widetilde{\beta}\frac{r^2}{2} - b_{12}(\widetilde{\gamma}_1 r^{\lambda_1} - \widetilde{\gamma}_2 r^{-\lambda_1} + \widetilde{\gamma}_3 + \widetilde{\beta}\lambda_2 r^2)$$
$$- b_{22}[\widetilde{\gamma}_1(\lambda_1 + 1)r^{\lambda_1} - \widetilde{\gamma}_2(1 - \lambda_1)r^{-\lambda_1} + \widetilde{\gamma}_3 + 3\widetilde{\beta}\lambda_2 r^2] - a_{23}\widetilde{\beta}r^2\}\cos\theta - \int_0^\theta \widetilde{\Gamma}_1(\theta)d\theta + \widetilde{\Gamma}_2(r), \tag{7.191}$$

where $\widetilde{\Gamma}_2(r)$ is a function to be determined. Substituting $\widetilde{U}$ and $\widetilde{V}$ in (7.189c) together with λ_1 and λ_2 of (7.187) yields

$$\widetilde{\gamma}_3\,(2b_{12}+b_{11}+b_{22}+b_{66})\cos\theta = -\int_0^\theta \widetilde{\Gamma}_1(\theta)\,d\theta - \widetilde{\Gamma}_{1,\theta}(\theta) - r\widetilde{\Gamma}_{2,r}(r) + \widetilde{\Gamma}_2(r). \qquad (7.192)$$

Differentiation of (7.192) with respect to θ yields

$$\widetilde{\Gamma}_{1,\theta\theta}(\theta) + \widetilde{\Gamma}_1(\theta) = \widetilde{\gamma}_3\,(2b_{12}+b_{11}+b_{22}+b_{66})\sin\theta. \qquad (7.193)$$

The solution of the above differential equation is given by

$$\widetilde{\Gamma}_1(\theta) = \widetilde{\gamma}_c\cos\theta + \widetilde{\gamma}_s\sin\theta - \frac{1}{2}\widetilde{\gamma}_3\,(2b_{12}+b_{11}+b_{22}+b_{66})\,(-\sin\theta + \underline{\theta\cos\theta)} \qquad (7.194)$$

where $\widetilde{\gamma}_s$ and $\widetilde{\gamma}_c$ are constants. The above underlined term shows that for the displacement $\widetilde{U}$ to remain single-valued, one must require $\widetilde{\gamma}_3 = 0$. Furthermore, the r.h.s. of (7.192) vanishes only for $\int_0^\theta \widetilde{\Gamma}_1(\theta)\,d\theta + \widetilde{\Gamma}_{1,\theta}(\theta) = \widetilde{\gamma}_0$ and $r\widetilde{\Gamma}_{2,r}(r) - \widetilde{\Gamma}_2(r) = -\widetilde{\gamma}_0$, where $\widetilde{\gamma}_0$ is a constant, and therefore, $\widetilde{\Gamma}_2(r) = \widetilde{\gamma}_0 + \widetilde{\gamma}_r r$, where $\widetilde{\gamma}_r$ is a constant as well. This means that one should substitute $\widetilde{\Gamma}_1(\theta) = \widetilde{\gamma}_c\cos\theta + \widetilde{\gamma}_s\sin\theta$ in (7.190) and $-\int_0^\theta \widetilde{\Gamma}_1(\theta)\,d\theta + \widetilde{\Gamma}_2(r) = -\widetilde{\gamma}_c\sin\theta + \widetilde{\gamma}_s\cos\theta + \widetilde{\gamma}_r r$ in (7.191). Overall, $\widetilde{\gamma}_c, \widetilde{\gamma}_s, \widetilde{\gamma}_r$ remain as parameters to be determined based on the boundary conditions of (7.170).

The boundary conditions require $\sigma_r = \tau_{r\theta} = 0$ over $r = R_1, R_2$, which may be written as

$$\widetilde{\gamma}_1 R_1^{\lambda_1-1} - \widetilde{\gamma}_2 R_1^{-\lambda_1-1} + \widetilde{\beta}\lambda_2 R_1 = 0, \qquad \widetilde{\gamma}_1 R_2^{\lambda_1-1} - \widetilde{\gamma}_2 R_2^{-\lambda_1-1} + \widetilde{\beta}\lambda_2 R_2 = 0. \qquad (7.195)$$

Solving the above system yields

$$\widetilde{\gamma}_1 = -\widetilde{\beta}\lambda_2\frac{R_2^{\lambda_1+2} - R_1^{\lambda_1+2}}{R_2^{2\lambda_1} - R_1^{2\lambda_1}}, \qquad \widetilde{\gamma}_2 = \widetilde{\beta}\lambda_2 R_1^{\lambda_1+2} R_2^{\lambda_1+2}\frac{R_2^{\lambda_1-2} - R_1^{\lambda_1-2}}{R_2^{2\lambda_1} - R_1^{2\lambda_1}}. \qquad (7.196)$$

The natural boundary conditions at the tip, namely, the second equation of (7.139) (while all other conditions are satisfied identically) yield $\widetilde{\beta} = \frac{M_x}{I^e_{yy}}$, where I^e_{yy} may be viewed as the "effective" cross-sectional moment (of inertia) of this case, since as indicated in Remark 7.8, the bending curvature may be written as $\kappa_y(z) = -a_{33}\frac{M_x}{I^e_{yy}}$, where

$$I^e_{yy} = \frac{\pi(R_2^4 - R_1^4)}{4} - \frac{\pi\lambda_2}{a_{33}}\Big[(a_{13}+3a_{23})\frac{(R_2^4 - R_1^4)}{4} - \frac{a_{13}+(1+\lambda_1)a_{23}}{\lambda_1+2}\frac{(R_2^{\lambda_1+2} - R_1^{\lambda_1+2})^2}{R_2^{2\lambda_1} - R_1^{2\lambda_1}}$$
$$- \frac{a_{13}+(1-\lambda_1)a_{23}}{\lambda_1-2}\frac{R_1^4 R_2^4 (R_2^{\lambda_1-2} - R_1^{\lambda_1-2})^2}{R_2^{2\lambda_1} - R_1^{2\lambda_1}}\Big]. \qquad (7.197)$$

The final stress components expressions become

$$\sigma_r\frac{I^e_{yy}}{M_x} = \lambda_2\Big[r - \frac{R_2^{\lambda_1+2} - R_1^{\lambda_1+2}}{R_2^{2\lambda_1} - R_1^{2\lambda_1}} r^{\lambda_1-1} - R_1^{\lambda_1+2} R_2^{\lambda_1+2}\frac{R_2^{\lambda_1-2} - R_1^{\lambda_1-2}}{R_2^{2\lambda_1} - R_1^{2\lambda_1}} r^{-\lambda_1-1}\Big]\sin\theta,$$

$$\sigma_\theta\frac{I^e_{yy}}{M_x} = \lambda_2\Big[3r - \frac{R_2^{\lambda_1+2} - R_1^{\lambda_1+2}}{R_2^{2\lambda_1} - R_1^{2\lambda_1}}(1+\lambda_1) r^{\lambda_1-1} - R_1^{\lambda_1+2} R_2^{\lambda_1+2}\frac{R_2^{\lambda_1-2} - R_1^{\lambda_1-2}}{R_2^{2\lambda_1} - R_1^{2\lambda_1}}(1-\lambda_1) r^{-\lambda_1-1}\Big]\sin\theta,$$

$$\tau_{r\theta}\frac{I^e_{yy}}{M_x} = -\lambda_2\Big[r - \frac{R_2^{\lambda_1+2} - R_1^{\lambda_1+2}}{R_2^{2\lambda_1} - R_1^{2\lambda_1}} r^{\lambda_1-1} - R_1^{\lambda_1+2} R_2^{\lambda_1+2}\frac{R_2^{\lambda_1-2} - R_1^{\lambda_1-2}}{R_2^{2\lambda_1} - R_1^{2\lambda_1}} r^{-\lambda_1-1}\Big]\cos\theta,$$

$$\sigma_z \frac{I^e_{yy}}{M_x} = r\sin\theta - \frac{\lambda_2}{a_{33}}\{(a_{13}+3a_{23})r - \frac{R_2^{\lambda_1+2}-R_1^{\lambda_1+2}}{R_2^{2\lambda_1}-R_1^{2\lambda_1}}[a_{13}+(1+\lambda_1)a_{23}]r^{\lambda_1-1}$$
$$-R_1^{\lambda_1+2}R_2^{\lambda_1+2}\frac{R_2^{\lambda_1-2}-R_1^{\lambda_1-2}}{R_2^{2\lambda_1}-R_1^{2\lambda_1}}[a_{13}+(1-\lambda_1)a_{23}]r^{-\lambda_1-1}\}\sin\theta,$$
$$\tau_{\theta z} = \tau_{rz} = 0. \tag{7.198}$$

Once the stress components are determined, the strain components are directly obtained from the constitutive relations for cylindrical MON13z anisotropy of (2.72). The displacements are then given by (7.169a–c) as

$$u = -\frac{z^2}{2}\widetilde{\beta}a_{33}\sin\theta + \widetilde{U}(r,\theta), \quad v = -\frac{z^2}{2}\widetilde{\beta}a_{33}\cos\theta + \widetilde{V}(r,\theta), \quad w = \widetilde{\beta}rza_{33}\sin\theta. \tag{7.199}$$

For a full circular cross-section (where $R_1 = 0$ and $R_2 = R$) and under the boundary conditions of (7.170), the above displacements become

$$u\frac{I^e_{yy}}{M_x} = \{-\frac{z^2}{2}a_{33} + \lambda_2(b_{11}+3b_{12})\frac{r^2}{2} - [b_{11}+(1+\lambda_1)b_{12}]\frac{\lambda_2}{\lambda_1}(\frac{r}{R})^{\lambda_1}R^2 + \frac{1}{2}r^2a_{13}\}\sin\theta,$$
$$v\frac{I^e_{yy}}{M_x} = \{-\frac{z^2}{2}a_{33} + \frac{1}{2}r^2a_{13} - r^2a_{23} + \lambda_2(b_{11}+3b_{12})\frac{r^2}{2} - [b_{11}+(1+\lambda_1)b_{12}]\frac{\lambda_2}{\lambda_1}(\frac{r}{R})^{\lambda_1}R^2$$
$$-[r^2-(\frac{r}{R})^{\lambda_1}R^2]\lambda_2b_{12} - [3r^2-(1+\lambda_1)(\frac{r}{R})^{\lambda_1}R^2]\lambda_2b_{22}\}\cos\theta,$$
$$w\frac{I^e_{yy}}{M_x} = rza_{33}\sin\theta. \tag{7.200}$$

Also, for a full circular cross-section, $\widetilde{\gamma}_2 = 0$ (see (7.196)), and therefore, when $\lambda_1 < 1$ the stresses become singular at $r = 0$ and only a solution of an annular cross-section is possible. For $\lambda_1 > 1$, the stresses are non-singular at $r = 0$, and a solution of both annular and circular cross-sections are possible. See illustrative examples in Fig. 7.17.

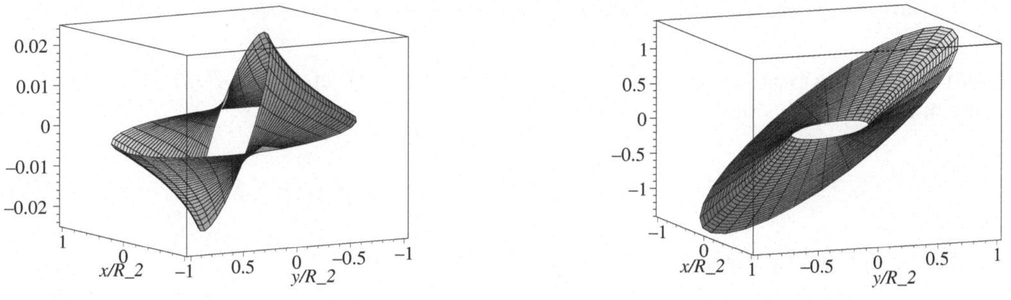

(a) $\frac{\sigma_r R_2^3}{M_x}$. (b) $\frac{\sigma_z R_2^3}{M_x}$.

Figure 7.17: Stresses in an annular cross-section of cylindrical anisotropy, $\lambda_1 = 0.67$, $\frac{R_1}{R_2} = 0.3$.

7.5.3.2 Stress Hypothesis Method

In this section we shall present a solution for an orthotropic beam under a tip force P_y while assuming $a_{13} = a_{23}$. The solution philosophy is founded on the St. Venant's semi-inverse method that has been extensively exploited throughout this chapter. For the present purpose we write

$$\sigma_r = \sigma_\theta = \tau_{r\theta} = 0,$$

$$\sigma_z = -\frac{P_y}{I_x}(l-z)r\sin\theta, \quad \tau_{\theta z} = \frac{P_y}{I_x a_{44}}(\frac{1}{r}\chi_{2,\theta} - \frac{2}{3}a_{13}r^2\cos\theta), \quad \tau_{rz} = \frac{P_y}{I_x a_{55}}\chi_{2,r}. \quad (7.201)$$

The components of strain are obtained from the constitutive relations for cylindrical MON13z anisotropy of (2.72). These components show that all compatibility equations (7.135a–f), are satisfied identically.

The equilibrium equations (7.137a,b) are satisfied identically as well. To fulfill the third equation of equilibrium, (7.137c), the function $\chi_2(r,\theta)$ must satisfy the following equation:

$$\nabla_6^{(2)}\chi_2 = -\frac{(3a_{44}+2a_{13})a_{55}}{3}r\sin\theta, \quad (7.202)$$

where $\nabla_6^{(2)}$ is the orthotropic version of the $\nabla_5^{(2)}$, see (3.200). The boundary conditions on the lateral (cylindrical) surface given by (7.138a, b) are satisfied identically. The third equation, (7.138c), then becomes

$$D_4^n\chi_2 = \frac{2a_{13}a_{55}}{3}r^2\cos\theta\cos(\bar{\mathbf{n}},\theta), \quad (7.203)$$

where D_4^n is the orthotropic version of the D_3^n, see (3.221). Hence, we have arrived at the Neumann BVP that should be solved for each cross-section geometry. In the general case, such a solution does not take into account the tip loads conditions that should be adjusted by superposition of suitable additional solutions.

Example 7.11 *Beam of Annular Cross-Section Under Transverse Tip Load.*
For the BVP of (7.202), (7.203) in an annular domain we shall seek a solution of the form

$$\chi_2(r,\theta) = f(r)\sin\theta, \quad (7.204)$$

where $f(r)$ is a function to be determined. Then, (7.202) may be written as

$$a_{44}(f_{,rr}\sin\theta + \frac{1}{r}f_{,r}\sin\theta) - a_{55}\frac{1}{r^2}f\sin\theta = -\frac{(3a_{44}+2a_{13})a_{55}}{3}r\sin\theta, \quad R_1 \le r \le R_2, \quad (7.205)$$

which should be solved under the boundary conditions $f_{,r} = 0$ at $r = R_1, R_2$. Equating coefficients of $\sin\theta$ on both sides of (7.205) yields

$$f_{,rr} + \frac{1}{r}f_{,r} - \frac{a_{55}}{a_{44}}\frac{1}{r^2}f = -\frac{(3a_{44}+2a_{13})a_{55}}{3a_{44}}r, \qquad R_1 \le r \le R_2. \quad (7.206)$$

The general solution of the above system is

$$f(r) = \widetilde{P}r^{\mu} + \widetilde{Q}r^{-\mu} + \widetilde{S}r^3, \qquad \mu = \sqrt{\frac{a_{55}}{a_{44}}}, \qquad \widetilde{S} = -\frac{(3a_{44}+2a_{13})a_{55}}{3(9a_{44}-a_{55})}, \quad (7.207)$$

where $\widetilde{P}$ and $\widetilde{Q}$ are constants to be determined. The above boundary conditions take the form

$$\mu\widetilde{P}R_1^{\mu-1} - \mu\widetilde{Q}R_1^{-\mu-1} + 3\widetilde{S}R_1^2 = 0, \qquad \mu\widetilde{P}R_2^{\mu-1} - \mu\widetilde{Q}R_2^{-\mu-1} + 3\widetilde{S}R_2^2 = 0. \quad (7.208)$$

Solving the above system yields

$$\widetilde{P} = -\frac{3\widetilde{S}(R_2^{\mu+3} - R_1^{\mu+3})}{\mu(R_2^{2\mu} - R_1^{2\mu})}, \qquad \widetilde{Q} = -\frac{3\widetilde{S}R_1^{\mu+1}R_2^{\mu+1}(R_1^{\mu-1}R_2^2 - R_2^{\mu-1}R_1^2)}{\mu(R_2^{2\mu} - R_1^{2\mu})}. \quad (7.209)$$

Thus, the stress components become

$$\sigma_r = \sigma_\theta = \tau_{r\theta} = 0, \qquad \sigma_z = -\frac{P_y(l-z)r\sin\theta}{I_x},$$

$$\tau_{\theta z} = \frac{P_y}{I_x}r^2\cos\theta + \frac{P_yR_2^2}{I_x a_{44}}3\widetilde{S}[\frac{3}{\mu^2}(\frac{r}{R_2})^2 - \frac{(1-c^{\mu+3})}{\mu(1-c^{2\mu})}\frac{r^{\mu-1}}{R_2^{\mu-1}} + \frac{(1-c^{\mu-3})}{\mu(1-c^{2\mu})}c^{\mu+3}\frac{r^{-\mu-1}}{R_2^{-\mu-1}}]\cos\theta,$$

$$\tau_{rz} = \frac{P_yR_2^2}{I_x}\frac{3\widetilde{S}}{a_{55}}[(\frac{r}{R_2})^2 - \frac{1-c^{\mu+3}}{1-c^{2\mu}}\frac{r^{\mu-1}}{R_2^{\mu-1}} - \frac{1-c^{\mu-3}}{1-c^{2\mu}}c^{\mu+3}\frac{r^{-\mu-1}}{R_2^{-\mu-1}}]\sin\theta, \tag{7.210}$$

where $c = \frac{R_1}{R_2}$ and $I_x = \frac{\pi(R_2^4 - R_1^4)}{4}$. Examination of the tip conditions (7.139) shows that they are all satisfied identically for $P_y \neq 0$ and $P_x = P_z = M_x = M_y = M_z = 0$, and therefore, no correction is required in this specific case. For an annular circular cross-section, the displacements are

$$u\frac{I_x}{P_y} = [-\frac{(l-z)r^2}{2}a_{13} + \frac{z^2}{2}(l-\frac{z}{3})a_{33}]\sin\theta, \quad v\frac{I_x}{P_y} = [\frac{(l-z)r^2}{2}a_{13} + \frac{z^2}{2}(l-\frac{z}{3})a_{33}]\cos\theta,$$

$$w\frac{I_x}{P_y} = -rz(l-\frac{1}{2}z)a_{33} + 3R_2^2\widetilde{S}[\frac{r^3}{3R_2^2}(1-\frac{1}{6\widetilde{S}}a_{13}) - \frac{1-c^{\mu+3}}{\mu(1-c^{2\mu})}(\frac{r^\mu}{R_2^{\mu-1}} + \frac{r^{-\mu}}{R_2^{-\mu-1}})]\sin\theta. \tag{7.211}$$

For a full circular cross-section, ($c = R_1 = 0$, $R_2 = R$, $I_x = \frac{\pi R^4}{4}$), the displacements are

$$u\frac{I_x}{P_y} = [-\frac{(l-z)r^2}{2}a_{13} + \frac{z^2}{2}(l-\frac{z}{3})a_{33}]\sin\theta, \tag{7.212a}$$

$$v\frac{I_x}{P_y} = [\frac{(l-z)r^2}{2}a_{13} + \frac{z^2}{2}(l-\frac{z}{3})a_{33}]\cos\theta, \tag{7.212b}$$

$$w\frac{I_x}{P_y} = -r[z(l-\frac{z}{2})a_{33} + \frac{3\widetilde{S}}{\mu}R^{3-\mu}r^\mu + (\frac{1}{6}a_{13} - \widetilde{S})r^2]\sin\theta. \tag{7.212c}$$

Also, for a full circular cross-section, $\widetilde{Q} = 0$. In such a case, when $\mu < 1$ the stresses become singular at $r = 0$ and only a solution in an annular cross-section is possible. For $\mu > 1$, the stresses are non-singular at $r = 0$ and solutions of both annular and circular cross-sections are possible, see examples in Fig. 7.18.

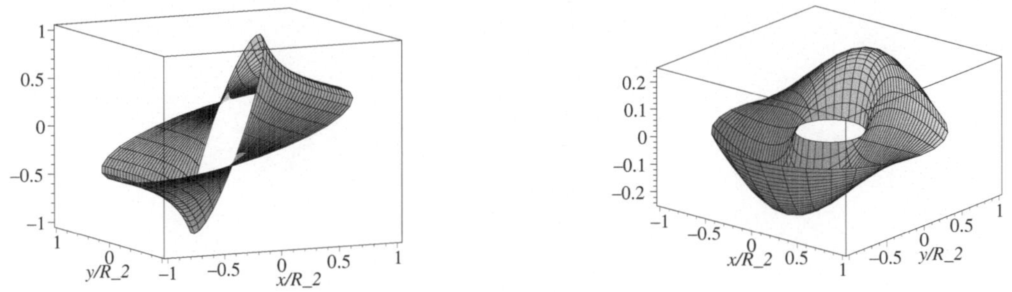

(a) $\frac{\tau_{\theta z}R_2^2}{P_y}$. (b) $\frac{\tau_{rz}R_2^2}{P_y}$.

Figure 7.18: Stresses in an annular cross-section of cylindrical anisotropy, $\mu = 0.88$, $\frac{R_1}{R_2} = 0.3$.

8

Non-Homogeneous Plane and Beam Analysis

This chapter is devoted to plane problems of non-homogeneous domain and to the analysis of beams with non-homogeneous cross-section that consist of various MON13z materials. For that purpose we adopt the domain topology definitions described in S.3.1.1 and the material constitutive relations of (7.1).

We first review formulations for BVPs in a plane non-homogeneous domain and also discuss their appearance in view of the complex potentials method (see S.4.4). We then continue the approach adopted in Chapter 7, and present separate formulations for non-homogeneous beams under tip and distributed loads.

In view of the generic non-homogeneous domain description of Fig. 3.1(b), the analysis presented in what follows applies to both the micro- and macro- problems. In the micro- case, and referring to a typical composite material where a fiber is surrounded by a matrix material, both the fiber and the matrix may be treated as made of transversely isotropic materials (which may be viewed as a "simpler" reduction of monoclinic materials, see S.2.6). In the macro- case, besides two-dimensional non-homogeneous problems, the plane analysis presented in what follows serves as a major component of the corresponding non-homogeneous beam analysis. In such cases, the "plane" notion is replaced by a "cross-section". A common example for such a non-homogeneous cross-section is the laminated domain shown in Fig. 3.1(c).

The origin of the solution method developed in this chapter for anisotropic beams is the collection of St. Venant's problem solutions for non-homogeneous beams presented by (Muskhelishvili, 1953). These solutions have also introduced the notion of principle axis of extension and principle plane of bending for non-homogeneous (or "compound") isotropic beams. The concept was later evolved by (Ruchadze, 1975) and (Ruchadze, 1978) for the case of anisotropic materials as well. Additional important contributors to the general methodology were (Sobolev, 1937) and (Sherman, 1943) who supplied mathematical proof for the solution existence of the auxiliary plane-strain BVPs employed by (Muskhelishvili, 1953).

In what follows, an *improved, unified and consistent* derivation of the above approach is presented, see also (Zivzivadze and Berekashvili, 1984), (Kazar (Kezerashvili) *et al.*, 2004) and (Grebshtein *et al.*, 2004). The derivation may be viewed as an extension to the case of a non-homogeneous cross-section of the tip loading effects derived in S.7.2, and the distributed loading effects derived in S.7.3.

8.1 Plane (Two-Dimensional) BVPs

In this section, we shall present generic versions of the Dirichlet/Neumann, the biharmonic and the Coupled-Plane BVPs described in S.3.2.6, S.3.3.5, S.3.4 for non-homogeneous domain.

The following discussion will make use of the non-homogeneous cross-sections schemes shown in Fig. 3.1. Such a domain, Ω, consists of $N \geq 1$ simply connected homogeneous domains $\Omega_{[j]}$ i.e., $\Omega = \bigcup_{j=1}^{N} \Omega_{[j]}$. The orientation of the normal $\bar{\mathbf{n}}$ over the dividing curves $\partial\Omega_{ij}$ deserves a special attention. First, for all operations that are carried out in a specific domain, $\Omega_{[j]}$, we preserve the definitions of (3.4) for both the (free) boundary contours and the dividing curves (i.e., $\bar{\mathbf{n}}$ is always oriented *outward of the domain*) and hence, the operation $\frac{d}{ds}\widetilde{A} = A_{,y}\cos(\bar{\mathbf{n}},x) - \widetilde{A}_{,x}\cos(\bar{\mathbf{n}},y)$ remains correct. Yet, when a subtraction of a function value over the dividing curves is required as shown by (3.1), *the same orientation of* $\bar{\mathbf{n}}$ *is assumed on both involved domains*. As a default, we shall assume that $\bar{\mathbf{n}}$ is oriented *outward of the domain of the lower domain index*, see Fig. 8.1.

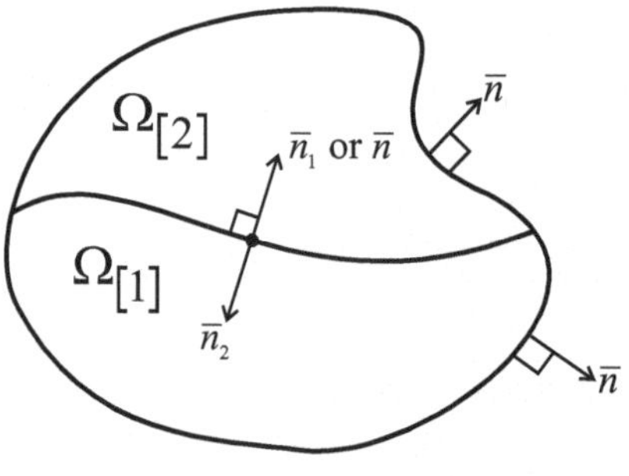

Figure 8.1: Normal orientation notation: $\bar{\mathbf{n}}_j$ is directed outward of domain $\Omega_{[j]}$ $(j = 1,2,\ldots)$. Over the dividing curves $\bar{\mathbf{n}}$ is selected to be equal to $\bar{\mathbf{n}}_1$.

It should also be noted that the $\frac{d}{ds}$-type of boundary conditions leave the values of the functions Λ, Φ and the derivatives $\Phi_{,x}$, $\Phi_{,y}$ undetermined by a constant. In other words, one may select arbitrarily the values of the above functions at specific points on $\partial\Omega_{ij}$ (i.e. a point for each one of them). As a general rule we shall select a point over the dividing curve, (say, $(x_1^{ij}, y_1^{ij}) \in \partial\Omega_{ij}$), where we shall force the functions to be equal for both domains, namely,

$$\{\Lambda,\, \Phi,\, \Phi_{,x},\, \Phi_{,y}\}(x_1^{ij}, y_1^{ij})_{[i]}^{[j]} = \{0,\, 0,\, 0,\, 0\}, \tag{8.1}$$

while in one of the domains, one may also select arbitrary values for these functions. For the sake of simplicity we also assume $\{\Lambda,\, \Phi,\, \Phi_{,x},\, \Phi_{,y}\}(0,0) = \{0,0,0,0\}$.

8.1.1 The Neumann BVP

Let F_0^Λ, F_3^Λ and $F^{\Lambda,ij}$, $F_3^{\Lambda,ij}$ be given functions on $\Omega_{[j]}$, $\partial\Omega_j$ and $\partial\Omega_{ij}$, respectively. Let Λ be an unknown function on Ω such that the restrictions $\Lambda^{[j]} = \Lambda_{|\Omega_{[j]}}$ are C^2-continuous functions. Then, the associated Neumann-type BVP becomes

$$\nabla_3^{(2)}\Lambda = F_0^\Lambda \qquad \text{over} \quad \Omega, \tag{8.2a}$$

$$D_1^n\Lambda = F_3^\Lambda \qquad \text{on} \quad \partial\Omega, \tag{8.2b}$$

$$\{\frac{1}{a_0}D_1^n\Lambda,\, \Lambda\}_{[i]}^{[j]} = \{F_3^{\Lambda,ij},\, F^{\Lambda,ij}\} \qquad \text{on} \quad \partial\Omega_{ij}. \tag{8.2c}$$

In the applications below $F_3^{\Lambda,ij} = [\frac{1}{a_0}F_3^{\Lambda}]_{[i]}^{[j]}$ and $F^{\Lambda,ij} = 0$ hold. Note that we have used the field and boundary operators that correspond to the plane-shear type of problems derived in S.3.3, although other (compatible) operators may be also employed. The existence condition of the Neumann problem solution in a homogeneous simply connected domain is given by (3.126) as $\oint_{\partial\Omega} D_1^n \Lambda = \iint_{\Omega} \nabla_3^{(2)} \Lambda$. By applying the same arguments to each domain, $\Omega_{[j]}$, we obtain the following necessary condition for the existence of the solution:

$$\sum_{j=1}^{N} \frac{1}{a_0} \oint_{\partial\Omega_j} F_3^{\Lambda} = \sum_{j=1}^{N} \frac{1}{a_0} \iint_{\Omega_{[j]}} F_0^{\Lambda}, \tag{8.3}$$

which in view of $F_3^{\Lambda,ij} = [\frac{1}{a_0}F_3^{\Lambda}]_{[i]}^{[j]}$ may also be written as

$$\oint_{\partial\Omega} \frac{1}{a_0} F_3^{\Lambda} + \sum_{ij} \int_{\partial\Omega_{ij}} F_3^{\Lambda,ij} = \iint_{\Omega} \frac{1}{a_0} F_0^{\Lambda}. \tag{8.4}$$

Note that (3.127a,b) hold in the non-homogeneous case as well.

Application of the above is presented in **P.8.1** while activated as a *harmonic solver* for a non-homogeneous rectangular domain, see Examples 8.3, 8.4.

8.1.2 The Dirichlet BVP

Let F_0^{Λ}, G_3 or F_3, G_3^{ij} or F_3^{ij} and G_w^{ij} be given functions on $\Omega_{[j]}$, $\partial\Omega_j$ and $\partial\Omega_{ij}$, respectively. Let Λ be an unknown function on Ω such that $\Lambda^{[j]} = \Lambda_{|\Omega_{[j]}}$ are C^2-continuous functions. Then, the associated Dirichlet BVP becomes

$$\nabla_3^{(2)} \Lambda = F_0^{\Lambda} \quad \text{over} \quad \Omega, \tag{8.5a}$$

$$\Lambda = G_3 \quad \text{or} \quad \frac{d}{ds}\Lambda = F_3 \quad \text{on} \quad \partial\Omega, \tag{8.5b}$$

$$\Lambda_{[i]}^{[j]} = G_3^{ij} \quad \text{or} \quad \frac{d}{ds}\Lambda_{[i]}^{[j]} = F_3^{ij} \quad \text{on} \quad \partial\Omega_{ij}, \tag{8.5c}$$

$$\widetilde{w}_{[i]}^{[j]} = G_w^{ij} \quad \text{on} \quad \partial\Omega_{ij}, \tag{8.5d}$$

where $\widetilde{w}$ is a functional depending on Λ and its partial derivatives, and the first condition discussed in (8.1) holds as well. In the applications below $F_3^{ij} = [F_3]_{[i]}^{[j]}$, $G_3^{ij} = [G_3]_{[i]}^{[j]}$ hold. Note again that we have used the field operator that corresponds to the plane-shear type of problems derived in S.3.3, although other operators may be employed. In view of (3.129) the following $\frac{d}{ds}$-case single-valued condition along a contour is necessary for the solution existence:

$$\oint_{\partial\Omega} F_3 + \sum_{ij} \int_{\partial\Omega_{ij}} F_3^{ij} = 0. \tag{8.6}$$

The weakening of the torsion problem in a laminated rectangle is discussed in Remark 8.1.

Remark 8.1 Let the system (8.5a–c) be written as (3.136a–c) for the torsion problem in a non-homogeneous domain, namely with $F_0^{\Psi} = -2\theta$, where $w(x,y)$ is the displacement in the z direction (i.e. the out-of-plane warping) and θ is a twist parameter.

We first note that (3.136c:a) may be written more conveniently using the identity $\frac{d}{ds}\psi_{|\partial\Omega} = \psi_{,y}\cos(\bar{\mathbf{n}},x) - \psi_{,x}\cos(\bar{\mathbf{n}},y)$, while in case of the horizontally laminated rectangular domain shown in Fig. 3.1(c), $\cos(\bar{\mathbf{n}},x) = 0$ and $\frac{d}{ds} = \pm\frac{d}{dx}$ along dividing curves.

Further on, it becomes clear that it is quite difficult to impose directly the condition $w_{[i]}^{[j]} = 0$ (see (3.136c:b)) as it requires the introduction of unknown integration constants into the formulation. Hence, we shall weaken this condition by replacing it with $\frac{d}{ds} w_{[i]}^{[j]} = 0$. From (3.114), (3.95a,b) one may write

$$w_{,y} = \gamma_{yz} - \theta x = a_{45}\psi_{,y} - a_{44}\psi_{,x} - \theta x, \qquad w_{,x} = \gamma_{xz} + \theta y = a_{55}\psi_{,y} - a_{45}\psi_{,x} + \theta y, \quad (8.7)$$

and by using the identity $\frac{d}{ds} w = w_{,y}\cos(\bar{\mathbf{n}},x) - w_{,x}\cos(\bar{\mathbf{n}},y)$, the condition $\frac{d}{ds} w_{[i]}^{[j]} = 0$ may be written explicitly in terms of the stress function ψ without any unknown constants.

Once a solution has been achieved, by integrating (8.7), one may find the integration constants for the displacement w at each region component, which will ensure continuous w displacement. This is done by equating the displacement w at selected points over the dividing curves (one on each dividing curve), say, $(x_0^{ij}, y_0^{ij}) \in \partial\Omega_{ij}$, namely, by requiring $w(x_0^{ij}, y_0^{ij})_{[i]}^{[j]} = 0$. This will ensure w continuity along the entire dividing curve since the condition $\frac{d}{ds} w_{[i]}^{[j]} = 0$ has been included in the BVP solution. Although not required for strain/stress analysis, one may also impose continuity of the stress function ψ at selected points over the dividing curves, say $(x_1^{ij}, y_1^{ij}) \in \partial\Omega_{ij}$, by fulfilling the requirement $\psi(x_1^{ij}, y_1^{ij})_{[i]}^{[j]} = 0$.

The twist parameter may be derived using the applied torsion moment, M_z, from the integral equality $M_z = 2\iint_\Omega \psi$ (see Remark 3.6).

8.1.3 The Biharmonic BVP

In this section we shall present a generic version of the biharmonic BVP in a non-homogeneous domain (the case of the biDirichlet BVP may be considered analogously).

Let F_0, $\{G_1, G_2, G_1^{ij}, G_2^{ij}\}$ or $\{F_1, F_2, F_1^{ij}, F_2^{ij}\}$ and G_u^{ij}, G_v^{ij} be given functions in $\Omega_{[j]}$, $\partial\Omega_j$ and $\partial\Omega_{ij}$, respectively. Also, let Φ be an unknown function on Ω such that $\Phi^{[j]} = \Phi_{|\Omega_{[j]}}$ are C^4-continuous functions. We shall express a generic biharmonic-type BVP for the unknown function Φ on Ω as

$$\nabla_n^{(4)}\Phi = F_0 \qquad \text{over} \quad \Omega, \quad (8.8a)$$

$$\{\Phi_{,x}, \Phi_{,y}\} = \{-G_1, G_2\} \quad \text{or} \quad \frac{d}{ds}\{\Phi_{,x}, \Phi_{,y}\} = \{-F_1, F_2\} \qquad \text{on} \quad \partial\Omega, \quad (8.8b)$$

$$\{\Phi_{,x}, \Phi_{,y}\}_{[i]}^{[j]} = \{-G_1^{ij}, G_2^{ij}\} \quad \text{or} \quad \frac{d}{ds}\{\Phi_{,x}, \Phi_{,y}\}_{[i]}^{[j]} = \{-F_1^{ij}, F_2^{ij}\} \quad \text{on} \quad \partial\Omega_{ij}, \quad (8.8c)$$

$$\{\tilde{u}, \tilde{v}\}_{[i]}^{[j]} = \{G_u^{ij}, G_v^{ij}\} \qquad \text{on} \quad \partial\Omega_{ij}, \quad (8.8d)$$

where $\tilde{u}$, $\tilde{v}$ are functionals depending on Φ and its partial derivatives, and the (last) three conditions discussed in (8.1) hold here as well. In the applications below we use a simpler definition $F_k^{ij} = [F_k]_{[i]}^{[j]}$ or $G_k^{ij} = [G_k]_{[i]}^{[j]}$. Note that no specific version is assigned to the biharmonic operator as the above system represents a generic problem.

The above formulation includes, in addition to the effect of external loads F_k, G_k over $\partial\Omega$, the effect of external loading F_k^{ij}, G_k^{ij}, that is "injected" directly over the dividing curves $\partial\Omega_{ij}$. Subsequently, the single-valued type boundary conditions are written as

$$\oint_{\partial\Omega}\{F_1, F_2, F_2 y - F_1 x\} + \sum_{ij}\int_{\partial\Omega_{ij}}\{F_1^{ij}, F_2^{ij}, F_2^{ij} y - F_1^{ij} x\} = \{0,0,0\}, \quad (8.9a)$$

$$\oint_{\partial\Omega}(G_2\cos(\bar{\mathbf{n}},x) + G_1\cos(\bar{\mathbf{n}},y)) + \sum_{ij}\int_{\partial\Omega_{ij}}(G_2^{ij}\cos(\bar{\mathbf{n}},x) + G_1^{ij}\cos(\bar{\mathbf{n}},y)) = 0. \quad (8.9b)$$

These equalities become clearer if the single-valued conditions are written first for each $\Omega_{[j]}$ domain component (see (3.72)), and then summed up and directly yields (8.9a,b).

Application of the above is presented in **P.8.2** while activated as a *biharmonic solver* for a laminated rectangular domain (see also Example 8.1).

Remark 8.2 discuss the weakening of the biharmonic BVP in a laminated rectangle.

Remark 8.2 Similar to Remark 8.1, the biharmonic BVP (8.8a–d) may also be weakened and subsequently simplified when the functionals $\tilde{u}, \tilde{v}$ are the actual displacements u, v as in the problem (3.88a–c) (where $F_1^{ij} = F_2^{ij} = 0$ and $G_u^{ij} = G_v^{ij} = 0$). In such a case, the displacements in each region component may be expressed as

$$u^{[j]} = U^{[j]}(x,y) - \omega_3^{0\,[j]} y + u^{0\,[j]}, \qquad v^{[j]} = V^{[j]}(x,y) + \omega_3^{0\,[j]} x + v^{0\,[j]}, \tag{8.10}$$

where $\omega_3^{0\,[j]}$, $u^{0\,[j]}$ and $v^{0\,[j]}$ are rigid rotation and displacements of $\Omega_{[j]}$ that are not known a priori, and therefore, direct substitution of (8.10) in (8.8d) prevents a unique solution of the system of (8.8a–d). We shall therefore use a different path by confining ourselves to the horizontally laminated domain such as the one shown in Fig. 3.1(c) (while noting that similar, but rather lengthy derivation, may also be obtained for more general cross-section geometries). In the present laminated case we write the displacement derivatives at each domain by omitting the specific $[j]$ indexing for all quantities as (see also (1.9a,b), (1.19), (3.4))

$$\frac{d}{ds}u = \underline{(\frac{\varepsilon_6}{2} - \omega_3)\cos(\bar{\mathbf{n}},x)} - \varepsilon_1 \cos(\bar{\mathbf{n}},y), \quad \frac{d}{ds}v = \underline{\varepsilon_2 \cos(\bar{\mathbf{n}},x)} - (\frac{\varepsilon_6}{2} + \omega_3)\cos(\bar{\mathbf{n}},y), \tag{8.11}$$

while the strain components are given as (see (3.27), (3.54))

$$\varepsilon_i = b_{i1}(\Phi_{,yy} + \overline{U}_1) - b_{i6}\Phi_{,xy} + b_{i2}(\Phi_{,xx} + \overline{U}_2), \qquad i = 1,2,6 \tag{8.12}$$

where $\overline{U}_i$ are known functions. Since for the case under discussion, $\frac{d}{ds} = \mp \frac{d}{dx}$ for $\cos(\bar{\mathbf{n}},y) = \pm 1$ and $\cos(\bar{\mathbf{n}},x) = 0$ hold for all dividing curves, the underlined terms in (8.11) vanish and $\frac{d}{ds}u$ does not contain the rotation and/or rigid body rotation components. Moreover, instead of $\frac{d}{ds}v$ one may use an even weaker form and write

$$\frac{d^2}{ds^2}v = v_{,xx} = \frac{1}{2}\varepsilon_{6,x} + \omega_{3,x}, \tag{8.13}$$

where according to (1.52), (1.53c) $\omega_{3,x} = \frac{1}{2}\varepsilon_{6,x} - \varepsilon_{1,y}$. We therefore replace (8.8d) with

$$\{u_{,x},\, v_{,xx}\}_{[i]}^{[j]} = \{0,0\} \quad \text{on} \quad \partial\Omega_{ij}, \tag{8.14}$$

which contains only strain and strain derivatives but no displacement or rotation terms.

Once the BVP is solved, and the stress and strain components are in hand, one may integrate the strain components into displacements (see S.1.2). In this process, we first select the rigid rotation and translations of one lamina, say $\Omega_{[1]}$, namely, $\omega_3^{0\,[1]}$, $u^{0\,[1]}$ and $v^{0\,[1]}$. Then, for an arbitrary point on the dividing contour with $\Omega_{[2]}$, say $(x_0^{12}, y_0^{12}) \in \partial\Omega_{12}$, one may determine $\omega_3^{0\,[2]}$, $u^{0\,[2]}$ and $v^{0\,[2]}$ from the three equations

$$\{u, v, v_{,x}\}(x_0^{12}, y_0^{12})_{[1]}^{[2]} = \{0,0,0\}, \tag{8.15}$$

which clearly ensure that $\{u,\, v\}_{[1]}^{[2]} = \{0,\, 0\}$ over the entire dividing curve since (8.14) is part of the BVP solution. Analogous process is then applied for the interface of $\Omega_{[2]}$ and $\Omega_{[3]}$, etc.

Example 8.1 *Deformation of a Non-Homogeneous Domain.*

By activating **P.8.2** with $ax\_problem = 0$, $NF = 2$, $k_\Phi = 10$, $d = h = 1$, $t_z = -30°, 30°$ we are able to define a loading of $F_2 = -\sigma_x = y^2 - \frac{1}{12}$ along the left vertical edge ($x = -0.5$) the integral of which vanishes. Therefore, the influence of this loading should vanish far from this edge (see St. Venant's Principle in S.5.1.3.1). The corresponding stress function, Φ, and the σ_x stress are presented in Fig. 8.2. As shown, the stress diminishes gradually as expected, and in this special case, the non-homogeneous cross-section behaves in a symmetric way similar to a homogeneous one.

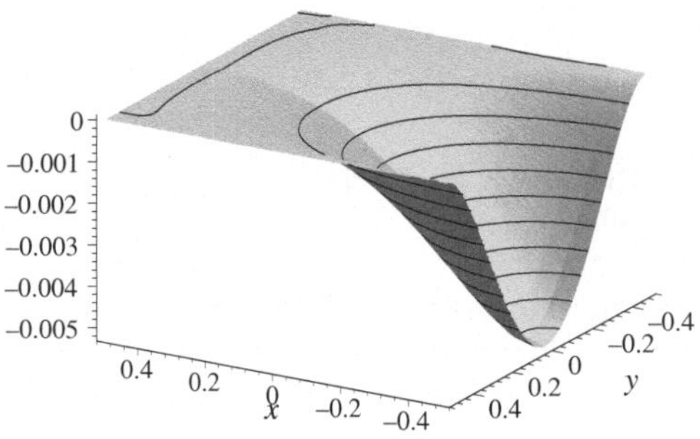

(a) $\Phi(x,y)$.

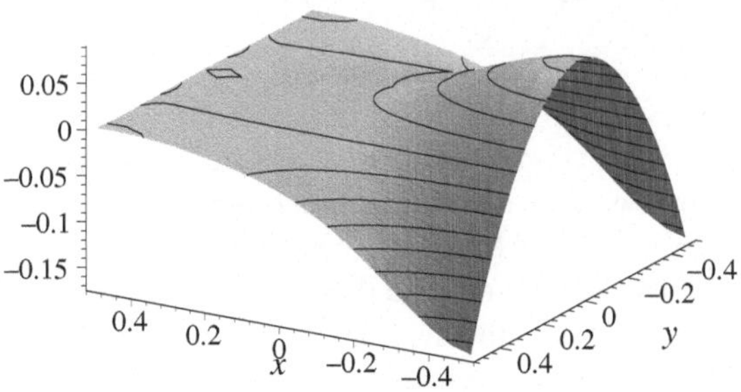

(b) $\sigma_x(x,y)$.

Figure 8.2: The stress function and a stress component in Example 8.1.

8.1.4 Coupled-Plane BVP

Let f, g, $\{G_k,\ G_k^{ij}\}$ or $\{F_k\ F_k^{ij}\}$ $(k = 1,2,3)$ and G_u^{ij}, G_v^{ij}, G_w^{ij} be given functions in $\Omega_{[j]}$, $\partial\Omega_j$ and $\partial\Omega_{ij}$, respectively. We also set Φ, Λ as unknown functions on Ω such that $\Phi^{[j]} = \Phi_{|\Omega_{[j]}}$ and $\Lambda^{[j]} = \Lambda_{|\Omega_{[j]}}$ are C^4- and C^3- continuous functions, respectively. Then, the Coupled-Plane BVP in a non-homogeneous domain becomes

$$\nabla_1^{(4)}\Phi + \nabla_1^{(3)}\Lambda = g, \qquad \nabla_1^{(3)}\Phi + \nabla_1^{(2)}\Lambda = f \qquad \text{over} \quad \Omega, \tag{8.16a}$$

$$\{\Phi_{,x},\ \Phi_{,y},\ \Lambda\} = \{-G_1, G_2, G_3\} \quad \text{or} \quad \frac{d}{ds}\{\Phi_{,x},\ \Phi_{,y},\ \Lambda\} = \{-F_1, F_2, F_3\} \quad \text{on} \quad \partial\Omega, \tag{8.16b}$$

$$\{\Phi_{,x}, \Phi_{,y}, \Lambda\}_{[i]}^{[j]} = \{-G_1^{ij}, G_2^{ij}, G_3^{ij}\} \ \text{or} \ \frac{d}{ds}\{\Phi_{,x}, \Phi_{,y}, \Lambda\}_{[i]}^{[j]} = \{-F_1^{ij}, F_2^{ij}, F_3^{ij}\} \text{ on } \partial\Omega_{ij}, \tag{8.16c}$$

$$\{\widetilde{u},\ \widetilde{v},\ \widetilde{w}\}_{[i]}^{[j]} = \{G_u^{ij},\ G_v^{ij},\ G_w^{ij}\} \quad \text{on} \quad \partial\Omega_{ij}, \tag{8.16d}$$

where $\widetilde{u}$, $\widetilde{v}$ and $\widetilde{w}$ are functionals of the functions Φ, Λ and their partial derivatives. In the applications below we use simpler notation $F_k^{ij} = [F_k]_{[i]}^{[j]}$ and $G_k^{ij} = [G_k]_{[i]}^{[j]}$. In view of (3.144), the single-valued type conditions (8.6), (8.9a) or (8.9b) are necessary for solution existence.

8.1.5 n-Coupled Dirichlet BVP

We extend below the BVP formulations of S.4.4, given for complex potentials, to the non-homogeneous domain Ω case. In this section, the indices k, m are specified for the values $1,\dots,n$, where integer $n > 0$ is the BVP level, while i, j are specified for the domain indices

$1,\ldots,N$. Usually, we shall omit the domain index for piecewise operators, functions, variables and elastic constants.

8.1.5.1 General Formulation

We define the *complex potentials* $\mathbf{H}_k(z_k) = \widetilde{\Lambda}_{1k}(z_k) + \mathbf{i}\widetilde{\Lambda}_{2k}(z_k)$ as piecewise functions of the complex variable $z_k = x_k + \mathbf{i}y_k$, and assume they are analytical in each homogeneous component $\Omega_{k[j]}$ of the Ω_k domain. Note that when the complex parameters μ_k are given, the non-homogeneous structure on Ω_k is induced from Ω by the affine isomorphism T_{μ_k}, namely, $\Omega_k = \bigcup_j T_{\mu_k}(\Omega_{k,[j]})$.

We now define the *n-Coupled Dirichlet BVP* for a non-homogeneous domain as follows: *For a given non-homogeneous domain Ω, piecewise functions F_m, F_m^{ij} or G_m, G_m^{ij}, and $G_{\Upsilon_m}^{ij}$ over $\partial\Omega$ and $\partial\Omega_{ij}$, respectively, sets of complex parameters μ_k $(\Im(\mu_k) > 0)$ and complex coefficients ν^{km}, η^{km}, find the complex potentials $\mathbf{H}_k(z_k)$, whose linear mixtures $\Lambda_m(x,y) = \Re[\sum_k \nu^{km}\mathbf{H}_k(x+\mu_k y)]$ and $\Upsilon_m(x,y) = \Re[\sum_k \eta^{km}\mathbf{H}_k(x+\mu_k y)]$ satisfy the following boundary conditions:*

$$\Lambda_m = G_m \quad \text{or} \quad \frac{d}{ds}\Lambda_m = F_m \quad \text{on} \quad \partial\Omega, \tag{8.17a}$$

$$\Lambda_m{}_{[i]}^{[j]} = G_m^{ij} \quad \text{or} \quad \frac{d}{ds}\Lambda_m{}_{[i]}^{[j]} = F_m^{ij} \quad \text{on} \quad \partial\Omega_{ij}, \tag{8.17b}$$

$$\Upsilon_m{}_{[i]}^{[j]} = G_{\Upsilon_m}^{ij} \quad \text{on} \quad \partial\Omega_{ij}. \tag{8.17c}$$

In the applications below $F_m^{ij} = [F_m]_{[i]}^{[j]}$, $G_m^{ij} = [G_m]_{[i]}^{[j]}$ hold. The sets of continuous functions $\Lambda_m^{[j]} = \Lambda_{m|\Omega_{[j]}}$, $\Upsilon_m^{[j]} = \Upsilon_{m|\Omega_{[j]}}$ present the piecewise functions Λ_m, Υ_m on each domain $\Omega_{[j]}$. One should also supplement the above $\frac{d}{ds}$-boundary conditions with "jumps" of Λ_m at special points $(x_1^{ij}, y_1^{ij}) \in \partial\Omega_{ij}$, as

$$\Lambda_m(x_1^{ij}, y_1^{ij})_{[i]}^{[j]} = \Lambda_m^{ji}. \tag{8.18}$$

Applying (4.96) to a non-homogeneous domain, one obtains the following single-valued conditions, which are necessary for the solution existence:

$$\oint_{\partial\Omega} F_m + \sum_{ij} \int_{\partial\Omega_{ij}} F_m^{ij} = 0. \tag{8.19}$$

Note that in the "diagonal" case where $\nu^{km} = \eta^{km} = 0$, $k \neq m$, (8.17a-c) are decomposed into n independent Dirichlet-type boundary conditions for certain harmonic functions in a non-homogeneous domain.

8.1.5.2 Application to Plane Problems

We present the Coupled-Plane BVP in a non-homogeneous domain, (8.16a–d), for $f(x,y) = g(x,y) = 0$, as the following 3-Coupled Dirichlet BVP, (8.17a,b),

$$\{\Lambda_1, \Lambda_2, \Lambda_3\} = \{G_1, G_2, G_3\} \quad \text{or} \quad \frac{d}{ds}\{\Lambda_1, \Lambda_2, \Lambda_3\} = \{F_1, F_2, F_3\} \quad \text{on} \quad \partial\Omega, \tag{8.20a}$$

$$\{\Lambda_1, \Lambda_2, \Lambda_3\}_{[i]}^{[j]} = \{G_1^{ij}, G_2^{ij}, G_3^{ij}\} \quad \text{or} \quad \frac{d}{ds}\{\Lambda_1, \Lambda_2, \Lambda_3\}_{[i]}^{[j]} = \{F_1^{ij}, F_2^{ij}, F_3^{ij}\} \quad \text{on} \quad \partial\Omega_{ij}, \tag{8.20b}$$

$$\{\Upsilon_1, \Upsilon_2, \Upsilon_3\}_{[i]}^{[j]} = \{G_u^{ij}, G_v^{ij}, G_w^{ij}\} \quad \text{on} \quad \partial\Omega_{ij}. \tag{8.20c}$$

We also assume $\Lambda_1 = -\Phi_{,x}$, $\Lambda_2 = \Phi_{,y}$, $\Lambda_3 = \Lambda$, $\Upsilon_1 = u$, $\Upsilon_2 = v$, $\Upsilon_3 = w$, where the r.h.s. expressions are given in terms of complex potentials, see (4.145), (4.147) and (4.150), (4.152). Note that in addition to (4.148), one should assume

$$\eta^{1k} = 2p_k, \qquad \eta^{2k} = 2q_k, \qquad \eta^{3k} = 2r_k. \tag{8.21}$$

One should supplement the $\frac{d}{ds}$-case of boundary conditions with continuity conditions for the Λ_m functions at special points $(x_1^{ij}, y_1^{ij}) \in \partial\Omega_{ij}$, i.e. (8.18) with $\Lambda_m^{ji} = 0$. The necessary conditions for solution existence in this $\frac{d}{ds}$-case are given by (8.19).

Note that the Dirichlet ($n = 1$) and biharmonic BVP ($n = 2$) are particular cases of the above formulation. For the Dirichlet BVP we set $\Lambda = \Lambda_3$, $\Upsilon_3 = w$ (and disregard all other components), while for the biharmonic BVP we set $\Lambda_1 = -\Phi_{,x}$, $\Lambda_2 = \Phi_{,y}$ and $\Upsilon_1 = u$, $\Upsilon_2 = v$ (and disregard all other components).

8.2 Uncoupled Beams Under Tip Loads

In this section we shall examine the beam behavior under tip loads only (i.e. the St. Venant's problem). As indicated, the following derivation may also be viewed as an extension (to the case of a non-homogeneous cross-section) of the tip loading effects derived in S.7.2.

In what follows we consider a "clamped-free" beam (i.e. a beam, which is clamped at its root ($z = 0$) and free at its tip ($z = l$)). Discussion of the above boundary conditions, including the integral manner in which they are applied, appears in S.5.1.3 (while the modifications required for other boundary conditions appear in S.5.1.3.2). Similar to S.7.2, and referring to Fig. 5.3(b), we shall consider here the effect of the three *tip forces* P_x, P_y, P_z, and the three *tip moments* M_x, M_y, M_z. We will also employ the St. Venant's semi-inverse method of solution discussed in S.1.6.3.

For all examples included in this section, it is important to reiterate and emphasize again, that according to the methods of solution developed in Chapter 4 and applied here, *the field equations are prescribed*. Therefore, the field equations are fulfilled *in an exact manner* (see S.4.3), and the solutions presented are "exact" in the sense that the only error included is the (relatively small) violation of the boundary conditions, while error indices, ε_{rel} (see S.4.3.4.3), were below 0.003 for all cases.

8.2.1 *General Aspects and Interlaminar Conditions*

Similar to the analysis presented in Chapter 7, we consider a uniform MON13z beam, but with non-homogeneous cross-section. Therefore, the derivation in what follows makes use of the formulation given in S.8.1. For the moment, the origin of the coordinate system is placed at an arbitrary point and is oriented at an arbitrary angle about the z-axis (which is perpendicular to all cross-sections, and therefore passes through the same point at each cross-section, see Fig. 5.1(a)). Later on, it will be shown that part of the analysis must be carried out while the coordinate system center is located at a specific point. Unless stated differently, all other general assumptions adopted in Chapter 7, including those related to the linear (small deformation) nature of the solution, apply here as well.

The governing equations for the problems under discussion are founded on the equilibrium equations (1.82a–c) and the compatibility equations (1.45a–f), which clearly apply in each domain component and depend on its specific material characteristics.

In the present non-homogeneous case, a special care should be addressed to the boundary conditions (5.16a–c) and other interlaminar conditions. First, these conditions should apply to the outer ("free") contours $\partial\Omega_j$. In addition, on the dividing contours, $\partial\Omega_{ij}$, stress continuity should be imposed. This requirement may be viewed as the "equilibrium conditions between regions", and it may be written using the notation of (3.1) as

$$[\sigma_x \cos(\bar{\mathbf{n}},x) + \tau_{xy} \cos(\bar{\mathbf{n}},y)]_{[i]}^{[j]} = 0, \tag{8.22a}$$

$$[\tau_{xy} \cos(\bar{\mathbf{n}},x) + \sigma_y \cos(\bar{\mathbf{n}},y)]_{[i]}^{[j]} = 0, \tag{8.22b}$$

$$[\tau_{xz} \cos(\bar{\mathbf{n}},x) + \tau_{yz} \cos(\bar{\mathbf{n}},y)]_{[i]}^{[j]} = 0, \tag{8.22c}$$

where $\bar{\mathbf{n}}$ is a normal to the contours $\partial\Omega_{ij}$, the orientation of which is explained in S.8.1 (note that some classical derivations consider (8.22a–c) as boundary conditions of each domain on the dividing curves, $\partial\Omega_{ij}$). As "compatibility conditions between regions", we shall require that the displacement components u, v, w will remain continuous, namely,

$$\{u, v, w\}_{[i]}^{[j]} = \{0, 0, 0\} \qquad \text{on} \quad \partial\Omega_{ij}. \tag{8.23}$$

8.2.2 *The Auxiliary Problems of Plane Deformation*

(Muskhelishvili, 1953) introduced the formulation of three *auxiliary problems of plane deformation* that fulfill the requirements of the St. Venant's problem for non-homogeneous beams, and presented some analytical solutions for extension, bending and torsion of beams composed from isotropic components.

Following the isotropic analysis of (Muskhelishvili, 1953), we shall now define three plane-strain (two-dimensional) problems in a non-homogeneous MON13z domain Ω, the importance of which will be clarified later on. The superscript $()^{(n)}$ in the following equations stands for the ordinal number of the auxiliary problem, i.e., $n = 1,2,3$. Subsequently, all other quantities related to the $n^{\underline{th}}$ problem will be denoted by a bar and a superscript as well (i.e., $\overline{()}^{(n)}$).

Within these auxiliary problems of plane deformation, we seek a solution for the non-homogeneous cross-section that satisfies the following conditions:

(a) The external stresses applied at the free contours $\partial\Omega_j \ \ (j = 1,\ldots,N)$ vanish. Hence (5.16a, b) show that

$$\overline{\sigma}_x^{(n)} \cos(\bar{\mathbf{n}},x) + \overline{\tau}_{xy}^{(n)} \cos(\bar{\mathbf{n}},y) = 0, \qquad \overline{\tau}_{xy}^{(n)} \cos(\bar{\mathbf{n}},x) + \overline{\sigma}_y^{(n)} \cos(\bar{\mathbf{n}},y) = 0. \tag{8.24}$$

(b) On the dividing contours between different materials, $\partial\Omega_{ij}$, one should require stress continuity in the form

$$[\overline{\sigma}_x^{(n)} \cos(\bar{\mathbf{n}},x) + \overline{\tau}_{xy}^{(n)} \cos(\bar{\mathbf{n}},y)]_{[i]}^{[j]} = 0, \qquad [\overline{\tau}_{xy}^{(n)} \cos(\bar{\mathbf{n}},x) + \overline{\sigma}_y^{(n)} \cos(\bar{\mathbf{n}},y)]_{[i]}^{[j]} = 0. \tag{8.25}$$

(c) The displacements $\overline{u}^{(n)}, \overline{v}^{(n)}$ on the dividing contours $\partial\Omega_{ij}$ differ by *known* functions $f_u^{(n)}(x,y)$ and $f_v^{(n)}(x,y)$.

We therefore reach three generic BVPs for plane-strain, where the unknown stress functions $\overline{\Phi}^{(n)}$ over the non-homogeneous domain Ω should be determined. Based on (8.8a–d), one should solve the following three auxiliary biharmonic BVPs:

$$\nabla_1^{(4)} \overline{\Phi}^{(n)} = 0 \qquad \text{over} \quad \Omega, \tag{8.26a}$$

$$\frac{d}{ds}\{\overline{\Phi}_{,x}^{(n)}, \overline{\Phi}_{,y}^{(n)}\} = \{0, 0\} \qquad \text{on} \quad \partial\Omega, \tag{8.26b}$$

$$\frac{d}{ds}\{\overline{\Phi}^{(n)}_{,x}, \overline{\Phi}^{(n)}_{,y}\}^{[j]}_{[i]} = \{0, 0\} \quad \text{on} \quad \partial\Omega_{ij}, \tag{8.26c}$$

$$\{\overline{u}^{(n)}, \overline{v}^{(n)}\}^{[j]}_{[i]} = \{f^{(n)}_u, f^{(n)}_v\} \quad \text{on} \quad \partial\Omega_{ij}, \tag{8.26d}$$

where for the sake of convenience, we assume $\overline{u}^{(n)}(0,0) = \overline{v}^{(n)}(0,0) = 0$.

Note that the single-valued conditions (8.9a) are satisfied as no external loading is imposed. Once $\overline{\Phi}^{(n)}$ $(n = 1,2,3)$ are determined, the corresponding stress components are found as $\overline{\sigma}^{(n)}_x = \overline{\Phi}^{(n)}_{,yy}$, $\overline{\sigma}^{(n)}_y = \overline{\Phi}^{(n)}_{,xx}$ and $\overline{\tau}^{(n)}_{xy} = -\overline{\Phi}^{(n)}_{,xy}$, and $\overline{\sigma}^{(n)}_z$ by (3.26a).

We shall present here the functions $f^{(n)}_u$ and $f^{(n)}_v$, which will be shown to be required throughout the derivation of the problem of non-homogeneous beams:

$$f^{(1)}_u = \left[\frac{a_{13}}{a_{33}}\right]^{[j]}_{[i]}\frac{x^2}{2} - \left[\frac{a_{23}}{a_{33}}\right]^{[j]}_{[i]}\frac{y^2}{2}, \qquad f^{(1)}_v = \left[\frac{a_{23}}{a_{33}}\right]^{[j]}_{[i]} xy + \left[\frac{a_{36}}{a_{33}}\right]^{[j]}_{[i]}\frac{x^2}{2}, \tag{8.27a}$$

$$f^{(2)}_u = \left[\frac{a_{13}}{a_{33}}\right]^{[j]}_{[i]} xy + \left[\frac{a_{36}}{a_{33}}\right]^{[j]}_{[i]}\frac{y^2}{2}, \qquad f^{(2)}_v = -\left[\frac{a_{13}}{a_{33}}\right]^{[j]}_{[i]}\frac{x^2}{2} + \left[\frac{a_{23}}{a_{33}}\right]^{[j]}_{[i]}\frac{y^2}{2}, \tag{8.27b}$$

$$f^{(3)}_u = -\left[\frac{a_{13}}{a_{33}}\right]^{[j]}_{[i]} x - \left[\frac{a_{36}}{a_{33}}\right]^{[j]}_{[i]}\frac{y}{2}, \qquad f^{(3)}_v = -\left[\frac{a_{36}}{a_{33}}\right]^{[j]}_{[i]}\frac{x}{2} - \left[\frac{a_{23}}{a_{33}}\right]^{[j]}_{[i]} y. \tag{8.27c}$$

It is important to note that the solutions of the above auxiliary problems depend on the specific selection of the system's origin placement and orientation. Further discussion in S.8.2.5.1 shows that (8.60), (8.61) should be satisfied. In the general case, these conditions enforce some iterative process for the determination of the system origin location as will be discussed within S.8.2.11. **P.8.2** produces the above auxiliary problems and generic biharmonic BVP solutions for a laminated rectangle, see Example 8.2.

Example 8.2 *The Auxiliary Problems of Plane Deformation in a Laminated Rectangle.*
In this example we present solutions of the three auxiliary problems of plane deformation for the non-homogeneous rectangle shown in Fig. 8.3(a). By activating **P.8.2** with *ax_problem* =

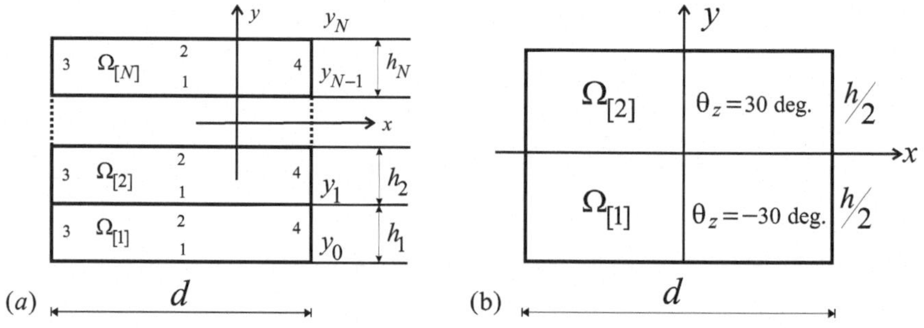

Figure 8.3: A non-homogeneous, rectangular cross-section. (a) Generic; (b) $N = 2$.

1,2,3, $NF = 2$, $k_\Phi = 25$, $d = h = 1$ and $t_z = -30°, 30°$, see Fig. 8.3(b), we determine the three auxiliary problems for the case under discussion. The above material selection creates two domains where all elastic moduli are identical except for a_{16}, a_{26}, a_{36} and a_{45} that are of identical magnitude but opposite signs. By (initially) placing the coordinate system at the cross-section midpoint (see Fig. 8.3(b)), equations (8.27a–c) are clearly simplified (as $y = 0$ on the interface line). In the present case, the stress functions $\overline{\Phi}^{(2)}, \overline{\Phi}^{(3)}$ for the second and third problems vanish. Therefore, $\overline{u}^{(2)} = \overline{v}^{(2)} = 0$ and $\overline{u}^{(3)} = \frac{a_{36}}{2a_{33}}y$, $\overline{v}^{(3)} = -\frac{a_{36}}{2a_{33}}x$ (i.e. piecewise linear functions). The resulting solutions are demonstrated in Figs. 8.4, 8.5 in terms of the deformation field. Note that these solutions create stress distributions that satisfy (8.60), (8.61)

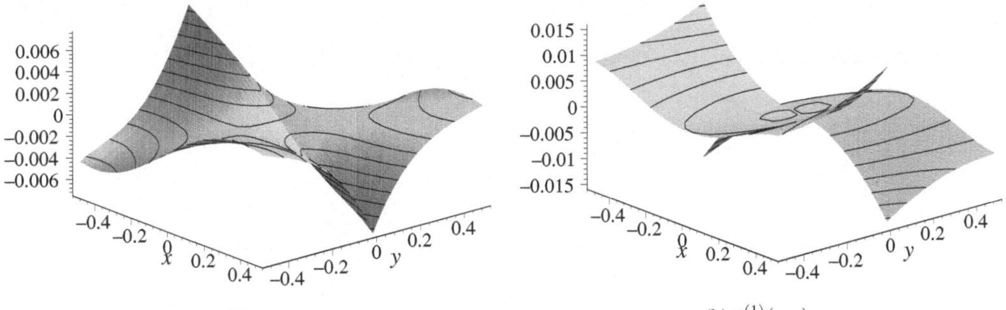

(a) $\overline{u}^{(1)}(x,y)$. (b) $\overline{v}^{(1)}(x,y)$.

Figure 8.4: Displacements of the first auxiliary problem in Example 8.2.

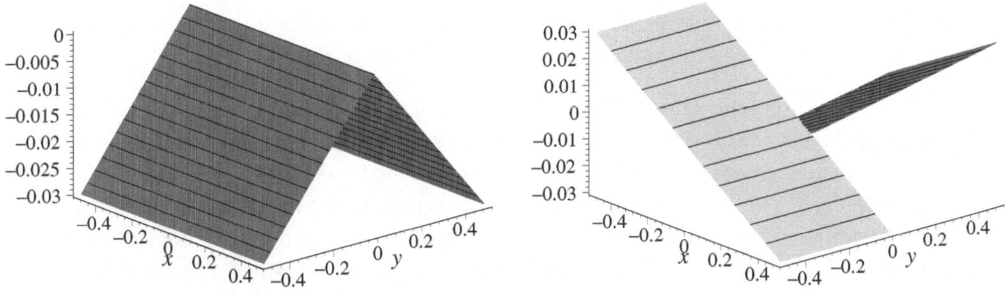

(a) $\overline{u}^{(3)}(x,y)$. (b) $\overline{v}^{(3)}(x,y)$.

Figure 8.5: Displacements of the third auxiliary problem in Example 8.2.

that will be shown to be required later on, and therefore, the initial assumption regarding the system origin location is correct.

8.2.2.1 The Auxiliary Problems in View of the Fundamental Problems of Elasticity

Since the above described auxiliary problems are defined in a non-standard way with prescribed displacements discontinuity, the purpose of this section is to show that these auxiliary problems can be presented as special cases of the three fundamental problems that are generally treated within the theory of elasticity, see S.1.6.1.

To facilitate the following discussion, we shall adopt the simplified non-homogeneous cross-section shown in Fig. 8.6, which consists of two materials, Material I and Material II, that occupy regions I, and II respectively. Suppose now that we define a contour parameter $t = \frac{2\pi}{L}s$ that replaces the arclength parameter $0 \le s \le L$, where L is the length of the inner contour (see also S.3.1.2). To this end, we solve two fundamental problems of the first kind (see S.1.6.1). Problem (1) treats domain I (without domain II) as shown in Fig. 8.6(b). The loads over the

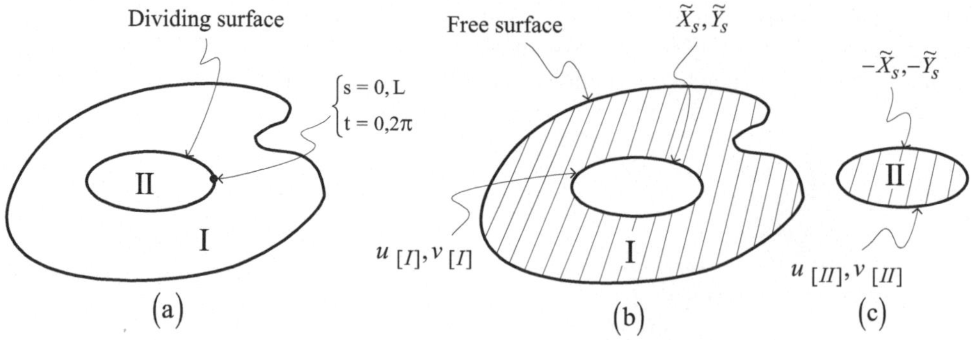

Figure 8.6: Illustrative domain compound of two sub-domains.

outer surface vanish, while in the inner surface we assume a generic distribution $X_s(t)$ and $Y_s(t)$. Problem (2) treats domain II (without domain I) as shown in Fig. 8.6(c), which is loaded by the opposite distributions $-X_s(t)$ and $-Y_s(t)$. From both problems we obtain a distribution of displacements over the dividing contour, which will be denoted $u_{[I]}(t)$, $v_{[I]}(t)$, $u_{[II]}(t)$, $v_{[II]}(t)$. These two problems are of the *first fundamental problem of elasticity* type, see S.1.6.1.

Subsequently, one may look for the two $\widetilde{X}(t)$ and $\widetilde{Y}(t)$ distributions that will yield the two prescribed $f_u(t) = u_{[I]}(t) - u_{[II]}(t)$ and $f_v(t) = v_{[I]}(t) - v_{[II]}(t)$ differences in displacements. Such a searching procedure is summarized by the l.h.s. chart of Fig. 8.7.

We shall now examine the same problem from the point of view of *the second and the third fundamental problems of elasticity*. In this case, when solving domain I we apply some generic distribution of displacements $u_{[I]}(t)$, $v_{[I]}(t)$, over the inner contour while keeping the outer contour traction-free, which is compatible with the definition of the third fundamental problem. When solving domain II, we apply $u_{[I]}(t) - f_u(t)$, $v_{[I]}(t) - f_u(t)$ over the inner contour, which is compatible with the definition of the second fundamental problem. Hence, the above solutions yield the surface loads $X_{s[I]}, Y_{s[I]}$ and $X_{s[II]}, Y_{s[II]}$ over the dividing contour.

In this case, one may look for the two $u_{[I]}(t)$, $v_{[I]}(t)$ distributions that will create equal traction by both solutions, i.e., $X_{s[I]} - X_{s[II]} = 0$ and $Y_{s[I]} - Y_{s[II]} = 0$. The above procedure is summarized by the r.h.s. chart of Fig. 8.7.

8.2.3 *Tip Axial Force*

To evaluate the effect of tip axial force, P_z, we first isolate and examine the relevant terms in the displacement expressions of a homogeneous beam as shown by (7.74a–c), namely,

$$u = \frac{P_z}{S_\Omega}(a_{13}x + \frac{1}{2}a_{36}y), \qquad v = \frac{P_z}{S_\Omega}(\frac{1}{2}a_{36}x + a_{23}y), \qquad w = \frac{P_z}{S_\Omega}a_{33}z. \tag{8.28}$$

Accordingly, the displacements in the non-homogeneous case are assumed to be of the form

$$u = \alpha_3(\frac{a_{13}}{a_{33}}x + \frac{a_{36}}{2a_{33}}y + \overline{u}^{(3)}), \qquad v = \alpha_3(\frac{a_{36}}{2a_{33}}x + \frac{a_{23}}{a_{33}}y + \overline{v}^{(3)}), \qquad w = \alpha_3 z, \tag{8.29}$$

where α_3 is a non-dimensional constant to be determined. Here $\overline{u}^{(3)}, \overline{v}^{(3)}$ are components of the solution of the third ($n = 3$) plane deformation auxiliary problem discussed in S.8.2.2. Clearly, the above selection of $f_u^{(3)}$ and $f_v^{(3)}$ in 8.27c ensures that the displacements in (8.29) will remain continuous over the dividing contours as required.

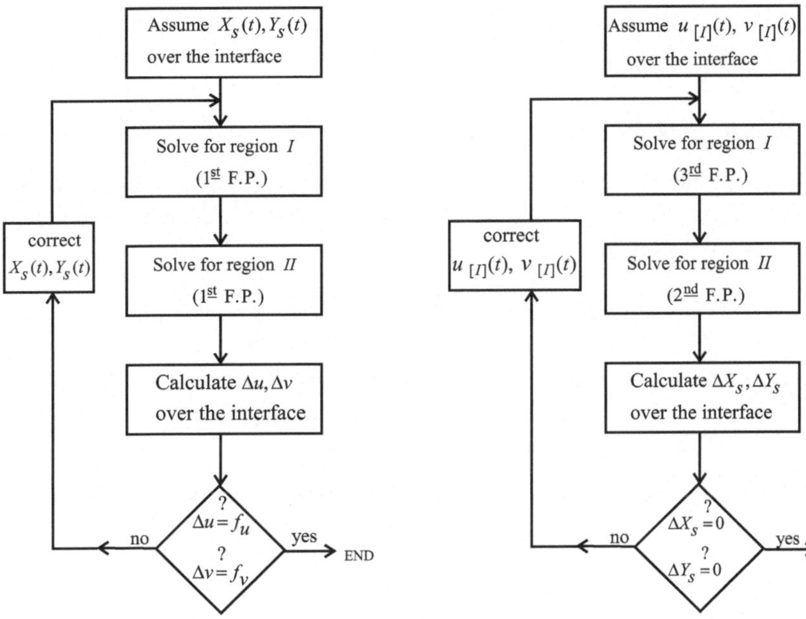

Figure 8.7: The auxiliary problems in view of the three fundamental problems (F.P.) of elasticity theory.

The strain components may be extracted from the above displacements as

$$\varepsilon_x = \alpha_3(\frac{a_{13}}{a_{33}} + \overline{\varepsilon}_x^{(3)}), \ \varepsilon_y = \alpha_3(\frac{a_{23}}{a_{33}} + \overline{\varepsilon}_y^{(3)}), \ \varepsilon_z = \alpha_3, \ \gamma_{yz} = \gamma_{xz} = 0, \ \gamma_{xy} = \alpha_3(\frac{a_{36}}{a_{33}} + \overline{\gamma}_{xy}^{(3)}). \quad (8.30)$$

The stress components may be obtained from the above strain components by employing the constitutive relations (7.1), or as a superposition of the solution for a homogeneous beam and the stress distribution of the above third auxiliary problem. In such a case, the homogeneous beam stress expressions are those derived in (7.72a–d) for the case where P_z is the only nonzero tip load, while $\frac{P_z}{S_\Omega}$ is replaced by $\frac{\alpha_3}{a_{33}}$, see (7.74a–c), (8.29). Hence,

$$\sigma_x = \alpha_3 \overline{\sigma}_x^{(3)}, \quad \sigma_y = \alpha_3 \overline{\sigma}_y^{(3)}, \quad \sigma_z = \alpha_3(\frac{1}{a_{33}} + \overline{\sigma}_z^{(3)}), \quad \tau_{yz} = \tau_{xz} = 0, \quad \tau_{xy} = \alpha_3 \overline{\tau}_{xy}^{(3)}. \quad (8.31)$$

Note that the stress resultants X_s and Y_s of (5.16a,b) are continuous across the dividing surfaces (when the same normal orientation is taken for both sides), while Z_s of (5.16c) vanishes.

Clearly, the above solution is consistent with all theory of elasticity requirements, as it is a superposition of two solutions that comply with the above requirements separately. To this end, the only aspect that has not been clarified are the magnitude and direction of the integral tip loads obtained by this solution. In the general case, the above stress components produce three tip load components: M_y, M_x and P_z. According to (5.7) these loads are given by

$$\{M_{y3}, M_{x3}, P_{z3}\} = \alpha_3\{\overline{I}_{13}, \overline{I}_{23}, \overline{I}_{33}\}, \quad (8.32)$$

where

$$\{\overline{I}_{13}, \overline{I}_{23}, \overline{I}_{33}\} = \iint_\Omega (\frac{1}{a_{33}} + \overline{\sigma}_z^{(3)})\{-x, y, 1\}. \quad (8.33)$$

It will become clearer later on that similar to the above way, in which this solution contributes not only to P_z but to M_x and M_y as well (since $\overline{I}_{23}$ and $\overline{I}_{13}$ do not vanish), other solutions will contribute to P_z. Hence, the requirement for a solution that produces P_z only is postponed at this stage until the effect of all tip loads components will be adequately derived.

8.2.4 Tip Bending Moments

The solution of the problem of bending by a tip moment, M_x, is derived in a similar way to the one described in S.8.2.3. We adopt again (7.74a–c) and express the displacement components in the non-homogeneous case as

$$u=\alpha_2(-\frac{a_{13}}{a_{33}}xy-\frac{a_{36}}{2a_{33}}y^2+\bar{u}^{(2)}),\ \ v=\alpha_2(\frac{a_{13}}{2a_{33}}x^2-\frac{a_{23}}{2a_{33}}y^2+\frac{z^2}{2}+\bar{v}^{(2)}),\ \ w=-\alpha_2 yz, \quad (8.34)$$

where α_2 is a constant to be determined, the dimension of which is $[\text{length}]^{-1}$. Here $\bar{u}^{(2)}, \bar{v}^{(2)}$ are the components of the solution of the second ($n=2$) auxiliary problem of plane deformation discussed in S.8.2.2. The strain components in this case are

$$\begin{aligned}
&\varepsilon_x=\alpha_2(-\frac{a_{13}}{a_{33}}y+\bar{\varepsilon}_x^{(2)}), \qquad \varepsilon_y=\alpha_2(-\frac{a_{23}}{a_{33}}y+\bar{\varepsilon}_y^{(2)}), \qquad \varepsilon_z=-\alpha_2 y,\\
&\gamma_{yz}=\gamma_{xz}=0, \qquad \gamma_{xy}=\alpha_2(-\frac{a_{36}}{a_{33}}y+\bar{\gamma}_{xy}^{(2)}),
\end{aligned} \quad (8.35)$$

and the corresponding stress components become

$$\sigma_x=\alpha_2\bar{\sigma}_x^{(2)}, \quad \sigma_y=\alpha_2\bar{\sigma}_y^{(2)}, \quad \sigma_z=-\alpha_2(\frac{y}{a_{33}}-\bar{\sigma}_z^{(2)}), \quad \tau_{yz}=\tau_{xz}=0, \quad \tau_{xy}=\alpha_2\bar{\tau}_{xy}^{(2)}. \quad (8.36)$$

In a similar way, one may derive the problem of bending by a tip moment M_y, for which the first auxiliary problem of plane deformation for a non-homogeneous beam is defined. The displacement components in this case are

$$u=\alpha_1(-\frac{a_{13}}{2a_{33}}x^2+\frac{a_{23}}{2a_{33}}y^2+\frac{z^2}{2}+\bar{u}^{(1)}),\ \ v=\alpha_1(-\frac{a_{36}}{2a_{33}}x^2-\frac{a_{23}}{a_{33}}xy+\bar{v}^{(1)}),\ \ w=-\alpha_1 xz \quad (8.37)$$

where α_1 is a constant to be determined, the dimension of which is $[\text{length}]^{-1}$. Here $\bar{u}^{(1)}, \bar{v}^{(1)}$ are the components of the solution of the first ($n=1$) auxiliary problem of plane deformation discussed in S.8.2.2. The components of strain for this case are

$$\begin{aligned}
&\varepsilon_x=\alpha_1(-\frac{a_{13}}{a_{33}}x+\bar{\varepsilon}_x^{(1)}), \qquad \varepsilon_y=\alpha_1(-\frac{a_{23}}{a_{33}}x+\bar{\varepsilon}_y^{(1)}), \qquad \varepsilon_z=-\alpha_1 x,\\
&\gamma_{yz}=\gamma_{xz}=0, \qquad \gamma_{xy}=\alpha_1(-\frac{a_{36}}{a_{33}}x+\bar{\gamma}_{xy}^{(1)}),
\end{aligned} \quad (8.38)$$

and for the components of stress we obtain

$$\sigma_x=\alpha_1\bar{\sigma}_x^{(1)}, \quad \sigma_y=\alpha_1\bar{\sigma}_y^{(1)}, \quad \sigma_z=-\alpha_1(\frac{x}{a_{33}}-\bar{\sigma}_z^{(1)}), \quad \tau_{yz}=\tau_{xz}=0, \quad \tau_{xy}=\alpha_1\bar{\tau}_{xy}^{(1)}. \quad (8.39)$$

The discussion regarding the X_s, Y_s continuity and the fulfillment of the integral tip boundary conditions discussed in S.8.2.3 apply in the present case of tip bending moments as well. The contributions of the above solutions to the tip loads are

$$\{M_{yi}, M_{xi}, P_{zi}\}=\alpha_i\{\bar{I}_{1i}, \bar{I}_{2i}, \bar{I}_{3i}\}, \qquad i=1,2 \quad (8.40)$$

where

$$\{\bar{I}_{12},\bar{I}_{22},\bar{I}_{32}\}=-\iint_\Omega(\frac{y}{a_{33}}-\bar{\sigma}_z^{(2)})\{-x,y,1\}, \quad (8.41a)$$

$$\{\bar{I}_{11},\bar{I}_{21},\bar{I}_{31}\}=-\iint_\Omega(\frac{x}{a_{33}}-\bar{\sigma}_z^{(1)})\{-x,y,1\}. \quad (8.41b)$$

As will be shown in S.8.2.5.1, the existence conditions for the bending functions require a selection of the coordinate system origin so that $\bar{I}_{32} = \bar{I}_{31} = 0$.

8.2.5 *Tip Bending Forces*

Following (7.74a–c), and (8.27a) it appears that one may express the displacement components created by P_x in the non-homogeneous case by using the solution of the first auxiliary problem discussed in S.8.2.2 as

$$u = \alpha_4[(l-z)(-\frac{a_{13}}{2a_{33}}x^2 + \frac{a_{23}}{2a_{33}}y^2 + \bar{u}^{(1)}) + \frac{z^2}{2}(l - \frac{z}{3})], \tag{8.42a}$$

$$v = \alpha_4(l-z)(-\frac{a_{36}}{2a_{33}}x^2 - \frac{a_{23}}{a_{33}}xy + \bar{v}^{(1)}), \tag{8.42b}$$

$$w = -\alpha_4[(l - \frac{z}{2})xz + xy^2 + \chi_1], \tag{8.42c}$$

where α_4 is a constant to be determined, the dimension of which is $[\text{length}]^{-2}$. Once the results of the first auxiliary problem are adopted, (8.42a–c) show that to enforce w displacement continuity one also needs to ensure that *the edgewise bending function,* $\chi_1(x,y)$, see S.7.2.2.1, is continuous as well. The strain components in this case are

$$\varepsilon_x = \alpha_4(l-z)(-\frac{a_{13}}{a_{33}}x + \bar{\varepsilon}_x^{(1)}), \qquad \varepsilon_y = \alpha_4(l-z)(-\frac{a_{23}}{a_{33}}x + \bar{\varepsilon}_y^{(1)}),$$

$$\gamma_{xy} = \alpha_4(l-z)(-\frac{a_{36}}{a_{33}}x + \bar{\gamma}_{xy}^{(1)}), \qquad \varepsilon_z = -\alpha_4(l-z)x,$$

$$\gamma_{yz} = \alpha_4[\frac{a_{36}x^2 + 2(a_{23} - 2a_{33})xy}{2a_{33}} - \chi_{1,y} - \bar{v}^{(1)}],$$

$$\gamma_{xz} = \alpha_4[\frac{a_{13}x^2 - (a_{23} + 2a_{33})y^2}{2a_{33}} - \chi_{1,x} - \bar{u}^{(1)}]. \tag{8.43}$$

The corresponding stress components take the form

$$\sigma_x = \alpha_4(l-z)\bar{\sigma}_x^{(1)}, \qquad \sigma_y = \alpha_4(l-z)\bar{\sigma}_y^{(1)}, \qquad \tau_{xy} = \alpha_4(l-z)\bar{\tau}_{xy}^{(1)}, \tag{8.44a}$$

$$\sigma_z = -\alpha_4(l-z)(\frac{x}{a_{33}} - \bar{\sigma}_z^{(1)}), \tag{8.44b}$$

$$\tau_{yz} = \frac{\alpha_4}{a_0}(Q^{\chi_1} + a_{45}\chi_{1,x} - a_{55}\chi_{1,y}), \qquad \tau_{xz} = \frac{\alpha_4}{a_0}(P^{\chi_1} - a_{44}\chi_{1,x} + a_{45}\chi_{1,y}), \tag{8.44c}$$

where $a_0 = a_{44}a_{55} - a_{45}^2$,

$$P^{\chi_1} = (P^{\chi_1})^* - (a_{44}\bar{u}^{(1)} - a_{45}\bar{v}^{(1)}), \qquad Q^{\chi_1} = (Q^{\chi_1})^* + (a_{45}\bar{u}^{(1)} - a_{55}\bar{v}^{(1)}), \tag{8.45}$$

and $(P^{\chi_1})^*$, $(Q^{\chi_1})^*$ are given in (7.54). The solution for bending by a transverse force P_y may be carried out analogously. We present its results only, as the logical steps are identical to the discussion regarding P_x. In this case, the second plane-strain problem is employed along with the *beamwise bending function*, χ_2, see S.7.2.2.2, and the displacements are

$$u = \alpha_5(l-z)(-\frac{a_{13}}{a_{33}}xy - \frac{a_{36}}{2a_{33}}y^2 + \bar{u}^{(2)}), \tag{8.46a}$$

$$v = \alpha_5[(l-z)(\frac{a_{13}}{2a_{33}}x^2 - \frac{a_{23}}{2a_{33}}y^2 + \bar{v}^{(2)}) + \frac{z^2}{2}(l - \frac{z}{3})], \tag{8.46b}$$

$$w = -\alpha_5[(l - \frac{z}{2})yz + x^2y + \chi_2], \tag{8.46c}$$

where α_5 is a constant to be determined, the dimension of which is $[\text{length}]^{-2}$. The resulting strain components are

$$\varepsilon_x = \alpha_5(l-z)(-\frac{a_{13}}{a_{33}}y + \bar{\varepsilon}_x^{(2)}), \qquad \varepsilon_y = \alpha_5(l-z)(-\frac{a_{23}}{a_{33}}y + \bar{\varepsilon}_y^{(2)}),$$
$$\gamma_{xy} = \alpha_5(l-z)(-\frac{a_{36}}{a_{33}}y + \bar{\gamma}_{xy}^{(2)}), \qquad \varepsilon_z = -\alpha_5(l-z)y,$$
$$\gamma_{yz} = \alpha_5[\frac{a_{23}y^2 - (a_{13}+2a_{33})x^2}{2a_{33}} - \chi_{2,y} - \bar{v}^{(2)}],$$
$$\gamma_{xz} = \alpha_5[\frac{a_{36}y^2 + 2(a_{13}-2a_{33})xy}{2a_{33}} - \chi_{2,x} - \bar{u}^{(2)}], \tag{8.47}$$

and for the stress components we obtain

$$\sigma_x = \alpha_5(l-z)\bar{\sigma}_x^{(2)}, \qquad \sigma_y = \alpha_5(l-z)\bar{\sigma}_y^{(2)}, \qquad \tau_{xy} = \alpha_5(l-z)\bar{\tau}_{xy}^{(2)}, \tag{8.48a}$$
$$\sigma_z = -\alpha_5(l-z)(\frac{y}{a_{33}} - \bar{\sigma}_z^{(2)}), \tag{8.48b}$$
$$\tau_{yz} = \frac{\alpha_5}{a_0}(Q^{\chi_2} - a_{55}\chi_{2,y} + a_{45}\chi_{2,x}), \qquad \tau_{xz} = \frac{\alpha_5}{a_0}(P^{\chi_2} - a_{44}\chi_{2,x} + a_{45}\chi_{2,y}) \tag{8.48c}$$

where

$$P^{\chi_2} = (P^{\chi_2})^* - (a_{44}\bar{u}^{(2)} - a_{45}\bar{v}^{(2)}), \qquad Q^{\chi_2} = (Q^{\chi_2})^* + (a_{45}\bar{u}^{(2)} - a_{55}\bar{v}^{(2)}), \tag{8.49}$$

and $(P^{\chi_2})^*$, $(Q^{\chi_2})^*$ are given in (7.64b). To derive the governing equations for χ_k, we employ the definition of the *generalized harmonic* operator $\nabla_3^{(2)}$ and the *generalized normal derivative operator* D_1^n, that appear in (3.97) and (3.118), respectively. For that purpose we note that the stress components of (8.44a–e) satisfy identically the equilibrium equations (1.82a,b). To satisfy the third equation of equilibrium, (1.82c), and similar to (7.56a), (7.63a), the functions $\chi_k(x,y)$ must fulfill the following requirement in each domain $\Omega_{[j]}$:

$$\nabla_3^{(2)}\chi_k = F_0^{\chi_k} \qquad \text{over} \quad \Omega, \tag{8.50}$$

where

$$F_0^{\chi_k} = (F_0^{\chi_k})^* - a_{44}\bar{\varepsilon}_x^{(k)} - a_{55}\bar{\varepsilon}_y^{(k)} + a_{45}\bar{\gamma}_{xy}^{(k)} - a_0\bar{\sigma}_z^{(k)}, \tag{8.51}$$

while the expressions for $(F_0^{\chi_k})^*$ are given by (7.57), (7.64a). The (only relevant) boundary condition $Z_s = 0$, see (5.16c), yields for the free boundaries (similar to (7.56b), (7.63b))

$$D_1^n\chi_k = P^{\chi_k}\cos(\bar{\mathbf{n}},x) + Q^{\chi_k}\cos(\bar{\mathbf{n}},y) \qquad \text{on} \quad \partial\Omega, \tag{8.52}$$

which for simplicity we write as $a_0 Z_s = 0$. Over the dividing contours, X_s and Y_s are continuous as earlier discussed, and the requirement for continuous Z_s, see (5.16c), yields

$$[\frac{1}{a_0}D_1^n\chi_k]_{[i]}^{[j]} = [\frac{1}{a_0}(P^{\chi_k}\cos(\bar{\mathbf{n}},x) + Q^{\chi_k}\cos(\bar{\mathbf{n}},y))]_{[i]}^{[j]} \quad \text{on} \quad \partial\Omega_{ij}. \tag{8.53}$$

As already indicated, from the condition for a continuous w displacement component over the dividing contours we write

$$\chi_k{}_{[i]}^{[j]} = 0 \quad \text{on} \quad \partial\Omega_{ij}. \tag{8.54}$$

As shown by (8.44b), (8.48b), over the tip cross-section, the normal stress vanishes, and thus, only P_x, P_y and M_z are created by the two sets of solutions presented in this section. The tip loads, which are proportional to α_4 and α_5, are given below by substituting $k = 4,5$, respectively, in the following equations:

$$\{P_{xk}, P_{yk}, M_{zk}\} = \alpha_k\{\bar{I}_{4k}, \bar{I}_{5k}, \bar{I}_{6k}\}, \tag{8.55}$$

where

$$\{\bar{I}_{4k}, \bar{I}_{5k}, \bar{I}_{6k}\} = \iint_{\Omega}\{\frac{\tau_{xz}}{\alpha_k}, \frac{\tau_{yz}}{\alpha_k}, x\frac{\tau_{yz}}{\alpha_k} - y\frac{\tau_{xz}}{\alpha_k}\}. \tag{8.56}$$

In the above, the expressions for $\frac{\tau_{yz}}{\alpha_4}$ and $\frac{\tau_{xz}}{\alpha_4}$ are taken from (8.44c), respectively, and the expressions for $\frac{\tau_{yz}}{\alpha_5}$ and $\frac{\tau_{xz}}{\alpha_5}$ are taken from (8.48c). For the above bending problems one may write

$$P^{\chi_1}_{,x} + Q^{\chi_1}_{,y} - F^{\chi_1}_0 = -a_0(\frac{x}{a_{33}} - \overline{\sigma}^{(1)}_z), \tag{8.57a}$$

$$P^{\chi_2}_{,x} + Q^{\chi_2}_{,y} - F^{\chi_2}_0 = -a_0(\frac{y}{a_{33}} - \overline{\sigma}^{(2)}_z). \tag{8.57b}$$

Substituting the above in the integral identities of (3.127a) (see also discussion in S.8.1.1) and using definitions of D_x, D_y in S.7.2.2, enables us to simplify (8.56) as

$$\{\bar{I}_{44}, \bar{I}_{54}, \bar{I}_{45}, \bar{I}_{55}, \bar{I}_{64}, \bar{I}_{65}\} = \{\bar{I}_{11}, -\bar{I}_{21}, \bar{I}_{12}, -\bar{I}_{22}, -D_x, -D_y\}. \tag{8.58}$$

8.2.5.1 Existence Conditions for the Bending Functions

To develop the existence condition of the Neumann BVP for χ_1 we employ (8.3) with $\Lambda = \chi_1$ and take $F^{\chi_1}_3$ to be equal to the r.h.s. of (8.52). By Green's Theorem we write for each Ω_j

$$\oint_{\partial\Omega_j}(\frac{1}{a_0}F^{\chi_1}_3) = \iint_{\Omega_j}\frac{1}{a_0}(P^{\chi_1}_{,x} + Q^{\chi_1}_{,y}). \tag{8.59}$$

Then, using (8.57a) yields the condition of existence for χ_1 as

$$\iint_{\Omega}(\frac{x}{a_{33}} \underline{- \overline{\sigma}^{(1)}_z}) = 0. \tag{8.60}$$

Analogously the condition of existence for χ_2 becomes

$$\iint_{\Omega}(\frac{y}{a_{33}} \underline{- \overline{\sigma}^{(2)}_z}) = 0. \tag{8.61}$$

The above conditions may be satisfied by suitable selection of the location of the coordinate system origin. Yet, in some cases, the resulting equations may not be explicit as discussed in S.8.2.11. It may also be shown that typically, the underlined terms in (8.60), (8.61) represent relatively "small corrections" in their respective expressions.

A solution for generalized harmonic functions in a rectangular non-homogeneous domain is documented in **P.8.1**.

Example 8.3 *Bending Functions in a Non-Homogeneous Rectangle.*

To present the bending functions in a non-homogeneous domain, we activate **P.8.1** for the cross-section of Fig. 8.3(b) and employ the auxiliary solutions of Example 8.2, namely $NF = 2$, $func = \chi_1, \chi_2$, $k_{La} = k_\Phi + 2 = 27$, $d = h = 1$, $t_z = -30°, 30°$. The results shown in Fig. 8.8 demonstrate the fulfillment of the continuity condition for these functions, while the fulfillment of the boundary conditions will be shown later on within the discussion of the resulting stress components of Example 8.7.

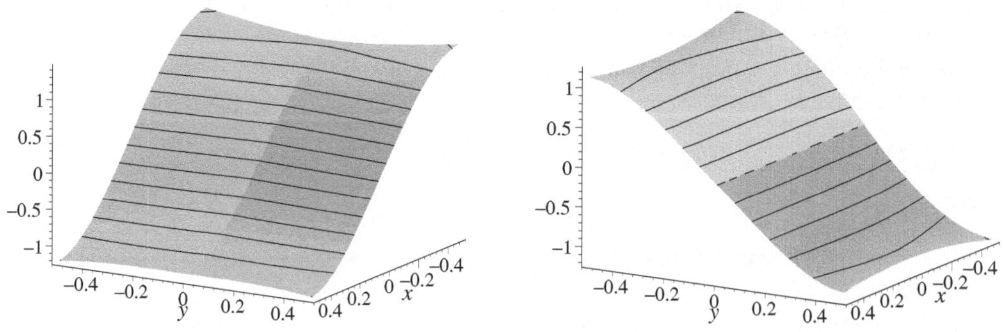

(a) $\chi_1(x,y)$. (b) $\chi_2(x,y)$.

Figure 8.8: Bending functions in Example 8.3.

8.2.6 *Torsional Moment*

As discussed in S.7.2.1, two basic methods may be employed for the present task. In the first method, the expressions are written for the warping function φ (see S.7.2.1.1), while in the second method, the stress function in torsion, ψ, is employed (see S.7.2.1.2). Solutions for some isotropic non-homogeneous domains are presented in (Muskhelishvili, 1953).

The expression for the torsional rigidity is the one given in (7.11) while clearly, the integration should be carried out over the entire cross-section. Employing (7.2a–c) with $M_z/D = \theta$, we assume that for a non-homogeneous beam, the symbolic expressions for the displacement components are identical to those of the homogeneous case, namely (rigid body parameters are not relevant for the discussion at this stage),

$$u = -\alpha_6 yz, \qquad v = \alpha_6 xz, \qquad w = \alpha_6\, \varphi(x,y), \tag{8.62}$$

where $\alpha_6 = \theta$ is the *twist* (angle per unit length), which is not known at this stage, and a warping function φ is subsequently defined for each domain $\Omega_{[j]}$. In contrast with the previous cases, the u and v deformation components *are not material-dependent*, and therefore, if adopted and found to be adequate for the non-homogeneous case, the above displacement hypothesis causes no discontinuity along the dividing contours. The continuity of the w component will be shortly discussed. The non-vanishing strain components in this case are given by (7.4) as

$$\gamma_{yz} = \alpha_6\,(\varphi_{,y} + x), \qquad \gamma_{xz} = \alpha_6\,(\varphi_{,x} - y), \tag{8.63}$$

and the only non-vanishing stress components are

$$\tau_{yz} = \frac{\alpha_6}{a_0}(-Q^{\varphi} - a_{45}\varphi_{,x} + a_{55}\varphi_{,y}), \qquad \tau_{xz} = \frac{\alpha_6}{a_0}(-P^{\varphi} + a_{44}\varphi_{,x} - a_{45}\varphi_{,y}) \tag{8.64}$$

while P^{φ} and Q^{φ} are given by (7.8). Similar to S.7.2.1.1, in order to maintain equilibrium, and analogously to (7.7a), the following condition should be imposed at each domain $\Omega_{[j]}$:

$$\nabla_3^{(2)}\,\varphi = 0 \qquad \text{over} \quad \Omega. \tag{8.65}$$

Over the free boundaries, and analogously to (7.7b), one should require

$$D_1^n\,\varphi = P^{\varphi}\cos(\bar{\mathbf{n}},x) + Q^{\varphi}\cos(\bar{\mathbf{n}},y) \qquad \text{on} \quad \partial\Omega. \tag{8.66}$$

The continuity condition of Z_s over the dividing contours becomes

$$[\frac{1}{a_0}D_1^n\varphi]_{[i]}^{[j]} = [\frac{1}{a_0}(P^\varphi \cos(\bar{\mathbf{n}},x) + Q^\varphi \cos(\bar{\mathbf{n}},y))]_{[i]}^{[j]} \qquad \text{on} \quad \partial\Omega_{ij}, \tag{8.67}$$

and the condition for the w displacement component to be continuous over the dividing contours is written as

$$\varphi_{[i]}^{[j]} = 0 \qquad \text{on} \quad \partial\Omega_{ij}. \tag{8.68}$$

One may verify that the existence condition of (8.4) with $\Lambda = \varphi$ is satisfied identically regardless of the location of the coordinate system.

Similar to (8.55), the contribution of the present solution to the tip loads is

$$\{P_{x6}, P_{y6}, M_{z6}\} = \alpha_6\{\bar{I}_{46}, \bar{I}_{56}, \bar{I}_{66}\}, \tag{8.69}$$

where

$$\{\bar{I}_{46}, \bar{I}_{56}, \bar{I}_{66}\} = \iint_\Omega \{\frac{\tau_{xz}}{\alpha_6}, \frac{\tau_{yz}}{\alpha_6}, x\frac{\tau_{yz}}{\alpha_6} - y\frac{\tau_{xz}}{\alpha_6}\}, \tag{8.70}$$

while the expressions for $\frac{\tau_{yz}}{\alpha_6}$ and $\frac{\tau_{xz}}{\alpha_6}$ are taken from (8.64). Using integral identities of S.3.3.3 and definition (7.11) of D, expressions (8.70) are simplified as

$$\{\bar{I}_{46},\bar{I}_{56},\bar{I}_{66}\} = \{0,0,D\}. \tag{8.71}$$

Remark 8.3 As a special case, one may present the torsion problem of a non-homogeneous beam with *continuously varying elastic moduli* (see also discussion in Remark 3.1). In such a case, the strain and the stress components may still be written as in (7.3) and (7.4), respectively. Likewise, the boundary condition of (7.6) holds and therefore, (7.7b) remains valid. However, the third equilibrium equation, (1.82c), should be written as

$$[\frac{a_{44}}{a_0}(\varphi_{,x} - y) - \frac{a_{45}}{a_0}(\varphi_{,y} + x)]_{,x} + [-\frac{a_{45}}{a_0}(\varphi_{,x} - y) + \frac{a_{55}}{a_0}(\varphi_{,y} + x)]_{,y} = 0 \quad \text{over} \quad \Omega, \tag{8.72}$$

or equivalently as

$$[A_{55}(\varphi_{,x} - y) + A_{45}(\varphi_{,y} + x)]_{,x} + [A_{45}(\varphi_{,x} - y) + A_{44}(\varphi_{,y} + x)]_{,y} = 0 \qquad \text{over} \quad \Omega, \tag{8.73}$$

which replaces the field equation of (7.7a). Elastic moduli distributions (i.e., $A_{ij}(x,y)$) that produce no warping in elliptical cross-sections were studied by (Ecsedi, 2004).

As another special case, we present the torsion problem of a non-homogeneous circular beam of cylindrical MON13z anisotropy with *continuously varying elastic moduli in the radial direction only*. The strain and the stress components may still be written as in (7.150) and (7.151a,b), respectively. Likewise, the boundary condition of (7.152:b) holds as well. However, the third equilibrium equation, (7.137c), and boundary equation, (7.138c) (with $\cos(\bar{\mathbf{n}},r) = 1$, $\cos(\bar{\mathbf{n}},\theta) = 0$), should be written as

$$(\frac{a_{44}}{a_0}\varphi_{,r} - \frac{a_{45}}{a_0}r)_{,r} + \frac{1}{r}(\frac{a_{44}}{a_0}\varphi_{,r} - \frac{a_{45}}{a_0}r) = 0 \qquad \text{for} \quad 0 \le r \le R, \tag{8.74a}$$

$$a_{44}\varphi_{,r}(R) = a_{45}R. \tag{8.74b}$$

Introducing a new function $\tilde{f}(r) = a_{44}\varphi_{,r} - a_{45}r$, we rewrite (8.74a,b) equivalently as

$$(\tilde{f})_{,r} + \frac{1}{r}\tilde{f} = 0, \qquad \tilde{f}(R) = 0. \tag{8.75}$$

The general solution of (8.75: a) is $\widetilde{f} = \frac{\widetilde{c}}{r}$, where $\widetilde{c}$ is a constant. In view of (8.75: b) we obtain $\widetilde{c} = 0$ and hence $\widetilde{f} \equiv 0$. The general solution of $a_{44}\varphi_{,r} - a_{45}r = 0$ is therefore

$$\varphi(r) = \int_0^r \frac{a_{45}}{a_{44}} r\,dr + \varphi(0). \tag{8.76}$$

For the case of $\frac{a_{45}}{a_{44}} = const.$, and assume $\varphi(0) = 0$, one obtains a quadratic function as in the homogeneous domain case, see S.7.5.2

$$\varphi = \frac{a_{45}}{2a_{44}} r^2. \tag{8.77}$$

Note that the boundary condition (8.74b) is satisfied for all $0 \le r \le R$, which indicates that $\tau_{rz} = 0$ for the entire domain. It is therefore clear that these solutions hold for thin-walled circular cross-sections as well. The torsional rigidity is finally expressed by (7.139:f) as

$$D = 2\pi \int_0^R \frac{1}{a_0}(-a_{45}\varphi_{,r} + a_{55}r)\, r^2\, dr = 2\pi \int_0^R \frac{1}{a_{44}} r^3\, dr. \tag{8.78}$$

Example 8.4 *Polynomial Based Torsion Function in a Non-Homogeneous Rectangle.*
To present an example for the torsion (warping) functions in a non-homogeneous cross-section, we activate **P.8.1** for the cross-section of Fig. 8.3(b) as in Example 8.3 but with $func = \varphi$. The resulting warping function is presented by Fig. 8.9. Due to the domain and material

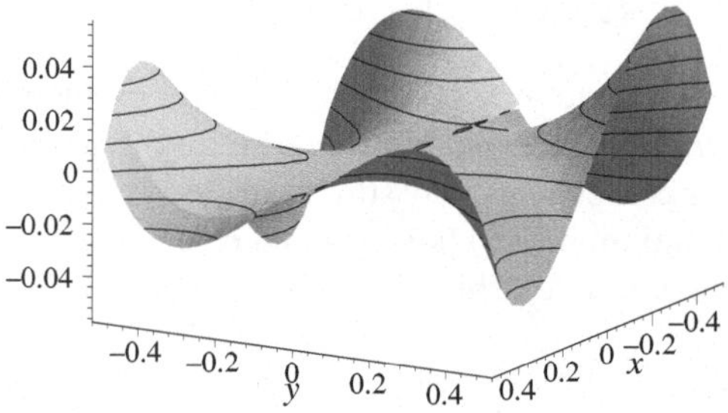

Figure 8.9: Torsion function, $\varphi(x,y)$ in Example 8.4.

"symmetries", we obtain $\varphi(x,-y) = \varphi(-x,y) = -\varphi(x,y)$, and hence $\varphi(x,0) = \varphi(0,y) \equiv 0$.

Example 8.5 *Fourier Series Based Torsion Function in a Non-Homogeneous Rectangle.*
Consider a non-homogeneous orthotropic beam the cross-section of which is a rectangle of width d and height $h = \sum_{i=1}^{N} h^{[i]}$, that consists of N different material domains $\Omega_{[j]}$ $(j = 1,\ldots,N)$ of heights $h^{[j]}$, see Fig. 8.3(a). For analytical convenience, the origin of the coordinate system is located at the (geometrical) centroid of the cross-section and throughout the derivation, we shall omit the index $[j]$ for piecewise parameters, functions and operations.

Since the materials are orthotropic the coefficients a_{44} and a_{55} are the only relevant material properties, so (8.64) become

$$\tau_{yz} = \frac{\theta}{a_{44}}(\varphi_{,y} + x), \qquad \tau_{xz} = \frac{\theta}{a_{55}}(\varphi_{,x} - y). \tag{8.79}$$

Also, in the present orthotropic case, the *generalized harmonic* operator and the *generalized normal derivative operator*, are simplified to $\nabla_4^{(2)}$ and D_2^n, see (3.196), (3.217). Hence, at each domain the field equation (8.65) becomes

$$\nabla_4^{(2)}\varphi = 0 \qquad \text{over} \quad \Omega, \tag{8.80}$$

while $P^{\varphi} = a_{44}y$ and $Q^{\varphi} = -a_{55}x$. The stress continuity conditions over the free boundaries and over the dividing contours are

$$D_2^n \varphi^{[j]} = [a_{44}y\cos(\bar{\mathbf{n}},x) - a_{55}x\cos(\bar{\mathbf{n}},y)]^{[j]}, \qquad 0 < j \leq N, \tag{8.81a}$$

$$[\frac{1}{a_0} D_2^n \varphi]_{[j]}^{[j+1]} = [\frac{1}{a_0}(a_{44}y\cos(\bar{\mathbf{n}},x) - a_{55}x\cos(\bar{\mathbf{n}},y))]_{[j]}^{[j+1]}, \qquad 0 < j < N. \tag{8.81b}$$

To solve the above BVP, we modify the Fourier series approach that was classically employed for solution of the torsion problem in an isotropic rectangular domain (see e.g. (Love, 1944), (Muskhelishvili, 1953)), (Timoshenko and Goodier, 1970). For that purpose, we introduce a piecewise harmonic function, $\Psi(x,y)$, as

$$\varphi = \frac{1}{\widetilde{a}}\Psi + xy, \tag{8.82}$$

where $\widetilde{a} = \sqrt{\frac{a_{44}}{a_{55}}}$ is a piecewise constant. $\Psi(x,y)$ is subsequently expanded as:

$$\Psi = \sum_{n=0}^{\infty} [A_n \sinh(m_n\widetilde{a}y) + B_n \cosh(m_n\widetilde{a}y)] \sin(m_n x), \tag{8.83}$$

where $m_n = \frac{(2n+1)\pi}{d}$ and A_n, B_n are piecewise constants. This selection shows that $\Psi_{,xx} = -m_n^2\Psi$ and $\Psi_{,yy} = m_n^2\widetilde{a}^2\Psi$, and thus, (8.80) identically fulfilled since

$$\nabla_4^{(2)}\Psi = 0 \qquad \text{over} \quad \Omega. \tag{8.84}$$

Subsequently, we only need to ensure that the boundary conditions are satisfied. The boundary condition (8.81a) on the free "vertical" lines is identically satisfied since it may be written as $a_{44}(\frac{1}{\widetilde{a}}\Psi_{,x} + y) = a_{44}y$, while $\Psi_{,x} = 0$ for $x = \pm\frac{d}{2}$. As discussed in what follows, the other conditions become

$$[\frac{1}{\widetilde{a}}\Psi_{,y}]^{[j]} = -2x, \qquad j \in \{1, N\} \tag{8.85a}$$

$$[\frac{1}{a_{44}\widetilde{a}}\Psi_{,y}]_{[j]}^{[j+1]} = -2x[\frac{1}{a_{44}}]_{[j]}^{[j+1]}, \qquad 0 < j < N \tag{8.85b}$$

$$[\frac{1}{\widetilde{a}}\Psi]_{[j]}^{[j+1]} = 0, \qquad 0 < j < N. \tag{8.85c}$$

Equation (8.85a) represents the boundary condition (8.81a), which is applied at the two free "horizontal" edges (i.e. two equations). Equation (8.85b) translates the interlaminar conditions of (8.81b), where the values $j = 1, \ldots, N-1$ are applied for each interface contour. The same values of j should be substituted in (8.85c) that are obtained from the interlaminar conditions (8.68).

To enable an analytical solution, the function "$2x$" in the interval $(-\frac{d}{2}, \frac{d}{2})$ may also be expanded as the following series (see (Muskhelishvili, 1953)):

$$2x = \sum_{n=0}^{\infty} m_n X_n \sin(m_n x), \qquad X_n = d^2\left(\frac{2}{\pi}\right)^3 \frac{(-1)^n}{(2n+1)^3}. \tag{8.86}$$

Thus, the two conditions of (8.85a) become for each $n \geq 0$,

$$A_n^{[1]} \cosh(m_n\widetilde{a}^{[1]}y^{[0]}) + B_n^{[1]} \sinh(m_n\widetilde{a}^{[1]}y^{[0]}) = -X_n, \tag{8.87a}$$

$$A_n^{[N]} \cosh(m_n\widetilde{a}^{[N]}y^{[N]}) + B_n^{[N]} \sinh(m_n\widetilde{a}^{[N]}y^{[N]}) = -X_n, \tag{8.87b}$$

where $y^{[j]} = \sum_{i=1}^{j} h^{[i]}$ (and $y^{[0]} = 0,\ y^{[N]} = h$). Also, (8.85b) gives for each $n \geq 0$ the $N-1$ conditions

$$[\frac{1}{a_{44}}(A_n \cosh(m_n\tilde{a}y) + B_n \sinh(m_n\tilde{a}y))]_{[j]}^{[j+1]} = -X_n[\frac{1}{a_{44}}]_{[j]}^{[j+1]}, \qquad 0 < j < N. \tag{8.88}$$

Finally, (8.85c) yields for each $n \geq 0$ the $N-1$ conditions

$$[\frac{1}{\tilde{a}}(A_n \sinh(m_n\tilde{a}y) + B_n \cosh(m_n\tilde{a}y))]_{[j]}^{[j+1]} = 0, \qquad 0 < j < N. \tag{8.89}$$

The system (8.87a,b), (8.88), (8.89) of $2N$ linear equations with $2N$ variables $A_n^{[j]}, B_n^{[j]},\ j \leq N$, may be solved by levels, for each n separately.

We shall now implement the above system for the $N = 2$ case. For each n, a system of four equations with the $A_n^{[1]}, A_n^{[2]}, B_n^{[1]}, B_n^{[2]}$ coefficients is created and can be analytically solved. Hence, one may write in a matrix form

$$\begin{bmatrix} \tilde{M}_{n11} & \tilde{M}_{n12} & 0 & 0 \\ \tilde{M}_{n21} & \tilde{M}_{n22} & \tilde{M}_{n23} & \tilde{M}_{n24} \\ \tilde{M}_{n31} & \tilde{M}_{n32} & \tilde{M}_{n33} & \tilde{M}_{n34} \\ 0 & 0 & \tilde{M}_{n43} & \tilde{M}_{n44} \end{bmatrix} \begin{Bmatrix} A_n^{[1]} \\ B_n^{[1]} \\ A_n^{[2]} \\ B_n^{[2]} \end{Bmatrix} = -X_n \begin{Bmatrix} 1 \\ [\frac{1}{a_{44}}]_{[1]}^{[2]} \\ 0 \\ 1 \end{Bmatrix}. \tag{8.90}$$

It is therefore shown that the above scheme constitutes a closed form solution for an infinite number of terms in (8.83). As another example, for the $N = 3$ case, we create for each n a system of six equations for the coefficients $A_n^{[j]}$ and $B_n^{[j]}$ ($j = 1,\ldots,3$), which is written in a matrix form as

$$\begin{bmatrix} \tilde{M}_{n11} & \tilde{M}_{n12} & 0 & 0 & 0 & 0 \\ \tilde{M}_{n21} & \tilde{M}_{n22} & \tilde{M}_{n23} & \tilde{M}_{n24} & 0 & 0 \\ \tilde{M}_{n31} & \tilde{M}_{n32} & \tilde{M}_{n33} & \tilde{M}_{n34} & 0 & 0 \\ 0 & 0 & \tilde{M}_{n43} & \tilde{M}_{n44} & \tilde{M}_{n45} & \tilde{M}_{n46} \\ 0 & 0 & \tilde{M}_{n53} & \tilde{M}_{n54} & \tilde{M}_{n55} & \tilde{M}_{n56} \\ 0 & 0 & 0 & 0 & \tilde{M}_{n65} & \tilde{M}_{n66} \end{bmatrix} \begin{Bmatrix} A_n^{[1]} \\ B_n^{[1]} \\ A_n^{[2]} \\ B_n^{[2]} \\ A_n^{[3]} \\ B_n^{[3]} \end{Bmatrix} = -X_n \begin{Bmatrix} 1 \\ [\frac{1}{a_{44}}]_{[1]}^{[2]} \\ 0 \\ [\frac{1}{a_{44}}]_{[2]}^{[3]} \\ 0 \\ 1 \end{Bmatrix}. \tag{8.91}$$

The above solution methodology is implemented by **P.8.3**, see illustrative results in Figs. 8.10, 8.11. To present some solution properties, we define (in addition to the $\tilde{a} = \sqrt{a_{44}/a_{55}}$) the non-dimensional parameters

$$s^{[j]} = \frac{a_{44}^{[j]}}{a_{44}^{0}}, \qquad t^{[j]} = \frac{h^{[j]}}{d}, \tag{8.92}$$

where a_{44}^0 is a reference typical value of a_{44}. Figure 8.10(a) presents the warping functions for the case of $N = 3$, $s^{[j]} = 1, \frac{3}{10}, 1$, $\tilde{a}^{[j]} = \frac{1}{2}, \frac{1}{3}, \frac{1}{2}$, $t^{[j]} = 2, 1, 2$, which is similar to Fig. 8.9. Figure 8.10(b) presents the stress field, $[\tau_{yz}, \tau_{xz}]$, for the case of $N = 2$, $s^{[j]} = 1, \frac{3}{10}$, $\tilde{a}^{[j]} = 1, 1$, $t^{[j]} = \frac{1}{2}, \frac{1}{2}$, and shows an interesting non-equal distribution of shear stresses over the two parts of the domain. This is due to the higher shear moduli of the lower part as reflected by the selection of $s^{[j]}$. The interlaminar stress in this torsion case is τ_{yz}. The stress component values along the interface line are clearly shown in Fig. 8.11(a) where $N = 3$, $s^{[j]} = 1, \frac{3}{10}, 1$, $\tilde{a}^{[j]} = \frac{1}{2}, \frac{1}{3}, \frac{1}{2}$, $t^{[j]} = 2, 1, 2$. For the same case, Fig. 8.11(b) shows that τ_{xz} is not continuous.

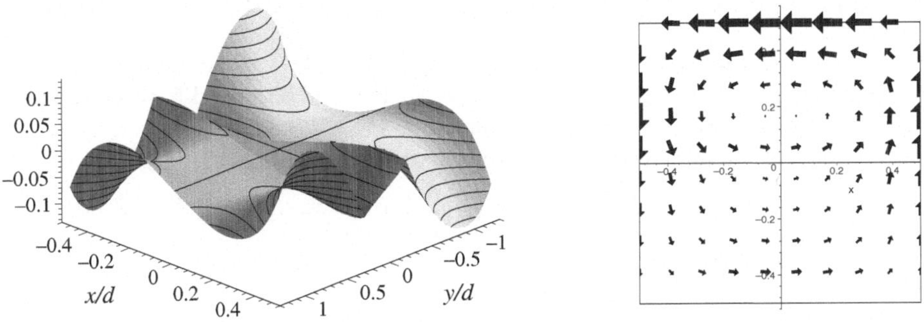

(a) φ.

(b) (τ_{xz}, τ_{yz})-vector.

Figure 8.10: (a) Warping function normalized by d^2, $N = 3$. (b) Stress vector field, $N = 2$.

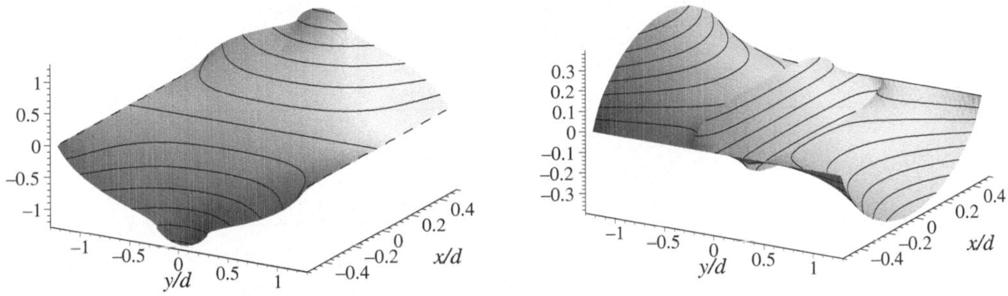

(a) τ_{yz}.

(b) τ_{xz}.

Figure 8.11: Stress components normalized by $\frac{\theta d}{a_{44}^0}$, $N = 3$.

8.2.7 Summarizing the Tip Loading Effects

We shall summarize here the so-called "St. Venant's Problem" in a non-homogeneous domain, namely, the effect of the three *tip forces* P_x, P_y, P_z, and three *tip moments* M_x, M_y, M_z, discussed in S.8.2.3–8.2.6. Verification of this solution appears in **P.8.4**.

The stress components are:

$$\begin{aligned}
\sigma_x &= \alpha_1\bar{\sigma}_x^{(1)} + \alpha_2\bar{\sigma}_x^{(2)} + \alpha_3\bar{\sigma}_x^{(3)} + \alpha_4(l-z)\bar{\sigma}_x^{(1)} + \alpha_5(l-z)\bar{\sigma}_x^{(2)}, \\
\sigma_y &= \alpha_1\bar{\sigma}_y^{(1)} + \alpha_2\bar{\sigma}_y^{(2)} + \alpha_3\bar{\sigma}_y^{(3)} + \alpha_4(l-z)\bar{\sigma}_y^{(1)} + \alpha_5(l-z)\bar{\sigma}_y^{(2)}, \\
\tau_{xy} &= \alpha_1\bar{\tau}_{xy}^{(1)} + \alpha_2\bar{\tau}_{xy}^{(2)} + \alpha_3\bar{\tau}_{xy}^{(3)} + \alpha_4(l-z)\bar{\tau}_{xy}^{(1)} + \alpha_5(l-z)\bar{\tau}_{xy}^{(2)}, \\
\sigma_z &= -\alpha_1\left(\frac{x}{a_{33}} - \bar{\sigma}_z^{(1)}\right) - \alpha_2\left(\frac{y}{a_{33}} - \bar{\sigma}_z^{(2)}\right) + \alpha_3\left(\frac{1}{a_{33}} + \bar{\sigma}_z^{(3)}\right) - \alpha_4(l-z)\left(\frac{x}{a_{33}} - \bar{\sigma}_z^{(1)}\right) \\
&\quad - \alpha_5(l-z)\left(\frac{y}{a_{33}} - \bar{\sigma}_z^{(2)}\right), \\
\tau_{yz} &= \frac{\alpha_4}{a_0}(Q^{\chi_1} + a_{45}\chi_{1,x} - a_{55}\chi_{1,y}) + \frac{\alpha_5}{a_0}(Q^{\chi_2} + a_{45}\chi_{2,x} - a_{55}\chi_{2,y}) - \frac{\alpha_6}{a_0}(Q^{\varphi} + a_{45}\varphi_{,x} - a_{55}\varphi_{,y}),
\end{aligned} \tag{8.93}$$

$$\tau_{xz} = \frac{\alpha_4}{a_0}(P^{\chi_1} - a_{44}\chi_{1,x} + a_{45}\chi_{1,y}) + \frac{\alpha_5}{a_0}(P^{\chi_2} - a_{44}\chi_{2,x} + a_{45}\chi_{2,y}) - \frac{\alpha_6}{a_0}(P^{\varphi} - a_{44}\varphi_{,x} + a_{45}\varphi_{,y}).$$

As shown, the above expressions are based on three non-trivial functions: the torsion function $\varphi(x,y)$, and the bending functions $\chi_1(x,y)$ and $\chi_2(x,y)$, that should be solved for each specific non-homogeneous cross-section geometry as discussed in S.8.2.5, S.8.2.6.

The strain components are obtained from the stress-strain relationships (7.1), as

$$\begin{aligned}
\varepsilon_x &= \alpha_1(-\frac{a_{13}}{a_{33}}x + \bar{\varepsilon}_x^{(1)}) + \alpha_2(-\frac{a_{13}}{a_{33}}y + \bar{\varepsilon}_x^{(2)}) + \alpha_3(\frac{a_{13}}{a_{33}} + \bar{\varepsilon}_x^{(3)}) - \alpha_4(l-z)(\frac{a_{13}}{a_{33}}x - \bar{\varepsilon}_x^{(1)}) \\
&\quad - \alpha_5(l-z)(\frac{a_{13}}{a_{33}}y - \bar{\varepsilon}_x^{(2)}), \\
\varepsilon_y &= \alpha_1(-\frac{a_{23}}{a_{33}}x + \bar{\varepsilon}_y^{(1)}) + \alpha_2(-\frac{a_{23}}{a_{33}}y + \bar{\varepsilon}_y^{(2)}) + \alpha_3(\frac{a_{23}}{a_{33}} + \bar{\varepsilon}_y^{(3)}) - \alpha_4(l-z)(\frac{a_{23}}{a_{33}}x - \bar{\varepsilon}_y^{(1)}) \\
&\quad - \alpha_5(l-z)(\frac{a_{23}}{a_{33}}y - \bar{\varepsilon}_y^{(2)}), \\
\gamma_{xy} &= \alpha_1(-\frac{a_{36}}{a_{33}}x + \bar{\gamma}_{xy}^{(1)}) + \alpha_2(-\frac{a_{36}}{a_{33}}y + \bar{\gamma}_{xy}^{(2)}) + \alpha_3(\frac{a_{36}}{a_{33}} + \bar{\gamma}_{xy}^{(3)}) - \alpha_4(l-z)(\frac{a_{36}}{a_{33}}x - \bar{\gamma}_{xy}^{(1)}) \\
&\quad - \alpha_5(l-z)(\frac{a_{36}}{a_{33}}y - \bar{\gamma}_{xy}^{(2)}), \\
\varepsilon_z &= \alpha_3 - \alpha_1 x - \alpha_2 y - \alpha_4(l-z)x - \alpha_5(l-z)y, \\
\gamma_{yz} &= \alpha_4[\frac{a_{36}x^2 + 2(a_{23} - 2a_{33})xy}{2a_{33}} - \chi_{1,y} - \bar{v}^{(1)}] + \alpha_5[\frac{a_{23}y^2 - (a_{13} + 2a_{33})x^2}{2a_{33}} - \chi_{2,y} - \bar{v}^{(2)}] \\
&\quad + \alpha_6(\varphi_{,y} + x), \\
\gamma_{xz} &= \alpha_4[\frac{a_{13}x^2 - (a_{23} + 2a_{33})y^2}{2a_{33}} - \chi_{1,x} - \bar{u}^{(1)}] + \alpha_5[\frac{a_{36}y^2 + 2(a_{13} - 2a_{33})xy}{2a_{33}} - \chi_{2,x} - \bar{u}^{(2)}] \\
&\quad + \alpha_6(\varphi_{,x} - y).
\end{aligned} \tag{8.94}$$

The corresponding displacements are

$$\begin{aligned}
u &= \alpha_1(-\frac{a_{13}}{2a_{33}}x^2 + \frac{a_{23}}{2a_{33}}y^2 + \frac{1}{2}z^2 + \bar{u}^{(1)}) + \alpha_2(-\frac{a_{13}}{a_{33}}xy - \frac{a_{36}}{2a_{33}}y^2 + \bar{u}^{(2)}) \\
&\quad + \alpha_3(\frac{a_{13}}{a_{33}}x + \frac{a_{36}}{2a_{33}}y + \bar{u}^{(3)}) + \alpha_4[(l-z)(-\frac{a_{13}}{2a_{33}}x^2 + \frac{a_{23}}{2a_{33}}y^2 + \bar{u}^{(1)}) + \frac{z^2}{2}(l - \frac{z}{3})] \\
&\quad + \alpha_5(l-z)(-\frac{a_{13}}{a_{33}}xy - \frac{a_{36}}{2a_{33}}y^2 + \bar{u}^{(2)}) - \alpha_6 yz + u^0 - \omega_z^0 y + \omega_y^0 z, \\
v &= \alpha_1(-\frac{a_{36}}{2a_{33}}x^2 - \frac{a_{23}}{a_{33}}xy + \bar{v}^{(1)}) + \alpha_2(\frac{a_{13}}{2a_{33}}x^2 - \frac{a_{23}}{2a_{33}}y^2 + \frac{1}{2}z^2 + \bar{v}^{(2)}) \\
&\quad + \alpha_3(\frac{a_{36}}{2a_{33}}x + \frac{a_{23}}{a_{33}}y + \bar{v}^{(3)}) + \alpha_4(l-z)(-\frac{a_{36}}{2a_{33}}x^2 - \frac{a_{23}}{a_{33}}xy + \bar{v}^{(1)}) \\
&\quad + \alpha_5[(l-z)(\frac{a_{13}}{2a_{33}}x^2 - \frac{a_{23}}{2a_{33}}y^2 + \bar{v}^{(2)}) + \frac{z^2}{2}(l - \frac{z}{3})] + \alpha_6 xz + v^0 - \omega_x^0 z + \omega_z^0 x, \\
w &= (\alpha_3 - \alpha_1 x - \alpha_2 y)z - \alpha_4[(l - \frac{z}{2})xz + xy^2 + \chi_1] - \alpha_5[(l - \frac{z}{2})yz + x^2 y + \chi_2] \\
&\quad + \alpha_6\varphi + w^0 - \omega_y^0 x + \omega_x^0 y,
\end{aligned} \tag{8.95}$$

where u^0, v^0, w^0, are rigid body displacements, and ω_x^0, ω_y^0, ω_z^0, are rigid body rotations.

The beam axis displacements and rotations are (excluding the rigid body terms)

$$\begin{aligned}
u_0 &= \frac{1}{2}\alpha_1 z^2 + \frac{1}{2}\alpha_4 z^2(l-z), \qquad v_0 = \frac{1}{2}\alpha_2 z^2 + \frac{1}{2}\alpha_5 z^2(l-z), \qquad w_0 = \alpha_3 z, \\
\omega_{x0} &= -\alpha_2 z + \alpha_5 z(\frac{3}{2}z - l), \qquad \omega_{y0} = \alpha_1 z - \alpha_4 z(\frac{3}{2}z - l), \qquad \omega_{z0} = \alpha_6 z.
\end{aligned} \tag{8.96}$$

Example 8.6 *A Non-Homogeneous Beam Under Hydrostatic Load.*

Similar to the above discussion of tip loads, one may adopt the solution presented in Example 3.5 for a prismatic body under hydrostatic load, $\widetilde{P}$. We *initially* assume that in the case of a non-homogeneous cross-section, the solution presented in (3.79) holds for each domain component separately. Therefore, stress continuity between domains will be maintained, but displacements continuity will not. To correct that, one should superimpose a new auxiliary (plane-strain) solution as described in S.8.2.2 (denoted by the superscript $()^{(p)}$), namely

$$u = -\widetilde{P}[(b_{11}+b_{12})x + \frac{1}{2}(b_{16}+b_{26})y + \overline{u}^{(p)}], \tag{8.97a}$$

$$v = -\widetilde{P}[(b_{22}+b_{12})y + \frac{1}{2}(b_{16}+b_{26})x + \overline{v}^{(p)}], \tag{8.97b}$$

while $w = 0$ in both (3.79) and the new auxiliary solution. The new auxiliary problem should be therefore based on

$$f_u^{(p)} = -[b_{11}+b_{12}]_{[i]}^{[j]} x - [b_{16}+b_{26}]_{[i]}^{[j]} \frac{y}{2}, \quad f_v^{(p)} = -[b_{16}+b_{26}]_{[i]}^{[j]} \frac{x}{2} - [b_{22}+b_{12}]_{[i]}^{[j]} y \tag{8.98}$$

which is essentially analogous to the third auxiliary problem, see (8.27c). The resulting nonzero stresses in this case are (see Example 3.5)

$$\sigma_x = -\widetilde{P}(1+\overline{\sigma}_x^{(p)}), \quad \sigma_y = -\widetilde{P}(1+\overline{\sigma}_y^{(p)}), \quad \tau_{xy} = -\widetilde{P}\overline{\tau}_{xy}^{(p)}, \quad \sigma_z = \widetilde{P}(\frac{a_{13}+a_{23}}{a_{33}} - \overline{\sigma}_z^{(p)}). \tag{8.99}$$

The above axial stress leaves the following tip resultants:

$$\{P_{zp}, M_{xp}, M_{yp}\} = \widetilde{P} \iint_{\Omega} (\frac{a_{13}+a_{23}}{a_{33}} - \overline{\sigma}_z^{(p)})\{1, y, -x\}. \tag{8.100}$$

Hence, to create a solution for a non-homogeneous beam under hydrostatic load with free tips, one needs to superimpose additional solutions for tip loads of magnitude $-P_{zp}$, $-M_{xp}$ and $-M_{yp}$ as summarized in S.8.2.7. To simulate the effect of "all-around uniform pressure" the above $-P_{zp}$ magnitude should be replaced by $-P_{zp} - \widetilde{P}S_{\Omega}$. Overall, in the general case, the present methodology shows that the hydrostatic problem requires a solution of four auxiliary (plane-strain) problems (three for the tip loads and the one discussed above).

8.2.8 *Fulfilling the Tip Conditions*

To fulfill the natural boundary conditions over the tip, one should determine the amplitudes α_i $(i = 1,\ldots,6)$ so that the integral conditions over the tip cross-section, namely, (5.7), will be satisfied. By adding up all contributions described above, we write

$$P_i = \sum_k P_{ik}, \qquad M_i = \sum_k M_{ik}, \qquad i = x, y, z. \tag{8.101}$$

Hence, one may construct the following two linear algebraic systems (see also (8.58)):

$$\begin{Bmatrix} M_y \\ M_x \\ P_z \end{Bmatrix} = \begin{bmatrix} \overline{I}_{11} & \overline{I}_{12} & \overline{I}_{13} \\ \overline{I}_{21} & \overline{I}_{22} & \overline{I}_{23} \\ \overline{I}_{31} & \overline{I}_{32} & \overline{I}_{33} \end{bmatrix} \begin{Bmatrix} \alpha_1 \\ \alpha_2 \\ \alpha_3 \end{Bmatrix}, \quad \begin{Bmatrix} P_x \\ P_y \\ M_z \end{Bmatrix} = \begin{bmatrix} \overline{I}_{11} & \overline{I}_{12} & 0 \\ -\overline{I}_{21} & -\overline{I}_{22} & 0 \\ -D_x & -D_y & D \end{bmatrix} \begin{Bmatrix} \alpha_4 \\ \alpha_5 \\ \alpha_6 \end{Bmatrix}. \tag{8.102}$$

The solution of the above systems yields the values of α_i $(i = 1,\ldots,6)$ for a given set of tip loads, P_x, P_y, P_z, M_x, M_y, M_z. As shown, the tip loads M_y, M_x and P_z will never induce any shear stresses, see (8.31), (8.36), and therefore will not influence the resultants P_x, P_y and M_z.

Similarly, the tip loads P_x, P_y and M_z will never induce any normal stress component σ_z at the tip cross-section ($z = l$), see (8.44b), (8.48b), and therefore will not influence the resultants M_y, M_x and P_z. Subsequently, although there are some mutual influences between loads in one direction and deformation in another direction (i.e. the off-diagonal terms do not necessarily vanish), similar to MON13z homogeneous beams discussed in Chapter 7, non-homogeneous MON13z beams are generally referred to as *uncoupled* as well.

When only components of the triad M_y, M_x and P_z are involved, $\alpha_4 = \alpha_5 = \alpha_6 = 0$ and the system origin location may be selected arbitrarily. However, when one of the P_x, P_y, M_z triad is involved, the system origin location should be selected so that (8.60), (8.61) are satisfied (and therefore $\bar{I}_{31}$ and $\bar{I}_{32}$ should vanish). Thus, in the general case, only the system orientation may be arbitrarily selected by convenience considerations. Typically, geometry simplicity dictates the system orientation, but there are other cases where simplification may be achieved by zeroing out an additional term in one of the matrices (with proper orientation selection).

The reader may verify that for the homogeneous cross-section case, the linear systems (8.102) become

$$\begin{Bmatrix} M_y \\ M_x \\ P_z \end{Bmatrix} = \begin{bmatrix} \frac{I_y}{a_{33}} & 0 & 0 \\ 0 & -\frac{I_x}{a_{33}} & 0 \\ 0 & 0 & \frac{S_\Omega}{a_{33}} \end{bmatrix} \begin{Bmatrix} \alpha_1 \\ \alpha_2 \\ \alpha_3 \end{Bmatrix}, \quad \begin{Bmatrix} P_x \\ P_y \\ M_z \end{Bmatrix} = \begin{bmatrix} \frac{I_y}{a_{33}} & 0 & 0 \\ 0 & \frac{I_x}{a_{33}} & 0 \\ -D_x & -D_y & D \end{bmatrix} \begin{Bmatrix} \alpha_4 \\ \alpha_5 \\ \alpha_6 \end{Bmatrix}, \tag{8.103}$$

and subsequently, (8.93) are reduced to (7.72a–d).

8.2.9 *Principal Axis of Extension and Principal Planes of Bending*

To facilitate the following derivation, we examine the inverse form of the first of (8.102) that relates M_y, M_x, P_z and $\alpha_1, \alpha_2, \alpha_3$. As shown by (8.96), this relation may also be written by the tip axial displacements and transverse rotations at the elastic axis, using inverse matrix $\tilde{I} (= \bar{I}^{-1})$, as

$$\begin{Bmatrix} \alpha_1 \\ \alpha_2 \\ \alpha_3 \end{Bmatrix} = \frac{1}{l} \begin{Bmatrix} \omega_{y0}(l) \\ \omega_{x0}(l) \\ w_0(l) \end{Bmatrix} = \begin{bmatrix} \tilde{I}_{11} & \tilde{I}_{12} & \tilde{I}_{13} \\ \tilde{I}_{21} & \tilde{I}_{22} & \tilde{I}_{23} \\ \tilde{I}_{31} & \tilde{I}_{32} & \tilde{I}_{33} \end{bmatrix} \begin{Bmatrix} M_y \\ M_x \\ P_z \end{Bmatrix}. \tag{8.104}$$

For a complete separation of the solutions involved (described in S.8.2.3 through S.8.2.4), the off-diagonal terms should vanish. Indeed, for a given geometry, one may try to find a location and orientation for the coordinate system that will cause (all or part of) the off-diagonal terms to vanish. By doing so, we define the *principal axis of extension and the principal planes of bending*.

We shall first look at the location and orientation of the coordinate system in which one may apply an axial force, P_z, that will induce zero lateral rotation of the elastic axis tip, namely, $\omega_{x0}(l) = \omega_{y0}(l) = 0$. Accordingly, one should require $\tilde{I}_{13} = \tilde{I}_{23} = 0$. In a similar manner, to ensure that M_x induces only $\omega_{x0}(l)$, one should require $\tilde{I}_{12} = \tilde{I}_{32} = 0$, and to ensure that M_y induces only $\omega_{y0}(l)$, one should require $\tilde{I}_{21} = \tilde{I}_{31} = 0$. Fundamentally, for a given geometry, (up to) three conditions may be imposed on the system origin location and orientation. Thus, each of the above three pairs of conditions can be implemented separately (with an extra arbitrary degree of freedom). In special cases, symmetry properties may cause the fulfillment of more than three conditions simultaneously.

It should also be noted that the conditions $\bar{I}_{31} = \bar{I}_{32} = 0$ are sufficient for providing the state of $\tilde{I}_{31} = \tilde{I}_{32} = 0$ discussed above.

8.2.10 Shear Center

By inverting the second matrix of (8.102) one may write (see also (8.96) with $\phi \equiv \omega_{z0}$),

$$\begin{Bmatrix} \alpha_4 \\ \alpha_5 \\ \alpha_6 \end{Bmatrix} = \frac{1}{l} \begin{Bmatrix} \frac{3}{l^2} u_0(l) \\ \frac{3}{l^2} v_0(l) \\ \phi(l) \end{Bmatrix} = \frac{1}{Det} \begin{bmatrix} -\bar{I}_{22} D & -\bar{I}_{12} D & 0 \\ \bar{I}_{21} D & \bar{I}_{11} D & 0 \\ \bar{I}_{21} D_y - \bar{I}_{22} D_x & \bar{I}_{11} D_y - \bar{I}_{12} D_x & \bar{I}_{12}\bar{I}_{21} - \bar{I}_{22}\bar{I}_{11} \end{bmatrix} \begin{Bmatrix} P_x \\ P_y \\ M_z \end{Bmatrix}, \quad (8.105)$$

where $Det = D(\bar{I}_{12}\bar{I}_{21} - \bar{I}_{22}\bar{I}_{11})$. In this case, the existence conditions for the bending functions derived in S.8.2.5.1 show that the system origin must be located at the point where (8.60), (8.61) are fulfilled.

It is first evident that *torsional moment does not induce any transverse bending*. Suppose now that we apply transverse loads P_x and P_y at $x = x_{sc}$ and $y = y_{sc}$. We are subsequently able to replace the above loads with two identical P_x and P_y loads that act at the origin along with additional torsional moment of magnitude $P_y x_{sc} - P_x y_{sc}$. The tip twist angle will therefore be

$$\frac{1}{l}\phi(l) = \frac{1}{Det}[(\bar{I}_{21} D_y - \bar{I}_{22} D_x)P_x + (\bar{I}_{11} D_y - \bar{I}_{12} D_x)P_y + (\bar{I}_{12}\bar{I}_{21} - \bar{I}_{22}\bar{I}_{11})(P_y x_{sc} - P_x y_{sc})], \quad (8.106)$$

which in the general case will vanish only if

$$y_{sc} = -\frac{\bar{I}_{22} D_x - \bar{I}_{21} D_y}{\bar{I}_{12}\bar{I}_{21} - \bar{I}_{22}\bar{I}_{11}}, \qquad x_{sc} = \frac{\bar{I}_{12} D_x - \bar{I}_{11} D_y}{\bar{I}_{12}\bar{I}_{21} - \bar{I}_{22}\bar{I}_{11}}. \quad (8.107)$$

These two coordinates comply with the classical definition of the *shear center*, as *the location in which one may apply transverse loads without creating a twist* ($\bar{I}_{12}\bar{I}_{21} - \bar{I}_{22}\bar{I}_{11}$ is always positive as it represents the inverse of the beam's torsional rigidity).

8.2.11 Solution Procedure

1. Extension and bending

To determine the effect of the P_z, M_x and M_y triad, the system origin location may be selected arbitrarily and the following procedure is adopted:

(1a) Solve the three auxiliary problems of plane deformation of S.8.2.2.

(1b) Calculate $\bar{I}_{ij}$ from (8.33), (8.41a,b).

(1c) Calculate $\alpha_1, \alpha_2, \alpha_3$ from (8.104).

(1d) Calculate the stress, strain and displacement components as shown in S.8.2.7 (while $\alpha_4 = \alpha_5 = \alpha_6 = 0$).

2. Shear and torsion

To determine the effect of the P_x, P_y, M_z triad, the following procedure is adopted:

(2a) Select system orientation (i.e. the axes direction in the x, y-plane).

(2b) Assume initial value ("guess") for the two coordinates of the system origin location. A reasonable guess is the one that emerges from (8.60), (8.61) by assuming $\overline{\sigma}_z^{(1)} = \overline{\sigma}_z^{(2)} = 0$.

(2c) Solve the three auxiliary problems of plane deformation of S.8.2.2.

Note that *for a y-"laminated"* rectangle, where $y = h_j$ on $\partial\Omega_{j+1,j}$, see e.g. Fig. 8.3(a), the BVP (and hence the stresses) in the second and third auxiliary problems depend on the y and x coordinates of the system origin, respectively. This is due to the fact that, as shown in (8.14) of Remark 8.2, one may weaken the BVP by differentiation of f_u and f_v, once and twice, respectively, with respect to x. Therefore, we replace (8.8d) with one of the following:

$$\{\overline{u}^{(1)}_{,x}, \overline{v}^{(1)}_{,xx}\}^{[j+1]}_{[j]} = \{\frac{a_{13}}{a_{33}}x, \frac{a_{36}}{a_{33}}\}^{[j+1]}_{[j]}, \qquad 0 < j < N, \tag{8.108a}$$

$$\{\overline{u}^{(2)}_{,x}, \overline{v}^{(2)}_{,xx}\}^{[j+1]}_{[j]} = [\frac{a_{13}}{a_{33}}]^{[j+1]}_{[j]}\{y, -1\}, \qquad 0 < j < N, \tag{8.108b}$$

$$\{\overline{u}^{(3)}_{,x}, \overline{v}^{(3)}_{,xx}\}^{[j+1]}_{[j]} = \{-\frac{a_{13}}{a_{33}}, 0\}^{[j+1]}_{[j]}, \qquad 0 < j < N. \tag{8.108c}$$

(2d) Modify the two coordinates of the system origin location so that (8.60), (8.61) will be satisfied.

(2e) Compare coordinates of the system origin obtained in (2d) with those used in (2c). If different, update the values and re-execute (2c) and (2d). Note that there are many cases where step (2c) does not affect the system origin coordinate, and no iterations are needed. The above updating procedure has to be determined for each case separately, although a simple replacement of the assumed coordinates with the calculated pair is probably sufficient in most cases. As already indicated, the $\overline{\sigma}^{(i)}_z$ terms in (8.60), (8.61) are relatively small.

(2f) Calculate D, D_x, D_y of (7.11), (7.60), (7.66), respectively, and $\overline{I}_{ij}$ of (8.41a,b).

(2g) Calculate $\alpha_4, \alpha_5, \alpha_6$ from (8.105).

(2h) Calculate the stress, strain and displacement components as shown in S.8.2.7 (while $\alpha_1 = \alpha_2 = \alpha_3 = 0$).

The above solution is illustrated by **P.8.5**.

8.3 Uncoupled Beam Under Axially Distributed Loads

In this section, we extend the derivation of S.7.3 for MON13z beams of non-homogeneous cross-sections under distributed loads. Special symbolic linking of the derivation to that of Chapter 7 enables us to present the following analysis in a relatively concise manner. However, it is highly recommended that the reader first become familiar with the derivation of S.7.3 and the previous parts of this chapter.

8.3.1 The Solution Hypothesis

To derive the solution hypothesis in this case we first adopt the notation of (6.1) to express the body and surface loads as $K^{\underline{th}}$-degree polynomials. We also adopt the solution methodology of S.7.3.1, and hence Table 7.1 and Fig. 7.5 are still valid with the exception of $\sigma_x, \sigma_y, \tau_{xy}$ the highest level of which is now $K+2$. The entire solution presented in what follows is *symbolically* verified in **P.8.6** and **P.8.7**.

The solution hypothesis contains an assumed form of the stress components. For that purpose, we use (7.81a–f) by denoting them with an asterisk, namely, $(\sigma_x)^*$, $(\sigma_y)^*$, etc. Note that *we only adopt the analytical expressions* while the involved terms and functions may be different in the present case. The solution hypothesis is founded on the following stress expressions:

$$\sigma_x = (\sigma_x)^* + \sum_{k=0}^{K} \frac{z^{k+1}}{k+1} [\frac{z}{k+2} (p_k \overline{\sigma}_x^{(1)} + q_k \overline{\sigma}_x^{(2)}) - d_k \overline{\sigma}_x^{(3)}], \tag{8.109a}$$

$$\sigma_y = (\sigma_y)^* + \sum_{k=0}^{K} \frac{z^{k+1}}{k+1} [\frac{z}{k+2} (p_k \overline{\sigma}_y^{(1)} + q_k \overline{\sigma}_y^{(2)}) - d_k \overline{\sigma}_y^{(3)}], \tag{8.109b}$$

$$\sigma_z = (\sigma_z)^* + \sum_{k=0}^{K} \frac{z^{k+1}}{k+1} [\frac{z}{k+2} (p_k \overline{\sigma}_z^{(1)} + q_k \overline{\sigma}_z^{(2)}) - d_k \overline{\sigma}_z^{(3)}], \tag{8.109c}$$

$$\tau_{yz} = (\tau_{yz})^*, \qquad \tau_{xz} = (\tau_{xz})^*, \tag{8.109d}$$

$$\tau_{xy} = (\tau_{xy})^* + \sum_{k=0}^{K} \frac{z^{k+1}}{k+1} [\frac{z}{k+2} (p_k \overline{\tau}_{xy}^{(1)} + q_k \overline{\tau}_{xy}^{(2)}) - d_k \overline{\tau}_{xy}^{(3)}]. \tag{8.109e}$$

To clarify this writing technique, as an example, the term p_k that appears in the stress expression for $(\sigma_z)^*$, see (7.81c) and in the above expressions, takes **different** values in the present non-homogeneous case, as will further be shown by (8.121a).

The body-force potentials $\overline{U}_1^{(k)}$, $\overline{U}_2^{(k)}$ that appear in the above $(\,)^*$ terms are defined for $k = 0, \ldots, K$ as in (7.78).

The expressions for $S_x^{(k)}$ and $S_y^{(k)}$ $(k = 0, \ldots, K+1)$ that are included in $(\sigma_x)^*$, $(\sigma_y)^*$, $(\sigma_z)^*$, $(\tau_{yz})^*$, $(\tau_{xz})^*$, see (7.81a–e), take the form

$$S_x^{(k)} = (S_x^{(k)})^* - d_k \int_0^x (a_{44} \overline{u}^{(3)} - a_{45} \overline{v}^{(3)}) dx, \tag{8.110a}$$

$$S_y^{(k)} = (S_y^{(k)})^* - d_k \int_0^y (a_{55} \overline{v}^{(3)} - a_{45} \overline{u}^{(3)}) dy. \tag{8.110b}$$

Here, the components denoted by asterisk are those of (7.83a,b). In these expressions, P^{χ_1}, Q^{χ_1} are those of (8.45) while P^{χ_2}, Q^{χ_2} are defined in (8.49).

Note that in (8.109a–f) and in all expressions derived in what follows, we replace $H^{(k)}$ by the r.h.s. of (7.88) as required by the single-value conditions for the present problem.

8.3.2 The Strain Components

Based on the above assumed stress expressions, the strain components are derived from the strain-stress relationships (see (7.1)) as

$$\varepsilon_x = (\varepsilon_x)^* + \sum_{k=0}^{K} \frac{z^{k+1}}{k+1} [\frac{z}{k+2} (p_k \overline{\varepsilon}_x^{(1)} + q_k \overline{\varepsilon}_x^{(2)}) - d_k \overline{\varepsilon}_x^{(3)}], \tag{8.111a}$$

$$\varepsilon_y = (\varepsilon_y)^* + \sum_{k=0}^{K} \frac{z^{k+1}}{k+1} [\frac{z}{k+2} (p_k \overline{\varepsilon}_y^{(1)} + q_k \overline{\varepsilon}_y^{(2)}) - d_k \overline{\varepsilon}_y^{(3)}], \tag{8.111b}$$

$$\varepsilon_z = (\varepsilon_z)^*, \tag{8.111c}$$

$$\gamma_{yz} = (\gamma_{yz})^* + \sum_{k=0}^{K} z^k [\frac{z}{k+1} (p_k \overline{v}^{(1)} + q_k \overline{v}^{(2)}) - d_k \overline{v}^{(3)}], \tag{8.111d}$$

$$\gamma_{xz} = (\gamma_{xz})^* + \sum_{k=0}^{K} z^k [\frac{z}{k+1} (p_k \overline{u}^{(1)} + q_k \overline{u}^{(2)}) - d_k \overline{u}^{(3)}], \tag{8.111e}$$

$$\gamma_{xy} = (\gamma_{xy})^* + \sum_{k=0}^{K} \frac{z^{k+1}}{k+1} [\frac{z}{k+2} (p_k \overline{\gamma}_{xy}^{(1)} + q_k \overline{\gamma}_{xy}^{(2)}) - d_k \overline{\gamma}_{xy}^{(3)}]. \tag{8.111f}$$

Note that in the above, the components denoted by asterisk are those of (7.104a–f). The equalities $\varepsilon_M^{(k)} = M_{,y}^{(k)}$, $\varepsilon_L^{(k)} = L_{,x}^{(k)}$ hold in this case as well, while $\gamma_{LM}^{(k)}$ that appear in $(\gamma_{xy})^*$ are defined by (7.106).

8.3.3 Displacements and Rotations

To this end, the integration scheme of S.1.2 may be used to determine the displacements via integration of (8.111a–f). The resulting displacement components are (the underlined terms are further discussed in S.8.3.8:**(d)**)

$$u = (u)^* + \sum_{k=0}^{K} \frac{z^{k+1}}{k+1}[\underline{\frac{z}{k+2}(p_k\bar{u}^{(1)} + q_k\bar{u}^{(2)})} - \underline{\underline{d_k\bar{u}^{(3)}}}], \tag{8.112a}$$

$$v = (v)^* + \sum_{k=0}^{K} \frac{z^{k+1}}{k+1}[\underline{\frac{z}{k+2}(p_k\bar{v}^{(1)} + q_k\bar{v}^{(2)})} - \underline{\underline{d_k\bar{v}^{(3)}}}], \tag{8.112b}$$

$$w = (w)^*. \tag{8.112c}$$

Here, the components denoted by asterisk are those of (7.107a–c). The rotations are

$$\omega_x = (\omega_x)^* - \frac{1}{2}\sum_{k=0}^{K} z^k[\frac{z}{k+1}(p_k\bar{v}^{(1)} + q_k\bar{v}^{(2)}) - d_k\bar{v}^{(3)}], \tag{8.113a}$$

$$\omega_y = (\omega_y)^* + \frac{1}{2}\sum_{k=0}^{K} z^k[\frac{z}{k+1}(p_k\bar{u}^{(1)} + q_k\bar{u}^{(2)}) - d_k\bar{u}^{(3)}], \tag{8.113b}$$

$$\omega_z = (\omega_z)^* + \sum_{k=0}^{K} \frac{z^{k+1}}{k+1}[\frac{z}{k+2}(p_k\bar{\omega}_z^{(1)} + q_k\bar{\omega}_z^{(2)}) - d_k\bar{\omega}_z^{(3)}], \tag{8.113c}$$

where the components denoted by asterisk are those of (7.109a–c).

Rigid body displacements and rotations are not included in the above expressions and are introduced by the tip loads correction discussed in S.8.3.9.

8.3.4 The Biharmonic Stress Functions

In the non-homogeneous case, $\Phi^{(k)}(x,y)$ $(k = 0,\ldots,K)$, are piecewise generalized biharmonic stress functions. Following S.8.1.3, the biharmonic BVP for each level is formulated as

$$\nabla_1^{(4)}\Phi^{(k)} = F_0^{(k)} \qquad \text{over} \quad \Omega, \tag{8.114a}$$

$$\frac{d}{ds}\{\Phi_{,x}^{(k)}, \Phi_{,y}^{(k)}\} = \{-F_1^{(k)}, F_2^{(k)}\} \qquad \text{on} \quad \partial\Omega, \tag{8.114b}$$

$$\frac{d}{ds}\{\Phi_{,x}^{(k)}, \Phi_{,y}^{(k)}\}_{[i]}^{[j]} = \{-F_1^{(k)}, F_2^{(k)}\}_{[i]}^{[j]} \qquad \text{on} \quad \partial\Omega_{ij}, \tag{8.114c}$$

$$\{L^{(k)}, M^{(k)}\}_{[i]}^{[j]} = \{0, 0\} \qquad \text{on} \quad \partial\Omega_{ij}. \tag{8.114d}$$

Here $F_i^{(k)} = (F_i^{(k)})^*$ for $i = 1,2,3$ and $k = 0,\ldots,K$, see (7.86), (7.101). For convenience we assume $\Phi^{(k)}(0,0) = \Phi_{,x}^{(k)}(0,0) = \Phi_{,y}^{(k)}(0,0) = 0$.

8.3.5 The Harmonic and Longitudinal Stress Functions

The longitudinal stress functions, $\omega^{(k)}(x,y)$ $(k = 0,\ldots,K)$, in the non-homogeneous domain are governed by

$$\nabla_3^{(2)}\omega^{(k)} = F_0^{\omega^{(k)}} \qquad \text{over} \quad \Omega, \tag{8.115a}$$

$$D_1^n\omega^{(k)} = P^{\omega^{(k)}}\cos(\bar{\mathbf{n}}, x) + Q^{\omega^{(k)}}\cos(\bar{\mathbf{n}}, y) \qquad \text{on} \quad \partial\Omega, \tag{8.115b}$$

$$[\frac{1}{a_0}D_1^n\omega^{(k)}]_{[i]}^{[j]} = [\frac{1}{a_0}(P^{\omega^{(k)}}\cos(\bar{\mathbf{n}}, x) + Q^{\omega^{(k)}}\cos(\bar{\mathbf{n}}, y))]_{[i]}^{[j]} \qquad \text{on} \quad \partial\Omega_{ij}, \tag{8.115c}$$

$$[\omega^{(k)}]_{[i]}^{[j]} = F^{\omega^{(k)},ij} \qquad \text{on} \quad \partial\Omega_{ij}, \tag{8.115d}$$

where

$$F_0^{\omega^{(k)}} = (F_0^{\omega^{(k)}})^* + d_k(a_{44}\bar{\varepsilon}_x^{(3)} + a_{55}\bar{\varepsilon}_y^{(3)} - a_{45}\bar{\gamma}_{xy}^{(3)} + a_0\bar{\sigma}_z^{(3)}), \tag{8.116}$$

and

$$P^{\omega^{(k)}} = (P^{\omega^{(k)}})^* + d_k(a_{44}\bar{u}^{(3)} - a_{45}\bar{v}^{(3)}), \quad Q^{\omega^{(k)}} = (Q^{\omega^{(k)}})^* + d_k(a_{55}\bar{v}^{(3)} - a_{45}\bar{u}^{(3)}). \tag{8.117}$$

The expressions for $(F_0^{\omega^{(k)}})^*$, $(P^{\omega^{(k)}})^*$, $(Q^{\omega^{(k)}})^*$ are given in (7.102a,b), (7.103a–c) and

$$F^{\omega^{(k)},ij} = -d_k\{\frac{1}{2a_{33}}[(a_{13}+\frac{a_{55}}{2})x^2 + (a_{45}+a_{36})xy + (a_{23}+\frac{a_{44}}{2})y^2]\}_{[i]}^{[j]}. \tag{8.118}$$

For convenience we assume $\omega^{(k)}(0,0) = 0$. Note that all additional (to the homogeneous case) expressions are proportional to d_k, and thus, similar to the homogeneous case (see S.7.3.3, 7.3.4), $\omega^{(k)} = 0$ for $k > K$. The harmonic stress functions, $H^{(k)}$, are then determined by (7.88).

8.3.6 The Auxiliary Functions

We define the functions as $L^{(k)}(x,y) = (L^{(k)})^*$, $M^{(k)}(x,y) = (M^{(k)})^*$ $(k = 0,\dots,K)$. Here, the components denoted by asterisk are those of (7.96a, b), while the functions $l^{(k)}(y)$ and $m^{(k)}(x)$ take in the present case the form

$$\begin{aligned} l^{(k)}(y) = (l^{(k)})^* + \{&\frac{b_{26}}{a_0}\int_0^y (a_{55}\bar{v}_k - a_{45}\bar{u}_k)dy - \frac{b_{22}}{a_0}\int_0^y(\int_0^y (a_{55}\bar{v}_k - a_{45}\bar{u}_k)_{,x}dy)\,dy \\ &- \frac{b_{12}}{a_0}\int_0^y (a_{44}\bar{u}_k - a_{45}\bar{v}_k)dy\}_{|x=0}, \end{aligned} \tag{8.119a}$$

$$\begin{aligned} m^{(k)}(x) = (m^{(k)})^* + \{&\frac{b_{16}}{a_0}\int_0^x (a_{44}\bar{u}_k - a_{45}\bar{v}_k)dx - \frac{b_{11}}{a_0}\int_0^x(\int_0^x (a_{44}\bar{u}_k - a_{45}\bar{v}_k)_{,y}dx)\,dx \\ &- \frac{b_{12}}{a_0}\int_0^x (a_{55}\bar{v}_k - a_{45}\bar{u}_k)dx\}_{|y=0}. \end{aligned} \tag{8.119b}$$

In the above, the components denoted by asterisk are those of (7.97a, b), and for efficient writing, we have introduced the notation

$$\bar{u}_k = p_k\bar{u}^{(1)} + q_k\bar{u}^{(2)} - (k+1)d_{k+1}\bar{u}^{(3)}, \qquad \bar{v}_k = p_k\bar{v}^{(1)} + q_k\bar{v}^{(2)} - (k+1)d_{k+1}\bar{v}^{(3)}. \tag{8.120}$$

8.3.7 The Loading Constants

The loading constants, p_k, q_k, τ_k, d_k $(k = 0,\dots,K)$, are defined by

$$\begin{aligned} \bar{I}_{11}p_k - \bar{I}_{12}q_k = \oint_{\partial\Omega} X_s^{(k)} + \iint_\Omega \{&X_b^{(k)} + \frac{k+1}{a_0}[a_{44}\omega_{,x}^{(k+1)} - a_{45}\omega_{,y}^{(k+1)} \\ &- (k+2)(a_{44}L^{(k+2)} - a_{45}M^{(k+2)}) + d_{k+1}(\frac{a_0x}{2a_{33}} - a_{44}\bar{u}^{(3)} + a_{45}\bar{v}^{(3)})]\}, \end{aligned} \tag{8.121a}$$

$$\begin{aligned} \bar{I}_{22}q_k - \bar{I}_{21}p_k = \oint_{\partial\Omega} Y_s^{(k)} + \iint_\Omega \{&Y_b^{(k)} + \frac{k+1}{a_0}[a_{55}\omega_{,y}^{(k+1)} - a_{45}\omega_{,x}^{(k+1)} \\ &- (k+2)(a_{55}M^{(k+2)} - a_{45}L^{(k+2)}) + d_{k+1}(\frac{a_0y}{2a_{33}} - a_{55}\bar{v}^{(3)} + a_{45}\bar{u}^{(3)})]\}, \end{aligned} \tag{8.121b}$$

$$\tau_k = (\tau_k)^* + (k+1)\frac{d_{k+1}}{D}\iint_\Omega \frac{1}{a_0}[(a_{44}\bar{u}^{(3)} - a_{45}\bar{v}^{(3)})y - (a_{55}\bar{v}^{(3)} - a_{45}\bar{u}^{(3)})x], \tag{8.121c}$$

$$\bar{I}_{33}d_k = \oint_{\partial\Omega} Z_s^{(k)} + \iint_{\Omega} \{Z_b^{(k)} + \frac{k+1}{a_{33}}[-a_{13}(\Phi_{,yy}^{(k+1)} + \overline{U}_1^{(k+1)}) - a_{23}(\Phi_{,xx}^{(k+1)} + \overline{U}_2^{(k+1)})$$
$$+ a_{36}\Phi_{,xy}^{(k+1)} + (p_{k+1}y + q_{k+1}x)xy + (k+2)(\frac{a_{13}a_{44} + a_{23}a_{55} - a_{36}a_{45} + a_0}{a_0}H^{(k+2)}$$
$$+ \frac{a_{13}}{a_0}S_x^{(k+2)} + \frac{a_{23}}{a_0}S_y^{(k+2)}) + (k+2)d_{k+2}\frac{a_{45} + a_{36}}{2}xy]\}, \qquad (8.121d)$$

where $\bar{I}_{33}$ is defined by (8.33), $\bar{I}_{22}, \bar{I}_{11}, \bar{I}_{12}$ and $\bar{I}_{21}$ are defined by (8.41a,b), and D is the torsional rigidity, see S.8.2.6. Note that in the general case, (8.121a,b) constitute a linearly system.

It should be stressed again that in the above writing, the quantities $(\tau_k)^*$ are symbolically equal to τ_k of (7.98).

8.3.8 *Verification of Solution Hypothesis*

The solution verification for the non-homogeneous case is similar to the one presented in S.7.3.5 for the homogeneous case. Since the expressions are given as a symbolic superposition of the "homogeneous case expressions" and "additional terms", only these additional terms will be discussed in what follows (except for the displacements where all terms should be considered simultaneously). Yet, the reader should realize that the notion "additional terms" includes *all* differences between the homogeneous and the non-homogeneous formulations. For example, the additional terms in σ_x are not just those that are added to $(\sigma_x)^*$ in (8.109a), but *all* additional terms that emerge from the different definition of the ingredients of $(\sigma_x)^*$.

(a) Equilibrium Equations. Considering the additional stress terms of (8.109e), and using the fact that the solutions of the auxiliary problems satisfy the equilibrium equations, one may verify that the equilibrium equations (1.82a, b), are satisfied identically,

From the third equilibrium equation, (1.82c), the additional (to the homogeneous case) terms in (8.51), and the additional terms to $F_0^{\omega(k)}$ $(k=0,\ldots,K)$ in (8.116), are extracted as products of the terms $p_k z^{k+1}$, $q_k z^{k+1}$ and a "free term" z^{k+1}, respectively, and yield the equations $\nabla_3^{(2)}\chi_1 = F_0^{\chi_1}$, $\nabla_3^{(2)}\chi_2 = F_0^{\chi_2}$, and $\nabla_3^{(2)}\omega^{(k)} = F_0^{\omega(k)}$, documented in (8.50), (8.115a).

(b) Boundary Conditions. From the outer contour condition of (5.16a) one may deduce the additional terms for the derivatives $\frac{d}{ds}\Phi_{,y}^{(k)}$ $(k=0,\ldots,K)$, over the contours $\partial\Omega$ and $\partial\Omega_{ij}$, see (8.114b). Similarly, from the outer contour condition of (5.16b), the additional terms for the derivatives $\frac{d}{ds}\Phi_{,x}^{(k)}$ are deduced. Interface conditions for the stress function derivatives, (8.114c), are simple corollary of the above.

From the third boundary condition, (5.16c), the additional terms for the normal derivatives $D_1^n\chi_1$ $D_1^n\chi_2$ and $D_1^n\omega^{(k)}$ derived in S.8.2.5, S.8.3.5 are obtained as the products of $p_k z^{k+1}$, $q_k z^{k+1}$ and z^{k+1}, respectively.

(c) Compatibility Equations. The additional terms of the strain components satisfy the compatibility equations due to the fact that the solutions of the auxiliary problems are consistent and inherently satisfy the compatibility equations.

(d) Displacements interface continuity.

The in-plane displacement components: Continuity for levels $k > K+2$, see (7.107a–c), follows from the fact that loading constants of S.8.3.7 are not domain dependent (i.e. constants over the entire non-homogeneous domain, Ω).

Displacements continuity for levels $k = K+2, K+1$ is achieved by the once and twice underlined terms of (8.112a,b), respectively, which cancel out the discontinuity of $(u)^*$, $(v)^*$.

For levels $k \leq K$ displacements continuity is part of the biharmonic BVP, since by explicit use of (8.112a,b) and (7.107a,b), the interface conditions of (8.114d) may also be written as

$$\{u^{(k)}, v^{(k)}\}_{[i]}^{[j]} = \{0, 0\} \qquad \text{on} \quad \partial\Omega_{ij}. \tag{8.122}$$

The out-of-plane displacement component:

Continuity of $w = (w)^*$ may be verified by examining (7.107c). This equation shows that the coefficients of z^{k+1}, z^{k+2} and z^{k+3} ($k = 0, \ldots, K$) do not contribute any discontinuity due to the fact that the loading constants are not domain dependent. For lower levels this continuity condition is imposed by (8.115d) as part of the Neumann BVP for $\omega^{(k)}$ when $H^{(k)}$ is replaced by the r.h.s. of (7.88). Note that this condition contains the functions χ_1, χ_2 and φ, which are all continuous along the interface.

(e) Singled-Value Conditions. We again exploit the single-valued conditions (8.9a) (actually (7.95) for each domain component) for $\Phi^{(k)}(x,y)$ and its first derivatives. These three conditions yield the expressions for the constants q_k, p_k and τ_k, respectively, shown in S.8.3.7.

(f) Existence of the longitudinal stress function. The existence condition of a generalized harmonic problem is given by (8.4) (actually (7.95) for each domain component) (3.125). Applying this condition to $\omega^{(k)}(x,y)$, namely, $\oint_{\partial\Omega} D^n \omega^{(k)} = \iint_{\Omega} \nabla_3^{(2)} \omega^{(k)}$, yields the definition of the constants d_k of S.8.3.7.

8.3.9 Solution Procedure

The solution procedure in this case is identical to the one presented in S.7.3.6.

Similar to the homogeneous case, the solution presented above does not ensure that the three forces and three moments at the beam tip vanish, see stress integral of (5.7). Hence, one needs to superimpose a series of solutions for tip loads (see S.8.2.7) in order to cancel out these resultants. These additional loads will clearly contribute to the strain and displacements as well and include rigid body effects.

8.4 Applications

Example 8.7 *The Effect of Tip Bending Forces.*

To determine the effect of transverse tip forces on a beam with the non-homogeneous cross-section shown in Fig. 8.3(b), we activate **P.8.5** and employ the auxiliary solutions of Example 8.2, and the bending and warping functions of Examples 8.3, 8.4. We therefore remain with the same configuration of $NF = 2$, $d = h = 1$, $t_z = -30°, 30°$, and in addition select $P_x = 1$.

The resulting σ_x and σ_y stress components are demonstrated in Figs. 8.12(a),(b), respectively. As shown, the boundary conditions are clearly satisfied as $\sigma_x = 0$ at $x = \pm\frac{d}{2}$ and $\sigma_y = 0$ at $y = \pm\frac{h}{2}$. Examination of the interlaminar line ($y = 0$) shows that σ_y rises towards the edges $x = \pm\frac{d}{2}$. The presented compressible stress will become a tensile one for $P_x = -1$, and thus, this type of free-edge interlaminar normal stress has the potential to cause an initiation of delamination phenomena. The above discussed phenomena are generally termed as "free edge stress singularity", see e.g. (Pagano and Schoeppner, 2000). To further explore this point, Figs. 8.13(a),(b) present the stress σ_y as a function of k_Φ (while $k_\Lambda = k_\Phi - 2$). As shown by Fig. 8.13(a), for a point that is not on the interface, a convergence to a stationary value (~ 0.038) is achieved. On the other hand, for a point that is on the interface, increasing values

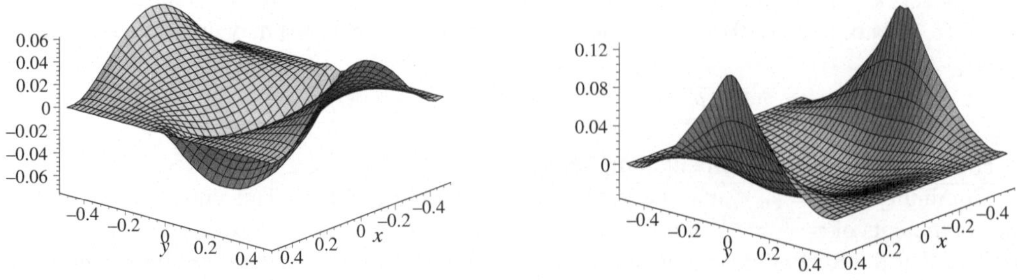

(a) $\sigma_x(x,y)$. (b) $\sigma_y(x,y)$.

Figure 8.12: Stress components due to P_x tip load in Example 8.7 ($k_\Phi = 25$, $l = 10$, $z = 5$).

are obtained. This tendency may be approximated in this case by a function that is proportional (for example) to $\sim k_\Phi^{0.064}$ as shown by the thick line in Fig. 8.13(b). Note that in most of the standard analyses, this singularity is presented as increasing value of stress for decreasing distance from the free edge. One may also examine the existence conditions of (8.60), (8.61)

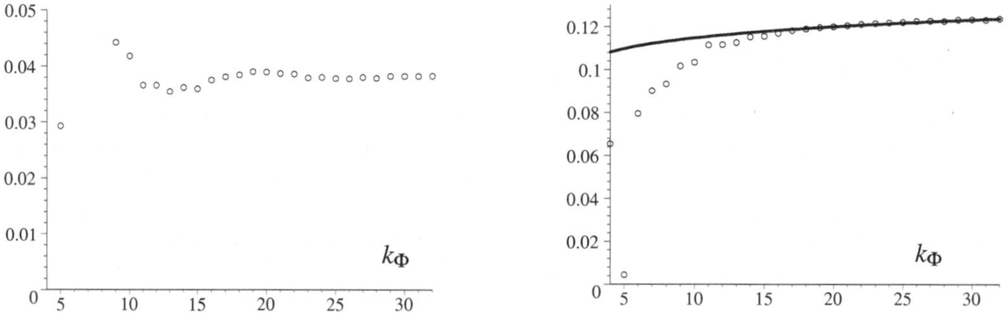

(a) $\sigma_y(\pm\frac{d}{2}, -\frac{h}{4})$. (b) $\sigma_y(\pm\frac{d}{2}, 0)$.

Figure 8.13: Stress at two edge locations vs. k_Φ in Example 8.7 ($l = 10$, $z = 5$).

and conclude that while (8.61) is identically satisfied, (8.60) leaves a residual for the existence condition for χ_1, which may be written by means of a residual offset $\widetilde{x}_0$ (of the system x-coordinate) defined as

$$\widetilde{x}_0 \iint_\Omega \frac{1}{a_{33}} = \iint_\Omega \left(\frac{x}{a_{33}} - \overline{\sigma}_z^{(1)}\right), \tag{8.123}$$

while clearly, one should requires $\widetilde{x}_0 = 0$. Fig. 8.14 presents $\widetilde{x}_0$ as a function of k_Φ, and demonstrates that $\widetilde{x}_0 \ll d (= 1)$ in any level of approximation and that it decreases (for example) as $\sim k_\Phi^{-2.3}$ (the thick line). Also, note that in the case under discussion, the stresses of the first auxiliary problem (and therefore also $\widetilde{x}_0$) do not depend on the location of the coordinate system. To establish that we examine (8.108a) that shows that one may replace (8.8d) with

$$\{\overline{u}^{(1)}_{,x}, \overline{v}^{(1)}_{,xx}\}^{[2]}_{[1]} = \{0, \frac{a_{36}}{a_{33}}\}^{[2]}_{[1]} \tag{8.124}$$

on the interface line. Therefore the resulting BVP and the stresses associated with it are independent of the location of the coordinates origin. Thus, an iterative process as suggested in (2e) of S.8.2.11 is not required here. The shear stresses τ_{yz} and τ_{xz} are presented in Figs. 8.15(a),(b),

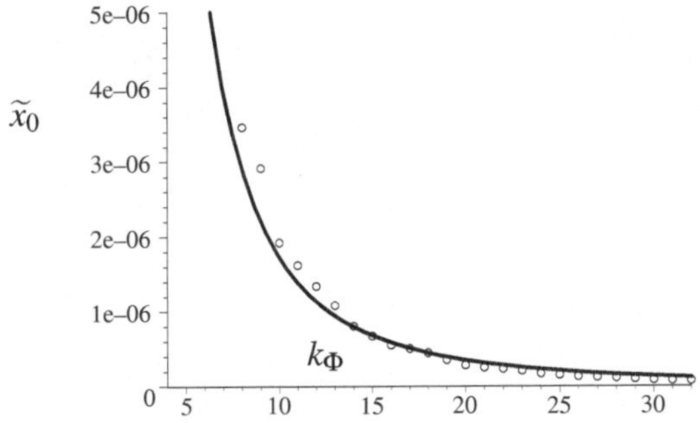

Figure 8.14: Fulfillment of the existence condition in Example 8.7.

respectively. Besides the clear fulfillment of the boundary conditions, note that τ_{yz} is smaller than τ_{xz} by an order of magnitude, and so the shear stress τ_{xy} which is presented in Fig. 8.16(a). In addition, Fig. 8.16(b) presents the displacement component $v(x,y)$. As far as P_y is con-

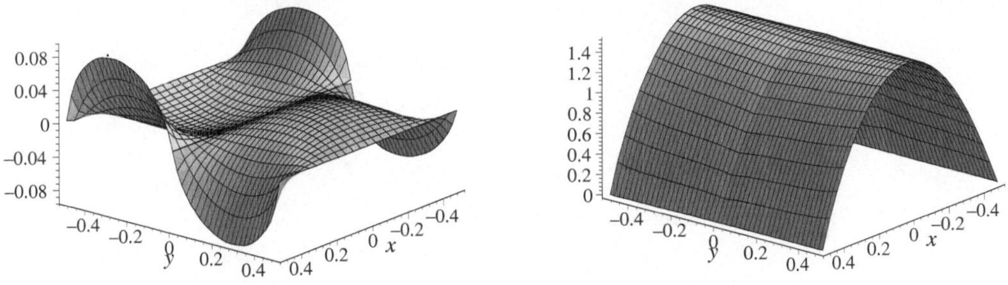

(a) $\tau_{yz}(x,y)$. (b) $\tau_{xz}(x,y)$.

Figure 8.15: Stress components in Example 8.7 ($k_\Phi = 25, l = 10, z = 5$).

cerned, the resulting shear stresses τ_{yz}, τ_{xz} are described in Figs. 8.17(a),(b), respectively. In this case τ_{yz} is much larger than τ_{xz} and unlike Fig. 8.15(b), its maximum value (which occurs along the interface line) is not constant. In addition, τ_{xz} is not continuous across the interface line.

Example 8.8 *A Non-Homogeneous Beam Under Constant Axial Body Force.*

Consider a non-homogeneous MON13z beam that undergoes a constant body force in the z direction only, namely $Z_b = Z_0 = const.$, while no other body, surface or tip loads are applied. In such a case, the loading constants of (8.121a–d) are $p_0 = q_0 = 0$ and

$$d_0 = Z_0\frac{S_\Omega}{\overline{I}_{33}}, \qquad \tau_0 = \frac{Z_0\, S_\Omega}{D\overline{I}_{33}} \iint_\Omega \frac{1}{a_0}[(a_{44}\overline{u}^{(3)} - a_{45}\overline{v}^{(3)})y - (a_{55}\overline{v}^{(3)} - a_{45}\overline{u}^{(3)})x], \quad (8.125)$$

where $D, \overline{I}_{33}$ are given by (7.11) and (8.33), respectively. Since $K = 0$, only the stress functions Φ, ω and the auxiliary functions L, M of level $k = 0$ should be considered, and hence, for the sake of convenience, in what follows we shall omit the index superscript.

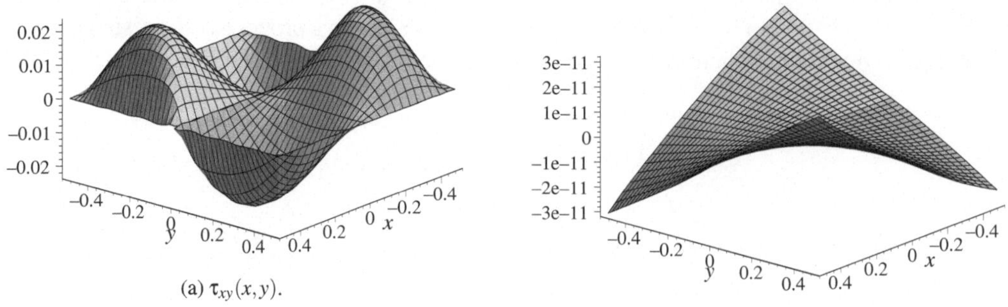

(a) $\tau_{xy}(x,y)$.

(b) $v(x,y)$.

Figure 8.16: Stress and displacement components in Example 8.7 ($k_\Phi = 25$, $l = 10$, $z = 5$).

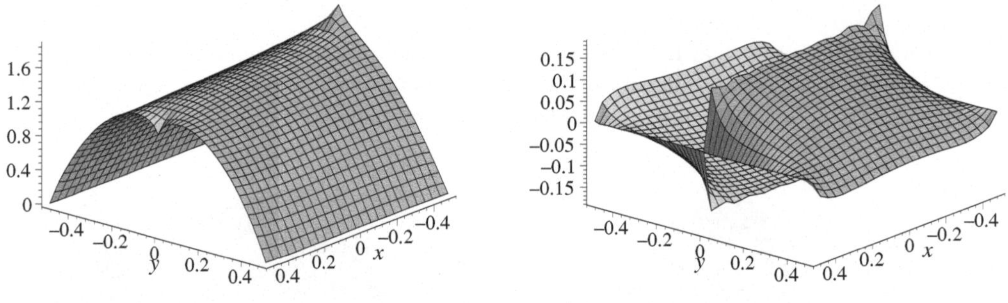

(a) $\tau_{yz}(x,y)$.

(b) $\tau_{xz}(x,y)$.

Figure 8.17: Stress components due to P_y tip load in Example 8.7 ($k_\Phi = 25$, $l = 10$, $z = 5$).

The Neumann-type BVP of (8.115a–d) for $\omega(x,y)$ should be written as

$$F_0^{\omega} = -a_0 Z_0 + d_0(a_{44}\bar{\varepsilon}_x^{(3)} + a_{55}\bar{\varepsilon}_y^{(3)} - a_{45}\bar{\gamma}_{xy}^{(3)} + a_0\bar{\sigma}_z^{(3)}),$$
$$P^{\omega} = d_0(-\frac{a_0}{2a_{33}}x + a_{44}\bar{u}^{(3)} - a_{45}\bar{v}^{(3)}), \quad Q^{\omega} = d_0(-\frac{a_0}{2a_{33}}y + a_{55}\bar{v}^{(3)} - a_{45}\bar{u}^{(3)}),$$
$$F^{\omega,ij} = -d_0\{\frac{1}{2a_{33}}[(a_{13} + \frac{a_{55}}{2})x^2 + (a_{45} + a_{36})xy + (a_{23} + \frac{a_{44}}{2})y^2]\}_{[i]}^{[j]}, \tag{8.126}$$

while for convenience we assume $\omega(0,0) = 0$ (compare with Example 7.4).
The biharmonic BVP (8.114a–d) for $\Phi(x,y)$ should be written with $F_0 = F_1 = F_2 = 0$ and

$$L = \tau_0\{\int_0^x [(\frac{b_{11}a_{44} + b_{12}a_{55} - b_{16}a_{45}}{a_0} - \frac{a_{13}}{a_{33}})\varphi - \frac{b_{11}}{a_0}\int_0^x P^{\varphi}dx - \frac{b_{12}}{a_0}\int_0^y Q^{\varphi}dy]dx$$
$$+ (b_{12}a_{44} - b_{22}a_{55} + b_{26}a_{45})\frac{y^3}{6a_0} - \int_0^y [(\frac{b_{12}a_{44} + b_{22}a_{55} - b_{26}a_{45}}{a_0} - \frac{a_{23}}{a_{33}})\int_0^y \varphi_{,x}dy$$
$$- (\frac{b_{16}a_{44} + b_{26}a_{55} - b_{66}a_{45}}{a_0} - \frac{a_{36}}{a_{33}})\varphi]|_{x=0}dy\} + b_{12}\Phi_{,x} - b_{16}\Phi_{,y} + b_{11}\int_0^x \Phi_{,yy}dx$$
$$- \int_0^y \{b_{22}\int_0^y \Phi_{,xxx}dy + b_{12}[2\Phi_{,xy} - \Phi_{,xy}(0,0)] - b_{16}[2\Phi_{,yy} - \Phi_{,yy}(0,0)]$$

$$- b_{26}[2\Phi_{,xx} - \Phi_{,xx}(0,0)] + b_{66}[\Phi_{,xy} - \frac{1}{2}\Phi_{,xy}(0,0)]\}_{|x=0}\, dy,$$
$$M = Sym(L). \tag{8.127}$$

The stresses then become

$$\sigma_x = \Phi_{,yy} - zd_0\overline{\sigma}_x^{(3)}, \qquad \sigma_y = \Phi_{,xx} - zd_0\overline{\sigma}_y^{(3)}, \qquad \tau_{xy} = -\Phi_{,xy} - zd_0\overline{\tau}_{xy}^{(3)}, \tag{8.128}$$
$$\sigma_z = \frac{1}{a_{33}}(a_{36}\Phi_{,xy} - a_{13}\Phi_{,yy} - a_{23}\Phi_{,xx}) - (\frac{1}{a_{33}} + \overline{\sigma}_z^{(3)})\, z d_0,$$
$$\tau_{yz} = \frac{1}{a_0}(a_{55}\omega_{,y} - a_{45}\omega_{,x}) + \frac{d_0}{2a_{33}}y - \frac{d_0}{a_0}(a_{55}\overline{v}^{(3)} - a_{45}\overline{u}^{(3)}) - \frac{z\tau_0}{a_0}(a_{55}\varphi_{,y} - a_{45}\varphi_{,x} - Q^{\varphi}),$$
$$\tau_{xz} = \frac{1}{a_0}(a_{44}\omega_{,x} - a_{45}\omega_{,y}) + \frac{d_0}{2a_{33}}x - \frac{d_0}{a_0}(a_{44}\overline{u}^{(3)} - a_{45}\overline{v}^{(3)}) - \frac{z\tau_0}{a_0}(a_{44}\varphi_{,x} - a_{45}\varphi_{,y} - P^{\varphi}).$$

Once Φ, ω are determined for a given geometry, one needs to superimpose suitable solutions of S.8.2.7 in order to cancel out the tip resultants that are induced by the above stresses.

Let a beam cross-section be geometrically symmetric about the y axis and consist of orthotropic material that is rotated about the z axis by $-30°$, $30°$ for the $y < 0$ and $y > 0$ regions. As discussed in Example 8.2, the elastic moduli are identical in two domains $\Omega_{[1]}$, $\Omega_{[2]}$ except for a_{16}, a_{26}, a_{36} and a_{45} that are of identical magnitude but opposite signs. The same is true for reduced elastic constants b_{ij}. By (initially) placing the coordinate system at the cross-section midpoint, show that the solution for the third auxiliary problem vanishes, i.e., $\overline{\Phi}^{(3)} = 0$, and $\overline{\varepsilon}_x^{(3)} = \overline{\varepsilon}_y^{(3)} = \overline{\gamma}_{xy}^{(3)} = \overline{\sigma}_z^{(3)} = 0$. Hence $\overline{I}_{33} = \frac{S_\Omega}{a_{33}}$ and the displacements become rigid body:

$$\overline{u}^{(3)} = \frac{a_{36}}{2a_{33}}y, \qquad \overline{v}^{(3)} = -\frac{a_{36}}{2a_{33}}x. \tag{8.129}$$

The loading constants are $p_0 = q_0 = 0$ and

$$d_0 = Z_0 a_{33}, \qquad \tau_0 = \frac{Z_0}{2Da_0}\iint_\Omega a_{36}(a_{44}y^2 + a_{55}x^2) = 0. \tag{8.130}$$

Since the biharmonic problem (8.114a–d) is homogeneous, $\Phi = 0$. Subsequently, by assuming that the harmonic function ω is of the form

$$\omega = -\frac{Z_0}{4}(a_{55}x^2 + 2a_{45}xy + a_{44}y^2) + \frac{Z_0}{2}a_{36}^{[1]}\overline{\varphi}(x,y), \tag{8.131}$$

where $\overline{\varphi}(x,y)$ is a generalized harmonic function to be determined by a Neumann-type BVP which may be written as (8.115a–d) with $F_0^{\overline{\varphi}} = F_{\overline{\varphi}}^{ij} = 0$, $P^{\overline{\varphi}} = \frac{a_{36}}{a_{36}^{[1]}}P^{\varphi}$, $Q^{\overline{\varphi}} = \frac{a_{36}}{a_{36}^{[1]}}Q^{\varphi}$. Then

$$\sigma_x = \sigma_y = \tau_{xy} = 0, \qquad \sigma_z = -Z_0 z, \tag{8.132a}$$
$$\tau_{yz} = \frac{Z_0 a_{36}^{[1]}}{2a_0}(a_{55}\overline{\varphi}_{,y} - a_{45}\overline{\varphi}_{,x} - \frac{a_{36}}{a_{36}^{[1]}}Q^{\varphi}), \quad \tau_{xz} = \frac{Z_0 a_{36}^{[1]}}{2a_0}(a_{44}\overline{\varphi}_{,x} - a_{45}\overline{\varphi}_{,y} - \frac{a_{36}}{a_{36}^{[1]}}P^{\varphi}). \tag{8.132b}$$

To create an illustrative example of $\overline{\varphi}(x,y)$, for the cross-section of Fig. 8.3(b), we activate **P.8.1** with $NF = 2$, $func = pphi$, $k_{La} = k_\Phi + 2 = 20$, $d = h = 1$, $t_z = -30°, 30°$. The resulting function $\overline{\varphi}$ has symmetries $\overline{\varphi}(x,-y) = -\overline{\varphi}(-x,y) = \overline{\varphi}(x,y)$ and is presented by Fig. 8.18. The corresponding shear stress components are described in Fig. 8.19. In this case, due to the symmetry $\overline{\varphi}(x,-y) = \overline{\varphi}(x,y)$, the stress components of (8.132a) induce only the tip resultant

$$\widetilde{P}_z = \iint_\Omega \sigma_z|_{z=l} = -Z_0 l S_\Omega. \tag{8.133}$$

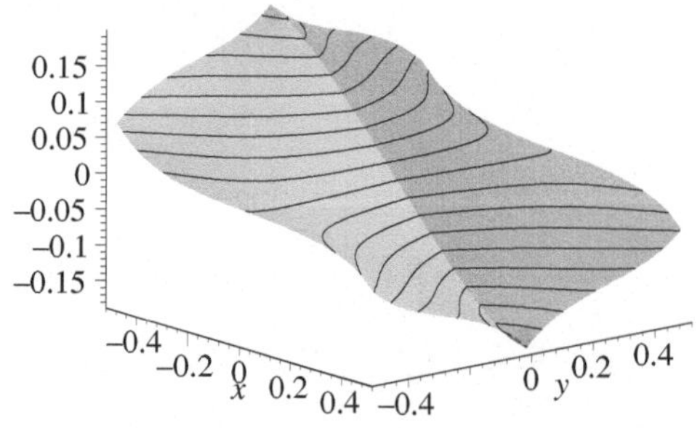

Figure 8.18: The function, $\overline{\varphi}(x,y)$ in Example 8.8.

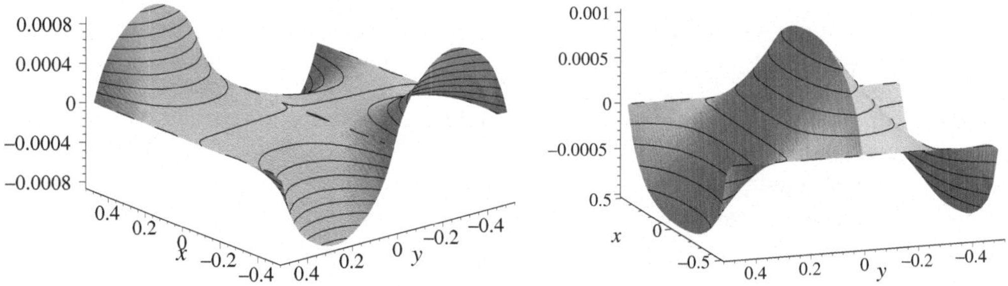

(a) τ_{yz}.

(b) τ_{xz}.

Figure 8.19: The shear stress components in Example 8.8.

To cancel out this resultant, one needs to superimpose suitable solutions for $P_z = -\widetilde{P}_z$. In the present case $\widetilde{I}_{13} = \widetilde{I}_{23} = 0$ and only a superposition of a solution with $\alpha_3 = Z_0\, l\, a_{33}$ is required. Referring to (7.108), (8.112c) the axis extension becomes $w_0 = Z_0\, a_{33}(lz - z^2/2)$.

9

Solid Coupled Monoclinic Beams

The objective of this chapter is twofold. First, it presents an *approximate* analytical model for *coupled* monoclinic beams. The model provides insight and fundamental understanding of the coupling mechanisms within anisotropic beams at the structural level. The derivation also supplies a simple but relatively accurate tool for quantitative estimation of coupled-beam behavior. In addition, an *exact* multilevel solution scheme, which is founded on an iterative process, that consists of the above analytical model and a standard plane-strain analysis of S.3.2.1, is derived.

Compared with the general and exact model for anisotropic beams developed in Chapter 6, the analysis presented in this chapter is considerably simpler, as the coupled solid beams considered in this chapter are made of MON13y material. As already discussed in S.2.4, such beams may also be viewed as made of orthotropic materials, the principal axes (of anisotropic) of which are not aligned with the beam axis, see Fig. 2.2. Depending on the specific lay-up mode, and as discussed in S.5.2.2, beams of MON13y material have the potential to induce coupling effects such as the "Beamwise Bending-Twist" coupling. Note that the analysis in S.5.2.2 was exact but yet confined to a homogeneous beam under tip axial load and transverse bending moments only, while with the solution methodology developed in this chapter, one may handle all other types of tip and distributed loads, together with a wider spectrum of cross-sections. Coupled beams of thin-walled cross-section are discussed in Chapter 10.

For further reading see (Rand, 1994b), (Rand, 1998a), (Rand, 1998b), (Rand, 1998c), (Rand, 2000c), (Rand, 2001a).

9.1 Background

9.1.1 Cross-Section Warping

Based on previous experience, it may be stated that adequate prediction of the elastic couplings must be founded on a detailed modeling of the axial distortion of the planes that were perpendicular to the beam axis before deformation. This deformation is usually referred to as the "out-of-plane warping". Since this warping is a local characteristic, numerical models that include a detailed warping deformation are typically based on a large number of warping degrees of freedom, see e.g. (Ochoa and Reddy, 1992) and (Fish *et al.*, 1994). Moreover, the warping deformation is usually very small compared with other deformation components, and thus, may pose some numerical difficulties in its prediction.

To cope with the above detailing challenge, a common simplification of numerical models is based on *assumed warping shape functions*, and many of them take into account only the torsional-related warping effects, which is in some cases accompanied by an additional linear distribution (that essentially accounts for constant transverse shear distribution only).

9.1.2 Approximate Analytical Solutions

Generally speaking, due to the large number of degrees of freedom involved, numerical and exact models that are capable of providing a detailed description of the beam behavior do not always supply enough insight into the structural mechanisms. Therefore, approximate analytical models have a clear and important role as part of any system of analysis tools. Such models have the potential to provide an analytical insight and a clear identification of the major parameters that control the beam behavior, and in particular, the resulting structural couplings.

Compared with the complete solution developed in Chapter 6, the present approach provides relatively high accuracy results with a tremendous saving in analytical derivation and complexity, and will therefore constitute an adequate tool for understanding the physical phenomena.

The approximate solutions documented in what follows are confined to the linear case and constitutes a "strength-of-materials type" set of fundamental solutions for anisotropic beams. As will be shown further on, these solutions may be generalized for much more extensive use than the one they were originally derived for. In addition, by supplying the exact out-of-plane warping deformation distribution for different basic cases, the present analytical model has also the potential to provide guidelines for numerical schemes that are based on various types of "built-in" warping shape functions.

The reader should also be aware of the fact that on many occasions, the term "beam" refers to a "slender thin plate" (or "Beam-Plate Models"), which are discussed in S.5.3.1. As far as the degrees of freedom involved, such models represent only a reduced version of the model developed in what follows.

9.1.3 Coupling Effects in Symmetric and Antisymmetric Solid Beams

Consider a MON13y material the constitutive relations of which are, see S.2.3,

$$\begin{Bmatrix} \varepsilon_x \\ \varepsilon_y \\ \varepsilon_z \\ \gamma_{yz} \\ \gamma_{xz} \\ \gamma_{xy} \end{Bmatrix} = \begin{bmatrix} a_{11} & a_{12} & a_{13} & 0 & a_{15} & 0 \\ & a_{22} & a_{23} & 0 & a_{25} & 0 \\ & & a_{33} & 0 & a_{35} & 0 \\ & & & a_{44} & 0 & a_{46} \\ & Sym. & & & a_{55} & 0 \\ & & & & & a_{66} \end{bmatrix} \begin{Bmatrix} \sigma_x \\ \sigma_y \\ \sigma_z \\ \tau_{yz} \\ \tau_{xz} \\ \tau_{xy} \end{Bmatrix}. \tag{9.1}$$

Examination of the above compliance matrix shows that ε_z is influenced by τ_{xz}, and likewise, γ_{xz} is influenced by σ_z. A similar conclusion is reached when the corresponding stiffness matrix is examined. In such a case, σ_z is influenced by γ_{xz}, and τ_{xz} is influenced by ε_z. Thus, as also shown in S.5.2.2, the above material has the potential to create coupling between global degrees of freedom, such as the Bending-Twist coupling. The above also explains the importance of accurate prediction of the out-of-plane warping, which is an important contributor to the γ_{xz} strain component, see (1.9b).

Even at this early stage, the coupling phenomena may be further illustrated qualitatively by the following descriptive explanation: Suppose that a given beam is loaded by a tip torsional moment, which clearly creates, among others, a distribution of the τ_{xz} stress component. Subsequently, ε_z will be induced, which is capable of introducing bending and/or extension deformation. While a quantitative analysis of such an effect is discussed in what follows, a general qualitative conclusion that emerges from the above basic observation is that the elastic couplings at the MON13y material level is a direct consequence of the couplings between normal stress and shear strain and between shear stress and normal strain at the material level.

So, one should distinguish between two basic configurations of solid anisotropic beams, which are commonly entitled "symmetric" and "antisymmetric". These definitions are quite generic and are demonstrated by the simplified cases shown in Fig. 9.1. In Fig. 9.1(a), a ho-

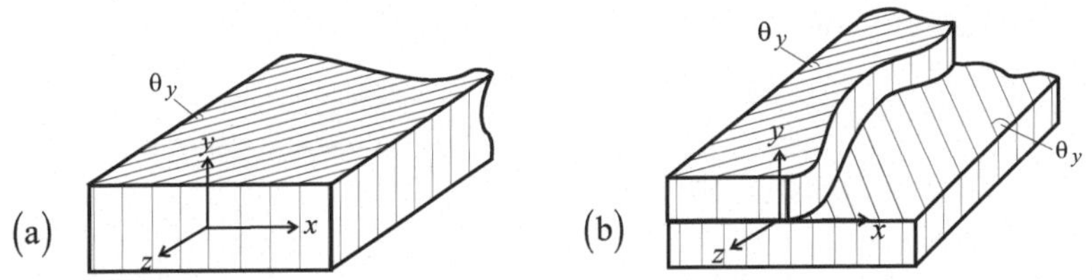

Figure 9.1: Solid beams made of MON13y orthotropic material: (a) symmetric, (b) antisymmetric.

mogeneous solid beam made of orthotropic material, which is oriented at an angle θ_y with respect to the beam axis, is presented (this configuration may also be viewed as a single-layer homogeneous lamina or as a laminated beam where all laminae are oriented in the same direction). Such a configuration is called "unbalanced" in some contexts, and will be denoted "symmetric" since the distribution of the "lamination angles" (θ_y) is symmetric (and in this special case even uniform) with respect to the y-axis. Figure 9.1(b) presents a solid beam made of two homogeneous parts of identical orthotropic material that are oriented at opposite $\pm\theta_y$ angles. This case is denoted "antisymmetric" (with respect to the y-axis). It should be emphasized again that the above two configurations are just simplified cases of symmetric and antisymmetric solid beams, while more realistic configurations consist of diversified shapes, various number of laminae and different lamination angles. Yet, the relatively simple models of Fig. 9.1 are useful when analytical solutions are pursued, as they enable clear physical and mathematical insight.

The involved coupling effects in solid symmetric and antisymmetric beams in the structural level are qualitatively summarized in Fig. 9.2 for both the material and the structural levels. The

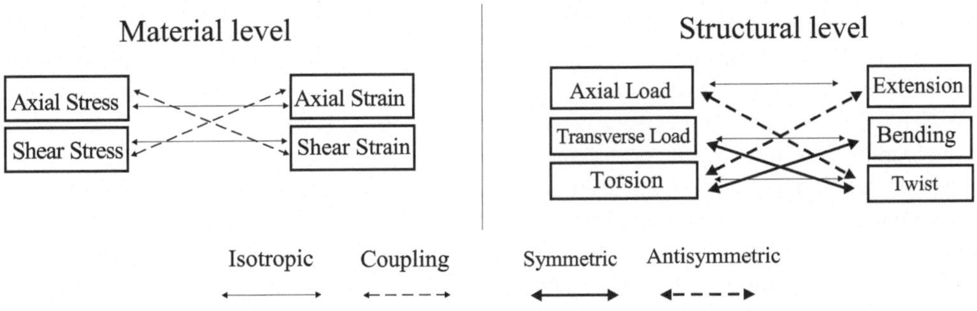

Figure 9.2: Material-level and structural-level coupling effects in MON13y beams.

thin-full arrows show isotropic relations. The thin-broken arrows show coupling at the material level. The thick-full arrows show coupling effects in symmetric beams, while the thick-broken arrows show coupling effects in antisymmetric beams.

In summary, the coupling effects in beams of solid cross-sections may be classified into two main categories: the Extension/Torsion coupling and the Bending/Torsion coupling. While in the former case one is interested in the coupling between axial extension and elastic twist, the latter coupling deals with the lateral bending and the elastic twist. In many engineering applications and mainly in aeronautical applications, the twist angle plays a major role (e.g. it directly influences the angle of attack in aircraft wings and helicopter blades), and therefore, the above elastic couplings are of primary importance. If modeled and tailored properly, such couplings may significantly contribute to the structural response of the beam and enable adequate tailoring of the beam structure to comply with specific requirements, see literature on *Optimization and Tailoring* of S.5.4.

9.2 An Approximate Analytical Model

Consider the untwisted, uniform anisotropic beam shown in Fig. 5.1(a). The z coordinate line coincides with the beam axis in the undeformed state. The beam length, l, is measured along the z coordinate line in the undeformed state. A solid, simply connected cross-section (domain) Ω, where $\partial\Omega$ stands for its contour (boundary), and the unit vector normal to the contour (and directed outwards), $\bar{\mathbf{n}}$, are presented in Fig. 5.1(b). All notation and definitions described in S.5.1, and in particular (5.1), hold throughout the following derivation.

The beam undergoes a distribution of loads (forces and moments) in the x, y and z directions, in addition to tip loads (i.e. tip forces and moments). The surface distributed loads are denoted $X_s(x,y,z)$, $Y_s(x,y,z)$, $Z_s(x,y,z)$, in x, y, z directions, respectively, and they act over the contour, i.e., $(x,y) \in \partial\Omega$. The body distributed loads are denoted $X_b(x,y,z)$, $Y_b(x,y,z)$, $Z_b(x,y,z)$ in the coordinate directions, and they act over each material point, i.e., $(x,y) \in \Omega$.

One may also define the "*equivalent loads per unit length*" that act along the beam. We denote the *force components* by $p_x(z)$, $p_y(z)$, and $p_z(z)$ (in the x, y, z directions, respectively), and the *torsional moment per unit length* by $m_z(z)$ (in the z direction), namely,

$$\{p_x, p_y, p_z, m_z\} = \oint_{\partial\Omega} \{X_s, Y_s, Z_s, xY_s - yX_s\} + \iint_{\Omega} \{X_b, Y_b, Z_b, xY_b - yX_b\}. \tag{9.2}$$

Formally, the *bending moments per unit length* m_x and m_y could also be defined. Yet, as shown in Remark 9.1, these components may be introduced by a proper Z_b distribution. Hence, the beam loading is fully described by the distributions p_x, p_y, p_z, m_z and Z_b.

For further clarification we distinguish between two cases: the case where the functions X_s, Y_s, Z_s and X_b, Y_b, Z_b are given, and the case where p_x, p_y, p_z, m_x, m_y, m_z are given. In the first case, we employ (9.2) to evaluate p_x, p_y, p_z and m_z, while in the second case we employ (9.7), (9.8) of Remark 9.1 to evaluate Z_b. In both cases, the subsequent analysis is carried out for the loading functions p_x, p_y, p_z, m_z and Z_b.

The deformation assumption is based on two types of displacements. First, since a "beam-like behavior" is expected (i.e. the beam is expected to deform like a slender body, where the most profound deformation components are those of the elastic axis), we introduce the one-dimensional functions $u_0(z)$, $v_0(z)$ and $w_0(z)$ for the displacements of the beam axis in the x, y, and z directions, respectively, and $\phi(z)$ for the twist angle. Since these components of the deformation are functions of z only, they represent "rigid" deformation of each cross-section that contains no in-plane cross-section warping (i.e. no deformation of the cross-section shape). To account for the out-of-plane warping (in the z direction), a three-dimensional warping function is superimposed upon the above mentioned displacements. This warping function is denoted $\Psi^w(x,y,z)$, and is assumed to be of zero average value over the cross-section area (i.e., $\iint_\Omega \Psi^w = 0$). Note that Ψ^w is the total out-of-plane warping that may originate from various sources (e.g. torsion, shear, bending, extension, etc.).

Since a linear formulation is under discussion, no distinction is made between deformed and undeformed direction, and the above assumed displacement functions may be directly used to construct the deformation and the resulting strain expressions without any involvement of coordinate transformation. The displacements are therefore written as

$$u = u_0(z) - y\phi(z), \quad v = v_0(z) + x\phi(z), \quad w = w_0(z) - xu_{0,z}(z) - yv_{0,z}(z) + \Psi^w(x,y,z). \tag{9.3}$$

Hence, the non-vanishing strain components, see (1.9a), are given by

$$\varepsilon_z = w_{0,z} - xu_{0,zz} - yv_{0,zz} + \Psi^w_{,z}, \tag{9.4a}$$

$$\gamma_{yz} = x\phi_{,z} + \Psi^w_{,y}, \qquad \gamma_{xz} = -y\phi_{,z} + \Psi^w_{,x}, \tag{9.4b}$$

while according to the reduced stress-strain relationships derived in what follows within S.9.2.1 (see (9.10)), the corresponding stress components are

$$\begin{Bmatrix} \sigma_z \\ \tau_{yz} \\ \tau_{xz} \end{Bmatrix} = \mathbf{Q} \cdot \begin{Bmatrix} \varepsilon_z \\ \gamma_{yz} \\ \gamma_{xz} \end{Bmatrix}. \tag{9.5}$$

Remark 9.1 The *bending moments per unit length* $m_x(z)$ and $m_y(z)$ are given by

$$\{m_x, m_y\} = \oint_{\partial\Omega} \{yZ_s, xZ_s\} + \iint_\Omega \{yZ_b, xZ_b\}. \tag{9.6}$$

Hence, for a given distribution of m_x and m_y, their influence may be introduced via *an additional body-force distribution* of the simplest type as

$$Z_b(x,y,z) = xZ_b^{(1)}(z) + yZ_b^{(2)}(z), \tag{9.7}$$

which affects the bending behavior but does not contribute to the $p_z(z)$ force component. Using (5.1) one obtains

$$Z_b^{(2)}(z) = \frac{m_x(z)}{I_x}, \qquad Z_b^{(1)}(z) = -\frac{m_y(z)}{I_y}. \tag{9.8}$$

Note that the above selection of linear distributions in (9.7) is arbitrary, since the present analysis is capable of incorporating a generic axial body-force distribution, which is more than required to simulate the distributions of $m_x(z)$ and $m_y(z)$.

9.2.1 *Reduced Stress-Strain Relationships*

We shall first develop a "reduced stress-strain relationship" for the material under discussion. For that purpose, we adopt the derivation and notation of (3.19–3.22) of S.3.2 by considering a MON13y material, see (9.1). For such a material, $a_{14} = a_{16} = a_{24} = a_{26} = a_{34} = a_{36} = a_{45} = a_{56} = 0$ and similarly, $A_{14} = A_{16} = A_{24} = A_{26} = A_{34} = A_{36} = A_{45} = A_{56} = 0$. Hence, the matrices $\mathbf{A}_\varepsilon$,$\mathbf{B}_\varepsilon$,$\mathbf{D}_\varepsilon$ and $\mathbf{A}_\sigma$,$\mathbf{B}_\sigma$,$\mathbf{D}_\sigma$, see (3.20), (3.22), become

$$\mathbf{A}_\varepsilon=\begin{bmatrix}a_{11}&a_{12}&0\\a_{12}&a_{22}&0\\0&0&a_{66}\end{bmatrix},\quad \mathbf{B}_\varepsilon=\mathbf{C}_\varepsilon{}^T=\begin{bmatrix}a_{13}&0&a_{15}\\a_{23}&0&a_{25}\\0&a_{46}&0\end{bmatrix},\quad \mathbf{D}_\varepsilon=\begin{bmatrix}a_{33}&0&a_{35}\\0&a_{44}&0\\a_{35}&0&a_{55}\end{bmatrix},$$
$$\mathbf{A}_\sigma=\begin{bmatrix}A_{11}&A_{12}&0\\A_{12}&A_{22}&0\\0&0&A_{66}\end{bmatrix},\quad \mathbf{B}_\sigma=\mathbf{C}_\sigma{}^T=\begin{bmatrix}A_{13}&0&A_{15}\\A_{23}&0&A_{25}\\0&A_{46}&0\end{bmatrix},\quad \mathbf{D}_\sigma=\begin{bmatrix}A_{33}&0&A_{35}\\0&A_{44}&0\\A_{35}&0&A_{55}\end{bmatrix}. \tag{9.9}$$

Due to the beam slenderness, and as shown by (9.5), the following analysis will mainly be focused on the stress and strain components that are created in planes perpendicular to the beam axis, namely, $\bar{\bar{\sigma}}$ $(=\sigma_z,\ \tau_{yz},\ \tau_{xz})$ and the corresponding strain components $\bar{\bar{\varepsilon}}$ $(=\varepsilon_z,\ \gamma_{yz},\ \gamma_{xz})$. We therefore define the reduced constitutive relations as $\bar{\bar{\varepsilon}}=\mathbf{q}\cdot\bar{\bar{\sigma}}$, $\bar{\bar{\sigma}}=\mathbf{Q}\cdot\bar{\bar{\varepsilon}}$, where $\mathbf{Q}=\mathbf{q}^{-1}$, and the above vectors and matrices are

$$\begin{Bmatrix}\varepsilon_z\\\gamma_{yz}\\\gamma_{xz}\end{Bmatrix}=\begin{bmatrix}q_{33}&0&q_{35}\\0&q_{44}&0\\q_{35}&0&q_{55}\end{bmatrix}\begin{Bmatrix}\sigma_z\\\tau_{yz}\\\tau_{xz}\end{Bmatrix},\qquad \begin{Bmatrix}\sigma_z\\\tau_{yz}\\\tau_{xz}\end{Bmatrix}=\begin{bmatrix}Q_{33}&0&Q_{35}\\0&Q_{44}&0\\Q_{35}&0&Q_{55}\end{bmatrix}\begin{Bmatrix}\varepsilon_z\\\gamma_{yz}\\\gamma_{xz}\end{Bmatrix}. \tag{9.10}$$

Since part of the stress and strain components ($\bar{\sigma}$ and $\bar{\varepsilon}$) do not appear in the above relations, some assumption has to be made regarding their role and the way their determination will be carried out. We continue the derivation by establishing two assumption versions that yield two different constitutive relations and therefore lead to two different solutions — see also Table 9.1.

Assumption → Application ↓	$\bar{\bar{\varepsilon}}=0$	$\bar{\bar{\sigma}}=0$	$\bar{\varepsilon}=0$	$\bar{\sigma}=0$
Plane-Strain (see S.3.2.1)	0			
Plane-Stress (see S.3.2.2)		0		
Rigid cross-section (Version I)			0	
Stress-Free cross-section (Version II)				0

Table 9.1: Summary of reduced constitutive relations schemes.

In "Version I" we assume an "*infinitely rigid cross-section*" (or in other words, neglect the in-plane strain) and therefore, $\bar{\varepsilon}=0$. Thus, $\mathbf{q}$ (and $\mathbf{Q}$) are derived by

$$\mathbf{q}=\mathbf{D}_\sigma{}^{-1}=\mathbf{D}_\varepsilon-\mathbf{C}_\varepsilon\cdot\mathbf{A}_\varepsilon{}^{-1}\cdot\mathbf{B}_\varepsilon. \tag{9.11}$$

Once $\bar{\bar{\sigma}}$ and $\bar{\bar{\varepsilon}}$ are obtained by a suitable solution, $\bar{\sigma}$ is determined as

$$\bar{\sigma}=\mathbf{B}_\sigma\cdot\bar{\bar{\varepsilon}}=-\mathbf{A}_\varepsilon{}^{-1}\cdot\mathbf{B}_\varepsilon\cdot\bar{\bar{\sigma}}. \tag{9.12}$$

In this case $Q_{ij}=A_{ij}$, and therefore

$$\begin{bmatrix}q_{33}&0&q_{35}\\0&q_{44}&0\\q_{35}&0&q_{55}\end{bmatrix}=\begin{bmatrix}\frac{A_{55}}{A_{33}A_{55}-A_{35}^2}&0&-\frac{A_{35}}{A_{33}A_{55}-A_{35}^2}\\0&\frac{1}{A_{44}}&0\\-\frac{A_{35}}{A_{33}A_{55}-A_{35}^2}&0&\frac{A_{33}}{A_{33}A_{55}-A_{35}^2}\end{bmatrix}. \tag{9.13}$$

Alternatively, (9.11) shows that

$$q_{33} = a_{33} - \frac{a_{13}^2 a_{22} - 2a_{13}a_{23}a_{12} + a_{23}^2 a_{11}}{a_{11}a_{22} - a_{12}^2}, \quad q_{55} = a_{55} - \frac{a_{15}^2 a_{22} - 2a_{15}a_{25}a_{12} + a_{25}^2 a_{11}}{a_{11}a_{22} - a_{12}^2},$$
$$q_{35} = a_{35} - \frac{a_{13}(a_{22}a_{15} - a_{12}a_{25}) + a_{23}(a_{11}a_{25} - a_{12}a_{15})}{a_{11}a_{22} - a_{12}^2}, \qquad q_{44} = a_{44} - \frac{a_{46}}{a_{66}}. \tag{9.14}$$

In "Version II" we assume an "*in-plane stress free cross-section*" (or, in other words, that the stress components σ_x, σ_y and σ_{xy} are much smaller than the other three components and may be neglected) and therefore, $\overline{\sigma} = 0$. This assumption is exact for isotropic beams and will be proved (by comparison with the exact solution of Chapter 6) to be excellent in many cases for the monoclinic material under discussion as well. In this case

$$\mathbf{Q} = \mathbf{D_\varepsilon}^{-1} = \mathbf{D_\sigma} - \mathbf{C_\sigma} \cdot \mathbf{A_\sigma}^{-1} \cdot \mathbf{B_\sigma}. \tag{9.15}$$

Once a solution is carried out and $\overline{\overline{\sigma}}$ and $\overline{\overline{\varepsilon}}$ are known, $\overline{\varepsilon}$ is directly obtained by

$$\overline{\varepsilon} = \mathbf{B_\varepsilon} \cdot \overline{\overline{\sigma}} = -\mathbf{A_\sigma}^{-1} \cdot \mathbf{B_\sigma} \cdot \overline{\overline{\varepsilon}}. \tag{9.16}$$

In this case $q_{ij} = a_{ij}$, and therefore

$$\begin{bmatrix} Q_{33} & 0 & Q_{35} \\ 0 & Q_{44} & 0 \\ Q_{35} & 0 & Q_{55} \end{bmatrix} = \begin{bmatrix} \frac{a_{55}}{a_{33}a_{55}-a_{35}^2} & 0 & -\frac{a_{35}}{a_{33}a_{55}-a_{35}^2} \\ 0 & \frac{1}{a_{44}} & 0 \\ -\frac{a_{35}}{a_{33}a_{55}-a_{35}^2} & 0 & \frac{a_{33}}{a_{33}a_{55}-a_{35}^2} \end{bmatrix}. \tag{9.17}$$

Alternatively, (9.15) shows that

$$Q_{33} = A_{33} - \frac{A_{13}^2 A_{22} - 2A_{13}A_{23}A_{12} + A_{23}^2 A_{11}}{A_{11}A_{22} - A_{12}^2}, \quad Q_{55} = A_{55} - \frac{A_{15}^2 A_{22} - 2A_{15}A_{25}A_{12} + A_{25}^2 A_{11}}{A_{11}A_{22} - A_{12}^2},$$
$$Q_{35} = A_{35} - \frac{A_{13}(A_{22}A_{15} - A_{12}A_{25}) + A_{23}(A_{11}A_{25} - A_{12}A_{15})}{A_{11}A_{22} - A_{12}^2}, \qquad Q_{44} = A_{44} - \frac{A_{46}}{A_{66}}. \tag{9.18}$$

In the (uncoupled) orthotropic case where $q_{35} = Q_{35} = 0$, the above compliance and stiffness matrices become diagonal and depend on a_{33}, a_{44}, and a_{55} or A_{33}, A_{44}, and A_{55} only. For further analytical convenience, we define $b_0 = q_{33}q_{55} - q_{35}^2 = (Q_{33}Q_{55} - Q_{35}^2)^{-1}$, so that $Q_{33} = \frac{q_{55}}{b_0}$, $Q_{35} = -\frac{q_{35}}{b_0}$ and $Q_{55} = \frac{q_{33}}{b_0}$. The above two sets of assumptions are complementary to the Plane-Strain and Plane-Stress sets discussed in S.3.2.1, S.3.2.2, see Table 9.1.

9.2.2 Equilibrium Equations

The approximate solution under discussion contains four "global" deformation parameters (i.e. the four cross-section displacements $u_0(z)$, $v_0(z)$, $w_0(z)$ and $\phi(z)$, which are common for the entire cross-section), and one "local" deformation parameter (i.e. the cross-section warping function $\Psi^w(x,y,z)$, which is a local deformation component). Subsequently, equilibrium is achieved by four integral equations and one differential equation. The integral equations are obtained by assuming that the beam is a slender structure, and thus, the exact way of introducing the external loads to each cross-section is immaterial. In other words, as far as the external loads are concerned, the beam is viewed as a "thin line" and as already indicated, only the distributed loads per unit length $p_x(z)$, $p_y(z)$, $p_z(z)$, $m_z(z)$ and the axial body-force distributions are taken into account. For that purpose, we first define resultant loads as the total loads acting at each cross-section. As will be shown shortly, this definition will serve as the integral form of the differential equilibrium equations, namely, (see also (3.102) and (6.9)),

$$\{\overline{P}_x, \overline{P}_y, \overline{P}_z, \overline{M}_z\}(z) = \iint_\Omega \{\tau_{xz}, \tau_{yz}, \sigma_z, x\tau_{yz} - y\tau_{xz}\}. \tag{9.19}$$

To derive formally the r.h.s. of (9.19), the differential equations of equilibrium (see (1.82a–c)) are first integrated in the following form:

$$\iint_\Omega (\sigma_{x,x} + \tau_{xz,z} + \tau_{xy,y} + X_b) = 0, \tag{9.20a}$$

$$\iint_\Omega (\sigma_{y,y} + \tau_{xy,x} + \tau_{yz,z} + Y_b) = 0, \tag{9.20b}$$

$$\iint_\Omega (\sigma_{z,z} + \tau_{xz,x} + \tau_{yz,y} + Z_b) = 0, \tag{9.20c}$$

$$\iint_\Omega (\sigma_{y,y} + \tau_{xy,x} + \tau_{yz,z} + Y_b)x - \iint_\Omega (\sigma_{x,x} + \tau_{xz,z} + \tau_{xy,y} + X_b)y = 0. \tag{9.20d}$$

By virtue of Remark 3.2, with the aid of (9.2), (9.19), (5.16a–c), one may write (9.20a–d) as

$$\{\overline{P}_x, \overline{P}_y, \overline{P}_z, \overline{M}_z\}_{,z} = -\{p_x, p_y, p_z, m_z\}. \tag{9.21}$$

Integrating the above equations for a clamped-free beam (at $z = 0$ and $z = l$) yields

$$\{\overline{P}_x, \overline{P}_y, \overline{P}_z, \overline{M}_z\} = \int_z^l \{p_x, p_y, p_z, m_z\}\, dz + \{P_x, P_y, P_z, M_z\}, \tag{9.22}$$

where P_x, P_y, P_z, M_z are the corresponding tip loads. Hence, equating the r.h.s. of (9.19) with the r.h.s. of (9.22) yields four integral equilibrium equations. Yet, for the sake of convenience, typically, the loads distributions $p_x(z)$, $p_y(z)$, $p_z(z)$, $m_z(z)$ are first integrated as shown in (9.22), and as indicated, (9.19) are referred to as the "integral equilibrium equations". These equations may also be viewed as the equilibrium conditions for the ("integral" or "global") cross-section unknowns $u_0(z)$, $v_0(z)$, $w_0(z)$ and $\phi(z)$. In addition to the above integral equilibrium equations, a differential equation of equilibrium is required for the out-of-plane warping $\Psi^w(x,y,z)$, which as already indicated, is a local deformation component. For that purpose, the differential equilibrium equation in the z direction is explicitly used, namely,

$$\sigma_{z,z} + \tau_{xz,x} + \tau_{yz,y} + Z_b = 0. \tag{9.23}$$

The coexistence of an integral and a differential equation in the z direction (i.e. (9.19:c), (9.23)) does not create any redundancy. This is due to the fact that the unknown deformation component w_0 and $\Psi^w(x,y,z)$ are also integral (i.e. a "mean" or "average" constant value over the cross-section) and differential (i.e. a distribution over the cross-section with zero average) components of the deformation in the z direction.

As generally discussed in S.5.1.3, the boundary conditions may be classified into two categories. The first category includes the "tip" boundary conditions at both ends. In the present analysis, a "clamped-free" beam will be under discussion, and therefore, geometrical boundary conditions are imposed at the root while natural boundary conditions are imposed at the tip. The second category includes the "outer surface" boundary conditions that account for the loading distribution over the beam outer surface, see Fig. 5.1(a). Analogous considerations may also be applied for other variants of the end boundary conditions, see S.5.1.3.2.

Overall, there are eight boundary conditions for the displacements $u_0(z)$, $v_0(z)$, $w_0(z)$ and $\phi(z)$ at the beam root and tip, and a "contour" boundary condition for the warping function,

Ψ^w, that should be satisfied over the entire contour of each cross-section. Within the present approximate model for a slender beam, the geometrical boundary conditions at the beam root cross-sections are imposed at the "centroid" only (i.e. the $x = y = 0$ point), which also may be viewed as an "averaged" or "global" condition. For a beam that is clamped at its "root" end ($z = 0$), these conditions are

$$u_0 = v_0 = w_0 = u_{0,x} = v_{0,x} = \phi = 0, \qquad \text{for} \quad x = y = z = 0. \tag{9.24}$$

The natural boundary conditions at the "tip" end cross-sections ($z = l$) are based on equating the tip transverse moments, M_x and M_y, to those obtained by integration of the stresses over the tip cross-section area, Ω^t, namely

$$\{M_x, M_y\} = \iint_{\Omega^t} \{y, -x\}\sigma_z. \tag{9.25}$$

The contour boundary condition is founded on the requirement of balancing between the internal stress components and the external traction. As far as the components of $\bar{\bar{\sigma}}$ are concerned, (5.16c) shows that along $\partial\Omega$,

$$\tau_{xz}\cos(\bar{\mathbf{n}}, x) + \tau_{yz}\cos(\bar{\mathbf{n}}, y) = Z_s. \tag{9.26}$$

In essence, one should also impose the boundary conditions (5.16a, b) for the components of $\bar{\sigma}$ $(= \sigma_x, \sigma_y, \sigma_{xy})$. However, such boundary conditions may not be imposed as the above stress components are not part of the present analysis. When Version I is employed, these components are given by (9.12), while for Version II, these components are zero (and therefore identically satisfy (5.16a, b) for the case of a "traction-free" outer surface, namely, $X_s = Y_s = 0$). Hence, in both cases, once a solution has been obtained, the level of fulfillment of the boundary conditions (5.16a, b) may be examined.

In summary, the present analytical model consists of four integral displacement components and four integral equilibrium equations, along with one local displacement component that is accompanied by a differential equilibrium equation. All equations are restricted by suitable boundary conditions.

As a general observation, the approximate analytical model under discussion sacrifices, for the sake of simplicity, some of the governing conditions or boundary conditions. As shown in Remark 9.2, the strain components, which are determined within Version II, may violate the compatibility conditions of (9.27). In addition, both versions have the potential to violate the contour boundary conditions (5.16a, b), however, Version II has a certain advantage as it identically satisfies the boundary condition for the common traction-free surface case.

There is no exact measure that may give an indication regarding the error induced by the above potential violation of the compatibility equations and/or boundary conditions. The exact version of the reduced constitutive relations should be selected according to the specific nature of the problem under discussion.

Remark 9.2 We shall discuss here some of the properties of the proposed approximate solution. The displacements assumption of (9.3) produces zero $\bar{\varepsilon}$ components, which are compatible with Version I of S.9.2.1. Yet, a solution that satisfies all equations and boundary conditions discussed above is still considered as "approximate" in the sense that it may violate the first two equilibrium equations, (1.82a,b), and the first two boundary conditions, (5.16a,b), since the present formulation does not explicitly impose these requirements.

On the other hand, if Version II is employed, the first two equilibrium equations are satisfied only if $\tau_{xy,z} = \tau_{xz,z} = 0$. In addition, since $\bar{\sigma} = 0$, the boundary conditions of (5.16a,b) are satisfied as well in the common case of a stress-free outer surface. However, the strain components of (9.4a–c) show that the compatibility equations (see (1.45a–f)) are violated unless

$$\varepsilon_{x,yy}+\varepsilon_{y,xx}-\gamma_{xy,xy}=0, \quad \varepsilon_{y,zz}=\varepsilon_{x,zz}=\gamma_{xy,zz}=0. \quad 2\varepsilon_{x,yz}=\gamma_{xy,xz}, \quad 2\varepsilon_{y,xz}=\gamma_{xy,yz}. \quad (9.27)$$

This potential violation of the compatibility equations that may be induced by Version II should be put in perspective. Note that (9.27) contain only the in-plane components ε_x, ε_y and γ_{xy}, and therefore, error may be induced over the in-plane deformation only, or in other words, the in-plane deformation may become non-integrable. This however, does not necessarily damage the quality of the prediction of the global beam behavior including the associated coupling effects.

9.2.3 *Analytical Solutions*

We shall present some analytical solutions for solid beams of homogeneous, simply connected cross-sections. The notion "homogeneous cross-section" stands for the case, where all laminae are identical and oriented at the same angle with respect to the z-axis (or the case of a "single lamina cross-section"). From an analytical point of view, such a configuration is simpler as all elastic moduli are constants over the cross-section domain (as discussed in S.9.1.3, such a configuration is usually denoted as *symmetric*). The solutions presented in what follows are for non-uniform axial force, tip edgewise and beamwise loads and torsional moment. Recall that the S.5.2.2 presents solutions for such beams under tip axial load and transverse tip bending moments only. Solutions for beams of non-homogeneous cross-section including *antisymmetric* beams are dealt with in S.9.3.

The solutions presented in what follows were generated using the well-established analytical technique generally termed to as "St. Venant's semi-inverse method of solution" (see S.1.6.3), which is based on a preliminary solution form "assumption" using unknown parameters. In later stages, these parameters are determined so that the governing equations are all fulfilled. Once such a procedure is successfully completed for any linear problem, the resulting solution is unique and exact. The reader should therefore realize that unless stated differently, all preliminary assumptions are of temporary nature only, as they convert into parts of the unique solution once one is successfully achieved.

It should also be emphasized that although we shall employ the notion "lamina" (and its derivatives), the analysis is *not* be confined to the case of thin solid beams, which are treated using the CLPT as "Beam-Plate Models" in S.5.3.1.

9.2.3.1 Non-Uniform Axial Force

A solution of a homogeneous beam of simply connected cross-section that undergoes axial load is discussed in this section. We consider a beam loaded by a tip tensile force, P_z, and a body-force distribution $Z_b(z)$ (i.e. uniform over the cross-section area). Following (9.2), (9.19:c) the axial resultant force, $\overline{P}_z(z)$, is obtained by integrating the equation $\overline{P}_{z,z}=-Z_bS_\Omega$ under the condition $\overline{P}_{z|z=l}=P_z$.

The solution for the case under discussion may be generated by assuming $v_0=\phi=0$ and a warping function of the form $\Psi^w=x\Psi_0(z)$, where $\Psi_0(z)$ is a (longitudinal) function to be determined. Substituting the above assumption in (9.4a,b) yields (see trail "B" of S.1.6.5.2)

$$\varepsilon_z=w_{0,z}-xu_{0,zz}+x\Psi_{0,z}, \qquad \gamma_{yz}=0, \qquad \gamma_{xz}=\Psi_0. \qquad (9.28)$$

Using (9.5), the stress components σ_z, τ_{yz} and τ_{xz} are obtained. Substituting the above stress components in two of the equilibrium equations, (9.19:a,c), for $\overline{P}_x(z)=0$ and $\overline{P}_z(z)\neq 0$, and using (5.1), yields two equations that may be solved for $w_{0,z}$ and Ψ_0 as

$$w_{0,z}=\frac{\overline{P}_z(z)}{S_\Omega}\cdot\frac{Q_{55}}{Q_{33}Q_{55}-Q_{35}^2}, \qquad \Psi_0=-\frac{\overline{P}_z(z)}{S_\Omega}\cdot\frac{Q_{35}}{Q_{33}Q_{55}-Q_{35}^2}, \qquad (9.29)$$

or

$$w_{0,z} = \frac{\overline{P}_z(z)}{S_\Omega} q_{33}, \qquad \Psi_0 = \frac{\overline{P}_z(z)}{S_\Omega} q_{35}. \tag{9.30}$$

Note that the equilibrium equations, (9.19:b,d) are identically satisfied in this case.

Fulfilling the tip conditions of (9.25) for zero tip moments requires $u_{0,zz} = \Psi_{0,z}$ at the tip, but this identity will be assumed to be valid for the entire beam (i.e. for any z). Hence, (9.28) shows that $\varepsilon_z = w_{0,z}$, and with the aid of (9.5) and (9.29) one obtains

$$\sigma_z = Q_{33} w_{0,z} + Q_{35} \Psi_0 = \frac{\overline{P}_z(z)}{S_\Omega}, \qquad \tau_{xz} = 0. \tag{9.31}$$

Therefore, the differential equilibrium equation in z direction, (9.23), is satisfied as well since $Z_b = -\overline{P}_{z,z}/S_\Omega$.

The above solution yields some interesting observations. First, the edgewise curvature $u_{0,zz}$ emerge solely from the warping longitudinal derivative $\Psi_{0,z}$. This is a coupling effect that is not observed in uniform extension (such as the result of applying a tip force only). Verification of the above coupling effect, in which lateral bending is associated with a non-uniform distribution of the axial resultant load, is represented by the general and exact analysis presented in Chapter 6 (see (6.124)). An additional important observation is gained by considering the simple case of a tip axial load (i.e., $Z_b = 0$ and $\overline{P}_z(z) = P_z$), where (9.30:a) indicates that the longitudinal strain variation is given by

$$w_{0,z} = \frac{\overline{P}_z}{S_\Omega} q_{33}, \tag{9.32}$$

which shows that $1/q_{33}$ plays the role of an effective extensional elastic modulus, see also Example 5.1 and Remark 9.3. It is therefore clear that the reduction in the axial stiffness for a homogeneous coupled beam is fully reflected by the variation of the value of q_{33}.

The present solution also indicates that the longitudinal variation of the warping is proportional to $x\overline{P}_z(z)$, and that the shear strain may be written as $\gamma_{xz} = \frac{q_{35}}{q_{33}} \varepsilon_z$ (or $\gamma_{xz} = -\frac{Q_{35}}{Q_{55}} \varepsilon_z$). Hence the shear strain induced by the coupling module q_{35} (or Q_{35}), reaches values that are on the order of the axial strain. For a "clamped-free" beam with "clamped-II" version of the geometrical boundary conditions at its root (see (5.13)), the displacements in the case under discussion are given by (see integration procedure of S.1.2)

$$u = \frac{q_{35}}{S_\Omega} \int_0^z \overline{P}(z) dz - z \frac{q_{35}}{2S_\Omega} \overline{P}(0), \qquad w = \frac{q_{33}}{S_\Omega} \int_0^z \overline{P}(z) dz + x \frac{q_{35}}{2S_\Omega} \overline{P}(0). \tag{9.33}$$

Finally, the Extension-Bending coupling presented above may be expressed as

$$\frac{u_{0,zz}}{\overline{P}_{z,z}} = \frac{q_{35}}{S_\Omega}. \tag{9.34}$$

The above equation is in agreement with the complete and exact model derived in Chapter 6, see (6.184).

Remark 9.3 The solution presented by (9.33) may be compared with the general exact solution in (5.25a–c), (5.27) for uniform tensile force (namely, $\overline{P}_z = P_z = const.$), which for the material under discussion may be written as

$$u = \frac{P_z}{S_\Omega} \left(a_{13} x + \frac{a_{35}}{2} z\right), \qquad v = \frac{P_z}{S_\Omega} a_{23} y, \qquad w = \frac{P_z}{S_\Omega} \left(\frac{a_{35}}{2} x + a_{33} z\right). \tag{9.35}$$

Here, Version II of the reduced stress-strain relationships should be adopted since, on one hand, in this specific case, it does not violate the compatibility equations, and, on the other hand, it fulfills the boundary condition of a free outer surface, since $\overline{\sigma} = 0$. Hence, by adopting Version II for which $q_{ij} = a_{ij}$, the present approximate model yields the expressions of (9.35), however, the underlined terms for u and v are missing in the present analysis, since fundamentally, the in-plane deformation of the cross-section is not captured by the model. In this specific case, which complies with all compatibility equations (and is therefore fully integrable), one may calculate the strain components

$$\varepsilon_x = \frac{P_z}{S_\Omega} a_{13}, \qquad \varepsilon_y = \frac{P_z}{S_\Omega} a_{23}, \qquad \gamma_{xy} = 0 \tag{9.36}$$

from (9.16), and integrate them to recover the missing terms. Note again that such a "post-analysis" is not always possible, since in the general case, Version II violates the compatibility equations.

Example 9.1 *A Beam Under Uniform Axial Body Force.*

Consider a clamped-free beam, which is oriented vertically, and extended axially by its own weight. In this case $Z_b = const.$ and $\overline{P}_z(z) = Z_b S_\Omega (l - z)$. For the reasons discussed in Remark 9.3, we select here Version II, and therefore, (9.33) hold as well, while u and w are given by

$$u = -\frac{Z_b}{2} a_{35} z (z - l), \qquad w = \frac{Z_b}{2} \left(a_{35} x l - a_{33} z^2 + 2 a_{33} l z \right). \tag{9.37}$$

Hence, the beam will ("parabolically") incline in the x direction.

9.2.3.2 Tip Edgewise Force

We shall derive an analytical solution for a beam of a rectangular cross-section that occupies the domain $|x| < \frac{\tilde{a}}{2}$, $|y| < \frac{\tilde{b}}{2}$. The beam undergoes a tip bending force in the x direction, P_x (no other surface or body forces are applied). The solution is derived by assuming $v_0 = \phi = 0$, and the following stress components (see trail "B" of S.1.6.5.2):

$$\sigma_z = -\frac{12 P_x}{\tilde{a}^3 \tilde{b}} x (l - z) + \frac{12 P_x}{\tilde{a}^3 \tilde{b}} \frac{q_{35}}{q_{33}} \left(x^2 - \frac{\tilde{a}^2}{12}\right), \qquad \tau_{yz} = 0, \qquad \tau_{xz} = -\frac{6 P_x}{\tilde{a}^3 \tilde{b}} [x^2 - (\frac{\tilde{a}}{2})^2]. \tag{9.38}$$

Substituting (9.38) in (9.19) shows that $\overline{P}_y = \overline{P}_z = \overline{M}_z = 0$ while $\overline{P}_x = P_x$. In addition, the tip conditions (9.25) are satisfied (i.e., $M_x = M_y = 0$) and so is the differential equilibrium equation (9.23). Moreover, the contour boundary condition (9.26) is satisfied for a free outer surface (namely, $Z_s = 0$) over both $x = \pm \tilde{a}/2$ and $y = \pm \tilde{b}/2$ edges. Hence this solution satisfies all requirements of the approximate model, and using (9.10) the corresponding strain components $\varepsilon_z, \gamma_{yz}, \gamma_{xz}$ may be obtained. Integration of these strain components while the other components are ignored (see general procedure in of S.1.2) under the root boundary condition of (9.24), yields the displacement components. Using the general format of (9.3) one may write

$$u_0 = \frac{2 P_x}{\tilde{a}^3 \tilde{b}} q_{33} z^2 (3l - z) + \underline{\frac{P_x}{2 \tilde{a} \tilde{b}} \left(-\frac{q_{35}^2}{q_{33}} + \frac{3}{2} q_{55}\right) z}, \qquad w_0 = \frac{P_x}{2 \tilde{a} \tilde{b}} q_{35} z, \tag{9.39}$$

where employing the underlined term converts the boundary conditions of the above result from "clamped-I" to "clamped-II" (see S.5.1.3). The out-of-plane warping takes the form

$$\Psi^w = \frac{P_x}{\tilde{a}^3 \tilde{b}} [2(2 \frac{q_{35}^2}{q_{33}} - q_{55}) x^3 - 6 q_{35} x^2 (l - z) + \tilde{a}^2 (-\frac{q_{35}^2}{q_{33}} + \frac{3}{2} q_{55}) x]. \tag{9.40}$$

The above solution shows that the material anisotropy induces coupling effects (which are reflected by $q_{35} \neq 0$) of linear axial extension due to lateral bending. Also note that the warping in this case has a cubic distribution in the x direction, which is also a function of z but is independent of y.

At this point, one should select a version of the constitutive relations. As shown (see (9.12)), if Version I is selected, the above solution violates the boundary conditions (5.16a, b) (for a free outer surface). If Version II is selected, the above boundary conditions are fulfilled in an exact manner, however, this yields $\varepsilon_x = a_{13}\sigma_z + a_{15}\tau_{xz}$, $\varepsilon_y = a_{23}\sigma_z + a_{25}\tau_{xz}$ and $\gamma_{xy} = a_{46}\tau_{xz}$, for which two out of the six compatibility equations ((9.27)) are violated, see also Remark 9.2. It is reasonable to assume that Version II will provide a better approximation in this case where the entire outer surface is free of traction. If Version II is selected, $q_{ij} = a_{ij}$, and the bending in the x direction takes a form similar to the isotropic case (where $1/a_{33}$ plays the role of Young's modulus).

The Bending-Extension coupling presented in this section may be summarized as

$$\frac{w_{0,z}}{P_x} = \frac{q_{35}}{2\tilde{a}\tilde{b}}. \tag{9.41}$$

The above solution for the influence of a tip edgewise force has been derived for a rectangular cross-section but it may be slightly modified to other cross-section shapes as far as the beam axis deformation is concerned. Such a generalization may be carried out by replacing $\tilde{a}^3\tilde{b}$ with $12I_y$ and $\tilde{a}\tilde{b}$ by S_Ω in (9.39). When applied to different solid cross-sections, this may serve as a first approximate estimation.

9.2.3.3 Tip Beamwise Force

We now consider the clamped-free beam discussed in S.9.2.3.2 that undergoes a tip beamwise force P_y. The solution in this case is initiated by assuming $u_0 = w_0 = 0$, and a warping function that is a sum of an "isotropic" warping function for a beam of a rectangular cross-section in bending, and an "isotropic" warping function for a beam of an elliptical cross-section in torsion, namely,

$$\Psi^w = -\frac{6P_y}{\tilde{a}\tilde{b}^3 Q_{44}}\left[\frac{y^3}{3} - \left(\frac{\tilde{b}}{2}\right)^2 y\right] - \phi_{,z}xy. \tag{9.42}$$

Using the strain expressions of (9.4a,b) and the constitutive relations of (9.5), the following stress components are obtained:

$$\sigma_z = -y[Q_{33}(v_{0,zz} + \underline{\phi_{,zz}x}) + 2Q_{35}\phi_{,z}], \tag{9.43a}$$

$$\tau_{yz} = -\frac{6P_y}{\tilde{a}\tilde{b}^3}\left(y^2 - \frac{\tilde{b}^2}{4}\right), \tag{9.43b}$$

$$\tau_{xz} = -y[Q_{35}(v_{0,zz} + \underline{\phi_{,zz}x}) + 2Q_{55}\phi_{,z}]. \tag{9.43c}$$

As shown, τ_{yz} satisfies the contour boundary condition at $y = \pm\tilde{b}/2$, see (9.26). To minimize the violation of the contour boundary condition at $x = \pm\tilde{a}/2$ we neglect the underlined terms in (9.43a–c) and equate τ_{xz} to zero, see Remark 9.4. This yields the very important twist-bending curvature relation

$$\frac{\phi_{,z}}{v_{0,zz}} = -\frac{Q_{35}}{2Q_{55}} = \frac{q_{35}}{2q_{33}}, \tag{9.44}$$

which derives the Bending-Torsion coupling mechanism in symmetric beams. From the illustrative magnitude of this coupling, shown in Fig. 5.7(b), it is clear that it is on the order of a unit (i.e., $\phi_{,z}$ is on the order of $v_{0,zz}$). In addition, by neglecting the underlined term in (9.43a,c) and using (9.44), the differential equilibrium equation (9.23) yields

$$v_{0,zzz} = -\frac{12P_y}{\tilde{a}\tilde{b}^3} \cdot \frac{Q_{55}}{Q_{33}Q_{55} - Q_{35}^2}. \tag{9.45}$$

The natural boundary conditions at the beam tip (9.25) show that $v_{0,zz}(l) = 0$ there. Carrying out the longitudinal integrations of $v_{0,zz}$, $v_{0,z}$, v_0 and ϕ for a clamped-free beam, yields the following bending and twist distributions:

$$v_0 = \frac{2P_y}{\tilde{a}\tilde{b}^3} q_{33} z^2 (3l - z) + \underline{\frac{3}{4} \frac{P_y}{\tilde{a}\tilde{b}} \frac{z}{q_{44}}}, \qquad \phi = \frac{3P_y}{\tilde{a}\tilde{b}^3} q_{35} z (2l - z), \tag{9.46}$$

where employing the underlined term converts the boundary conditions of the above result from "clamped-I" to "clamped-II", see S.5.1.3. The expression of (9.46) for ϕ demonstrates the three-dimensional nature of the warping, due to the second term in (9.42).

Re-substitution of the above results in the following complete stress expressions (with no neglect of the $\phi_{,zz}xy$ term) yields:

$$\sigma_z = -\frac{12P_y}{\tilde{a}\tilde{b}^3} y (l - z) + \frac{6P_x}{\tilde{a}\tilde{b}^3} \cdot \frac{q_{35}q_{55}}{q_{35}q_{55} - q_{35}^2} xy, \tag{9.47a}$$

$$\tau_{yz} = -\frac{6P_y}{\tilde{a}\tilde{b}^3} \left(y^2 - \frac{\tilde{b}^2}{4}\right), \qquad \tau_{xz} = -\frac{6P_y}{\tilde{a}\tilde{b}^3} \cdot \frac{q_{35}^2}{q_{35}q_{55} - q_{35}^2} xy. \tag{9.47b}$$

Selection of Version II for the constitutive relations seems to be natural in this case since the entire outer surface is traction-free. In addition, no compatibility equations are violated. The Bending-Torsion coupling mechanism described by (9.44) may be verified by the complete and exact analysis of Chapter 6, see (6.162).

Remark 9.4 The neglect of the underlined terms in (9.43a–c) is founded on the fully justified assumption $|\Psi^w_{,z}| \ll |y v_{0,zz}|$, that may be phrased as *assuming that the longitudinal warping derivative is small compared with the longitudinal derivative of the axial displacements induced by the beam rotation*, see also Fig. 5.3(a).

Using the expressions of (9.47a,b) to re-evaluate the governing equations and boundary conditions show that they are all satisfied except for the differential equilibrium equation and the boundary condition for τ_{xz} at $x = \pm\tilde{a}/2$. This violation may be corrected by a superposition of a two-dimensional solution that consists of two stress components, τ^0_{yz} and τ^0_{xz} only. This solution has to fulfill the equilibrium equation

$$\tau^0_{xz,x} + \tau^0_{yz,y} = -\tau_{xz,x} \tag{9.48}$$

under the boundary conditions $\tau^0_{xz} = -\tau_{xz}$ for $x = \pm\tilde{a}/2$ and $\tau^0_{yz} = 0$ for $y = \pm\tilde{b}/2$, while producing zero resultant forces $\overline{P}_x$, $\overline{P}_y$ and moment $\overline{M}_z$. A solution of such a plane problem is discussed in S.3.3 (see S.3.3.1, S.3.3.2 where τ_{xz} of (9.48) is denoted τ^P_{xz}), and ought to be relatively small and of minor importance.

Remark 9.5 As shown above, the twist, $\theta \equiv \phi_{,z}$, is a linear function of z, see (9.46). Therefore, one may not employ the technique used in MON13z (or simpler) beam analyses, to determine the location of a "shear center" (see Remark 7.6). This concept may be presented as looking for the distance, say x_{sc}, by which one may shift the tip load along the x-axis in order to cancel out the twist. This distance may be determined by $\frac{P_y x_{sc}}{D} = -\theta$ (where D is the beam torsional rigidity that will be derived in S.9.2.3.4), which yields

$$x_{sc} = -\frac{6D}{\widetilde{a}\widetilde{b}^3} q_{35}(l-z). \tag{9.49}$$

The above may be interpreted as *the offset required to zero out twist at z*, and clearly the twist angle rate, θ, will not vanish for all z values and some twist angle (ϕ) will be created. To zero out the total twist angle at z, x_{sc} should be selected so that

$$\frac{3}{\widetilde{a}\widetilde{b}^3} q_{35} z(2l-z) + \frac{x_{sc}}{D} z = 0. \tag{9.50}$$

We therefore conclude again that since the twist induced by bending due to the coupling mechanism in symmetric beams may not be simulated by an offset of the tip load, no "equivalent shear center" may be defined.

9.2.3.4 Tip Torsional Moment

When a tip torsional moment is applied, a more complicated analytical solution should be considered. In this case $\overline{M}_z(z) = M_z$ and all other tip loads vanish. As a first step, for given material properties, Q_{33}, Q_{44}, Q_{55} and Q_{35}, we shall solve the "uncoupled" case where $Q_{35} = 0$ (while Q_{33}, Q_{44} and Q_{55} remain unchanged, see also Remark 9.6). We denote the warping function in this solution $\varphi(x,y)$, where the warping itself is given by $\phi_{,z}\varphi$. As previously indicated, $\phi_{,z}$ is the beam twist (i.e. the rotation angle per unit length) given by $\phi_{,z} = M_z/D$, while D is the torsional rigidity in this uncoupled case. The solution for such an uncoupled case is relatively simple (see S.7.2.1 or (Sokolnikoff, 1983)) as the only nonzero stress components are

$$\tau_{xz} = \frac{M_z}{D} Q_{55}(\varphi_{,x} - y), \qquad \tau_{yz} = \frac{M_z}{D} Q_{44}(\varphi_{,y} + x). \tag{9.51}$$

Having $\varphi(x,y)$ and D of the uncoupled case, we adopt trail "B" of S.1.6.5.2 and assume that in the coupled case,

$$\phi_{,z} = \frac{M_z}{D} + \widetilde{\alpha}, \tag{9.52a}$$

$$\Psi^w = \frac{M_z}{D}\varphi - \widetilde{\alpha}xy + \widetilde{\beta}x^2 + \widetilde{\gamma}x + \widetilde{\delta} \tag{9.52b}$$

where $\widetilde{\alpha}$, $\widetilde{\beta}$, $\widetilde{\gamma}$, and $\widetilde{\delta}$ are coefficients to be determined. All of the above coefficients should vanish in the uncoupled case, and therefore, must be functions of Q_{35} that vanish for $Q_{35} = 0$. As will be shown later on, none of the above coefficients depends on z, and thus, $\Psi^w_{,z} = 0$. Therefore, the shear stress components in the coupled case become

$$\tau_{yz} = Q_{44}\left(x\phi_{,z} + \Psi^w_{,y}\right), \qquad \tau_{xz} = Q_{35}\left(w_{0,z} - xu_{0,zz} - yv_{0,zz}\right) + Q_{55}\left(-y\phi_{,z} + \Psi^w_{,x}\right). \tag{9.53}$$

We now suggest the hypothesis (which will be proved to be exact later on), that the stress components in the coupled case ((9.53)), should be equal to those of the uncoupled case, namely, (9.51), respectively, see also Remark 9.7. This occurs only if

$$\widetilde{\alpha} = -\frac{Q_{35}}{2Q_{55}} v_{0,zz}, \qquad \widetilde{\beta} = \frac{Q_{35}}{2Q_{55}} u_{0,zz}, \qquad \widetilde{\gamma} = -\frac{Q_{35}}{Q_{55}} w_{0,z}. \tag{9.54}$$

Using (9.4a,b), (9.5), the axial stress becomes

$$\sigma_z = \left(w_{0,z} - xu_{0,zz} - yv_{0,zz}\right)\left(Q_{33} - \frac{Q_{35}^2}{Q_{55}}\right) + Q_{35}\frac{M_z}{D}\left(\varphi_{,x} - y\right). \tag{9.55}$$

We now substitute the above stress in (9.19:c) for $\overline{P}_z = 0$, and in (9.25) for $M_x = M_y = 0$. These substitutions yield three equations to be solved for $u_{0,zz}$, $v_{0,zz}$ and $w_{0,z}$, which by employing (5.1) become

$$\{u_{0,zz}, v_{0,zz}, w_{0,z}\} = \{\frac{I_{\varphi x}}{I_y}, \frac{I_{\varphi y} - I_x}{I_x}, -\frac{I_\varphi}{S_\Omega}\}\frac{M_z Q_{35} Q_{55}}{D(Q_{33}Q_{55} - Q_{35}^2)}, \tag{9.56}$$

where

$$\{I_\varphi, I_{\varphi x}, I_{\varphi y}\} = \iint_\Omega \{1, x, y\}\varphi_{,x}. \tag{9.57}$$

The reader should note that (9.19:a,b,d), (9.26) are all satisfied as the stresses involved are identical to those of the uncoupled case. The same is true for the differential equation of equilibrium (9.23) since $\sigma_{z,z} = 0$.

We now integrate the first of (9.51) over the cross-section area and equate it to zero (since $\overline{P}_x = 0$). This immediately shows that $I_\varphi = 0$ for any cross-section. Hence, $w_{0,z} = 0$.

We next exploit some basic characteristics of the shear stresses in torsion. The discussion for MON13z material in S.7.2.1 is applicable in the present context as the uncoupled case may be viewed as the orthotropic case as well. As shown there, $\tau_{xz} = \psi_{,y}$, and by substituting $\psi = \Lambda$ in (3.9a:a) of Remark 3.2, one obtains $\iint_\Omega \tau_{xz} x = 0$. Hence, (9.51:a) shows that $I_{\varphi x} = 0$ for any cross-section as well, and therefore, $u_{0,zz} = 0$.

Multiplying (9.51:a) by y and noting that its contribution to the torsional moment is always half of the total moment (see also S.7.2.1.2, (7.29)), yields $\frac{M_z}{2} = -\frac{M_z}{D}Q_{55}(I_{\varphi y} - I_x)$, and by rearranging

$$I_{\varphi y} = I_x - \frac{D}{2Q_{55}}. \tag{9.58}$$

Substituting the above in the second of (9.56) yields

$$v_{0,zz} = -\frac{M_z}{2I_x} \cdot \frac{Q_{35}}{Q_{33}Q_{55} - Q_{35}^2}. \tag{9.59}$$

As shown, the coupling effect emerges in the form of a constant bending curvature (which reflects parabolic bending). The above may also be verified by the results of the exact and general analysis of Chapter 6. Using (9.52a), (9.54) one obtains

$$\phi_{,z} = \frac{M_z}{D}(1 + \frac{D}{4I_x Q_{55}} \cdot \frac{Q_{35}^2}{Q_{33}Q_{55} - Q_{35}^2}). \tag{9.60}$$

Hence, the ratio of the additional twist (compared with the uncoupled case) to the induced bending curvature is identical to the one presented in (9.44), namely

$$\frac{\Delta\phi_{,z}}{v_{0,zz}} = -\frac{Q_{35}}{2Q_{55}} = \frac{q_{35}}{2q_{33}}, \tag{9.61}$$

see also Remark 9.8. The *equivalent torsional rigidity*, $D_{eq} = M_z/\phi_{,z}$, in the coupled case may be written as

$$\frac{D_{eq}}{D} = (1 + \frac{D}{D_0})^{-1}, \qquad D_0 = 4I_x Q_{55}\frac{Q_{33}Q_{55} - Q_{35}^2}{Q_{35}^2} = 4I_x\frac{q_{33}}{q_{35}^2}. \tag{9.62}$$

The above result may also be explained in terms of the general formulation of Chapter 6 (see (6.137)). The D/D_0 represents the ratio between the torsional rigidity, D, and the bending stiffness, $\frac{I_x}{q_{33}}$, since it may be written as

$$\frac{D}{D_0} = \frac{D}{4}\left(\frac{q_{35}}{q_{33}}\right)^2\left(\frac{q_{33}}{I_x}\right)^{-1}. \tag{9.63}$$

As schematically shown in Fig. 9.3, in the coupled case, there is a loss of torsional rigidity for increasing values of D/D_0.

To further discuss the above result, we assume that an orthotropic material is in hand, and therefore, one may vary the material orientation angle, say θ_y (about the y-axis) in order to obtain a family of MON13y materials, which are currently under discussion (see Fig. 2.2). While changing the above angle, two different mechanisms take place. First, the shear elastic moduli Q_{44} and Q_{55} are substantially changed (typically increase up to a maximum value and then decrease again as θ_y approaches 90°), and the "uncoupled torsional rigidity", D, behaves in a similar way. Yet, the "uncoupled torsional rigidity" does not represent any physical quantity as it is a mathematical definition of a hypothetical problem (in which the elastic moduli are identical except for Q_{35} that is set to zero). The second mechanism is the reduction of the actual torsional rigidity, D_{eq}, which according to (9.62) and Fig. 9.3 is always smaller than D. Thus, in general, the variation of the effective torsional rigidity will not be monotonic (this effect is also shown in Fig. 9.4(a) for the case of an elliptical cross-section).

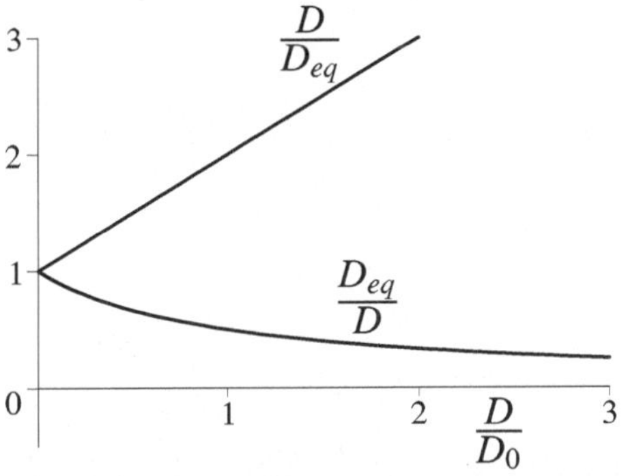

Figure 9.3: The ratio of the coupled torsional rigidity to the uncoupled one (and its inverse).

Remark 9.6 The so-called "uncoupled" case is characterized by Q_{33}, Q_{44} and Q_{55} of the actual problem while Q_{35} is set to zero. Therefore, quantities that were determined using q_{ij} for such an uncoupled case should be adopted by substituting $q_{33} = 1/Q_{33}$, $q_{44} = 1/Q_{44}$ and $q_{55} = 1/Q_{55}$.

Remark 9.7 The shear stress resultants in the "uncoupled" and in the "coupled" cases should be identical as they must add up to the same torsional moment and (zero) transverse forces. Yet, since the moment at the tip is applied in an integral manner, it is not retroactively clear that the shear stress distributions are identical.

Remark 9.8 For a very thin rectangular cross-section (of width $\widetilde{d}$ and thickness $\widetilde{h}$ where $\widetilde{h} \ll \widetilde{d}$), (10.69) shows that $D = \frac{1}{3}\widetilde{d}\widetilde{h}^3 Q_{55}$, while $I_x = \frac{1}{12}\widetilde{d}\widetilde{h}^3$, and therefore, $D = 4I_x Q_{55}$. Then, from (9.59), (9.60) follow

$$\frac{v_{0,zz}}{\phi_{,z}} = \frac{2q_{35}}{q_{55}}, \tag{9.64}$$

which stands for the Torsion-Bending coupling in thin homogeneous lamina (see also S.5.3.1, Example 5.3).

Example 9.2 *Torsional Rigidity of Elliptical Beam.*

To illustrate the above derivation, we adopt the solution for a beam of elliptical cross-section (of semi-axes $\widetilde{a}$ and $\widetilde{b}$) in the uncoupled case presented in S.7.2.1 and Example 7.1. In this case,

D is given by (7.38), and should be written as

$$D = \frac{\pi \tilde{a}^3 \tilde{b}^3}{q_{55}\tilde{a}^2 + q_{44}\tilde{b}^2}, \tag{9.65}$$

while D_0 is determined with $I_x = \frac{1}{4}\pi\tilde{a}\tilde{b}^3$. Hence, by (9.62) the ratio of the coupled torsional rigidity to the uncoupled one is

$$\frac{D_{eq}}{D} = \frac{q_{33}(q_{55}\tilde{a}^2 + q_{44}\tilde{b}^2)}{q_{33}(q_{55}\tilde{a}^2 + q_{44}\tilde{b}^2) + \tilde{a}^2 a_{35}^2}. \tag{9.66}$$

By adopting the uncoupled warping function φ of (7.39), we obtain from (9.52b), (9.59), (9.60)

$$v_{0,zz} = \frac{2M_z}{\pi\tilde{a}\tilde{b}^3} q_{35}, \tag{9.67a}$$

$$\phi_{,z} = \frac{M_z}{\pi\tilde{a}^3\tilde{b}^3}(\tilde{b}^2 q_{44} + \tilde{a}^2 q_{55} + \tilde{a}^2 \frac{q_{35}^2}{q_{33}}), \quad \Psi^w = \frac{M_z}{\pi\tilde{a}^3\tilde{b}^3}(\tilde{b}^2 q_{44} - \tilde{a}^2 q_{55} - \tilde{a}^2 \frac{q_{35}^2}{q_{33}})xy. \tag{9.67b}$$

By exploiting the fact that τ_{xz} and τ_{yz} are identical to those of the uncoupled case given by (9.51), and by using (9.10), (9.55) one may write

$$\varepsilon_z = y\frac{2M_z}{\pi\tilde{a}\tilde{b}^3}[\frac{q_{35}{}^3}{q_{33}\, q_{55} - q_{35}{}^2} - q_{35}], \tag{9.68a}$$

$$\gamma_{yz} = x\frac{2M_z}{\pi\tilde{a}^3\tilde{b}} q_{44}, \qquad \gamma_{xz} = y\frac{2M_z}{\pi\tilde{a}\tilde{b}^3}[\frac{q_{35}^2}{q_{33}}(\frac{q_{33}\, q_{55}}{q_{33}\, q_{55} - q_{35}{}^2} - 1) - q_{55}]. \tag{9.68b}$$

For Version II of the constitutive relations, one should set $q_{ij} = a_{ij}$ in the above strain expressions and use (9.16) to write

$$\varepsilon_x = y\frac{2M_z}{\pi\tilde{a}\tilde{b}^3}[a_{15} - \frac{a_{13}\, a_{35}{}^3}{a_{33}\,(a_{33}\, a_{55} - a_{35}{}^2)}], \quad \varepsilon_y = y\frac{2M_z}{\pi\tilde{a}\tilde{b}^3}[a_{25} - \frac{a_{23}\, a_{35}{}^3}{a_{33}\,(a_{33}\, a_{55} - a_{35}{}^2)}], \tag{9.69a}$$

$$\gamma_{xy} = x\frac{2M_z}{\pi\tilde{a}^3\tilde{b}} a_{46}, \tag{9.69b}$$

which clearly satisfy the conditions on the traction-free outer surface and do not violate compatibility. Deformation is determined by the standard scheme of S.1.2. The above solution may be employed to determine the torsional rigidity characteristic of a coupled elliptical beam in two ways. First, as already discussed, one may exploit an orthotropic material and study the influence of its lamination angle, θ_y, which is reflected via its influence on the elastic moduli, see D, D_{eq} and their ratio in Fig. 9.4 (that has been drawn for ellipse with $\tilde{a}/\tilde{b} = 2$ made of typical Graphite/Epoxy orthotropic lamina).

In addition, it is interesting to explore the influence of the "beam-thickness ratio", $\tilde{a}/\tilde{b}$, on D_{eq}. In the isotropic case, $D_{iso} = G\frac{\pi\tilde{a}^3\tilde{b}^3}{\tilde{a}^2+\tilde{b}^2}$, and therefore the non-dimensional torsional rigidity $\overline{D} = \frac{D_{eq}}{I_x G}$ becomes $\overline{D} = \frac{4\tilde{a}^2}{\tilde{a}^2+\tilde{b}^2}$. In the anisotropic case, we define $\overline{D}$ by the elastic shear modulus G_{23} as $\overline{D} = \frac{D_{eq}}{I_x G_{23}}$. Figure 9.5 presents the above values for typical Graphite/Epoxy orthotropic lamina. It indicates that $\overline{D}$ varies between ~ 1.5 to ~ 3 (as opposed to a value between 2 and 4 in the isotropic case). Figure 9.5 therefore supplies a quick estimation of the torsional rigidity.

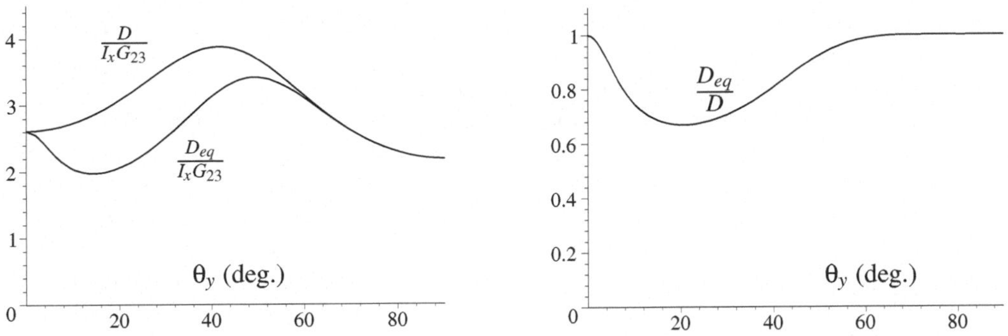

(a) The coupled and uncoupled torsional rigidity.

(b) The ratio of coupled to uncoupled torsional rigidity.

Figure 9.4: Torsional rigidity as a function of the lamination angle for an elliptical cross-section.

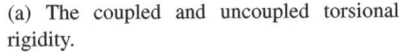

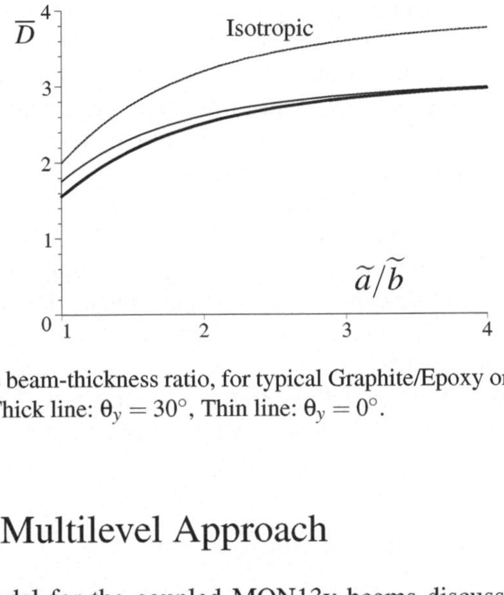

Figure 9.5: $\overline{D}$ vs. ellipse beam-thickness ratio, for typical Graphite/Epoxy orthotropic lamina, compared with the isotropic case. Thick line: $\theta_y = 30°$, Thin line: $\theta_y = 0°$.

9.3 An Exact Multilevel Approach

In what follows, a model for the coupled MON13y beams discussed in S.9.2 that contains no approximations or neglected components, but is based on a multilevel iterative solution procedure, is presented. In general, the analysis employs a series of solution levels that are properly interconnected, and employed one at a time.

The analysis proposed here deals with three levels of solutions. Each solution level is focused on a different aspect of the structural response. The level hierarchy emerges from the fact that all "upwards" data (from a lower level to an upper one) is of a primary importance for the upper level solution, while all "downwards" data (from an upper level to a lower one) is of a secondary (or minor) importance for the lower level. As a result, qualitatively speaking, all downwards information may be viewed as a "small correction" for the lower level, which ensures excellent convergence properties of the entire procedure.

To derive a multilevel analysis, one first needs to identify the basic physical phenomena involved. Then, a way needs to be found to *gradually include* the above phenomena according to their overall importance and the changes they might induce on other physical phenomena.

When dealing with anisotropic beams, the extension, the two global lateral bending curvature components and the twist are the primary group of parameters that apply. Then, the cross-section in-plane stress components constitute a set of additional parameters that should be accounted for. Finally, the interlaminar effects have to be considered.

As mentioned above, solutions of lower levels dramatically influence solutions of higher levels but not vice versa. Yet, all levels are consistently and fully coupled. Hence, once the overall multilevel process converges, all equations of equilibrium and compatibility and all boundary conditions are satisfied simultaneously, and therefore constitute the (unique) three-dimensional solution of this linear problem. Practically, the present methodology enables modeling and solution of the primary phenomena at each level while ignoring irrelevant data from other levels. On top of the resulting efficiency, this feature enables a clear insight into the phenomena associated with each solution level.

To clarify the contribution of the proposed methodology, it should be mentioned that the literature contains *numerical methodologies* that are titled as "multiscale", "global-local" or "hierarchical" approaches (e.g. (Fish *et al.*, 1994), (Fish and Markolefas, 1994), (Ghosh *et al.*, 1995), (Srinivasan *et al.*, 1997), (Mitchell and Reddy, 1998)). Such analyses are based on numerical techniques, usually on finite-element based solution schemes, that allow various levels of meshing refinement at various locations over the structure. Subsequently, these solutions employ detailed meshing of areas of interest, which enables the capturing of fine local effects that are of smaller scale and could not be adequately modeled by the lower level (coarse) analysis, although all physical phenomena were modeled in both levels. Generally speaking, the main focus of such analyses is the tailoring of the different numerical meshing used at different levels. In contrast with the above class of methods, and although one may adopt some numerical techniques in order to implement the present analysis as well, the essence of the present approach is *independent of any numerical aspect and level of discretization.* In fact, the present approach will be demonstrated by some *analytic* examples in what follows. To make this point even clearer, it should be mentioned that the multilevel strategy offered in what follows is not a "grid correction" but rather a hierarchy of physical phenomena, in which higher level phenomena are determined in the background posed by lower level phenomena (and subsequently hardly influence this background state).

The reader is also referred to the model reported in (Savoia *et al.*, 1993) which has a relevancy to the present derivation as it also employs an iterative variational approach for a laminated beam model.

9.3.1 Displacements and Stress-Strain Relationships

As already indicated, the proposed model is exact and therefore includes all displacement components. However, for convenience purposes, instead of employing generic $u(x,y,z)$, $v(x,y,z)$ and $w(x,y,z)$ displacement functions, the following form (which is a natural extension of (9.3)) is adopted:

$$u = u_0 - y\phi + \Psi^u, \qquad v = v_0 + x\phi + \Psi^v, \qquad w = w_0 - xu_{0,z} - yv_{0,z} + \Psi^w. \tag{9.70}$$

Here, $u_0(z)$, $v_0(z)$, $w_0(z)$ are the beam axis displacement components in the x, y and z directions, respectively and $\phi(z)$ is the cross-section twist angle. In addition, $\Psi^u(x,y,z)$ and $\Psi^v(x,y,z)$ are the components of the in-plane warping in the x and y directions, respectively, while $\Psi^w(x,y,z)$ is the out-of-plane warping. These three generic warping functions are superimposed upon the above mentioned u_0, v_0, w_0 and ϕ_0 deformation components and are assumed to be of zero average value over the cross-section area (i.e., $\iint_\Omega \Psi^u = \iint_\Omega \Psi^v = \iint_\Omega \Psi^w = 0$). In this linear case, no distinction is made between the deformed and the undeformed directions and the order of superposition of the deformation components is therefore immaterial.

The above separation of the displacements into cross-section components (u_0, v_0, w_0 and ϕ) and local warping components (Ψ^u, Ψ^v and Ψ^w) is convenient due to the fact that the beam slenderness induces cross-section displacements, which are orders of magnitude higher than the warping components. Yet, these warping deformation components play a key role in the involved physical phenomena.

In the case under discussion we employ the exact and full stiffness matrix for MON13y materials, which according to (9.1) is written as

$$\begin{Bmatrix} \sigma_x \\ \sigma_y \\ \sigma_z \\ \tau_{yz} \\ \tau_{xz} \\ \tau_{xy} \end{Bmatrix} = \begin{bmatrix} A_{11} & A_{12} & A_{13} & 0 & A_{15} & 0 \\ & A_{22} & A_{23} & 0 & A_{25} & 0 \\ & & A_{33} & 0 & A_{35} & 0 \\ & & & A_{44} & 0 & A_{46} \\ & Sym. & & & A_{55} & 0 \\ & & & & & A_{66} \end{bmatrix} \begin{Bmatrix} \varepsilon_x \\ \varepsilon_y \\ \varepsilon_z \\ \gamma_{yz} \\ \gamma_{xz} \\ \gamma_{xy} \end{Bmatrix}. \tag{9.71}$$

We shall also use the sub-matrices notation of (9.9).

9.3.2 Definition of Solution Levels

The problem is divided into three solution levels while as stated before, each level is responsible for the modeling of a different physical phenomenon.

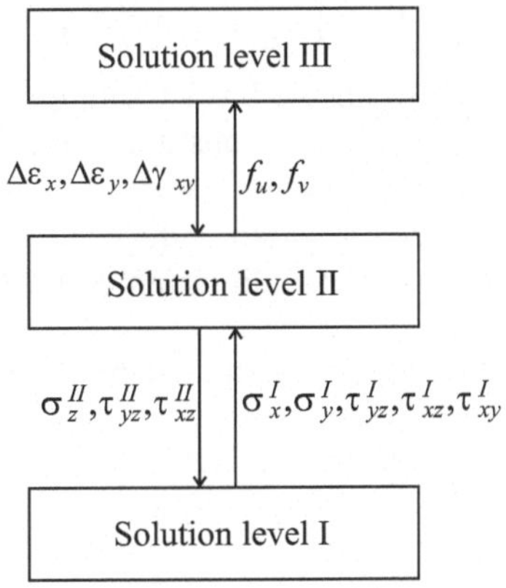

Figure 9.6: Hierarchy of solution levels and their mutual influences.

Figure 9.6 describes the solution levels and their mutual influences. The first (lowest) level is denoted “level I” and includes the displacements u_0, v_0, w_0 and ϕ and the out-of-plane warping $\Psi^w(x,y,z)$. This analysis level ignores the in-plane warping and any of the interlaminar effects related to these components, and is similar to the analysis in S.9.2 with a selection of Version I of the constitutive relations. Yet, since this level contains the out-of-plane warping, it does take care of the interlaminar effects related to it in a non-homogeneous cross-section. Subsequently, a (higher) “level II” solution is aimed towards the determination of the in-plane stress components, namely, the components σ_x, σ_y, τ_{xy} and the in-plane warping components Ψ^v and Ψ^w, while in the case of a non-homogeneous cross-section, the requirements for deformation and loads continuity along the interface line are ignored. Further on, the (highest) “level III” solution is carried out to refine the solution by adjusting deformation and loads along the interlaminar lines. All of the above solutions levels are interconnected (see Fig. 9.6), or in other words, all constraints and effects produced by higher levels are fed downwards and affect the lower level solutions and vice versa.

The following discussion contains a detailed description of each level analysis, and its interaction with other levels, along with a description of their execution manner.

9.3.2.1 Level I

As already indicated, a level I solution includes the cross-section displacements u_0, v_0, w_0 and ϕ and the out-of-plane warping Ψ^w only. Thus, the influence of the in-plane warping components is accounted for by the data obtained from solution levels II in the generic form of the stress distributions σ_z^{II}, τ_{yz}^{II}, τ_{xz}^{II}, see Fig. 9.6. In essence, a level I solution (which is similar to the solution in S.9.2) contains only the normal strain ε_z, and the shear strains γ_{yz} and γ_{xz}, which are given by (9.4a,b).

Based on (9.71) and the above discussion, one may define a "level problem", the stress components of which are denoted by asterisks as

$$\begin{Bmatrix} \sigma_z^* \\ \tau_{yz}^* \\ \tau_{xz}^* \end{Bmatrix} = \mathbf{D}_\sigma \cdot \begin{Bmatrix} \varepsilon_z \\ \gamma_{yz} \\ \gamma_{xz} \end{Bmatrix}, \tag{9.72}$$

where $\mathbf{D}_\sigma$ is given by (9.9), and the stress vector on the l.h.s. is

$$\begin{Bmatrix} \sigma_z^* \\ \tau_{yz}^* \\ \tau_{xz}^* \end{Bmatrix} = \begin{Bmatrix} \sigma_z \\ \tau_{yz} \\ \tau_{xz} \end{Bmatrix} - \begin{Bmatrix} \sigma_z^{II} \\ \tau_{yz}^{II} \\ \tau_{xz}^{II} \end{Bmatrix}. \tag{9.73}$$

Note that despite the similarities between the level I solution and the solution presented in S.9.2, we do not have here different versions of constitutive relations as we directly employ all elastic moduli.

For this level I problem, the global deformation of the beam discussed in S.9.2 is adopted by writing the integral form of the equations of equilibrium derived for this case as (see (9.19))

$$\{\overline{P}_x^*, \overline{P}_y^*, \overline{P}_z^*\} = \iint_\Omega \{\varepsilon_z A_{35} + \gamma_{xz} A_{55},\ \gamma_{yz} A_{44},\ \varepsilon_z A_{33} + \gamma_{xz} A_{35}\}, \tag{9.74a}$$

$$\overline{M}_z^* = \iint_\Omega x\gamma_{yz} A_{44} - y\,(\varepsilon_z A_{35} + \gamma_{xz} A_{55}), \tag{9.74b}$$

where

$$\{\overline{P}_x^*, \overline{P}_y^*, \overline{P}_z^*, \overline{M}_z^*\} = \{\overline{P}_x, \overline{P}_y, \overline{P}_z, \overline{M}_z\} - \iint_\Omega \{\tau_{xz}^{II}, \tau_{yz}^{II}, \sigma_z^{II}, x\tau_{yz}^{II} - y\tau_{xz}^{II}\}. \tag{9.75}$$

Further on, the differential equilibrium equation in the z direction given by (9.23) may be directly used as

$$\sigma_{z,z}^* + \tau_{xz,x}^* + \tau_{yz,y}^* + Z_b{}^* = 0 \tag{9.76}$$

where

$$Z_b{}^* = Z_b - \sigma_{z,z}^{II} - \tau_{xz,x}^{II} - \tau_{yz,y}^{II}. \tag{9.77}$$

The geometrical boundary conditions at a clamped ("root") end given by (9.24) are directly adopted as well, while the natural boundary conditions at the beam tip (which are based on equating the transverse moments, M_x and M_y, to those obtained by integration of the stresses over the tip cross-section area, Ω^t), take the form (see (9.25))

$$M_x - \iint_{\Omega^t} y\sigma_z^{II} = \iint_{\Omega^t} y\,(\varepsilon_z A_{33} + \gamma_{xz} A_{35}), \tag{9.78a}$$

$$M_y + \iint_{\Omega^t} x\sigma_z^{II} = -\iint_{\Omega^t} x\,(\varepsilon_z A_{33} + \gamma_{xz} A_{35}). \tag{9.78b}$$

The boundary conditions on $\partial\Omega$ are similar to those of (9.26),

$$\tau^*_{xz}\cos(\bar{\mathbf{n}},x)+\tau^*_{yz}\cos(\bar{\mathbf{n}},y)=Z^*_s, \tag{9.79}$$

where

$$Z^*_s=Z_s-\tau^{II}_{xz}\cos(\bar{\mathbf{n}},x)-\tau^{II}_{yz}\cos(\bar{\mathbf{n}},y). \tag{9.80}$$

As indicated by the above discussion, the solution in level I, which is formulated in terms of the $()^*$ parameters has to satisfy the integral equilibrium equations (9.74a–d), the differential equilibrium equation (9.76), the tip conditions (9.78a,b), and the contour boundary condition (9.79). However, when a non-homogeneous cross-section is under discussion, we adopt the notation of Fig. 3.1 and (3.1), and thus, in addition to the above requirements, one has to fulfill the interlaminar condition

$$[\tau^*_{xz}\cos(\bar{\mathbf{n}},x)+\tau^*_{yz}\cos(\bar{\mathbf{n}},y)]^{[j]}_{[i]}=-[\tau^{II}_{xz}\cos(\bar{\mathbf{n}},x)+\tau^{II}_{yz}\cos(\bar{\mathbf{n}},y)]^{[j]}_{[i]}, \tag{9.81}$$

which is created by the $[Z_s]^{[j]}_{[i]}=0$ requirement. On top of that, to ensure w continuity along the interlaminar surfaces, the values of the out-of-plane warping Ψ^w should remain continuous, namely,

$$[\Psi^w]^{[j]}_{[i]}=0. \tag{9.82}$$

For convenience, in all cases, the averaged out-of-plane warping is set to zero.

Within the iterative process, σ_z, τ_{yz} and τ_{xz} as may be derived from (9.73) and ε_z, γ_{yz} and γ_{xz} are *the net additional contributions of each level I iteration* to the respective stress and strain components. These values accumulate as iterations progress until convergence is achieved.

For a level II solution, the following stress components are transferred (except for σ^I_z, which is provided below only for the sake of completeness), see also (9.9):

$$\begin{Bmatrix}\sigma^I_z\\ \tau^I_{yz}\\ \tau^I_{xz}\end{Bmatrix}=\mathbf{D}_\sigma\cdot\begin{Bmatrix}\varepsilon^I_z\\ \gamma^I_{yz}\\ \gamma^I_{xz}\end{Bmatrix},\qquad \begin{Bmatrix}\sigma^I_x\\ \sigma^I_y\\ \tau^I_{xy}\end{Bmatrix}=\mathbf{B}_\sigma\cdot\begin{Bmatrix}\varepsilon^I_z\\ \gamma^I_{yz}\\ \gamma^I_{xz}\end{Bmatrix}, \tag{9.83}$$

where $\varepsilon^I_z,\gamma^I_{yz},\gamma^I_{xz}$ are the total values accumulated up to and including the current iteration.

9.3.2.2 Level II

The level II solution deals with the in-plane analysis of each cross-section. The associated strain components are, see (1.9a,b), (9.70),

$$\varepsilon_x=\Psi^u_{,x},\qquad \varepsilon_y=\Psi^v_{,y},\qquad \gamma_{xy}=\Psi^u_{,y}+\Psi^v_{,x}. \tag{9.84}$$

The solution in level II is driven by the stresses of level I, and by the body forces and surface loads in the x and y directions. Thus, the stress components of this level problem are given by (see also (9.9)):

$$\begin{Bmatrix}\sigma^*_x\\ \sigma^*_y\\ \tau^*_{xy}\end{Bmatrix}=\mathbf{A}_\sigma\cdot\begin{Bmatrix}\varepsilon_x\\ \varepsilon_y\\ \gamma_{xy}\end{Bmatrix}, \tag{9.85}$$

where

$$\begin{Bmatrix}\sigma^*_x\\ \sigma^*_y\\ \tau^*_{xy}\end{Bmatrix}=\begin{Bmatrix}\sigma_x\\ \sigma_y\\ \tau_{xy}\end{Bmatrix}-\begin{Bmatrix}\sigma^I_x\\ \sigma^I_y\\ \tau^I_{xy}\end{Bmatrix}. \tag{9.86}$$

Equilibrium within this solution is maintained by the two differential equations in the x and y directions, namely,

$$\sigma_{x,x}+\tau_{xy,y}+\tau^I_{xz,z}+X_b=0, \qquad \sigma_{y,y}+\tau_{xy,x}+\tau^I_{yz,z}+Y_b=0, \tag{9.87}$$

which may be written as

$$\sigma^*_{x,x}+\tau^*_{xy,y}+X_b{}^*=0, \qquad \sigma^*_{y,y}+\tau^*_{xy,x}+Y_b{}^*=0, \tag{9.88}$$

where

$$X_b{}^*=X_b+\tau^I_{xz,z}+\sigma^I_{x,x}+\tau^I_{xy,y}, \tag{9.89a}$$
$$Y_b{}^*=Y_b+\tau^I_{yz,z}+\sigma^I_{y,y}+\tau^I_{xy,x}. \tag{9.89b}$$

The associated boundary conditions along $\partial\Omega$,

$$\sigma_x\cos(\bar{\mathbf{n}},x)+\tau_{xy}\cos(\bar{\mathbf{n}},y)=X_s, \qquad \tau_{xy}\cos(\bar{\mathbf{n}},x)+\sigma_y\cos(\bar{\mathbf{n}},y)=Y_s, \tag{9.90}$$

may be written as

$$\sigma^*_x\cos(\bar{\mathbf{n}},x)+\tau^*_{xy}\cos(\bar{\mathbf{n}},y)=X^*_s, \qquad \tau^*_{xy}\cos(\bar{\mathbf{n}},x)+\sigma^*_y\cos(\bar{\mathbf{n}},y)=Y^*_s, \tag{9.91}$$

where

$$X^*_s=X_s-\sigma^I_x\cos(\bar{\mathbf{n}},x)-\tau^I_{xy}\cos(\bar{\mathbf{n}},y), \tag{9.92a}$$
$$Y^*_s=Y_s-\tau^I_{xy}\cos(\bar{\mathbf{n}},x)-\sigma^I_y\cos(\bar{\mathbf{n}},y). \tag{9.92b}$$

As shown, the solution in level II constitutes a two-dimensional plane-strain problem for *orthotropic material*. Following S.3.2.1, S.3.2.4, S.3.2.5 we therefore write

$$\nabla^{(4)}_2\Phi=F_0 \qquad \text{over} \quad \Omega, \tag{9.93a}$$
$$\frac{d}{ds}\{\Phi_{,x},\Phi_{,y}\}=\{-F_1,F_2\} \qquad \text{on} \quad \partial\Omega. \tag{9.93b}$$

For these equations we employ the orthotropic operator of (3.203), where the stress function definition of (3.54) is adopted. $F_0(x,y)$ is written with the aid of (3.55), (3.57). F_1 and F_2 are written with the aid of (3.65).

When a non-homogeneous cross-section is under discussion, we adopt the notation of Fig. 3.1 and the formulation in S.8.1.3, and seek a solution of (9.93a,b), which also satisfies

$$\frac{d}{ds}\{\Phi_{,x},\Phi_{,y}\}^{[j]}_{[i]}=\{-F^{ij}_1,F^{ij}_2\} \qquad \text{on} \quad \partial\Omega_{ij}, \tag{9.94a}$$
$$\{u,v\}^{[j]}_{[i]}=\{0,0\} \qquad \text{on} \quad \partial\Omega_{ij}, \tag{9.94b}$$

where F^{ij}_1 and F^{ij}_2 are the net forces in y and x directions, which are "injected" to the interface, and according to (9.92a,b), are given by

$$F^{ij}_1=-\left[\tau^I_{xy}\cos(\bar{\mathbf{n}},x)+\sigma^I_y\cos(\bar{\mathbf{n}},y)\right]_{[i]}-\left[\tau^I_{xy}\cos(\bar{\mathbf{n}},x)+\sigma^I_y\cos(\bar{\mathbf{n}},y)\right]_{[j]}, \tag{9.95a}$$
$$F^{ij}_2=-\left[\sigma^I_x\cos(\bar{\mathbf{n}},x)+\tau^I_{xy}\cos(\bar{\mathbf{n}},y)\right]_{[i]}-\left[\sigma^I_x\cos(\bar{\mathbf{n}},x)+\tau^I_{xy}\cos(\bar{\mathbf{n}},y)\right]_{[j]}, \tag{9.95b}$$

where here, $\bar{\mathbf{n}}$ is the *outwards normal* of the respective domain. Hence, the two-dimensional solution methodologies described in Chapter 4 may be adopted to carry out this level solution as well.

The reader may verify that the single-value conditions of (3.73a,b), (3.74) are fulfilled for this level problem, namely,

$$\oint_{\partial\Omega} Y_s^* + \iint_{\Omega} Y_b^* = 0, \tag{9.96a}$$

$$\oint_{\partial\Omega} X_s^* + \iint_{\Omega} X_b^* = 0, \tag{9.96b}$$

$$\oint_{\partial\Omega} (Y_s^* x - X_s^* y) + \iint_{\Omega} (Y_b^* x - X_b{}^* y) = 0. \tag{9.96c}$$

Considering (9.96a) as an example, by virtue of Green's Theorem, (9.89b), (9.92b) show that

$$\oint_{\partial\Omega} Y_s^* + \iint_{\Omega} Y_b^* = \oint_{\partial\Omega} Y_s + \iint_{\Omega} Y_b + \iint_{\Omega} \tau_{yz,z}^I. \tag{9.97}$$

The above r.h.s. vanishes since (9.2:b) shows that $p_y = \oint_{\partial\Omega} Y_s + \iint_{\Omega} Y_b$, while according to (9.19:b), (9.21:b), $\overline{P}_{y,z} = \iint_{\Omega} \tau_{yz,z} = -p_y$.

To this end, within the present multilevel approach one may adopt two alternative ways:

(a) To solve the BVP of (9.93a,b), (9.94a,b) as is. This will ensure that the displacement components over the interface lines will remain continuous. In such a case, a level III solution is not required.

(b) To ignore (9.94b), namely, to solve the BVP of (9.93a,b), (9.94a). Such a solution is generally termed a "stress solution", as it does not consider the requirement of deformation continuity along the interlaminar surfaces. In general, a stress solution is much simpler and more analytically achievable, while supplying an adequate approximation compared with the complete one, and therefore selection of this option is advocated in the present context. However, to correct this selection a level III solution is required as well.

For the solution in level III, one needs to derive the difference between the displacement components along the interface lines. This is done by integrating $\varepsilon_x, \varepsilon_y, \gamma_{xy}$ for the displacement functions $u(x,y)$ and $v(x,y)$ in each domain, $\Omega_{[i]}$. We then adopt the notation of (3.1), and note that the displacements over the dividing surface differ by f_u and f_v, namely,

$$\{u, v\}_{[i]}^{[j]} = \{f_u(x,y), f_v(x,y)\} \qquad \text{on} \quad \partial\Omega_{ij}. \tag{9.98}$$

Hence, f_u and f_v are the differences (or the "jumps") in the corresponding displacement components along the interface lines. When the solution in level II ignores (9.94b) and accounts for stress continuity only, these differences do not vanish in the general case, as schematically shown in Fig. 9.7, and therefore the above stress solution is not exact. When a level III solution is activated, its strain distributions $\Delta\varepsilon_x$, $\Delta\varepsilon_y$ and $\Delta\gamma_{xy}$ are added to the strain components derived in level II, see Fig. 9.6.

Overall, in each iteration, σ_x, σ_y and τ_{xy} as obtained by (9.86) and ε_x, ε_y and γ_{xy} (that include $\Delta\varepsilon_x$, $\Delta\varepsilon_y$ and $\Delta\gamma_{xy}$ if a level III solution has been employed) are the net additional contributions of each level II iteration to the respective stress and strain components. These values accumulate as iteration progresses until convergence is achieved.

For the solution in level I (see (9.71)) we write

$$\begin{Bmatrix} \sigma_z^{II} \\ \tau_{yz}^{II} \\ \tau_{xz}^{II} \end{Bmatrix} = \mathbf{C}_\sigma \cdot \begin{Bmatrix} \varepsilon_x^{II} \\ \varepsilon_y^{II} \\ \gamma_{xy}^{II} \end{Bmatrix}, \tag{9.99}$$

where $\varepsilon_x^{II}, \varepsilon_y^{II}, \gamma_{xy}^{II}$ are the total values of strain components accumulated in level II up to (and including) the current iteration.

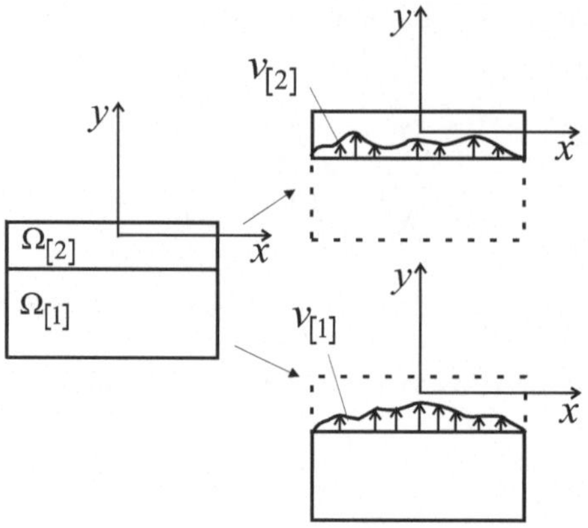

Figure 9.7: A non-homogeneous cross-section made of two materials and the resulting $v_{[j]}$ distributions derived in level II when (9.94b) is ignored.

Remark 9.9 In general, the expressions for the body forces X_b^*, Y_b^* depend on the axial coordinate z. Hence, the solution in level II is parametric with respect to z. In a "discrete" terminology, this two-dimensional solution needs to be carried out for each cross-section along the beam axis.

Remark 9.10 Considering the single-value conditions of the solution in level II for a non-homogeneous cross-section, we write for each domain, Ω_i, similar to (9.97),

$$\oint_{\partial\Omega_i} Y_s^* + \sum_{j(i)} \int_{\partial\Omega_{ij}} \left[\tau_{xy}^* \cos(\bar{\mathbf{n}},x) + \sigma_y^* \cos(\bar{\mathbf{n}},y)\right] + \iint_{\Omega_i} Y_b^* = 0, \tag{9.100}$$

where $\bar{\mathbf{n}}$ is directed outwards to the domain, $\partial\Omega_i$ are the free contours of the domain, and $\partial\Omega_{ij}$ are the interlaminar interface line between the domain and its surrounding domains, Ω_j (i.e. the above summation is carried out only over the interfaces $j(i)$ between the $i^{\underline{th}}$ domain and its surrounding domains). By employing (9.86), (9.89b), (9.92b) we write

$$\oint_{\partial\Omega_i} [Y_s - \underline{\tau_{xy}^I \cos(\bar{\mathbf{n}},x)} - \underline{\sigma_y^I \cos(\bar{\mathbf{n}},y)}] - \sum_j \int_{\partial\Omega_{ij}} [\underline{\tau_{xy}^I \cos(\bar{\mathbf{n}},x)} + \underline{\sigma_y^I \cos(\bar{\mathbf{n}},y)}]$$
$$+ \iint_{\Omega_i} [Y_b + \tau_{yz,z}^I + \underline{\sigma_{y,y}^I} + \underline{\tau_{xy,x}^I}] = -\sum_j \int_{\partial\Omega_{ij}} [\tau_{xy} \cos(\bar{\mathbf{n}},x) + \sigma_y \cos(\bar{\mathbf{n}},y)]. \tag{9.101}$$

The (once and twice) underlined terms are cancelled out by applying Green's Theorem. Adding up all domains cancels out the net effect of the terms on the r.h.s.of (9.101), which leads again to the r.h.s. of (9.97). Returning to (9.100), by adding all domains, the single-value condition for a level II BVP becomes

$$\oint_{\partial\Omega} Y_s^* + \iint_{\Omega} Y_b^* = -\sum_{j(i)} \int_{\partial\Omega_{ij}} \left[\tau_{xy}^* \cos(\bar{\mathbf{n}},x) + \sigma_y^* \cos(\bar{\mathbf{n}},y)\right], \tag{9.102}$$

where the summation is carried out for all interface lines. In the general case, the above r.h.s. does not vanish (see e.g. Example 9.5). Equation (9.102) replaces (9.96a) for the non-homogeneous case, and similar conditions may be written for (9.96b,c) as well.

9.3.2.3 Level III

As already indicated, the solution in level III is required only when a non-homogeneous cross-section is under discussion and when (9.94b) is ignored within a level II solution. A level III solution exhibits high similarity with the auxiliary problems of plane deformation discussed within S.8.2.2 (while only the functions $f_u(x,y)$ and $f_v(x,y)$ are different). This level solution handles the displacements differences in the interface domains by producing an (orthotropic) *plane-strain*, two-dimensional solution, see S.3.2, that has to fulfill the following conditions:

(a) The external stresses applied at the free contours $\partial\Omega_j$ $(j = 1,\ldots,N)$ vanish. Hence (9.90) show that

$$\sigma_x^{III}\cos(\bar{\mathbf{n}},x) + \tau_{xy}^{III}\cos(\bar{\mathbf{n}},y) = 0, \qquad \tau_{xy}^{III}\cos(\bar{\mathbf{n}},x) + \sigma_y^{III}\cos(\bar{\mathbf{n}},y) = 0, \tag{9.103}$$

where the normal vector to the contour, $\bar{\mathbf{n}}$, is taken as the outer normal to $\Omega_{[j]}$.

(b) On the dividing surfaces, $\partial\Omega_{ij}$, stress continuity in the following form is applied (see (9.90)):

$$\left[\sigma_x^{III}\cos(\bar{\mathbf{n}},x) + \tau_{xy}^{III}\cos(\bar{\mathbf{n}},y)\right]_{[i]}^{[j]} = 0, \qquad \left[\tau_{xy}^{III}\cos(\bar{\mathbf{n}},x) + \sigma_y^{III}\cos(\bar{\mathbf{n}},y)\right]_{[i]}^{[j]} = 0 \tag{9.104}$$

where $\bar{\mathbf{n}}$ is a normal to the contours $\partial\Omega_{ij}$, the orientation of which is immaterial, however, *the same orientation is used on both sides of the equilibrium sign*, see Fig. 8.1. Clearly, in the special case when a dividing contour is characterized by $y = const.$, the above conditions yield $[\tau_{xy}^{III}]^{[i]} = [\tau_{xy}^{III}]^{[i]}$ and $[\sigma_y^{III}]^{[i]} = [\sigma_y^{III}]^{[i]}$, namely, the two stress components should be equal on both sides of the dividing plane.

(c) The displacements on the dividing surface differ by the same functions obtained in (9.98) with the opposite sign, namely,

$$\{u^{III}, v^{III}\}_{[i]}^{[j]} = -\{f_u(x,y), f_v(x,y)\} \qquad \text{on} \quad \partial\Omega_{ij}. \tag{9.105}$$

It is therefore clear that adding the solution of level III to the one obtained in level II yields a refined plane-strain solution of level II, while stress resultant and displacement compatibility is maintained along the interface lines.

As already indicated, the strain distributions in this level are denoted $\Delta\varepsilon_x$, $\Delta\varepsilon_y$ and $\Delta\gamma_{xy}$ and are added to those of level II.

9.3.2.4 Summary

The solution levels described above may be executed in various iterative manners. Referring to Fig. 9.6, the simplest way of executing the three solution levels may be described as follows. The process is initiated by level I (with no data from level II), continues on to level II (with data from level I). Then, depending on the version selection in S.9.3.2.2, a solution of level III may be required, in which case its results are superimposed on those of level II. The level I solution is then activated again with the updated data from level II, and this procedure is repeated until global convergence is achieved. Note again that the solution in each level is "independent" and may be executed by various analytical and/or numerical techniques.

Once a convergence has been achieved, the resulting stress components are the accumulated σ_z, τ_{yz}, and τ_{xz} from level I and the accumulated σ_x, σ_y and τ_{xy} from level II.

9.3.3 Examples

Example 9.3 *An Isotropic Homogeneous Beam Under Bending Tip Load.*

The above multilevel concept will be first demonstrated for the case of a uniform isotropic beam of a rectangular cross-section $\Omega = \{|x| \le \frac{\tilde{a}}{2},\ |y| \le \frac{\tilde{b}}{2}\}$, and undergoes a constant beamwise bending tip load P_y.

Level I: The solution in level I is divided into two parts. The first part consists of the stress analysis of the elementary solution, which assumes that all stress components vanish except for

$$\sigma_z = -\frac{P_y}{I_x} y\,(l-z)\,, \qquad \tau_{yz} = -\frac{P_y}{2I_x}(y^2 - \frac{\widetilde{b}^2}{4}), \tag{9.106}$$

and the only deformation component is $v_0 = \frac{P_y}{6EI_x} z^2(3l-z)$. The reader may verify that this solution satisfies all required integral and differential equilibrium equations and boundary conditions, but does not satisfy the compatibility equations. We therefore seek an additional solution consisting of two stress components only, $\Delta\tau_{yz}$, $\Delta\tau_{xz}$, that are superimposed on the first one so that compatibility is fulfilled. We subsequently reach a problem that is equivalent to the two-dimensional problem described in S.3.3, where the above elementary solution serves as the particular solution, see also Example 3.10. Hence according to S.3.3, one needs to solve the following Poisson's equation ($a_{44} = a_{55} = 2(1+\nu)/E$ and $a_{45} = 0$ in this case):

$$\nabla^{(2)}\psi = -\frac{\nu}{1+\nu}\frac{P_y}{I_x}x + \widetilde{c}, \tag{9.107}$$

under the boundary condition $\psi_{|\partial\Omega} = 0$ (and $\widetilde{c} = 0$ for zero rotation angle). This solution will supply the additional stress components

$$\Delta\tau_{xz} = \psi_{,y}, \qquad \Delta\tau_{yz} = -\psi_{,x}, \tag{9.108}$$

which will ensure satisfying all necessary conditions in level I (note that according to S.3.3, these additional components will make no contribution to P_x, P_y and M_z). Exact solution of (9.107) appears in (Timoshenko and Goodier, 1970). Detailed description of this relatively small correction is not required here since in this isotropic case, only ε_z influences the following steps.

For level II one obtains $\varepsilon_z = \sigma_z^*/A_{33}$ (see (9.72)), namely,

$$\varepsilon_z = -\frac{P_y}{I_x}\cdot\frac{(1+\nu)\,(1-2\nu)}{(1-\nu)\,E}y\,(l-z)\,, \tag{9.109}$$

and therefore, since in this first iteration, $\varepsilon_z^I = \varepsilon_z$,

$$\sigma_x^I = \sigma_y^I = -\frac{P_y}{I_x}\cdot\frac{\nu}{1-\nu}y(l-z)\,, \qquad \tau_{xy}^I = 0. \tag{9.110}$$

The exact expressions of the above τ_{yz}^I and τ_{xz}^I are immaterial since they are not functions of z. As indicated in Remark 9.9, z is a parameter in (9.110), and therefore, in a "discrete analysis language", a level II solution should be carried out "for each cross-section".

Level II: The solution in level II is derived by the body forces, see (9.89a,b),

$$X_b{}^* = 0, \qquad Y_b{}^* = -\frac{P_y}{I_x}\cdot\frac{\nu}{1-\nu}\,(l-z)\,. \tag{9.111}$$

The surface loads that are induced from a level I solution (see (9.92a,b)), are

$$X_s^{*(R)} = -X_s^{*(L)} = -\frac{P_y}{I_x}\cdot\frac{\nu}{1-\nu}y(l-z), \qquad Y_s^{*(U)} = Y_s^{*(D)} = \frac{b}{2}\frac{P_y}{I_x}\cdot\frac{\nu}{1-\nu}(l-z), \tag{9.112}$$

where the superscripts (R), (L), (U) and (D) stand for the right, left upper, and lower edges of the cross-section, respectively, see Fig. 9.8. The reader may verify that all single-value

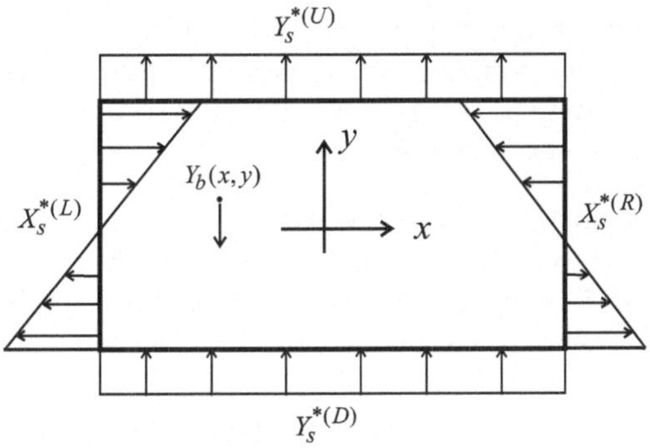

Figure 9.8: In-plane stress distribution for a level II solution of Example 9.3.

conditions of (9.96a–c) are satisfied, and therefore, this plane-strain problem is well posed. The solution of this level problem is simply

$$\sigma_x^* = \sigma_y^* = \frac{P_y}{I_x} \cdot \frac{\nu}{1-\nu} y(l-z), \qquad \tau_{xy}^* = 0. \tag{9.113}$$

Note that since $\sigma_x^* = -\sigma_x^I$, the total stress added at this stage is $\sigma_x = \sigma_x^I + \sigma_x^* = 0$ (similarly, no σ_y is added). The strain components in level II are

$$\varepsilon_x = \varepsilon_y = \frac{P_y}{I_x} \cdot \frac{\nu(1+\nu)(1-2\nu)}{(1-\nu)E} y(l-z), \qquad \gamma_{xy} = 0. \tag{9.114}$$

For level I, the following stress is calculated, see (9.99):

$$\sigma_z^{II} = \frac{2P_y}{I_x} \cdot \frac{\nu^2}{1-\nu} y(l-z). \tag{9.115}$$

Hence, as shown by (9.77), (9.80), the only loading that should be fed back to level I is $Z_b{}^* = \sigma_{z,z}^{II}$, which is a linear function of y. There are no induced Z_s^* as $\tau_{yz}^{II} = \tau_{xz}^{II} = 0$ in this isotropic case.

The iterative process: We should now carry out an additional level I solution. The solution here will simply be

$$\sigma_z^* = Z_b{}^*(l-z), \qquad \tau_{yz}^* = \tau_{xz}^* = 0. \tag{9.116}$$

Since $\sigma_z^* = -\sigma_z^{II}$, the total stress added at this stage is $\sigma_z = \sigma_z^{II} + \sigma_z^* = 0$. Hence, the process moves on again to level II, which is derived by the new σ_z^*. We therefore see two important features of this process:

(a) Level I always produces σ_z^* that are linear functions of y. By comparing $\sigma_z^* = -\sigma_z^{II}$ of (9.115) with σ_z^* of (9.106), one may conclude that these stress components produce (an infinite) geometrical series with the constant common ratio $q_\sigma = \frac{2\nu^2}{1-\nu}$.

(b) No stresses are added during the iterations to any of the stress components.

Subsequently, the iterations continue without modifying the stresses (and they therefore preserve their values of the first level I solution and $\sigma_x = \sigma_y = \tau_{xy} = 0$), while the strain components will converge to the value obtained in the first iteration divided by $1 - q_\sigma$. Hence,

$$\varepsilon_z = -\frac{P_y}{EI_x} y(l-z), \qquad \varepsilon_x = \varepsilon_y = \frac{\nu P_y}{EI_x} y(l-z), \tag{9.117}$$

which coincides with the exact solution of this case. The typical value of $q_\sigma \cong 0.26$ (for $\nu = 0.3$) in this example demonstrates the nominal convergence rate of the scheme in this isotropic case. At this stage, displacements may be obtained by integration. No level III solution is required, as a homogeneous cross-section is under discussion.

Example 9.4 *A Coupled Beam Under Tip Moment.*

We shall discuss here the case of a coupled MON13y homogeneous beam (of generic cross-section) that undergoes a uniform bending moment, M_x, which is applied at the tips by a linear axial stress distribution $\sigma_z = \frac{M_x}{I_x} y$. Since a homogeneous beam is under discussion, only level I and II are derived below.

The solution in Level I is based on the assumption (which will be proved to be exact later on) that the data from the level II solution is of the form $\sigma_z^{II} = \widetilde{D}y$, $\tau_{xz}^{II} = \widetilde{E}y$, and $\tau_{yz}^{II} = 0$ where $\widetilde{D}$ and $\widetilde{E}$ are constants, and clearly, for the first iteration $\widetilde{D} = \widetilde{E} = 0$ will be used. In addition, w and u are set to zero, $v_{0,zz}$ and $\phi_{,z}$ are assumed to be constants, and the following out-of-plane warping form is used:

$$\Psi^w = -\phi_{,z} xy. \tag{9.118}$$

Substituting the above initial assumptions in (9.4a,b), (9.72) shows that the strains are given by $\varepsilon_z = -y v_{0,zz}$, $\gamma_{yz} = 0$ and $\gamma_{xz} = -2y\phi_{,z}$, and thus,

$$\sigma_z^* = \sigma_z - \widetilde{D}y = -(A_{33} v_{0,zz} + 2A_{35}\phi_{,z})\, y, \tag{9.119a}$$

$$\tau_{yz}^* = \tau_{yz} = 0, \tag{9.119b}$$

$$\tau_{xz}^* = \tau_{xz} - \widetilde{E}y = -(A_{35} v_{0,zz} + 2A_{55}\phi_{,z})\, y. \tag{9.119c}$$

One may now determine $v_{0,zz}$ and $\phi_{,z}$ from (9.119a,c) by the requirements $\sigma_z = yM_x/I_x$ and $\tau_{xz} = 0$, which yield

$$\begin{bmatrix} A_{35} & A_{33} \\ A_{55} & A_{35} \end{bmatrix} \cdot \begin{Bmatrix} 2\phi_{,z} \\ v_{0,zz} \end{Bmatrix} = \begin{Bmatrix} \widetilde{D} - \frac{M_x}{I_x} \\ \widetilde{E} \end{Bmatrix}. \tag{9.120}$$

The reader may verify that the above values for $v_{0,zz}$, $\phi_{,z}$ and the expression for Ψ^w satisfy all equations of equilibrium (9.74a,b), (9.76) and boundary conditions (9.24), (9.78a,b), (9.79) of level I. This concludes the solution in this level, while for the solution in level II, the following stress components are transferred, see (9.83):

$$\begin{Bmatrix} \tau_{xz}^I \\ \sigma_x^I \\ \sigma_y^I \end{Bmatrix} = -y \begin{bmatrix} A_{55} & A_{35} \\ A_{15} & A_{13} \\ A_{25} & A_{23} \end{bmatrix} \cdot \begin{Bmatrix} 2\phi_{,z} \\ v_{0,zz} \end{Bmatrix}, \qquad \tau_{yz}^I = \tau_{xy}^I = 0. \tag{9.121}$$

The solution in Level II is based on the assumption that the in-plane warping components are of the form $\Psi^u = \widetilde{F}xy$ and $\Psi^v = \widetilde{G}x^2 + \widetilde{H}y^2$, where $\widetilde{F}$, $\widetilde{G}$ and $\widetilde{H}$ are constants. According to (9.84)

$$\varepsilon_x = \Psi^u_{,x} = \widetilde{F}y, \qquad \varepsilon_y = \Psi^v_{,y} = 2\widetilde{H}y, \qquad \gamma_{xy} = \Psi^u_{,y} + \Psi^v_{,x} = (\widetilde{F} + 2\widetilde{G})x, \tag{9.122}$$

and thus, (9.85) shows that

$$\sigma_x^* = (A_{11}\widetilde{F} + 2A_{12}\widetilde{H})\, y, \qquad \sigma_y^* = (A_{12}\widetilde{F} + 2A_{22}\widetilde{H})\, y, \qquad \tau_{yz}^* = A_{66}(\widetilde{F} + 2\widetilde{G})x. \tag{9.123}$$

The requirement for $\sigma_x = \sigma_y = \tau_{xy} = 0$ (namely, $\sigma_x^* = -\sigma_x^I$, $\sigma_y^* = -\sigma_y^I$, $\tau_{xy}^* = -\tau_{xy}^I$) shows that $\widetilde{G} = -\widetilde{F}/2$ while $\widetilde{F}$ and $\widetilde{H}$ are obtained from

$$\begin{bmatrix} A_{11} & A_{12} \\ A_{12} & A_{22} \end{bmatrix} \cdot \begin{Bmatrix} \widetilde{F} \\ 2\widetilde{H} \end{Bmatrix} = \begin{bmatrix} A_{15} & A_{13} \\ A_{25} & A_{23} \end{bmatrix} \cdot \begin{Bmatrix} 2\phi_{,x} \\ v_{,x} \end{Bmatrix}. \tag{9.124}$$

The solution of the above system of equations ensures that both the differential equations, (9.87), and the boundary conditions, (9.90), are satisfied. This concludes the solution in level II while (9.99) is used to construct the stress components $\sigma_z^{II} = \widetilde{D}y$ and $\tau_{xz}^{II} = \widetilde{E}y$, which are fed back to level I, as

$$\begin{Bmatrix} \widetilde{D} \\ \widetilde{E} \end{Bmatrix} = \begin{bmatrix} A_{13} & A_{23} \\ A_{15} & A_{25} \end{bmatrix} \cdot \begin{Bmatrix} \widetilde{F} \\ 2\widetilde{H} \end{Bmatrix}, \tag{9.125}$$

while as previously assumed $\tau_{yz}^{II} = 0$. With these values the iterative process returns to level I, and the above steps are repeated until convergence is achieved.

The above multilevel solution includes two deformation parameters, $\phi_{,z}$ and $v_{0,zz}$ that are transferred from level I to level II, and two deformation parameters, $\widetilde{D}$ and $\widetilde{E}$ that are fed back from level II to level I. Subsequently, an iteration matrix may be formulated for each one of these two sets of deformation parameters and a derivation of its spectral number may be carried out. To derive the iteration matrix for the $\widetilde{D}$ and $\widetilde{E}$ parameters, (9.120) is substituted in (9.124), which is then substituted in (9.125). This enables us to write

$$\begin{bmatrix} \widetilde{D} \\ \widetilde{E} \end{bmatrix}_{i+1} = [\widetilde{\alpha}] \cdot \begin{bmatrix} \widetilde{D} \\ \widetilde{E} \end{bmatrix}_{i} + [\widetilde{\beta}], \tag{9.126}$$

where the subscript i represents the iteration number. The iteration matrix components, $[\widetilde{\alpha}]$ and $[\widetilde{\beta}]$ are

$$[\widetilde{\alpha}] = \begin{bmatrix} A_{13} & A_{23} \\ A_{15} & A_{25} \end{bmatrix} \begin{bmatrix} A_{11} & A_{12} \\ A_{12} & A_{22} \end{bmatrix}^{-1} \begin{bmatrix} A_{15} & A_{13} \\ A_{25} & A_{23} \end{bmatrix} \begin{bmatrix} A_{35} & A_{33} \\ A_{55} & A_{35} \end{bmatrix}^{-1}, \quad [\widetilde{\beta}] = [\widetilde{\alpha}] \cdot \begin{Bmatrix} -\frac{M_x}{I_x} \\ 0 \end{Bmatrix}. \tag{9.127}$$

Introducing elastic moduli of typical Graphite/Epoxy orthotropic lamina shows that the *spectral radius* of $[\widetilde{\alpha}]$ (defined as the maximal absolute value of its eigenvalues) is $\rho_\alpha = 0.44$, while the condition $\rho_\alpha \leq 1$ ensures stability and convergence of the scheme.

The above closed-form solution and the determination of the spectral radius of the overall multilevel iteration process demonstrate the feasibility and applicability of the method to laminated anisotropic beams as well.

As already stated, no general proof for generic configurations may be developed in the present context, since the proposed multilevel analysis is not confined to a specific solution methodology at each level. Figure 9.9(a) demonstrates the non-dimensional values of $\widetilde{D}$ and $\widetilde{E}$ as obtained by a sequential execution of the above solution levels for typical Graphite/Epoxy orthotropic lamina with a lamination angle $\theta_y = 30°$, see Fig. 2.2. As shown, convergence is achieved for six iterations. In the initial state, where only a level I analysis has been carried out, $\widetilde{D} = \widetilde{E} = 0$, and (9.120) shows that

$$\phi_{,z} = \frac{A_{35}}{A_{33}A_{55} - A_{35}^2} \cdot \frac{M_x}{2I_x}, \qquad v_{0,zz} = -\frac{A_{55}}{A_{33}A_{55} - A_{35}^2} \cdot \frac{M_x}{I_x}, \tag{9.128}$$

which is an approximate solution that would have come out of the analysis of S.9.1 with material properties of Version I. However, once a convergence is achieved, $\widetilde{D} \cong -0.661$ and $\widetilde{E} \cong 0.276$, which is compatible with the exact solution $\phi_{,z} = -\frac{M_x}{I_x}\frac{a_{35}}{2}$ and $v_{0,zz} = -\frac{M_x}{I_x}a_{33}$, see Remark 9.11. Figure 9.9(b) shows the values of $v_{0,zz}$ and $\phi_{,z}$ as they are modified by the in-plane warping. Throughout the iterative process, the absolute value of $\phi_{,z}$ is changed by 10% and the absolute value of $v_{0,zz}$ is changed by 26%.

Remark 9.11 The exact solution for $v_{0,zz}$ and $\phi_{,z}$ is given by (5.25b), (5.26c), respectively, by substituting

$$\phi_{,z} = \omega_{z0,z}(z), \qquad v_{0,zz} = v_{,zz}(0,0,z), \qquad \widetilde{a} = \widetilde{b} = 0, \qquad \widetilde{c} = \frac{M_x}{I_x}. \tag{9.129}$$

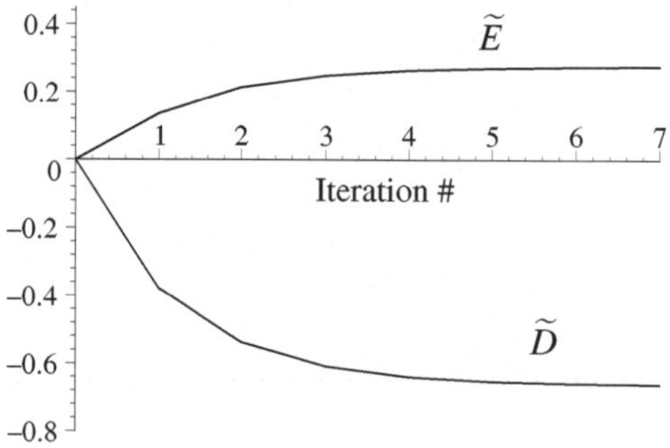

(a) $\widetilde{D}$ and $\widetilde{E}$.

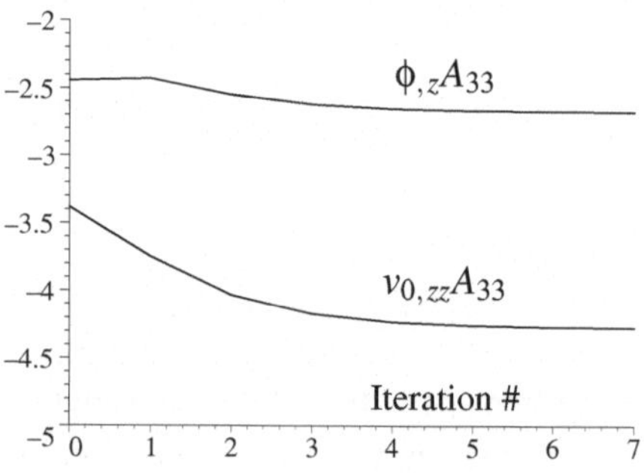

(b) Normalized twist ($\phi_{,z}A_{33}$) and bending curvature ($v_{0,zz}A_{33}$).

Figure 9.9: The iterative process of Example 9.4 ($M_x/I_x = 1$).

It is also worth mentioning that if a level I solution is carried out for Version II assumptions (see S.9.2.1), i.e. with A_{ij} replaced by Q_{ij}, the same exact solution would have been obtained without any iterative process. This is due to the fact that the assumption on which Version II is founded (i.e. the in-plane stress components vanish) does coincide with the present case boundary conditions.

Example 9.5 *Level I Solution for a Non-Homogeneous Isotropic Beam.*
The case of a non-homogeneous isotropic beam the cross-section of which consists of two materials as shown in Fig. 9.10 is discussed in this example. The beam is assumed to undergo a tip loading P_y and we therefore set $u_0 = w_0 = \phi_0 = 0$. To enable a linear strain distribution

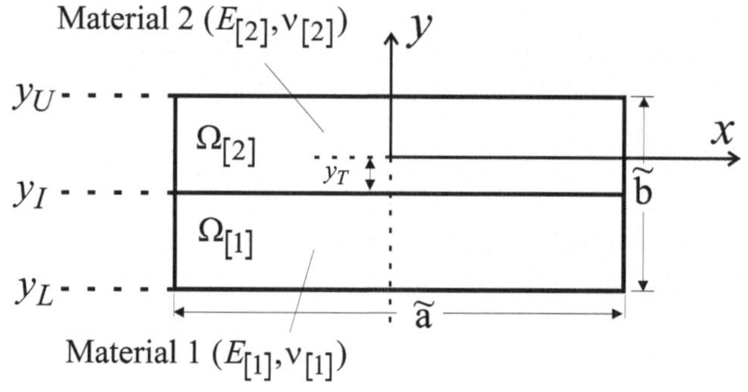

Figure 9.10: A non-homogeneous isotropic beam of Example 9.5.

that induces no axial load, we shall first place the origin of the coordinate system at the center of tension, which is given by

$$y_T = \frac{\iint_\Omega E y}{\iint_\Omega E} = \frac{\widetilde{b}}{4} \cdot \frac{E_{[1]} - E_{[2]}}{E_{[1]} + E_{[2]}}. \tag{9.130}$$

Subsequently, the normal stress is a linear function (different within each material), namely,

$$\sigma_z^{*[i]} = -y v_{0,zz} E_{[i]}, \quad i = 1, 2. \tag{9.131}$$

By equating the resulting M_x moment to $-P_y(l-z)$ one obtains

$$v_{0,zz} = P_y(l-z)\widetilde{C}, \qquad \widetilde{C} = \frac{96}{\widetilde{a}\widetilde{b}^3} \cdot \frac{E_{[1]} + E_{[2]}}{E_{[1]}^2 + 14E_{[1]}E_{[2]} + E_{[2]}^2}. \tag{9.132}$$

Based on the definitions of Fig. 9.10, $y_L = -\frac{\widetilde{b}}{2} - y_T$, $y_U = \frac{\widetilde{b}}{2} - y_T$, $y_I = -y_T$, and the corresponding τ_{yz} shear stresses are integrated so as to satisfy the equilibrium equation (9.23)

$$\tau_{yz}^{*[i]} = -\left(y^2 - y_U^2\right)\frac{P_y E_{[i]}}{2}\widetilde{C}, \qquad i = 1,2. \tag{9.133}$$

The above axial and shear stress components are shown in Fig. 9.11 for $\frac{E_{[2]}}{E_{[1]}} = 4$, $\widetilde{a} = 2$, $\widetilde{b} = 1$.

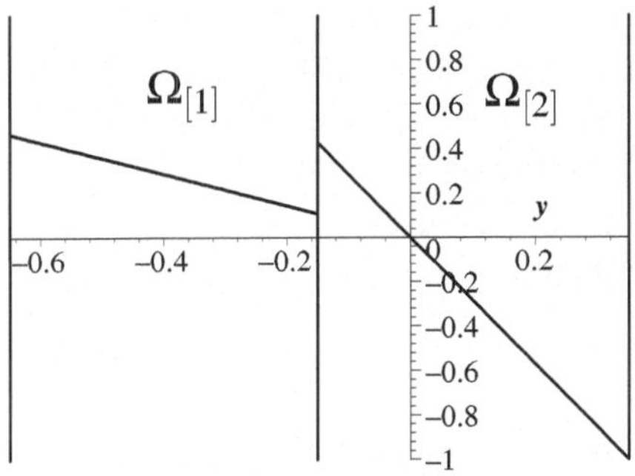

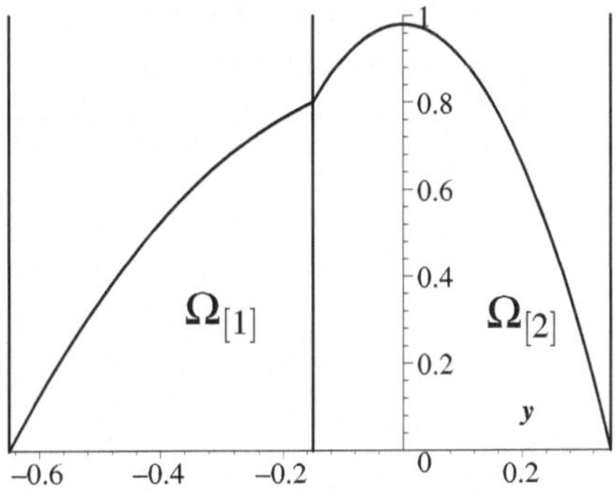

Figure 9.11: Stress distributions for the non-homogeneous cross-section of Example 9.5 (non-dimensional, normalized differently).

The reader may verify that the integral equilibrium equations (9.19) are satisfied, and in particular,

$$\iint_{\Omega_{[1]}} \{\sigma_z^{I[1]}, \tau_{yz}^{I[1]}\} + \iint_{\Omega_{[2]}} \{\sigma_z^{I[2]}, \tau_{yz}^{I[2]}\} = \{0, P_y\}. \tag{9.134}$$

The out-of-plane warping is obtained by integrating $\Psi_{,y}^{w[i]} = 2\tau_{yz}^{*[i]}\left(1+\nu_{[i]}\right)/E_{[i]}$ with respect to y. Hence,

$$\Psi^{w[i]} = -\left(\frac{y^3}{3} - y_U^2 y\right)P_y\left(1+\nu_{[i]}\right)\widetilde{C} + c^{[i]}, \qquad i = 1,2, \tag{9.135}$$

where the coefficients $c^{[1]}$ and $c^{[2]}$ are determined from the conditions of out-of-plane warping continuity (see (9.82)) and zero average value, namely,

$$\Psi^{w[1]} = \Psi^{w[2]} \quad (y = y_I), \qquad \iint_{\Omega_{[1]}} \Psi^{w[1]} + \iint_{\Omega_{[2]}} \Psi^{w[2]} = 0. \tag{9.136}$$

The warping distribution is presented in Fig. 9.12.

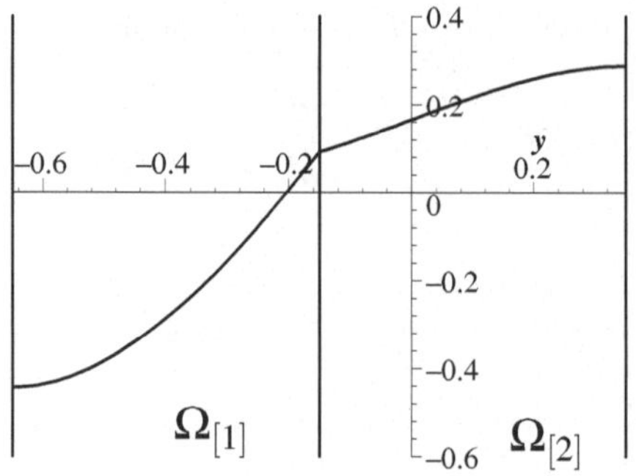

Figure 9.12: The out-of-plane warping in Example 9.5.

The above stress distributions fulfill all required conditions except for compatibility, similar to the case in Example 9.3. We therefore need to superimpose a solution, the role of which is to supply additional $\Delta\tau_{yz}^{*[1]}$, $\Delta\tau_{xz}^{*[1]}$, $\Delta\tau_{yz}^{*[2]}$, $\Delta\tau_{xz}^{*[2]}$ so that compatibility will be satisfied. This additional solution should be determined by the two-dimensional scheme of S.3.3.1 — see details in Remark 9.12. The exact shape of the above corrections is of minor importance since the stress components, which are transferred to level II, are functions of $\varepsilon_z^{I[i]} = \sigma_z^{*[i]}/A_{33}^{[i]}$ only.

For level II, the following stress components are determined:

$$\sigma_x^{I[i]} = \sigma_y^{I[i]} = -\frac{\nu_{[i]}}{1-\nu_{[i]}} E_{[i]} P_y \widetilde{C} y\,(l-z)\,, \qquad \tau_{xy}^{I[i]} = 0. \tag{9.137}$$

At this stage, one may employ (9.89a,b) and (9.92a,b) to determine the body forces $X_b{}^*$, $Y_b{}^*$ and the surface loads $X_s{}^*$, $Y_s{}^*$ that derive the solution in Level II. Note that τ_{yz}^{I} τ_{xz}^{I} are of no importance for level II in this specific example as their derivatives with respect to z vanish. Level II solution for the present case is essentially a solution the non-homogeneous plane-strain BVP of (9.94a,b), see also general aspects of this problem in S.8.1.3.

Remark 9.12 To determine the additional solution required in the above example (see S.3.3.1, and analogous discussion in Example 9.3), and by adopting the formulation of (8.5a–d), we write the BVP conditions (while $i=1$, $j=2$) as

$$\nabla^{(2)}\psi = -\frac{E\nu}{1+\nu} P_y \widetilde{C} x, \qquad \text{over} \quad \Omega, \tag{9.138a}$$

$$\frac{d}{ds}\psi = 0 \qquad \text{on} \quad \partial\Omega, \tag{9.138b}$$

$$\frac{d}{ds}\{\psi, \Psi^w\}_{[i]}^{[j]} = \{0,0\} \qquad \text{on} \quad \partial\Omega_{ij}. \tag{9.138c}$$

Regarding the $\frac{d}{ds}\Psi^{w[i]}$ condition required above, we employ the weakening approach of Remark 8.1, write $\frac{d}{ds}\Psi^{w[i]} = \underline{\Psi_{,y}^{w[i]}\cos(\bar{\mathbf{n}},x)} - \Psi_{,x}^{w[i]}\cos(\bar{\mathbf{n}},y)$ and set the underlined term to zero. Then, according to (9.4b), $\gamma_{xz} = \Psi_{,x}^{w[i]}$ (since $\phi_{,z}=0$ in the present case), and thus in the isotropic case $\Psi_{,x}^{w[i]} = 2\psi_{,y}^{[i]}\frac{1+\nu_{[i]}}{E_{[i]}}$. We therefore replace (9.138c) by

$$\{\psi_{,x}, 2\psi_{,xy}\frac{1+\nu}{E}\}_{[i]}^{[j]} = \{0,0\} \qquad \text{on} \quad \partial\Omega_{ij}. \tag{9.139}$$

For continuous $\Psi^{w[i]}$ over the interface lines, we also require that $\Psi^{w[i]}$ will be continuous at specific points over the interface line. The required stress components are then expressed as $\Delta\tau_{yz}^{*[i]} = -\psi_{,x}^{[i]}$ and $\Delta\tau_{yz}^{*[i]} = \psi_{,y}^{[i]}$.

10
Thin-Walled Coupled Monoclinic Beams

Thin-walled structures constitute a very important ingredient in many advanced applications where the specific strength (i.e. "strength to weight" ratio) plays a major role. Representatives of such applications are aircraft wings and helicopter blades that are typically built and modeled as thin-walled beams. The analysis of thin-walled anisotropic beams originally emerged from the theories for isotropic beams and tubes which were gradually modified to include the effects induced by anisotropic materials, see literature on *Thin-Walled Beams* in S.5.4.

Similar to solid anisotropic beams that were discussed in Chapter 9, thin-walled anisotropic beams offer *structural (elastic) couplings* as well. Such couplings emerge from the stress-strain coupling characteristics at the material level, as shown by Fig. 9.2 which is also applicable for the present case of thin-walled beams.

10.1 Background

In what follows, we shall first derive and explore the behavior of thin-walled beams, a cross-section of which may be characterized as a *multiply connected domain*. Such cross-sections are usually denoted as "closed" thin-walled cross-sections and are further classified as either "single-cell" or "multi-cell" geometries, see e.g., Fig. 10.1(a),(b), respectively. Then, thin-walled cross-sections, which are characterized as *simply connected domains* or "open" cross-sections will be discussed, see e.g., Fig. 10.1(c). There are also cases of "mixed" ("open" and "closed") geometries as shown in Fig. 10.1(d). By definition, a cross-section is considered as "thin-walled" if *all* its components may be viewed as thin (either straight or curved) panels of typical thickness, t, which is much smaller than a characteristic dimension of the cross-section (e.g. its overall width or height), while common practice calls for a thickness that is at least 10 to 15 times smaller than a typical width or height. More rigorous criteria were developed by (Rappel and Rand, 2000).

Due to the specific strength characteristics of thin-walled anisotropic beams, the literature contains a vast theoretical and experimental research effort in this area, most of which is focused on closed cross-sections since their torsional rigidity is profoundly higher. Yet, in view of modern applications where soft (controllable) torsional rigidity may be proved as a useful

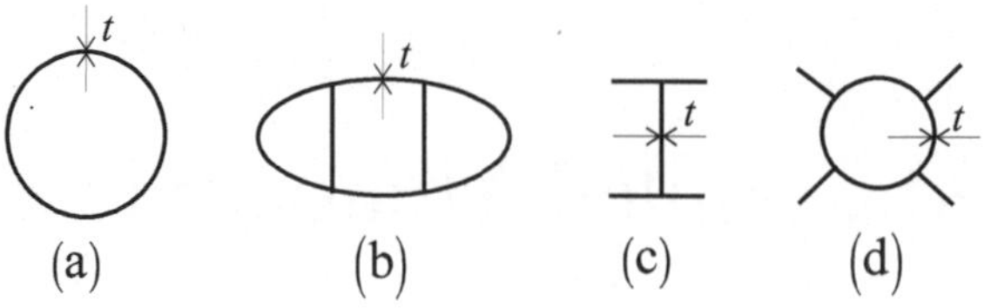

Figure 10.1: A scheme of various closed and open thin-walled cross-section.

characteristic as well, the discussion in this book will also be devoted to thin-walled beams of open cross-sections.

The warping characteristics of open thin-walled beams provide substantially different coupling behavior compared with similar beams of closed cross-section. For example, the torsion of beams of open cross-sections produces very large out-of-plane warping deformation compared with closed thin-walled geometries. To determine such a response, we shall invoke analysis tools that are similar to those utilized for solid simply connected domains within Chapter 9.

Similar to Chapter 9, in this chapter we shall seek a "strength of materials" type of solutions, and thus, the derivation presented in what follows is focused on relatively simple closed-form analytical solutions that will provide numerous advantages. On top of their ability to furnish a simple estimation of the structural beam behavior due to various loading modes, they supply a clear physical explanation for the structural behavior and insight into the coupling mechanisms. They may also be used for identifying critical parameters of various new configurations and subsequently, contribute to the efficiency of detailed analyses as well.

10.2 Multiply Connected Domain

10.2.1 The Elastic Coupling Effects

Simple examples of symmetric and antisymmetric closed, single-cell, thin-walled beams are shown in Fig. 10.2. In this figure, rectangles ("box-beams") are shown. The walls are orthotropic laminae, which are oriented at various θ_y angles with respect to the beam axis, and therefore effectively create various MON13y materials, see S.5.2.2.

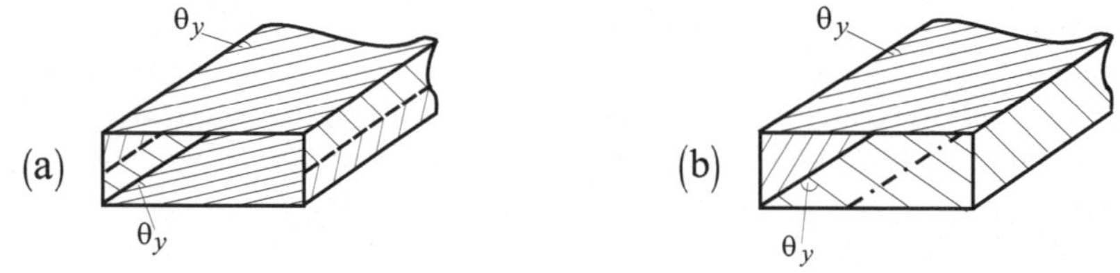

Figure 10.2: Thin-walled beams made of orthotropic laminae. (a) symmetric, (b) antisymmetric.

Referring to Fig. 10.2, in the first case, (a), the upper and the lower horizontal walls are laid at the same angle with respect to the z-axis to create a "symmetric" cross-section. In the second case, (b), the upper and the lower horizontal walls are laid at opposite angles with respect to the z-axis to create an "antisymmetric" cross-section (note that practically, an antisymmetric beam may be obtained by wrapping a continuous lamina along its contour). The various coupling effects in symmetric and antisymmetric thin-walled beams are similar to those discussed in S.9.1.3 and shown in Fig. 9.2. These coupling effects will be demonstrated by the collection of solutions that will be presented in what follows.

10.2.2 An Approximate Analytical Model

10.2.2.1 Kinematics

The analysis presented below is for a straight and untwisted single cell thin-walled beam as schematically shown in Fig. 10.3(a). The beam length, l, is measured along the z coordinate line while the x and y coordinates define the cross-section planes. A scheme of a single-cell thin-walled cross-section is presented in Fig. 10.3(b). The wall is assumed to consist of MON13y

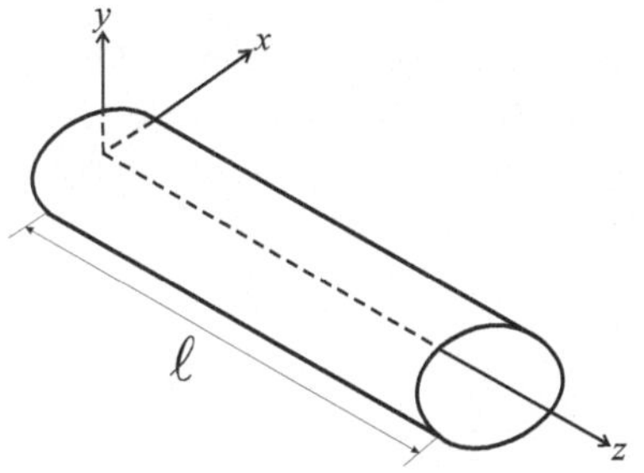

(a) Undeformed thin-walled beam.

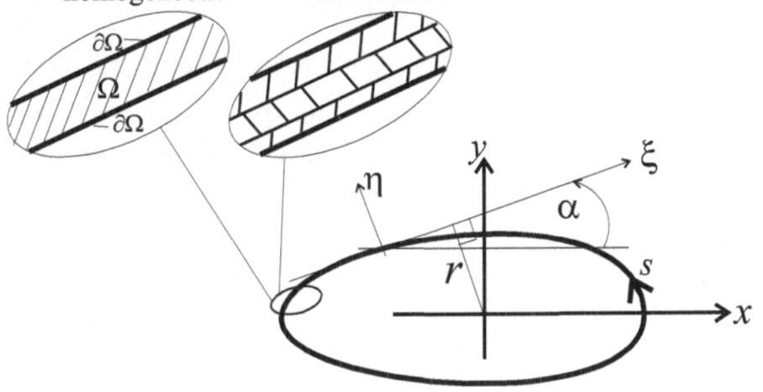

(b) Thin-walled cross-section.

Figure 10.3: Thin-walled beam notation.

material, which is laid "tangent to the wall" as will be discussed shortly, and it may be either homogeneous or laminated. Note that in this case, Ω and $\partial\Omega$ represent the "material region" and its "outer surface", respectively. All notation and definitions described in S.5.1 hold throughout the following derivation as well.

To describe the deformation of the beam presented in Fig. 10.3, a local coordinate system ξ-η-ζ is fixed at each point along the cross-section walls so that ζ is "parallel" to the z-axis, ξ coincides with the local tangent to the wall, and η is perpendicular to ζ and ξ. The angle between the x and the ξ axes is denoted α. The η-ξ origin is located at the wall middle plane. With this notation it is clear that an element of area may be written as $d\Omega = (d\xi - \eta d\alpha)d\eta$ where $d\alpha = \alpha_{,\xi} d\xi$, and therefore, integration over the cross-section area should be carried out as

$$\iint_{\Omega} \tilde{\mathrm{H}} = \oint_{s} [\int_{-t/2}^{t/2} \tilde{\mathrm{H}}\,(1 - \eta\alpha_{,\xi})\, d\eta] d\xi, \tag{10.1}$$

where $t(\xi)$ is the local wall thickness.

The fundamental deformation assumptions that will be used in the present derivation are similar to those presented in S.9.2 for solid beams, and are based on two types of displacements. First, the cross-section displacements $u_0(z)$, $v_0(z)$, and $w_0(z)$ in the x, y, and z directions, respectively, and a twist angle, $\phi(z)$, about the z direction are defined. These components of the deformation are functions of z only, and therefore, they represent "rigid body" displacements of each cross-section that contains no warping.

To account for the out-of-plane warping, a three-dimensional warping function is superimposed (in the z direction) upon the above mentioned displacements. This warping function is denoted Ψ^w, and is a function of η and s, where s is a circumferential coordinate along the wall directed counterclockwise (i.e. in the positive z direction), see Fig. 10.3(b). Ψ^w is assumed to be of zero average value over the cross-section area (i.e., $\iint_\Omega \Psi^w = 0$). Similar to the case of solid beams, the out-of-plane warping may originate from various sources (e.g. torsion, bending, extension, etc.).

As a general rule, although z and ζ directions are identical, a quantity that is a function of z only will be denoted as $\mathrm{H}(z)$, while a quantity that is a function of (z,s) will be denoted as $\mathrm{H}(\zeta,s)$. Hence, although $\frac{\partial}{\partial z} = \frac{\partial}{\partial \zeta}$ for the global cross-section quantities (such as u_0, v_0, w_0, and ϕ), derivatives with respect to z will be retained, while for the local characteristics (such as Ψ^w), derivatives with respect to ζ will be used. In addition, although both s and ξ stand for the circumferential direction (while $\frac{\partial}{\partial \xi} = -\frac{\partial}{\partial s}$), the notation $\mathrm{H}(\zeta,s)$ will be kept.

10.2.2.2 Constitutive Relations

Since the material is assumed to align with the wall direction (i.e. the ξ coordinate line is tangent to the local wall, see Fig. 10.3(b)), we adopt here the constitutive relations derived in S.9.2.1 (either Version I or Version II). For the present case of a thin-walled cross-section, these constitutive relations are applicable in the local ξ,η,ζ system directions, namely,

$$\left\{ \begin{array}{c} \sigma_\zeta \\ \tau_{\eta\zeta} \\ \tau_{\xi\zeta} \end{array} \right\} = \mathbf{Q} \cdot \left\{ \begin{array}{c} \varepsilon_\zeta \\ \gamma_{\eta\zeta} \\ \gamma_{\xi\zeta} \end{array} \right\}. \tag{10.2}$$

To derive the strain components in the above ξ,η,ζ directions we employ the strain definition of (9.4a) and the strain transformation rules derived in S.1.3.4 (where the rotation angle about the z direction is α) to obtain

$$\left\{ \begin{array}{c} \varepsilon_\zeta \\ \gamma_{\eta\zeta} \\ \gamma_{\xi\zeta} \end{array} \right\} = \begin{bmatrix} 1 & 0 & 0 \\ 0 & \cos\alpha & -\sin\alpha \\ 0 & \sin\alpha & \cos\alpha \end{bmatrix} \cdot \left\{ \begin{array}{c} \varepsilon_{zz} \\ \gamma_{yz} \\ \gamma_{xz} \end{array} \right\}, \tag{10.3}$$

which by employing the identities (note that $\frac{\partial}{\partial \xi} = -\frac{\partial}{\partial s}$)

$$\cos\alpha = x_{,\xi} = y_{,\eta}, \qquad \sin\alpha = y_{,\xi} = -x_{,\eta}, \tag{10.4}$$

yields

$$\varepsilon_\zeta = w_{0,z} - x u_{0,zz} - y v_{0,zz} + \Psi^w_{,\zeta}, \tag{10.5a}$$

$$\gamma_{\eta\zeta} = \phi_{,z}(x\cos\alpha + y\sin\alpha) + \Psi^w_{,\eta}, \tag{10.5b}$$

$$\gamma_{\xi\zeta} = \phi_{,z}(x\sin\alpha - y\cos\alpha) + \Psi^w_{,\xi}. \tag{10.5c}$$

As shown in Fig. 10.4, the normal distance from the x,y-plane origin to the tangent line is given by $r = y\cos\alpha - x\sin\alpha$, while the distance along the tangent line between the point under discussion and the closest point to the origin is $r' = x\cos\alpha + y\sin\alpha$. Note that the above interpretation of r as the normal distance forces it to be a positive number while the quantity $(y\cos\alpha - x\sin\alpha)$ may change sign. More generally, by defining an infinitesimal element along the tangent line as $d\vec{s} = \hat{x}dx + \hat{y}dy$, and the radius vector to the point under discussion as $\vec{R} = x\hat{x} + y\hat{y}$ (where $\hat{x}$ and $\hat{y}$ are unit vectors in the x and y directions, respectively), one may determine the normal distance, r, as

$$r = \frac{\left| d\vec{s} \times \vec{R} \right|}{\left| d\vec{s} \right|} = \left| -y\cos\alpha + x\sin\alpha \right|. \tag{10.6}$$

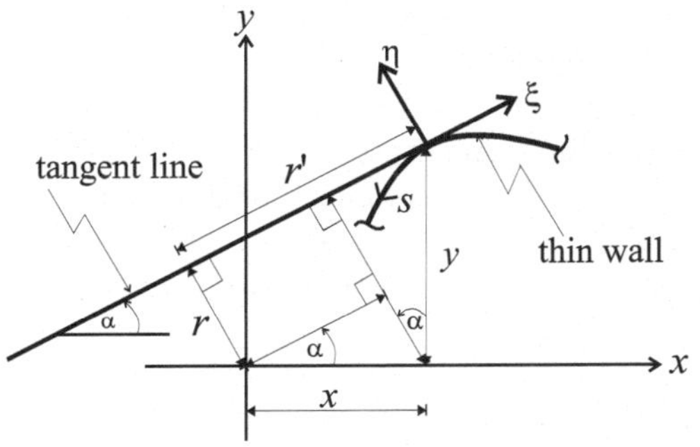

Figure 10.4: Geometrical interpretation of r and r'.

Hence, the quantity $(-y\cos\alpha + x\sin\alpha)$ will be denoted as $\pm r$, where the plus sign stands for cases where $d\vec{s}\times\vec{R}$ is oriented in the positive z direction. With the above definitions, (10.5a–c) may be written as

$$\varepsilon_\zeta = w_{0,z} - xu_{0,zz} - yv_{0,zz} + \Psi^w_{,\zeta}, \qquad \gamma_{\eta\zeta} = r'\phi_{,z} + \Psi^w_{,\eta}, \qquad \gamma_{\xi\zeta} = \pm r\phi_{,z} + \Psi^w_{,\xi}. \tag{10.7}$$

At this stage, we shall make a further simplification in the constitutive relations by ignoring $\gamma_{\eta\zeta}$ and $\tau_{\eta\zeta}$. To justify this neglect, we first notice that (10.5b) shows that for all cases where no twist takes place, one may simply assume $\Psi^w_{,\eta} = 0$ (i.e. constant warping through the thickness) to obtain $\gamma_{\eta\zeta} = \tau_{\eta\zeta} = 0$. Yet, when $\phi_{,z} \neq 0$, $\gamma_{\eta\zeta}$ does not vanish anymore, and indeed, Ψ^w becomes a function of η. In such a case, we set

$$\Psi^w(s,\eta,\zeta) = \psi^w(s,\zeta) - \eta\phi_{,z}r'(s,\zeta). \tag{10.8}$$

Then, $\Psi^w_{,\eta} = -\phi_{,z}r'$, which yields $\gamma_{\eta\zeta} = 0$. In classical "isotropic" terminology, $\psi^w(s,\zeta)$ should be denoted as "contour warping" while the $-\eta\phi_{,z}r'$ term should be denoted as (linear) "thickness warping". This notation is related to Vlasov and Kirchhoff assumptions that were originally applied to torsion and flexure problems of thin-walled open isotropic beams, see e.g. (Gjelsvik, 1981). Thus, using the above hypothesis, we arrive at the simplified expressions

$$\varepsilon_\zeta = w_{0,z} - x\,u_{0,zz} - y\,v_{0,zz} + \psi^w_{,\zeta}, \tag{10.9a}$$
$$\gamma_{\eta\zeta} = 0, \tag{10.9b}$$
$$\gamma_{\xi\zeta} = \mp r\phi_{,z} + \psi^w_{,\xi}, \tag{10.9c}$$

provided that we further neglect $|\eta\,\phi_{,z}\frac{\partial r'}{\partial\xi}|$ compared with $|\psi^w_{,\xi}|$. This neglect is perfectly true for $\eta = 0$, and thus, from this point on, $\psi^w(s,\zeta)$ will be referred to as the *middle-plane warping*, while we keep in mind that the actual warping is given by (10.8). The reduced stress-strain relations in the present case are therefore

$$\begin{Bmatrix} \sigma_\zeta \\ \tau_{\xi\zeta} \end{Bmatrix} = \begin{bmatrix} Q_{33} & Q_{35} \\ Q_{35} & Q_{55} \end{bmatrix} \cdot \begin{Bmatrix} \varepsilon_\zeta \\ \gamma_{\xi\zeta} \end{Bmatrix}. \tag{10.10}$$

In summarizing this section, we note that the present formulation is founded on the neglect of five strain/stress components, i.e. the $\tau_{\eta\zeta}$, $\gamma_{\eta\zeta}$ components, and the triad ε_ξ, ε_η, $\gamma_{\xi\eta}$ for the reduced constitutive relations of Version I, or the triad σ_ξ, σ_η, $\tau_{\xi\eta}$ for the reduced constitutive relations of Version II (see S.9.2.1).

Remark 10.1 Note that, similar to the case of solid cross-sections, Version II is more suitable to the case of a traction-free outer surface.

An additional supporting argument for using Version II in the case of traction-free outer surface is common in thin plate analyses, and is usually referred to as a "plane-stress" state within the thin wall, see S.3.2.2. Here, one may argue that due to the fact that the stress components vanish over $\partial\Omega$, and the relatively small wall thickness (compared with other cross-section dimensions), the corresponding $\tau_{\xi\eta}$, σ_η and $\tau_{\eta\zeta}$ stress components ought to remain small over the entire wall thickness.

10.2.2.3 Governing Equations and Boundary Conditions

The definition of "equivalent loads per unit length" of (9.2) holds in the present case as well. Thus, similar to (9.19), equilibrium is achieved by four integral equations and one differential equation. The integral equations equate the cross-section resultants $\overline{P}_x$ (in the x direction), $\overline{P}_y$ (in the y direction), $\overline{P}_z$ (in the z direction) and the moment resultant $\overline{M}_z$ (in the z direction) to the corresponding loads obtained by stress integrations. The cross-section equilibrium equations may be therefore written for a thin-walled cross-section as

$$\{\overline{P}_x, \overline{P}_y, \overline{P}_z, \overline{M}_z\} = \iint_\Omega \{\tau_{\xi\zeta}\cos\alpha, \tau_{\xi\zeta}\sin\alpha, \sigma_\zeta, -r\tau_{\xi\zeta}\}. \tag{10.11}$$

To derive the differential equilibrium equation in the z-direction for this case (similar to (9.23)), one may employ the corresponding differential equilibrium equations for cylindrical coordinates derived in S.1.3.2, namely,

$$\frac{1}{\rho}\left(\sigma_{\theta z,\theta} + \sigma_{\rho z}\right) + \sigma_{\rho z,\rho} + \sigma_{z,\zeta} + Z_b = 0, \tag{10.12}$$

where Z_b is the body-force component in the ζ (or z) direction. This equation should be implemented by applying the following replacements: $\theta \to \xi$, $\rho \to \eta$, $z \to \zeta$, $\frac{\partial}{\partial\xi} = \frac{1}{\rho}\frac{\partial}{\partial\theta}$ and $\frac{1}{\rho} \to \kappa$ where κ stands for the local curvature. This yields

$$\sigma_{\zeta,\zeta} + \tau_{\xi\zeta,\xi} + \tau_{\eta\zeta,\eta} + \kappa\tau_{\eta\zeta} + Z_b = 0. \tag{10.13}$$

In addition, by neglecting $\tau_{\eta\zeta}$ as previously discussed, one obtains the following differential equilibrium equation:

$$\sigma_{\zeta,\zeta} + \tau_{\xi\zeta,\xi} + Z_b = 0. \tag{10.14}$$

Similar to the case of a solid cross-section, there are eight beam-type boundary conditions at the beam root and tip. For a "clamped-free" beam, six of them are the geometrical boundary conditions given by (9.24). The remaining two natural boundary conditions at the beam tip are based on equating the external transverse tip moments, M_x, M_y, to those obtained by stress integrations over the tip cross-section area, Ω^t. These conditions are given by (9.25), namely,

$$\{M_x, M_y\} = \iint_{\Omega^t} \{y, -x\}\sigma_\zeta. \tag{10.15}$$

No contour boundary conditions are required for a thin-walled cross-section since $\tau_{\eta\zeta}$ is neglected.

In what follows we shall present some solutions that comply with the above formulation and illustrate the behavior of closed thin-walled coupled monoclinic beams. In these solutions, it will be assumed that the wall thickness is homogeneous. In terms of composite laminae, this assumption means that we confine these solutions either to the case of a single-lamina wall, or to the case where all laminae are identical and oriented at the same angle with respect to the z-axis, see Fig. 10.3(b) and its discussion. From an analytical point of view, this simplification is useful since the elastic moduli are all constants throughout the thickness. As a further *approximation*, the solutions derived in what follows may be used for laminated walls as well while employing the laminate's "effective" or "smeared" properties.

Remark 10.2 Apparently, the neglect of the stress components $\tau_{\xi\eta}$, σ_η and $\tau_{\eta\zeta}$ "contradicts" the fact that in the general case, the wall outer surface is loaded. This contradiction may be removed by noting that the "net" or "integral" effect of these loads is accounted for by the integral equilibrium equations since these loads contribute to the "equivalent loads per unit length" p_x, p_y, p_z and m_z, which determine the variation (with respect to z) of $\overline{P}_x$ $\overline{P}_y$, $\overline{P}_z$ and $\overline{M}_z$, respectively, see S.9.2.2 and in particular, (9.22).

10.2.2.4 Antisymmetric Beam Under Tip Torsional Moment and Axial Force

The solution described in this section is derived for a single cell cross-section of arbitrary geometry, constant wall thickness and constant elastic moduli. As already discussed, such a cross-section should be classified as *antisymmetric*, see Fig. 10.2(b). In this case, we assume that the origin is located inside the cross-section and therefore, equation (10.9c) is written with the minus sign for the entire cross-section. When such a beam is subjected to a tip torsion moment, M_z, and a tip axial force, P_z, an exact analytical solution may be generated by assuming $u_0 = v_0 = 0$ and constant strains ε_ζ and $\gamma_{\xi\zeta}$ over each cross-section and along the beam (see trail "B" of S.1.6.5.2). Consequently, σ_ζ and $\tau_{\xi\zeta}$ are also constants, and ψ^w is a function of s only. Then, (10.11) shows that

$$P_z = pt\sigma_\zeta, \qquad M_z = -2A_m t\tau_{\xi\zeta}, \tag{10.16}$$

where p is the cross-section circumference and A_m is the area enclosed by the median line. Since σ_ζ and $\tau_{\xi\zeta}$ are constants, the differential equilibrium equation (10.14) is satisfied. Substituting (10.9a,b), (10.10) in (10.16: b) yields two equations for the unknowns $w_{0,z}$ and $\gamma_{\xi\zeta}$. In this system, $\gamma_{\xi\zeta}$ is replaced by $\phi_{,z}$ by integrating (10.9c) under the condition $\oint_{\partial\Omega}\psi^w_{,s}\,ds = 0$, which yields $\gamma_{\xi\zeta} = -2A_m\phi_{,z}/p$. Solving these equations yields

$$\begin{Bmatrix} w_{0,z} \\ \phi_{,z} \end{Bmatrix} = \frac{1}{Q_{33}Q_{55} - Q_{35}^2}\begin{bmatrix} \frac{Q_{55}}{pt} & \frac{Q_{35}}{2A_m t} \\ \frac{Q_{35}}{2A_m t} & \frac{Q_{33}p}{4A_m^2 t} \end{bmatrix} \cdot \begin{Bmatrix} P_z \\ M_z \end{Bmatrix}. \tag{10.17}$$

Having $w_{0,z}$ and $\phi_{,z}$, the variation of ψ^w as a function of s is obtained by integrating (10.9c) (i.e., $\psi^w_{,s} = -r\phi_{,z} - \gamma_{\xi\zeta}$) under the condition $\oint_{\partial\Omega}\psi^w\,ds = 0$.

Clearly, when Version II of the constitutive relations is adopted (see S.9.2.1) for the present case where all strain components are constants, no violation of the compatibility equations occur, while all traction-free boundary conditions are fully satisfied (i.e., $\tau_{\eta\eta}$ and $\tau_{\xi\eta}$ vanish as well).

The Extension-Twist coupling in thin-walled antisymmetric beams is represented by the off-diagonal terms in the matrix of (10.17). As shown, $\phi_{,z} = \frac{p}{2A_m}\frac{Q_{35}}{Q_{55}}w_{0,z}$ in the case of a tip axial force, and $w_{0,z} = \frac{2A_m}{p}\frac{Q_{35}}{Q_{33}}\phi_{,z}$ in the case of a tip torsion moment. In both cases, reciprocity shows that

$$\phi_{,z\,|P_z=1} = w_{0,z\,|M_z=1}. \tag{10.18}$$

For a circular closed cross-section of radius R, (10.17) shows that

$$\frac{w_{0,z}}{R\phi_{,z}}_{|M_z} = \frac{Q_{35}}{Q_{33}}, \qquad \frac{R\phi_{,z}}{w_{0,z}}_{|P_z} = \frac{Q_{35}}{Q_{55}}. \tag{10.19}$$

Figure 10.5 presents the above couplings as obtained for typical Graphite/Epoxy orthotropic material that is laid at various lamination angles with respect to the beam axis (see Fig. 10.2(b)).

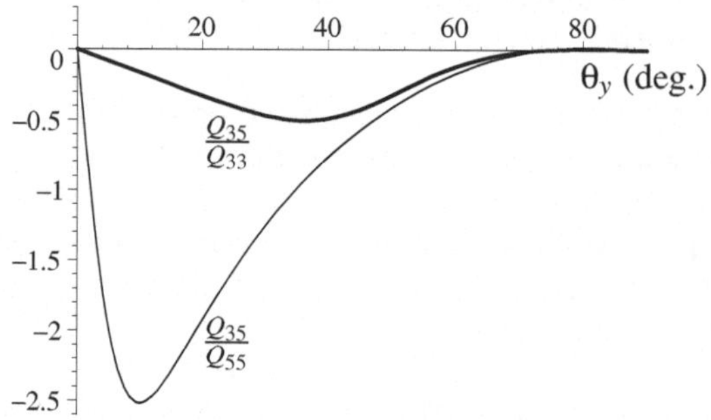

Figure 10.5: Typical Torsion-Extension (thick line) and Extension-Torsion (thin line) coupling magnitude in a circular cross-section, as functions of the lamination angle.

Similar to the case of a homogeneous solid cross-section, for a given set of elastic moduli, it is possible to define an "uncoupled" case where $Q_{35} = 0$ and the values of Q_{33} and Q_{55} are kept unchanged (Q_{44} is not relevant in this case — see (10.10)). Equation (10.17) shows that the "uncoupled" torsional rigidity, D, the "coupled" torsional rigidity, D_{eq} and their ratio are

$$D = Q_{55} \cdot \frac{4A_m^2 t}{p}, \qquad D_{eq} = \frac{Q_{33}Q_{55} - Q_{35}^2}{Q_{33}} \cdot \frac{4A_m^2 t}{p}, \qquad \frac{D_{eq}}{D} = 1 - \frac{Q_{35}^2}{Q_{33}Q_{55}}. \tag{10.20}$$

These quantities are presented in Fig. 10.6 for a thin-walled cross-section made of typical Graphite/Epoxy orthotropic lamina (all values are normalized with their respective $\theta_y = 0$ values), where they are also compared with the results obtained in S.9.2.3.4 for solid cross-sections.

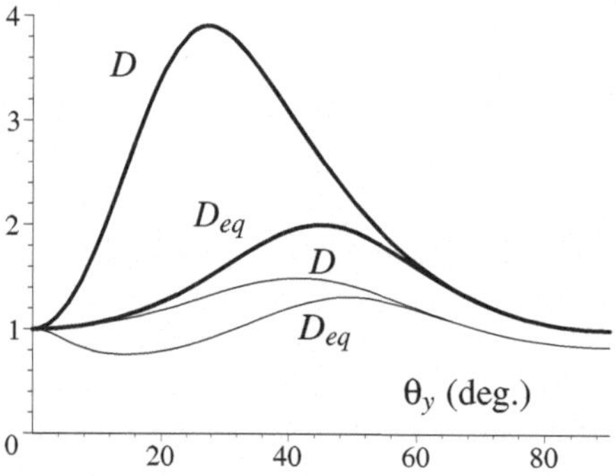

(a) The "coupled" and "uncoupled" torsional rigidity.

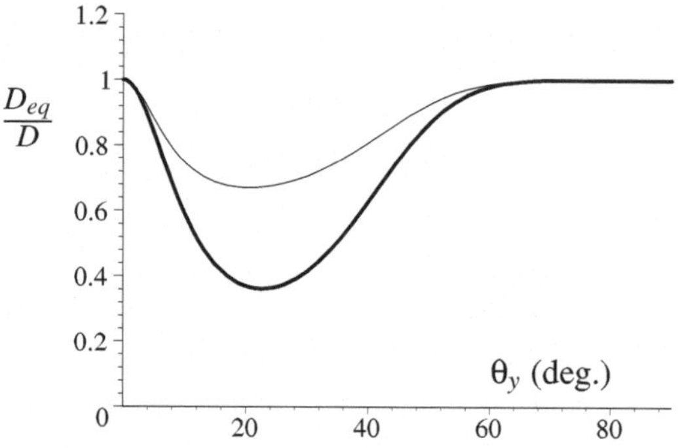

(b) The ratio of "coupled" to "uncoupled" torsional rigidity.

Figure 10.6: Thick lines: Torsional rigidity as a function of the lay-up angle. Thin lines: The values shown in Fig. 9.4 for a solid elliptic cross-section (of $\tilde{a}/\tilde{b} = 2$).

Note that unlike the solid beam case, both the "uncoupled" and "coupled" torsional rigidities of (10.20) may be viewed as products of the geometrical factor $\frac{4A_m^2 t}{p}$ and an elastic moduli (or an "effective" one). In Fig. 10.6, the "uncoupled" case reflects the variations due to the changes in the shear elastic modulus, Q_{55}, while the "coupled" case reflects also the coupling effects. Within this comparison, one should also note that the above discussed coupling is of Twist-Bending type for solid beams, while in the present case is of Twist-Extension type.

As far as the beam behavior in extension is concerned, namely, the effect of a tip axial load, P_z, on the mean extensional strain $w_{0,z}$, it is interesting to see that the result of (10.17) is identical to (9.29), which has been obtained for a symmetric solid cross-section (by substituting $S_\Omega = pt$ and considering the case of a uniform axial load, i.e., $\overline{P}_z = P_z$).

The solution presented by (10.17) may be put in a more enlightening form. Defining the cross-sectional "moment (of inertia) for torsion" in this case as $J = 4A_m^2 t/p$ and noting that the cross-section area is $S_\Omega = pt$, one may write

$$\begin{Bmatrix} w_{0,z} \\ \phi_{,z} \end{Bmatrix} = \frac{1}{Q_{33}Q_{55} - Q_{35}^2} \begin{bmatrix} \frac{Q_{55}}{S_\Omega} & \frac{Q_{35}}{\sqrt{S_\Omega J}} \\ \frac{Q_{35}}{\sqrt{S_\Omega J}} & \frac{Q_{33}}{J} \end{bmatrix} \cdot \begin{Bmatrix} P_z \\ M_z \end{Bmatrix}, \tag{10.21}$$

which clearly exposes the role of the quantities S_Ω, J and $\sqrt{S_\Omega J}$ that represent "axial", "torsional", and "coupled axial-torsional" stiffness elements as expected in this problem. This formulation may serve as a first approximation and generalization for cases of non-uniform thickness distribution, where J and S_Ω should be replaced by $4A_m^2 / \oint_{\partial\Omega} \frac{ds}{t}$ and $\oint_{\partial\Omega} t\, ds$, respectively.

Example 10.1 *Torsional Rigidity of Thin-Walled Orthotropic Beam.*

It is interesting to use the above results to examine the torsional rigidity of a thin-walled orthotropic beam, and to compare it with the case of a thin-walled cross-section with Cartesian anisotropy.

For that purpose, we first adopt the solution presented in Example 7.1 for which a stress function has been calculated for a solid elliptical cross-section of semi-axes of $\widetilde{a}, \widetilde{b}$. The corresponding stress function, ψ, and torsional rigidity, D, are given by (7.41) and (7.38), respectively. Clearly, the contour of an ellipse of semi-axes $\widetilde{a}k, \widetilde{b}k$ is a stress line in the solid ellipse for which ψ is constant, and therefore, the boundary condition $\tau_N = 0$ is satisfied over it (see (7.28a)). Hence, the above stress function, ψ, applies also to the cross-section for which S_Ω is defined as the region between the contours $x^2/\widetilde{a}^2 + y^2/\widetilde{b}^2 = 1$ and $x^2/\widetilde{a}^2 + y^2/\widetilde{b}^2 = k^2$, where $k < 1$. In such a case, a thin-walled cross-section is obtained for $k \to 1$. The torsional rigidity of such a thin-walled cross-section is therefore

$$D_{\text{Cartesian}} = \frac{\pi \widetilde{a}^3 \widetilde{b}^3}{a_{55}\widetilde{a}^2 + a_{44}\widetilde{b}^2} \left(1 - k^4\right). \tag{10.22}$$

Returning to the result of (10.17), for the above thin elliptical cross-section and $1 \le \widetilde{a}/\widetilde{b} < 1.5$, one may employ the following approximations: $A_m \cong \pi\widetilde{a}\widetilde{b}$, $t \cong (1-k)(\widetilde{a}+\widetilde{b})/2$, and $p \cong \pi\sqrt{2(\widetilde{a}^2+\widetilde{b}^2)}$. Hence, for the orthotropic case under discussion, one obtains (see (10.20:a))

$$D = Q_{55}\frac{4A_m^2 t}{p} = \frac{4\pi\widetilde{a}^3\widetilde{b}^3}{a_{55}(\widetilde{a}^2+\widetilde{b}^2)}(1-k)\widetilde{f}, \qquad \widetilde{f} = \frac{1}{2\sqrt{2}}(1+\frac{\widetilde{b}}{\widetilde{a}})\sqrt{1+\frac{\widetilde{a}^2}{\widetilde{b}^2}}. \tag{10.23}$$

This quantity stands for the case where the material direction is "parallel" to the local wall direction as shown in Fig. 10.7(b), while that of (10.22) stands for Cartesian material, see Fig. 10.7(a). Therefore (10.22) depends on both a_{44} and a_{55}, while (10.23) does not include a_{44}. A similar effect in solid beams is discussed in S.7.5.2.

It is interesting to note that (10.22) may be written for $k \to 1$ as

$$D_{\text{Cartesian}} \cong \frac{\pi\widetilde{a}^3\widetilde{b}^3}{a_{55}\widetilde{a}^2 + a_{44}\widetilde{b}^2}\left(1+k^2\right)(1+k)(1-k) \cong \frac{4\pi\widetilde{a}^3\widetilde{b}^3}{a_{55}\widetilde{a}^2 + a_{44}\widetilde{b}^2}(1-k), \tag{10.24}$$

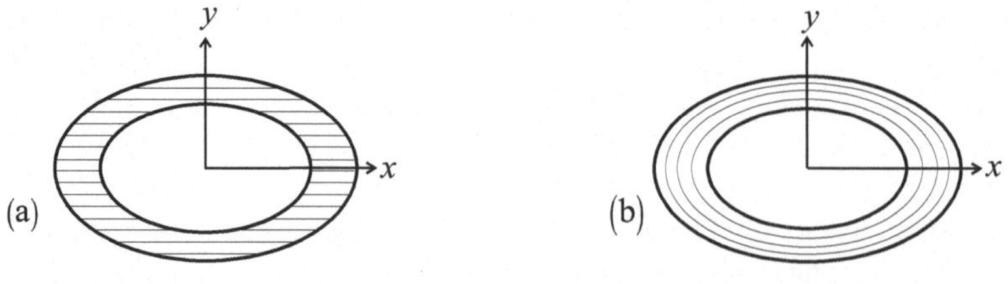

Figure 10.7: A thin-walled cross-section made of two confocal ellipses; material laminae are laid (a) in Cartesian coordinates, (b) "parallel" to the local wall direction.

which coincides with (10.23) for $a_{44} = a_{55}$ (and $\widetilde{f} \cong 1$, which is essentially the range for which the approximations made above in the expressions for A_m, t and p are valid). As a specific example consider a circular cross-section of radius $\widetilde{R}$ (where $\widetilde{f} = 1$) for which

$$D_{\text{Cartesian}} \cong \frac{4\pi\widetilde{R}^4}{a_{44}+a_{55}}(1-k), \qquad D = \frac{4\pi\widetilde{R}^4}{2a_{55}}(1-k). \tag{10.25}$$

It is therefore shown again that a_{55} "plays also the role" of a_{44} in the torsion mechanism of thin-walled beams.

10.2.2.5 Antisymmetric Beam Under Bending Moments

To study the effect of tip bending moments M_x and M_y, we shall consider an arbitrary antisymmetric thin-walled cross-section of constant elastic moduli similar to the one discussed in S.10.2.2.4. Since no other loads are applied we set $\overline{M}_x = M_x$ and $\overline{M}_y = M_y$.

For the strain expressions of (10.9a,b) we assume $\psi^w = \psi^w(s)$, $\phi = w_0 = 0$, and

$$v_{0,zz} = -\widetilde{Q}\frac{M_x}{I_x}, \qquad u_{0,zz} = \widetilde{Q}\frac{M_y}{I_y}, \qquad \widetilde{Q} = \frac{Q_{55}}{Q_{33}Q_{55}-Q_{35}^2}. \tag{10.26}$$

The above yields

$$\varepsilon_\zeta = \widetilde{Q}\frac{M_x}{I_x}y - \widetilde{Q}\frac{M_y}{I_y}x, \qquad \gamma_{\xi\zeta} = -\psi^w_{,s} = -\frac{Q_{35}}{Q_{55}}\varepsilon_\zeta, \tag{10.27}$$

and hence,

$$\sigma_\zeta = \frac{M_x}{I_x}y - \frac{M_y}{I_y}x, \qquad \tau_{\xi\zeta} = 0. \tag{10.28}$$

In view of (5.1), the above stress components clearly satisfy the integral equilibrium equations (10.11), the differential equilibrium equation (10.14), and the tip conditions (10.15).

We therefore see that for given values of M_x and M_y, ε_ζ and $\gamma_{\xi\zeta}$ are directly determined by (10.27). Subsequently, $\psi^w(s)$ is obtained by integrating the second equation of (10.27). Note that (10.28) are identical to those presented for the general case in (5.23).

10.2.2.6 Symmetric Beam Under Tip Axial Force

To form an illustrative symmetric thin-walled beam, a single-cell rectangular cross-section that consists of two horizontal walls at $y = \pm\widetilde{b}/2$ and two vertical walls at $x = \pm\widetilde{a}/2$ is employed — see Fig. 10.8. The wall thickness, t, is constant. The upper half of this cross-section (i.e. the part for which $y > 0$) is assumed to be characterized by the elastic moduli Q_{33}, Q_{55} and Q_{35},

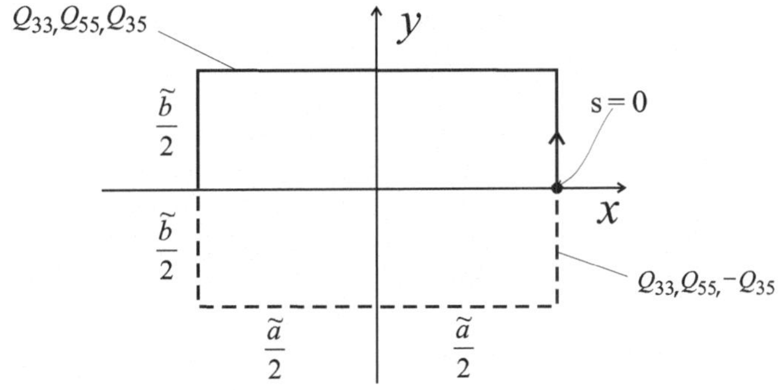

Figure 10.8: Simplified symmetric single-cell rectangular cross-section.

while the lower half (i.e. the part for which $y < 0$) is characterized by Q_{33}, Q_{55} and $-Q_{35}$. As already discussed, such a cross-section should be classified as *symmetric*, see Fig. 10.2(a).

For vanishing values of Q_{35}, it is clear that the results of S.10.2.2.4 are applicable and yield $u_0 = v_0 = \phi = \psi^w = 0$ and $w_{0,z} = P_z/2Q_{33}(\widetilde{a}+\widetilde{b})t$, where P_z is the tip axial force. When a nonzero value of Q_{35} is considered, "additional" strains $\Delta\varepsilon_\zeta = \Delta w_{0,z}$ and $\Delta\gamma_{\xi\zeta} = -\psi^w_{,s}$ should appear in order to provide the same stress distribution (and therefore, to ensure that the fulfillment of both equilibrium equations and boundary conditions is maintained). To generate an exact closed-form analytical solution for this case, ψ^w is assumed to be given by

$$\psi^w(s) = \begin{cases} \widetilde{\alpha}[s - \frac{1}{2}(\widetilde{a}+\widetilde{b})], & \text{for} \quad 0 \le s \le \widetilde{a}+\widetilde{b}, \\ \widetilde{\alpha}[\frac{3}{2}(\widetilde{a}+\widetilde{b}) - s], & \text{for} \quad \widetilde{a}+\widetilde{b} \le s \le 2(\widetilde{a}+\widetilde{b}), \end{cases} \tag{10.29}$$

where $\widetilde{\alpha}$ is a constant to be determined and the origin of the circumferential coordinate s is located at $x = \widetilde{a}/2, y = 0$ — see Fig. 10.8. The requirement for zero additional stresses yields (see (10.10)):

$$\Delta\sigma_\zeta = Q_{33}\Delta w_{0,z} - Q_{35}\widetilde{\alpha} = 0, \qquad \Delta\tau_{\xi\zeta} = Q_{35}(w_{0,z} + \Delta w_{0,z}) - Q_{55}\widetilde{\alpha} = 0. \tag{10.30}$$

It should be emphasized that since the signs of both the elastic modulus Q_{35} and $\psi^w_{,s}$ are identical to the sign of the y coordinate for each point over the cross-section, (10.30) hold for the entire beam. Solving (10.30) yields the values of $\Delta w_{0,z}$ and $\widetilde{\alpha}$ as

$$\Delta w_{0,z} = \frac{Q_{35}^2}{Q_{33}Q_{55} - Q_{35}^2} w_{0,z}, \qquad \widetilde{\alpha} = \frac{Q_{33}Q_{35}}{Q_{33}Q_{55} - Q_{35}^2} w_{0,z}. \tag{10.31}$$

Thus, similar to the case of a symmetric solid beam, the coupling in the present case induces additional axial strain and a warping distribution. It is also interesting to note that the total $w_{0,z} + \Delta w_{0,z}$ value obtained here is in complete agreement with (9.29) (for $S_\Omega = 2(\widetilde{a}+\widetilde{b})t$ and a uniform axial load, i.e., $\overline{P}_z = P_z$) although they were derived for different symmetric solid and thin-walled beams. Also, (9.28), (9.29) show warping, which is linear with respect to x. Likewise, in the present case the warping over the horizontal walls varies linearly with respect to x as well (as $\psi^w_{,s} = \pm\widetilde{\alpha}$ for $y = \pm\widetilde{b}/2$), which is identical to the rate of change of the warping given in S.9.2.3.1.

Since all strain components remain constant, no violation of the compatibility equations occurs when Version II is adopted for the present case, while all traction-free boundary conditions are fully satisfied,

10.3 Simply Connected Domain

The characteristics of beams of simply connected or "open" cross-section (i.e. cross-sections that include no closed cells) are under discussion in this section. Classical analysis of torsion and flexure of thin-walled open isotropic beams may be found in (Gjelsvik, 1981).

The analysis of open anisotropic thin-walled beams under torsion is quite different from the one required when such beams undergo transverse bending and axial loads. This is due to the fact that in the former case, the stress components significantly vary throughout the wall thickness, while in the latter case, they are nearly constants (throughout the thickness). In addition, in some cases, the torsion problem may be confined to a two-dimensional problem where the local values of the twist, bending curvature and axial strain are determined based on the local resultants. On the other hand, the analysis of an anisotropic beam under bending loads poses a relatively complex three-dimensional problem as will be described later on.

In view of the above, the following discussion will be divided into two main parts. First, the analysis of an open thin-walled cross-section under transverse and axial loading will be addressed. Then, the beam behavior under torsional moment will be discussed. The analysis is introduced by its fundamental steps in order to allow a clear treatment of the associated simplifying assumptions, and to provide insight into the associated physical phenomena.

As far as the definition of symmetric and antisymmetric configurations of open beams is concerned, Fig. 10.9 presents illustrative cross-sections made of orthotropic material, which is laid at an angle θ_y with respect to the beam's z-axis. As shown, the lamination angles of the vertical walls in Fig. 10.9(a) are symmetric with respect to a vertical plane, which is parallel to the y,z-plane that passes through the middle of the horizontal wall, while the lamination angles of the vertical walls in Fig. 10.9(b) are antisymmetric with respect to the same plane. It should be noted that the above definitions are related to the edgewise bending kinematics that consists of the transverse displacement in the x direction and rotation about the y direction, and we therefore expect symmetric or antisymmetric behavior for phenomena associated with this edgewise bending for the cross-sections shown in Fig. 10.9.

Note that such definitions are possible only in cross-sections that pose certain geometrical requirements (for example, no clear definitions of "symmetry" or "antisymmetric" lamination with respect to planes that are parallel to the x,z-plane may be related to the cross-sections shown in Fig. 10.9).

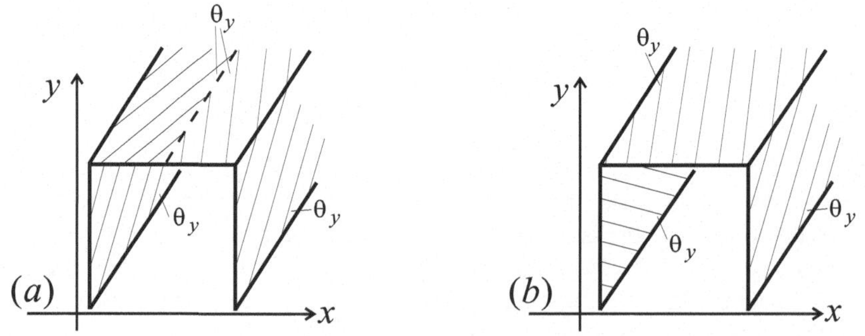

Figure 10.9: Illustrative thin-walled beams of open cross-section: (a) symmetric, (b) antisymmetric.

10.3.1 The Transverse and Axial Loads Effect

The prediction of the structural behavior of anisotropic beams of open cross-section geometry due to transverse loading (i.e. the action of either distributed and/or concentrated transverse forces and bending moments) and axial loading is dealt with in what follows. As already mentioned, in this case, the thin-walled geometry produces nearly uniform distribution of the normal and the shear stresses throughout the wall thickness.

In addition, typically, bending and extension loads produce relatively complex three-dimensional effects. Nevertheless, the beam response due to the torsional moment distribution (that will be derived within S.10.3.2) will be invoked to augment the present derivation by superposition, as will be discussed later on.

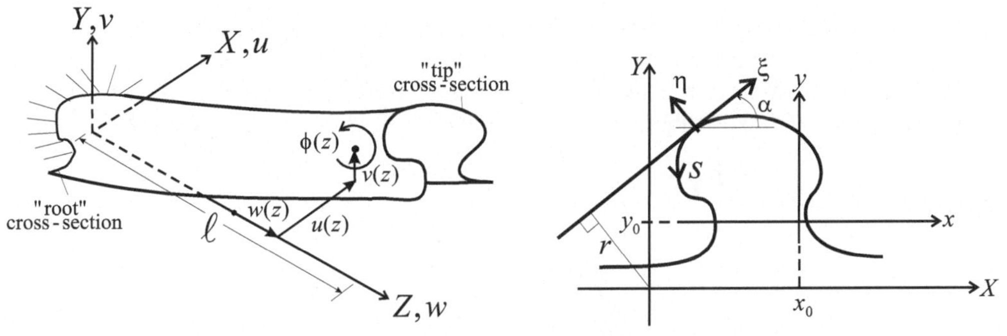

(a) The cross-section deformation components (the out-of-plane warping, $\psi^w(\xi,\zeta)$, is not shown).

(b) A generic open cross-section geometry.

Figure 10.10: A scheme of a thin-walled anisotropic beam of open cross-section geometry.

For the sake of clarity and computational efficiency, the following discussion is presented by sequential steps, although a "one step" presentation of the theory is also possible. The analysis is derived for the linear (small deformation) case where all loads are independent of the deformation. Figure 10.10(a) presents an open thin-walled beam, the axis of which is parallel (before deformation) to the z coordinate line. Figure 10.10(b) presents a scheme of an "open" thin-walled cross-section. The beam is defined in the domain between its "root" and the "tip" cross-sections, and its length is denoted l. Similar to S.10.2, we discuss here the case of slender beams, and unless stated differently, all relevant notation and definitions will be adopted here as well. As will become clearer later on, it is convenient to define the cross-section geometry by an additional x, y coordinate system, so that $X(\xi) = x_0 + x(s)$ and $Y(\xi) = y_0 + y(s)$, where x_0 and y_0 are constants — see Fig. 10.10(b).

As shown by (10.9a,c), the above kinematics yields the following axial strain, ε_ζ (in the ζ direction), and shear strain, $\gamma_{\xi\zeta}$ (in the wall direction)

$$\varepsilon_\zeta = w_{0,z} - X u_{0,zz} - Y v_{0,zz} + \psi^w_{,\zeta}, \qquad \gamma_{\xi\zeta} = \underline{-r\phi_{,z}} + \psi^w_{,\xi}, \tag{10.32}$$

while as previously discussed, the sign of the underlined term should be changed when $d\vec{s} \times \vec{R}$ is oriented in the positive z direction. The constitutive relations of (10.10) hold here as well. We also adopt the first three integral equilibrium equations of (10.11), and the differential equilibrium equation, (10.14).

As a first step, the analysis will be derived for a general location of the origin of the x, y-plane (x_0, y_0) — see Fig. 10.10(b). In addition, the twist induced shear deformation is assumed to be embedded in a temporary warping function $\widetilde{\psi}^w(\xi,\zeta)$, which is obtained by (see (10.32:b))

$$\widetilde{\psi}^w_{,\xi} = \psi^w_{,\xi} - r\phi_{,z}, \tag{10.33}$$

and therefore, $\phi = 0$ is temporarily assumed. Later on, the evaluation of ϕ for this case (i.e. due to the transverse and the axial loads) will be discussed. In addition, according to the above discussion, all torsional moments are ignored at this stage of the analysis.

To derive the governing equations, we first express the stress components in terms of displacements, as shown in (10.9a,c),

$$\sigma_{\zeta\zeta} = Q_{33}(w_{0,z} - Xu_{0,zz} - Yv_{0,zz} + \widetilde{\psi}^w_{,\zeta}) + Q_{35}(\widetilde{\psi}^w_{,\xi}), \tag{10.34a}$$

$$\tau_{\xi\zeta} = Q_{35}(w_{0,z} - Xu_{0,zz} - Yv_{0,zz} + \widetilde{\psi}^w_{,\zeta}) + Q_{55}(\widetilde{\psi}^w_{,\xi}). \tag{10.34b}$$

The boundary conditions for the above differential equation will be discussed using the scheme presented in Fig. 10.11 for a non-homogeneous open cross-section. As shown, a realistic case is one where the cross-section consists of segments of constant elastic moduli rather than a continuous elastic moduli distribution (note that in the present thin-walled case the notion "seg-

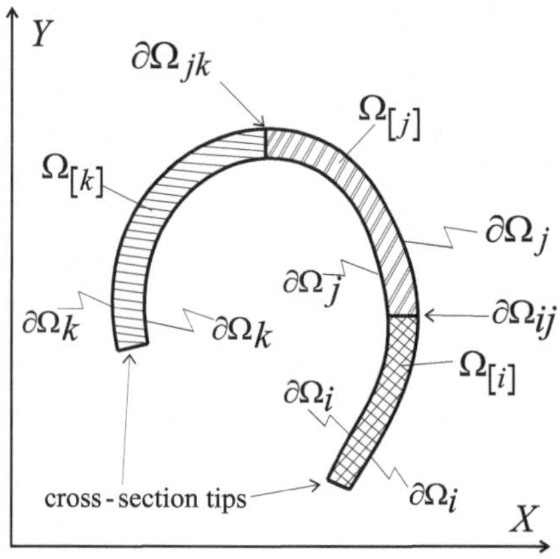

Figure 10.11: A scheme of a non-homogeneous, open cross-section.

ment" replaces the notion "region" or "domain" used for solid cross-sections). At the (assumed traction-free) cross-section tips (see Fig. 10.11), the boundary condition requires $\tau_{\xi\zeta} = 0$. This condition implies that at these tips (see (10.34b)):

$$\widetilde{\psi}^w_{,\xi} = -\frac{Q_{35}}{Q_{55}}(w_{0,z} - Xu_{0,zz} - Yv_{0,zz} + \widetilde{\psi}^w_{,\zeta}). \tag{10.35}$$

Note that the above condition together with $\tau_{\zeta\eta} = 0$ yield a traction-free outer surface over the entire beam, see also Remark 10.1. Similarly, between two segments (of different elastic moduli), the shear stress should remain continuous. Therefore, the shear strain $\widetilde{\psi}^w_{,\xi}$ exhibits a "step change" between segments, which is obtained from

$$[Q_{35}(w_{0,z} - Xu_{0,zz} - Yv_{0,zz} + \widetilde{\psi}^w_{,\zeta}) + Q_{55}\widetilde{\psi}^w_{,\xi}]^{[i]}_{[j]} = 0. \tag{10.36}$$

To this end, for each cross-section, a system of equations is solved. This system consists of the three first equations of (10.11) and the differential equilibrium equation, (10.14), while considering the segment-free tips conditions and the jumps between segments as indicated above. Such a solution yields the $u_0(z)$, $v_0(z)$, $w_0(z)$ and $\widetilde{\psi}^w(s)$ distributions along the beam. Further discussion of this process for general cases is provided in Remark 10.3.

So far, the analysis assumed that x_0 and y_0 are known. Subsequently, in order to comply with the previously discussed assumptions of vanishing torsion moment, the above solution for each cross-section is repeated and solved for different origin locations (x_0,y_0) in order to find the one that will provide zero resultant moment at each cross-section, namely,

$$\iint_{\Omega} r\tau_{\xi\zeta} = 0. \tag{10.37}$$

Once an (x_0,y_0) location that satisfies (10.37) is found as a function of z (or in words "for each cross-section"), it is possible to "perturb" the displacements expressions obtained at the previous stage by adding arbitrary modifications to x_0 and y_0 that will be denoted Δx_0 and Δy_0, respectively. These modifications are required in order to locate the origin *of all cross-sections* along the same line (parallel to the z-axis). A practical selection is $\Delta x_0 = -x_0$ and $\Delta y_0 = -y_0$, which maintains the $X-Y-Z$ system as the global reference system. To cancel out the changes, which are induced by the above Δx_0 and Δy_0 modifications, two additional steps need to be carried out. First, as previously indicated, no torsional moment is induced when the origin is located at (x_0,y_0). Yet, when the origin location is modified, a torsional moment will be created. This moment should be determined and saved for further consideration of the beam behavior under torsion discussed in S.10.3.2. In addition, in order to induce no changes in the axial or shear strain components (see (10.34a,b)), one has to introduce an additional "cross-section" axial deformation, $\Delta w_{0,z}$, given by

$$\Delta w_{0,z} = \Delta x_0 u_{0,zz} + \Delta y_0 v_{0,zz}. \tag{10.38}$$

At this stage, one may introduce the twist angle, $\phi_{,z}$, into the formulation. Again, this is the *bending and extension induced twist angle* while the *torsion induced angle* will be determined separately. The determination of this component of the twist angle is carried out by the determination of the warping function, ψ^w, using $\psi^w_{,\xi} = \widetilde{\psi}^w_{,\xi} + r\phi_{,z}$ (see (10.33)), namely

$$\psi^w(\xi,\zeta) = \widetilde{\psi}^w(z) + \phi_{,z}(z)\int_0^{\xi} r(\xi,\zeta)d\xi + \widetilde{c}(z), \tag{10.39}$$

where the constant term, $\widetilde{c}(z)$, is selected so that $\iint_{\Omega}\psi^w(\xi,\zeta) = 0$ for each z. The value of $\phi_{,z}$ is then selected as the one that minimizes the linear component of the warping distribution at the root, namely,

$$\frac{\partial}{\partial\phi_{,z}}[(\iint_{\Omega^r}\psi^w X)^2 + (\iint_{\Omega^r}\psi^w Y)^2] = 0. \tag{10.40}$$

The above criterion for selecting $\phi_{,z}$ emerges from boundary condition considerations at the beam root and/or tip ends. To clarify this point, one should recall that the axial displacement is given by $w = w_0 - Xu_{0,z} - Yv_{0,z} + \psi^w$ (see (10.32)). Hence, by considering the case where a complete restraint of the axial deformation is required at the root, it is clear that the requirements of $u_{0,z} = 0$ and $v_{0,z} = 0$ there (see (9.24)), force the warping to contain minimal (and if possible, zero) components of linear variations in the X and Y directions (otherwise, nonzero values of $u_{0,z}$ and $v_{0,z}$ will be required to minimize the axial displacements at the root). Note that in the general case, a fulfillment of (10.40) does not provide a complete axial displacement restraint. Such a complete restraint of the axial deformation may require additional distribution of axial displacements. Yet, (10.40) ensures that the additional axial displacement components required for a complete restraint will contain zero (or minimal) linear components, and therefore, their effect will be of a "local nature" only.

Finally, to conclude the analysis stages for the effect of transverse and axial loading, the values of $w_0(z)$ should be integrated per (10.38), and likewise, $\phi_{,z}$ should be integrated to provide the $\phi(z)$ distribution (see boundary conditions in (9.24)). Note that for tip transverse loads, P_x, P_y, if the (x_0, y_0) location obtained by the above criteria yields no twist angle, it may be considered as the "shear center" location.

As indicated, Remark 10.3 discusses the solution procedure in general cases, while simple study cases are analytically derived within S.10.3.1.1, S.10.3.1.2.

Remark 10.3 In the general case it is convenient to conceptually consider a discrete model for the distribution of the warping $\widetilde{\psi}^w$ along the wall, which may be represented by the values $\widetilde{\psi}^w(1)\cdots\widetilde{\psi}^w(N)$ at $s(1)\cdots s(N)$, respectively. Note that since the involved equations contain warping derivatives with respect to ξ, one should adopt a "finite-difference type" of technique. However, this does not lessen the generality of the present discussion since alternatively, the warping could have been expanded into a series of some admissible complete set of shape functions (of s) the coefficient of which could replace the above discrete values of the warping. When such a technique is adopted, the warping derivatives may be determined analytically. Subsequently, once the derivatives with respect to ξ are carried out, the resulting system of equations (i.e. the three first equations of (10.11) and the differential equilibrium equation, (10.14)) may be put as

$$\mathbf{S}\cdot\begin{Bmatrix} w_{0,z} \\ u_{0,zzz} \\ v_{0,zzz} \\ \widetilde{\psi}^w(1) \\ \vdots \\ \widetilde{\psi}^w(N) \end{Bmatrix} = \begin{Bmatrix} P_x+\widetilde{R}_x \\ P_y+\widetilde{R}_y \\ P_z+\widetilde{R}_z \\ Z_b(1)+\widetilde{B}(1) \\ \vdots \\ Z_b(N)+\widetilde{B}(N) \end{Bmatrix} \tag{10.41}$$

where $\mathbf{S}$ is a constant square matrix, $\widetilde{R}_x$, $\widetilde{R}_y$ and $\widetilde{R}_z$ are functions of $u_{0,zz}$, $v_{0,zz}$ and $\widetilde{\psi}^w_{,z}(i)$, while $\widetilde{B}(i)$ are functions of $u_{0,zz}$, $v_{0,zz}$ $w_{0,zz}$, $\widetilde{\psi}^w_{,z}(i)$ and $\widetilde{\psi}^w_{,zz}(i)$. Note that the coordinate z is a parameter in the above system of equations, where the boundary conditions of (10.35), (10.36) are already embedded. The condition of zero average warping is not implemented here as it discussed in connection with (10.39). Thus, before solving this system, one of the warping unknowns, say $\widetilde{\psi}^w(1)$, is assumed to vanish, and therefore, effectively, only the "warping derivatives" (or differences between warping values at discrete locations) are determined at this stage. As shown, although the system is linear, in the general case, an implicit solution may be required. Therefore, if an analytical solution is not available, the following iterative scheme should be adopted:

(a) The iterative scheme is initiated by some deformation assumption. Then, the resultant external loads at the discrete cross-sections along the beam are evaluated. Subsequently, (10.41), is solved and the unknown vector is obtained *for each cross-section.*

(b) The natural boundary conditions at the beam tip (10.15) are used to obtain the values of $u_{0,zz}, v_{0,zz}$ there. These values are integrated along the beam and the distributions of $u_{0,zz}(z)$ and $v_{0,zz}(z)$ are obtained. With the aid of the geometric boundary conditions at the root (9.24), the distributions of $u_{0,x}$, $v_{0,x}$, w_0, u_0 and v_0 along the beam are also determined.

(c) Using the above new estimation of the deformation, the r.h.s. vector of (10.41) at each cross-section is updated (since it is a function of the longitudinal derivatives of the unknowns), and the iterative process is repeated until convergence is achieved.

Study of typical cases has shown that the above quasi-explicit scheme exhibits excellent convergence characteristics. To complete the solution, one should determine Δx_0, Δy_0, ϕ and ψ^w per the above discussion.

10.3.1.1 The Extension-Torsion Coupling

To demonstrate the Extension-Torsion coupling in thin-walled open cross-sections, a beam that undergoes a tip axial load (i.e. a uniform axial resultant load, P_z) is considered. To simplify the analytical solution, a special case of an anisotropic beam of open circular cross-section is examined. Figure 10.12(a)(1) presents the general notation while for the present discussion, the case of a "half-tube" shown in Fig. 10.12(a)(2) is adopted (by selecting $\alpha_0 = 90°$ and $\alpha_E = -90°$). The elastic moduli are assumed to be constants (i.e. not functions of ξ and/or

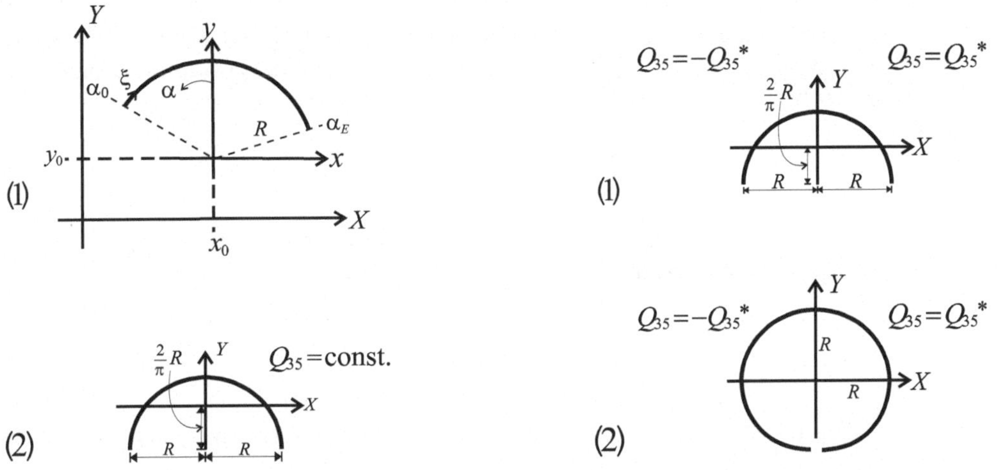

(a) (1) Notation; (2) Antisymmetric (uniform) "half-tube" cross-section.

(b) Symmetric (non-uniform) circular open cross-sections. (1) A "half-tube"; (2) A "full-tube".

Figure 10.12: Schemes of a circular thin-walled cross-sections.

η), and therefore, this lay-up configuration is generally referred to as "antisymmetric" (see Fig. 10.9), and is expected to produce an Extension-Torsion coupling.

By neglecting the warping derivatives in the z direction due to the anticipated uniform longitudinal behavior (which will be proved to be valid in what follows), the differential equilibrium equation becomes (see (10.14), (10.34a,b))

$$Q_{33}(w_{0,zz} - Xu_{0,zzz} - Yv_{0,zzz}) - Q_{35}\left(x_{,\xi}u_{0,zz} + y_{,\xi}v_{0,zz}\right) + Q_{55}\widetilde{\psi}^w_{,\xi} = 0. \tag{10.42}$$

To satisfy this equation, the following temporary warping distribution is selected:

$$\widetilde{\psi}^w = a_c \cos\alpha + a_s \sin\alpha + a_2\xi^2 + a_1\xi + a_0, \tag{10.43}$$

where a_c, a_s, a_2, a_1 and a_0 are constants to be determined. Substituting (10.43) in (10.42) yields

$$a_c = -\frac{1}{\alpha^2_{,\xi}Q_{33}}\left(Q_{33}Rv_{0,zzz} + Q_{35}u_{0,zz}\right), \qquad a_s = \frac{1}{\alpha^2_{,\xi}Q_{55}}\left(Q_{33}Ru_{,zzz} - Q_{35}v_{0,zz}\right), \tag{10.44a}$$

$$a_2 = -\frac{Q_{33}}{2Q_{55}}\left(w_{0,zz} - x_0 u_{,zzz} - y_0 v_{,zzz}\right). \tag{10.44b}$$

As already mentioned, (10.35) should be implemented at the cross-section (assumed traction-free) tips. This condition is implemented by equating the following first warping derivatives as obtained from (10.35) and (10.43), respectively:

$$\widetilde{\psi}^w_{,\xi} = -\frac{Q_{35}}{Q_{55}}\left[w_{0,z} - X u_{0,zz} - Y v_{0,zz}\right], \tag{10.45a}$$

$$\widetilde{\psi}^w_{,\xi} = +\alpha_{,\xi}(a_s \cos\alpha - a_c \sin\alpha) + 2a_2\xi + a_1. \tag{10.45b}$$

When a_c, a_s and a_2 of (10.44a,b) are used, equations (10.45a,b) yield a single equation for a_1. When the tip values of x, y α, $\alpha_{,\xi}$ and ξ are substituted in (10.45a,b), a value of $a_1^{(0)}$ is obtained for a_1 at one tip (say, $\alpha = \alpha_0$), and similarly, a value $a_1^{(E)}$ is derived for a_1 at the other tip (say $\alpha = \alpha_E$). Thus, in the general case, the requirement $a_1^{(0)} = a_1^{(E)}$ should also be imposed. However, by assuming $u_0(z) = v_0(z) = 0$ and constant $w_{0,z}$ (as will be proved to be valid later on), a_c, a_s and a_2 become zero and the following condition creates no shear stresses over the entire cross-section:

$$\widetilde{\psi}^w_{,\xi} = a_1 = -\frac{Q_{35}}{Q_{55}} w_{0,z}, \tag{10.46}$$

which clearly takes care of the above $a_1^{(0)} = a_1^{(E)}$ requirement as well. Then, by substituting (10.46) in (10.34a) and using the third equation of (10.11), the axial strain becomes

$$w_{0,z} = \frac{P_z}{S_\Omega} \cdot \frac{1}{Q_{33} - \frac{Q_{35}^2}{Q_{55}}}, \tag{10.47}$$

which shows that indeed, $w_{0,zz} = 0$, and that $Q_{33} - Q_{35}^2/Q_{55}$ is the effective axial modulus in this case.

The boundary conditions of (10.15) show that to ensure $u_{0,zz} = 0$ and $v_{0,zz} = 0$ at the tip end of the beam (which is essential to support the above $u_0(x) = v_0(x) = 0$ assumption), one should select $x_0 = 0$ and $y_0 = -2R/\pi$. Such a selection locates the X, Y coordinates origin at the cross-section area center — see Fig. 10.12(a)(2) (in other words, unless this location is selected, the above obtained axial stress distribution,which was found to be uniform over the cross-section will produce non-vanishing transverse tip moments).

The solution described so far replaces the general procedure described in Remark 10.3 for each cross-section as it yields $u_0(z)$, $v_0(z)$, $w_0(z)$ and $\widetilde{\psi}^w_{,\xi}$. Also, (10.37) is satisfied, and thus, no $\Delta w_{0,z}$ is required, see (10.38). For the cross-section under discussion, $r = R(1 - 2/\pi \cos\alpha)$, and subsequently, in order to determine $\phi_{,z}$ using (10.40), $\psi^w_{,\xi}$ takes the form

$$\psi^w_{,\xi} = -\frac{Q_{35}}{Q_{55}} w_{0,z} + R\phi_{,z}(1 - \frac{2}{\pi}\cos\alpha). \tag{10.48}$$

Integrating the above distribution yields

$$\psi^w = \frac{\alpha}{\alpha_{,\xi}}(-\frac{Q_{35}}{Q_{55}} w_{0,z} + R\phi_{,z}) - \frac{2R}{\pi\alpha_{,\xi}}\phi_{,z}\sin\alpha + \widetilde{c}. \tag{10.49}$$

The requirement $\iint_\Omega \psi^w = 0$ shows that $\widetilde{c} = 0$, and (10.40) is satisfied provided that

$$R\phi_{,z}\left(1 - \frac{2}{\pi}\frac{\iint_\Omega X\sin\alpha}{\iint_\Omega X\alpha}\right) = \frac{Q_{35}}{Q_{55}} w_{0,z}, \tag{10.50}$$

where the integrals containing the Y coordinate vanish due to symmetry considerations and were therefore omitted from (10.50). Carrying out the involved integrations shows that

$$\phi_{,z} = \frac{2}{R}\frac{Q_{35}}{Q_{55}} w_{0,z}. \tag{10.51}$$

For a Version II material (see S.9.2.1), the above Extension-Torsion coupling may also be written as

$$\frac{R\phi_{,z}}{w_{0,z}} = -2\frac{a_{35}}{a_{33}}. \tag{10.52}$$

Figure 5.7(b) of S.5.2.2 presents the ratio a_{35}/a_{33} for a typical Graphite/Epoxy orthotropic material as a function of its lamination angle, from which it is evident that its maximal magnitude is around a value of 2.5 for a lamination angle of about $\theta_y = 15°$. Substituting (10.51) in (10.49) yields:

$$\frac{\psi^w}{R} = \frac{Q_{35}}{Q_{55}} w_{0,z}(\frac{4}{\pi}\sin\alpha - \alpha) \tag{10.53}$$

which shows that the warping magnitude is derived by $\frac{Q_{35}}{Q_{55}} w_{0,z}$. The above warping distribution is presented in Fig. 10.13(a). Clearly, similar to the pure isotropic case, the case of zero ply angle ($Q_{35} = 0$) contains no warping. As the ply angle is increased, a significant warping distribution on the order of the uniform axial extension appears. Note that as shown by (10.47), for a given P_z, $w_{0,z}$ is a function of Q_{33}, Q_{35} and Q_{55} as well.

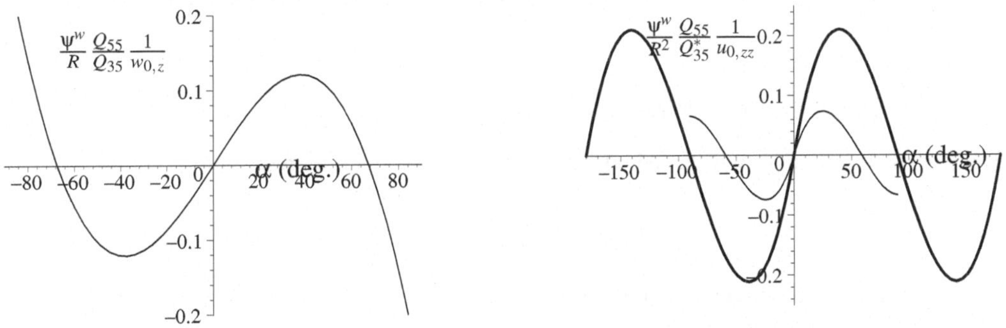

(a) Warping due to a tip axial force for a "half-tube" cross-section (see (10.53)).

(b) Warping due to a tip bending moment. The thin and the thick lines represent "half-tube" and "full-tube" cross-sections, respectively.

Figure 10.13: Out-of-plane non-dimensional warping distributions.

Repeating the above analysis for a full circle (namely, $\alpha_0 = 180°$ and $\alpha_E = -180°$) shows that (10.46), (10.47) still hold, while the X,Y coordinates origin should be located at the center, and $r = R$. Therefore,

$$\psi^w_{,\xi} = -\frac{Q_{35}}{Q_{55}} w_{0,z} + R\phi_{,z}. \tag{10.54}$$

Integrating by requiring $\iint_\Omega \psi^w = 0$ shows that $\psi^w = 0$, and

$$\frac{R\phi_{,z}}{w_{0,z}}_{|F} = \frac{R\phi_{,z}}{2w_{0,z}}_{|H} \tag{10.55}$$

where the indices "H" and "F" stand for half and full tubes, respectively. The above solution for a full "open" circle is identical to the one obtained by (10.17) for a full "closed" circle (when $P_z \neq 0$, $M_z = 0$).

10.3.1.2 The Bending-Torsion Coupling

A closed-form solution that demonstrates the Bending-Torsion coupling in open thin-walled beams is presented in this section. To produce a simplified model, a circular ("half-tube") cross-section is used again with a system location as in Fig. 10.12(a)(2). To create the desired Bending-Torsion coupling, the ply angle of the l.h.s. of the cross-section (i.e. the arc for $X < 0$ and $\alpha > 0$), is assumed to be of equal magnitude but of opposite sign to the ply angle of the r.h.s. of the cross-section (i.e. for $X > 0$ and $\alpha < 0$). Such an arrangement creates a "symmetric" configurations where all elastic moduli are constants along the circular wall except for Q_{35}, which takes the value $Q_{35} = Q^*_{35}$ for $X > 0$ and, $Q_{35} = -Q^*_{35}$ for $X < 0$ — see Fig. 10.12(b)(1). The above characteristics of the elastic moduli become clearer in light of the symmetric distribution (with respect to the ply angle) of Q_{33} and Q_{55}, and the antisymmetric distribution of the Q_{35}. Note that Q^*_{35} in this case is the absolute value of Q_{35} obtained for the ply angle under consideration.

The beam is assumed to undergo a tip moment, M_y, and therefore, a uniform resultant bending moment $\overline{M}_y = M_y$ is induced at each cross-section. In addition, it is assumed that $v_0(z) = w_0(z) = 0$, and that $u_{0,zz}$ is a constant. To obtain the desired solution, $\widetilde{\psi}^w_{,\xi}$ is assumed to be equal to $-\tilde{a}X$ and $+\tilde{a}X$ over the l.h.s. and the r.h.s. of the cross-section, respectively, where $\tilde{a}$ is a constant. All of the above initial assumptions will be shown to be valid in what follows. Subsequently, (10.34b) shows that

$$\tau_{\xi\zeta} = \begin{cases} -Q^*_{35}(-Xu_{0,zz}) - Q_{55}\tilde{a}X, & X < 0, \\ Q^*_{35}(-Xu_{0,zz}) + Q_{55}\tilde{a}X, & X \geq 0. \end{cases} \tag{10.56}$$

Hence, to assure that $\tau_{\xi\zeta}$ will vanish along the *entire* cross-section, one should require

$$\tilde{a} = \frac{Q^*_{35}}{Q_{55}} u_{0,zz}, \tag{10.57}$$

which also satisfies the first two equations of (10.11) and (10.35)–(10.37). Subsequently, from (10.34a), (10.57) one may write:

$$\sigma_{\zeta\zeta} = -(Q_{33} - \frac{Q^{*2}_{35}}{Q_{55}})Xu_{0,zz}, \tag{10.58}$$

which satisfies the third equation of (10.11), and with the aid of (10.15) shows that $M_x = 0$ and

$$M_y = u_{0,zz}(Q_{33} - \frac{Q^{*2}_{35}}{Q_{55}})I_y. \tag{10.59}$$

Similar to (10.47), $Q_{33} - Q^{*2}_{35}/Q_{55}$ becomes the effective bending modulus, and (10.59) may be used for the determination of $u_{0,zz}$ per given M_y. The warping is determined using (10.39) as

$$\psi^w_{,\xi} = \pm\tilde{a}X + R\phi_{,z}(1 - \frac{2}{\pi}\cos\alpha), \tag{10.60}$$

where the upper and lower signs represent the $X \geq 0$ and $X < 0$ regions, respectively. By substituting $X = -R\sin\alpha$, (10.60) is integrated and with the aid of the requirement $\iint_\Omega \psi^w = 0$, ψ^w takes the form

$$\frac{\psi^w}{R^2 u_{0,zz}} = \frac{Q^*_{35}}{Q_{55}}[\mp\cos\alpha \pm 1 + \tilde{b}(\frac{2}{\pi}\sin\alpha - \alpha)]. \tag{10.61}$$

$\widetilde{b}$ is defined as $\phi_{,z}/\widetilde{a}$ so that it reflects the geometrical coupling factor that determines the amount of twist per unit bending curvature, since according to the above definition and (10.57)

$$\frac{\phi_{,z}}{u_{0,zz}} = \widetilde{b}\frac{Q_{35}^*}{Q_{55}}. \tag{10.62}$$

Application of (10.40) shows that $\widetilde{b}$ is given by

$$\widetilde{b} = \frac{\iint_\Omega (\mp\cos\alpha \pm 1)X}{\iint_\Omega (\alpha - \frac{2}{\pi}\sin\alpha)X}, \tag{10.63}$$

where again, all integrals containing Y vanish and were therefore omitted. Carrying out the integrations for the present configuration yields $\widetilde{b} = -1$. Thus, a negative twist is induced by a lateral bending curvature.

The thin line in Fig. 10.13(b) presents the out-of-plane warping distribution in this case as a function of α. As shown, the warping at the cross-section free tips, namely, at the $\alpha = \pm 90°$ locations is equal in magnitude but opposite in sign. It is also interesting to note that the coupling magnitude in a *solid* symmetric homogeneous cross-section (see (5.40)) may be expressed by (10.62) with $\widetilde{b} = -1/2$.

To produce a similar solution for a "full-tube" cross-section as shown in Fig. 10.12(b)(2), the X, Y coordinates origin is located at the circle center, and the above discussion is still valid, however, one should use $r = R$ and ignore the term $2/\pi\cos\alpha$ in (10.60) and the terms $(2/\pi)\sin\alpha$ in (10.61), (10.63) (while clearly, the integration is carried out for the entire "full-tube" cross-section). The above calculation shows that $\widetilde{b} = -2/\pi$ in this case. Thus,

$$\left[\frac{\phi_{,z}}{u_{0,zz}}\right]_{|F} = \frac{2}{\pi}\left[\frac{\phi_{,z}}{u_{0,zz}}\right]_{|H} \tag{10.64}$$

or in words: a "half-tube" symmetric open cross-section produces a Bending-Torsion coupling magnitude, which is 57% larger than a symmetric "full-tube" cross-section.

The thick line in Fig. 10.13(b) presents the out-of-plane warping distribution for a "full-tube" cross-section. Note that the distribution is continuous (at $\alpha = \pm 180°$) and contains a warping magnitude that is larger than the one obtained for a "half-tube" cross-section.

10.3.2 The Torsional Moment Effect

As already discussed, due to their substantially different warping characteristic, anisotropic beams of open cross-sections under torsion produce different coupling characteristics compared with similar beams of a closed cross-section. Torsion of beams of open cross-sections results in relatively large out-of-plane warping deformation. Such a response should be evaluated using tools that are similar to those utilized in the analysis of simply connected domains, which are known to be relatively complex.

We shall first examine the orthotropic rectangular cross-section of length p and width t shown in Fig. 10.14(a). To determine the torsional rigidity and the stress/strain distributions in this orthotropic cross-section we shall exploit the transformation of Remark 7.2. For reasons that will clarified later on, we shall denote this solution as the "uncoupled case" and add to its components a superscript "UC". According to Remark 7.2, the associated transformation is carried out by $\xi' = \xi\sqrt{a_{55}/a_{44}}$, and $\eta' = \eta$. Hence, the above cross-section is transformed into an isotropic rectangle of length $p' = p\sqrt{a_{55}/a_{44}}$ and width t for which it is well known

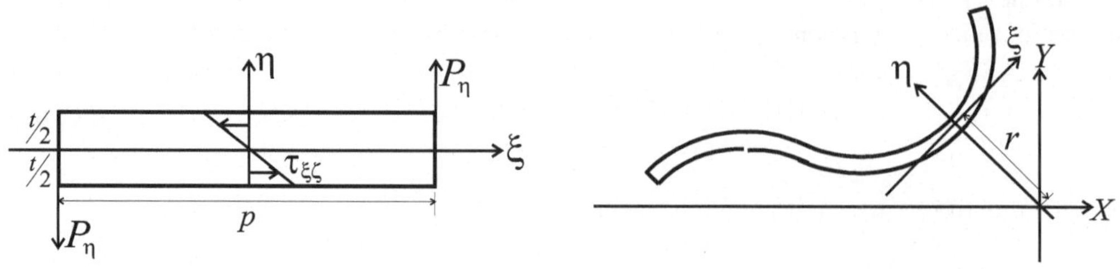

(a) A straight orthotropic cross-section, $Q_{35} = 0$, $r = 0$.

(b) A curved anisotropic cross-section, $Q_{35} \neq 0$, $r \neq 0$.

Figure 10.14: Simply connected ("open") thin-walled cross-sections.

(see e.g. (Timoshenko and Goodier, 1970)) that the torsional rigidity is

$$\frac{D'}{G} = \frac{1}{3}p't^3[1 - \frac{192}{\pi^5}\frac{t}{p'}\sum_{n=1,3,5...}^{\infty}\frac{1}{n^5}\tanh(\frac{n\pi p'}{2t})], \tag{10.65}$$

and that $\tau_{\xi\zeta}$ is given by

$$\frac{\tau'_{\xi\zeta}}{G} = -\phi_{,z}\frac{8b}{\pi^2}\sum_{n=1,3,...}\frac{1}{n^2}(-1)^{\frac{n-1}{2}}\left[1 - \frac{\cosh(\frac{n\pi\xi'}{t})}{\cosh(\frac{n\pi p'}{2t})}\right]\sin\left(\frac{n\pi\eta'}{t}\right). \tag{10.66}$$

Following (7.24) (by replacing a_{44} with $1/Q_{44}$ and a_{55} with $1/Q_{55}$), the torsional rigidity in the uncoupled case is

$$D^{UC} = \frac{D'}{G}\sqrt{\frac{Q_{55}}{Q_{44}}}Q_{55}. \tag{10.67}$$

Clearly, in the present orthotropic case where $a_{45} = 0$ one may write $\tau_{yz} = \gamma_{yz}/a_{44}$ $\tau_{xz} = \gamma_{xz}/a_{55}$. Also, (7.22:b) shows that $\tau_{xz} = \tau'_{xz}/(a_{55}G)$. Therefore, one of the strain components in the uncoupled case becomes

$$\gamma_{\xi\zeta}^{UC} = -\phi_{,z}\frac{8t}{\pi^2}\sum_{n=1,3,5...}\frac{1}{n^2}(-1)^{\frac{n-1}{2}}\left[1 - \frac{\cosh(n\pi\sqrt{\frac{Q_{44}}{Q_{55}}}\cdot\frac{\xi}{t})}{\cosh(\frac{n\pi}{2}\sqrt{\frac{Q_{44}}{Q_{55}}}\cdot\frac{p}{t})}\right]\sin\left(\frac{n\pi\eta}{t}\right). \tag{10.68}$$

At this stage it is possible to examine the limiting case where the ratio p/t is very large. Equations (10.65), (10.67) show that the torsional rigidity in this case becomes (see also Remark 10.4)

$$D^{tw} = D^{UC}|_{\frac{t}{p}\to 0} \cong \frac{1}{3}pt^3Q_{55}. \tag{10.69}$$

Also, for very large p/t ratio, the expressions of $\gamma_{\zeta\eta}^{UC}$ given in (10.68) take the form

$$\gamma_{\xi\zeta}^{tw} = \gamma_{\xi\zeta}^{UC}|_{\frac{t}{p}\to 0} \cong -2\phi_{,z}\eta. \tag{10.70}$$

For the present thin-walled case (i.e. high p/t ratio), the shear strain component $\gamma_{\zeta\eta}^{tw}$ is negligible over the entire cross-section except for the tip regions. At these regions, the net effect of the $\tau^{tw}_{\zeta\eta}$ stress component may be expressed by two concentrated loads, P_η, see Fig. 10.14(a), given by

$$P_\eta = \frac{1}{2p} D^{tw} \phi_{,z}. \tag{10.71}$$

The above is based on the fact that the moment induced by the stress $\tau^{tw}_{\xi\zeta} = Q_{55}\gamma^{tw}_{\xi\zeta}$ is half of the total torsional moment $D^{tw}\phi_{,z}$ (see (3.108) and its discussion). The other half is therefore given by $P_\eta p$.

From this point and on, we consider the function $\phi(z)$ as known and given "locally" by the local resultant moment, $\overline{M}_z$ as $\phi_{,z}(z) = \overline{M}_z(z)/D^{tw}$.

The next extension of the above formulation consists of applying it to the case of a curved thin-walled segment as shown in Fig. 10.14(b), where ξ is stretched along the middle surface of the curved segment and η is perpendicular to ξ (and ζ) at each point along the wall. It is then assumed that due to the fact that the wall is very thin, the distribution of stresses is independent of the segment curvature in the x, y-plane (a similar assumption is traditionally adopted in the isotropic case — see e.g. (Gjelsvik, 1981)). Obviously, and also as implied by (10.1), such a distribution will maintain the same torsional rigidity (given by (10.69)).

The purpose of the following derivation is therefore to determine the additional deformation that is required to maintain the above stress distribution in cases where the segment is not straight and/or cases where the elastic modulus Q_{35} does not vanish. Subsequently, additional degrees of freedom should be considered. These degrees of freedom are the cross-section deformation components in the X, Y and Z directions, which are denoted u, v and w, respectively, and an additional warping distribution, $\Delta\Psi^w$. In other words, the task of the present extension is to find the appropriate values of the above degrees of freedom that will assure local stress components, which are identical to the $\tau^w_{\xi\zeta} = -2\phi_{,z}\eta Q_{55}$ component (see (10.70)), and the $\tau^w_{\zeta\eta}$ component as reflected by the P_η forces. Note that the above $\Delta\Psi^w$ distribution is the warping that needs to be added to the already exist warping distribution, which is embedded in the $\tau^w_{\xi\zeta}$ and $\tau^w_{\zeta\eta}$ distributions.

Similar to the case of closed thin-walled cross-sections, we also set $\Delta\Psi^w(\xi,\eta,\zeta) = \psi^w(\xi,\zeta) -\eta\phi_{,z}r'$. Then, $\Delta\Psi^w_{,\eta} = -\phi_{,z}r'$. Hence overall, we are looking for the warping function $\psi^w(\xi,\zeta)$ (i.e. uniform over the thickness) and the displacement derivatives $u_{0,zz}$, $v_{0,zz}$ and $w_{0,z}$.

To determine the above additional degrees of freedom, we first express the changes in the axial and the shear stresses, which are induced by the fact that the segment has now a non-vanishing value of the normal distance (r, see Fig. 10.14(b)) and elastic modulus Q_{35}, as

$$\Delta\varepsilon_{\zeta\zeta} = w_{0,z} - Xu_{0,zz} - Yv_{0,zz} + \psi^w_{,\zeta}, \qquad \Delta\gamma_{\eta\zeta} = 0, \qquad \Delta\gamma_{\xi\zeta} = -r\phi_{,z} + \psi^w_{,\xi}. \tag{10.72}$$

The corresponding changes in stresses are

$$\Delta\sigma_{\zeta\zeta} = Q_{33}\Delta\varepsilon_{\zeta\zeta} + Q_{35}(\gamma^{tw}_{\xi\zeta} + \Delta\gamma_{\zeta\xi}), \quad \Delta\tau_{\eta\zeta} = 0, \quad \Delta\tau_{\xi\zeta} = Q_{35}\Delta\varepsilon_{\zeta\zeta} + Q_{55}(\Delta\gamma_{\xi\zeta}). \tag{10.73}$$

It is now possible to cancel the additional $\Delta\tau_{\xi\zeta}$ shear stress by requiring that

$$\psi^w_{,\xi} = -\frac{Q_{35}}{Q_{55}}[w_{0,z} - Xu_{0,zz} - Yv_{0,zz} + \psi^w_{,\zeta}] + r\phi_{,z}, \tag{10.74}$$

which also shows that the additional resultant forces in the X and Y directions will vanish (as required by the present pure torsion case), and so will the additional torsional moment, since they are all integrals of this shear stress.

At this stage, the generalization of St. Venant's classical theory assumption will be adopted. Accordingly, the warping is viewed as a product of a warping shape function, $\overline{\psi}^w(\xi)$, and a warping amplitude $\phi_{,z}(z)$, namely: $\psi^w(\xi,\zeta) = \phi_{,z}(z)\cdot\overline{\psi}^w(\xi)$. This assumption appears also in Vlasov's theory (see e.g. (Gjelsvik, 1981)) for isotropic open thin-walled cross-sections. Subsequently it is possible to write $\psi^w_{,z} = \phi_{,zz}\overline{\psi}^w(\xi)$ or

$$\psi^w_{,z} = \frac{\phi_{,zz}}{\phi_{,z}}\psi^w. \tag{10.75}$$

It should be noted that the above assumption does not imply that the warping is induced only by the twist. The entire discussion in this section is focused on the effect of torsional moment only, and therefore, the warping under discussion is the *torsional induced warping*, which is one component of the overall warping distribution.

Substitution (10.75) in (10.74) yields the following linear, ordinary, first-order differential equation for ψ^w:

$$\psi^w_{,\xi} + \mu\frac{\psi^w}{l} = r\phi_{,z} - \frac{Q_{35}}{Q_{55}}(w_{0,z} - Xu_{0,zz} - Yv_{0,zz}), \tag{10.76}$$

where μ is a non-dimensional parameter that combines a measure of material coupling and longitudinal twist variation magnitude

$$\mu = \frac{Q_{35}}{Q_{55}}\frac{\phi_{,zz}}{\phi_{,z}}l. \tag{10.77}$$

Equation (10.76) should be solved under the condition of $\iint_\Omega \psi^w = 0$. Once ψ^w is determined, the additional axial stress becomes (see (10.73), (10.75)),

$$\Delta\sigma_\zeta = (Q_{33} - \frac{Q_{35}^2}{Q_{55}})(w_{0,z} - Xu_{0,zz} - Yv_{0,zz} + \frac{\phi_{,zz}}{\phi_{,z}}\psi^w) + Q_{35}\gamma^{tw}_{\xi\zeta}. \tag{10.78}$$

To assure that the above additional axial stress induces no net resultant axial force and transverse bending moments as required by the present pure torsion case, one should require that the following three integrals will vanish (see (5.7)):

$$\iint_\Omega \Delta\sigma_{\zeta\zeta}\{1, X, Y\} = \{0, 0, 0\}. \tag{10.79}$$

The above requirement may be put as

$$\mathbf{M}_\sigma\cdot\begin{Bmatrix} w_{0,z} \\ u_{0,zz} \\ v_{0,zz}\end{Bmatrix}\frac{1}{\phi_{,z}} = -\frac{\phi_{,zz}}{(\phi_{,z})^2}\iint_\Omega (Q_{33} - \frac{Q_{35}^2}{Q_{55}})\begin{Bmatrix} 1 \\ Y \\ -X\end{Bmatrix}\psi^w - \frac{1}{\phi_{,z}}\iint_\Omega Q_{35}\begin{Bmatrix} 1 \\ Y \\ -X\end{Bmatrix}\gamma^{tw}_{\xi\zeta}, \tag{10.80}$$

where a matrix $\mathbf{M}_\sigma$ is given by

$$\mathbf{M}_\sigma = \iint_\Omega (Q_{33} - \frac{Q_{35}^2}{Q_{55}})\begin{bmatrix} 1 & -X & -Y \\ Y & -XY & -Y^2 \\ -X & X^2 & YX\end{bmatrix}. \tag{10.81}$$

The integrations have to be executed carefully according to (10.1), which enables us to write the second vector on the r.h.s. of (10.80) as

$$-\frac{1}{\phi_{,z}}\iint_\Omega Q_{35}\begin{Bmatrix} 1 \\ Y \\ -X\end{Bmatrix}\gamma^{tw}_{\xi\zeta} = -\frac{1}{6}\int_0^p t(\xi)^3 Q_{35}\alpha_{,\xi}\begin{Bmatrix} 1 \\ Y \\ -X\end{Bmatrix}d\xi. \tag{10.82}$$

In general, (10.76) and (10.80) are coupled and should be solved simultaneously. Such a solution yields the desired additional values of $w_{0,z}/\phi_{,z}$, $u_{0,zz}/\phi_{,z}$, $v_{0,zz}/\phi_{,z}$, and the warping distribution, $\psi^w/\phi_{,z}$.

So far, the origin location (x_0, y_0) was assumed to be known. However, for cross-sections of complex geometries and ply angle distributions, the determination of the above location values is not trivial. In general it should be noted that on one hand, in the present case, the torsional moment may be consider as a "free vector" the location of which is immaterial, but on the other hand, different (x_0, y_0) sets may induce different values of $w_{0,z}$, $u_{0,zz}$, $v_{0,zz}$ and ψ^w. Hence, in order to establish suitable criteria for selecting x_0 and y_0, one should invoke the arguments that were previously discussed in the context of transverse bending and led to the requirement that at the root cross-section, the warping will be orthogonal to linear variations in the X and Y directions, see S.10.3.1. In establishing the mathematical expressions for these requirements, one may not preclude geometries where the above requirement may not be fulfilled. In such cases, (x_0, y_0) should be selected as the set that minimizes the magnitude of the linear variation of the axial distortion at the root cross-section. Thus, *among all solutions that simultaneously satisfy (10.76) and (10.80), the physical solution is the one that also satisfies the following conditions*:

$$[(\iint_{\Omega^r} \psi^w X)^2 + (\iint_{\Omega^r} \psi^w Y)^2]_{,x_0} = 0, \qquad [(\iint_{\Omega^r} \psi^w X)^2 + (\iint_{\Omega^r} \psi^w Y)^2]_{,y_0} = 0. \quad (10.83)$$

For a uniform twist (i.e. the case of a tip torsional moment M_z, where $\phi_{,zz} = 0$), if the (x_0, y_0) location obtained by the above criteria yields $u_{0,zz} = v_{0,zz} = 0$, it may be considered as the "shear center" location.

Based on the above discussion, the analysis of a non-homogeneous cross-section such as the one shown in Fig. 10.15 may be executed by adding up the contribution of each segment so that

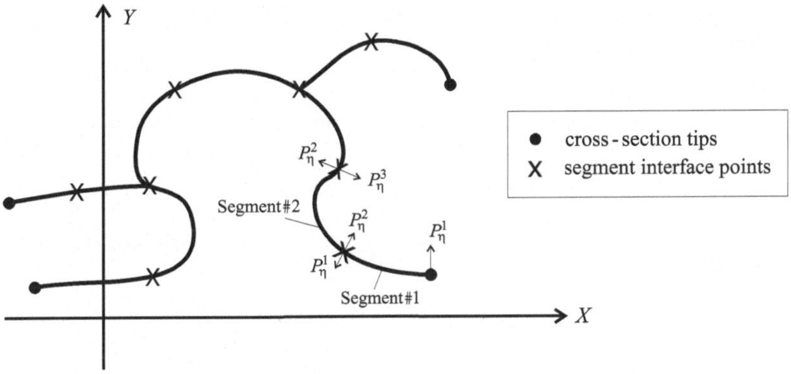

Figure 10.15: A scheme of a non-homogeneous, multi-branch, open cross-section.

the total torsional rigidity will be the sum of the torsional rigidity of all segments. Equation (10.70) and the relation $\tau_{\xi\zeta}^{tw} = \gamma_{\xi\zeta}^{tw} Q_{55}$ will still be valid locally based on the local Q_{55} value. Subsequently, concatenated forces will appear at the interface points between segments that will reflect changes in the corresponding P_η^i values and directions. In this case, the r.h.s. of (10.76) will be different for different segments and clearly, the integrals in (10.80) should be taken for the entire cross-section. Hence, by adding suitable constants to ψ^w at each segment, one should also be able to make sure that its continuity between segments is maintained, and that $\iint_\Omega \psi^w = 0$ is fulfilled.

Remark 10.4 As shown by the $\Delta\tau_{\xi\zeta}=0$ requirement that led to (10.74), the torsional rigidity shown in (10.69) is the one that holds for open cross-sections even when they are not straight and/or their elastic module Q_{35} does not vanish. This rigidity is substantially smaller than the one obtained for closed cross-sections by (10.20). To demonstrate that we consider a generic closed cross-section of circumference p and effective torsional rigidity D_{eq}, as discussed in (10.20). If this cross-section is opened at a point, the torsional rigidity is given by D^{tw} of (10.69). Their ratio is

$$\frac{D^{tw}}{D_{eq}}=\frac{p^2t^2}{12A_m^2}\cdot\frac{Q_{33}Q_{55}}{Q_{33}Q_{55}-Q_{35}^2}. \tag{10.84}$$

For a generic cross-section, one may define, $\widetilde{d}$, as a typical cross-section length dimension, and therefore write $A_m=\widetilde{c}_1\widetilde{d}^2$, $p=\widetilde{c}_2\widetilde{d}$ where $\widetilde{c}_1$ and $\widetilde{c}_2$ are constants. Hence,

$$\frac{D^{tw}}{D_{eq}}=\frac{1}{12}\left(\frac{\widetilde{c}_2}{\widetilde{c}_1}\right)^2\left(\frac{t}{\widetilde{d}}\right)^2\frac{Q_{33}Q_{55}}{Q_{33}Q_{55}-Q_{35}^2}. \tag{10.85}$$

The above equation reflects the fact that in general, the torsional rigidity of a thin-walled cross-section is smaller by a factor that is dominated by the square of the thickness ratio $t/\widetilde{d}$ (for a thin-walled circular isotropic cross-section of radius $\widetilde{R}$, a value of $D^{tw}/D_{eq}=\frac{1}{3}(t/\widetilde{R})^2$ is obtained).

10.3.2.1 The Torsion-Extension Coupling

To analytically demonstrate the application of the above derivation and the associated Torsion-Extension coupling in open cross-sections, the circular cross-section shown in Fig. 10.12(a)(1) will be re-examined. The cross-section is assumed to consist of constant elastic moduli distributions. In this case, $\xi=(\alpha_0-\alpha)R$ and the perpendicular distance is $r=R+y_0\cos\alpha-x_0\sin\alpha$. Also, $x=-R\sin\alpha$, $y=R\cos\alpha$ and $\alpha_{,\xi}=-1/R$. To clarify the involved phenomena, the cases of $\mu=0$ and $\mu\neq 0$ are dealt with separately.

The case of $\mu=0$: For this case, the general solution of (10.76), is given by

$$\psi^w=\widetilde{a}_0+\widetilde{a}_c\cos\alpha+\widetilde{a}_s\sin\alpha+\widetilde{a}_1\alpha, \tag{10.86}$$

where $\widetilde{a}_0$, $\widetilde{a}_c$, $\widetilde{a}_s$ and $\widetilde{a}_1$ are constant coefficients to be determined. Balancing terms of $\cos\alpha$, $\sin\alpha$, and α, yields

$$\alpha_{,\xi}\frac{\widetilde{a}_c}{\phi_{,z}}=\frac{Q_{35}}{Q_{55}}R\frac{u_{0,zz}}{\phi_{,z}}-x_0,\qquad \alpha_{,\xi}\frac{\widetilde{a}_s}{\phi_{,z}}=\frac{Q_{35}}{Q_{55}}R\frac{v_{0,zz}}{\phi_{,x}}+y_0, \tag{10.87a}$$

$$\alpha_{,\xi}\frac{\widetilde{a}_1}{\phi_{,z}}=-\frac{Q_{35}}{Q_{55}}\left(\frac{w_{0,z}}{\phi_{,z}}-x_0\frac{u_{0,zz}}{\phi_{,x}}-y_0\frac{v_{0,zz}}{\phi_{,z}}\right)+R. \tag{10.87b}$$

The remaining $\widetilde{a}_0$ coefficient is determined to ensure $\iint_\Omega\psi^w=0$, and for that purpose, it is expressed as $\widetilde{a}_0=\widetilde{f}_c\widetilde{a}_c+\widetilde{f}_s\widetilde{a}_s+\widetilde{f}_1\widetilde{a}_1$ where the associated coefficients are given by

$$\{\widetilde{f}_c,\widetilde{f}_s,\widetilde{f}_1\}=-\frac{1}{S_\Omega}\iint_\Omega\{\cos\alpha,\sin\alpha,\alpha\}. \tag{10.88}$$

Two separate sub-cases yield $\mu=0$, namely, the case of $\phi_{,zz}=0$ and the case of $Q_{35}=0$ (see (10.77)). In the first sub-case, direct evaluation of $w_{0,z}/\phi_{,z}$, $u_{0,zz}/\phi_{,x}$ and $v_{0,zz}/\phi_{,z}$ from (10.80) is possible. Then, the coefficient $\widetilde{a}_c$, $\widetilde{a}_s$, $\widetilde{a}_1$ and $\widetilde{a}_0$ may be determined by (10.87a,b), (10.88). In the second sub-case, $\widetilde{a}_c$, $\widetilde{a}_s$, $\widetilde{a}_1$ and $\widetilde{a}_0$ are directly obtained from (10.87a,b), (10.88), and subsequently, (10.80), (10.82) are used to obtain $w_{0,z}/\phi_{,z}$, $u_{0,zz}/\phi_{,z}$ and $v_{0,zz}/\phi_{,z}$. In both cases, x_0 and y_0 are determined so that (10.83) are satisfied.

Note that if both $\phi_{,zz}$ and Q_{35} are zero, the solution returns to the fundamental model while $w_{0,z}$, $u_{0,zz}$, $v_{0,zz}$ are all zero, and (10.76) shows that the additional warping distribution is $\psi^w_{,\xi} = r\phi_{,z}$, similar to the classical isotropic case.

Example 10.2 *Half-Tube Isotropic Cross-Section Under Tip Torsional Moment.*
As an example we consider the isotropic cross-section of Fig. 10.16(a). In this case $\phi_{,zz}0$,

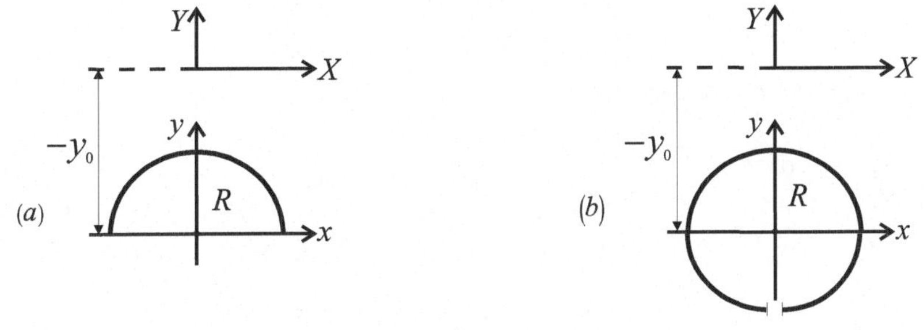

Figure 10.16: Open circular cross-sections: (a) A "half tube"; (b) A "full tube".

$Q_{35} = 0$, and (10.80) shows that $w_{0,z} = u_{0,zz} = v_{0,zz} = 0$. We also assume $x_0 = 0$ while y_0 remains unknown. Equations (10.87a,b), (10.88) then show that

$$\frac{\tilde{a}_c}{\phi_{,z}} = Rx_0, \qquad \frac{\tilde{a}_s}{\phi_{,z}} = -Ry_0, \qquad \frac{\tilde{a}_1}{\phi_{,z}} = -R^2, \qquad \frac{\tilde{a}_0}{\phi_{,z}} = 0. \tag{10.89}$$

Thus,

$$\frac{\psi^w}{\phi_{,z}} = Rx_0 \cos\alpha - Ry_0 \sin\alpha - R^2\alpha. \tag{10.90}$$

We therefore write the square brackets in (10.83) as

$$\left(\iint_\Omega \psi^w X\right)^2 + \left(\iint_\Omega \psi^w Y\right)^2 = R^8 t\phi_{,z}\left[\left(\frac{\pi}{2}\frac{y_0}{R} + 2\right)^2 + \left(\frac{\pi}{2}\frac{x_0}{R} + 2\frac{x_0}{R}\frac{y_0}{R}\right)^2\right] \tag{10.91}$$

while the solution of (10.83) yields $y_0 = -\frac{4R}{\pi}$, which is a classical result.

<u>**The case of $\mu \neq 0$:**</u> In this case, the general solution of (10.76) becomes

$$\psi^w = \tilde{a}_0 + \tilde{a}_c \cos\alpha + \tilde{a}_s \sin\alpha + \tilde{a}_1 e^{-\frac{\mu}{p\alpha_{,\xi}}\alpha}, \tag{10.92}$$

where the constant coefficients $\tilde{a}_0$, $\tilde{a}_c$ and $\tilde{a}_s$ are obtained from

$$\begin{bmatrix} \mu/p & 0 & 0 \\ 0 & \mu/p & \alpha_{,\xi} \\ 0 & -\alpha_{,\xi} & \mu/p \end{bmatrix} \cdot \begin{Bmatrix} \tilde{a}_0/\phi_{,z} \\ \tilde{a}_c/\phi_{,z} \\ \tilde{a}_s/\phi_{,z} \end{Bmatrix} = \frac{Q_{35}}{Q_{55}} \begin{bmatrix} -1 & x_0 & y_0 \\ 0 & 0 & R \\ 0 & -R & 0 \end{bmatrix} \cdot \begin{Bmatrix} w_{0,z}/\phi_{,z} \\ u_{0,zz}/\phi_{,z} \\ v_{0,zz}/\phi_{,x} \end{Bmatrix} + \begin{Bmatrix} R \\ y_0 \\ -x_0 \end{Bmatrix}, \tag{10.93}$$

and $\tilde{a}_1$ is determined to ensure $\iint_\Omega \psi^w = 0$. For that purpose, $\tilde{a}_1$ may be expressed as

$$\tilde{a}_1 = \tilde{f}_0\tilde{a}_0 + \tilde{f}_c\tilde{a}_c + \tilde{f}_s\tilde{a}_s, \tag{10.94}$$

where

$$\{\widetilde{f}_0, \widetilde{f}_c, \widetilde{f}_s\} = -\frac{1}{\iint_\Omega e^{-\frac{\mu}{p\alpha_{,\xi}}\alpha}} \iint_\Omega \{1, \cos\alpha, \sin\alpha\}. \tag{10.95}$$

Substituting the above expression for ψ^w in (10.80) enables us to express the first vector of its r.h.s. in terms of the coefficient $\widetilde{a}_0$, $\widetilde{a}_c$ and $\widetilde{a}_s$ as

$$-\frac{\phi_{,zz}}{(\phi_{,z})^2}(Q_{33} - \frac{Q_{35}^2}{Q_{55}}) \iint_\Omega \begin{Bmatrix} 1 \\ Y \\ -X \end{Bmatrix} \psi^w = -\frac{\phi_{,zz}}{\phi_{,z}}(Q_{33} - \frac{Q_{35}^2}{Q_{55}})(\psi_1^w + \psi_2^w) \begin{Bmatrix} \widetilde{a}_0 \\ \widetilde{a}_c \\ \widetilde{a}_s \end{Bmatrix} \frac{1}{\phi_{,z}}, \tag{10.96}$$

where the matrices ψ_1^w and ψ_2^w are

$$\psi_1^w = \iint_\Omega \begin{bmatrix} 1 & \cos\alpha & \sin\alpha \\ Y & Y\cos\alpha & Y\sin\alpha \\ -X & -X\cos\alpha & -X\cos\alpha \end{bmatrix}, \quad \psi_2^w = \iint_\Omega e^{-\frac{\mu}{\alpha_{,\xi}}\alpha} \begin{bmatrix} \widetilde{f}_0 & \widetilde{f}_c & \widetilde{f}_s \\ Y\widetilde{f}_0 & Y\widetilde{f}_c & Y\widetilde{f}_s \\ -X\widetilde{f}_0 & -X\widetilde{f}_c & -X\widetilde{f}_s \end{bmatrix}. \tag{10.97}$$

It is now possible to substitute the coefficients $\widetilde{a}_0, \widetilde{a}_c, \widetilde{a}_s$ from (10.93) in (10.96). Further substitution of the resulting equation in (10.80) (and use of (10.82)) yields a system of equations for $w_{0,z}/\phi_{,z}$, $u_{0,zz}/\phi_{,z}$ and $v_{0,zz}/\phi_{,z}$. Then, additional application of (10.93) yields $\widetilde{a}_0$, $\widetilde{a}_c$, $\widetilde{a}_s$, which subsequently may be used to determine $\widetilde{a}_1$ (10.94) and the entire warping distribution is revealed. Note again, the above scheme is founded on a known set of x_0 and y_0 values, which have to be determined using (10.83).

Example 10.3 *Torsion-Extension Coupling in Tubes.*
To analytically demonstrate the Torsion-Extension coupling effects in the antisymmetric thin-walled circular cross-section under discussion (see Fig. 10.16(a)), it is worth examining the case of a uniform beam, which is subjected to a tip moment, M_z (and therefore $\phi_{,z}$ is constant and $\phi_{,zz} = 0$). We subsequently set $\alpha_{,\xi} = -1/R$, $\alpha_E = -\pi/2$, $\alpha_0 = \pi/2$, and due to the symmetry assume $x_0 = 0$. Hence, (10.80–10.82) show that

$$(Q_{33} - \frac{Q_{35}^2}{Q_{55}})Rt \begin{bmatrix} \pi & 0 & -2R - \pi y_0 \\ 2R + \pi y_0 & 0 & -\frac{\pi}{2}R^2 - 4y_0R - \pi y_0^2 \\ 0 & \frac{\pi}{2}R^2 & 0 \end{bmatrix} \cdot \begin{Bmatrix} w_{0,z}/\phi_{,z} \\ u_{0,zz}/\phi_{,z} \\ v_{0,zz}/\phi_{,z} \end{Bmatrix} = \frac{t^3}{6} Q_{35} \begin{Bmatrix} \pi \\ 2R + \pi y_0 \\ 0 \end{Bmatrix}. \tag{10.98}$$

The solution of the above system is clearly $u_{0,zz} = v_{0,zz} = 0$, and

$$\frac{w_{0,z}}{\phi_{,z}} = \frac{t^2}{6R} \cdot \frac{Q_{55}Q_{35}}{Q_{33}Q_{55} - Q_{35}^2}, \tag{10.99}$$

which represents the Torsion-Extension coupling of the present cross-section, namely, the ratio of the induced extension to the applied twist. Note that this solution also shows that the normal stress integral over the thickness vanishes for any ξ, namely, (see (10.1))

$$\int_{-t/2}^{t/2} \Delta\sigma_{\zeta\zeta} \left(1 + \frac{\eta}{R}\right) d\eta = 0, \tag{10.100}$$

since in this case, see (10.70), (10.78)

$$\Delta\sigma_{\zeta\zeta} = (Q_{33} - \frac{Q_{35}^2}{Q_{55}}) w_{0,z} + Q_{35}(-2\phi_{,z}\eta) = Q_{35}(\frac{t^2}{6R} - 2\eta). \tag{10.101}$$

Furthermore, (10.87a,b), (10.88) show that $\widetilde{a}_c = \widetilde{a}_0 = 0$ and the warping becomes $\psi^w = \widetilde{a}_s \sin\alpha + \widetilde{a}_1\alpha$, where

$$\widetilde{a}_s = -\phi_{,z}Ry_0, \qquad \widetilde{a}_1 = -\phi_{,z}R^2 f_{35}, \qquad \widetilde{f}_{35} = 1 - \frac{1}{6}\left(\frac{t}{R}\right)^2 \frac{Q_{35}^2}{Q_{33}Q_{55} - Q_{35}^2}. \tag{10.102}$$

Practically, $\widetilde{f}_{35} \cong 1$ and stands for the small correction due to the effect of Q_{35}. We therefore write the square brackets in (10.83) for this case as

$$(\iint_\Omega \psi^w X)^2 + (\iint_\Omega \psi^w Y)^2 = R^8 t\phi_{,z}\left(\frac{\pi}{2}\frac{y_0}{R} + 2\widetilde{f}_{35}\right)^2, \tag{10.103}$$

while the solution of (10.83) yields $y_0 = -\frac{4R}{\pi}\widetilde{f}_{35}$. The difference in warping between the two cross-section tips is, see Fig. 10.17,

$$\psi^w(\frac{\pi}{2}) - \psi^w(-\frac{\pi}{2}) = -\phi_{,z}R^2(\pi - \frac{8}{\pi})\widetilde{f}_{35}. \tag{10.104}$$

Hence, a relatively large amount of warping is induced in this case (as a quantitative reference, simple calculation shows that for a twist of *one degree per length of a radius* the difference in warping between the two cross-section tips is about *one percent of a radius*). Also note that $\frac{\psi^w}{\phi_{,z}}\frac{1}{\widetilde{f}_{35}}\frac{1}{R^2}$ in this case is identical to $\frac{\psi^w}{w_{0,z}}\frac{Q_{55}}{Q_{35}}\frac{1}{R}$ of (10.53). This stresses the tremendous difference between the two phenomena. While warping due to extension is a pure consequence of the anisotropy characteristics (and therefore vanishes for $Q_{35} = 0$), warping due to twist is mainly a geometrical phenomena and is hardly affected by anisotropy characteristics (and therefore Q_{35} plays a minor role in its influence on $\widetilde{f}_{35}$).

Since $u_{0,zz} = v_{0,zz} = 0$, and (10.99) is valid for any case where $\alpha_E = -\alpha_0$ and t, Q_{ij} are constants, the same warping function is valid for a full-tube open cross-section where $\alpha_E = -\pi$, $\alpha_0 = \pi$ as shown in Fig. 10.16(b). In such a case $y_0 = -2R\widetilde{f}_{35}$ is obtained, and

$$\psi^w(\pi) - \psi^w(-\pi) = -\phi_{,z}R^2 2\pi\widetilde{f}_{35}, \tag{10.105}$$

which is about 10 times larger than the difference shown by (10.104), see comparison in Fig. 10.17.

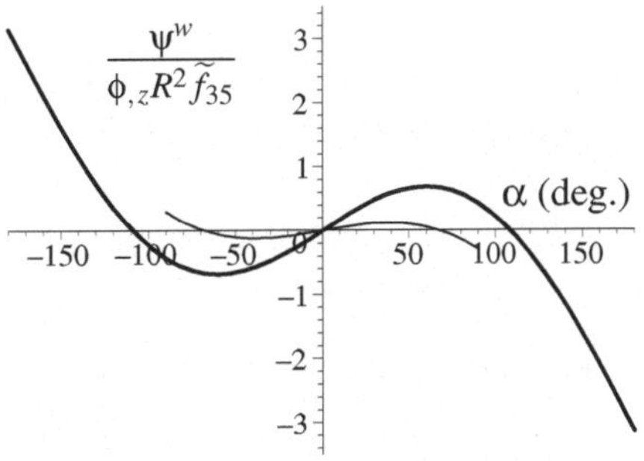

Figure 10.17: Out-of-plane warping distributions in Example 10.3 for half-tube (thin line) and full-tube (thick line) cross-sections.

Substituting $\phi_{,z} = M_z/D^{tw}$ in (10.99) and using (10.69) yields the amount of axial strain per unit torsional moment as

$$\frac{w_{0,z}}{M_z} = \frac{1}{2ptR}\frac{Q_{35}}{Q_{33}Q_{55} - Q_{35}^2}, \tag{10.106}$$

which is valid for any $\alpha_E = -\alpha_0$ values. See Remark 10.5 for comparison of this result with the one obtained for closed cross-sections.

We shall now compare the above Torsion-Extension coupling with the Extension-Torsion coupling obtained in (10.51), (10.55) for the "half-tube" (H) cross-section of Fig. 10.12(a)(2) and a similar "full" (F) cross-section. The corresponding equations are

$$\left[\frac{w_{0,z}}{R\phi_{,z}}\right]_{|M_z} = \left(\frac{t}{R}\right)^2 \underline{\frac{1}{6}\cdot\frac{Q_{55}Q_{35}}{Q_{33}Q_{55}-Q_{35}^2}}, \quad \left[\frac{R\phi_{,z}}{w_{0,z}}\right]_{|H,P_z} = 2\underline{\frac{Q_{35}}{Q_{55}}}, \quad \left[\frac{R\phi_{,z}}{w_{0,z}}\right]_{|F,P_z} = \underline{\frac{Q_{35}}{Q_{55}}}. \tag{10.107}$$

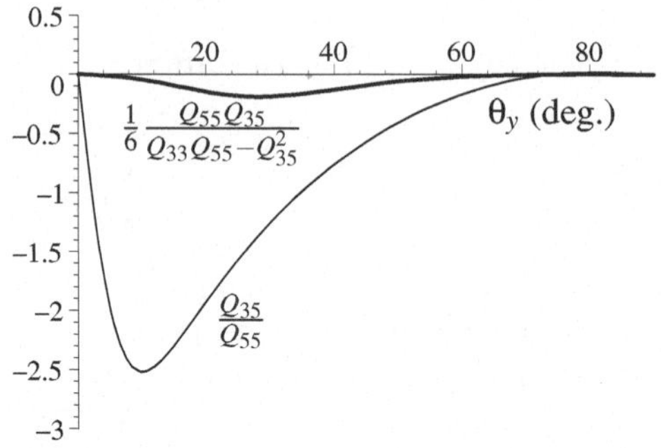

Figure 10.18: Material-related quantities that determine the Torsion-Extension (thick line) and Extension-Torsion (thin line) coupling magnitude in Example 10.3.

Figure 10.18 presents the underlined quantities in (10.107) for a typical Graphite/Epoxy orthotropic material as a function of its lamination angle. As shown, in view of the fact that $\left(\frac{t}{R}\right)^2 \ll 1$, the amount of axial extension that is obtained per unit twist in the case of tip torsional moment is substantially lower than the amount of twist that is obtained per unit axial extension in the case of tip axial force. The above results may be compared with those of a closed "full-tube" cross-section shown by Fig. 10.5 and (10.19), that indicate that in contrast with (10.107), in closed thin-walled cross-sections, the above coupling effects are of the same order.

Remark 10.5 Since (10.106) is valid for any $\alpha_E = -\alpha_0$ values, it may be compared with the quantity obtained for a closed cross-section by (10.17), namely,

$$\frac{w_{0,z}}{M_z} = \frac{Q_{35}}{Q_{33}Q_{55}-Q_{35}^2}\frac{1}{2A_m t}. \tag{10.108}$$

Substituting $A_m = \pi R^2$ in (10.108) and $p = 2\pi R$ in (10.106) shows that

$$\left[\frac{w_{0,z}}{M_z}\right]_{|C} = 2\left[\frac{w_{0,z}}{M_z}\right]_{|O} \tag{10.109}$$

where"C" and "O" stand for closed and open "full-circle" cross-sections, respectively. An important conclusion that emerges from the above discussion is that although much larger twist angles are created in open cross-sections per a given torsional moment (see also Remark 10.4), the amount of axial strain obtained per unit torsional moment in open cross-sections is of the same order as in closed cross-sections.

Remark 10.6 The model presented in S.10.3.1 serves as an adequate tool for analyzing cross-sections that consist of closed loops with additional "open segments" and undergo axial and transverse bending, see Fig. 10.15. However, in view of Remark 10.4, for the determination of the effect of the torsional moment derived in S.10.3.1.1, the open segments may be neglected and a model for thin-walled closed cross-sections should be invoked.

Remark 10.7 For a uniform beam having the cross-section of Fig. 10.12(a)(2), equations (10.47), (10.51) supply the amount of twist per unit force, $\phi_{,z}/P_z$, obtained in the case of a tip axial force, which may be compared with (10.106) that gives the amount of axial extension per unit moment, $w_{0,z}/M_z$, in the case of tip torsional moment. Such a comparison shows that due to the unique coupling mechanisms, the above ratios are not identical as obtained in the case of closed cross-sections (see (10.18)), but are related by: $(\phi_{,z})_{P_z=1} = 4\,(w_{0,z})_{M_z=1}$. It is therefore interesting to verify that the above solutions do comply with the Reciprocal Theorem of Betti and Rayleigh, see (1.14). For that purpose, we write the associated work balance in terms of the involved stress and strain components using integrations over the tip cross-section (since the loads are applied there, although the same result holds for any cross-section) as

$$\iint_\Omega \sigma_{\zeta\zeta}^{P_z}\varepsilon_{\zeta\zeta}^{M_z} + \iint_\Omega \tau_{\xi\zeta}^{P_z}\gamma_{\xi\zeta}^{M_z} = \iint_\Omega \sigma_{\zeta\zeta}^{M_z}\varepsilon_{\zeta\zeta}^{P_z} + \iint_\Omega \tau_{\xi\zeta}^{M_z}\gamma_{\xi\zeta}^{P_z} \tag{10.110}$$

where the superscripts P_z and M_z indicate axial tip load and the tip torsional moment, respectively. The solutions presented in S.10.3.1.1 and S.10.3.2.1 show that for a circular cross-section

$$\sigma_{\zeta\zeta}^{P_z} = \frac{P_z}{pt}, \qquad \varepsilon_{\zeta\zeta}^{P_z} = \frac{Q_{55}}{Q_{33}Q_{55}-Q_{35}^2}\frac{P_z}{pt}, \qquad \gamma_{\xi\zeta}^{P_z} = -\frac{Q_{35}}{Q_{33}Q_{55}-Q_{35}^2}\frac{P_z}{pt}, \tag{10.111a}$$

$$\sigma_{\zeta\zeta}^{M_z} = \frac{M_z}{pt}\frac{Q_{35}}{Q_{55}}\left(\frac{1}{2R}-\frac{6}{t^2}\eta\right), \quad \tau_{\xi\zeta}^{M_z} = -M_z\frac{6}{pt^3}\eta, \quad \varepsilon_{\zeta}^{M_z} = \frac{M_z}{2ptR}\frac{Q_{35}}{Q_{33}Q_{55}-Q_{35}^2}, \tag{10.111b}$$

while $\tau_{\xi\zeta}^{P_z} = 0$ (and therefore $\gamma_{\xi\zeta}^{M_z} = 0$ is absent). By using (10.1) one may show that (10.110) is fulfilled.

11
Program Descriptions

This chapter contains the instructions for the symbolic `Maple` programs and their applications as described throughout the book. Section P.1 consists of the programs **P.1.1 — P.1.17** of Chapter 1, Section P.2 consists of the programs **P.2.1 — P.2.13** of Chapter 2, etc. Additional technical comments and instructions appear in the programs. For further reading on *computer algebra programs and tools* see e.g. (Grabmeier *et al.*, 2003), (Heck, 2003), (Maple, 2003) and (Rovenski, 2000).

All programs are included in the enclosed CD which may be used in two ways:

Viewing the programs (without `Maple`):

- Select "Book_html" folder.
- Open "book_contents.html".
- Select the chapter that contains the program you wish to view.
- Select the desired program.

Activating the programs (with `Maple` 8,9 or a later version):

- Copy the folder "book_tmp" to your D hard drive (if another hard drive is selected, the string "currentdir("d:/book_tmp")" that appears at the top of each program should be changed as well).
- Open "book_contents.mws".
- Select the desired program.

Note that part of the programs read data and partial codes from suitable text files. Therefore, to run the programs, in addition to the `Maple` program files (*.mws), the folder "book_tmp" contains also a system of *.dat and *.txt files.

A word of precaution: The programs documented in this chapter were written for demonstration and illustration of specific cases and analytical techniques. They are not meant to be a "software package" nor "generic solvers" for problems in anisotropic elasticity. Their value is mainly educational.

P.1 Programs for Chapter 1

P.1.1 Strain Tensor in Space

This program derives the strain tensor in curvilinear orthogonal coordinates of space as discussed in S.1.1.2, S.1.1.3. The program may incorporate five built-in coordinate systems. Additional systems should be first introduced to **P.1.16**.

Selected Parameters:

- s – An index for a three-dimensional coordinate system, see Example 1.9:

$$s = \begin{cases} 0 & - \text{ Cartesian coordinates } (\alpha_1 = x,\ \alpha_2 = y,\ \alpha_3 = z), \\ 1 & - \text{ Cylindrical coordinates } (\alpha_1 = \rho,\ \alpha_2 = \theta^c,\ \alpha_3 = z), \\ 2 & - \text{ Spherical coordinates } (\alpha_1 = \rho,\ \alpha_2 = \theta^s,\ \alpha_3 = \phi^s), \\ 3 & - \text{ Elliptical-cylindrical coordinates } (\alpha_3 = z), \\ 4 & - \text{ Bipolar-cylindrical coordinates } (\alpha_3 = z). \end{cases}$$

P.1.2 Strain Tensor in the Plane

This program derives the strain tensor in curvilinear orthogonal coordinates of the plane as discussed in S.1.1.2, S.1.1.3. The program may incorporate four built-in coordinate systems. Additional systems should be first introduced to **P.1.17**.

Selected Parameters:

- s – An index for a two-dimensional coordinate system, see Example 1.9:

$$s = \begin{cases} 0 & - \text{ Cartesian coordinates } (\alpha_1 = x,\ \alpha_2 = y), \\ 1 & - \text{ Polar coordinates } (\alpha_1 = \rho,\ \alpha_2 = \theta^c), \\ 2 & - \text{ Elliptical coordinates}, \\ 3 & - \text{ Bipolar coordinates}. \end{cases}$$

P.1.3 Compatibility Equations in Space

This program derives the compatibility equations in curvilinear orthogonal coordinates of space as discussed in S.1.2.1.

The program contains two methods to derive the required expressions. The first method exploits the `tensor` package tools of `Maple`, while the second method uses linear algebra tools (such as vectors, matrices, quadratic forms, etc.). The results are compared to show that both methods are identical.

The program reads the output of **P.1.16** for five built-in coordinate systems. Additional systems should be therefore introduced first to **P.1.16**.

Selected Parameters:

- s – see **P.1.1**.

P.1.4 Compatibility Equations in the Plane

This program derives the compatibility equation in curvilinear orthogonal coordinates of the plane as discussed in S.1.1.3.

The program contains two methods to derive the required expressions. The first method exploits the tensor package tools of `Maple`, while the second method uses linear algebra tools (such as vectors, matrices, quadratic forms, etc.). The results are compared to show that both methods are identical.

The program reads the output of **P.1.17** for four built-in coordinate systems. Additional coordinate systems should be therefore introduced first to **P.1.17**.

Selected Parameters:

- s – see **P.1.2**.

P.1.5 Displacements by Strain Integration in Space

This program performs strain integration and yields displacements as discussed in S.1.2.2 for Cartesian coordinates in space. For clarification purposes, the program uses a direct integration method (and not a `Maple` command). If a "Compatibility is not satisfied" message is obtained, the system is not consistent and integration is not possible.

Selected Parameters:

- x_0, y_0, z_0 – A reference point where the rigid body parameters are defined.
- $u_0, v_0, w_0, \omega_1, \omega_2, \omega_3$ – Rigid body displacements and rotations, in the x, y, z directions, respectively (close these entries for symbolic rigid body terms).
- $Eps_i,\ i = 1,\ldots,6$ – Consistent set of strain components, see (1.21).

P.1.6 Displacements by Strain Integration in the Plane

This program performs strain integration and yields displacements as discussed in S.1.2.2 for Cartesian coordinates in the plane. For clarification purposes, the program uses a direct integration method (and not a `Maple` command). If a "Compatibility is not satisfied" message is obtained, the system is not consistent and integration is not possible.

Selected Parameters:

- x_0, y_0 – A reference point where the rigid body parameters are defined.
- u_0, v_0, ω_3 – Rigid body displacements in the x and y directions, respectively, and rotation in the z direction (close these entries for symbolic rigid body terms).
- $Eps_i,\ i = 1,2,6$ – Consistent set of strain components, see (1.21).

P.1.7 Equilibrium Equations in Space

This program derives the equilibrium equations for curvilinear orthogonal coordinates in space as discussed in S.1.3.2. The program reads the output of **P.1.16** for five built-in coordinate systems. Additional coordinate systems should be therefore introduced first to **P.1.16**.

Selected Parameters:

- s – see **P.1.1**.

P.1.8 Equilibrium Equations in the Plane

This program derives the equilibrium equations for curvilinear orthogonal coordinates in the plane as discussed in S.1.3.2. The program reads the output of **P.1.17** for four built-in coordinate systems. Additional coordinate systems should be therefore introduced first to **P.1.17**.

Selected Parameters:

- s – see **P.1.2**.

P.1.9 Stress/Strain Tensor Transformation due to Coordinate System Rotation

This program demonstrates generic symbolic transformations of a stress (or strain) tensor under space coordinates rotation that are defined by the three Euler's angles ϕ, θ, ψ as discussed in S.1.3.3, S.1.7 (see order of rotations in (1.206) and its discussion).

The program contains two methods to derive the required expressions. The first method exploits the tensor package tools of `Maple`, while the second method uses linear algebra tools (such as vectors, matrices, quadratic forms, etc.). The results are compared to show that both methods are identical.

The program uses the general notation "σ" for the tensor, its components and the related quantities, see S.1.3.3. These may technically be replaced by "ε" when a strain tensor is under discussion, see S.1.3.4.

P.1.10 Application of Stress/Strain Tensor Transformation

This program demonstrates the transformations of stress/strain tensor components of **P.1.9** for specific Euler's angles ϕ, θ, ψ as discussed in S.1.3.3, S.1.7 (see order of rotations in (1.206)).

The program uses the general notation "σ" for the tensor, its components and the related quantities, see S.1.3.3. These may technically be replaced by "ε" when a strain tensor is under discussion, see S.1.3.4.

Selected Parameters:

- $t_x \equiv \phi$ – The rotation angle about the x axis (degrees).
- $t_y \equiv \theta$ – The rotation angle about the y axis (degrees).
- $t_z \equiv \psi$ – The rotation angle about the z axis (degrees).

P.1.11 Stress/Strain Tensor Transformations from Cartesian to Curvilinear Coordinates

This program demonstrates transformations of a stress/strain tensor defined in Cartesian coordinates to various curvilinear coordinates. As discussed in S.1.3.3, the output of this program is written in terms of the curvilinear coordinates $\alpha_1, \alpha_2, \alpha_3$ (and *not* in terms of the transformation rotation angles), and is therefore a function of the specific point.

The program uses the general notation "σ" for the tensor, its components and the related quantities, see S.1.3.3. These may technically be replaced by "ε" when a strain tensor is under discussion, see S.1.3.4.

Selected Parameters:

- s – see **P.1.1**.
- *Point* – The point under discussion given by the three values of α_1, α_2 and α_3.

P.1.12 Stress/Strain Visualization

This program reproduces Example 1.1 and Figs. 1.7, 1.8, 1.9(a),(b), 1.10(a),(b), 1.11. The program uses the general notation "σ" for the tensor, its components and the related quantities, see S.1.3.3. These may technically be replaced by "ε" when a strain tensor is under discussion, see S.1.3.4.

Selected Parameters:

- σ – Stress tensor in Cartesian coordinates at a point.

P.1.13 Euler's Equation for a Functional Based on a Function of One or Two Variables

This program implements the functional of (1.165) to produce Euler's equation (1.167), and the functional of (1.170) to produce Euler's equation (1.171), see S.1.5.3.

Selected Parameters:

- Ex – Example # (1,2,3,4,5,6).

$$Ex = \begin{cases} 1,2 & \text{– the results of Example 1.4,} \\ 3,4 & \text{– the results of Example 1.5,} \\ 5 & \text{– the result of Example 1.6,} \\ 6 & \text{– the result of Example 1.2.} \end{cases}$$

P.1.14 Elastica

This program produces the results of Example 1.3.

Selected Parameters (*for Euler's spiral*):

- L – Rod's length.
- θ_e – Rod's end angle.
- $\tilde{g}$ – The constraint parameter.

P.1.15 Rotation Matrix in Space

This program derives a space rotation matrix using Euler's angles, see (1.206).

P.1.16 Curvilinear Coordinates in Space

This program derives some characteristics of orthogonal curvilinear coordinates in space.

The programs contains two methods to derive the required expressions. The first method exploits the tensor package tools of `Maple`, while the second method uses linear algebra tools (such as vectors, matrices, quadratic forms, etc.). The results are compared to show that both methods are identical.

Additional coordinate systems may be introduced by defining their functions $f_i(\alpha_1,\alpha_2,\alpha_3)$, $i=1,2,3$, see (1.210).

Selected Parameters:

- s – see **P.1.1**.

P.1.17 Curvilinear Coordinates in the Plane

This program derives some characteristics of orthogonal curvilinear coordinates in the plane. Additional coordinate systems may be introduced by defining their functions $f_i(\alpha_1,\alpha_2)$, $i=1,2$, see (1.210).

Selected Parameters:

- s – see **P.1.2**.

P.2 Programs for Chapter 2

P.2.1 Compliance and Stiffness Matrices Presentation

This program (symbolically) presents the compliance and stiffness matrices of various materials.

Selected Parameters:

- *material* – Material type (the number in the code stands for the number of independent unknowns).

$$
material = \begin{cases} 0 - \text{General anisotropic, GEN21,} \\ 1 - \text{Orthotropic, ORT9,} \\ 2 - \text{Isotropic, ISO2,} \\ 3 - \text{Cubic, CUB3,} \\ 11,12,13 - \text{Monoclinic: MON13x, MON13y, MON13z, respectively,} \\ 21,22,23 - \text{Transversely Isotropic: TI5x, TI5y, TI5z, respectively,} \\ 31,32,33 - \text{Tetragonal: TRG6x, TRG6y, TRG6z, respectively.} \end{cases}
$$

P.2.2 Material Data by Compliance Matrix

This program introduces numerical values to the compliance matrix. It then calculates the stiffness matrix. The user has to update only the material that is relevant to his needs (see list of materials in **P.2.1**).

Selected Parameters:

- a_{ij} – Elements of the compliance matrix.

P.2.3 Material Data by Stiffness Matrix

This program introduces numerical values to the stiffness matrix. It then calculates the compliance matrix. The user has to update only the material that is relevant to his needs (see list of materials in **P.2.1**).

Selected Parameters:

- A_{ij} – Elements of the stiffness matrix.

P.2.4 Compliance Matrix Positiveness

This program develops the inequalities required for positive-definite stress-strain law of various materials, as discussed in S.2.10.

P.2.5 Generic Compliance Matrix Transformation due to Coordinate System Rotation

This program demonstrates the transformation of a generic compliance matrix of Cartesian anisotropy between two coordinate systems in terms of the three Euler's angles ϕ, θ, ψ as discussed in S.1.7, S.2.12 (see order of rotations in (1.206)).

Close all inputs to get a complete symbolic expression. Set a numerical value only for angles that should not appear symbolically.

Selected Parameters:

- ϕ, θ, ψ – Rotation angles of orthotropic material, see **P.1.10**. To change the orthotropic material characteristics, modify and activate first **P.2.2** or **P.2.3**.

P.2.6 Application of the Compliance Matrix Transformation

This program demonstrates the transformation of **P.2.5**. Prior to this code, **P.2.2** or **P.2.3** should be activated.

Selected Parameters:

- t_x , t_y, t_z – see **P.2.5**.
- *material* – Material type when $t_x = t_y = t_z = 0$, see **P.2.1**.

P.2.7 Compliance Matrix Transformation due to Coordinate System Rotation

The program produces symbolic transformation of a specific compliance matrix between two coordinate systems in terms of the three Euler's angles ϕ, θ, ψ as discussed in S.1.7, S.2.12 (see order of rotations in (1.206)).

Close all inputs to get a complete symbolic expression. Set a numerical value only for angles that should not appear symbolically.

Selected Parameters:

- ϕ, θ, ψ – see **P.2.5**.
- *material* – Material type when $\phi = \theta = \psi = 0$, see **P.2.1**.

P.2.8 Visualization of a Compliance Matrix Transformation

This program demonstrates graphs in Cartesian and spherical coordinates for elements of the transformed compliance matrix. The material and rotation angles are those of the last run of **P.2.6**.

Selected Parameters:

- i_0, j_0 – Indices to plot a_{i_0,j_0}.

P.2.9 Generic Stiffness Matrix Transformation due to Coordinate System Rotation

This symbolic program demonstrates the transformation of a generic stiffness matrix of Cartesian anisotropy between two coordinate systems in terms of the three Euler's angles ϕ, θ, ψ as discussed in S.1.7, S.2.13 (see order of rotations in (1.206)).

Close all inputs to get a complete symbolic expression. Set a numerical value only for angles that should not appear symbolically.

See selected parameters and details in **P.2.5**.

P.2.10 Stiffness Matrix Transformation due to Coordinate System Rotation

The program produces symbolic transformation of a specific stiffness matrix between two coordinate systems in terms of the three Euler's angles ϕ, θ, ψ as discussed in S.1.7, S.2.13 (see order of rotations in (1.206)).

Close all inputs to get a complete symbolic expression. Set a numerical value only for angles that should not appear symbolically.

See selected parameters and details in **P.2.7**.

P.2.11 Application of the Stiffness Matrix Transformation

This program demonstrates the transformation of **P.2.9**. Prior to this code, **P.2.2** or **P.2.3** should be activated.

See selected parameters in **P.2.6**.

P.2.12 Compliance Matrix Transformation from Cartesian to Curvilinear Coordinates

This program executes transformations of a compliance matrix given in Cartesian coordinates into its form in various orthogonal curvilinear coordinates.

Selected Parameters:

- s – see **P.1.1**.
- *material* – Material type, see **P.2.1**.
- *Point* – The point under discussion given by the three values of α_1, α_2, α_3.

P.2.13 Principal Directions of Anisotropy

This program produces the transformation of a stiffness matrix to the principal directions of anisotropy for the material and rotation angles defined in **P.2.11**.

P.3 Programs for Chapter 3

Table P.1 summarizes all programs that present prescribed solutions of field equations.

"Field" Equation	Programs
Poisson: $\nabla_3^{(2)}\Lambda = \cdots$	**P.3.7** **P.3.8** (A)
Biharmonic: $\nabla_j^{(4)}\Phi = \cdots,\ (j = 1,2)$	**P.3.5** **P.3.6** (A)
Coupled-Plane: $\nabla_1^{(4)}\Phi + \nabla_1^{(3)}\Lambda = \cdots$ $\nabla_1^{(3)}\Phi + \nabla_1^{(2)}\Lambda = \cdots$	**P.3.10** **P.3.11** (A)

Table P.1: Prescribed solution programs. "(A)" stands for an application to a rectangular domain.

P.3.1 Illustrative Parametrizations

This program examines a given prescribed parametrization, see S.3.1.2, S.3.1.3. It contains three types of examples based on Cartesian coordinates, polar coordinates, and complex functions.

Selected Parameters:

- *Part 1:* $x(t), y(t)$ – Explicit periodic functions $x(t), y(t)$ for $t \in [0, 2\pi]$.
- *Part 2:* $\rho(t)$ – Explicit periodic functions $\rho(t)$ and/or $\theta(t)$ for $t \in [0, 2\pi]$.
- *Part 3:* $w(z)$ – Explicit complex function $w(z)$, $z = \cos(t) + \mathbf{i}\sin(t)$, for $t \in [0, 2\pi]$.

P.3.2 Regular Polygon Parametrization Using the Schwarz-Christoffel Integral

This program illustrates the parametrization of a regular polygon using single-valued conformal mapping. See S.3.1.2 and Example 3.2.

Selected Parameters:

- n – Number of vertices (or edges) of the polygon.
- ii – An integer that controls the number of terms in the series expansion. The series are evaluated up to the power of $ii \times n$.
- AA, BB – Complex numbers (the constants $\widetilde{A}$, $\widetilde{B}$ of (3.10)) for "stretching" and "rigid" rotation and translation of the domain.

P.3.3 Generic Polygon Parametrization Using the Schwarz-Christoffel Integral

This program illustrates the parametrization of a generic polygon using single-valued conformal mapping. See S.3.1.4 and Example 3.3.

Selected Parameters:

- *n, ii, AA, BB* – see **P.3.2**.
- φ_i, $i = 1,\ldots,n-1$ – Interior angles of the polygon. The last angle is determined by the program.
- P_i, $i = 1,\ldots,n-2$ – Points on the unit circle that correspond to the vertices. The last two points are determined by the program.

P.3.4 Fourier Series Parametrization of a Polygon

This program illustrates the parametrization of a polygon using Fourier series, see S.3.1.4 and Example 3.3. The area centroid is also determined.

Selected Parameters:

- $case = \begin{cases} 1 & - \text{ Each edge occupies an equal length } \Delta t = \frac{2\pi}{n} \text{ (Method B)}, \\ 2 & - \text{ The variation of } t \text{ devoted for each edge is proportional to its length.} \\ & \quad \text{Therefore, } \Delta s = \Delta t \frac{l}{2\pi}, \text{ where } l \text{ is the circumferential (Method A).} \end{cases}$
- *xx, yy* – Arrays containing the coordinates of the n vertices of a generic polygon. The resulting continuous functions $x(t)$ and $y(t)$ are piecewise linear, where t is linearly spread between two adjoint vertices so that $-\pi \le t \le \pi$.
- $N(= k_p)$ – Number of harmonics in the final Fourier series expansions of $x(t)$ and $y(t)$.

P.3.5 Prescribed Polynomial Solution of the Biharmonic Equation

This program derives prescribed polynomial solutions (including displacements integration) of the biharmonic equation for MON13z and isotropic materials, see S.3.2.3 and Example 3.4. Φ is assumed as the following homogeneous polynomial of degree k_Φ:

$$\Phi = \sum_{i=0}^{k_\Phi} \overline{B}_{i,k_\Phi-i} x^i y^{k_\Phi-i}.$$

Selected Parameters:

- k_Φ – The polynomial degree.
- S – An array of four numbers, which are the first indices of the $\overline{B}_{i,k_\Phi-i}$ coefficients that are left as free parameters (not applicable for isotropic material).
- $case = \begin{cases} case_a & - \text{ Plane- stress solution, } (\nabla_3^{(4)}, \text{ see (3.206))}, \\ case_b & - \text{ Plane- strain solution, } (\nabla_1^{(4)}, \text{ see (3.202))}. \end{cases}$
- $iso = \begin{cases} 0 & - \text{ MON13z material}, \\ 1 & - \text{ Isotropic material}. \end{cases}$

P.3.6 Application of Prescribed Polynomial Solution of the Biharmonic Equation

This program illustrates the prescribed polynomial solutions of **P.3.5** in a rectangular domain, see S.3.2.3, Example 3.4.

Selected Parameters:

- All parameters of **P.3.5**. To change the isotropic material characteristics, modify and activate first **P.2.2** or **P.2.3**.
- k_0 – An integer number $= 1,2,3,4$ that determines the only nonzero coefficient between the four coefficients selected as free parameters. For example, when $k_0 = 2$ the second coefficient is set to be equal 1 (unit) and all other three coefficients are set to zero.
- d, h – Semi-width and height, respectively, of the rectangular domain.
- t_z – The rotation angle of an orthotropic material about the z-axis (in degrees, valid for $iso = 0$ only). This rotation creates a MON13z material. For $t_z = 0$, the material remains orthotropic. To change the orthotropic material characteristics, modify and activate first **P.2.2** or **P.2.3**.

P.3.7 Prescribed Polynomial Solution of Laplace's Equation

This program derives prescribed polynomial solutions of Laplace's equation for MON13z and isotropic materials. Λ is assumed as the following homogeneous polynomial of degree k_Λ:

$$\Lambda = \sum_{i=0}^{k_\Lambda} \overline{H}_{i,k_\Lambda - i} x^i y^{k_\Lambda - i}.$$

Selected Parameters:

- k_Λ – The polynomial degree.
- S – An array of two numbers, which are the first indices of the two $\overline{H}_{i,k_\Lambda - i}$ coefficients that are left as free parameters (not applicable for isotropic material).
- iso – see **P.3.5**.

P.3.8 Application of Prescribed Polynomial Solution of Laplace's Equation

This program illustrates the prescribed polynomial solution of **P.3.7** in a rectangle.

Selected Parameters:

- All parameters of **P.3.7**.
- k_0 – An integer number $= 1,2$ that determines the only nonzero coefficient between the two coefficients selected as free parameters. When $k_0 = 1$ the first coefficient is set to be equal to 1 (unit) and the other is set to zero, and vice versa for $k_0 = 2$.
- d, h, t_z – see **P.3.6**.
- a_{ij} – Compliance matrices for MON13z and ISO2 materials.

P.3.9 Affine Transformation

This program illustrates the affine transformation of (3.131) for given $\widetilde{a}$ and $\widetilde{b}$ parameters.

Selected Parameters:

- a_0, b_0 – The coefficients $\widetilde{a}, \widetilde{b}$, respectively, in S.3.3.4.
- $r(\theta)$ – The original contour in the x, y-plane given in polar coordinates. The introduction of generic polygons is also demonstrated.

P.3.10 Prescribed Polynomial Solution of Coupled-Plane Equations

This program illustrates prescribed polynomial solutions for Coupled-Plane equations with MON13z material.

Selected Parameters:

- k_Φ – The polynomial degree of Φ ($k_\Lambda = k_\Phi - 1$).
- S_1 – An array of four numbers, which are the first indices of the two $\overline{B}_{i,k_\Phi - i}$ coefficients of Φ that are left as free parameters. See **P.3.5**.
- S_2 – An array of two numbers, which are the first indices of the two $\overline{H}_{i,k_\Lambda - i}$ coefficients of Λ that are left as free parameters, see **P.3.7**.

P.3.11 Application of Prescribed Polynomial Solution of Coupled-Plane Equations

This program illustrates the prescribed polynomial solutions of Coupled-Plane equations in a rectangle, see **P.3.10**.

Selected Parameters:

- All parameters of **P.3.10**.
- k_1 – An integer number $= 1, 2, 3, 4$ that determines the only nonzero coefficient among the above four coefficients selected for Φ.
- k_2 – An integer number $= 1, 2$ that determines the only nonzero coefficient among the above four coefficients selected for Λ.
- d, h, t_z – see **P.3.6**.
- a_{ij} – Compliance matrix for MON13y material.

P.3.12 Ellipticity of the Differential Operators

This program illustrates the differential operators described in S.3.6.5. A positive-definite quadratic form of energy is built and special cases of the stress vector are shown to coincide with the above operators' characteristic polynomials. The program also presents numerical values for the polynomial roots.

Selected Parameters:

- t_x, t_y, t_z – see **P.2.5**.

P.4 Programs for Chapter 4

Table P.2 lists all programs for particular solutions of the field equations described in S.4.2. Table P.3 summarizes all programs for homogeneous polynomial BVP solutions described in S.4.3. Table P.4 summarizes programs for circular and annular isotropic domains discussed in S.4.6.

Type of equation(s)	"Field" Equation	Elliptical domain	Generic domain
Laplace	$\nabla_3^{(2)}\Lambda = F_0^\Lambda$	**P.4.4**	**P.4.3**
Biharmonic	$\nabla_j^{(4)}\Phi = F_0,\ (j=1,2)$	**P.4.1**	**P.4.2**
Coupled-plane	$\nabla_1^{(4)}\Phi + \nabla_1^{(3)}\Lambda = g$ $\nabla_1^{(3)}\Phi + \nabla_1^{(2)}\Lambda = f$	**P.4.6**	**P.4.5**

Table P.2: Particular solution programs.

Homogeneous BVP	Elliptical domain	Generic domain	Application
$\nabla_2^{(2)}\Lambda = 0,\ \frac{d}{ds}\Lambda = \cdots,\ D^n\Lambda = \cdots$	**P.4.8**	**P.4.13**	**P.4.18**
$\nabla_j^{(4)}\Phi = 0,\ (j=1,2)$ $\frac{d}{ds}\Phi_{,x} = \cdots,\ \frac{d}{ds}\Phi_{,y} = \cdots$	**P.4.10**	**P.4.14**	**P.4.16** **P.4.17**
$\nabla_1^{(4)}\Phi + \nabla_1^{(3)}\Lambda = 0$ $\nabla_1^{(3)}\Phi + \nabla_1^{(2)}\Lambda = 0$ $\frac{d}{ds}\Lambda = \cdots,\ \frac{d}{ds}\Phi_{,x} = \cdots,\ \frac{d}{ds}\Phi_{,y} = \cdots$	**P.4.12**	**P.4.15**	

Table P.3: Homogeneous BVP solution programs.

P.4.1 Particular Polynomial Solution of the Biharmonic Equation in an Ellipse

This program determines a particular solution, $\Phi_p^{(k)}$, of a biharmonic equation in an elliptical domain as discussed in S.4.2.1.

Selected Parameters:

- *case* – see **P.3.5**.
- a_0, b_0 – Ellipse semi-axes, ($\widetilde{a}$, $\widetilde{b}$ in S.4.2.1).

BVP	Circular domain	Annular domain
Dirichlet & Neumann	**P.4.23**	**P.4.24**
Biharmonic	**P.4.25**	**P.4.26**

Table P.4: Solution programs for circular and annular isotropic domains.

- k_0 – The value of k in S.4.2.1.
- a_{ij}, b_{ij} – The a_{ij}, b_{ij} matrices elements. Due to the complexity of the symbolic solution, it is more convenient to work with numerical data. The data supplied is for a typical Graphite/Epoxy orthotropic material, which is rotated by 30° about the z-axis (i.e. for a MON13z material but it can be replaced by another matrix of MON13z material). Rational numbers are used to demonstrate solution exactness.

Note: To obtain a symbolic output, the data lines for a_0, b_0 and a_{ij}, b_{ij} should be closed. Working with low values of k_0 is recommended in such a case.

P.4.2 Particular Polynomial Solution of the Biharmonic Equation

This program determines a particular solution, $\Phi_p^{(k)}$, of a biharmonic equation as discussed in S.4.2.1. The program sets to zero the four coefficients with the highest power of x (i.e. the four highest first indices).

Selected Parameters:

- k_0 – The value of k in S.4.2.1.

P.4.3 Particular Polynomial Solution of Poisson's Equation

This program determines a particular solution, $\Lambda_p^{(k)}$, for Poisson's equation as discussed in S.4.2.2. The program sets to zero the two coefficients with the highest power of x (i.e. the two highest first indices).

Selected Parameters:

- k_0 – The value of k in S.4.2.2.

P.4.4 Particular Polynomial Solution of Poisson's Equation in an Ellipse

This program determines a particular polynomial solution, $\Lambda_p^{(k)}$, for Poisson's equation in an elliptical domain as discussed in S.4.2.2.

Selected Parameters:

- a_0, b_0, a_{ij} – see **P.4.1**.
- k_0 – see **P.4.3**.

Note: To obtain a symbolic output, the data lines for a_0, b_0 and a_{ij} should be closed. Working with low values of k_0 is recommended in such a case.

P.4.5 Particular Polynomial Solution of Coupled-Plane Equations

This program determines a particular polynomial solution, $\Phi_p^{(k)}$ and $\Lambda_p^{(k+1)}$, of the coupled-plane BVP as discussed in S.4.2.3. The program sets to zero the coefficients with the highest power of x, i.e. the highest first indices (four for $\Phi_p^{(k)}$ and two for $\Lambda_p^{(k+1)}$).

Selected Parameters:

- k_g – The value of k in S.4.2.3 (i.e. the polynomial degree of g). $k_g = -1$ represents $g = 0$ and $f = \overline{f}_{00} = const.$, $k_g = 0$ represents $g = const.$ and $f = \overline{f}_{10}x + \overline{f}_{01}y$, etc.

P.4.6 Particular Polynomial Solution of Coupled-Plane Equations in an Ellipse

This program determines particular solutions, $\Phi_p^{(k)}$ and $\Lambda_p^{(k+1)}$, for coupled-plane BVPs in an elliptical domain as discussed in S.4.2.3.

Selected Parameters:

- a_0, b_0, b_{ij} – see **P.4.1**.

- k_g – see **P.4.5**.

P.4.7 Prescribed Polynomial Boundary Functions

This program determines symbolically (with numerical examples) prescribed polynomials that satisfy the required existence and uniqueness conditions for various cross-sections, see S.4.3.1.

Selected Parameters:

- $$\text{BVP}=\begin{cases} -1 & - \text{ determines } F_3^{\Lambda} \text{ so that } \oint_{\partial\Omega} F_3^{\Lambda} = 0, \text{ see (4.2)}, \\ 1 & - \text{ determines } F_3 \text{ so that } \oint_{\partial\Omega} F_3 = 0, \text{ see (4.2)}, \\ 2 & - \text{ determines } F_1,\ F_2 \text{ so that (4.4) are satisfied}, \\ 3 & - \text{ determines } F_1,\ F_2,\ F_3 \text{ so that the first set (4.11) is satisfied.} \end{cases}$$

- k_{lower}, k_{upper} – Lower and upper polynomial levels.

- $$n=\begin{cases} 3,4,5,6,\ldots & - \text{ Domain definition as a regular n-vertices polygon}, \\ -4,-5,-6,-7,-8 & - \text{ Domain definition as a built-in non-convex polygon.} \end{cases}$$

- $ii(= k_p)$ – An integer $(1,2,\ldots)$ that controls the number of terms in the contour series expansion. Select $ii = 1$ for an elliptical domain (n has no influence in this case).

- R_0 – Magnification constant for domain geometry (when $n > 0$ is selected).

- m_0 – When an elliptical domain is selected by $ii = 1$, m_0 stands for the ellipse slenderness, namely, $m_0 = \widetilde{a}/\widetilde{b}$ (where $\widetilde{a} > \widetilde{b} > 0$ are the semi-axes along the x, y directions, respectively).

P.4.8 Exact/Conditional Polynomial Solution of Dirichlet/Neumann Homogeneous BVPs

When conditional solution is required, the set of equations entitled "data relations set #2" represents the internal relations that have to be fulfilled by the data in order to constitute an exact polynomial solution. See discussion in S.4.3.3.1.

Selected Parameters:

- $\text{BVP}=\begin{cases} 0 & - \text{ The Dirichlet problem (4.13a), (4.1b),} \\ 1 & - \text{ The Dirichlet problem (4.13a), (4.38a),} \\ -1 & - \text{ The Neumann problem (4.13a), (4.38b).} \end{cases}$

  For BVP $=\pm 1$, (4.39) must be satisfied. Otherwise, some of the coefficients are eliminated by the program so that the above conditions are fulfilled. The above coefficients elimination is indicated as the set of equations entitled "data relations set #1".

- *iso* - see **P.3.5**.

- $k_3(=k_Q)$ – Polynomial degree of G_3 in (4.1b) or P_j and Q_j in (4.38a, b).

- k_0 – Degree of the lower homogeneous polynomial that participates in G_3 or P_j and Q_j (i.e. $k_0 = 0, 1, \ldots, k_3$).

- t_z – see **P.3.6**. Valid only for $iso = 0$. Close marked lines for symbolic output.

- $x(t), y(t), k_p$ – The parametrization functions and level. These functions may include symbolic parameters that will appear in the output as well.

P.4.9 Symbolic Verification of the Neumann BVP Solution

This program proves that the polynomial solution presented in Example 4.3 is an identical and concise version of the formal one obtained by **P.4.8**.

No Selected Parameters.

P.4.10 Exact/Conditional Polynomial Solution of Homogeneous Biharmonic BVPs

When conditional solution is required, the set of equations entitled "data relations set #2" represents the internal relations that have to be fulfilled by the data in order to constitute an exact polynomial solution. See discussion in S.4.3.3.2.

Selected Parameters:

- *case, iso* – see **P.3.5**.

- $k_1(=k_Q)$ – Polynomial degree of P_j and Q_j in (4.40).

- k_0 – Degree of the lower homogeneous polynomial that participates in P_j and Q_j (i.e. $k_0 = 0, 1, \ldots, k_3$). Note that (4.41) must be satisfied. Otherwise, some of the coefficients are eliminated by the program so that the above conditions are fulfilled. The above coefficients elimination is indicated as the set of equations entitled "data relations set #1".

- t_z, $x(t), y(t)$, k_p – see **P.4.8**.
- Ex=rot – Open this line for the case of an elliptical domain under centrifugal loading as described in Example 4.12.

P.4.11 Symbolic Verification of the Biharmonic BVP Solution

This program proves that the polynomial solution presented in Example 4.7 is an identical and concise version of the formal one obtained by **P.4.10**.

No Selected Parameters.

P.4.12 Exact/Conditional Polynomial Solution of Homogeneous Coupled-Plane BVPs

This program supports the discussion in S.4.3.3, S.4.3.3.3. When conditional solution is required, the set of equations entitled "data relations set #2" represents the internal relations that have to be fulfilled by the data in order to constitute an exact polynomial solution.

Selected Parameters:

- $k_1 (= k_Q)$ – Polynomial degree of P_j and Q_j in (4.40).
- k_0 – Degree of the lower homogeneous polynomial that participates in P_j and Q_j (i.e. $k_0 = 0, 1, \ldots, k_3$). Note that (4.39), (4.41) must be satisfied. Otherwise, some of the coefficients are eliminated by the program so that the above conditions are fulfilled. The above coefficients elimination is indicated as the set of equations entitled "data relations set #1".
- t_x, t_y, t_z – see **P.2.6**. Close marked lines for symbolic output.
- $x(t), y(t)$, k_p – see **P.4.8**.

P.4.13 Approximate Polynomial Solution of Homogeneous Dirichlet/Neumann BVPs

This program supports the discussion in S.4.3, S.4.3.4. Rational numbers are used to ensure exact single-value conditions and exact prescribed solution. The minimization process is executed with rational numbers for isotropic materials and by floating point arithmetic for anisotropic materials. The program also contains comparisons with exact solutions for torsion of an anisotropic ellipse and an isotropic triangle.

Selected Parameters:

- BVP, t_z – see **P.4.8**.
- *iso* – see **P.3.5**.
- $n = 3, 4, 5, 6, \ldots$ – Domain definition as a regular n-vertices polygon.
- $method\_ = \begin{cases} residual, \\ contour. \end{cases}$

For $n > 0$ the parametrization is carried out as described in S.3.1.3. The minimization process is carried out by either the *residual* method of S.4.3.4.1, or by the *contour* method of S.4.3.4.2.

- n, ii, R_0, m_0 – see **P.4.7**. Note that when $ii = 1$, the solution becomes exact, see also **P.4.8**.

- Open the indicated lines for random boundary data.

- P_3, Q_3, G_3 – User defined boundary polynomials (used only when random data is not selected). For BVP$= \pm 1$, make sure that the "Boundary Data Verification" section yields "0". Otherwise, the boundary polynomials must be corrected.

- $k_3\,(=k_Q)$ – Polynomial degree of P_3, Q_3 or G_3 in (4.1b).

- $k_{0,\mathrm{rand}}\,(\leq k_3)$ – Degree of the lower homogeneous polynomial that participates in P_3, Q_3 or G_3. Valid only for random boundary data.

- $kk\,(=k_a)$ – The amount of solution degree augmentation discussed in S.4.3.4.

- dd – Material matrix scaling. The entire matrix is multiplied by 10^{dd}.

P.4.14 Approximate Polynomial Solution of Homogeneous Biharmonic BVPs

This program supports the discussion in S.4.3.4. Rational numbers are used to ensure exact single-value conditions and exact prescribed solution, see S.4.3, S.4.3.4. The minimization process is executed with rational numbers.

Selected Parameters:

- *case* – see **P.3.5**.

- *iso*, method_, n, ii, R_0, m_0, t_z, dd, kk – see **P.4.13**.

- $k_1\,(=k_Q)$ – Polynomial degree of P_j and Q_j, $j = 1,2$ in (4.40).

- P_1, Q_1, P_2, Q_2 – User defined boundary polynomials (used only when random data is not selected). Note that (4.41) must be satisfied. For that purpose, make sure that the "Boundary Data Verification" section yields "[0, 0, 0]". Otherwise, the boundary polynomials must be corrected.

- Open the indicated lines for random boundary data.

- $k_{0,\mathrm{rand}}\,(\leq k_1)$ – Degree of the lower homogeneous polynomial that participates in P_1, Q_1, P_2, Q_2. Valid only for random boundary data.

- corr=true/false – Set "corr=true" to add a linear function to Φ (i.e., $\Delta\Phi = \widetilde{A}x + \widetilde{B}y + \widetilde{C}$) to ensure $\Phi = 0$ for the points of $t = 0, \frac{2\pi}{3}, \frac{4\pi}{3}$.

P.4.15 Approximate Polynomial Solution of Homogeneous Coupled-Plane BVPs

This program supports the discussion in S.4.3.4.

Selected Parameters:

- method_, n, ii, R_0, m_0, dd – see **P.4.13**.
- $P_1, Q_1, P_2, Q_2, P_3, Q_3$ – User defined boundary polynomials (used only when random data is not selected). Note that (4.39), (4.41) must be satisfied. For that purpose, make sure that the "Boundary Data Verification" section yields "[0, 0, 0, 0]". Otherwise, the boundary polynomials must be corrected.
- $k_j,\ j = 1,3$ – Polynomial degree of $P_j, Q_j,\ j = 1,3$ in (4.38a), (4.40).
- $kk_L\,(= k_a)$ – The amount of solution degree augmentation for Λ discussed in S.4.3.4.
- $kk_\Phi\,(= k_a)$ – The amount of solution degree augmentation for Φ discussed in S.4.3.4.
- Open the indicated lines for random boundary data.
- $k_{0,\text{rand}}\,(\leq k_3)$ – Degree of the lower homogeneous polynomial that participates in $P_j, Q_j,\ j = 1,3$. Valid only for random boundary data.
- w_Φ – The relative weight for Φ in the error minimization process, while the relative weight for Λ is a unit. This value should be adjusted for each specific case, and mainly when Φ and Λ differ by orders of magnitude.
- t_x, t_y, t_z – see **P.2.6**.
- corr – see **P.4.14**.

P.4.16 Rotating Plate: Application of the Biharmonic BVP Solution

This program presents an application of **P.4.14** to the specific case of a rotating plane domain for $\gamma = 1$, see (4.88) of Example 4.12.

Selected Parameters: see **P.4.14**.

P.4.17 Bending of Thin Plates: Application of the Biharmonic BVP Solution

This program presents an application of **P.4.14** to the specific case of bending of anisotropic plates discussed in S.3.5.2 and Example 4.13.

Selected Parameters:

- iso, ii, R_0, m_0, t_z, kk – see **P.4.14**.
- $n = 3, 4, 5, \ldots$ – Domain definition as a regular n-vertices polygon.
- h – Plate thickness.
- k_{min}, k_0 – The lowest and the highest polynomial level of the loading, respectively.

- Define the loading by $S_1(i,j)$ as

$$q(x,y) = \sum_{s=k_{min}}^{k_0} \sum_{i=0}^{s} S_1(i,s-i)x^i y^{s-i}.$$

- $bc = \begin{cases} 1 & - \ \textit{clamped} \text{ edge boundary conditions, see (3.184)}, \\ 2 & - \ \textit{simply supported} \text{ edge boundary conditions, see (3.185)}. \end{cases}$

P.4.18 Approximate Polynomial Solution of Homogeneous n-Coupled Dirichlet BVPs

This program presents a solution for the n-Coupled Dirichlet BVP by complex potentials as shown in S.4.4.1.

Selected Parameters:

- $BVP = \begin{cases} 0 & - \ \text{for the first case presented by (4.95)}, \\ 1 & - \ \text{for the second case presented by (4.95)}. \end{cases}$
- $NF = 1,2,3$ – Number of functions Λ_k (n in S.4.4.1).
- method_, n, ii, R_0, m_0, kk – see **P.4.13**.
- μ_k – The corresponding roots (complex numbers $= \alpha_k + \mathbf{i}\beta_k$, $k = 1,\ldots,NF$).
- P_m, Q_m or G_m – User defined boundary polynomials (used only when random data is not selected).

  For BVP=1: $F_m = P_m \cos(\bar{\mathbf{n}},x) + Q_m \cos(\bar{\mathbf{n}},y)$, $m = 1,2,3$. Make sure that the "Boundary Data Verification" section yields "Verify=[0]". Otherwise, the boundary polynomials must be corrected.

- $k_3\ (= k_Q)$ – Polynomial degree of P_m, Q_m or G_m in (4.95).
- Open the indicated lines for random boundary data.
- $k_{0,\text{rand}}\ (\leq k_3)$ – Degree of the lower homogeneous polynomial that participates in P_m, Q_m or G_m. Valid only for random boundary data.
- ν^{ij} – The coefficients $\nu^{ij} = \nu_1^{ij} + \mathbf{i}\nu_2^{ij}$ (complex numbers), see (4.91).

P.4.19 Fourier Series Solution of Homogeneous Dirichlet BVPs in a Rectangle

This program solves the first problem presented in S.4.4.2, in its non-homogeneous version, namely

$$\nabla_2^{(2)}\Psi = -2\theta \qquad \text{over} \quad \Omega, \qquad \Psi = 0 \qquad \text{on} \quad \partial\Omega.$$

The program introduces a particular solution $\Psi_p = -y(y-h)$ and corrects the boundary conditions accordingly.

Selected Parameters:

- d,h – Rectangle dimensions, see Fig. 4.15(a).

- β_3 – Imaginary part of μ_3.
- N – Number of terms in Fourier series.
- $\widetilde{G}_3$ – The boundary function over the edges.

P.4.20 Fourier Series Solution of Homogeneous Coupled-Plane BVPs in a Rectangle

This program presents a solution for the homogeneous coupled-plane BVP discussed in S.4.4.5.

Selected Parameters:

- N, h, d, β_3 – see **P.4.19**.
- $P_1, Q_1, P_2, Q_2, P_3, Q_3$ – User defined boundary polynomials (used only when random data is not selected). Note that (4.39), (4.41) must be satisfied.

  The program is set to solve the torsion problem ($\overline{f}_{00} = -2$) in three versions:

$$Ex = \begin{cases} 1 & - \quad \Lambda_p = \frac{\overline{f}_{00}}{2b_{44}} x(x-d), \\ 2 & - \quad \Lambda_p = \frac{\overline{f}_{00}}{2b_{55}} y(x-y), \\ 3 & - \quad \Lambda_p = \frac{\overline{f}_{00}}{4} [\frac{1}{b_{44}} x(x-d) + \frac{1}{b_{55}} y(x-y)]. \end{cases}$$

- $iso = \begin{cases} 0 & - \quad \text{Anisotropic material,} \\ 1 & - \quad \text{Isotropic material.} \end{cases}$
- t_x, t_y, t_z – see **P.2.6**.

P.4.21 Equilibrium Equations in Terms of Displacements

This program demonstrates a verification of (4.182) and produces analogous formulas for partial differential operators of various orthogonal coordinates.

Selected Parameters:

- s – An index for a three-dimensional coordinate system, see Example 1.9 and **P.1.1**.

P.4.22 Prescribed Solutions in an Isotropic Parallelepiped

This program demonstrates the solution scheme of S.4.5.2.

Selected Parameters:

- NN – Number of terms in Fourier series (maximal value of n or m).
- d, h, L – The dimensions d, h, l, respectively, in S.4.5.2.
- λ, μ – Lamé's constants for isotropic materials, see Remark 2.2.
- f_0 – An $NN \times NN$ matrix. $f_0(n,m)$ is the f_0 coefficient described in (4.200a). The program sums up sll solutions for $n, m \in [1, \ldots, NN]$.

P.4.23 Fourier Series Solution of the Dirichlet/Neumann BVPs in an Isotropic Circle

This program supports the discussion in S.4.6.2, S.4.6.3.

Selected Parameters:

- BVP= $\begin{cases} 0 & - \text{ The Dirichlet problem for } \Lambda_{|\partial\Omega} = F_3(\theta) \text{ boundary condition,} \\ 1 & - \text{ The Dirichlet problem for } \frac{d}{d\theta}\Lambda_{|\partial\Omega} = F_3(\theta) \text{ boundary condition,} \\ -1 & - \text{ The Neumann problem for } \frac{d}{dr}\Lambda_{|\partial\Omega} = F_3(\theta) \text{ boundary condition.} \end{cases}$

- F_3 – 2π-periodic boundary function of θ. For BVP$=\pm 1$, F_3 must satisfy the condition $\oint F_3\, d\theta = 0$ and a_0 is set to zero in these cases.

- N – The Fourier series degree. Depending on $F_3(\theta)$, a sufficiently large number should be selected. A convergence study should be performed.

- R – The circle radius.

P.4.24 Fourier Series Solution of the Dirichlet/Neumann BVPs in an Isotropic Circular Ring

This program supports the discussion in S.4.6.4, S.4.6.5. The program also contains the symbolic calculation of the coefficients involved in the above sections.

Selected Parameters:

- BVP – see **P.4.23**.

- F_{31}, F_{32} – 2π-periodic boundary functions of θ.

  For BVP$=1$, F_{3i} must satisfy $\oint_{\partial\Omega} F_{3i}\, d\theta = 0$, $i = 1,2$, and therefore, A_{i0} are set to zero. a_0 and d_0 are also set to zero in this case.

  For BVP$=-1$, F_{3i} must satisfy (4.229). Otherwise, A_{20} is corrected to satisfy (4.230). a_0 is set to zero in this case.

- N – see **P.4.23**. Depending on $F_3(\theta)$, a sufficiently large number should be selected.

- $0 < R_1 < R_2$ – The annular radii.

P.4.25 Fourier Series Solution of the Biharmonic BVP in an Isotropic Circle

This program supports the discussion in S.4.6.6. The program also contains the symbolic calculation of the coefficients involved in the above sections.

Selected Parameters:

- G_1, G_2 – 2π-periodic boundary functions of θ.

- N – see **P.4.23**. Depending on G_1, G_2, a sufficiently large number should be selected.

- R – The circle radius.

P.4.26 Fourier Series Solution of the Biharmonic BVP in an Isotropic Circular Ring

This program supports the discussion in S.4.6.7.

Selected Parameters:

- $G_{ij},\ i,j=1,2$ – 2π-periodic boundary value functions of θ, see (4.237).
- N – see **P.4.23**.
- $0<R_1<R_2$ – The annular radii.

P.5 Programs for Chapter 5

P.5.1 Elementary "Strength-of-Materials" Isotropic Beam Analysis

This program carries out the elementary "Strength-of-Materials" isotropic beam analysis that is documented in S.5.3.4 for an elliptical cross-section. The program employs the torsion induced warping function and the torsional rigidity derived in Chapter 7, see Example 7.1.

Selected Parameters:

- a_0, b_0 – Ellipse semi-axes.
- l – Beam's length.
- $P_{x_{tip}}, P_{y_{tip}}, P_{z_{tip}}$ – Tip (concentrated) forces in the x, y and z directions.
- $M_{x_{tip}}, M_{y_{tip}}, M_{z_{tip}}$ – Tip (concentrated) moments in the x, y and z directions.
- $k_p, M(k_p+1,6)$ – Distributed loads input according to the following definition, see also S.5.3.4:

$$\{p_x,p_y,p_z\}=\sum\nolimits_{j=0}^{k_p}\{M(j+1,1),M(j+1,2),M(j+1,3)\}z^j,$$
$$\{m_x,m_y,m_z\}=\sum\nolimits_{j=0}^{k_p}\{M(j+1,4),M(j+1,5),M(j+1,6)\}z^j.$$

P.6 Programs for Chapter 6

P.6.1 An Anisotropic Beam of Elliptical Cross-Section

This program demonstrates a complete analytical solution for an elliptical beam of general anisotropy. The code is founded on the analysis of Chapter 6, and other techniques for solution of the Coupled-Plane BVP discussed in Chapters 3 4.

To ensure compatibility with other programs, the stress function ψ notation is kept as Λ.

Selected Parameters:

- a_0, b_0 – see **P.5.1**.
- l – Beam's length.

- *material* – The material type, see **P.2.2**, **P.2.3**.

- t_x, t_y, t_z – Material rotation angles (degrees) about coordinate axes. See **P.1.10**.

- $K\ (\geq -1)$ – Solution level, see Chapter 6.

- $P_{\alpha_t}, M_{\alpha_t},\ \alpha = x,y,z$ – Tip (concentrated) forces and moments.

- $P_\alpha^{(k)}(x,y), Q_\alpha^{(k)}(x,y),\ \alpha = x,y,z$ – Surface loading according to the following definitions, see (6.1):

$$X_s^{(k)} = P_x^{(k)}(x,y)\cos(\bar{\mathbf{n}},x) + Q_x^{(k)}(x,y)\cos(\bar{\mathbf{n}},y),$$
$$Y_s^{(k)} = P_y^{(k)}(x,y)\cos(\bar{\mathbf{n}},x) + Q_y^{(k)}(x,y)\cos(\bar{\mathbf{n}},y),$$
$$Z_s^{(k)} = P_z^{(k)}(x,y)\cos(\bar{\mathbf{n}},x) + Q_z^{(k)}(x,y)\cos(\bar{\mathbf{n}},y).$$

- $X_b^{(k)}, Y_b^{(k)}, Z_b^{(k)}$ – Body-force loading according to the definitions of (6.1).

- $P_1^{(0)}, P_2^{(0)}, P_3^{(0)}, M_1^{(0)}, M_2^{(0)}, M_3^{(0)}$ – Root loads. Should be calculated ("off-line") and introduced as input according to Remark 6.1.

P.7 Programs for Chapter 7

P.7.1 Tip Loads Effect in a Monoclinic Beam

This program verifies the tip loads solutions of S.7.2.3 for a MON13z beam. The program contains the associated stress, strain and displacement expressions.

P.7.2 Auxiliary Harmonic Functions for Elliptical Monoclinic Cross-Sections

This program symbolically derives the harmonic functions φ, χ_k and ω documented in Examples 7.1, 7.3, 7.4 for MON13z beam. ω is determined for a constant body force $Z_b = Z_0$.

- Ex_h – Harmonic function selection:

$$Ex_h = 1 \text{ for } \varphi, \quad Ex_h = 2 \text{ for } \chi_1, \quad Ex_h = 3 \text{ for } \chi_2, \quad Ex_h = 6 \text{ for } \omega/Z_0.$$

P.7.3 A Monoclinic Beam Under Axially Non-Uniform Distributed Loads (I)

This program documents and verifies the expressions for MON13z beam under axially non-uniform distributed loads derived in S.7.3. The program verifies the equilibrium equations, the boundary conditions, and the compatibility conditions.

Selected Parameters:

- K – Solution level, see S.7.3.

P.7.4 A Monoclinic Beam Under Axially Non-Uniform Distributed Loads (II)

This program documents and verifies the expressions for a MON13z beam under axially non-uniform distributed loads derived in S.7.3. The program verifies the single-value conditions of the biharmonic function.

Selected Parameters:

- K – Solution level, see S.7.3.

P.7.5 Solution for an Elliptical Monoclinic Beam Under Constant Longitudinal Loading

This program prepares the symbolic expressions for the polynomial-based solutions for a homogeneous MON13z beam of elliptical cross-section presented in Examples 7.4, 7.6, 7.7.

P.7.6 Solution Implementation for an Elliptical Monoclinic Beam Under Constant Longitudinal Loading

This program presents a numerical implementation for the symbolic expressions derived in **P.7.5** for a clamped-free beam.

- $material$ – Select material type, see **P.2.1** (excluding 0,11,12). To change the material characteristics, modify and activate first **P.2.2** or **P.2.3**.
- t_z – see **P.3.6**.
- a_0, b_0 – see **P.5.1**.
- X_0, Y_0, Z_0 – Values of the constant body forces (in the x, y and z directions, respectively).
- γ_1, HH – γ_1 and H of Example 7.7.
- l_0 – Beam's length.
- z_{plot} – Axial coordinate of the cross-section for which the results are plotted.

P.8 Programs for Chapter 8

P.8.1 Auxiliary Harmonic Functions in a Non-Homogeneous Rectangle

This program serves as a *harmonic solver*, and as a demonstrator of the auxiliary torsion and bending functions φ, χ_1, χ_2 in a non-homogeneous (and homogeneous, $N = 1$) rectangle shown in Fig. 8.3(a), see S.8.2.5.

- Problem selection: $func = \begin{cases} 0 & for\ a\ harmonic\ solver, \\ chi1 & for\ \chi_1, \\ chi2 & for\ \chi_2, \\ phi & for\ \varphi, \\ pphi & for\ \bar{\varphi}. \end{cases}$

- For $func = 0$, define loading functions $P_{[j]}$ and $Q_{[j]}$ and displacement jumps $W^{[j]}$ over the $j^{\underline{th}}$ interface line. $j = 1,\ldots,NF$ stands for the domain number.
- NF, d,h, $t_z^{[j]}$ – see **P.8.2**. When set to calculate φ, χ_1, χ_2 or $\bar{\varphi}$, the program reads the auxiliary problems solutions from the output of **P.8.2**. In such a case these data items must be identical to those used in **P.8.2**.
- k_{La} – Polynomial degree selection. The actual degree is determined by the program (as a function of the auxiliary problem solution level) and shown as k_Λ.

P.8.2 Plane Deformation and the Auxiliary Biharmonic Problems

This program serves as a *biharmonic solver*, and as a demonstrator of the auxiliary problems of plane deformation in a non-homogeneous rectangular domain, see S.8.2.2.

- *ax_problem* – Problem selection:
  $= 0$ for a biharmonic solver,
  $= 1,2,3$ for the first, second or third auxiliary problem, respectively.
- For *ax_problem* $= 0$, define loading functions $F_1^{[ji]}$ and $F_2^{[ji]}$ and displacements jumps $U^{[j]}$ and $V^{[j]}$ over the $j^{\underline{th}}$ interface line. $j = 1,\ldots,NF$ stands for the domain number, $i = 1,2,3,4$ stands for the edge number, see Fig. 8.3(a).
- NF (≥ 1) – Number of domains, see N in Fig. 8.3(a). $NF = 1$ stands for homogeneous rectangular cross-section.
- $k_\Phi > 3$ – Polynomial degree of Φ.
- d,h – Width and (total) height of a cross-section, see Fig. 8.3(a).
- $t_z^{[j]}$ – The rotation angles (degrees) of the NF orthotropic materials about the z-axis. These different rotation angle create different MON13z materials. When $t_z^{[j]} = 0$, the material remains orthotropic. To change the orthotropic material characteristics, modify and activate first **P.2.2** or **P.2.3**.

P.8.3 Fourier Series Based Torsion Function in a Non-Homogeneous Orthotropic Rectangle

This program implements the solution methodology presented in Example 8.5.

Selected Parameters:

- NL – Number of layers (N in Example 8.5).
- $h^{[j]}$, $a_0^{[j]}$, $s^{[j]}$ $(j = 1,\ldots,NL)$ – The parameters $t^{[j]}$, $\tilde{a}^{[j]}$, $s^{[j]}$ in Example 8.5.
- K – Series truncating value (replaces ∞ in (8.83)).

P.8.4 A Non-Homogeneous Monoclinic Beam Under Tip Loads

This program documents and verifies the expressions for the non-homogeneous MON13z beam tip loads derived in S.8.2. The program contains the associated stress, strain and displacement expressions.

P.8.5 A Non-Homogeneous Beam of Rectangular Cross-Section Under Tip Loads

This program determines the behavior of a beam of non-homogeneous rectangular cross-section under the tip force components P_x, P_y, P_z and tip moment components M_x, M_y, M_z, see Examples 8.3, 8.4, 8.7 and S.8.2.7.

The program reads the auxiliary problems solutions from the output of **P.8.2**, and the torsion and bending functions φ, χ_1, χ_2 from the output of **P.8.1**.

- P_x, P_y, P_z, M_x, M_y, M_z – The tip loads.

P.8.6 A Monoclinic Non-Homogeneous Beam Under Axially Distributed Non-Uniform Loads (I)

This program documents and verifies the expressions for a non-homogeneous MON13z beam under axially non-uniform distributed loads derived in S.7.3.3. The program verifies the equilibrium equations, the boundary conditions, and the compatibility conditions.

Selected Parameters:

- K – Solution level, see S.7.3.3.

P.8.7 A Monoclinic Non-Homogeneous Beam Under Axially Distributed Non-Uniform Loads (II)

This program documents and verifies the expressions for a non-homogeneous MON13z beam under axially non-uniform distributed loads derived in S.7.3.3. The program verifies the single-value conditions of the biharmonic function.

Selected Parameters:

- K – Solution level, see S.7.3.3.

References

Abdelnaser, A. S., and Singh, M. P. 1993. Random vibrations of cantilevered composite beams with torsion-bending coupling. *Probabilistic Engineering Mechanics*, **8**(3,4), 143–151.

Abramovich, H. 1992. Shear deformation and rotary inertia effects of vibrating composite beams. *Composite Structures*, **20**(3), 165–173.

Abramovich, H., and Livshits, A. 1994. Free vibrations of nonsymmetric cross-ply laminated composite beams. *Journal of Sound and Vibration*, **176**(5), 597–612.

Ackermann, J. A., and Kozik, T. J. 1995. End effects in laminated anisotropic beams — part I. *Journal of Energy Resources Technology, Transactions of the ASME*, **117**(4), 279–284.

Adams, D. F, Carlsson, L. A., and Pipes, R. B. 2002. *Experimental characterization of advanced composite materials*. Third edn. CRC Press.

Akovali, G. 2001. *Handbook of composite fabrication*. Shawbury, UK: Rapra Technology Ltd.

Al-Amery, R. I. M., and Roberts, T. M. 1990. Nonlinear finite difference analysis of composite beams with partial interaction. *Computers and Structures*, **35**(1), 81–87.

Alkahe, J., and Rand, O. 2000. Analytic extraction of the elastic coupling mechanisms in composite blades. *Composite Structures*, **49**(4), 399–413.

Almansi, E. 1901. Sopra la deformazione dei cilindri sollecitati lateralmente. *Atti della Academia Nazionale dei Lincei*, **10**, Nota I: 333–338, Nota II: 400–408.

Ambartsumyan, S. A. 1970. *Theory of anisotropic plates*. Progress in Material Science, vol. II. Lancaster, PA, USA: Technomic Publishing.

Ambartsumyan, S. A. 1974. *The general theory of anisotropic shells*. Moscow, USSR: Nauka.

Armanios, E. A., and Badir, A. M. 1995. Free vibration analysis of anisotropic thin-walled closed-section beams. *AIAA Journal*, **33**(10), 1905–1910.

Ascione, L., and Fraternali, F. 1992. A penalty model for the analysis of curved composite beams. *Computers and Structures*, **45**(5–6), 985–999.

Ashbee, K. H. G. 1993. *Fundamental principles of fiber reinforced composites*. Second edn. Lancaster, PA , USA: Technomic Publishing.

Atanackovic, T. M., and Guran, A. 2000. *Theory of elasticity for scientists and engineers*. Cambridge, MA, USA: Birkhäuser Boston.

Atanasoff, H., and Vizzini, A. J. 1992. Foam tool manufacture of stiffness-coupled composite box beams. *SAMPE Quarterly*, **23**(Apr.), 37–42.

Atilgan, A. R., and Hodges, D. H. 1991. Unified nonlinear analysis for non-homogeneous anisotropic beams with closed cross sections. *AIAA Journal*, **29**(11), 1990–1999.

Baida, E. N. 1959. The problem of elastic state in orthotropic and isotropic parallelepiped. *Izv. VUZ, Stroitelstvo i Mehanika*, Jun., 28–42. (in Russian).

Banerjee, J. R., and Williams, F. W. 1995. Free vibration of composite beams — an exact method using symbolic computation. *Journal of Aircraft*, **32**(3), 636–642.

Bank, L. C. 1990. Modifications to beam theory for bending and twisting of open-section composite beams. *Composite Structures*, **15**(2), 93–114.

Bank, L. C., and Cofie, E. 1993. Coupled deflection and rotation of anisotropic open-section composite stiffeners. *Journal of Aircraft*, **30**(1), 139–141.

Bank, L. C., and Cofie, E. 1994. Hybrid force/stiffness matrix method for the analysis of thin-walled composite frames. *Composite Structures*, **28**(4), 391–404.

Barber, J. R. 1992. *Elasticity*. Dordrecht, The Netherlands: Kluwer Academic Publisher.

Barbero, E. J., Lopez-anido, R., and Davalos, J. F. 1993. On the mechanics of thin-walled laminated composite beams. *Journal of Composite Materials*, **27**(8), 806–829.

Barkai, S. M., and Rand, O. 1998. The Influence of Composite Induced Couplings on Tiltrotor Whirl Flutter Stability in Forward Flight. *Journal of The American Helicopter Society*, **43**(2), 133–145.

Bassiouni, A. S., Gad-Elrab, R. M., and Elmahdy, T. H. 1999. Dynamic analysis for laminated composite beams. *Composite Structures*, **44**(2–3), 81–87.

Bauld, N. R., and Tzeng, L. S. 1984. A Vlasov theory for fiber-reinforced beams with thin-walled open cross-sections. *International Journal of Solids and Structures*, **20**(3), 277–297.

Beltzer, A. I. 1990. *Variational and finite element methods, a symbolic computation approach*. New York, NY, USA: Springer-Verlag.

Berdichevsky, V., Armanios, E., and Badir, A. 1992. Theory of anisotropic thin-walled closed cross-section beams. *Composite Engineering*, **2**(5–7), 411–432.

Bhaskar, K., and Librescu, L. 1995. A geometrically non-linear theory for laminated anisotropic thin-walled beams. *International Journal of Engineering Science*, **33**(9), 1331–1344.

Bogomolov, A. V., and Borisenko, V. A. 1994. Improvement in the accuracy of determining shear moduli for orthotropic bars with twisting. *Strength of Materials*, **26**(3), 213–215.

Boresi, A. P., and Chong, K. P. 1999. *Elasticity in engineering mechanics*. Second edn. New York, NY, USA: John Wiley & Sons.

Brazier, L. G. 1927. On the flexure of thin cylindrical shells and other "thin" sections. *Proceedings of the Royal Society of London*, **A**(116), 104–114.

Breivik, N. L., Gurdal, Z., and Griffin, O. H. Jr. 1993. Compression of laminated composite beams with initial damage. *Journal of Reinforced Plastics and Composites*, **12**(7), 813–824.

Brush, D. O., and Almroth, B. O. 1975. *Buckling of bars, plates, and shells*. International student edn. New York, NY, USA: McGraw-Hill Companies.

Bull, J. W. (ed). 1995. *Numerical analysis and modelling of composite materials*. Dordrecht, The Netherlands: Kluwer Academic Publisher.

Calladine, C. R. 1983. *Theory of shell structures*. Cambridge University Press.

Cardoso, J. B., Sousa, L. G., Castro, J. A., and Valido, A. J. 2002. Optimal design of laminated composite beam structures. *Structural and Multidisciplinary Optimization*, **24**(3), 205–211.

Cartan, E. 1946. *Leçons sur la geometrie des espaces de Riemann*. Second edn. Cahiers scientifiques; fasc. 2. Paris, France: Gauthier–Villars.

Cesnik, C. E. S., and Hodges, D. H. 1997. VABS: A new concept for composite rotor blade cross-sectional modeling. *Journal of The American Helicopter Society*, **42**(1), 27–38.

Cesnik, C. E. S., Hodges, D. H., and Sutyrin, V. G. 1996. Cross-sectional analysis of composite beams including large initial twist and curvature effects. *AIAA Journal*, **34**(9), 1913–1920.

Chandiramani, N. K., Librescu, L., and Shete, C. D. 2002. On the free-vibration of rotating composite beams using a higher-order shear formulation. *Aerospace Science and Technology*, **6**(8), 545–561.

Chandra, R., and Chopra, I. 1991. Experimental and theoretical analysis of composite I-beam with elastic couplings. *AIAA Journal*, **29**(12), 2197–2206.

Chandra, R., and Chopra, I. 1992a. Experimental-theoretical investigation of the vibration characteristics of rotating composite box beams. *Journal of Aircraft*, **29**(4), 657–664.

Chandra, R., and Chopra, I. 1992b. Structural response of composite beams and blades with elastic couplings. *Composite Engineering*, **2**(5–7), 347–374.

Chandra, R., and Chopra, I. 1993. Analytical-experimental investigation of free-vibration characteristics of rotating composite I-beams. *Journal of Aircraft*, **30**(6), 927–934.

Chandra, R., Stemple, A. D., and Chopra, I. 1990. Thin-walled composite beams under bending, torsional, and extensional loads. *Journal of Aircraft*, **27**(7), 619–626.

Chandrashekhara, K., and Bangera, K. M. 1993a. Linear and geometrically non-linear analysis of composite beams under transverse loading. *Composite Science and Technology*, **47**(4), 339–347.

Chandrashekhara, K., and Bangera, K. M. 1993b. Vibration of symmetrically laminated clamped-free beam with a mass at the free end. *Journal of Sound and Vibration*, **160**(1), 93–101.

Chandrashekhara, K., Krishnamurthy, K., and Roy, S. 1990. Free vibration of composite beams including rotary inertia and shear deformation. *Composite Structures*, **14**(4), 269–279.

Chen, L., Ifju, P. G., and Sankar, B. V. 2001. A novel double cantilever beam test for stitched composite laminates. *Journal of Composite Materials*, **35**(13), 1137–1149.

Chen, W. F., and Saleeb, A. F. 1994. *Constitutive equations for engineering materials*. Vol. I. Amsterdam, The Netherlands: Elsevier.

Chen, W. H., and Gibson, R. F. 1998. Property distribution determination for nonuniform composite beams from vibration response measurements and Galerkin's method. *Journal of Applied Mechanics*, **65**(1), 127–133.

Chernykh, K. F., and Kulman, N. K. 1998. *An introduction to modern anisotropic elasticity*. New York, NY, USA: Begell House.

Cheung, V. K., and Zhou, D. 1999. The free vibrations of rectangular composite plates with point-supports using static beam functions. *Composite Structures*, **44**(2–3), 145–154.

Chou, P. C., and Pagano, N. J. 1992. *Elasticity: tensor, dyadic, and engineering approaches*. New York, NY, USA: Dover Publications.

Ciarlet, P. G. 1988. *Mathematical elasticity volume I: three-dimensional elasticity*. Vol. I. Amsterdam, The Netherlands: North–Holland.

Ciarlet, P. G. 1997. *Mathematical elasticity volume II: theory of plates*. Vol. II. Amsterdam, The Netherlands: North–Holland.

Conceicao Antonio, C. A., Trigo Barbosa, J., and Simas Dinis, L. 2000. Optimal design of beam reinforced composite structures under elasto-plastic loading conditions. *Structural and Multidisciplinary Optimization*, **19**(1), 50–63.

Cortinez, V. H., and Machado, S. 2001. A two-dimensional analysis of anisotropic vibrating beams. *Journal of Sound and Vibration*, **242**(3), 553–558.

Courant, R., and Hilbert, D. 1989. *Methods in mathematical physics*. Vol. I & II. New York, NY, USA: John Wiley & Sons.

Creaghan, S. G., and Palazotto, A. N. 1994. Nonlinear large displacement and moderate rotational characteristics of composite beams incorporating transverse shear strain. *Computers and Structures*, **51**(4), 357–371.

Dancila, D. S., Armanios, E. A., and Lentz, W. K. 1999. Free vibration of thin-walled closed-section composite beams with optimum and near-optimum coupling. *Journal of Thermoplastic Composite Materials*, **12**(1), 2–12.

Daniel, I. M., and Ishai, O. 1994. *Engineering mechanics of composite materials*. New York, NY, USA: Oxford University Press.

Davalos, J. F., Kim, Y., and Barbero, E. J. 1994. Analysis of laminated beams with a layer-wise constant shear theory. *Composite Structures*, **28**(3), 241–253.

Davi, G., and Milazzo, A. 1999. Bending stress fields in composite laminate beams by a boundary integral formulation. *Computers and Structures*, **71**(3), 267–276.

Demakos, C. B. 2003. Stress fields in fiber reinforced laminate beams due to bending and torsion moments. *Journal of Reinforced Plastics and Composites*, **22**(5), 399–418.

Devries, F., Dumontet, H., Duvaut, G., and Lene, F. 1989. Homogenization and damage for composite structures. *International Journal for Numerical Methods in Engineering*, **27**, 285–298.

Doghri, I. 2000. *Mechanics of deformable solids*. Berlin, Germany: Springer-Verlag.

Dostal, C. A. 1987. *Engineered materials handbook*. Vol. 1. Metals Park, Ohio, USA: ASM International.

Dudchenko, A., Lurie, S., and Obraztsov, I. 1984. *Anisotropic many-layered plates and shells, reviews of sciences and technics*. Mechanics of solids, vol. 15. Moscow, USSR: VINITI.

Dzhanelidze, G. Yu. 1960. The Almansi Problem. *Proceedings of Leningrad Polytechnic Institute*, **210**, 25–38. (in Russian).

Ecsedi, I. 2004. Elliptic cross section without warping under torsion. *Mechanics Research Communications*, **31**(2), 147–150.

Eisenberger, M., Abramovich, H., and Shulepov, O. 1995. Dynamic stiffness analysis of laminated beams using a first order shear deformation theory. *Composite Structures*, **31**(4), 265–271.

Evrard, T., Butler, R., Hughes, S. W., and Banerjee, J. R. 2000. Ply angle optimization of nonuniform composite beams subject to aeroelastic constraints. *AIAA Journal*, **38**(10), 1992–1994.

Filonenko-Borodich, M. 1965. *Theory of elasticity*. New York, NY, USA: Dover Publications.

Fish, J., and Markolefas, S. 1994. Adaptive global-local refinement strategy based on the interior error estimates of the *h*-method. *International Journal for Numerical Methods in Engineering*, **37**, 827–838.

Fish, J., Markolefas, S., Guttal, R., and Nayah, P. 1994. On adaptive multilevel superposition of finite element meshes for linear elastostatics. *Applied Numerical Mathematics*, **14**, 135–164.

Floros, M., and Smith, E. C. 1997. Finite element modeling of open-section composite beams with warping restraint effects. *AIAA Journal*, **35**(8), 1341–1347.

Frisch-Fay, R. 1962. *Flexible bars*. Washington, D.C., USA: Butterworth & Co. Publishers.

Gadelrab, R. M. 1996. The effect of delamination on the natural frequencies of a laminated composite beam. *Journal of Sound and Vibration*, **197**(3), 283–292.

Gandhi, F., and Lee, S. W. 1992. A composite beam finite element model with *p*-version assumed warping displacement. *Composite Engineering*, **2**(5–7), 329–345.

Ghosh, S., Lee, K., and Moorthy, S. 1995. Multiple scale analysis of heterogeneous elastic structures using homogenization theory and Voronoi cell finite element method. *International Journal of Solids and Structures*, **32**(1), 27–62.

Ghugal, Y. M., and Shimpi, R. P. 2001. A review of refined shear deformation theories for isotropic and anisotropic laminated beams. *Journal of Reinforced Plastics and Composites*, **20**(3), 255–272.

Gibson, R. F. 1994. *Principles of composite material mechanics.* New York, NY, USA: McGraw-Hill Companies.

Giurgiutiu, V., and Reifsnider, K. L. 1994. Development of strength theories for random fiber composites. *Journal of Composites Technology and Research*, **16**(2), 103–114.

Gjelsvik, A. 1981. *The theory of thin walled bars.* New York, NY, USA: John Wiley & Sons.

Gould, P. L. 1994. *Introduction to linear elasticity.* Second edn. Berlin, Germany: Springer-Verlag.

Grabmeier, J., Kaltofen, E., and Weispfenning, V. (eds). 2003. *Computer algebra handbook. Foundations, applications, systems.* Berlin/Heidelberg, Germany: Springer-Verlag.

Grebshtein, M., Kazar (Kezerashvili), M., Rovenski, V., and Rand, O. 2004. Non-homogeneous uncoupled beam under axially distributed loads. *Science and Engineering of Composite Materials (in Press).* Also: Proceedings of *The 24th Congress of the International Council of the Aeronautical Sciences*, 29 August - 3 September 2004, Yokohama, Japan.

Green, A. E., and Zerna, W. 1992. *Theoretical elasticity.* Second edn. New York, NY, USA: Dover Publications.

Halpin, J. C. 1992. *Primer on composite materials analysis.* Second edn. Lancaster, PA, USA: Technomic Publishing.

Harbert, S. J., and Hogan, H. A. 1992. An analysis of curvature and layup effects on delamination in notched composite beams. *Journal of Reinforced Plastics and Composites*, **11**(4), 443–457.

Harrison, C., and Butler, R. 2001. Locating delaminations in composite beams using gradient techniques and a genetic algorithm. *AIAA Journal*, **39**(7), 1383–1389.

Hearmon, R. F. S. 1961. *An introduction to applied anisotropic elasticity.* London, UK: Oxford University Press.

Heck, A. 2003. *Introduction to Maple.* Third edn. New York, NY, USA: Springer-Verlag.

Heredia, F. E., He, M. Y., and Evans, A. G. 1996. Mechanical performance of ceramic matrix composite I-beams. *Composites Part A*, **27**(12), 1157–1167.

Hodges, D. H. 2003. Geometrically exact, intrinsic theory for dynamics of curved and twisted anisotropic beams. *AIAA Journal*, **41**(6), 1131–1137.

Hodges, D. H., Shang, X., and Cesnik, C. E. S. 1996. Finite element solution of nonlinear intrinsic equations for curved composite beams. *Journal of The American Helicopter Society*, **41**(4), 313–321.

Horgan, C. O. 1972. On Saint-Venant's principle in anisotropic elasticity. *Journal of Elasticity*, **2**(3), 169–180.

Hull, D., and Clyne, T. W. 1996. *An introduction to composite materials*. New York, NY, USA: Cambridge University Press.

Icardi, U., Di Sciuva, M., and Librescu, L. 2000. Dynamic response of adaptive cross-ply cantilevers featuring interlaminar bonding imperfections. *AIAA Journal*, **38**(3), 499–506.

Iesan, D. 1987. *Saint-Venant's problem*. Lecture notes in mathematics. Berlin/Heidelberg, Germany: Springer-Verlag.

Ip, K. H., and Tse, P. C. 2001. Determination of dynamic flexural and shear moduli of thick composite beams using natural frequencies. *Journal of Composite Materials*, **35**(17), 1553–1569.

Jenkins, C. H. 1998. *Manual on experimental methods of mechanical testing of composites*. Second edn. Lilburn, GA, USA: The Fairmont Press.

Jeon, S. M., Cho, M. H., and Lee, I. 1995. Static and dynamic analysis of composite box beams using large deflection theory. *Computers and Structures*, **57**(4), 635–642.

Johnson, E. R., Vasiliev, V. V., and Vasiliev, D. V. 2001. Anisotropic thin-walled beams with closed cross-sectional contours. *AIAA Journal*, **39**(12), 2389–2393.

Jones, R. M. 1999. *Mechanics of composite materials*. Second edn. Materials science and engineering series. Philadelphia, PA, USA: Taylor & Francis Group.

Jung, S. N., Nagaraj, V. T., and Chopra, I. 2001. Refined structural dynamics model for composite rotor blades. *AIAA Journal*, **39**(2), 339–348.

Kalamkarov, A. L., and Kolpakov, A. G. 1997. *Analysis, design and optimization of composite structures*. London, UK: John Wiley & Sons.

Kalfon, J. P., and Rand, O. 1993. Nonlinear analysis of composite thin-walled helicopter blades. *Computers and Structures*, **48**(1), 51–61.

Karczmarzyk, S. 1996. An exact elastodynamic solution to vibration problems of a composite structure in the plane stress state. *Journal of Sound and Vibration*, **196**(1), 85–96.

Kardomateas, G. A. 1991. End force loading of generally anisotropic curved beams with linearly varying elastic constants. *International Journal of Solids and Structures*, **27**(1), 59–71.

Kazar (Kezerashvili), M., Rovenski, V., Grebshtein, M., and Rand, O. 2004. Nonhomogeneous uncoupled beam under tip loads. *Science and Engineering of Composite Materials (in Press)*. Also: Proceedings of *The 59th Annual Forum of the American Helicopter Society*, 6-8 May, 2003, Arizona, USA.

Kelly, A., and Zweben, C. (eds). 2000. *Comprehensive composite materials*. Vol. 1–6. Amsterdam, The Netherlands: Elsevier.

Kezerashvili, M. I. 1986. On the generalized Neumann problem for an homogeneous anisotropic bar with elliptic cross-section. *Transactions of Georgian Polytechnical Institute*, **6**(303), 74–76.

Khdeir, A. A., and Reddy, J. N. 1994. Free vibration of cross-ply laminated beams with arbitrary boundary conditions. *International Journal of Engineering Science*, **32**(12), 1971–1980.

Kim, C., and White, S. R. 1997. Constrained warping of thin-walled hollow composite beams. *AIAA Journal*, **35**(6), 1082–1084.

King, J. L. 1991. An improved theory for the free transverse vibrations of anisotropic beams. *Journal of Sound and Vibration*, **148**(3), 493–506.

King, Y. F., and Chan, W. S. 1993. Sensitivity analysis of delamination characterization in composite tapered beams. *Journal of Reinforced Plastics and Composites*, **12**(4), 386–403.

Ko, K. E., and Kim, J. H. 2003. Thermally induced vibrations of spinning thin-walled composite beam. *AIAA Journal*, **41**(2), 296–303.

Kocks, U. F., Tomé, C. N., and Wenk, H. R. 1998. *Texture and anisotropy: preferred orientations in polycrystals and their effect on materials properties*. Cambridge, UK: Cambridge University Press.

Kollár, L. P., and Pluzsik, A. 2002. Analysis of thin-walled composite beams with arbitrary layup. *Journal of Reinforced Plastics and Composites*, **21**(16), 1423–1465.

Kollár, L. P., and Springer, G. S. 2003. *Mechanics of composite structures*. New York, NY, USA: Cambridge University Press.

Kolpakov, A. G. 2004. *Stressed composite structures homogenized models for thin-walled nonhomogeneous structures with initial stresses*. Foundations of engineering mechanics. Berlin, Heidelberg, New York: Springer-Verlag.

Koo, J. S., and Kwak, B. M. 1994. A laminated composite beam element separately interpolated for the bending and shearing deflections without increase in nodal DOF. *Computers and Structures*, **53**(5), 1091–1098.

Kosmodamianskii, A. S. 1956. Bending of anisotropic beam under generic distributed load. *Engineering Bulletin, Institute of Mechanics, Academy of Sciences, SSSR*, **24**, 114–126. (in Russian).

Krawczuk, M., Ostachowicz, W., and Zak, A. 1996. Analysis of natural frequencies of delaminated composite beams based on finite element method. *Structural Engineering and Mechanics*, **4**(3), 243–255.

Krishnaswamy, S., Chandrashekhara, K., and Wu, W. Z. B. 1992. Analytical solutions to vibration of generally layered composite beams. *Journal of Sound and Vibration*, **159**(1), 85–99.

Krylov, V., and Sorokin, S. V. 1997. Dynamics of elastic beams with controlled distributed stiffness parameters. *Smart Materials and Structures*, **6**(5), 573–582.

Lai, H. Y., Fallahi, B., Gupta, R., and Bo, S. 1993. Dynamic characterization of thin-walled laminated channel beams with warping. *Computers and Structures*, **48**(6), 1137–1151.

Lakes, R. S. 1987. Foam structures with a negative Poisson's ratio. *Science*, **235**, 1038–1040.

Landau, L. D., and Lifschitz, E. M. 1986. *Theory of elasticity*. Third edn. Vol. 7. Oxford, UK: Pergamon Press.

Langhaar, H. L. 1962. *Energy methods in applied mechanics*. New York, NY, USA: John Wiley & Sons.

Lee, C. Y., Liu, D., and Lu, X. 1992. Static and vibration analysis of laminated composite beams with an interlaminar shear stress continuity theory. *International Journal for Numerical Methods in Engineering*, **33**(2), 409–424.

Lekhnitskii, S. G. 1968. *Anisotropic plates*. New York, NY, USA: Gordon and Breach, Science Publishers.

Lekhnitskii, S. G. 1981. *Theory of elasticity of an anisotropic body*. Moscow, USSR: Mir Publishers.

Lentz, W. K., and Armanios, E. A. 1998. Optimum coupling in thin-walled, closed-section composite beams. *Journal of Aerospace Engineering*, **11**(3), 81–89.

Li, W. J., and Chen, J. Y. 1993. Optimum design of composite structures subjected to multiple displacement constraints. *Journal of Strain Analysis for Engineering Design*, **28**(2), 135–141.

Liao, K., Tenek, L. H., and Reifsnider, K. L. 1992. Detection of stress fields and delaminations in vibrating composite beams by a thermoelastic technique. *Journal of Composites Technology and Research*, **14**(3), 176–181.

Liao, Y. S. 1993. A generalized method for the optimal design of beams under flexural vibration. *Journal of Sound and Vibration*, **167**(2), 193–202.

Librescu, L., and Song, O. 1991. Behavior of thin-walled beams made of advanced composite materials and incorporating non-classical effects. *Applied Mechanics Reviews*, **44**(11/2), 174–180.

Librescu, L., and Song, O. 1992. On the static aeroelastic tailoring of composite aircraft swept wings modelled as thin-walled beam structures. *Composite Engineering*, **2**(5–7), 497–512.

Liebenson, L. S. 1940. Some problems of the bending and torsion of the anisotropic prisms. *Uch. Zapiski Moskovskogo Universiteta*, **46**, 139–160.

Liu, S., Kutlu, Z., and Chang, F. K. 1993. Matrix cracking and delamination in laminated composite beams subjected to a transverse concentrated line load. *Journal of Composite Materials*, **27**(5), 436–470.

Loughlan, J., and Ata, M. 1997. The Constrained torsional characteristics of some carbon fibre composite box-beams. *Thin–Walled Structures*, **28**(3–4), 233–252.

Love, A. E. H. 1944. *A treatise on the mathematical theory of elasticity*. Fourth edn. New York, NY, USA: Dover Publications.

Lu, J. K. 1995. *Complex variable methods in plane elasticity*. Singapore: World Scientific Publ.

Lurie, S. A., and Vasiliev, V. V. 1995. *Biharmonic problem of the theory of elasticity*. London, UK: Gordon and Breach, Science Publishers.

Mahajan, P. 1998. Contact behavior of an orthotropic laminated beam indented by a rigid cylinder. *Composite Science and Technology*, **58**(3–4), 505–513.

Mahapatra, D. R., Gopalakrishnan, S., and Sankar, T. S. 2000. Spectral element based solutions for wave propagation analysis of multiply connected unsymmetric laminated composite beams. *Journal of Sound and Vibration*, **237**(5), 819–836.

Mahapatra, D. R., Nag, A., and Gopalakrishnan, S. 2002. Identification of delamination in composite beams using spectral estimation and a genetic algorithm. *Smart Materials and Structures*, **11**(6), 899–908.

Mallick, P. K. 1993. *Fiber-reinforced composites*. New York, NY, USA: Marcel Dekker.

Mallick, P. K. 1997. *Composites engineering*. New York, NY, USA: Marcel Dekker.

Manne, P. M., and Tsai, S. W. 1998. Design optimization of composite plates: Part II — structural optimization by plydrop tapering. *Journal of Composite Materials*, **32**(6), 572–578.

Maple, 9. 2003. *Learning guide*. Waterloo, Canada: Waterloo Maple Inc.

Massa, J. C., and Barbero, E. J. 1998. Strength of materials formulation for thin walled composite beams with torsion. *Journal of Composite Materials*, **32**(17), 1560–1594.

Mazor, D., and Rand, O. 2000. The influence of the in-plane warping on the behavior of thin-walled beams. *Thin–Walled Structures*, **37**(4), 363–390.

McCarthy, T. R., and Chattopadhyay, A. 1998. Investigation of composite box beam dynamics using a higher-order theory. *Composite Structures*, **41**(3–4), 273–284.

Michell, J. H. 1901. The theory of uniformly loaded beams. *Journal of Mathematics*, **32**, 28–42.

Milazzo, A. 2000. Interlaminar stresses in laminated composite beam-type structures under shear/bending. *AIAA Journal*, **38**(4), 687–694.

Miller, D. A., and Palazotto, A. N. 1995. Nonlinear finite element analysis of composite beams and arches using a large rotation theory. *Finite Elements in Analysis and Design*, **19**(3), 131–152.

Milne-Thomson, L. M. 1960. *Plane elastic systems*. Berlin, Germany: Springer-Verlag.

Miravete, A. 1996. *Optimisation of composite structures design*. Cambridge, UK: Woodhead Publishing.

Mitchell, J. A., and Reddy, J. N. 1998. A multilevel hierarchical preconditioner for thin elastic solids. *International Journal for Numerical Methods in Engineering*, **43**, 1383–1400.

Mohamed Nabi, S., and Ganesan, N. 1994. A generalized element for the free vibration analysis of composite beams. *Computers and Structures*, **51**(5), 607–610.

Murakami, H., and Yamakawa, J. 1996. On approximate solutions for the deformation of plane anisotropic beams. *Composites Part B*, **27B**(5), 493–504.

Murakami, H., Reissner, E., and Yamakawa, J. 1996. Anisotropic beam theories with shear deformation. *Journal of Applied Mechanics*, **63**(3), 660–668.

Muravskii, G. B. 2001. *Mechanics of non-homogeneous and anisotropic foundations.* New York, NY, USA: Springer-Verlag.

Murray, N. W. 1986. *Introduction to the theory of thin-walled structures.* New York, NY, USA: Clarendon Press.

Muskhelishvili, N. I. 1953. *Some basic problems of the mathematical theory of elasticity.* Groningen, The Netherlands: P. Noordhoff.

Nag, A., Mahapatra, D. R., and Gopalakrishnan, S. 2002. Identification of delamination in composite beams using spectral estimation and a genetic algorithm. *Smart Materials and Structures*, **11**(6), 899–908.

Nagaraj, V. T. 1996. Approximate formula for the frequencies of a rotating Timoshenko beam. *Journal of Aircraft*, **33**(3), 637–639.

Naik, N. K., and Ganesh, V. K. 1993. Optimum design studies on FRP beams with holes. *Composite Structures*, **24**(1), 59–66.

Nouri, T., and Gay, D. 1994. Shear stresses in orthotropic composite beams. *International Journal of Engineering Science*, **32**(10), 1647–1667.

Novozhilov, V. V. 1961. *Theory of elasticity.* London, UK: Pergamon Press.

Ochoa, O. O., and Reddy, J. N. 1992. *Finite element analysis of composite laminates.* Dordrecht, The Netherlands: Kluwer Academic Publisher.

Omidvar, B. 1998. Shear coefficient in orthotropic thin-walled composite beams. *Journal of Composites for Construction*, **2**(1), 46–56.

Oprea, J. 1997. *Differential geometry and its applications.* New Jersey, USA: Prentice Hall.

Ozdil, F., and Carlsson, L. A. 1999. Beam analysis of angle-ply laminate DCB specimens. *Composite Science and Technology*, **59**(2), 305–315.

Pagano, N. J., and Schoeppner, G. A. 2000. *Comprehensive Composite Materials.* Vol. 2. Elsevier. Chap. Delamination of polymer matrix composites: problems and assessment, pages 443–528.

Pal, N. C., and Ghosh, A. K. 1993. An experimental investigation on impact response of laminated composite beams. *Experimental Mechanics*, **33**(2), 159–163.

Pardini, L. C., Neto, F. L., and McEnaney, B. 2000. Modelling of mechanical properties of CRFC composites under flexure loading. *Journal of the Brazilian Society of Mechanical Sciences*, **22**(2), 272–291.

Parton, V. Z., and Kudryavtsev, B. A. 1993. *Engineering mechanics of composite structures.* Boca Raton, Florida, USA: CRC Press.

Parton, V. Z., and Perlin, P. I. 1984. *Mathematical methods of the theory of elasticity.* Vol. I & II. Moscow, USSR: Mir Publishers.

Pinheiro, M. A., and Sankar, B. V. 2000. Beam finite element for analyzing free edge delaminations. *Journal of Thermoplastic Composite Materials*, **13**(4), 272–291.

Pluzsik, A., and Kollár, L. P. 2002. Effects of shear deformation and restrained warping on the displacements of composite beams. *Journal of Reinforced Plastics and Composites*, **21**(17), 1517–1541.

Pollock, G. D., Zak, A. R., Hilton, H. H., and Ahmad, M. F. 1995. Shear center for elastic thin-walled composite beams. *Structural Engineering and Mechanics*, **3**(1), 91–103.

Ponte-Castaneda, P. 1996. Exact second-order estimates for the effective mechanical properties of nonlinear composite materials. *Journal of the Mechanics and Physics of Solids*, **44**(6), 827–862.

Ponte-Castaneda, P., and Willis, J. R. 1995. The effect of spatial distribution on the effective behavior of composite materials and cracked media. *Journal of the Mechanics and Physics of Solids*, **43**(12), 1919–1951.

Prathap, G., Vinayak, R. U., and Naganarayana, B. P. 1996. Beam elements based on a higher order theory — II. Boundary layer sensitivity and stress oscillations. *Computers and Structures*, **58**(4), 791–796.

Puspita, G., Barrau, J. J., and Gay, D. 1993. Computation of flexural and torsional homogeneous properties and stresses in composite beams with orthotropic phases. *Composite Structures*, **24**(1), 43–49.

Qatu, M. S. 1992. In-plane vibration of slightly curved laminated composite beams. *Journal of Sound and Vibration*, **159**(2), 327–338.

Qiao, P., Barbero, E. J., and Davalos, J. F. 2000. On the linear viscoelasticity of thin-walled laminated composite Beams. *Journal of Composite Materials*, **34**(1), 39–68.

Qin, Z., and Librescu, L. 2001. Static and dynamic validations of a refined thin-walled composite beam model. *AIAA Journal*, **39**(12), 2422–2424.

Qin, Z., and Librescu, L. 2003. Dynamic aeroelastic response of aircraft wings modeled as anisotropic thin-walled beams. *Journal of Aircraft*, **40**(3), 532–543.

Ramtekkar, G. S., M., Desai Y., and H., Shah A. 2002. Natural vibrations of laminated composite beams by using mixed finite element modelling. *Journal of Sound and Vibration*, **257**(4), 635–651.

Rand, O. 1991a. Experimental investigation of periodically excited rotating composite rotor blades. *Journal of Aircraft*, **28**(12), 876–884.

Rand, O. 1991b. Periodic response of thin-walled composite helicopter rotor blades. *Journal of The American Helicopter Society*, **36**(4), 3–11.

Rand, O. 1992. Exact solution for general torsion problems using boundary singularities. *Journal of Engineering Mechanics*, **118**(10), 2141–2147.

Rand, O. 1994a. Free vibration of thin-walled composite blades. *Composite Structures*, **28**(2), 169–180.

Rand, O. 1994b. Nonlinear analysis of orthotropic beams of solid cross-sections. *Composite Structures*, **29**(1), 27–45.

Rand, O. 1995. Experimental study of the natural frequencies of rotating thin-walled composite blades. *Thin–Walled Structures*, **21**(2), 191–207.

Rand, O. 1998a. Fundamental closed-form solutions for solid and thin-walled composite beams including a complete out-of-plane warping model. *International Journal of Solids and Structures*, **35**(21), 2775–2793.

Rand, O. 1998b. Interlaminar shear stresses in solid composite beams using a complete out-of-plane shear deformation model. *Computers and Structures*, **66**(6), 713–723.

Rand, O. 1998c. Similarities between solid and thin-walled composite beams by analytic approach. *Journal of Aircraft*, **35**(4), 604–615.

Rand, O. 2000a. Bending and extension of thin-walled composite beams of open cross-sectional geometry. *Journal of Applied Mechanics*, **67**(4), 813–818.

Rand, O. 2000b. In-plane warping effects in thin-walled box-beams. *AIAA Journal*, **38**(3), 542–544.

Rand, O. 2000c. On the importance of cross-sectional warping in solid composite beams. *Composite Structures*, **49**(4), 393–397.

Rand, O. 2001a. A multilevel analysis of solid laminated composite beams. *International Journal of Solids and Structures*, **38**(22–23), 4017–4043.

Rand, O. 2001b. Nonlinear in-plane warping deformation in elastically coupled open thin-walled beams. *Computers and Structures*, **79**(3), 281–291.

Rand, O., and Barkai, S. M. 1997. Dynamic characterization of thin-walled laminated channel beams with warping. *Composite Structures*, **39**(1–2), 39–54.

Rappel, O., and Rand, O. 2000. Analysis of elastically coupled thick-walled composite blades. *International Journal of Solids and Structures*, **37**(7), 1019–1043.

Ratcliffe, C. P., and Bagaria, W. J. 1998. Vibration technique for locating delamination in a composite beam. *AIAA Journal*, **36**(6), 1074–1077.

Ray, C., and Satsangi, S. K. 1999. Laminated stiffened plate — a first ply failure analysis. *Journal of Reinforced Plastics and Composites*, **18**(12), 1061–1076.

Reddy, J. N. 1999. *Theory and analysis of elastic plates*. Philadelphia, PA, USA: Taylor & Francis Group.

Reifsnider, K. L., and Case, S. W. 2002. *Damage tolerance and durability of material systems*. New York, NY, USA: John Wiley & Sons.

Reismann, H., and Pawlik, P. S. 1991. *Elasticity: theory and applications*. Vol. I. Melbourne: Krieger Publishing.

Rovenski, V. 2000. *Geometry of curves and surfaces with Maple*. Cambridge, MA, USA: Birkhäuser Boston.

Rovenski, V., and Rand, O. 2001. Analysis of anisotropic beams: an analytic approach. *Journal of Applied Mechanics*, **68**(4), 674–678.

Rovenski, V., and Rand, O. 2003. *Beams of general anisotropy with axially distributed loads*. TAE Report, no. 945. Haifa, Israel: Technion - Israel Institute of Technology.

Ruchadze, A. K. 1975. On one problem of elastic equilibrium of homogeneous isotropic prismatic bar. *Transactions of Georgian Polytechnical Institute*, **3**(176), 208–218. (in Russian).

Ruchadze, A. K. 1978. On the centre of bending of composite prismatic bars of different anisotropic material. *Bulletin of the Academy of Sciences of the Georgian SSR*, **90**(2), 317–320. (in Russian).

Ruchadze, A. K., and Berekashvili, R. A. 1980. On one generalized problem of Almansi. *Bulletin of the Academy of Sciences of the Georgian SSR*, **100**(3), 561–564. (in Russian).

Saada, A. S. 1993. *Elasticity Theory and Applications*. Second edn. Malabar, Florida, USA: Krieger Publishing Company.

Sagan, H. 1969. *Introduction to the calculus of variations*. New York, NY, USA: Dover Publications.

Sankar, B. V. 1991. A finite element for modeling delaminations in composite beams. *Computers and Structures*, **38**(2), 239–246.

Sankar, B. V., and Hu, S. 1991. Dynamic delamination propagation in composite beams. *Journal of Composite Materials*, **25**(11), 1414–1426.

Sankar, B. V., and Marrey, R. V. 1993. A unit-cell model of textile composite beams for predicting stiffness properties. *Composite Science and Technology*, **49**(1), 61–69.

Savoia, M. 1996. On the accuracy of one-dimensional models for multilayered composite beams. *International Journal of Solids and Structures*, **33**(4), 521–544.

Savoia, M., Laudiero, F., and Tralli, A. 1993. A refined theory for laminated beams: part I: new high order approach & part II:an iterative variational approach. *Meccanica*, **28**(Oct.), 39–51, 217–255.

Segura, J. M. 1990. An approximate method of determination of shear stresses due to flexure in composite beams. *International Journal of Engineering Science*, **28**(8), 735–750.

Sherman, D. I. 1943. Plane deformation in isotropic, inhomogeneous media. *Prikladnaya Matematika i Mehanika*, **7**, 301–309. (in Russian).

Shi, G., Lam, K. Y., and Tay, T. E. 1998. On efficient finite element modeling of composite beams and plates using higher-order theories and an accurate composite beam element. *Composite Structures*, **41**(2), 159–165.

Shi, Y. B., and Hull, D. 1992. Fracture of delaminated unidirectional composite beams. *Journal of Composite Materials*, **26**(15), 2172–2195.

Shi, Y. B., and Yee, A. F. 1994. Intraply crack and delamination interaction in laminate beams under transverse loading. *Composite Structures*, **29**(3), 287–297.

Shimpi, R. P., and Ghugal, Y. M. 1999. Layerwise trigonometric shear deformation theory for two layered cross-ply laminated beams. *Journal of Reinforced Plastics and Composites*, **18**(16), 1516–1543.

Shokrieh, M. M., and Rezaei, D. 2003. Analysis and optimization of a composite leaf spring. *Composite Structures*, **60**(3), 317–325.

Sierakowski, R. L., and Chaturvedi, S. K. 1997. *Dynamic loading and characterization of fiber-reinforced composites*. New York, NY, USA: John Wiley & Sons.

Singh, G., Rao, G. V., and Iyengar, N. G. R. 1991. Analysis of the nonlinear vibrations of unsymmetrically laminated composite beams. *AIAA Journal*, **29**(Oct.), 1727–1735.

Singh, G., Rao, G. V., and Iyengar, N. G. R. 1992. Nonlinear bending of thin and thick unsymmetrically laminated composite beams using refined finite element model. *Computers and Structures*, **42**(4), 471–479.

Slaughter, W. S. 2001. *Linearized theory of elasticity*. Basel: Birkhäuser-Verlag.

Smith, S. J., and Bank, L. C. 1992. Modifications to beam theory for bending and twisting of open-section composite beams — experimental verification. *Composite Structures*, **22**(3), 169–177.

Sobolev, S. L. 1937. On a regional problem for polyharmonic equations. *Matematiczeski Sbornik*, **2(44)**(3), 465–499. (in Russian).

Sokolnikoff, I. S. 1983. *Mathematical theory of elasticity*. Second edn. New York, NY, USA: McGraw-Hill Companies.

Soldatos, K. P., and Watson, P. 1997. A general theory for the accurate stress analysis of homogeneous and laminated composite beams. *International Journal of Solids and Structures*, **34**(22), 2857–2885.

Song, O., and Librescu, L. 1993. Free vibration of anisotropic composite thin-walled beams of closed cross-section contour. *Journal of Sound and Vibration*, **167**(1), 129–147.

Song, O., and Librescu, L. 1995. Dynamic theory of open cross-section thin-walled beams composed of advanced composite materials. *Journal of Thermoplastic Composite Materials*, **8**(2), 225–238.

Song, O., and Librescu, L. 1997a. Anisotropy and structural coupling on vibration and instability of spinning thin-walled beams. *Journal of Sound and Vibration*, **204**(3), 477–494.

Song, O., and Librescu, L. 1997b. Structural modeling and free vibration analysis of rotating composite thin-walled beams. *Journal of The American Helicopter Society*, **42**(4), 358–369.

Soutas-Little, R. W. 1999. *Elasticity*. New York, NY, USA: Dover Publications.

Srinivasan, S., Biggers, S. B., and Latour, R. A. 1997. Analysis of composite structures using the 3-D global/3-D local method. *Journal of Reinforced Plastics and Composites*, **16**(4), 353–371.

Steeds, J. W. 1973. *Introduction to anisotropic elasticity theory of dislocations*. Oxford, UK: Oxford University Press.

Steif, P. S., and Trojnacki, A. 1994. Bend strength versus tensile strength of fiber-reinforced ceramics. *Journal of the American Ceramic Society*, **77**(1), 221–229.

Stronge, W. J., and Yu, T. X. 1993. *Dynamic models for structural plasticity*. London, UK: Springer-Verlag.

Stuart, L. M. 1996. *International encyclopedia of composites*. Second edn. Vol. 1–6. John Wiley & Sons.

Subrahmanyam, K. B. 1993. Analysis of thin-walled composite beams by energy method. *Journal of Reinforced Plastics and Composites*, **12**(6), 642–669.

Suresh, J. K., and Nagaraj, V. T. 1996. Higher-order shear deformation theory for thin-walled composite beams. *Journal of Aircraft*, **33**(5), 978–986.

Suresh, J. K., Venkatesan, C., and Ramamurti, V. 1990. Structural dynamic analysis of composite beams. *Journal of Sound and Vibration*, **143**(3), 503–519.

Suresh, R., and Malhotra, S. K. 1997. Some studies on static analysis of composite thin-walled box beam. *Computers and Structures*, **62**(4), 625–634.

Suzuki, S. 1990. On the strength of a composite beam. *Journal of Sound and Vibration*, **136**(2), 315–321.

Tanghe Carrier, F. 2000. Remarkable property of the Saint-Venant warping function for orthotropic composite beams. *Mechanics Research Communications*, **27**(2), 143–148.

Tarnopol'skii, Y. M., Zhigun, I. G., and Polyakov, V. A. 1992. *Spatially reinforced composites*. Lancaster, PA, USA: Technomic Publishing.

Taylor, J. M., and Butler, R. 1997. Optimum design and validation of flat composite beams subject to frequency constraints. *AIAA Journal*, **35**(3), 540–545.

Timoshenko, S. P., and Goodier, J. N. 1970. *Theory of elasticity*. Third edn. Tokyo, Japan: McGraw-Hill Companies.

Ting, T. C. T. 1996. *Anisotropic elasticity: theory and applications*. Oxford, UK: Oxford University Press.

Todoroki, A., Tanaka, M., and Shimamura, Y. 2003. High performance estimations of delamination of graphite/epoxy laminates with electric resistance change method. *Composite Science and Technology*, **63**(13), 1911–1920.

Tripathy, A. K., Patel, H. J., and Pang, S. S. 1994. Bending analysis of laminated composite box beams. *Journal of Engineering Materials and Technology, Transactions of the ASME*, **116**(1), 121–129.

Tseng, Y. P., Huang, C. S., and Kao, M. S. 2000. In-plane vibration of laminated curved beams with variable curvature by dynamic stiffness analysis. *Composite Structures*, **50**(2), 103–114.

Ugural, A. C., and Fenster, S. K. 2003. *Advanced strength and applied elasticity*. Fourth edn. New Jersey, USA: Prentice Hall.

Vashakmadze, T. S. 1999. *The theory of anisotropic elastic plates*. Mathematics and Its Applications, vol. 476. Dordrecht, The Netherlands: Kluwer Academic Publisher.

Vaze, S. P., and Corona, E. 1997. Response and stability of square tubes under bending. *Journal of Applied Mechanics*, **64**(3), 649–657.

Vinayak, R. U., Prathap, G., and Naganarayana, B. P. 1996. Beam elements based on a higher order theory — I. Formulation and analysis of performance. *Computers and Structures*, **58**(4), 775–789.

Vinson, J. R., and Chou, T. 1975. *Composite materials and their use in structures*. Essex, UK: Elsevier.

Vinson, J. R., and Sierakowski, R. L. 2002. *The behavior of structures composed of composite materials*. Second edn. Dordrecht, The Netherlands: Kluwer Academic Publisher.

Vishnyakov, V. V., and Beresnev, A. N. 1964. Solution of the first and second fundamental problems of theory of elasticity for rectangular prism. *Pages 300–301 of: Reports of the third Siberian conference on mathematics and mechanics*. (in Russian).

Vlot, A. D., and Gunnink, J. W. 2001. *Fibre metal laminates: an introduction*. Dordrecht, The Netherlands: Kluwer Academic Publisher.

Volovoi, V. V., and Hodges, D. H. 2000. Theory of anisotropic thin-walled beams. *Journal of Applied Mechanics*, **67**(3), 453–459.

Volovoi, V. V., and Hodges, D. H. 2002. Single and multicelled composite thin-walled beams. *AIAA Journal*, **40**(5), 960–965.

Volovoi, V. V., Hodges, D. H., Berdichevsky, V. L., and Sutyrin, V. G. 1998. Dynamic dispersion curves for non-homogeneous, anisotropic beams with cross-sections of arbitrary geometry. *Journal of Sound and Vibration*, **215**(5), 1101–1120.

Volovoi, V. V., Hodges, D. H., Cesnik, C. E. S., and Popescu, B. 2001. Assessment of beam modeling methods for rotor blade applications. *Mathematical and Computer Modelling*, **33**, 1099–1112.

Whitney, J. M. 1987. *Structural analysis of laminated anisotropic plates*. Lancaster, PA, USA: Technomic Publishing.

Whitney, J. M. 1995. Cylindrical bending versus beam theory in the analysis of composite laminates. *Composites*, **26**(5), 395–398.

Wisnom, M. R., and Jones, M. I. 1996. Size effects on interlaminar tensile and shear strength of unidirectional glass fibre/epoxy. *Journal of Reinforced Plastics and Composites*, **15**(1), 2–15.

Wu, J. J., Barnett, D. M., and Ting, T. C. T. 1992. *Modern theory of anisotropic elasticity and applications*. Philadelphia, PA, USA: Society for Industrial & Applied Mathematics.

Wu, X. X., and Sun, C. T. 1991. Vibration analysis of laminated composite thin-walled beams using finite elements. *AIAA Journal*, **29**(5), 736–742.

Wu, X. X., and Sun, C. T. 1992. Simplified theory for composite thin-walled beams. *AIAA Journal*, **30**(12), 2945–2951.

Xie, M., and Adams, D. F. 1995. Contact finite element modeling of the short beam shear test for composite materials. *Computers and Structures*, **57**(2), 183–191.

Yildirim, V. K., and Erhan, S. E. 1999. Free vibration analysis of symmetric cross-ply laminated composite beams with the help of the transfer matrix approach. *Communications in Numerical Methods in Engineering*, **15**(9), 651–660.

Yu, H. 1994. Higher-order finite element for analysis of composite laminated structures. *Composite Structures*, **28**(4), 375–383.

Yu, W., Volovoi, V. V., Hodges, D. H., and Hong, X. 2002. Validation of the variational asymptotic beam sectional analysis. *AIAA Journal*, **40**(10), 2105–2112.

Zhang, Y., and Sikar skie, D. L. 1999. Strength prediction of random fiber composite beams using a phenomenological-mechanistic model. *Journal of Composite Materials*, **33**(20), 1882–1896.

Zivzivadze, R. T., and Berekashvili, R. A. 1984. Generalization of Almansi problem for compound anisotropic cylindrical beams. *Transactions of Georgian Polytechnical Institute*, **9**(279), 130–135.

Index